CW01024383

SNOW LEOPARDS

This is a Volume in the series

Biodiversity of the World:
Conservation from Genes to Landscapes
Edited By Philip J. Nyhus

SNOW LEOPARDS

BIODIVERSITY OF THE WORLD: CONSERVATION FROM GENES TO LANDSCAPES

Series editor

PHILIP J. NYHUS
Environmental Studies Program
Colby College, Waterville, ME, USA

Volume editors

THOMAS McCARTHY
Snow Leopard Program
Panthera, New York, NY, USA

DAVID MALLON
Division of Biology and Conservation Ecology
Manchester Metropolitan University
Manchester, UK

AMSTERDAM • BOSTON • HEIDELBERG • LONDON
NEW YORK • OXFORD • PARIS • SAN DIEGO
SAN FRANCISCO • SINGAPORE • SYDNEY • TOKYO

Academic Press is an imprint of Elsevier

Academic Press is an imprint of Elsevier
125 London Wall, London EC2Y 5AS, United Kingdom
525 B Street, Suite 1800, San Diego, CA 92101-4495, United States
50 Hampshire Street, 5th Floor, Cambridge, MA 02139, United States
The Boulevard, Langford Lane, Kidlington, Oxford OX5 1GB, UK

British Library Cataloguing-in-Publication Data
A catalogue record for this book is available from the British Library

Library of Congress Cataloging-in-Publication Data
A catalog record for this book is available from the Library of Congress

ISBN: 978-0-12-802213-9

For information on all Academic Press publications
visit our website at https://www.elsevier.com/

Publisher: Sara Tenney
Acquisition Editor: Kristi Gomez
Editorial Project Manager: Pat Gonzalez
Production Project Manager: Lucía Pérez
Designer: Matthew Limbert

Typeset by Thomson Digital

We dedicate this book to those heroes of snow leopard conservation who have gone before us. Tireless, dedicated, and visionary – we were privileged to call them friends and mentors.

Helen Freeman
Jachliin Tserendeleg
Chering Nurbu
Chandra Gurung
Mingma Sherpa
Rinchen Wangchuk
Pralad Yonzen
Lkhagvasumberel Tumursukh

And to the memory of Peter Matthiessen, whose book Snow Leopard *brought the mysterious nature of the cat and its realm to readers around the world.*

Contents

35. South Asia: Bhutan

TSHEWANG R. WANGCHUK, LHENDUP THARCHEN

36. South Asia: India

YASH VEER BHATNAGAR, VINOD BIHARI MATHUR, SAMBANDAM SATHYAKUMAR, ABHISHEK GHOSHAL, RISHI KUMAR SHARMA, AJAY BIJOOR, RANGASWAMY RAGHUNATH, RADHIKA TIMBADIA, PANNA LAL

37. South Asia: Nepal

SOM ALE, KARAN B. SHAH, RODNEY M. JACKSON

38. South Asia: Pakistan

38.1 Snow Leopard Conservation in Pakistan: A Historical Perspective

AHMAD KHAN

38.2 The Current State of Snow Leopard Conservation in Pakistan

JAFFAR UD DIN, HUSSAIN ALI, MUHAMMAD ALI NAWAZ

39. Northern Range: Mongolia

BARIUSHAA MUNKHTSOG, LKHAGVAJAV PUREVJAV, THOMAS McCARTHY, RANA BAYRAKÇISMITH

40. Northern Range: Russia

MIKHAIL PALTSYN, ANDREY POYARKOV, SERGEI SPITSYN, ALEXANDER KUKSIN, SERGEI ISTOMOV, JAMES P. GIBBS, RODNEY M. JACKSON, JENNIFER CASTNER, SVETLANA KOZLOVA, ALEXANDER KARNAUKHOV, SERGEI MALYKH, MIROSLAV KORABLEV, ELENA ZVYCHAINAYA, VYACHESLAV ROZHNOV

41. China: The Tibetan Plateau, Sanjiangyuan Region

YANLIN LIU, BYRON WECKWORTH, JUAN LI, LINGYUN XIAO, XIANG ZHAO, ZHI LU

42. China: Current State of Snow Leopard Conservation in China

PHILIP RIORDAN, KUN SHI

VII

THE FUTURE OF SNOW LEOPARDS

43. Sharing the Conservation Message

RANA BAYRAKÇISMITH, SIBYLLE NORAS, HEATHER HEMMINGMOORE

44. Global Strategies for Snow Leopard Conservation: A Synthesis

ERIC W. SANDERSON, DAVID MALLON, THOMAS McCARTHY, PETER ZAHLER, KIM FISHER

45. The Global Snow Leopard and Ecosystem Protection Program

ANDREW ZAKHARENKA, KOUSTUBH SHARMA, CHYNGYZ KOCHOROV, BRAD RUTHERFORD, KESHAV VARMA, ANAND SETH, ANDREY KUSHLIN, SUSAN LUMPKIN, JOHN SEIDENSTICKER, BRUNO LAPORTE, BORIS TICHOMIROW, RODNEY M. JACKSON, CHARUDUTT MISHRA, BAKHTIYAR ABDIEV, ABDUL WALI MODAQIQ, SONAM WANGCHUK, ZHANG ZHONGTIAN, SHAKTI KANT KHANDURI, BAKYTBEK DUISEKEYEV, BATBOLD DORJGURKHEM, MEGH BAHADUR PANDEY, SYED MAHMOOD NASIR, MUHAMMAD ALI NAWAZ, IRINA FOMINYKH, NURALI SAIDOV, NODIRJON YUNUSOV

46. Joining up the Spots: Aligning Approaches to Big Cat Conservation from Policy to the Field

URS BREITENMOSER, TABEA LANZ, ROLAND BÜRKI, CHRISTINE BREITENMOSER-WÜRSTEN

47. Future Prospects for Snow Leopard Survival

DAVID MALLON, THOMAS McCARTHY

List of Contributors

Bakhtiyar Abdiev State Agency on Environment Protection and Forestry, Bishkek, Kyrgyz Republic

Bayarjargal Agvaantseren Mongolian Snow Leopard Conservation Foundation, Ulaanbaatar, Mongolia

Som Ale Biological Sciences, University of Illinois, Chicago, IL; Snow Leopard Conservancy, Sonoma, CA, USA

Hussain Ali Snow Leopard Foundation, Islamabad, Pakistan

Priscilla Allen Independent, Seattle, WA, USA

Chagat Almashev Foundation for Sustainable Development of the Altai, Gorno-Altaisk, Altai Republic, Russia

George Amato American Museum of Natural History, Sackler Institute for Comparative Genomics, New York, NY, USA

Sukh Amgalanbaatar President, Argali Wildlife Research Center, Ulaanbaatar, Mongolia

Zayiniddin Amirov Institute of Zoology and Parasitology, Academy of Sciences, Dushanbe, Tajikistan

Maksat Anarbaev National Center for Mountain Regions Development; Snow Leopard Program, Panthera, Bishkek, Kyrgyz Republic

Rafael Arenas Wildlife Conservation Society, New York, NY, USA

Bakhtyor Aromov Gissar State Nature Reserve, Gosbiokontrol, Tashkent, Uzbekistan

Roland Bürki KORA, Bern, Switzerland

Rana Bayrakçısmith Snow Leopard Program, Panthera, New York, NY; Snow Leopard Network, Seattle, WA, USA

Nick Beale Panthera, London, UK

Yash Veer Bhatnagar Snow Leopard Trust and Nature Conservation Foundation, Mysore, Karnataka, India; Snow Leopard Trust, Seattle, WA, USA

Ajay Bijoor Snow Leopard Trust and Nature Conservation Foundation, Mysore, Karnataka, India; Snow Leopard Trust, Seattle, WA, USA

Ekaterina Yu. Blidchenko A.N. Severtsov Institute of Ecology and Evolution, Russian Academy of Sciences, Moscow, Russia

Leif Blomqvist Nordens Ark Åby Säteri, Hunnebostrand, Sweden

Jordi Boixader Iberian lynx *Ex Situ* Conservation Program, Andalusia, Spain

Ioana Botezatu INTERPOL, Lyon, France

Urs Breitenmoser IUCN Cat Specialist Group, Bern, Switzerland

Christine Breitenmoser-Würsten IUCN Cat Specialist Group, Bern, Switzerland

Hafeez Buzdar Snow Leopard Foundation, Islamabad, Pakistan

Elena Bykova Institute of the Gene Pool of Plants and Animals, Uzbek Academy of Sciences, Tashkent, Uzbekistan

Anthony Caragiulo American Museum of Natural History, Sackler Institute for Comparative Genomics, New York, NY, USA

Jennifer Castner The Altai Project, East Lansing, MI, USA

Raghunandan S. Chundawat Bagh Aap Aur Van (BAAVAN), New Delhi, India

Apela Colorado Worldwide Indigenous Science Network, Lahaina, HI, USA

Marc Criffield Florida Fish and Wildlife Conservation Commission, Naples, FL, USA

Jigmet Dadul Snow Leopard Conservancy India Trust, Leh, Ladakh, Jammu & Kashmir, India

Unurzul Dashzeveg Mongolian Snow Leopard Conservation Foundation, Ulaanbaatar, Mongolia

Galbadrakh Davaa Mongolia Country Program, The Nature Conservancy, Ulaanbaatar, Mongolia

Askar Davletbakov Institute for Biology and Soil Sciences, National Academy of Sciences of the Kyrgyz Republic, Bishkek, Kyrgyz Republic

Jaffar Ud Din Snow Leopard Foundation, Islamabad, Pakistan

Batbold Dorjgurkhem International Cooperation Division, Ministry of Environment and Green Development, Ulaanbaatar, Mongolia

Carlos A. Driscoll WWF Chair in Conservation Genetics, Faculty of Wildlife Sciences, Wildlife Institute of India, Dehradun, Uttarakhand, India

Bakytbek Duisekeyev Wildlife Department, Ministry of Agriculture, Astana, Kazakhstan

Alexander Esipov Institute of the Gene Pool of Plants and Animals, Uzbek Academy of Sciences, Tashkent

John D. Farrington World Wildlife Fund, Thimphu, Bhutan

Leonardo Fernández The Conservation Land Trust, Scalabrini Ortiz, Argentina

Kim Fisher WCS Asia Program, Wildlife Conservation Society, New York, NY, USA

Irina Fominykh Department of International Cooperation, Ministry of Natural Resources and Environment, Moscow, Russian Federation

Joseph L. Fox Natural & Environmental Sciences Department, Western State Colorado University, Gunnison, CO, USA

Maribel García-Tardío The Conservation Land Trust, Scalabrini Ortiz, Argentina

Germán Garrote The Conservation Land Trust, Scalabrini Ortiz, Argentina

Abhishek Ghoshal Nature Conservation Foundation, Mysore, Karnataka; Wildlife Institute of India, Dehradun, Uttarakhand, India

James P. Gibbs Department of Environmental and Forest Biology, State University of New York, College of Environmental Science and Forestry, Syracuse, NY, USA

Martin Gilbert Wildlife Health & Health Policy Program, Wildlife Conservation Society, Bronx, NY, USA

Richard B. Harris Game Division, Wildlife Program, Washington Department of Fish and Wildlife, Olympia, WA, USA

Michael Heiner Conservation Lands, Fort Collins, The Nature Conservancy, CO, USA

Heather Hemmingmoore Snow Leopard Network, London, UK

Jose A. Hernandez-Blanco A.N. Severtsov Institute of Ecology and Evolution, Russian Academy of Sciences, Moscow, Russia

Darla Hillard Snow Leopard Conservancy, Sonoma, CA, USA

Paul Hotham Fauna & Flora International, Cambridge, UK

Pippa Howard Fauna & Flora International, Cambridge, UK

Don Hunter Rocky Mountain Cat Conservancy, Fort Collins, CO, USA

Shafqat Hussain Department of Anthropology, Trinity College, Hartford, CT, USA

Sergei Istomov Science Department, Sayano-Shushensky State Nature Biosphere Reserve, Shushenskoe, Russia

Rodney M. Jackson Snow Leopard Conservancy, Sonoma, CA, USA

Ignacio Ignacio Jiménez-Peréz The Conservation Land Trust, Scalabrini Ortiz, Argentina

Örjan Johansson Grimsö Wildlife Research Station, Swedish University of Agricultural Sciences, Riddarhyttan, Sweden

Shannon Kachel School of Environmental and Forest Resources, University of Washington, Seattle, WA, USA

Khalil Karimov Tajikistan Snow Leopard Program, Panthera, Khorog, GBAO, Tajikistan

Alexander Karnaukhov Laboratory of Behavior and Behavioral Ecology of Mammals, A.N. Severtsov Institute of Ecology and Evolution, Russian Academy of Sciences, Moscow, Russia

Joseph Kiesecker Conservation Lands, Fort Collins, The Nature Conservancy, CO, USA

Ahmad Khan Department of Geographical Sciences, University of Maryland, College Park, MD, USA

Ashiq Ahmad Khan Snow Leopard Foundation, Islamabad, Pakistan

Shakti Kant Khanduri Ministry of Environment and Forests, New Delhi, India

Ambika Khatiwada National Trust for Nature Conservation, Sauraha, Chitwan, Nepal

Tungalagtuya Khuukhenduu Nomadic Nature Conservation, Ulaanbaatar, Mongolia

Andrew C. Kitchener Department of Natural Sciences, National Museums Scotland; Institute of Geography, University of Edinburgh, Edinburgh, UK

Chyngyz Kochorov GSLEP Secretariat, Bishkek, Kyrgyz Republic

Fred W. Koontz Vice President of Field Conservation, Woodland Park Zoo, Seattle, WA, USA

Miroslav Korablev Laboratory of Behavior and Behavioral Ecology of Mammals, A.N. Severtsov Institute of Ecology and Evolution, Russian Academy of Sciences, Moscow, Russia

Svetlana Kozlova Independent Consultant on Results-Based Management, Syracuse, NY, USA

Zairbek Kubanychbekov Snow Leopard Program, Panthera; Panthera Foundation Kyrgyzstan, Bishkek, Kyrgyz Republic

Kyran Kunkel Conservation Science Collaborative and Wildlife Biology Program, University of Montana, Missoula, MT, USA

Alexander Kuksin Science Department, Ubsunurskaya Kotlovina State Nature Biosphere Reserve, Kyzyl, Russia

Andrey Kushlin Independent Advisor, Washington, DC, USA

Panna Lal Wildlife Institute of India, Dehradun, Uttarakhand, India

Wendy Brewer Lama KarmaQuest Ecotourism and Adventure Travel, Half Moon Bay, CA, USA

Tabea Lanz IUCN Cat Specialist Group, Bern, Switzerland

Bruno Laporte Leadership, Knowledge, Learning, LLC, Washington, DC, USA

Juan Li Center for Nature and Society, College of Life Sciences, Peking University, Beijing, China; Department of Environmental Science, Policy and Management, University of California, Berkeley, CA, USA

Thomas Lind Kolmården Wildlife Park, Kolmården, Sweden

Yanlin Liu Shan Shui Conservation Center, Beijing, China

Oleg Loginov Snow Leopard Fund, Ust-Kamenogorsk, Kazakhstan

Irina Loginova Snow Leopard Fund, Ust Kamenogorsk, Kazakhstan

Guillermo López LIFE+ Iberlince project: Recovery of the Historic Distribution Range of the Iberian Lynx, Andalusia, Spain

Marcos López-Parra The Conservation Land Trust, Scalabrini Ortiz, Argentina

Sandro Lovari Ev.-K2-C.N.R., Bergamo, and Department of Life Sciences, University of Siena, Siena, Italy

Zhi Lu Shan Shui Conservation Center; Center for Nature and Society, College of Life Sciences, Peking University, Beijing, China

Susan Lumpkin Independent Advisor, Washington, DC, USA

Aishwarya Maheshwari Independent Researcher, Saurashtra University and Wildlife Institute India, Ghaziabad, Uttar Pradesh, India

Bipasha Majumder Independent, Mumbai, Maharashtra, India

David Mallon Division of Biology and Conservation Ecology, Manchester Metropolitan University, Manchester, UK

Sergei Malykh Laboratory of Behavior and Behavioral Ecology of Mammals, A.N. Severtsov Institute of Ecology and Evolution, Russian Academy of Sciences, Moscow, Russia

Vinod Bihari Mathur Wildlife Institute of India, Dehradun, Uttarakhand, India

Kyle McCarthy Department of Entomology and Wildlife Ecology, University of Delaware, Newark, DE, USA

Thomas McCarthy Snow Leopard Program, Panthera, New York, NY, USA

Stefan Michel Denver Zoological Foundation, Khorog, GBAO, Tajikistan

Tserenadmid Mijiddorj Mongolian Snow Leopard Conservation Foundation, Ulaanbaatar, Mongolia

Dale G. Miquelle Wildlife Conservation Society, New York, NY, USA

Charudutt Mishra Snow Leopard Trust and Nature Conservation Foundation, Mysore, Karnataka, India

John Mock Secretary and Trustee-at-Large, American Institute of Afghanistan Studies, Boston University, Boston, MA, USA

Abdul Wali Modaqiq National Environmental Protection Agency, Kabul, Afghanistan

Ghulam Mohammad Baltistan Wildlife and Conservation Development Organization (BWCDO), Skardu, Gilgit-Baltistan, Pakistan

Zalmai Moheb Afghanistan Program, Wildlife Conservation Society, Kabul, Afghanistan

Sayed Naqibullah Mostafawi Wildlife Conservation Society, Kabul, Afghanistan

Bariushaa Munkhtsog Institute of General and Experimental Biology, Mongolian Academy of Sciences, Irbis Mongolian Centre, Ulaanbaatar, Mongolia

Ranjini Murali Snow Leopard Trust and Nature Conservation Foundation, Mysore, Karnataka, India

Sergey V. Naidenko A.N. Severtsov Institute of Ecology and Evolution, Russian Academy of Sciences, Moscow, Russia

Tsewang Namgail Snow Leopard Conservancy India Trust, Leh, Ladakh, Jammu & Kashmir, India

Syed Mahmood Nasir Ministry of Climate Change, Islamabad, Pakistan

Muhammad Ali Nawaz Department of Animal Sciences, Quaid-i-Azam University; Snow Leopard Foundation, Islamabad, Pakistan

Sibylle Noras Snow Leopard Network, Melbourne, Australia

Helen Nyul Fauna & Flora International, Cambridge, UK

James Oakleaf Conservation Lands, Fort Collins, The Nature Conservancy, CO, USA

Dave Onorato Florida Fish and Wildlife Conservation Commission, Naples, FL, USA

Stéphane Ostrowski Wildlife Health & Health Policy Program, Wildlife Conservation Society, Bronx, NY, USA

Sujatha Padmanabhan Kalpavriksh Environment Action Group, Pune, Maharashtra, India

Richard Paley Afghanistan Program, Wildlife Conservation Society, Kabul, Afghanistan

Mikhail Paltsyn Department of Environmental and Forest Biology, State University of New York, College of Environmental Science and Forestry, Syracuse, NY, USA

Megh Bahadur Pandey Department of National Parks and Wildlife Conservation, Ministry of Forest and Soil Conservation, Kathmandu, Nepal

Andrey Poyarkov Laboratory of Behavior and Behavioral Ecology of Mammals, A.N. Severtsov Institute of Ecology and Evolution, Russian Academy of Sciences, Moscow, Russia

Herbert H.T. Prins Environmental Sciences Group – Resource Ecology, Wageningen University, Wageningen, The Netherlands

Yelizaveta Protas Independent, Jackson Heights, New York, NY, USA

Lkhagvajav Purevjav Snow Leopard Conservation Fund, Ulaanbaatar, Mongolia

Betsy Gaines Quammen Yellowstone Theological Institute, Bozeman, MT, USA

Rangaswamy Raghunath Nature Conservation Foundation, Mysore, Karnataka, India

Richard P. Reading Department of Biological Sciences & Graduate School of Social Work, University of Denver, Denver, CO, USA

Stephen R. Redpath Institute of Biological & Environmental Sciences, Aberdeen University, Aberdeen, UK

Teresa del Rey The Conservation Land Trust, Scalabrini Ortiz, Argentina

Philip Riordan Department of Zoology, University of Oxford, Oxford, UK

Hugh S. Robinson Landscape Analysis Laboratory, Panthera, New York; College of Forestry and Conservation, University of Montana, Missoula, MT, USA

Tatjana Rosen Tajikistan and Kyrgyzstan Snow Leopard Programs, Panthera, Khorog, Gorno-Badakhshan, Tajikistan

Viatcheslav V. Rozhnov A.N. Severtsov Institute of Ecology and Evolution, Russian Academy of Sciences, Moscow, Russia

Vyacheslav Rozhnov Laboratory of Behavior and Behavioral Ecology of Mammals, A.N. Severtsov Institute of Ecology and Evolution, Russian Academy of Sciences, Moscow, Russia

Gema Ruiz The Conservation Land Trust, Scalabrini Ortiz, Argentina

Jennifer Snell Rullman Snow Leopard Trust, Seattle, WA, USA

Brad Rutherford Snow Leopard Trust, Seattle, WA, USA

Nargiza Ryskulova Worldwide Indigenous Science Central Asian Consultant, Bishkek, Kyrgyzstan

Abdusattor Saidov Academy of Sciences, Dushanbe, Tajikistan

Nurali Saidov State Agency of Natural Protected Areas, Dushanbe, Tajikistan

Jarkyn Samanchina Fauna & Flora International, Bishkek, Kyrgyz Republic

Eric W. Sanderson WCS Asia Program, Wildlife Conservation Society, New York, NY, USA

Sambandam Sathyakumar Wildlife Institute of India, Dehradun, Uttarakhand, India

John Seidensticker Smithsonian Conservation Biology Institute, Washington, DC, USA

Anand Seth Independent Advisor, Washington, DC, USA

Karan B. Shah Natural History Museum, Tribhuvan University, Swoyambhu, Kathmandu, Nepal

Safdar Ali Shah Wildlife Department, Khyber Pakhtunkhwa, Peshawar

Koustubh Sharma GSLEP Secretariat, Bishkek, Kyrgyz Republic

Rishi Kumar Sharma World Wide Fund for Nature – India, New Delhi, India

Wasim Shehzad Institute of Biochemistry & Biotechnology, University of Veterinary & Animal Sciences, Lahore, Pakistan

Kun Shi The Wildlife Institute, Beijing Forestry University, Beijing, China

Miguel A. Simón Consejería de Medio Ambiente de la Junta de Andalucía, Jaén, Spain

Anthony Simms Wildlife Conservation Society, Bronx, NY, USA

Alexander Sliwa Cologne Zoo, Cologne, Germany

Karma Sonam Snow Leopard Trust and Nature Conservation Foundation, Mysore, Karnataka, India

Pavel A. Sorokin A.N. Severtsov Institute of Ecology and Evolution, Russian Academy of Sciences, Moscow, Russia

Sergei Spitsyn Science Department, Altaiskiy State Nature Biosphere Reserve, Yailu, Russia

Kulbhushansingh R. Suryawanshi Snow Leopard Trust and Nature Conservation Foundation, Mysore, Karnataka, India

Jay Tetzloff Miller Park Zoo, Bloomington, IL, USA

Lhendup Tharchen Jigme Dorji National Park, Damji, Bhutan

Tanzin Thinley Snow Leopard Trust and Nature Conservation Foundation, Mysore, Karnataka, India

Patrick R. Thomas Vice President and General Curator, Wildlife Conservation Society and Associate Director, Bronx Zoo, New York, NY, USA

Boris Tichomirow Middle Asia Program, NABU, Berlin, Germany

Radhika Timbadia Nature Conservation Foundation, Mysore, Karnataka, India; Snow Leopard Trust, Seattle, WA, USA

Pranav Trivedi Snow Leopard Trust and Nature Conservation Foundation, Mysore, Karnataka, India

Kuban Jumabai uulu Snow Leopard Trust/Snow Leopard Foundation in Kyrgyzstan, Bishkek, Kyrgyz Republic

Keshav Varma GTI Council, New Delhi, India

Stephanie von Meibom TRAFFIC – Europe, Frankfurt, Germany

Sonam Wangchuk Ministry of Agriculture and Forest, Thimpu, Bhutan

Tshewang R. Wangchuk Bhutan Foundation, Washington, DC, USA

Byron Weckworth Snow Leopard Program, Panthera, New York, NY, USA

Mike Weddle Jane Goodall Environmental Middle School, Salem, OR, USA

Per Wegge Department of Ecology and Natural Resource Management, Norwegian University of Life Sciences, Ås, Norway

Tony Whitten Fauna & Flora International, Cambridge, UK

Wai-Ming Wong Department of Field Programs, Panthera, New York, NY, USA

Lingyun Xiao Center for Nature and Society, College of Life Sciences, Peking University, Beijing, China

Anna A. Yachmennikova A.N. Severtsov Institute of Ecology and Evolution, Russian Academy of Sciences, Moscow, Russia

Bayarjargal Yunden Mongolia Country Program, The Nature Conservancy, Ulaanbaatar, Mongolia

Hang Yin Shan Shui Conservation Center, Beijing, China

Nobuyuki Yamaguchi Department of Biological and Environmental Sciences, College of Arts and Sciences, Qatar University, Doha, Qatar

Nodirjon Yunusov International Relations Department, State Committee for Nature Protection, Tashkent, Uzbekistan

Peter Zahler WCS Asia Program, Wildlife Conservation Society, New York, NY, USA

Andrew Zakharenka Global Tiger Initiative Secretariat, World Bank, Washington, DC, USA

Xiang Zhao Shan Shui Conservation Center, Beijing, China

Zhang Zhongtian Department of International Cooperation, State Forestry Administration, Beijing, China

Elena Zvychainaya Laboratory of Behavior and Behavioral Ecology of Mammals, A.N. Severtsov Institute of Ecology and Evolution, Russian Academy of Sciences, Moscow, Russia

Foreword

George B. Schaller

Panthera and Wildlife Conservation Society, New York, NY, USA

I am a true son of snowy mountains
Watching over solitude, persisting
Through all temporal stages
Crowded among hardened waves of boulders
I stand guard here—

From "I, Snow Leopard" by Jidi Majia[1]

In its gaunt high terrain and among the harsh wind-swept peaks, the snow leopard endures in 12 countries of Central Asia. The wild remoteness of the habitat has so far assured its survival. And so, too, has its smoky gray coat speckled with rosettes that make it almost invisible in its rocky realm, a ghost, a phantom.

Once, in December 1970, I watched a female snow leopard with a small cub for several days as they fed on a kill near their den. Tucked in my sleeping bag, I remained near them during bitter Pakistani nights, the mountains glowing in moonlight, the landscape belonging wholly to them, the spirit cats. I have sought to relive these hours in other ranges, in Afghanistan, Tajikistan, and China. But only in Mongolia, where I radio-collared a male snow leopard in 1990, was I able to locate and spend leisurely hours again with one of these cats. Peter Matthiessen (1978) wrote in his classic book *The Snow Leopard* "*that the snow leopard is, that it is here, that its frosty eyes watch us from the mountains—that is enough.*" Not anymore.

We still have too little information to assure the snow leopard a future. This book, *Snow Leopards*, a volume of the series *Biodiversity of the World: Conservation from Genes to Landscapes*, offers a superb overview of knowledge by experts from various countries, and also notes the gaps in our understanding of the snow leopard's needs. I have roamed Asia's mountains for several decades, and here add a few observations and comments.

There are about two million km^2 of suitable snow leopard habitat. The mountain ranges retain remarkable connectivity, many of them merging at some point. An adventurous snow leopard could readily make an arduous trek from the Himalaya in Bhutan or Nepal west and north for more than 2000 km on obscure trails to Mongolia. Only a fraction of the total range has been designated officially as reserves, many too small to sustain a viable snow leopard population. Most of them also have people living in them, grazing livestock, tilling small fields when possible, collecting fuelwood. Far more snow leopards live outside of reserves than in them. The cat needs landscapes, parts of which consist of protected core areas for all species, and the rest devoted to achieving a measure of ecological harmony between habitat, wildlife, and communities with their livestock. The Sanjiangyuan region in Qinghai Province of China is an excellent example of this landscape approach (Li et al., 2014). However, with climate change, considerable snow leopard habitat may be lost because of an upward shift in the tree line and concomitant loss of the alpine zone (Forrest et al., 2012).

Snow leopards are borderland cats in many places. This provides an opportunity to expand

[1]This is a fragment of a long poem by Jidi Majia, a poet of the traditional Yi culture in China, published in 2015.

landscapes across international borders through cooperative ventures and at the same time perhaps help defuse transboundary tensions. The adjoining Sagarmatha National Park in Nepal and the Qomolangma Nature Reserve in China are one example of such a transfrontier landscape. Others in need of consideration include the Siachen glacier area on the Pakistan–India border and the Pamir Mountains where the frontiers of China, Afghanistan, Pakistan, and Tajikistan converge.

Recent research on snow leopards has been fragmentary and usually brief, lasting a few weeks or months. In 1982, Jackson et al. (1990) began their 3.5-year research in Nepal using radio-telemetry, the first long-term effort, followed by Tom McCarthy's 4-year study in the Gobi desert of Mongolia during the 1990s, and the South Gobi program undertaken by Panthera and the Snow Leopard Trust, using GPS satellite collars, mainly from 2008 to 2013. In recent years, camera traps have not only provided solid data on population size and density (generally 1–5/100 km^2) but also revealed the snow leopard's beauty and secrets of behavior in stunning photographs.

Radio-collars have delineated home range size which may range from 20 to over 400 km^2 depending partly on prey density (McCarthy and Chapron, 2003). But how many snow leopards survive in the wild, a species that was listed as endangered in 1972 and banned from international trade in 1975? Published estimates, often little more than guesses, range from 3000 to 8000, with over half of the cats in China. Such nebulous estimates are not surprising because many mountain ranges have been sampled at only a site or two, and no long-term monitoring of populations, lasting a decade or more, has been done to measure such dynamics as their birth and death rates.

It is also important to know more about the snow leopard's role within its ecosystem, a top predator yet part of the interdependent natural processes. Studies have mainly focused on the snow leopard and its prey such as blue sheep, ibex, and livestock. Yet a whole carnivore guild of wolf, brown bear, lynx, and fox compete for the same prey and partly rely on scavenging snow leopard kills, as do vultures and various raptor species. Such resource competition could be especially marked between the snow leopard and the slightly larger common leopard which are not only sympatric along the Himalaya but also kill prey of similar size (Lovari et al., 2013). Further interactions between carnivore species were recorded by Li et al. (2013a) at a snow leopard signpost beneath a rock overhang which attracted other carnivores and even blue sheep and domestic yak, the species apparently exchanging information at the site. Free-roaming Tibetan mastiff dogs also checked the signpost. Such dogs, either feral or belonging to households or monasteries, often hunt in packs and kill blue sheep and even a snow leopard on occasion. The canine distemper virus could easily be transmitted from such dogs to snow leopards and other carnivores, an issue which has not been addressed so far.

Livestock penetrates snow leopard habitat into all but the steepest terrain where it has contact with blue sheep, ibex, and others. Pastures are often seriously overgrazed, the biomass of domestic animals many times that of wild herbivores. Being opportunists, snow leopards tend to prey on livestock proportionately more in such situations, although households readily admit that they may lose more animals to disease than to predation. The transmission of disease, such as hoof-and-mouth and sarcoptic mange, from domestic to wild ungulates is an ever-present threat. To create a sustainable landscape as a legacy to the future will entail a broad spectrum of activities, from implementing rangeland management and controlling disease to passing on conservation values to communities.

Camera traps, satellite collars, GIS, and other recent innovations are invaluable tools for gaining statistically valid data. Technology is seductive. However, it may come at a loss of time spent in the field by an investigator, a loss of reality of knowing a species intimately, of seeing the landscape as an animal does, the musky odor of a snow leopard scent mark, the circling griffon vultures that signal a kill, the scrape on a mountain pass.

The snow leopard is often viewed as an icon, a guardian deity, or, in academic terminology, as a flagship species symbolizing the high-altitude ecosystem. This, however, has little relevance to poor local families who depend for subsistence mainly on livestock, and lose animals to predation by snow leopard and wolf. Annual losses to livestock in a community are usually about 1–2%, as studies show, but with as much as 12%, wolves generally killing more than do snow leopards. Even the loss of a few animals may cause hardship to families who own only 20–30 animals, as is often the case in Pakistan and Nepal. To eliminate predators in retaliation by shooting, trapping, or poisoning is an obvious option for a family. Legally protected in all its range countries, the snow leopard continues to be killed in retribution with the added incentive of profiting from the sale of hide and body parts (Koshkarev and Vyrypaev, 2000; Li and Lu, 2014). To raise awareness about snow leopards, disseminate information about their status, and induce countries to increase their protection, international conferences have been held in Seattle (2002), Beijing (2008), and the Kyrgyz Republic (2013), among others. Publications such as the *Snow Leopard Survival Strategy* (McCarthy and Chapron, 2003; Snow Leopard Network, 2014) and this book are valuable sources of information, as are the intermittent news items circulated by the Snow Leopard Network. Publicity of this kind will, one hopes, stifle forever an impulse such as the one shown by Norsultan Nazarbaev, the President of Kazakhstan, when he publicly presented King Juan Carlos of Spain with a snow leopard coat in 1999.

A basic challenge is how best to apply scientific information to achieve conservation goals, to address the conflicting demands of protecting the environment and its species and promoting the welfare of communities. This is especially true when confronting an economic issue such as snow leopard–livestock conflict. For example, in the Wakhan Corridor of Afghanistan the Wildlife Conservation Society initiated a comprehensive community program which includes construction of predator-proof corrals, environmental education in schools, promotion of tourism, handicraft manufacture, veterinary services, compensation payments for losses to predators, and the establishment of a Wakhan-Pamir Association to integrate activities of the Wakhi and Kyrgyz communities (Simms et al., 2011). Such innovative projects have also been implemented in India, China, Pakistan, and Mongolia, to name four others, and these also include livestock insurance, handicrafts, tourism, and education to give communities a sense of ownership of programs (Jackson et al., 2010; Li et al., 2014; Mishra et al., 2003; Rosen et al., 2012).

Not all goes smoothly, however, with such efforts to improve a local quality of life. Almost half of the persons interviewed by Oli et al. (1994) about predation on livestock in Nepal felt that the eradication of snow leopards is the only option. A trophy-hunting program for ibex and markhor in Pakistan from which communities received a hefty 80% of the license fees from the government had problems when locals demanded the removal of snow leopards because they killed potential trophy animals (Rosen et al., 2012).

I was struck on reading the literature on compensation payments by how little is often done to address a main cause of the depredation, and how little projects may demand in reciprocity, such as in labor or cooperation from the local community. Much predation on livestock is due to poor herding practices, as I have observed over many years. Sheep and goats may have a herder nearby, even if a wholly inattentive one, whereas cattle, yaks, and horses are often left unattended on high pastures for weeks. If herders are in short supply, as when children are in school, cooperative communal herding is an option if there is the requisite social will. Households are, for example, fully aware of the need to secure their livestock at night yet rely all too often on luck that a predator will not invade a corral. Partly this is due to financial constraints, and now donor organizations often provide construction materials and designs for predator-proof corrals to poor families. However, as Oli et al. (1994) have noted, communities are often resistant to change in that *"closer guarding or the construction of enclosures for night-time corralling were considered by herdsmen to be unacceptable."* The same lack of initiative is evident in training dogs to guard livestock. Most households on the Tibetan Plateau have at least two to three dogs, primarily mastiffs, which are either tied up on short chains or stray in the countryside. They are essentially ineffective in protecting livestock. An obvious solution would be to keep the dogs from a very young age throughout the day and night with a sheep herd in the corral and on the range. Improved herding practices, including optimum livestock densities, can be suggested but may be difficult to implement unless culturally acceptable.

Snow leopards represent a complex mixture of cultural ideas, facts, and illusions. Outsiders arriving at a community to offer assistance, to reach across cultures, tend to view the landscape and its wildlife in terms of collecting data leading to management, and of offering economic assistance and incentives to achieve their goals. Hussain (2002) emphasized that such a utilitarian view of nature is not sufficient in resolving conservation problems. Spiritual, ethical, and aesthetic values are part of every culture. And so are music, art, and poetry. Nature and culture are one. An appeal to the beauty and mystery of the snow leopard touches the emotions, it reaches the heart. In northern Pakistan the most revered animal in the spirit realm is the snow leopard; in Mongolia, shamans ride on snow leopard pelts to reach the upper spirits (Hussain, 2002).

The world's major religions have been negligent in promoting their moral principles on behalf of the environment. Islam and Buddhism encompass much of the snow leopard's range, the former mainly in the western part, and the latter over much of the rest. Working with imams, the religious leaders in Pakistan, Sheikh (2006) found that according to their creed "all creatures are living beings like us, and worthy of respect and protection." Buddhism has as its basic tenet the love, respect, and compassion toward all living beings. Taking advantage of the fact that most Tibetan Buddhist monasteries on the Tibetan Plateau are located near snow leopard habitat, the Center for Nature and Society at Peking University began in 2009 a cooperative project with several monasteries. Every monastery has a sacred land, averaging 75 km^2 in size (Li et al., 2014), in which the killing of wildlife, grazing of livestock, and other destructive activities are forbidden. Monasteries vary considerably in adherence to these rules (see Schaller, 2012). The Tarthung monastery, however, is a glowing example for all religious communities to emulate. Monks monitor and study wildlife, conduct conservation classes in local schools, patrol their land, and involve communities in their Niabuayuze Environmental Protection Association, named for a holy mountain. In July 2013, the Association helped to organize a convocation of 31 monasteries at a holy lake to discuss and promote conservation in the region.

If we accept the moral responsibility of saving the snow leopard, we have to use all our knowledge, empathy, passion, persistence, and collaboration to stimulate governments and communities to resolve ecological issues. And this includes spiritual values to help lead us to a healthy and harmonious mountain environment guarded by the snow leopard, forever.

References

Forrest, J., Wikramanayake, E., Shrestha, R., Areendran, G., Gyeltshen, K., Maheshwari, A., Mazumdar, S., Naidoo, R., Thapa, G., Thapa, K., 2012. Conservation and climate change: assessing the vulnerability of snow leopard habitat to treeline shift in the Himalaya. Biol. Cons. 150, 129–135.

Hussain, S., 2002. Nature and human nature: conservation, values, and snow leopard. Contributed Papers to the Snow Leopard Survival Strategy, Summit, International Snow Leopard Trust, Seattle, pp. 65–72.

Jackson, R., Ahlborn, G., Shah, K.B., 1990. Capture and immobilization of wild snow leopards. In: Blomqvist, L. (Ed.), International Pedigree Book of Snow Leopards. Helsinki Zoo, Finland, pp. 93–102.

Jackson, R., Mishra, C., McCarthy, T., Ale, S., 2010. Snow leopards: conflict and conservation. In: Macdonald, D., Loveridge, A. (Eds.), Biology and Conservation of Wild Felids. Oxford University Press, New York, pp. 417–430.

Koshkarev, E., Vyrypaev, V., 2000. What happened to the snow leopard since the break-up of the Soviet Union. Snow Line 16, 1–2, 7–8.

Li, J., Schaller, G., McCarthy, T., et al., 2013. A communal sign post of snow leopards (*Panthera uncia*) and other species on the Tibetan Plateau, China. Int. J. Biodivers. 2013, 8, ID 370905.

Li, J., Wang, D., Yin, H., et al., 2014. Role of Tibetan Buddhist monasteries in snow leopard conservation. Conserv. Biol. 28 (1), 87–94.

Li, J., Lu, Z., 2014. Snow leopard poaching and trade in China 2000–2013. Biol. Cons. 176, 207–211.

Lovari, S., Ventimiglia, M., Minder, I., 2013. Food habits of two leopard species, competition, climate change and upper tree line: a way to the devrease of an endangered species. Ethol. Ecol. Evol. 25 (4), 305–318.

Matthiessen, P., 1978. The Snow Leopard. Viking Press, New York.

McCarthy, T., Chapron, G., 2003. Snow leopard survival strategy. International Snow Leopard Trust and Snow Leopard Network, Seattle.

Mishra, C., Allen, P., McCarthy, T., et al., 2003. The role of incentive programs in conserving the snow leopard. Cons. Biol. 17 (6), 1512–1520.

Oli, M., Taylor, I., Rogers, M., 1994. Snow leopard *Panthera uncia* predation on livestock: an assessment of local perceptions in the Annapurna Conservation Area, Nepal. Biol. Cons. 68, 63–68.

Rosen, T., Hussain, S., Mohammad, G., et al., 2012. Reconciling sustainable development of mountain communities with large carnivore conservation. Mt. Res. Dev. 32 (3), 286–293.

Schaller, G., 2012. Tibet Wild. Island Press, Washington, DC.

Sheikh, K., 2006. Involving religious leaders in conservation education in the Western Karakoram, Pakistan. Mt. Res. Dev. 26 (4), 312–322.

Simms, A., Moheb, Z., Salahudin, et al., 2011. Saving threatened species in Afghanistan: snow leopards in the Wakhan Corridor. Int. J. Environ. Stud. 63 (3), 299–312.

Snow Leopard Network, 2014. Snow leopard survival strategy, Revised Version 2014.1. Snow Leopard Network. Available from: www.snowleopardnetwork.org.

Preface

Biological diversity is a defining feature of planet Earth, and this diversity has numerous biological, economic, cultural, and ethical values. One species, *Homo sapiens*, has been a disproportionately effective competitor whose interactions with the environment are causing dramatic biological, chemical, and geological changes. Over seven billion humans are now consuming Earth's biological and natural resources, modifying its climate, and transforming virtually every terrestrial, freshwater, and marine ecosystem from the tropics to the Arctic and Antarctic. One impact of this growing human ecological footprint is an expanding number of species facing population declines and extinctions at a rate as great as any in Earth's history.

The challenge of protecting biological diversity has spawned an unprecedented wave of innovation and creativity as a growing number of scholars, practitioners, activists, and policy makers seek to understand and staunch biodiversity loss. Advances in conservation science, theory, technology, and our understanding of policy and human behavior are revolutionizing how we protect the world's biodiversity. The cumulative result of these conservation activities is a global renaissance in the field of conservation that is leading to more and more positive stories about the recovery and protection of threatened and endangered species, populations, communities, and ecosystems.

To address these challenges and opportunities, Academic Press, an imprint of Elsevier, has created a new series, *Biodiversity of the World: Conservation from Genes to Landscapes*. The first books in this series focus on specific taxa of broad conservation interest, with a perspective that crosses disciplinary boundaries from molecular biology to landscape ecology, and including human interactions and impacts. This series aims to provide readers with comprehensive, authoritative, and innovative coverage of interdisciplinary topics including biodiversity, behavior, evolution, ecology, conservation, and management that is accessible and affordable. The books are edited and written by prominent scholars and practitioners to illuminate and emphasize advances in conservation science and sustainable solutions. Academic Press has a distinguished 70-year legacy; Elsevier, one of the world's largest scientific publishers, is a global leader in science, technology, and health publishing.

The idea for this series had its genesis from eight books on Animal Behavior, Ecology, Conservation, and Management published by Noyes in the 1980s. This series ended with the closure of Noyes and was briefly restarted by William Andrew Publishers, which was subsequently acquired by Elsevier in 2009. Three books were published by Elsevier under the Academic Press imprint. One of these is a book I coedited with Ronald Tilson, *Tigers of the World: The Science, Politics, and Conservation of Panthera tigris*, second ed. (2010); this book has been a model for this new series.

Another inspiration for the series was an encounter I had some years ago with the eminent biologist and naturalist E.O. Wilson, Pellegrino University Professor Emeritus at Harvard University. In a meeting with my students to discuss one of his books, Professor Wilson lamented to the class that many of the university positions held by field naturalists were being redefined to hire geneticists and molecular biologists, resulting in increasing disciplinary specialization and loss of interdisciplinary perspectives. He told my students that he believed an effective method to learn about biology and conservation would be to study and develop an understanding of specific species or taxa, as he did with ants, from small to large scales across disciplinary boundaries. It is in this spirit that we launch our new series.

Snow Leopards, edited by Tom McCarthy and David Mallon, is the first volume in the series. *Snow Leopards* includes contributions from the world's authorities on the science and conservation of snow leopards and is destined to become the definitive volume on this extraordinary species. One goal for the series is to publish timely books that will be recognized internationally for their comprehensive, multidisciplinary, and authoritative treatment of specific species and taxonomic groups, particularly those of high conservation concern, and to create a distinctive repertoire by inviting scholars and practitioners with international reputations and stellar accomplishments to serve as volume editors. I cannot imagine a more appropriate volume or two better editors than Tom and David, and I am very pleased that *Snow Leopards* is our inaugural book in the series.

This series would not have been possible without the support and enthusiasm of Kristi Gomez, Senior Acquisition Editor for Life Sciences and Patricia Gonzalez, Senior Editorial Project Manager for Animal and Plant Sciences, and their colleagues at Elsevier. There is a pipeline of books scheduled to follow *Snow Leopards* that focus on similar extraordinary threatened taxa, written and edited by internationally recognized experts. I hope you share our enthusiastic anticipation for these next volumes exploring the science and conservation of the world's biological diversity from genes to landscapes.

Philip J. Nyhus, Series Editor
Biodiversity of the World: Conservation from Genes to Landscapes

Director, Environmental Studies Program,
Colby College, Waterville, ME, USA

Acknowledgments

In late 2013, we were approached by Philip Nyhus about the possibility of editing a book on snow leopards for a new series, *Biodiversity of the World: Conservation from Genes to Landscapes*, to be published by Elsevier under the Academic Press imprint. Recognizing that this would not only be the first book of the new series, but the first ever comprehensive book on snow leopards, we took little time to consider and accept the offer. As the series editor, Philip expertly guided us through the process of developing an outline of the book and a list of potential contributors to present to the publishers. It was during that early developmental stage that we realized our combined half-century of snow leopard research and conservation experience still left us far short of the knowledge required to fill a 600 page book on the species. But those years in the field had allowed us to meet and work with snow leopard experts from around the globe. It was to them, our colleagues, mentors, and friends, that we turned for assistance on this extensive undertaking. It was heartening that not one of them hesitated to join us and freely contribute their vast collective knowledge and thoughts on the animal that has been the focus of their professional lives, their degree work, and their passions. It is to them, nearly 200 contributing authors of the 47 chapters in this book, that we must express our heartfelt gratitude.

Once the book concept was accepted by Elsevier, it quickly became clear that any snow leopard knowledge deficit we had was dwarfed by our lack of savvy regarding book publishing. Thankfully we found ourselves in the good hands of Kristi Gomez, Pat Gonzalez, Lucía Pérez, and others on the staff at Elsevier who helped us at every turn. From chapter organization to editing, proofing, and final layout, their professionalism and patience made an otherwise daunting task for two field biologists, a manageable and enjoyable experience.

We have to give much credit and thanks to Rana Bayrakçısmith, Snow Leopard Program Manager with Panthera, for her meticulous copyediting of nearly the entire book. A tedious task, she sought out stray commas, ill-placed hyphens, missing citations, and erroneous cross references. The book reads better for it.

We had many professional images of snow leopards, or camera trap pictures, to choose from when planning the cover of the book. We think it most fitting that a local photographer from Leh, Ladakh, Mr Cheetah Dorjey, provided the stunning image that graces our cover and captures perfectly the mystic nature of the cat and its mountain realm.

We feel profoundly fortunate to have had this opportunity bring all that is known about snow leopards into one volume. But it must be seen as the collective work of cadre of dedicated and passionate individuals who believe the snow leopard deserves forever to be a living icon of Asia's high peaks.

Thomas McCarthy
David Mallon

SECTION I

Defining the Snow Leopard

CHAPTER

1

What is a Snow Leopard? Taxonomy, Morphology, and Phylogeny

Andrew C. Kitchener[*,**], *Carlos A. Driscoll*[†], *Nobuyuki Yamaguchi*[‡]

*Department of Natural Sciences, National Museums Scotland, Edinburgh, UK
**Institute of Geography, University of Edinburgh, Edinburgh, UK
[†]WWF Chair in Conservation Genetics, Faculty of Wildlife Sciences, Wildlife Institute of India, Dehradun, Uttarakhand, India
[‡]Department of Biological and Environmental Sciences, College of Arts and Sciences, Qatar University, Doha, Qatar

INTRODUCTION

The snow leopard is the smallest of the so-called big cats of the genus *Panthera* with a head and body length of 1–1.3 m, tail length 0.8–1.1 m, and a weight of 20–50 kg. The snow leopard is adapted to montane habitats in Central Asia, including principally the Altai, Tian Shan, Kun Lun, Pamir, Hindu Kush, Karakorum, and Himalayan mountain ranges, where it preys on ungulates, particularly blue sheep (*Pseudois* spp.), goats and ibex (*Capra* spp.), marmots (*Marmota* spp.), and lagomorphs. In order to survive in this often hostile environment, the snow leopard has evolved a suite of adaptations for combating low temperatures and hunting on steep heterogeneous slopes at high altitude. These adaptations have resulted in the snow leopard being treated as taxonomically and phylogenetically distinct from other big cats (*Panthera* spp.) in the past, but recent molecular data have confirmed its inclusion within *Panthera* as a sister species to the tiger (*P. tigris*). In this chapter we explore the morphological characteristics of the snow leopard and its adaptations in relation to its montane habitats, and we also explain its taxonomic and phylogenetic history to provide a clearer understanding of its position within the Felidae.

Snow Leopards
http://dx.doi.org/10.1016/B978-0-12-802213-9.00001-8

TAXONOMIC HISTORY AND GEOGRAPHICAL VARIATION

The snow leopard was first described by Buffon (1761) as l'Once. The snow leopard was initially described as either a felid related to the lynx or a small panther, with long hairs and long tail, occurring from North Africa through Arabia to southern Asia (Buffon, 1761). Clearly there was considerable confusion with leopards and lynxes. Schreber (1775) copied Buffon's figure of l'Once, but gave it a scientific name, *Felis uncia*. The type locality was subsequently fixed by Pocock (1930) as the Altai Mountains, although Ognev (1962) stated that it was the southern slope of the Kopet Dagh Mountains. However, this mountain range is outside the geographical distribution of the snow leopard, so that the Altai Mountains are regarded as the type locality.

The snow leopard was at first placed in *Felis* with all other cats, but Pocock (1916b) followed Gray (1854) and placed it in the genus *Uncia*, which Gray had created for *Felis irbis*, a junior synonym of *F. uncia*. However, more recently based on phylogenetic data, the snow leopard has been placed in the genus *Panthera* (see below). Few other scientific names have been proposed for the snow leopard; Ehrenberg (1830) described *F. irbis* from the Altai Mountains, but this is synonymous with *P. uncia*, while Horsfield (1855) described *F. uncioides* from Nepal, which was recognized as a subspecies, *F. uncia uncioides* by Stroganov (1962) on the basis of apparent differences in pelage, including lighter coloration and a reduction of spots compared with snow leopards from Central Asia. Hemmer (1972) found that these differences were inconsistent and considered that the snow leopard is monotypic. However, there have been no detailed molecular or craniometric analyses to investigate geographical variation.

More recently Zukowsky (1950) described *Uncia uncia schneideri* from Sikkim based on an aberrant individual, while Medvedev (2000) described *Panthera baikalensis-romanii* from the spurs of the Malkhan range in the Petrovsk-Zabaikal region, Chita Region of the River Ungo, which is said to be darker and browner and lacking rosettes, except in the lumbar region, compared with Central Asian snow leopards. However, Wozencraft (2005) included it within the synonymy of *P. uncia* without comment.

Fox (1994) highlighted the gap in distribution between the main southern "Tibetan" (i.e., Himalayan) population and the northern population in Russia and Mongolia. He suggested that these two populations may differ from each other. However, snow leopards are able to travel across more than 50 km of open steppes, not an optimal habitat for them, between isolated massifs, which suggests that fragmented populations, seemingly separated by distance and unsuitable habitat, may not be totally disconnected (McCarthy and Chapron, 2003). Further research is clearly required for understanding the intraspecific phylogeny and biogeography of snow leopards.

FOSSIL RECORD

Few fossil snow leopard remains have been found and even fewer have been dated to any degree of accuracy. Fossil remains were reported by Brandt (1871) and Tscherski (1892) from Upper Pleistocene in caves in the Altai. Other fragmentary fossils from locality 3, Choukoutien, China, have been variously identified as leopard, *Panthera pardus* or *P. uncia* (Hemmer, 1972). There is a mandibular ramus from the Siwaliks in the Natural History Museum London (Register no. 16537a), which appears to be from a snow leopard (Dennell et al., 2005). Other putative snow leopard fossils, from locality 1 at Choukoutien, China; Stranská Skála, Russia; and Woldrich, Austria, are misidentifications (Hemmer, 1972).

Dennell et al. (2005) have described a *Panthera* cf. *uncia* dated to 1.2–1.4 Mya from locality 73 in the Pabbi Hills, Pakistan (Upper Siwaliks).

Deng et al. (2011) mention an almost complete skull of an apparently ancestral snow leopard, *Panthera* sp., from the Zanda fauna of Tibet. This skull is dated to ca. 4.4 Mya (but could be as much as 5.95 Mya) and is the oldest known snow-leopard-like fossil, but it is about 10% smaller than today's animals. This specimen has recently been described as a new species, *Panthera blytheae* (Tseng et al., 2014). Although sharing characters with *P. uncia*, such as canines with an almost circular cross-section, a weakly inclined mandibular symphysis, a smooth transition between the mandibular rami and symphysis, the presence of a frontonasal depression, a narrow distance between the anterior edge of the bullae and the glenoid ridge, a sharp-turning ventral premaxilla–maxilla border at the canines, and straight and symmetrical P4 cusp alignment, *P. blytheae* can be distinguished from it and other *Panthera* by uniquely having a small labial cusp on the posterior cingulum of P3 and converging ridges on the labial surface of P4. Phylogenetic analyses of skull characters place *P. blytheae* as a sister species to *P. uncia* in the same clade as the tiger, *P. tigris* (Tseng et al., 2014), although with the information currently available, we cannot reject the possibility that *P. blytheae* is an earlier species of the genus *Panthera* without a close phylogenetic relationship to *P. uncia* in particular.

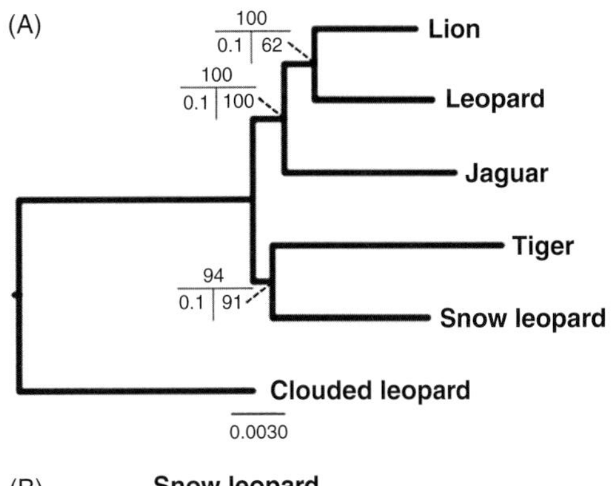

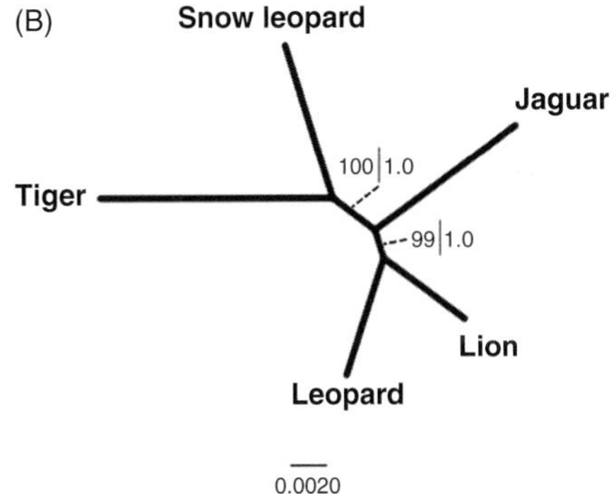

FIGURE 1.1 **Maximum likelihood (ML) tree based on analysis of the complete supermatrix.** (A) Rooted with clouded leopard as outgroup. 1000 ML bootstrap replicate percentages depicted on the top, Bayesian posterior probabilities (BPP) on the bottom left, and BEST posterior probabilities on the bottom right. (B) Unrooted topology with ML bootstrap percentages on the left and BPP on the right (Davis et al., 2010; reproduced with permission).

PHYLOGENY

In the most comprehensive genetic phylogeny to date, Davis et al. (2010) inferred the relationships among *Panthera*, including clouded leopard (*Neofelis nebulosa*), snow leopard, tiger, jaguar (*P. onca*), leopard (*P. pardus*), and lion (*P. leo*), using 39 Y chromosome segments, 3 autosomal genes, and 4 mitochondrial genes in a supermatrix phylogenetic analysis (Fig. 1.1). An independent inference, using the major urinary protein, transthyretin, recapitulated an identical topology. Transthyretin is putatively involved in male scent marking and is speculated to be a speciation protein.

Molecular dating, using a Bayesian relaxed clock approach, indicates that the snow leopard and the tiger diverged from each other 2.70–3.70 Mya, which is prior to the divergence of the jaguar (2.56–3.66 Mya) from the lion/leopard ancestor, and of that ancestor into those two species (1.95–3.10 Mya). Thus, in this analysis, the snow leopard is a sister species to the tiger, however distantly related, with the caveat that the tiger is the closest relative among extant taxa used in this analysis (i.e., excluding *P. blytheae*, *P. palaeosinensis*; see Christiansen, 2007). Based on the assumption that *P. blytheae* and *P. uncia* are sister taxa, Tseng et al. (2014) suggest that the divergence between the snow leopard and tiger lineages took place 4.86–5.13 Mya, making this the oldest divergence date estimated so far.

MORPHOLOGICAL ADAPTATIONS

The snow leopard has a number of morphological adaptations for living and hunting at high elevations in montane habitats. These are reviewed next.

Pelage

The snow leopard has the longest and densest pelage of any *Panthera*, with 4000 hairs per square centimeter, and a ratio of 8 underfur hairs to every guard hair (Heptner and Sludskii, 1992). The dense underfur is long, 43 mm cf. guard hairs 50 mm long. Hemmer (1972) compared summer and winter fur lengths (Table 1.1). The long guard hairs and thick underfur are effective at trapping a layer of air close to the skin for insulation.

The coat pattern of the snow leopard (rosettes) differs from that of the sister taxon, the tiger (stripe), in comparison to the other closely related taxa within the genus *Panthera* (lion – usually juveniles only – jaguar and leopard), all of which have rosettes. This may indicate that the basal primitive coat pattern of the family Felidae is flecks (i.e., dots), from which nearly all other patterns have developed within a relatively short period without involving any non-flecked patterns in between (Allen et al., 2011; Werdelin and Olsson, 1997). The dorsal pelage ranges from pale gray to creamy smoke gray, often tinged with brown or even reddish brown, and is marked with a pattern of solid brownish black spots on the head, neck, and lower limbs, while large rosettes or rings (50–90 mm diameter), often with a few small spots inside, are found on the flanks and tail, but the density of rosettes is lower than in the pelages of leopards and jaguars (Hemmer, 1972). A row of elongated spots and two lateral rows of elongated rings run along the midline of the back to the base of the tail (Hemmer, 1972). The underparts are whitish from the chin to the anus. The basic ground color, coupled with the disruptive effect of the markings, helps match snow leopards to their rocky environment and break up their outline for effective stalking and hunting.

TABLE 1.1 Fur Lengths (Millimeters) in Summer and Winter on Different Parts of the Body of Snow Leopards

Body parts	Summer	Winter
Flanks	25	50
Tail	50	60
Belly	50	120
Back	–	30–55

From Hemmer, H., 1972. Uncia uncia. *Mammal. Species 20, 1–5.*

Skull

The snow leopard's skull is typical for cats, with short jaws for a powerful killing bite and a large cranium for the attachment of large temporalis muscles. However, in comparison to other *Panthera* species, the skull of the snow leopard is easily distinguishable (Fig. 1.2) and it is also the smallest; the greatest length of skulls for males is 174.0–196.1 mm in males ($n = 9$) and 175.0–178.8 mm in females ($n = 6$) (Heptner and Sludskii, 1992; NY, unpublished data). The skull is broader, shorter, and more vaulted than in other *Panthera*, and particularly elevated between postorbital processes (Haltenorth, 1937; Pocock, 1916b) because of the inflated nasal cavity and broad nasal bones. The large nasal cavity probably allows for efficient countercurrent warming of inhaled air and cooling of exhaled air when breathing. Schauenberg's index (Schauenberg, 1969; greatest length of skull/cranial volume) suggests that *P. uncia* has a relatively large brain along with the other four *Panthera* species, *Puma concolor*, *Acinonyx jubatus*, and *Lynx lynx* amongst 38 species of felids that were investigated (Schauenberg, 1971).

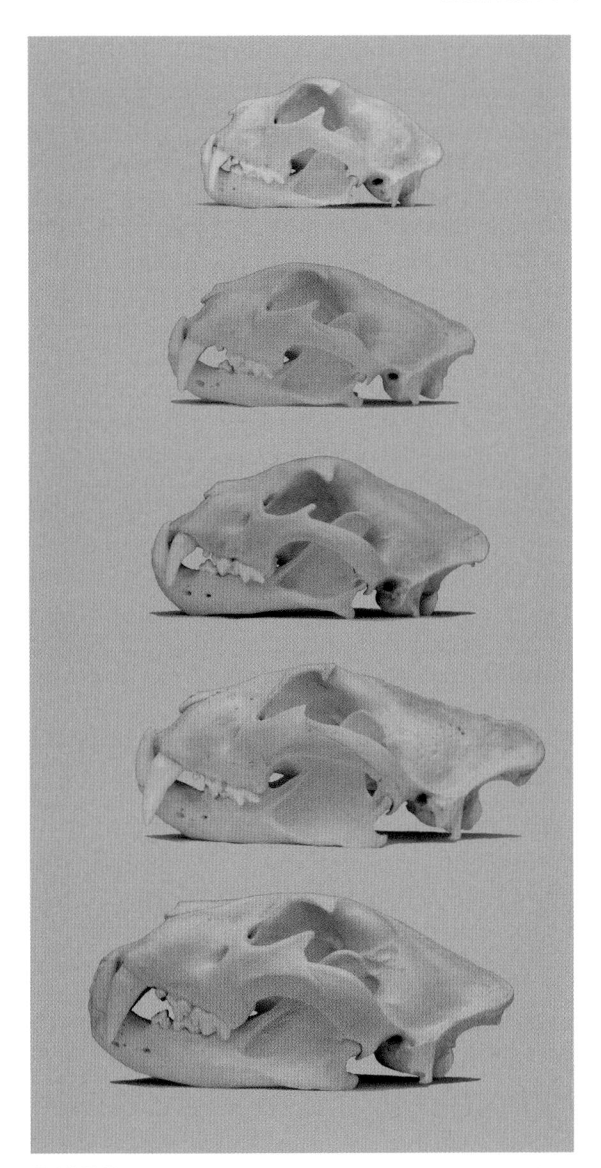

FIGURE 1.2 **A comparison of the skulls of *Panthera* cats, all lateral views, from the top: snow leopard, leopard, jaguar, tiger, lion.** Note the small, highly vaulted skull of the snow leopard (N. Yamaguchi).

In comparison with other cats, excluding the cheetah (*A. jubatus*), the snow leopard has a larger than expected nasal aperture relative to skull length and palate width (Torregrosa et al., 2010), which allows not only for a greater volume and density of turbinals for warming and humidifying inhaled cold dry air, but also for extracting as much oxygen as possible by increasing the volume of each breath (Hemmer, 1972; Torregrosa et al., 2010). However, in relation to body mass, nasal aperture size is not greater than expected so that skull length and palate width are reduced, or skull width is increased to allow for a larger nasal cavity (Hemmer, 1972; Torregrosa et al., 2010).

Teeth and Jaws

The snow leopard has significantly more slender canines along the anteroposterior axis than the other *Panthera* species, except the leopard (Christiansen, 2007). However, its canines are less blade-like, and hence, they have a rounder cross-section along the entire crown, similar to the jaguar's, but the middle of the crown is less rounded compared to that of the lion. Christiansen (2007) estimated average bite forces at the canine tip for various large cats (Fig. 1.3; Table 1.2).

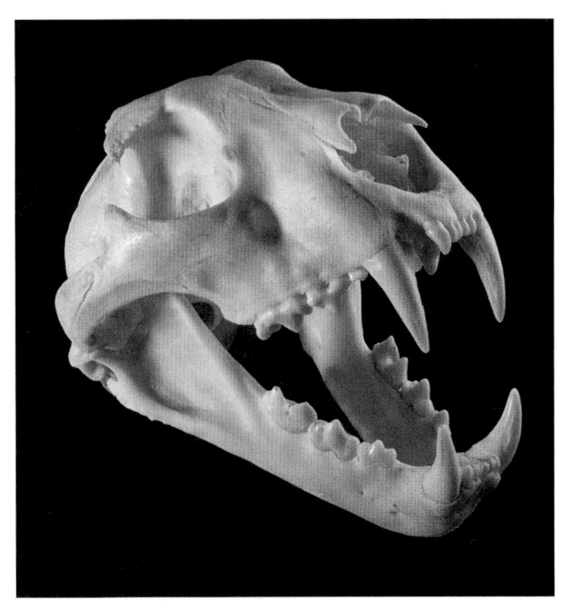

FIGURE 1.3 **Skull of a snow leopard showing dentition (National Museums Scotland).**

TABLE 1.2 Average Estimated Bite Forces at the Canine Tips of Large Cats

Species	*n*	Average bite force (*N*)
Snow leopard	9	363.0
Tiger	14	1234.3
Lion	10	1198.6
Jaguar	9	879.5
Leopard	8	558.6
Clouded leopard	12	344.2
Puma	10	499.6

From Christiansen, P., 2007. Canine morphology in the larger Felidae: implications for feeding ecology. Biol. J. Linn. Soc. 91, 573–592.

The average estimated bite force is the lowest for any *Panthera* and reflects the smaller body size of this species, but its bite force is much lower than the similar-sized puma (*P. concolor*) and equivalent to that of the smaller clouded leopard. The snow leopard's moderate canines are appropriate for killing small to medium-sized prey, although it is not apparent why they should have a more robust circular cross-section, given that a throat bite is typically used for killing larger prey, such as goats and blue sheep (Sunquist and Sunquist, 2002). Circular cross-sections suggest that forces may act from any direction; perhaps the difficulties of dealing with prey on steep heterogeneous slopes means that the direction of forces acting on the canine is more unpredictable than for the killing bites of other *Panthera*.

Compared with other *Panthera* cats, the snow leopard has a jaw gape of more than 70°, only slightly less than the clouded leopard's when measured from their skulls (Christiansen and Adolfsen, 2005), although it is possible that they can achieve an even wider gape than this in life. It is unclear why this is so, but perhaps the relatively large prey of snow leopards (i.e., montane ungulates), which have a wide throat or nape for killing bites, requires the snow leopard to have a wider gape for a throat bite.

Limbs and Vertebral Column

In terms of limb proportions, the snow leopard most resembles the cheetah, which is an open-country pursuit predator (Gonyea, 1976). The humeroradial index (94.6%) is only slightly less than that of the lion (98.3%) and the cheetah (103.3%), while the femorotibial index (105%) matches that of the cheetah, indicating longer lower limbs for a longer stride and potentially higher running speeds. The intermembral index is only 84.7% and falls within values for other large cats. The snow leopard's hunting behavior has been recorded on film in recent years and indicates that from an ambush it can display rapid acceleration and pursuit of bovid prey, with long leaps and sharp turns. The relatively longer tibiae would allow for more effective leaping, which is also supported by the relatively long thoracic (42.4% presacral vertebral length) and lumbar (35.6%) segments of the vertebral column, which ranked second among those of all large felids, thus allowing for more flexibility in leaping and turning.

Rieger (1984) mentioned a muscle, the musculus endopectoralis (= pectoralis major), which runs from the posterior sternum to the distal humerus, and apparently acts as a "spring" when a jumping mammal lands. Among felids the pectoralis major has the highest relative weight, emphasizing its importance in absorbing energy when landing after leaping. Snow leopards have apparently been recorded leaping as far as 15 m across a gorge (Ognev, 1962).

Tail

The snow leopard also has a long tail (75–90% of head-and-body length; Hemmer, 1972; mean 83% in 13 males and mean 82.2% in 15 females, ACK, unpublished data), which acts as a balancing organ (Rieger, 1984), when leaping between rocks and ascending or descending steep slopes, especially while moving rapidly in pursuit of prey. The tail is also used as a muffler to insulate paws and head from the cold at high altitude when resting (Rieger, 1984).

Laryngeal Anatomy

The genus *Panthera*, including the snow leopard, is characterized by the epihyal bone of the hyoid complex being replaced by an elastic ligament that allows the larynx to move away from the pharynx and, hence, permits roaring and other loud vocalizations. This anatomical feature is usually cited as being of taxonomic significance (Pocock, 1916a) and explains why *Panthera* cats roar and other cats cannot. Peters and Hast (1994) showed that most *Panthera* cats have large vocal folds with large fibroelastic pads, which, they hypothesized, vibrate to produce low-frequency sounds that are amplified by the long larynx, and bell-shaped pharynx and mouth, resulting in loud low-frequency vocalizations, including roars. The elastic epihyal allows the larynx to lower the formant frequencies of vocalizations and does not affect a cat's ability to purr, as thought previously (Weissengruber et al., 2002). However, snow leopards have the small pointed vocal folds of smaller cats and are unable to roar, although they produce long moaning calls to partners, grunts and moans, and they can apparently purr when inhaling and exhaling like smaller cats (Hemmer, 1972).

Physiological Adaptations

Living at high elevations, it would be expected that snow leopards should show physiological adaptations for breathing air with low levels of oxygen. However, surprisingly little information is available. Marma and Yunchis (1968) found that like other montane mammals, they have small red blood cells (RBCs) (mean 5.5 μm diameter, range 4.73–6.15 μm; cf. tiger 7.3 ± 0.45 μm; Shrivastav and Singh, 2012), a high concentration of hemoglobin (16.4 g%), high hematocrit value (47%; relative volume of RBCs), and a large number of RBCs (14.1–16.8 million/mm^3). A high concentration of small RBCs with a high surface-area-to-volume ratio probably helps the snow leopard to extract sufficient oxygen while breathing at high elevations. Recently it has been shown in studies of the human genome that two loci, EGLN1 (Egl nine homolog 1) and EPAS1 (endothelial PAS domain-containing protein 1) are involved in mediating physiological adaptation to high altitude (Cho et al., 2013). A recent comparison of whole genomes between members of the *Panthera* showed that the snow leopard has a specific genetic determinant in EGLN1 (Met39 (nonpolar)→Lys39; positively charged), which is probably also associated with physiological adaptation to high altitude (Cho et al., 2013). EGLN1 is typically highly conserved in mammals, so this change in the snow leopard genome may alter protein function. Also, two changes specific to snow leopards have been recorded in EPAS1 (Ile663 and Arg794); the latter was predicted to bring about a functional change of this protein (Cho et al., 2013). Further studies are required to determine how these changes to the snow leopard's genome provide adaptation to high altitude. However, recently Janecka et al. (2015) have investigated the ability of snow leopard hemoglobin to bind oxygen. Typically cats have hemoglobin with a low oxygen-binding affinity and reduced sensitivity to the allosteric cofactor 2,3–diphosphoglycerate (DPG), and the snow leopard is surprisingly no exception. Further studies are required to determine compensatory mechanisms in the oxygen transport pathway that allow snow leopards to show extreme hypoxia intolerance at altitudes of up to 6,000 meters or more.

CONCLUSIONS

Owing to its rarity in the wild and in museum collections, the snow leopard's anatomy and physiology have not been well studied in comparison with those of most other *Panthera* cats. The current large captive population offers good opportunities for studying these aspects from studies of living and dead animals. However, from what we do know, the snow leopard is adapted for leaping and turning to capture ungulate prey on steep mountain slopes, it is well insulated and camouflaged to survive in this cold environment and it has anatomical and physiological adaptations that allow it to maximize oxygen extraction from the low levels at high elevations while conserving heat energy. We recognize the potential for further areas of fruitful research, including relative lung volume in relation to low oxygen levels, eye anatomy in relation to the snow leopard's diurnal behavior, and whether there is any significant geographical variation in morphology and genetics, which could affect the snow leopard's conservation.

References

Allen, W.L., Cuthill, I.C., Scott-Samuel, N.E., Baddeley, R., 2011. Why the leopard got its spots: relating pattern development to ecology in felids. Proc. Royal Soc. B 278, 1373–1380.

Brandt, F., 1871. Neue Untersuchungen über die in den altaischen Höhlen aufgefundenen Säugethierreste, ein Betrag zue quaternären Fauna des Russischen Reiches. Bull. Acad. Imp. Sci. St. Pétersb. 15, 147–205.

Buffon, G.-L., 1761. La panthére, l'once et le léopard. In: Buffon, G.-L., Comte de, L.C., Daubenton, L.J.M. (Eds.), Histoire Naturelle, Générale et Particuliére avec la Description du Cabinet du Roi, vol. 9, De l'Imprimière Royale, Paris, pp. 151–172, pl. 13.

Cho, Y.S., Hu, L., Hou, H., Lee, H., Xu, J., Kwon, S. et al., 2013. The tiger genome and comparative analysis with lion and snow leopard genomes. Nat. Commun. 4 (2433), 1–7.

Christiansen, P., 2007. Canine morphology in the larger Felidae: implications for feeding ecology. Biol. J. Linn. Soc. 91, 573–592.

Christiansen, P., Adolfsen, J.S., 2005. Bite forces, canine strength and skull allometry in carnivores (Mammalia, Carnivora). J. Zool. (London) 266, 133–151.

Davis, B.W., Li, G., Murphy, W.J., 2010. Supermatrix and species tree methods resolve phylogenetic relationships within the big cats, *Panthera* (Carnivora: Felidae). Mol. Phylogenet. Evol. 56 (1), 64–76.

Deng, T., Wang, X., Fortelius, M., Li, Q., Wang, Y., Tseng, Z.J., Takeuchi, G.T., Saylor, J.E., Säilä, L.K., Xie, G., 2011. Out of Tibet: Pliocene woolly rhino suggests high-plateau origin of Ice Age megaherbivores. Science 333 (6047), 1285–1288.

Dennell, R.W., Turner, A., Coard, R., Beech, M., Anwar, M., 2005. Two Upper Siwalik (Pinjor Stage) fossil accumulations from localities 73 and 362 in the Pabbi Hills, Pakistan. J. Palaeontol. Soc. (India) 50, 101–111.

Ehrenberg, M.C.G., 1830. Observations et données nouvelles sur le tigre du nord et la panthère du nord, recueillies dans le voyage de Sibérie fait par M.A. de Humboldt, en l'année 1829. Ann. Sci. Nat. 21, 387–412.

Fox, J.L., 1994. Snow leopard conservation in the wild – a comprehensive perspective on a low density and highly fragmented population. In: Fox, J.L., Jizeng, D. (Eds.), Proceedings of the Seventh International Snow Leopard Symposium. July 25–30, 1992, International Snow Leopard Trust, Seattle, Xining, Qinghai, China, pp. 3–15.

Gonyea, W.J., 1976. Adaptive differences in the body proportions of large felids. Acta Anat. 96, 81–96.

Gray, J.E., 1854. The ounces. Ann. Mag. Nat. Hist. Ser. 2 14, 394.

Haltenorth, T., 1937. Die Verwandtschaftiliche Stellung der Großkatzen zueinander VII. Zeitschrift für Säugetierkunde 12, 7–240.

Hemmer, H., 1972. *Uncia uncia*. Mammal. Species 20, 1–5.

Heptner, V.G., Sludskii, A.A., 1992. Mammals of the Soviet Union. Part 2 Carnivora (hyaenas and cats), vol. II, E.J. Brill, Leiden, Netherlands.

Horsfield, T., 1855. Brief notices of several new or little-known species of Mammalia, lately discovered and collected in Nepal, by Brian Houghton Hodgson. Ann. Mag. Nat. Hist. Ser. 2 16, 101–114.

Janecka, J., Nielsen, S.S.E., Andersen, S.D., Hoffmann, F.G., Weber, R.E., Anderson, T., et al., 2015. Genetically based low oxygen affinities of felid hemoglobins: lack of biochemical adaptation to high-altitude hypoxia in the snow leopard. J. Exp. Biol. 218, 2402–2409.

Marma, B.B., Yunchis, V.V., 1968. Observation on the breeding, management and physiology of snow leopards, *Panthera u. uncia*, at Kaunas Zoo from 1962 to 1967. Int. Zoo Yearb. 8, 66–73.

McCarthy, T.M., Chapron, G. (Eds.), 2003. Snow Leopard Survival Strategy. International Snow Leopard Trust and Snow Leopard Network, Seattle.

Medvedev, D.G., 2000. Morfologicheskie otlichiya irbisa iz Yuzhnogo Zabaikalia. Vestnik Irkutskoi Gosudarstvennoi sel'skokhozyaistvennoi akademyi, vypusk 20, 20–30.

Ognev, S.I., 1962. Mammals of the USSR and adjacent countries, vol. III Carnivora (Fissipedia and Pinnipedia). Israel Program Scientific Translations, Jerusalem.

Peters, G., Hast, M.H., 1994. Hyoid structure, laryngeal anatomy, and vocalization in felids (Mammalia: Carnivora; Felidae). Zeitschrift für Säugetierkunde 59 (2), 87–104.

Pocock, R.I., 1916a. On the hyoidean apparatus of the lion (*F. leo*) and related species of Felidae. Ann. Mag. Nat. Hist. Ser. 8 18, 222–229.

Pocock, R.I., 1916b. On the tooth-change, cranial characters and classification of the snow-leopard or ounce (*Felis uncia*). Ann. Mag. Nat. Hist. Ser. 8 18, 306–316.

Pocock, R.I., 1930. The panthers and ounces of Asia. Part II. J. Bombay Nat. Hist. Soc. 34, 307–336, pls. VII–XIII.

Rieger, I., 1984. Tail functions in ounces, *Uncia uncia*. Blomqvist, L. (Ed.), International Pedigree Book of Snow Leopards, *Panthera uncia*, 4, Helsinki Zoo, Helsinki, pp. 85–97.

Schauenberg, P., 1969. L'identification du chat forestier d'Europe *Felis s. silvestris* Schreber 1777 par une méthode ostéométrique. Revue Suisse de Zoologie 76, 433–441.

Schauenberg, P., 1971. L'indice crânien des Félidés. Revue Suisse de Zoologie 78, 317–320.

Schreber J.C.D., 1775. Die Säugethiere in Abbildungen nach der Natur mit Beschreibungen, vol. 2(14), Wolfgang Walther, Erlangen.

Shrivastav, A.B., Singh, K.P., 2012. Tigers blood: Haematological and biochemical studies. In: Moschandreou, T.E. (Ed.). Blood Cell – An Overview of Studies in Hematology, pp. 229–242. INTECH Open Access Publisher. http://dx.doi.org/10.5772/50360.

Stroganov S.S.U., 1962. Zveri Sibiri: Khishchnye. Akademiya Nauk SSSR, Moscow, (In Russian).

Sunquist, M., Sunquist, F., 2002. Wild Cats of the World. Chicago University Press, Chicago.

Torregrosa, V., Petrucci, M., Pérez-Claros, J.A., Palmqvist, P., 2010. Nasal aperture area and body mass in felids: ecophysiological implications and paleobiological inferences. Geobios 43 (6), 653–661.

Tscherski, J.D., 1892. Wissenschaftliche Resultate der von der kaiserlichen Akademie der Wissenschaften zur Erforschung des Janalandes unde der Neusibirischen Inslen in den Jahren 1885 und 1886 ausgesandten Expedition. IV.: Beschreibung der Sammlung posttertiärer Säeugetiere. Mém. l'Acad. Imp. Sci. St. Pétersbourg, 7th ser. 40 (1), 1–511.

Tseng, Z.J., Wang, X., Slater, G.J., Takeuchi, G.T., Li, Q., Liu, J., Xie, G., 2014. Himalayan fossils of the oldest known pantherine establish ancient origin of big cats. Proc. Royal Soc. B 281 (1774), , 20132686.

Weissengruber, G.E., Gerhard Forstenpointner, G., Peters, G., Kübber-Heiss, A., Fitch, W.T., 2002. Hyoid apparatus and pharynx in the lion (*Panthera leo*), jaguar (*Panthera onca*), tiger (*Panthera tigris*), cheetah (*Acinonyx jubatus*) and domestic cat (*Felis silvestris* f. *catus*). J. Anat. 201, 195–209.

Werdelin, L., Olsson, L., 1997. How the leopard got its spots: a phylogenetic view of the evolution of felid coat patterns. Biol. J. Linn. Soc. 62, 383–400.

Wozencraft, W.C., 2005. Order Carnivora. In: Wilson, D.E., Reeder, D.M. (Eds.), Mammal Species of the World. A Taxonomic and Geographic Reference, vol. 1, third ed. Johns Hopkins University Press, Baltimore, MD, pp. 532–722.

Zukowsky, L., 1950. Grossäuger, die Hagenbeck entdeckte. Der Zoologischer Garten (N.F.) 17, 211–221.

CHAPTER

2

What is a Snow Leopard? Behavior and Ecology

Joseph L. Fox, Raghunandan S. Chundawat***

*Natural & Environmental Sciences Department, Western State Colorado University, Gunnison, CO, USA

**Bagh Aap Aur Van (BAAVAN), New Delhi, India

INTRODUCTION

From the first author's initial encounter with snow leopard sign in Nepal in the early 1970s, through surveys and research on the species by both authors in India in the 1980s and 1990s, and a compilation of the first review of snow leopard (*Panthera uncia*) ecology (Fox, 1989), the development of our knowledge of snow leopard behavior and ecology over the past 25 years has grown substantially. Given the logistical challenges of the snow leopard's rugged and remote habitat, as well as its secretive habits, progress is slow but impressive. Recent research encompassing satellite telemetry (see Chapter 26) and noninvasive methods (see Chapters 27.1 and 27.2) has contributed immensely, and within the past decade or so such techniques have resulted in dramatic progress. This chapter updates the original 1989 review, presenting current information on snow leopard behavior and ecology in light of previous work. Zoo-based studies from the 1960s and 1970s still provide our basic knowledge of many snow leopard behavioral and ontogenetic characteristics. Because snow leopards are not subjected to consistent direct observation in the wild, research has continued to be focused on population characteristics, ecological parameters, and diet. Our review here encompasses the first sign-based surveys and anecdotal observations from the 1970s, to the first radio telemetry study in the early 1980s, through the development of satellite telemetry, camera-trapping techniques, and genetic analyses, overall illustrating some 50 years' work on the behavior and ecology of this elusive large predator.

GENERAL PHYSICAL CHARACTERISTICS

Adult snow leopards measure 100–130 cm from nose to tail, with a tail length of 80–100 cm (ca. 80% of body length), the longest relative to body size of any felid (Nowak, 1999). Average shoulder height is about 60 cm. Whereas Hemmer (1972)

Snow Leopards
http://dx.doi.org/10.1016/B978-0-12-802213-9.00002-X

reported an overall weight range of 25–75 kg (presumably captive animals), subsequent reports of healthy wild animals place the adult (>3 years) range at 36–52 kg (with only males reaching the 43–52 kg range), and subadults (2–3 years) at 21–40 kg (Chundawat, 1992; Jackson et al., 1990; Johansson et al., 2013; McCarthy et al., 2005; Oli, 1994). Note that McCarthy et al. (2005) monitored for several years an old (9–11 years) adult female in poor condition (worn/broken teeth and other physical injuries) whose weight was only ca. 21 kg. The size of adult foot pads, as reflected in pug-mark dimensions, is 9–10 cm in length and a 7–8 cm width, with front paws slightly larger than rear ones (Fox, 1989). Two cubs in captivity averaged pugmark sizes of 5 × 5 cm (length x width) at age 4 months, 7 × 6 cm at 9 months, and 8 × 7 cm at 12 months (Fox, 1989). At present the limited data set available does not demonstrate a significant latitudinal difference in snow leopard size/weight or appendage size, as could be expected from Bergmann's Rule or Allen's Rule, over the species' ca. 3000 km north–south range. In any case, the altitudinal differences in habitat along this gradient (discussed later) may compensate for latitudinal climate differences.

The coat of the snow leopard is white to cream-yellow in background color, mottled with gray to black spots and rosettes. Compact elongated spots and two lateral rows of elongated rings are frequently consolidated into solid black stripes in young animals, which break up into large spots as the animal grows (Hemmer, 1972). Spots on the head and neck are solid, whereas larger rings or rosettes, most enclosing smaller spots, occur on the body and tail. Snow leopards can be recognized individually by their facial spot patterns (Blomqvist and Nystrom, 1980), and such identification from facial and other body parts using remote camera captures has recently begun to be used to assess population characteristics in the wild (Jackson et al., 2006; Sharma et al., 2014). The snow leopard's coat is long and thick (longer hair present in winter) with molt occurring twice yearly (Novikov, 1956), which can influence spot patterns and individual recognition (Jackson et al., 2006). In summer, belly and tail hair length is about 50 mm and hair on the flanks is about 25 mm; in winter hair on the back is 30–55 mm, on the sides about 50 mm, on the belly up to 120 mm, and on the tail up to 60 mm (Hemmer, 1972).

BEHAVIOR

Group Composition and Spatial Distribution

Adult snow leopards are generally solitary, although groups of 2–4 may form during the breeding season or with the birth of cubs (Fox et al., 1988; Jackson and Ahlborn, 1988; McCarthy et al., 2005; Schaller, 1977). Adult females and cubs stay together for about 1–2 years (Novikov, 1956). Early ground-based radio telemetry research suggested a lack of territoriality in snow leopards (Jackson and Ahlborn, 1988; McCarthy et al., 2005), but subsequent work using satellite telemetry and larger samples have shown strong within-sex territoriality, with overlap occurring primarily between adult males and adult females and for subadults (O. Johansson, Swedish University of Agricultural Sciences, unpublished data). Individuals apparently maintain separation by actively avoiding each other, probably facilitated through scent-marking, scraping, and the deposition of other sign, with such marking densest in overlapping core activity areas.

Movement and Aggressive Behavior

Snow leopard walking, running, leaping, and climbing movements are as in the closely related members of the *Pantherinae,* with an especially highly developed jumping ability (Hemmer, 1972; Ognev, 1935). The long tail appears to be important for balance in rugged terrain, in thermoregulation, and in a number of communication functions that indicate current mood to conspecifics (Rieger, 1984). Zoo-based study has shown that aggressive threats and attack postures of the snow leopard are similar to those in other large Felidae (Leyhausen, 1979).

Early ground-based tracking research in western Nepal and Ladakh reported daily snow leopard movements of up to 7 km and average straight-line distance movements of 0.8 km per day (Chundawat, 1992; Jackson and Ahlborn, 1988). In contrast, in a less rugged study area in Mongolia, movements were substantially greater, commonly over 12 km daily (McCarthy et al., 2005), and satellite telemetry from the latter and more recent work (O. Johansson, Swedish University of Agricultural Sciences, unpublished data) indicates even greater average movement and subsequent home range size (see later sections).

Vocalizations

The vocal repertoire of the snow leopard is similar to other Felidae. Vocalizations include the nonaggressive prusten (a puffing sound emitted through the nostrils), mew calls, the mew/main call, copulatory hissing, growling (females) and a loud cry (males), and agonistic spitting, hissing, growling, and screaming/roaring (Hemmer 1972; Peters, 1980). Snow leopards, however, do not roar like other Pantherinae due to differences in their larynx morphology (Nowak, 1999). The mew/main call, associated primarily with the breeding period, and the copulatory cry or yowl are probably the vocalizations that can most easily be heard in the wild. In western Nepal, Jackson and Ahlborn (1988) reported having occasionally heard the characteristic mew/main call during the January–March breeding season, usually in late evening.

Marking Behavior

Marking behavior includes scraping, claw raking, spraying (squirting urine and scent), and cheek/head rubbing (Ahlborn and Jackson, 1988; Blomqvist and Sten, 1982; Chundawat, 1992; Freeman, 1983; Rieger, 1978, 1980; Wemmer and Scow, 1977). Snow leopards scrape their hind feet over horizontal surfaces (usually of loose material), sometimes urinating on the pile of material formed behind the scrape (Rieger, 1978), and these scrapes are made at sites with specific topographical characteristics (usually near cliffs or boulders) along well-used travel routes (see next paragraph). Where trees are present, isolated trees or ones along travel routes or crossroads are sometimes marked by snow leopards raking their claws vertically along the trunks (Ahlborn and Jackson, 1988). Snow leopards spray mark by elevating their tails vertically upright and squirting urine and scent backwards and up against near-vertical surfaces, the underside of rock overhangs, or bushes. Captive males spray-marked more than females, and both sexes scraped equally often (Rieger, 1978), but we do not know about such sex-related differences in marking frequency in the wild. Snow leopards rub their cheeks/heads against odoriferous surfaces such as spray marks, meat, or plants (Rieger, 1978), sometimes leaving behind hairs that can be used for genetic identification.

In wild snow leopards, both sexes (age > 1.5 years) commonly make scrapes and spray marks, scraping being more common, and such marking occurs most frequently during January–March (Ahlborn and Jackson, 1988). Scrape dimensions averaged 36 cm total length, 20 cm pit length, 19 cm pit width, 5 cm pit depth, and 6 cm height of scraped up material (Fox et al., 1988). Detectable urine marking of scrapings was present at about 19% of scrape sites (Ahlborn and Jackson, 1988). In western Nepal, scrape sites were frequently reused; about 50% of scrape sites were single, but only 15% of all scrapes occurred at these sites (Ahlborn and Jackson, 1988). New scraping occurred on about 86% of snow leopard visits along study transects, and fresh snow leopard scrapes occurred with a frequency of 0.7–1.0/km along travel routes during late winter (Ahlborn and Jackson, 1988; Fox et al., 1988). On well-used snow leopard travel routes in west Nepal, the maximum frequency of scrapes was 235/km. In long-distance surveys of snow leopard signs in northwest India, the frequency of scrapes ranged as high as 37/km of valley bottom transects (Fox et al., 1991). Most snow leopard feces were found at scrape sites, predominantly at well-used locations, but sympatric carnivores such as foxes and wolves also deposit feces at such "signpost" sites (Janečka et al., 2008; Li et al., 2013). Fresh feces were found at a frequency of 0.08/km along snow leopard travel routes (Ahlborn and Jackson, 1988). But again note that recent genetic identification of feces has shown that, even with experienced observers, misidentification of scats by up to 40% can occur, most frequently misidentifying fox scats as those of snow leopard (Jackson et al., 2006). Frequency of all markings depends on snow leopard population density, age/sex composition, as well as terrain characteristics (Ahlborn and Jackson, 1988; Fox et al., 1991).

Activity, Social, and Sexual Behavior Patterns

Captive male and female snow leopard pairs were found to be inactive about 70% of daylight hours (Freeman, 1982, 1983). About 15% of the time was spent walking, 6% grooming or other active social contact, 4% sniffing and flehmen on either objects or mate, and the remaining 5% in marking activities, vocalizations (primarily prusten), or back-rolling (Freeman, 1982, 1983). Aggressive behaviors accounted for only a trace of observation time. Males performed social grooming and marking behavior (head rubbing, scraping, spraying) more frequently than females, whereas females did more back-rolling, autogrooming, and prusten vocalizing (Freeman, 1983). Increases in active social contact for both sexes, back-rolling in females, and a decrease in marking by females were all associated with the female's estrous period (Freeman, 1983). In the wild in a southwestern Mongolia study site, year-round data based on animals with activity-sensing collars revealed more activity at night (51% of the time) than during the day (35%), and a tendency toward crepuscular maximums (McCarthy et al., 2005).

Snow leopards copulate in both ventral/dorsal and dorsal/dorsal postures, although the latter is more common (Freeman, 1983). Copulation occurs over a 3–6 day period. The male usually grips the fur on the female's neck when he mounts, and at copulatory climax he gives a loud piercing yowl (Freeman, 1983). The male only irregularly performs the characteristic felid copulatory bite, but usually utters the yowl (Rieger and Peters, 1981).

Food Procurement Behavior

Snow leopards stalk and kill their prey as in other large solitary felids, and large prey items may be killed with either a nape bite or suffocation associated with a throat bite (Fox and Chundawat, 1988; Schaller, 1977). Subadult snow leopards (estimated 20 kg) can kill adult blue sheep weighing more than 55 kg (Jackson and Hillard, 1986). Schaller (1977) reported that large prey are opened and eaten beginning with the chest and forelegs or lower abdomen, leaving the digestive tract intact, but Fox and Chundawat (1988) report initial feeding on the viscera of a domestic goat. The snow leopard eats in either a squatting or lying position (Shaposhnikov, 1956), but the crouched position is more common (Fox and Chundawat, 1988; Hemmer, 1972), and interestingly is more characteristic of smaller cats. No evidence has been found that snow leopards cover the remains of their kills (Jackson and Ahlborn, 1988). Snow leopards appear to remain around kills for longer periods than do other large cats (Fox and Chundawat, 1988; Schaller, 1977), a trait making them relatively vulnerable to revenge killings by humans upset over livestock losses.

ECOLOGY AND HABITAT

Vegetation

Snow leopards inhabit the high mountains of central Asia, with a large portion of their range being predominantly treeless due to either alpine or desert conditions, at elevations of about 600–4000 m in the northern part of their range to 1800–5800 m in the southern portions (Dang, 1967; Fox et al., 1991; Jackson and Ahlborn, 1988; McCarthy et al., 2005; Ognev, 1935; Stroganov, 1962). Vegetation in snow leopard range varies from treeless alpine zones in most mountain systems, through dense forest-alpine ecotones (southern slopes of the Himalayas, western and eastern edges of the Tibetan Plateau in China and the former Soviet Republics, and in adjacent parts of southwestern Russia and northern Mongolia), to open forests and woodland habitats in relatively dry parts of many mountain systems, and scrubland and desert mountain habitats in the central portion of snow leopard range as one approaches the Gobi and Taklamakan Desert regions (Koshkarev, 1984; Mallon, 1984; Novikov, 1956; Schaller, 1977).

Terrain

Snow leopards typically occupy rugged mountainous terrain (Fox et al., 1988; Jackson and Ahlborn, 1988; Koshkarev, 1984; Mallon, 1984; McCarthy et al., 2005; Schaller, 1977), although in the central portion of their range, they are known to make some use of less rugged desert mountain foothills and to travel over flat expanses to access isolated rugged mountain masses (Johansson et al., 2015; McCarthy et al., 2005; Zhirnov and Ilyinsky, 1986). In Ladakh, India, snow leopard winter travel routes occurred on terrain averaging 24° in slope angle, and 35 m from steep cliffs or other sharp breaks in terrain; 50% of all travel route locations were within 5 m of such cliffs (Fox et al., 1988). Snow leopards in western Nepal showed preference for cliffs, areas with slopes in excess of 40°, and areas within 25 m of edges such as cliffs. Preferred bedding sites were situated on or near ridges, cliffs, and other sites with good visibility. Snow leopards preferred to move, bed, and mark along linear topographic features such as major ridgelines, bluff edges, gullies, and the base or crest of broken cliffs (Jackson and Ahlborn, 1988). In southwestern Mongolia, snow leopards used slopes of >20° much more than expected and traveled closer to terrain edges than random expectations would dictate (McCarthy et al., 2005). In areas where movements are made over relatively flat terrain, overall habitat use was still overwhelmingly in rugged terrain, and home range algorithms that account for the distribution of such terrain differences are best predictors of actual habitat use (O. Johansson, Swedish University of Agricultural Sciences, unpublished data).

Marking Site Characteristics

Scrape markings are the most abundant type of sign left by snow leopards in the wild (Schaller, 1977), concentrated in areas where snow leopards travel most, and are generally associated with rugged terrain features (Ahlborn and Jackson, 1988; Fox et al., 1991). In western Nepal and Ladakh, these areas were consistently associated with the confluence of river or stream gorges, more so along valley bottoms but also along ridgelines (Ahlborn and Jackson, 1988; Fox et al., 1988). In Ladakh, 80% of scrapes were found at the base of cliffs, the remainder being found near free boulders, trees, or slight breaks in terrain (e.g., streambanks and trailsides). The scrapes were usually found 80–100 cm from the base of cliffs, boulders, or trees. Scrape substrates are usually loose materials such as snow and sandy or gravelly soil (Fox et al., 1988). Boulders, rock outcrops, and cliff-faces comprised 91–100% of the features which were spray-marked by snow leopards, with trees also occasionally being so marked

(Ahlborn and Jackson, 1988; Fox et al., 1988). The spray marks were 80–100 cm above ground level and the orientation of the surface sprayed (usually an overhang) averaged an angle of 166° (Fox et al., 1988). Scrape sites were found at the base of 73% of spray-marked rocks, and odor on spray sites sometimes could be detected more than 60 days following spraying (Ahlborn and Jackson, 1988).

Density and Home Range

Early sign-based surveys of snow leopard presence provided estimates of densities over large areas of anywhere from less than 1 to about 3/100 km^2 (Bold and Dorzhzunduy, 1976; Koshkarev, 1984; Schaller et al., 1988). The first intensive study that compared radio telemetry results with sign abundance (Ahlborn and Jackson, 1988; Jackson and Ahlborn, 1989) provided a possible basis for determining relationships between the two. However, the first attempt, in India's Ladakh region, to estimate density on this basis (Fox et al., 1991) gave relatively low results (e.g., 1–2/100 km^2) compared to later more intensive studies in the same general area, according to estimates of 4–5/100 km^2 by both Chundawat (1992) and Jackson et al. (2006). Mallon (1988), using intensive searches for tracks after snowfall, gave what turned out to be more realistic estimates of 2–4 snow leopards per 100 km^2 in good areas of the same region. Sign density is affected by factors other than just animal abundance (e.g., terrain characteristics and season), and its relationship to snow leopard density should be interpreted with caution.

Reported snow leopard population densities from small-scale intensive studies have varied from 5–6/100 km^2 (Jackson and Ahlborn, 1989) in west Nepal, and 4–5/100 km^2 in an area of good habitat in Ladakh, India (Chundawat, 1992), to 1–2/100 km^2 in large-scale surveys and/or overall less productive habitat (Fox et al., 1988; Mallon, 1984; Schaller et al., 1988). A recent intensive 4-year study of snow leopards in a desert mountain ecosystem in southern Mongolia documented a stable snow leopard population density of 1.1–1.2 snow leopards per 100 km^2 (0.7–0.8 adults/100 km^2) (Sharma et al., 2014). Notably, this study demonstrated substantial variation in age-sex composition during this period of overall population stability (Sharma et al., 2014).

Initial ground-based radiotelemetry studies of snow leopards, in the early 1980s to mid-1990s, in the southern part of their range reported both male and female home ranges of well less than 100 km^2 (Chundawat, 1992; Jackson and Ahlborn, 1989; Oli, 1997). By the late 1990s the first such study in more marginal desert habitat of southwest Mongolia (that included isolated mountain blocks) reported home ranges of less than 150 km^2, but a satellite-collared adult male in this study showed a much larger range of 1590 km^2 and suggestions of a range of possibly up to 4500 km^2 (McCarthy et al., 2005). Recent satellite telemetry-based studies in southern Mongolia, after comparing various home range estimators to provide the most meaningful measure of this geographical variable, show an average annual adult male home range of 207 km^2 and an average adult female home range of 124 km^2 (Johansson et al., 2015).

Prey Species

The primary prey of snow leopards is wild sheep and goats whose typical habitat is the rugged terrain of mountainous regions. Smaller mammalian species (e.g., marmots, *Marmota* spp.; hares, *Lepus* spp.; pikas, *Ochotona* spp.) and various birds have been reported in snow leopard diet. Domestic livestock, primarily sheep and goats, comprise a significant component of snow leopard diet in many areas, and occasionally horses, yaks, and cattle are also taken. A comprehensive review of snow leopard prey and diet is presented in Chapter 4.

Energy/Food Requirements and Prey Choice

Early estimates of snow leopard food requirements, derived to elucidate the species' possible reintroduction requirements/consequences (Wemmer and Sunquist, 1988), and general ecology (Fox, 1989; Jackson and Ahlborn, 1984) tended to simplify prey requirements (i.e., emphasizing adult ungulate prey selection) and therefore underestimate both diet breadth and prey kill frequency. Wegge et al. (2011) refined this list of requirements to include on-site diet information and age/sex composition of ungulate prey individuals taken, to conclude kill rates of about once every 10 days in an area where livestock comprised 35–40% of snow leopard prey (ca. 4% loss of total domestic livestock numbers in the area). Recent satellite telemetry of snow leopards in an area in southern Mongolia with significant livestock presence and minimal alternative nonungulate prey shows a 73% wild versus 27% domestic ungulate prey choice and, with significant seasonal variation favoring increased livestock predation in winter (Johansson et al., 2015). The average kill frequency of every 8–9 days reflects predation on a significant number of juvenile wild ungulates and domestic animals, especially in winter (Johansson et al., 2015). Marmots (*Marmota* spp.) are reported to comprise a significant portion of snow leopard diet on the eastern Tibetan Plateau in China (Schaller et al., 1988) and in eastern Kyrgyzstan (Jumabay-Uulu et al., 2014), and need to be considered in assessing wild and domestic prey requirements in some areas. Nevertheless, some of the older diet studies may need to be reassessed in light of recent documentation of the common misidentification of scats. An array of influences on snow leopard predation choice and kill rate that affect management issues related to livestock predation (e.g., Bagchi and Mishra 2006; Johansson et al., 2015; Wegge et al., 2011) clearly need to be considered in addressing conservation issues.

Reproduction and Ontogeny

In the wild and in captivity snow leopards breed during late winter (January–March), commonly with 2–3 (rarely 1 or 4–5) cubs born in April–June following a 90–105-day gestation period. The female's estrous period lasts 2–8 days (Rieger, 1984). Captive-born snow leopard cubs weigh 0.3–0.6 kg at birth, 1–2 kg after 25 days, 3–5 kg after 50 days, and continue to gain about 1 kg every 2 weeks until they weigh 25–30 kg at age 1 year (Gaughan and Doherty, 1982; Juncys, 1964; Kitchener et al., 1975; Marma and Yunchis, 1968; O'Conner and Freeman, 1982). Young open their eyes after about 1 week, their ears after 2 weeks, walk at 2.5 weeks, retract claws at 3.5 weeks, eat voluntarily at 7 weeks, eat solid food and actively play at 8 weeks, and follow their mother at 12 weeks (Frueh, 1968; Juncys, 1964; O'Conner and Freeman, 1982). Young are weaned at about 5 months (Petzsch, 1968). The first teeth appear at about 3 weeks (Freeman and Hutchins, 1978; Frueh, 1968; Novikov, 1956), and continue to erupt in the following order (upper and lower case initials refer to upper and lower teeth): I1 or i1, I2 or i2, m1, P2, M1, I3, C or c, P4, P3, p3, p4 (Marma and Yunchis, 1968; Pocock, 1916). Sexual maturity is reached at the age of 2–3 years (Koivisto et al., 1977; Petzsch, 1968; Rieger, 1980), and females in captivity have successfully bred until about 15 years. In captivity snow leopards have lived up to 21 years (Wharton and Freeman, 1988), whereas in the wild the oldest adult recorded to date was 11 years (McCarthy et al., 2005).

Disease and Mortality

Reported snow leopard mortality in the wild has been primarily attributed to human trapping and hunting, but disease and natural mortality causes are not well-documented from wild populations (see Chapter 9). Significant mortality losses to both adult and subadult age classes have recently been reported in a stable snow leopard population in Mongolia (Sharma et al., 2014), however, again the only documented mortality cause (1 of 4) was human related. In the wild, diseased or injured snow leopards have been suggested as being more likely to prey on domestic livestock (e.g., Fox and Chundawat, 1988), but note the widespread predation on livestock reported in Nepal (Wegge et al., 2011) and southern Mongolia (Johansson et al., 2015), in the latter case showing relatively high instances of such predation by females and subadults.

References

Ahlborn, G., Jackson, R.M., 1988. Marking in free-ranging snow leopards in west Nepal: a preliminary assessment. Proc. Intl. Snow Leopard Symp. 5, 24–49.

Bagchi, S., Mishra, C., 2006. Living with large carnivores: predation on livestock by the snow leopard (*Uncia uncia*). J. Zool. (London) 268, 217–224.

Blomqvist, L., Nystrom, V., 1980. On identifying snow leopards, *Panthera uncia*, by their facial markings. Int. Ped. Book of Snow Leopards 2, 159–167.

Blomqvist, L., Sten, I., 1982. Reproductive biology of the snow leopard, *Panthera uncia*. Int. Ped. Book of Snow Leopards 3, 71–79.

Bold, A., Dorzhzunduy, S., 1976. Information on the snow leopard in the southern Gobi-Altai. Trudi Inst. Obschei I Experimentalnoi Biologii, Ulan Bator 11, 27–43, In Mongolian, Russian Summary.

Chundawat, R.S., 1992. Ecological studies on snow leopard and its associated species in Hemis National Park, Ladakh. PhD Thesis, University of Rajasthan.

Dang, H., 1967. The snow leopard and its prey. Cheetal 10 (1), 72–84.

Fox, J.L., 1989. A Review of the Status and Ecology of the Snow Leopard. International Snow Leopard Trust, Seattle.

Fox, J.L., Chundawat, R.S., 1988. Observations of snow leopard stalking, killing, and feeding behavior. Mammalia 52, 137–140.

Fox, J.L., Sinha, S.P., Chundawat, R.S., Das, P.K., 1988. A field survey of snow leopard in northwestern India. Proc. Intl. Symp. 5, 99–111.

Fox, J.L., Sinha, P., Chundawat, Das, P., 1991. Status of snow leopard *Panthera uncia* in northwest India. Biol. Conserv. 55, 283–298.

Fox, J.L., Sinha, S.P., Chundawat, R.S., 1991. The mountain ungulates of Ladakh. Biol. Conserv. 58, 167–190.

Freeman, H., 1982. Characteristics of the social behavior of the snow leopard. Int. Ped. Book of Snow Leopards 3, 117–120.

Freeman, H., 1983. Behavior in adult pairs of captive snow leopards (*Panthera uncia*). Zoo Biol. 2, 1–22.

Freeman, H., Hutchins, M., 1978. Captive management of snow leopard cubs. D. Zool. Garten (N.F.) 48, 49–62.

Frueh, R., 1968. A note on breeding snow leopards at St. Louis Zoo. Int. Zoo Yearb. 8, 74–76.

Gaughan, M.M., Doherty, J.G., 1982. Snow leopard rearing: infant development with particular emphasis on play behavior. Int. Ped. Book of Snow Leopards 3, 121–126.

Hemmer, H., 1972. *Uncia uncia*. Mammalian Species, Amer. Soc. Mammal. 20, 1–5.

Jackson, R.M., Ahlborn, G., 1984. A preliminary habitat suitability model for the snow leopard *Panthera uncia* in west Nepal. Int. Ped. Book of Snow Leopards 4, 43–52.

Jackson, R.M., Ahlborn, G., 1988. Observations on the ecology of snow leopard (*Panthera uncia*) in west Nepal. Proc. Int. Snow Leopard Symp. 5, 65–87.

Jackson, R.M., Ahlborn, G., 1989. Snow leopards (*Panthera uncia*) in Nepal – home range and movements. Natl. Geogr. Res.Explor. 5, 161–175.

Jackson, R.M., Hillard, D., 1986. Tracking the elusive snow leopard. Natl. Geogr. 169, 792–809.

Jackson, R.M., Ahlborn, G., Shah, K.B., 1990. Capture and immobilization of wild snow leopards. Int. Ped. Book of Snow Leopards 6, 93–102.

Jackson, R.M., Roe, J.O., Wangchuk, R., Hunter, D.O., 2006. Estimating snow leopard population abundance using photography and capture-recapture techniques. Wildl. Soc. Bull. 34, 772–781.

Janečka, J.E., Jackson, R., Yuguang, Z., Diqiang, L., Munkhtsog, B., Buckley-Beason, V., Murphy, W.J., 2008. Population monitoring of snow leopards using noninvasive collection of scat samples: a pilot study. Animal Conserv. 11, 401–411.

Johansson, O., Malmsten, J., Mishra, C., Lkhagvajav, P., Mc Carthy, T., 2013. Reversible immobilization of free-ranging snow leopards (*Panthera uncia*) with a combination of Medetomidine and Tiletamine-zolazepam. J. Wildl. Dis. 49, 338–349.

Johansson, O., McCarthy, T., Samelius, G., Andrén, H., Tumursukh, L., Mishra, C., 2015. Snow leopard predation in a livestock dominated landscape in Mongolia. Biol. Conserv. 184, 251–258.

Jumabay-Uulu, K., Wegge, P., Mishra, C., Sharma, K., 2014. Large carnivores and low diversity of optimal prey: a comparison of the diets of snow leopards *Panthera uncia* and wolves *Canis lupus* in Sarychat-Ertash Reserve in Kyrgyzstan. Oryx 48, 529–535.

Juncys, V., 1964. Zur fortpflanzung des schneelleoparden (*Uncia uncia*) im zoologischen garten. D. Zool. Garten (N.F.) 26, 303–306.

Kitchener, S.L., Merritt, D.A., Rosenthal, M.A., 1975. Observations on the breeding and husbandry of snow leopards (*Panthera uncia*) at Lincoln Park Zoo, Chicago. Int. Zoo Yearb. 15, 212–217.

Koivisto, I., Wahlberg, C., Muuronen, P., 1977. Breeding the snow leopard, *Panthera uncia*, at Helsinki Zoo 1967–1976. Int. Zoo Yearb. 17, 39–44.

Koshkarev, E.P., 1984. Characteristics of snow leopard (*Uncia uncia*) movements in the Tien Shan. Int. Ped. Book of Snow Leopards 4, 15–21.

Leyhausen, P., 1979. Cat Behavior (Transl. B.A. Tonkin). Garland STPM Press, New York.

Li, J., Schaller, G.B., McCarthy, T.M., Wang, D., Jiagong, D., Cai, Z., Basang, P., Lu, Z., 2013. A communal sign post of snow leopards (*Panthera uncia*) and other species on the Tibetan Plateau. Int. J. Biodivers. 2013, 370905.

Mallon, D., 1984. The snow leopard in Ladakh. Int. Ped. Book of Snow Leopards 4, 3–10.

Mallon, D., 1988. A further report on snow leopard in Ladakh. Proc. Int. Snow Leopard Symp. 5, 89–97.

Marma, B.B., Yunchis, V.V., 1968. Observations on the breeding, management and physiology of snow leopards, *Panthera u. uncia* at Kaunas Zoo from 1962 to 1967. Int. Zoo Yearb. 8, 66–74.

McCarthy, T.M., Fuller, T.K., Munkhtsog, B., 2005. Movements and activities of snow leopards in southwestern Mongolia. Biol. Conserv. 124, 527–537.

Novikov, G.A., 1956. Carnivorous mammals of the fauna of the USSR. Zool. Inst. Acad. Sci. USSR. (Eng. Trans. 1962. Israel Program for Scientific Translations, 284 p).

Nowak, R.M., 1999, sixth ed. Walker's Mammals of the World vol. I Johns Hopkins University Press, Baltimore.

O'Conner, T., Freeman, H., 1982. Maternal behavior and behavioral development in the captive snow leopard (*Panthera uncia*). Int. Ped. Book of Snow Leopards 3, 103–110.

Ognev, S.K., 1935. Mammals of USSR and adjacent countries. Vol. III. Carnivora. (Eng. trans. 1962, Israel Program for Scientific Translations, 656 p.).

Oli, M., 1994. Snow leopards and blue sheep in Nepal: densities and predator:prey ratio. J. Mammal. 75, 998–1004.

Oli, M., 1997. Winter home range of snow leopards in Nepal. Mammalia 61, 355–360.

Peters, G., 1980. The vocal repertoire of the snow leopard (*Uncia uncia,* Schreber 1775). Int. Ped. Book of Snow Leopards 2, 137–158.

Petzsch, H., 1968. Die Katzen. Urania-Verlag, Leipzig.

Pocock, R.I., 1916. On the tooth-change, cranial characteristics and classification of the snow leopard or ounce (*Felis uncia*). Ann. Mag. Nat. Hist. Ser. 8 18, 306–316.

Rieger, I., 1978. Scent marking behavior of ounces, *Uncia uncia*. Int. Ped. Book of Snow Leopards 1, 50–70.

Rieger, I., 1980. Some aspects of the history of ounce knowledge. Int. Ped. Book of Snow Leopards 2, 1–36.

Rieger, I., 1984. Oestrous timing in ounces, *Uncia uncia*, Schreger (1775). Int. Ped. Book of Snow Leopards 4, 99–103.

Rieger, I., Peters, G., 1981. Einige beobachtungen zumpaarungs und lautgebundsverhalten von irbissen, *Uncia uncia*, im zoologischen. Zeitschr. Saugetierk. 46, 35–48.

Schaller, G.B., 1977. Mountain monarchs: wild sheep and goats of the Himalaya. University of Chicago Press, Chicago.

Schaller, G.B., Ren Junrang, Qiu Mingjiang, 1988. Status of snow leopard *Panthera uncia* in Qinghai and Gansu Provinces, China. Biol. Conserv. 45, 179–194.

Shaposhnikov, F.D., 1956. The snow leopard in the western Tien Shan. Priroda 7, 113–114, (Russian).

Sharma, K., Bayrakcismith, R., Tumursukh, L., Johansson, O., Sevger, P., McCarthy, T., Mishra, C., 2014. Vigorous dynamics underlie a stable population of the endangered snow leopard *Panthera uncia* in Tost Mountains, South Gobi, Mongolia. PLoS ONE 9 (7), e101319.

Stroganov S.U., 1962. Carnivorous mammals of Siberia. Biol. Inst. Acad. Sci. USSR, Siberian Branch. (Eng. Transl., 1969, Israel Program for Scientific Translations).

Wegge, P., Shrestha, R., Flagstad, Ø., 2011. Snow leopard *Panthera uncia* predation on livestock and wild prey in a mountain valley in northern Nepal: implications for conservation management. Wildl. Biol. 18, 131–141.

Wemmer, C., Scow, K., 1977. Communication in the Felidae with emphasis on scent marking and contact patterns. How Animals Communicate. Indiana University Press, Bloomington, IN, p. 749-L760.

Wemmer, C., Sunquist, M., 1988. Snow leopard reintroductions: economic and energetic considerations. Proc. Int. Snow Leopard Symp. 5, 193–205.

Wharton, D., Freeman, H., 1988. The snow leopard in North America: captive breeding under the Species Survival Plan. Proc. Int. Snow Leopard Symp. 5, 131–136.

Zhirnov, L., Ilyinsky, V., 1986. The Great Gobi National Park – A Refuge for Rare Animals of the Central Asian Deserts. GKNT, Moscow.

CHAPTER

3

What is a Snow Leopard? Biogeography and Status Overview

*Thomas McCarthy**, *David Mallon***, *Eric W. Sanderson*[†], *Peter Zahler*[†], *Kim Fisher*[†]

***Snow Leopard Program, Panthera, New York, NY, USA**

****Division of Biology and Conservation Ecology, Manchester Metropolitan University, Manchester, UK**

[†]WCS Asia Program, Wildlife Conservation Society, New York, NY, USA

INTRODUCTION

The dearth of paleontological records provides little evidence regarding the evolution of the snow leopard (*Panthera uncia*) or its historic range, although a single fossil specimen from the Zanda Basin in Tibet may be from an ancestral snow leopard living there some 4.4–5.9 Mya (Tseng et al., 2014) (see also Chapter 1). Beyond a few Late Pleistocene remains found in caves in the Altai (Brandt, 1871; Tscherski, 1892), and samples from upper Siwalik deposits in Northern Pakistan dated to 1.2–1.4 Mya (Dennell et al., 2005), no other confirmed snow leopard fossils have been recorded. Several remains of *Panthera pardus* have been misidentified as snow leopard from sites as divergent as China to Austria (Hemmer, 1972; Thenius, 1969). This leaves us with no clear picture of snow leopard distribution prior to modern records.

The snow leopard was described to science by Schreber (1775) on the basis of a single drawing by Buffon (1761). Again, misidentification of common leopards and even lynx as snow leopards put the species' range as extending from Northern Africa to Southern Asia (see Chapter 1). It was not until the latter half of the twentieth century that the first range map was published (Fig. 3.1) that reflected a somewhat accurate estimate of the species' distribution (Hemmer, 1972), although that map still includes several "uncertain records" that we now know to be erroneous. It is the lack of accurate historic range maps that makes it impossible for

Snow Leopards
http://dx.doi.org/10.1016/B978-0-12-802213-9.00003-1

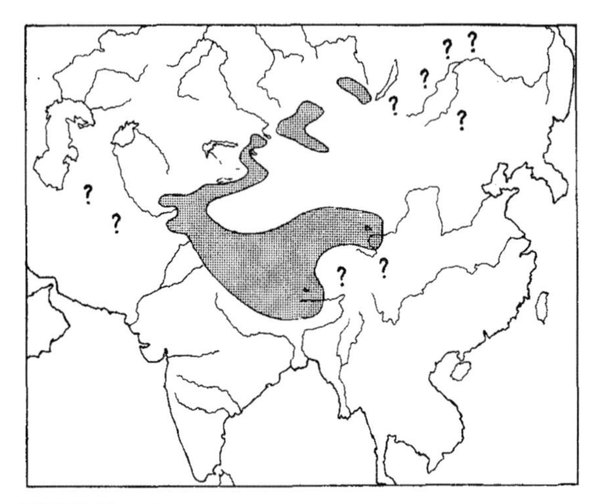

FIGURE 3.1 Map of snow leopard distribution where "?" indicate uncertain records, from Hemmer (1972).

us to estimate the extent of range loss for the species with any accuracy.

Today we know the range of the snow leopard extends from southern Siberia in the north across a broad arc including the mountains of Central Asia, the Tibetan Plateau and ending in the Himalayas in the south and east. They occur in the Sayan, Altai, Tien Shan, Kunlun, Pamir, Hindu Kush, Karakoram, and Himalayan ranges. They are known from 12 countries including Afghanistan, Bhutan, China, India, Kazakhstan, Kyrgyzstan, Mongolia, Nepal, Pakistan, Russia, Tajikistan, and Uzbekistan. Northern Myanmar may potentially have resident snow leopards, but current presence has yet to be confirmed.

Citing the needs for better maps to guide both research and conservation, Hunter and Jackson (1997) (see Chapter 25) used geographic information system (GIS) tools to model potential snow leopard habitat from map-based suitability criteria. Using 1:1,000,000 paper maps, hand-drawn polygons around the major mountain ranges of Asia were digitized. Unsuitable habitat was excluded using elevational upper and lower bounds. The lower elevation limit followed a north–south gradient (e.g., 1220 m in Mongolia and 3350 m in parts of Nepal), with the upper limit set at 5180 m except on the Tibetan Plateau where it was extended to 5485 m to include known snow leopard habitat. Permanent ice and water bodies were also excluded. Slope was used as an indicator of ruggedness to delineate "fair" (slope 0–30°) and "good" (slope > 30°) habitat. The resultant map indicated potential range of about 3,025,000 km^2, much greater than previous estimates (Fox, 1989, 1994), and identified two areas that had not previously been identified as potential snow leopard habitat – northern Myanmar and parts of Yunnan province in China. The overall importance of China was made clear, with more than 60% of potential range determined to be within that country. The map provided a vital resource for snow leopard conservation for more than a decade.

The first attempt to collect snow leopard observational data and compare them to existing maps (Williams, 2006) seemed to validate the Hunter-Jackson model. Of 1,496 expert-validated snow leopard observations (sightings or sign), 88% fell within the Hunter-Jackson–modeled potential range. With the exception of five likely erroneous observations well outside known range to the east of Lake Baikal in Russia, most observations were close enough to potential range polygons to be the result of slight errors by experts placing observational marks on 1:100,000,000 maps.

The next update to snow leopard range maps came in 2008 when several nongovernmental conservation organizations, including Panthera, the Snow Leopard Trust (SLT), the Snow Leopard Network (SLN), and the Wildlife Conservation Society (WCS), initiated a geographically based, range-wide assessment and conservation planning exercise for snow leopards.

RANGE-WIDE ASSESSMENT MEETING (BEIJING, CHINA, 2008)

The 2008 mapping and planning effort employed a similar methodology to the one used to assess range-wide priorities for jaguars

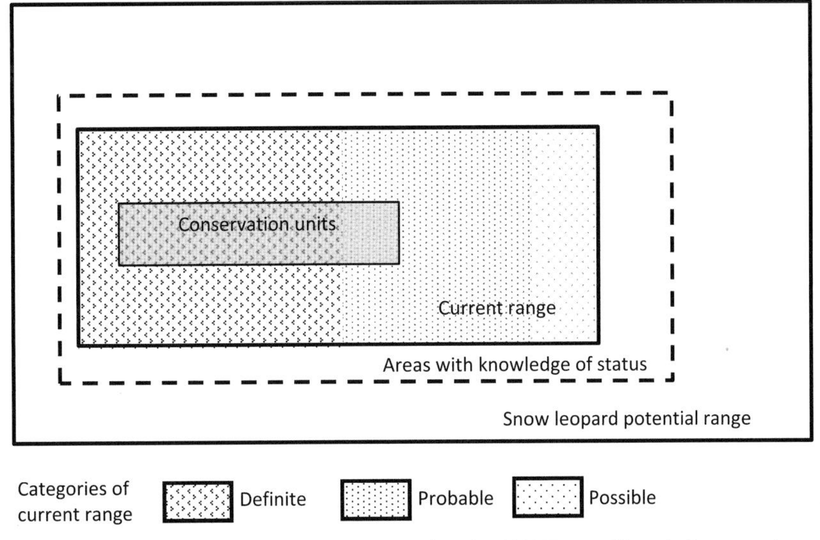

FIGURE 3.2 Graphical depiction of the nested data sets used in the 2008 Expert Knowledge mapping process.

(Sanderson et al., 2002; Zeller, 2007), tigers (Dinerstein et al., 2007; Sanderson et al., 2008), lions (IUCN/SSC Cat Specialist Group, 2006), cheetahs (IUCN/SSC, 2007), and other species (Altrichter et al., 2012; Sanderson et al., 2008; Thorbjarnarson et al., 2006). Species conservation planning guidelines from the IUCN suggest an analogous procedure (IUCN/SSC, 2008).

Through the range-wide assessment process, 22 experts on snow leopard distribution and status, including representatives from 11 of the 12 snow leopard range states (excluding Kazakhstan), worked together to establish the basic biogeography of the species, including: the potential range of the species (hereafter "potential range"), parts of the potential range where snow leopard status and distribution is known ("known potential range"), where the species is thought to currently be found ("current range"), and areas of importance to its long-term conservation ("snow leopard conservation units [SLCUs]"). Experts worked collaboratively in three regional groups on hard-copy maps at 1:1,000,000–1:3,000,000 scale, which were then digitized in a GIS. Two iterations of the data were completed, each time improving and synthesizing the data, and providing contextual information on questionnaires matched to each feature in the GIS. Collaborative work encouraged cooperation and critical review of snow leopard geography during the mapping process.

These datasets fit together in a logically consistent structure as shown in Fig. 3.2. The most extensive data represent the potential distribution of the species. The potential distribution of the species was defined by known habitat features important to snow leopards, particularly rugged, mountainous terrain, and the distribution of contemporary and historical snow leopard observations, based on expert opinion. After the workshop, water bodies and permanent snow and ice fields were masked out as shown in the GlobCover 2009 global land cover dataset (Bontemps et al., 2011). Survey locations between 1983 and 2008 were recorded as point features across this potential distribution. Experts were asked to summarize all observations within a 10 km radius of the location marked on the map and to provide ancillary information about the data of the survey, the identity of the surveyor, survey techniques, and whether evidence of snow leopard presence was uncovered.

Within the potential distribution, all areas were marked as known or unknown based on the experts' collective judgment. "Knowledge" meant any member of the expert group being able to competently comment on the status of snow leopards within the area, in the terms described in the following list. Within the known areas, all areas with evidence of snow leopard within the last 5 years were indicated. Areas where snow leopards are thought to occur (i.e., "current range") were further characterized in terms of the probability of presence, as described. Recognizing that extrapolating from survey locations about the distribution of snow leopards is difficult in the absence of detailed ranging data, each range polygon was classified by the probability that snow leopards currently occur according to the following definitions:

- *Definitive extant range* – the species is certain to occur in this area based on definitive recent observations of the species within the last year (i.e., 2007–2008).
- *Probable extant range* – the species likely occurs in this area based on an assessment of existing habitat, prey base, and connectivity, and there exist some recent but nondefinitive information or more definitive but older information about the species within the last 5 years (i.e., 2003–2008).
- *Possible extant range* – the species may occur in this area based on an assessment of existing habitat and possible connectivity to known areas where the species occurs, but there is no specific information about the species in this area within the last 5 years.

Finally within the known, current range areas, three types of SLCUs were identified. They were:

- Type I – This unit contains enough resident snow leopards to be self-sustaining over the next 100 years
- Type II – This unit contains fewer snow leopards than Type I, but adequate habitat for them to increase if threats are removed
- Type III – This unit contains fewer snow leopards than Type I, but would require large investments to make the population viable over the long term because the area is small or the habitat inadequate.

Potential SLCUs in areas currently thought to be lacking snow leopards were labeled Type R, according to the following definition:

Type R – An area with no snow leopards currently (because of past persecution) but where threats are currently low and there is adequate habitat and prey base to support a snow leopard population (potentially Type I) in the future.

Next, experts were asked to estimate the snow leopard population size (adults and large subadults) and trend in each SLCU and to indicate the basis for the assessment (e.g., camera trap survey, sign count, informed estimate, guess, or other). Each SLCU was then characterized in terms of the following factors thought to be related to the long-term survival for the species:

- Connectivity (in terms of dispersal to/from the SLCU and other SLCUs)
- Habitat quality
- Prey availability
- Direct killing of snow leopards

Lastly, experts were instructed to estimate the percentage of each SLCU under different land tenure systems and how effective the actual protection (as opposed to paper protection) of the area is for long-term survival of snow leopards.

Major ecological divisions across the range were then mapped by experts with regard to snow leopard ecology based on differences in prey base, snow leopard density, and habitat use. These snow leopard significant "ecological settings" were used to develop ecologically representative goals for snow leopard conservation (see Chapter 44).

All spatial and attribute data were compiled in a GIS (ArcGIS 10.1, ESRI, Redlands, CA) and transformed to Albers equal area conic map projection with the following parameters (central meridian at 95° east longitude; first and second standard parallels and latitude of origin at 15, 65, and 30° north latitude, respectively, based on the WGS84 datum). All areas were reported to the nearest 1 km^2 as defined on this coordinate system.

OUTPUTS

Potential Range and Ecological Settings

Experts identified a total of 3,256,841 km^2 of potential snow leopard range (Fig. 3.4A, Table 3.1), or areas where snow leopards could have occurred within the past 100 years during the 2008 exercise. Potential range as determined in this process was about 7.6% larger than the 3,025,000 km^2 estimate of Hunter and Jackson (1997). Nine ecological settings were identified as relevant to snow leopards within potential range, which included: Altai-Sayan, Trans-Altai Alashan Gobi, Tian Shan, Pamirs, Hindu-Kush, Karakoram, Himalayas, Hengduan, and Tibetan Plateau (Fig. 3.3). Notably, about 45% of all potential range was in the Tibetan Plateau ecological setting (Table 3.2).

Area of Expert Knowledge

The combined knowledge of the expert group pertaining to snow leopard status covered 91% of potential range (Table 3.1). Only in the

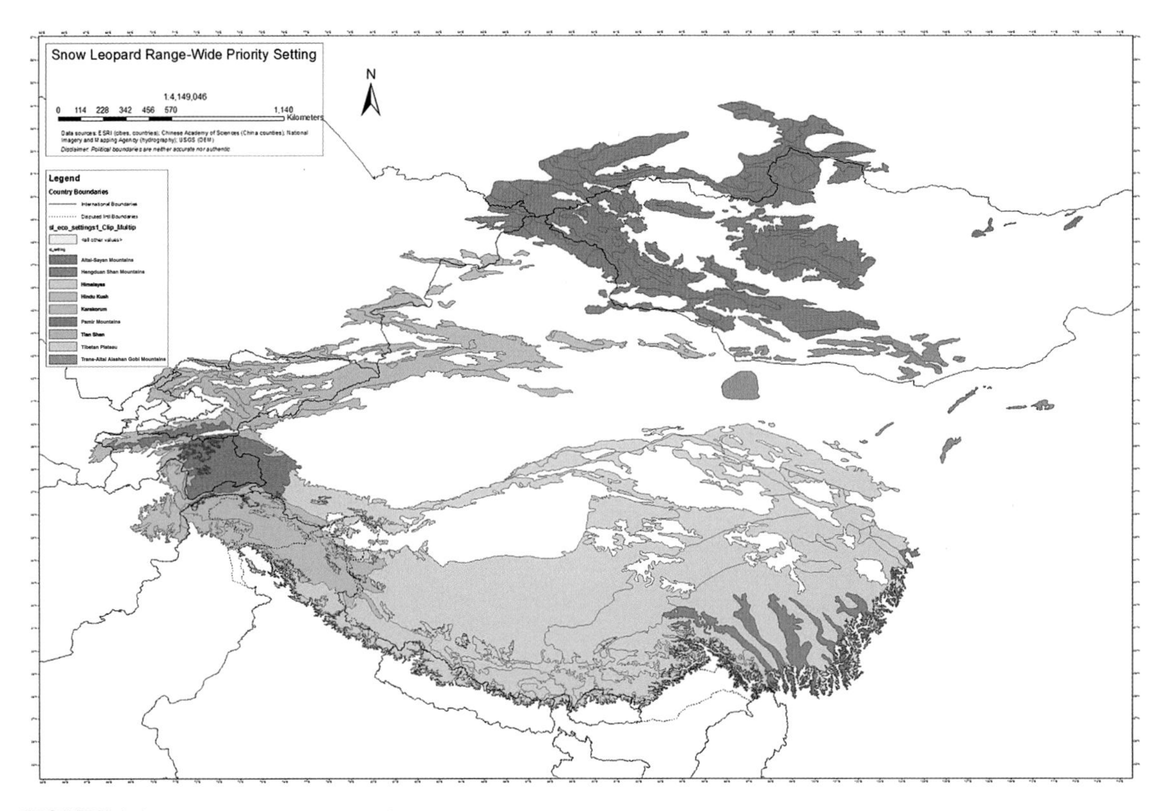

FIGURE 3.3 Snow leopard ecological settings.

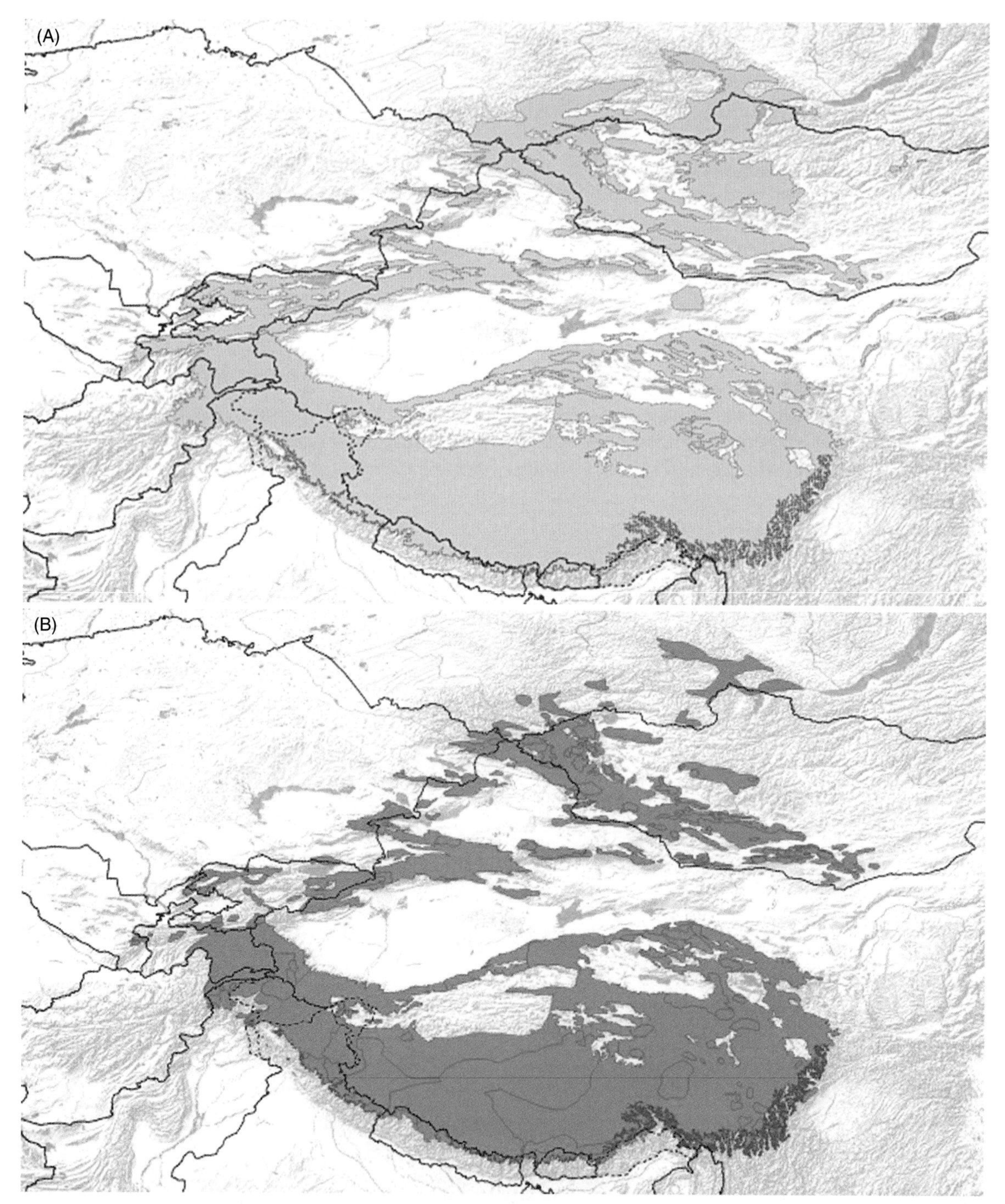

FIGURE 3.4 Spatial distribution of snow leopard data: (A) extent of potential range, (B) extent of current range, (C) probability of snow leopard occurrence in current range, (D) distribution of SLCUs.

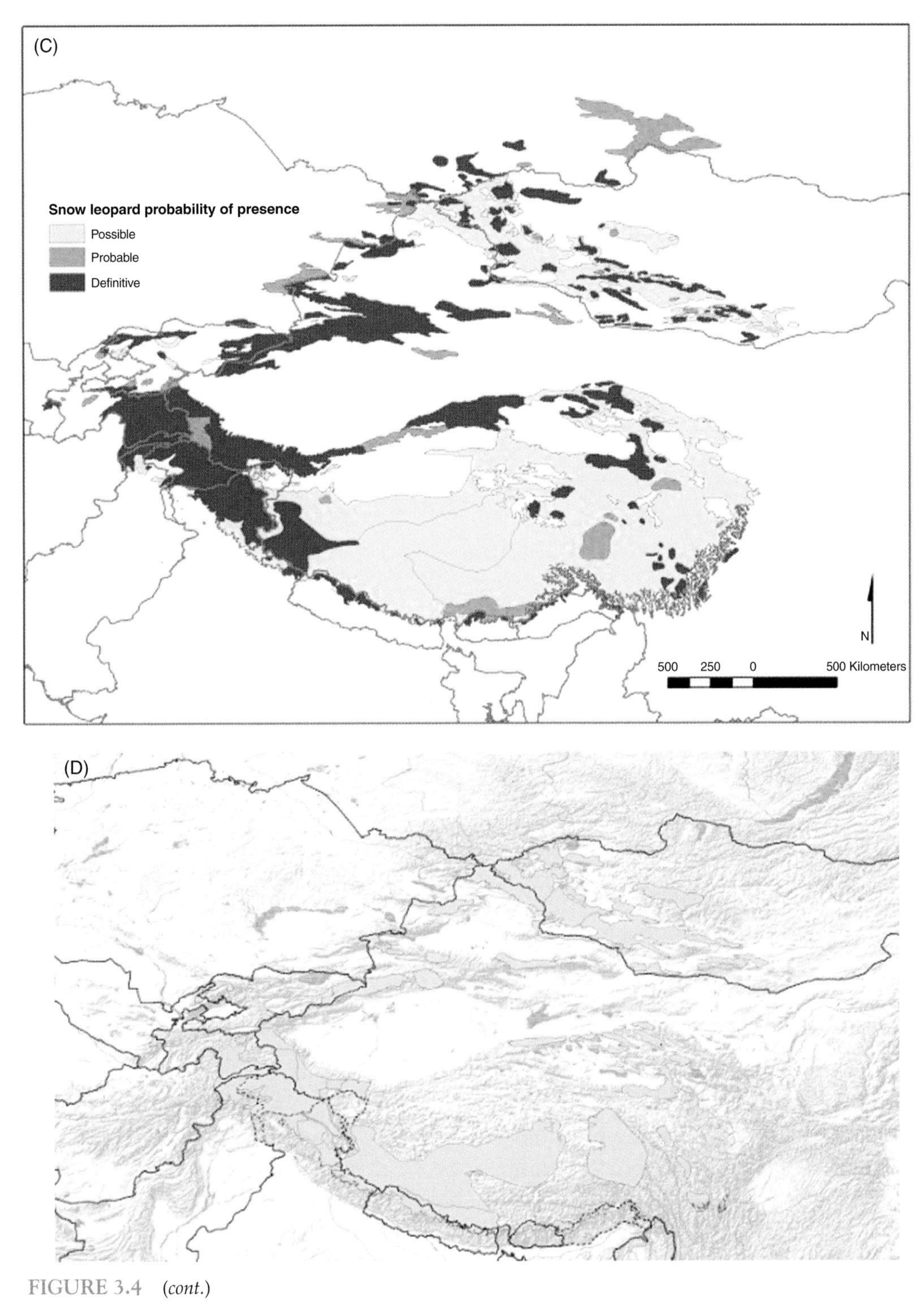

FIGURE 3.4 (*cont.*)

TABLE 3.1 Characterization of Snow Leopard Range

Categories of snow leopard range	Area (km^2)	Number of distinct areas*	Potential range (%)	% Known range	% Current range	% SLCUs
Potential snow leopard range	3,256,841	271	100			
Potential range where knowledge is lacking	276,793		8			
Knowledge area of potential range	2,980,048		92	100%		
Current range	2,778,309		85	93%	100%	
Definitive extant range	892,435	129	27	30%	32%	
Probable extant range	214,969	38	7	7%	8%	
Possible extant range	1,670,906	23	51	56%	60%	
Conservation units (SLCUs)	1,222,073	69	38	41%	44%	100%
Type I	835,755	35	26	28%	30%	68%
Type II	364,192	30	11	12%	13%	30%
Type III	22,127	4	1	1%	1%	2%
Type R	–	–	–	–	–	–
Global snow leopard landscapes (2013)	589,876	20	18	20%	21%	48%

* *Counts only include areas more than or equal to 12 km^2 (approximate smallest snow leopard home range in good habitat).*

TABLE 3.2 Summary of Snow Leopard Range by Ecological Setting

Ecological setting	Potential range (km^2)	Extent of knowledge (%)	Current range (%)	SLCU (%)
Altai-Sayan	632,062	65	61	38
Hengduan Shan	114,283	100	100	22
Himalayas	270,283	92	91	59
Hindu Kush	53,814	23	18	10
Karakoram	206,735	100	91	74
Pamirs	120,996	100	92	51
Tian Shan	397,409	100	74	29
Tibetan Plateau	1,560,978	100	98	36
Trans-Altai Alashan Gobi	82,348	100	70	36
Total	3,438,909	91	86	39

Altay-Sayan and Hindu Kush ecological settings could experts not confidently comment on snow leopard status in at least 90% of the area. There, the status of snow leopards was unknown in 35 and 78% of the areas, respectively.

Current Snow Leopard Range

Most of potential range (85.3%) is thought to have been occupied by snow leopards within the previous 5 years, meaning that 2,778,309 km^2 met the definition of current range (Fig. 3.4B, Table 3.1). Within current range, the presence of snow leopards was considered definitive in only 32% of the area, probable in 8%, and possible in 60% (Fig. 3.4C, Table 3.1). That means that the expert group could only say that snow leopard occurrence was certain in 27% of the vast 3.26 million km^2 of potential range.

No attempt was made to determine what portion of their range snow leopards may have been extirpated from because that would entail having a valid depiction of their historic range, which is unavailable for the species.

Snow Leopard Conservation Units

Snow leopard conservation units (SLCUs) were identified in all ecological settings (Fig. 3.4D, Table 3.1). The percentage of each ecological setting that fell into an SLCU ranged from a low of 10% in the Hindu Kush to a high of 74% in the Karakoram. Within the Tibetan Plateau ecological setting, where nearly half of all potential range is found, 36% of the area fell within an SLCU, 89% of which was Type I, or self-sustaining over the long term.

A total of 69 SLCUs were delineated with a combined area of about 1.2 million km^2, or about 38% of potential range (Table 3.1). Most SLCUs were classified as Type I (68%), representing areas considered to contain populations that are potentially self-sustaining over the next 100 years. Just 30% of SLCUs were considered of Type II, where habitat was adequate and threat removal would yield population increases. Only 2% of SLCUs were considered Type III, where threats, habitat quality, or prey base was so adverse that snow leopard populations were not considered viable without extensive investments. Type III SLCUs only occurred in the Altay-Sayan and the Himalayas ecological settings.

At least one SLCU was identified in each range country (Table 3.3). Thirty were identified in China, which is consistent with the fact that ~68% of current range is found in that country. SLCU size ranged from 405 km^2 (North Sichuan) to the vast Tibetan Plateau SLCU at just over 370,000 km^2, with 90% of the units being between 1,300 km^2 and 86,000 km^2.

Experts had adequate information to characterize the habitat quality and prey availability in 83 and 86% of SLCUs, respectively (Table 3.3), with values falling in the good or medium category in 78% of units for habitat quality and 75% for prey availability. Experts said information on connectivity between SLCUs was lacking in 54% of the units, although where known, dispersal between units was thought to be infrequent or none in 47%, versus frequent in 53% of the SLCUs.

Direct killing of snow leopards was thought to occur in 47 (68%) of SLCUs, but in only one unit was it considered to be frequent ("Lots") as opposed to occasional ("Some"). In 16% of SLCUs, there was thought to be no killing of snow leopards.

The snow leopard population size was estimated in 56 of the 69 SLCUs, while inadequate information was available for experts to comment on the remaining units (Table 3.3). Estimates were given either as a range (e.g., 1–49, 50–100, etc.) or as discrete numbers. Summing the high and low estimates yielded an overall population size of between 4,678 and 8,745 within 56 SLCUs totaling 875,818 km^2. Snow leopard density is most often reported as adult cats per 100 km^2. Using the lower bound of the

TABLE 3.3 Snow Leopard Conservation Units (SLCU) and Their Attributes by Country

Country	SLCU	Type	Area (km^2)	Habitat quality	Connectivity	Prey availability	Snow leopard population	Population trend	Killing of snow leopards	Source of population estimate
Afghanistan	Wakhan-Nuristan	I	14,555	Medium	Unknown	Medium	~50–100	Unknown	Some	Sign frequency, livestock attacks
Bhutan	Bhutan	I	8,762	High	Infrequent	High	~50–100	Unknown	Some	Informed estimate
China	Bayinbuluke	I	13,458	Medium	Frequent	Medium	~140	Decreasing	Some	Faunal surveys, questionnaire
China	Dulan county	I	4,619	High	Unknown	Medium	~1–49	Increasing	None	Camera trap
China	Kunlun	I	10,910	Medium	Frequent	Medium	~255	Decreasing	Some	Faunal surveys, questionnaire
China	North Sichuan	I	405	High	Unknown	High	Unknown	Stable	None	Interview, research, sign frequency
China	N. Central Tien-Shan	I	5,980	Medium	Frequent	Medium	~140	Decreasing	Some	Faunal surveys, questionnaire
China	Northeast Sichuan	I	6,230	High	Unknown	High	~251–500	Stable	None	Interview, research, sign frequency
China	Pahir West Kunlun	I	24,902	Unknown	Unknown	Unknown	~1–49	Unknown	Unknown	Informed estimate
China	Pamir	I	15,595	Medium	Infrequent	Medium	~251–500	Stable	Some	Faunal surveys
China	S. Central Sichuan	I	5,354	High	Unknown	High	~251–500	Stable	None	Interview, research, sign frequency
China	S. Sichuan	I	11,334	High	Unknown	High	~251–500	Stable	None	Interview, research, sign frequency
China	SW Sichuan	I	2,577	High	Unknown	High	~251–500	Stable	None	Interview, research, sign frequency
China	Taxkurgan	I	19,797	Unknown	Unknown	Unknown	~101–250	Unknown	Unknown	Informed estimate

China	Tibetan Plateau	I	370,282	Unknown	Unknown	Unknown	Unknown	Unknown	Unknown	Unknown
China	Tomur	I	19,984	Medium	Frequent	Medium	~101–250	Stable	Some	Camera trap, sign frequency
China	Tramkar Rasan	I	120,162	High	Unknown	High	~1–49	Stable	None	Informed estimate
China	W. Kunlun (b)	I	5,311	Medium	Infrequent	Medium	~251–500	Stable	Some	Faunal surveys
China	Bogda	II	13,310	Medium	Frequent	Medium	~90	Decreasing	Some	Sign, surveys, questionnaire
China	Gonghe county	II	24,843	Medium	Unknown	Medium	~1–49	Decreasing	None	Observed by local people
China	Qilian Mtns. 1	II	4,836	No data	Unknown	Unknown	Unknown	Unknown	Unknown	Unknown
China	Qilian Mtns. 2 (a)	II	3,200	No data	Unknown	Unknown	Unknown	Unknown	Unknown	Unknown
China	Qilian Mtns 2 (b)	II	1,478	No data	Unknown	Unknown	Unknown	Unknown	Unknown	Unknown
China	Qilian Mtns 2 (c)	II	5,627	No data	Unknown	Unknown	Unknown	Unknown	Unknown	Unknown
China	Qilian Mtns 2 (d)	II	3,372	No data	Unknown	Unknown	Unknown	Unknown	Unknown	Unknown
China	Qilian Mtns 2 (e)	II	7,367	No data	Unknown	Unknown	Unknown	Unknown	Unknown	Unknown
China	Tarbahetai	II	14,629	Medium	Infrequent	Medium	~70	Decreasing	Some	Faunal surveys, questionnaire
China	Tianjun county	II	1,746	Medium	Unknown	Medium	~1–49	Decreasing	None	Interview, general research
China	Wenquan	II	10,603	Poor	Infrequent	Poor	~40	Decreasing	Some	Faunal surveys, questionnaire
China	W. Altai	II	29,652	Medium	Infrequent	Poor	~101–250	Decreasing	Some	Sign, surveys, questionnaire
China	W. Kunlun (a)	II	12,312	Medium	Frequent	Medium	~255	Decreasing	Some	Faunal surveys, questionnaire
China	Baitag	III	7,645	Poor	None	Poor	~15	Decreasing	Some	Sign, surveys, questionnaire

(Continued)

TABLE 3.3 Snow Leopard Conservation Units (SLCU) and Their Attributes by Country (*Cont.*)

Country	SLCU	Type	Area (km^2)	Habitat quality	Connectivity	Prey availability	Snow leopard population	Population trend	Killing of snow leopards	Source of population estimate
India	Eastern Himalayas	I	5,670	High	None	High	Unknown	Unknown	Some	Sign frequency
India	Lahul-Spiti	I	14,789	High	Frequent	High	~1–49	Stable	Some	Sign frequency
India	North Sikkim	I	1,684	Medium	Frequent	High	~1–49	Stable	Some	Sign frequency
India	Uttarakhand	I	14,793	High	Frequent	Medium	~1–49	Stable	Some	Sign frequency
India	W. Tibetan Himalayas	I	46,371	High	None	High	Unknown	Unknown	Some	Sign frequency
India	Zanskar (S. of Indus)	I	17,994	High	Unknown	High	~6	Stable	Some	Camera trap, occupancy surveys
India	Dran and Sonamore	II	9,503	Medium	Frequent	Medium	Unknown	Stable	Lots	Sign frequency
India	Nubra Region	II	16,443	Medium	Frequent	High	Unknown	Stable	Some	Sign frequency
India	W. Arunachal	II	7,177	Unknown	Unknown	Poor	~1–49	Unknown	Some	Informed estimate
India	E. Arunachal	III	5,268	Unknown	Unknown	Poor	~1–49	Unknown	Some	Informed estimate
Kazakhstan	Talgar Junction	II	3,542	High	None	High	~60	Increasing	Some	Sign frequency
Kyrgyzstan	Kyrgyz Range	I	4,264	High	Unknown	Medium	~1–49	Increasing	None	Informed estimate
Kyrgyzstan	Syrts	I	13,819	High	Unknown	Medium	~50–100	Stable	Some	Camera trap, sign frequency
Kyrgyzstan	Alay-Kuu	II	2,015	Medium	Unknown	Medium	~9	Stable	Some	Informed estimate
Kyrgyzstan	West Tien-Shan	II	3,862	Medium	Unknown	Medium	~1–49	Decreasing	Some	Informed estimate
Mongolia	NW Altai	I	13,404	High	Frequent	Medium	~101–250	Unknown	Some	Camera trap, sign frequency
Mongolia	South Gobi	I	17,603	High	Infrequent	Medium	~101–250	Decreasing	Some	Sign frequency
Mongolia	Central Altai	II	77,373	Medium	Frequent	Medium	~150	Unknown	Some	Telemetry, sign frequency

Mongolia	Khangai	II	23,953	Medium	Infrequent	Medium	~1–49	Decreasing	Some	Sign frequency, oral information
Mongolia	Trans Altai	II	8,825	Medium	Infrequent	Medium	~1–49	Decreasing	None	Sign frequency
Mongolia	W. Altai (a)	II	36,412	High	Frequent	Medium	~101–250	Unknown	Some	Sign frequency
Mongolia	W. Altai (b)	II	39,988	Medium	Frequent	Medium	~101–250	Unknown	Some	Sign frequency
Mongolia	Khankhokhii	III	7,443	Poor	Infrequent	Poor	~1–49	Unknown	Unknown	Sign frequency
Nepal	GSHL—C. Nepal	I	15,024	High	Unknown	Medium	~251–500	Increasing	Some	Informed estimate
Nepal	GSHL—E. Nepal	II	8,129	High	Unknown	Medium	~50–100	Increasing	Some	Informed estimate
Nepal	GSHL—W. Nepal	III	5,981	Unknown	Unknown	Unknown	Unknown	Unknown	Unknown	Unknown
Pakistan	Central Karakoram	I	86,384	Medium	Frequent	Medium	~250	Stable	Some	Sign, surveys, habitat quality
Russia	Argut River Basin	I	1,819	High	Frequent	High	~20–25	Stable	Some	Winter track census
Russia	Sayana–Shushensky	I	1,527	High	Unknown	High	~10–15	Stable	Some	Winter track census
Russia	Chkhachev Ridge	II	1,305	Medium	Infrequent	Medium	~5–6	Stable	Some	Winter track census
Russia	Sengelen Ridge	II	2,140	Medium	Infrequent	Medium	~13	Stable	Some	Sign frequency
Russia	South Altai Ridge	II	1,569	Medium	Frequent	Medium	~12–15	Stable	Some	Winter track census
Russia	Tsagan-Shibetu	II	2,210	Medium	Infrequent	Medium	~8–10	Stable	Some	Winter track census
Tajikistan	Badakhan	I	27,368	High	Unknown	Medium	~180	Stable	Some	Sign frequency
Tajikistan	Balandkiik	I	2,029	High	Unknown	High	~18	Stable	Some	Sign frequency
Uzbekistan	Hissar and Tupalang	I	2,602	High	Frequent	Medium	~1–49	Stable	Some	Informed estimate
Uzbekistan	Ugam-Chatkal NP	II	5,234	High	Frequent	Poor	~1–49	Stable	Some	Informed estimate

population estimate for each SLCU, the mean density was 0.92 leopards per 100 km^2, while the high estimate leads to a mean density of 1.8 leopards per 100 km^2. The population trend was considered to be stable or increasing in 33 (48%) of SLCUs and decreasing in 15 (22%). No information on trend was reported by experts in 11 (30%) of the units.

Land tenure and the effectiveness of protection in each SLCU are summarized in Table 3.4. Tenure type varies dramatically across different SLCUs, though most have some kind of public land tenure, whether in communal lands, protected areas, or military controlled designations, the latter speaking to the need for close collaboration with military entities on conservation issues. Not surprisingly, only a small amount of private land occurs within the SLCUs. Effectiveness of protection also varies dramatically across the range, but experts felt that few SLCUs received full protection, and none did so outside of China. Interestingly, in some SLCUs with several prominent protected areas, experts felt protection was still highly ineffective over a substantial portion of the unit.

DISCUSSION

Recognizing that conservation decision making is inherently sensitive to the accuracy of the underlying data, the 2008 range-wide assessment and mapping process relied on the proven expert-driven methodology that had been developed for tigers (Wikramanayake et al., 1998) and adapted for jaguars (Sanderson et al., 2002) and other species (Sanderson et al., 2008; Thorbjarnarson et al., 2006) to evaluate the biogeography of the snow leopard. That said, the quality of the summary is only as good as the information on which it was based. Fortunately nearly all of the active community of snow leopard researchers contributed.

Snow leopard potential range, as determined by the expert knowledge process of 2008, covers some 3.26 million km^2, which compares favorably to the only other effort that used GIS tools and expert review to derive a "potential range" estimate (3.025 million km^2, Hunter and Jackson, 1997). Across that vast rugged and sparsely populated area, stretching north to south from Siberia to the Himalayas, with an even greater east-to-west expanse from central China to Uzbekistan, experts felt at least somewhat knowledgeable about the status of snow leopards in more than 90% of the potential range. This is an extraordinary increase in the level of knowledge over the past two decades, when previously it was difficult even to develop a range map for the species.

Within that knowledge area, the experts identified about 2.8 million km^2 of current range. Current, occupied, or inhabited range has been variously reported in the literature and ranges from about 1.2–1.8 million km^2 (Fox, 1994; GSLEP, 2013; Jackson et al., 2010). Comparing those estimates to those of the 2008 process is not straightforward, because in most cases previous studies lack a clear definition of "range" that was used, and lack details on methods for synthesis and analysis. Some, such as the Global Snow Leopard Ecosystem Protection Program (GSLEP, 2013), rely on outdated estimates at the range country level, some dating back to the mid-1990s. The expert process used in Beijing in 2008 was clear, transparent, relied on current knowledge and was guided by explicit and unambiguous definitions, and yielded, arguably, the most defensible depiction of snow leopard range to date. That said, much work remains to be done to document the distribution of snow leopards and their movements.

It is promising that nearly 836,000 km^2, or 30% of current range, fell within 35 distinct Type I SLCUs, where resident snow leopard populations are considered large enough and with access to an adequate prey base to be self-sustaining over the next 100 years. Of those, 17 SLCUs were thought to contain more than 100 snow leopards. One may wish to consider this when

TABLE 3.4 Percentage of Area by Land Tenure Type and by Assessed Level of the Actual Effectiveness of Protection Efforts for Snow Leopards in the Snow Leopard Conservation Units (SLCUs) in 2008

		Land tenure*						Protection effectiveness*			
Country	SLCU	Public—not formally protected	Communal ownership or use	Protected area	Border control or military	Private	Balance of land tenure**	Full	Partial	Ineffective	Unknown
Afghanistan	Wakhan-Nuristan	65	5	15	8	5	2		90		10
Bhutan	Bhutan	30	5	60	0	5	0		80	20	
China	Baitag	88	8	0	3	5	−4		100		
China	Bayinbuluke	60	8	25	0	8	−1		100		
China	Bogda	60	8	25	8	5	−6		98		8
China	Dulan county	100	0	0	0	0	0	100			
China	Gonghe county	100	0	0	0	0	0	100			
China	Kunlun	65	8	15	8	8	−4		96		8
China	North Sichuan	26	0	26	0	0	48		42	10	48
China	N. Central Tien-Shan	60	8	25	0	8	−1		100		
China	Northeast Sichuan	26	0	26	0	0	48		42	10	48
China	Pahir West Kunlun	100	0	0	0	0	0				100
China	Pamir	65	15	25	8	8	−21		113		8
China	Qilian Mtns. 1	0	0	0	0	0	100				100
China	Qilian Mtns. 2 (a)	0	0	0	0	0	100				100
China	Qilian Mtns. 2 (b)	0	0	0	0	0	100				100
China	Qilian Mtns. 2 (c)	0	0	0	0	0	100				100
China	Qilian Mtns. 2 (d)	0	0	0	0	0	100				100
China	Qilian Mtns. 2 (e)	0	0	0	0	0	100				100
China	S. Central Sichuan	26	0	26	0	0	48		42	10	48
China	S. Sichuan	26	0	26	0	0	48		42	10	48
China	SW Sichuan	26	0	26	0	0	48		42	10	48
China	Tarbahetai	85	3	0	8	3	1		91		9

(Continued)

TABLE 3.4 Percentage of Area by Land Tenure Type and by Assessed Level of the Actual Effectiveness of Protection Efforts for Snow Leopards in the Snow Leopard Conservation Units (SLCUs) in 2008 (*Cont.*)

		Land tenure*						Protection effectiveness*			
Country	**SLCU**	**Public—not formally protected**	**Communal ownership or use**	**Protected area**	**Border control or military**	**Private**	**Balance of land tenure****	**Full**	**Partial**	**Ineffective**	**Unknown**
China	Taxkurgan	60	0	40	0	0	0		40		60
China	Tianjun county	100	0	0	0	0	0	100			
China	Tibetan Plateau	0	0	0	0	0	100				100
China	Tomur	55	10	35	8	8	−16		108		8
China	Tramkar Rasan	0	0	100	0	0	0	40	60		
China	Wenquan	78	13	15	5	5	−16		111		5
China	W. Altai	72	0	28	0	0	0				100
China	W. Kunlun (a)	65	8	15	8	8	−4		96		8
China	W. Kunlun (b)	65	15	25	8	8	−21		113		8
India	Dran and Sonamore	0	0	20	0	0	80	5	15		80
India	E. Arunchal Pradesh	0	90	10	0	0	0				100
India	Eastern Himalayas	60	0	40	0	0	0				100
India	Lahul-Spiti	0	85	20	0	0	−5		20	85	
India	North Sikkum	40	40	20	0	0	0		60		40
India	Nubra Region	0	0	100	0	0	0			100	
India	Uttarakhand	70	0	30	0	0	0		10		90
India	W. Arunachal	0	0	0	0	0	100				100
India	W. Tibetan Himalayas	60	0	40	0	0	0				100
India	Zanskar (S. of Indus)	90	90	20	0	10	−110				100
Kazakhstan	Talgar Junction	10	90	0	0	0	0		100		
Kyrgyzstan	Alay-Kuu	50	0	0	50	0	0			50	50

Kyrgyzstan	Kyrgyz Range	72	0	8	0	20	0	5	23	72	
Kyrgyzstan	Syrts	70	0	8	20	2	0	5	5	70	20
Kyrgyzstan	Western Tien-Shan	25	0	75	0	0	0		60	25	15
Mongolia	Central Altai	80	0	20	0	0	0	20		80	
Mongolia	Khangai	60	0	40	0	0	0		40	60	
Mongolia	Khankhokhii	55	0	45	0	0	0				100
Mongolia	NW Altai	50	0	50	0	0	0				100
Mongolia	South Gobi	40	0	60	0	0	0			40	60
Mongolia	Trans Altai	0	0	100	0	0	0				100
Mongolia	W. Altai (a)	85	0	15	0	0	0				100
Mongolia	W. Altai (b)	5	95	0	50	0	−50				100
Nepal	GSHL—C. Nepal	0	0	0	0	0	100				100
Nepal	GSHL—E. Nepal	0	0	0	0	0	100				100
Nepal	GSHL—W. Nepal	0	0	0	0	0	100				100
Pakistan	Central Kara-koram	15	45	10	0	30	0	10	15	75	
Russia	Argut River Basin	10	0	90	0	0	0		90	10	
Russia	Chkhachev Ridge	50	0	50	0	0	0		50	50	
Russia	Sayana–Shushensky	0	0	100	0	0	0		50		50
Russia	Sengelen Ridge	100	0	0	0	0	0			100	
Russia	South Altai Ridge	0	0	100	0	0	0		100		
Russia	Tsagan-Shibetu	100	0	0	0	0	0			100	
Tajikistan	Badakhan	65	0	35	0	0	0		100		
Tajikistan	Balandkiik	0	0	100	0	0	0		100		
Uzbekistan	Hissar and Tu-palang	0	0	0	0	0	100				100
Uzbekistan	Ugam-Chatkal NP	0	0	0	0	0	100				100

** Percentages may not add to 100% because of overlapping land tenure types; see subsequent note.*

*** Negative land tenure balances indicate overlapping land tenure classes (e.g. formal protection over communal lands or military control of public lands or some other combination); positive balance numbers indicate unknown.*

looking at the goal of the GSLEP (see Chapter 45) of "Securing at least 20 snow leopard landscapes across the cat's range by 2020," where secure snow leopard landscapes are defined as those that contain at least 100 breeding-age snow leopards conserved with the involvement of local communities, support adequate and secure prey populations, and have functional connectivity to other snow leopard landscapes. Perhaps the expert-driven process of 2008 illuminates how close we already were to achieving the goal of the GSLEP. It is important to note, however, that considerable work is still required to define and monitor key snow leopard populations to keep them from falling below this population target level.

The snow leopard population estimates reported here are only for the portion of the range that falls in an SLCU, or about 44% of current range. Yet the estimate of 4678–8745 exceeds most of the published population estimates for the entire range, such as 3920–6390 (GSLEP, 2013), 4080–6590 (McCarthy and Chapron, 2003), and 4500–7500 (Jackson et al., 2010). In terms of density, the estimate we report here (0.9–1.8 snow leopards per 100 km^2) is not unreasonable and is on the low end of what has been reported regionally in fair to good range (McCarthy and Chapron, 2003). The higher population estimate is also consistent with another recent expert assessment completed in China in 2012, where snow leopard numbers for that country alone were put at 4500 or more (see Chapter 42), nearly double past estimates. Lastly, snow leopard population figures are often labeled little more than "guesstimates," and that is probably an accurate statement for many historical estimates. However, Table 3.3 documents the source of the estimate in 2008, and most are derived from field studies involving camera traps, questionnaires, sign frequency counts, faunal surveys, winter track counts, and other research visits to snow leopard habitat. As more research is undertaken in snow leopard range and more technologies such as camera traps and noninvasive genetics are employed, it is becoming clear that there are likely more snow leopards than previously thought.

One aspect of the SLCU characterization where expert knowledge was often lacking was the connectivity between SLCUs (46% unknown). Indeed, a component of snow leopard biogeography that has received little attention, or at least formal study, is that of connectivity. On cursory examination, the current range map for the species seems to depict a fairly fragmented distribution, particularly in the north and west portions of the range. However, all of our range maps have been constructed under the premise that occupied habitat is only found in rugged mountainous areas, and anything lying in between mountain ranges is generally not included. Yet snow leopards are known to cross large expanses of flat low-lying terrain to reach adjacent mountain systems in Mongolia (McCarthy et al., 2005; Sharma et al., 2014), so use of nonmountainous habitat as travel corridors may partially negate the effects of their otherwise patchy distribution. Unlike some large cats, snow leopard range overall is generally contiguous, although a break between the northern and southern part of the range at the Dzungarian Basin in northern Xinjiang Province, China, has been discussed by Fox (1994) and more recently by Riordan et al. (2015) who used resistant kernel modeling to assess snow leopard population connectivity across its range. To date there have been no studies that investigate genetic differences between the supposed northern and southern distributions. Much more research into connectivity across the cat's range is clearly needed.

Snow leopards are rare, secretive, and sparsely distributed across vast tracts of remote, nearly inaccessible habitat, and research into the species biogeography and basic biology has been slow in coming. Yet it has been only 40 years since the first generally accurate but rudimentary depiction of snow leopard range was published (Hemmer, 1972), and in the intervening years we have advanced from simple VHF to GPS-satellite collars that have added substantially to our knowledge of snow leopard movements and habitat use (see Chapter 26). Camera traps, hand-built for studies in the 1990s (McCarthy, 2000), are now used extensively across snow leopard range to document occurrence and numbers (see Chapter 28), with one NGO alone (Panthera) currently deploying >500 cameras in seven range states. Noninvasive fecal genetic methods (see Chapter 27.1) help us validate presence and estimate numbers and sex ratios. Collectively, these and other less intensive studies are adding to our body of knowledge on the biogeography and status of the species.

The expert-led 2008 assessment process yielded a robust and geographically explicit picture of snow leopard distribution and conservation status and provides a sound basis to build from as additional information becomes available. Although it is the best we have now, soon the 2008 study will be a decade old. Going forward it will be critical to continue to survey and aggregate information on snow leopard distribution, status, and population numbers so that the community writ large – policy makers, local communities, and scientists – can measure progress toward successful snow leopard conservation.

References

Altrichter, M., Taber, A., Beck, H., Reyna-Hurtado, R., Lizarraga, L., Keuroghlian, A., Sanderson, E.W., 2012. Range-wide declines of a key neotropical ecosystem architect, the near threatened white-lipped peccary *Tayassu pecari*. Oryx 46, 87–98.

Bontemps, S., Defourny, P., Bogaert, E., Arino, O., Kalogirou, V., Perez, J., 2011. GLOBCOVER 2009 Products Description and Validation Report. European Space Agency and Université c=Catholique du Lovain, Louvain-la-Neuve, Belgium, http://due.esrin.esa.int/globcover/LandCover2009/GLOBCOVER2009_Validation_Report_2.2.pdf (accessed January 27, 2015).

Brandt, F., 1871. Neue Untersuchungen über die in den altaischen Höhlen aufgefundenen Säugethierreste, ein Betrag zue quaternären Fauna des Russischen Reiches. Bulletin de l'Académie Impériale des Sciences de St. Pétersbourg 15, 147–205.

Buffon, Comte de, G.-L.L., 1761. La panthére, l'once et le léopard. Buffon, G.-L., Daubenton, L.J.M. (Eds.), Histoire Naturelle, générale et particuliére avec la description du Cabinet du Roi, Vol. 9, De l'Imprimière Royale, Paris, pp. 151–172, pl. 13.

Dennell, R.W., Turner, A., Coard, R., Beech, M., Anwar, M., 2005. Two Upper Siwalik (Pinjor Stage) fossil accumulations from localities 73 and 362 in the Pabbi Hills, Pakistan. J. Palaeontol. Soc. India 50, 101–111.

Dinerstein, E., Loucks, C., Wikramanayake, E., Ginsberg, J., Sanderson, E., Seidensticker, J., Forrest, J., Bryja, G., Heydlauff, A., Klenzendorf, S., Leimgruber, P., Mills, J., O'brien, T.G., Shrestha, M., Simons, R., Songer, M., 2007. The fate of wild tigers. BioScience 57, 508–514.

Fox, J.L., 1989. A Review of the Status and Ecology of the Snow Leopard (*Panthera uncia*). International Snow Leopard Trust, Seattle, WA, USA.

Fox, J.L., 1994. Snow leopard conservation in the wild – a comprehensive perspective on a low density and highly fragmented population. In: Fox, J.L., Jizeng, D. (Eds.), Proceedings of the Seventh International Snow Leopard Symposium, Xining, Qinghai, China, July 25–30, 1992, pp. 3–15. International Snow Leopard Trust, Seattle.

GSLEP, 2013. Global Snow Leopard & Ecosystem Protection Program: A New International Effort to Save the Snow Leopard and Conserve High-Mountain Ecosystems. Snow Leopard Working Secretariat, Bishkek, Dyrgyz Republic. http://akilbirs.com/files/final_gslep_web_11_%2014_%2013.pdf (accessed January 27, 2015).

Hemmer, H., 1972. *Uncia uncia*. Mammalian Species No. 20, 1–5.

Hunter, D.O., Jackson, R., 1997. A range-wide model of potential snow leopard habitat. In: Jackson, R., Ahmad, A. (Eds.). Proceedings of the Eighth International Snow Leopard Symposium, Islamabad, Pakistan. International Snow Leopard Trust, Seattle, USA and World Wildlife Fund-Pakistan, Islamabad, Pakistan, pp. 51–56.

IUCN/SSC Cat Specialist Group, 2006. Conservation strategy for the lion in eastern and southern Africa. IUCN/SSC, Gland, Switzerland. http://rocal-lion.org/documents/english/LionESweb.pdf.(accessed January 27, 2015).

IUCN/SSC, 2007. Regional conservation strategy for cheetah and wild dog in eastern Africa. IUCN/SSC, Gland, Switzerland. http://www.cheetahandwilddog.org/documents/ (accessed January 27, 2015).

IUCN / SSC, 2008. Strategic Planning for Species Conservation. IUCN, Gland, Switzerland. data.iucn.org/dbtw-wpd/edocs/2008-048.pdf.(accessed January 27, 2015).

Jackson, R.M., Mishra, C., McCarthy, T., Ale, S.B., 2010. Snow leopards, conservation and conflict. In: MacDonald, D., Loveridge, A. (Eds.), The Biology and Conservation of Wild Felids. Oxford University Press, UK, pp. 417–430.

McCarthy, T.M., 2000. Ecology and Conservation of Snow Leopards, Gobi Brown Bears, and Wild Bactrian Camels in Mongolia. PhD thesis, University of Massachusetts. 134 pp.

McCarthy, T.M., Chapron, G. (Eds.), 2003. Snow Leopard Survival Strategy. International Snow Leopard Trust and Snow Leopard Network, Seattle, WA, USA.

McCarthy, T.M., Fuller, T.K., Munkhtsog, B., 2005. Movements and activities of snow leopards in Southwestern Mongolia. Biol. Conserv. 124, 527–537.

Riordan, P., Cushman, S.A., Mallon, D., Shi, K., Hughes, J., 2015. Predicting global population connectivity and targeting conservation action for snow leopard across its range. Ecography 38, 001–008.

Sanderson, E.W., Forrest, J., Loucks, C., Ginsberg, J., Dinerstein, E., Seidensticker, J., Leimgruber, P., Songer, M., Heydlauff, A., O'Brien, T., Bryja, G., Klenzendorf, S., Sanderson, E.W., Redford, K.H., Chetkiewicz, C.L.B., Medellin, R.A., Rabinowitz, A.R., Robinson, J.G., Taber, A.B., 2002. Planning to save a species: the jaguar as a model. Conser. Biol. 16, 58–72.

Sanderson, E.W., Redford, K.H., Weber, B., Aune, K., Baldes, D., Berger, J., Carter, D., Curtin, C., Derr, J.N., Dobrott, S., Fearn, E., Fleener, C., Forrest, S., Gerlach, C., Gates, C.C., Gross, J.E., Gogan, P., Grassel, S., Hilty, J.A., Jensen, M., Kunkel, K., Lammers, D., List, R., Minkowski, K., Olson, T., Pague, C., Robertson, P.B., Stephenson, B., 2008. The ecological future of the North American Bison: conceiving long-term, large-scale conservation of wildlife. Conserv. Biol. 22, 252–266.

Schreber, J.C.D., 1775. Die Säugethiere in Abbildungen nach der Natur mit Beschreibungen, vol. 2(14). Wolfgang Walther, Erlangen.

Sharma, K., Bayrakcismith, R., Tumursukh, L., Johansson, O., Sevger, P., McCarthy, T., Mishra, C., 2014. Vigorous dynamics underlie a stable population of the endangered snow leopard *Panthera uncia* in Tost Mountains, South Gobi, Mongolia. PLOS One DOI: 10.1371/journal.pone.0101319.

Thenius, E., 1969. Über das Vorkommen fossiler Schneeleoparden (Subgenus *Uncia*, Carnivora, Mammalia). Säugetierk Mitt. 17, 234–242.

Thorbjarnarson, J., et al., 2006. Regional habitat conservation priorities for the American crocodile. Biol. Conserv. 128, 25–36.

Tscherski, J.D., 1892. Beschreibung der Sammlung posttertiairer Saugethiere. Mem. Acad. Imp. Sci. St. Petersbourg Ser. 7 40 (1), 1–511, pls. 1–6.

Tseng, Z.J., Wang, X., Slater, G.J., Takeuchi, G.T., Li, Q., Liu, J., Xie, G., 2014. Himalayan fossils of the oldest known pantherine establish ancient origin of big cats. Proc. Royal Soc. B 281 (1774), .

Wikramanayake, E.D., Dinerstein, E., Robinson, J.G., Karanth, U., Rabinowitz, A., Olson, D., Mathew, T., Hedao, P., Conner, M., Hemley, G., Bolze, D., 1998. An ecology-based method for defining priorities for large mammal conservation: the tiger as case study. Conserv. Biol. 12, 865–878.

Williams, P.A., 2006. A GIS assessment of snow leopard potential range and protected areas throughout inner Asia; and the development of an internet mapping service for snow leopard protection. Master's Thesis, University of Montana, 101 pp.

Zeller, K.A., 2007. Jaguars in the New Millennium Data Set Update: the State of the Jaguar in 2006. Wildlife Conservation Society, Bronx, NY.

CHAPTER

4

Snow Leopard Prey and Diet

*David Mallon**, *Richard B. Harris***, *Per Wegge*[†]

*Division of Biology and Conservation Ecology,
Manchester Metropolitan University, Manchester, UK
**Game Division, Wildlife Program,
Washington Department of Fish and Wildlife, Olympia, WA, USA
[†]Department of Ecology and Natural Resource Management,
Norwegian University of Life Sciences, Ås, Norway

INTRODUCTION

Snow leopards (*Panthera uncia*) have long been reported as preying mainly on mountain ungulates and domestic livestock, supplemented by smaller mammals and gamebirds (e.g., Bannikov, 1954; Blanford, 1888–1891; Heptner and Sludskii, 1972; Roberts, 1977; Schaller, 1977). Many subsequent reports have detailed individual prey species and aspects of dietary composition. For a summary see Sunquist and Sunquist (2002, and references therein).

A quite wide prey spectrum has been reported in the literature (Table 4.1). These prey items have been identified through direct observation by researchers, local reports, analysis of stomach contents or prey remains in scats, and lately by remote cameras. In a few cases, species may have been inferred as prey because their distribution overlaps that of the snow leopard so are essentially speculative.

PREY SPECIES

Mountain Ungulates (Caprinae)

The two most frequently reported prey species of the snow leopard are the bharal (or blue sheep/naur, *Pseudois nayaur*) and Siberian ibex (*Capra sibirica*). The ranges of these two species together cover virtually the entire range of the snow leopard. Bharal and Siberian ibex have largely allopatric distributions, with some small zones areas of overlap. Other mountain ungulates comprise a smaller element of the diet but are locally important.

Bharal (or Blue Sheep/Naur) **(P. nayaur)**

This species occurs across the Qinghai-Tibet Plateau and the Himalayas and Trans-Himalayas, from Yunnan in the southeast through northeast India, Bhutan, Sikkim, Nepal, and northwest India, just extending into the Karakoram range in the Shimshal valley of northern Pakistan.

Snow Leopards
http://dx.doi.org/10.1016/B978-0-12-802213-9.00004-3

TABLE 4.1 Species Recorded in Snow Leopard Diet

Prey species	Reported by:
Siberian ibex *Capra ibex*	Anwar et al. (2011); Shehzad et al. (2012a,b)
Markhor *C. falconeri*	Schaller (1977); Anwar et al. (2011)
Blue sheep *P. nayaur*	Devkota et al. (2013)
Himalayan tahr *H. jemlahicus*	Lovari et al. (2009)
Argali *O. ammon*	Heptner and Sludskii (1972); Shehzad et al. (2012a,b)
Ladakh urial *O. orientalis/vignei*	Chundawat and Rawat (1994)
Red deer *C. elaphus*	Lhagvasuren and Munkhtsog (2000); Zhiryakov and Baidavletov (2002)
White-lipped deer *Przewakskium albirostris*	Schaller (1998)
Siberian roe deer *C. pygargus*	Heptner and Sludskii (1972); Lhagvasuren and Munkhtsog (2000)
Musk deer *Moschus* spp.	Schaller (1998); Ferretti et al. (2014)
Moose *A. alces*	Zhiryakov and Baidavletov (2002)
Goitered gazelle *G. subgutturosa*	Bannikov (1954); Dash et al. (1977); Heptner and Sludskii (1972); Lhagvasuren and Munkhtsog (2000)
Wild ass *E. hemionus*	Bannikov (1954); Dash et al. (1977)
Wild boar *S. scrofa*	Zhiryakov and Baidavletov (2002)
Siberian marmot *M. sibirica*	Bannikov (1954)
Himalayan marmot *M. himalayana*	Devkota et al. (2013); Oli et al. (1993)
Grey marmot *M. baibacina*	Zhiryakov and Baidavletov (2002)
Long-tailed marmot *M. caudata*	Jumabay-Uulu et al. (2014)
Menzbier's marmot *M. menzbieri*	(see Chapter 34; Esipov et al.)
Red squirrel *S. vulgaris*	Zhiryakov and Baidavletov (2002)
Tien Shan ground squirrel *S. relictus*	Heptner and Sludskii (1972)
Royle's vole *A. roylei*	Oli et al. (1993)
Pika *Ochotona* spp.	Chundawat and Rawat (1994)
Royle's pika *O. roylei*	Oli et al. (1993)
Tolai hare *L. tolai*	Bold and Dorjzunduy (1976); Schaller (1998); Zhiryakov and Baidavletov (2002)
Arctic hare *L. timidus*	Zhiryakov and Baidavletov (2002)
Woolly hare *L. oiostolus*	Chundawat and Rawat (1994)
Red fox *V. vulpes*	Oli et al. (1993); Chundawat and Rawat (1994); Zhiryakov and Baidavletov (2002)
Stone marten *M. foina*	Oli et al. (1993)
Yellow-throated marten *M. flavigula*	Lyngdoh et al. (2014)
Least weasel *M. nivalis*	Oli et al. (1993)
Common palm civet *P. hermaphroditus*	Lyngdoh et al. (2014)
Eurasian badger *M. meles*	Zhiryakov and Baidavletov (2002)
Brown bear *U. arctos*	Heptner and Sludskii (1972)
Snowcock *Tetraogallus* spp.	Bannikov, (1954); Heptner and Sludskii (1972); Zhiryakov and Baidavletov (2002)
Rock ptarmigan *L. mutus*	Zhiryakov and Baidavletov (2002)

TABLE 4.1 Species Recorded in Snow Leopard Diet (*cont.*)

Prey species	Reported by:
Chukar *Alectoris chukar*	Shehzad et al. (2012a,b)
Pheasants	Dang (1967); Heptner and Sludskii (1972);
Pigeons	Bold and Dorjzunduy (1976); Heptner and Sludskii (1972)
Birds	Anwar et al. (2011); Devkota et al. (2013)
Livestock	
Horse	Devkota et al. (2013); Heptner and Sludskii (1972)
Donkey	Bagchi and Mishra (2006)
Yak and hybrid	Devkota et al. 2013; Anwar et al. (2011)
cattle	Anwar et al. (2011); Heptner and Sludskii, 1972
Sheep	Devkota et al. (2013); Anwar et al. (2011) Shehzad et al. (2012a,b)
Goat	Devkota et al. (2013); Anwar et al. (2011) Shehzad et al. (2012a,b)
Dog	Sunquist and Sunquist (2002)
Camel	Johansson et al. (2015)
Insects	Bold and Dorjzunduy (1976) Lhagvasuren and Munkhtsog (2000)
Vegetation	Bold and Dorjzunduy (1976); Schaller (1977); Mallon (1991); Wegge et al. (2012)

The northeastern limit of the distribution is in the Helan Shan range on the border between Ningxia Hui Autonomous Region and Inner Mongolia in China (Harris, 2014; Wang and Hoffmann, 1987). Bharal occur at elevations of 2500–5500 m (Schaller, 1977; Wang and Hoffmann, 1987). Harris (2014) estimated the total population size at somewhere between 47,000 (a conservative estimate) and 414,000 (probably an overestimate). It apparently remains fairly abundant in much of its range among the ranges of the Qinghai-Tibet Plateau, but the population trend is unknown; it is assessed on the IUCN Red List with a status of Least Concern (Harris, 2014). Males weigh 60–75 kg, females 35–55 kg (Wang and Hoffmann, 1987). It is reported as the main prey of snow leopard in Shey in northwest Nepal (Schaller, 1977); Annapurna area (Oli et al., 1993; Wegge et al., 2012), sites in Gansu, Qinghai, and Xinjiang, China (Schaller et al., 1987, 1988), and Ladakh, India (Chundawat and Rawat, 1994).

Siberian or Himalayan Ibex (Capra sibirica)

This species ranges from the Hindu Kush in Afghanistan, northeast through the Tien Shan and Pamir mountains of Central Asia, to the Altai and Sayan ranges in Mongolia and southern Siberia. It also occurs in the Karakoram and northwest Himalayas as far east as the Sutlej River in India (Fedosenko and Blank, 2001; Schaller, 1977; Shackleton, 1997). Some local declines have been reported, but the population trend is unknown and it is assessed with a status of Least Concern on the IUCN Red List (Reading and Shank, 2008). No overall population estimate is available. In India, population estimates include a minimum of 6000 in Ladakh and perhaps 4000 on the south side of the main Himalayas in Jammu & Kashmir and Himachal Pradesh (Fox et al., 1991, 1992). According to Shackleton (1997), there were an estimated 100,000–110,000 ibex in the former Soviet Union; most in Kyrgyzstan and Tajikistan (ca. 70,000) and Kazakhstan (ca. 17,000); 8,000–9,000 in southern Siberia and 2,400 in Uzbekistan. Males may reach a weight of 130 kg (Fedosenko and Blank, 2001); it is reported to be the main snow leopard prey in Mongolia (Bannikov, 1954; McCarthy, 2000), Russia and the Central Asian countries (Heptner and Sludskii, 1972) and Chinese Tien Shan (Schaller et al., 1988).

Markhor (Capra falconeri)

Markhor have a relatively limited distribution in the Pir Panjal range of Jammu

& Kashmir, northwest India, northern Pakistan, northeastern Afghanistan, and southern Tajikistan. Other populations in Baluchistan (Pakistan), Uzbekistan, and Turkmenistan lie outside snow leopard range (Shackleton, 1997). Markhor occupy elevations between 600 m and 3600 m at and around the treeline and are found in scrub forests made up primarily of oak (*Quercus ilex*), pine (*Pinus gerardiana*), and juniper (*Juniperus macropoda*) (Heptner and Sludskii, 1972; Roberts, 1977; Schaller, 1977). Males weigh 80–86 kg and females 41 kg (Heptner and Sludskii, 1972). Total numbers have been recently estimated at ca. 9700 (including the populations outside snow leopard range) with an increasing trend, and it has been reassessed as Near Threatened on the IUCN Red List, down from Endangered (Michel and Rosen Michel, 2015). In April 2012, 1018 were counted in Tajikistan (Michel et al., 2015). The population in Pir Panjal numbered 350–375 in 2004–2006 (Bhatnagar et al., 2009). The interface with snow leopards is much less than with the two preceding species, but markhor may represent an important prey item in Chitral and Gilgit, northern Pakistan (Roberts, 1977; Schaller, 1977) and perhaps a few other places.

Himalayan Tahr (Hemitragus jemlahicus)

This species occurs along the Himalayan range from southern Jammu & Kashmir in India, eastwards through Nepal to Sikkim, and in a few places on the Chinese side of the border (Bhatnagar and Lovari, 2008; Shackleton, 1997), but not recorded in Bhutan. Himalayan tahr inhabit steep rocky mountain sides, around the treeline, forest edges, and rhododendron scrub, between 2500 m and 4000 m (Bhatnagar and Lovari, 2008; Smith and Xie, 2008). There is no overall population estimate available, but the species is considered to be declining, and it is assessed as Near Threatened on the IUCN Red List (Bhatnagar and Lovari, 2008). The overlap with snow leopard distribution is fairly limited, though Himalayan tahr has been shown to be an important prey species in Sagarmatha NP in Nepal (Ferretti et al., 2014; Lovari et al., 2009) and may be so elsewhere along the Himalayas.

Argali (Ovis ammon)

Argali has an extensive range across Central Asia, the Qinghai-Tibet Plateau, and the Altai and Sayan ranges of Mongolia and Russia (Harris and Reading, 2008; Mallon et al., 2014). Argali is assessed as Near Threatened and declining on the IUCN Red List (Harris and Reading, 2008). No robust population estimate is available. Argali distribution coincides with much of that of the snow leopard, but as is typical of wild sheep, argali prefer less precipitous and more open, rolling terrain, and these differences in habitat preferences may therefore reduce the overlap in some localities. Jumabay-Uulu et al. (2014) recorded that argali made up more than 50%, of snow leopard diet in Sarychat-Ertash State Reserve in Kyrgyzstan. Shehzad et al. (2012a,b) recorded argali in 8.6% of 81 scats in Tost Uul, southern Mongolia (compared to 70.4% for Siberian ibex).

Urial (Ovis orientalis)

Urial occur in the northwest Himalayas, Karakoram, Hindu Kush, and southwest Pamir, as well as outside snow leopard range in western Central Asia, Iran, the Caucasus, and Turkey (Shackleton, 1997). Several subspecies have been named and the eastern forms are separated by some authors as *Ovis vignei*. The eastern limit of urial range is in the Indus Valley of Ladakh, India. The species is assessed as Vulnerable and declining on the IUCN Red List (Valdez 2008), but no global population estimate is available. Urial generally occur at lower elevations than the other mountain ungulates in the region, and they also prefer more open terrain. Both these factors reduce the extent of overlap with the snow leopard. Chundawat and Rawat (1994) recorded urial in 0.4% of 173 scats collected in Hemis NP, Ladakh, India.

Deer

Several species of deer (Cervidae and Moschidae) occur within snow leopard range.

Red Deer (Cervus elaphus)

This largest species comprises several named subspecies, some of which are sometimes regarded as separate species. Red deer occur widely, but patchily in Central Asia, the Altai mountains, parts of the Qinghai-Tibet Plateau, southeast Tibet, and Jammu & Kashmir (India), mainly in areas of forest and alpine meadows (Lovari et al., 2008; Smith and Xie, 2008). The species is classified as Least Concern with an increasing trend globally (Lovari et al., 2008), though many populations within snow leopard range may be under threat from poaching; it is reported as prey in Kazakhstan, Mongolia, China, and Russia (Heptner and Sludskii, 1972; Lhagvasuren and Munkhtsog, 2000; Schaller, 1998, Zhiryakov and Baidavletov, 2002).

White-Lipped Deer (Cervus albirostris)

The species previously ranged across much of the eastern Tibetan Plateau and currently occurs in fragmented populations in Gansu, Qinghai, eastern Tibet, western Sichuan, and northwest Yunnan (Koizumi et al., 1993; Schaller 1998). White-lipped deer inhabit conifer forest, rhododendron and willow scrub as well as alpine grasslands up to 5100 m (Koizumi et al., 1993). It is listed as Vulnerable on the IUCN Red List, with an unknown population trend (Harris, 2015). White-lipped deer comprised 4.5% of the diet in one sample of snow leopard scats from southern Qinghai (Schaller, 1998).

Siberian Roe Deer (Capreolus pygargus)

Roe deer occur in the Tien Shan, Altai, and Sayan and eastern parts of the Qinghai-Tibet Plateau; it is listed as Least Concern but decreasing on the IUCN Red List (Gonzalez and Tsytsulina, 2008). Roe deer remains were found in 11.5% and 31% of snow leopard scats from two sites in Kazakhstan (Zhiryakov and Baidavletov, 2002).

Musk Deer (Moschus spp.)

Six species occur across many parts of snow leopard range. All are mainly forest and alpine/subalpine scrub dwelling species, so habitat overlap with the snow leopard is to some extent limited. Five species are considered Endangered on the IUCN Red List and one is Vulnerable (IUCN, 2015). Musk deer made up 15.3% of the diet in one sample from Zadoi, southern Qinghai (Schaller, 1998) and 19% in Sagarmatha NP, Nepal (Ferretti et al., 2014). Musk deer are also recorded as prey in Russia (Heptner and Sludskii, 1972) and northern Pakistan (see Chapter 38.2).

Other Ungulates

Wild boar (*Sus scrofa*) has been recorded as prey in Central Asia (Heptner and Sludskii, 1972) and Zhiryakov and Baidavletov (2002) found wild boar remains in 63.6% of one small sample of scats from northern Kazakhstan. Moose (*Alces alces*) has also been recorded as prey in Kazakhstan (Zhiryakov and Baidavletov, 2002). Other species include goitered gazelle (*Gazella subgutturosa*) (Lhagvasuren and Munkhtsog, 2000; Novikov, 1956); wild camel (*Camelus ferus*), and wild ass (*Equus hemionus*) in Mongolia (Dash et al., 1977; Tulgat and Schaller, 1992); and goral (*Nemorhaedus* spp.), serow (*Capricornis* spp.), and takin (*Budorcas taxicolor*) (Dang, 1967), though it is unclear if all these species were confirmed through dietary analysis or inferred on the basis of local reports or range overlaps.

Small and Medium-Sized Mammals

Marmots

Five species (*Marmota sibirica, Marmota himalayana, Marmota baibacina, Marmota caudata,*

Marmota menzbieri) occur across the global range of the snow leopard. *M. menzbieri* has a restricted range in the western Tien Shan and is listed as Vulnerable on the IUCN Red List (Tsytsulina, 2008), while the others are more widely distributed and assessed as Least Concern. Marmots are absent from some arid areas such as the Gobi and other localities. Marmots hibernate for 5–7 months of the year, depending on species, latitude, and altitude, and even where they occur, are thus available for only about half the year. Marmots weigh 4–6.5 kg and may be an important supplementary prey in some localities. Oli et al. (1993) reported marmot remains in 22.9–27.9% of 213 scats from spring to fall in Nepal. Marmot was recorded in 36.5–65.3% of the samples from four sites across Qinghai, China (Schaller, 1998); 15–20% in Mongolia (McCarthy, 2000); and 8% in Sarychat-Ertash Reserve in Kyrgyzstan (Jumabay-Uulu et al., 2014).

Hares

Four species (*Lepus oiostolus, Lepus tibetanus, Lepus timidus, Lepus tolai*) occur across almost the whole of snow leopard range. All are listed as Least Concern (IUCN, 2015). *L. tolai* has been recorded in the diet in South Gobi of Mongolia (Bold and Dorjzunduy, 1976); *L. tolai* and *L. timidus* recorded in Kazakhstan (Zhiryakov and Baidavletov, 2002). *L. tolai* remains made up 1% of 72 scats analyzed from Taxkorgan, Xinjiang, China, (Schaller et al., 1987) and 4.5% of the diet in the Shule Nanshan, Qinghai (Schaller, 1998). *L. oiostolus* remains were found in 3.1% of 172 scats from Hemis NP in Ladakh, India (Chundawat and Rawat, 1994).

Other Mammals

Pikas (*Ochotona* spp.) are small members of the Lagomorpha. Around 26 species are distributed within and around snow leopard range (Hoffmann and Smith, 2005). *Ochotona roylei* has been reported in a few studies. Pikas weigh up to ca. 350 g so are unlikely to be a significant component in snow leopard diet; the same point applies to small rodents such as voles (Microtinae, e.g., *Alticola roylei*; Oli et al., 1993). Red squirrel (*Sciurus vulgaris*) was reported in scats in Kazakhstan (Zhiryakov and Baidavletov, 2002) and relict ground squirrel (*Spermophilus relictus*) by Heptner and Sludskii (1972).

Several carnivores have been reported as prey, including red fox (*Vulpes vulpes*) (Chundawat and Rawat 1994; Oli et al., 1993), stone marten (*Martes foina*) and least weasel (*Mustela nivalis*) (Oli et al., 1993); and Eurasian badger (*Meles meles*) (Zhiryakov and Baidavletov, 2002). A snow leopard killed and partially ate a 2-year-old brown bear (*Ursus arctos*) in the Tien Shan (Heptner and Sludskii, 1972). Common palm civet (*Paradoxurus hermaphroditus*) was reported in 16.8% of scats in a review by Lyngdoh et al. (2014), a surprising result since this species occurs only up to 2400 m along the Himalayas (Duckworth et al., 2008) and thus has a limited overlap with the snow leopard.

Birds

Birds, mainly Galliformes, have been reported as prey by several authors (e.g., Anwar et al., 2011; Heptner and Sludskii, 1972), usually in small quantities. Three species of snowcock (*Tetraogallus himalayensis, Tetraogallus tibetanus, Tetraogallus altaicus*) occur at high elevations across most of snow leopard range; all are assessed as Least Concern (BirdLife International 2012). Their remains have been found in scats in Kazakhstan (Zhiryakov and Baidavletov, 2002) and Mongolia (Lhagvasuren and Munkhtsog, 2000). Rock ptarmigan (*Lagopus mutus*) was also recorded in the diet in Kazakhstan (Zhiryakov and Baidavletov, 2002). Chukar (*Alectoris chukar*) made up 1.2% of scats from Tost Uul in the South Gobi of Mongolia (Shehzad et al., 2012a,b) and blue hill pigeon (*Columba rupestris*) remains occurred in scats from the same location (Bold and Dorjzunduy, 1976). Birds and eggshell were found in scats in Manang, Nepal (Wegge et al., 2012).

Invertebrates

Bold and Dorjzunduy (1976) reported grasshoppers (Orthoptera) in scats collected in South Gobi, and Lhagvasuren and Munkhtsog (2000) also found insects in scats from Mongolia.

Vegetation

One interesting facet of snow leopard diet is the frequency with which vegetation occurs in snow leopard scats, in contrast to other big cats (*Panthera* spp.). Bold and Dorjzunduy (1976) found feather grass (*Stipa* spp.) in 18% of 50 scats from South Gobi, Mongolia, and observed that it was "*apparently consumed to a significant extent in autumn.*" Schaller (1977) found *Rheum emodi*, *Polygonum* sp. and grass in his sample from northern Pakistan. Vegetation occurred in 0.7, 9.1, and 11% of scats in three samples from Kazakhstan (Zhiryakov and Baidavletov, 2002). In northern Nepal, grass, twigs, or leaves occurred in 19.3% of 213 scats, with six mainly composed of plant material (Oli et al., 1993) and similar plant material was also found in their sample by Devkota et al. (2013). Plant material occurred in 62% of scats sampled from Phu Valley, Manang, in Nepal, and often dominated the contents (Wegge et al., 2012).

Scats containing twigs of tamarisk (*Myricaria* spp.) have been reported in Ladakh: 29 of 50 scats, 4 wholly composed of twigs (Mallon 1984, 1991); and 41% of 173 scats, 25 wholly composed of twigs (Chundawat and Rawat, 1994); and also Kyrgyzstan (in 45% of scats; Jumabay-Uulu et al., 2014). Schaller (1998) said that vegetation made up 2.2–11% of scats from four sites in Qinghai, China, and included twigs of *Tamarix*, *Salsola arbuscula*, and *Sibiraea angustata*. Chundawat and Rawat (1994) actually observed one snow leopard feeding on a *Myricaria* bush after it had fed on a kill and reported further signs of feeding on these bushes, especially during the mating season.

Snow leopards are not physiologically adapted to digest plant cellulose so the reasons for consuming vegetation are not clear. Since *Myricaria* is clearly sought out or preferred, at least in Ladakh, perhaps the most likely explanation is that it contains secondary compounds that possess some dietary value; it may perhaps act as a vermifuge. Chemical analysis of the plant would be necessary to investigate this. Other potential reasons include acting as a scour, to bind material to be expelled, or help to keep the digestive system functioning in some way.

Livestock

Predation on domestic livestock, (sheep and goats, yaks, cattle and yak-cattle hybrids, horses, donkeys, camels, and dogs) has been reported from most parts of snow leopard range. The proportion of livestock consumed ranges from zero where unavailable (e.g., Jackson and Ahlborn, 1988; Jumabay-Uulu et al., 2014) upward, but is generally around 15–30% (Bagchi and Mishra, 2006; Chundawat and Rawat, 1994; Oli et al., 1993; Schaller et al., 1988; Shehzad et al., 2012a,b; Wegge et al., 2012) but can reach 70% (Anwar et al., 2011). The number and type of domestic animals killed and frequency of predation events depend on several factors, including herding and guarding practices, availability of other prey, season, terrain, and other factors. The role of livestock in snow leopard diet is described in more detail in Chapter 6. Livestock are attacked on open pastures, especially when left to roam free but on occasion, a snow leopard may gain entry to a night time corral housing flocks of sheep and goats and kill many at one time (Jackson et al., 2010; Mallon, 1984; Sunquist and Sunquist, 2002). These "surplus killing" events cause heavy economic losses and are greatly resented by herders. Remedial measures to improve the security of nighttime corrals in order to reduce this threat are described in Chapter 14.1.

DIETARY COMPOSITION

The proportions of different categories of prey and of individual species in snow leopard diet vary, as illustrated by the figures cited in the preceding species summaries. These in turn reflect variations in availability and density of prey across snow leopard range, seasonal effects, and site-specific factors such as elevation, vegetation, and terrain type. Variation at the level of the individual snow leopard is also evident. Lyngdoh et al. (2014) reviewed 25 dietary studies (14 published and 11 unpublished) from across the range, then used cluster analysis to identify four regional zones that contained unique prey assemblages and characteristics.

The results overall, do however coalesce around the point that wild ungulates and/or livestock comprise the bulk of the diet and are supplemented by medium and small mammals and birds, with vegetation a minor but widespread component, for reasons still to be ascertained. Making comparisons among and between studies is not straightforward because results may be reported variously as frequency of occurrence, total scat content, or estimated biomass; sample sizes are often small; they are rarely accompanied by robust information on prey availability, and sampling biases may influence the results to an unknown extent.

A further, important caveat is that information on diet has until recently largely been garnered from microscopic analysis of hair in scats, and identified through patterns of the medulla and cross-section compared to reference collections (see e.g., Oli, 1993). This methodology is widely established, but accurate results depend completely on the scats analyzed being assigned correctly to the target species.

Snow leopard scats have traditionally been identified by eye. Recent studies using more advanced techniques of fecal DNA analysis have demonstrated that visual identification of snow leopard scats is not reliable, even where care has been taken to sample only scats that were judged to be snow leopards based on their appearance and/or in association with other field signs, such as scrapes, tracks, and scent marks. Error rates for misidentified snow leopard scats were 30.5% in Manang, Nepal (Wegge et al., 2012); 34% in Sarychat-Ertash reserve, Kyrgyzstan (Jumabay-Uulu et al., 2014); 35% in Ladakh and 51% in Mongolia (Janečka et al., 2008); 41% in China and Kyrgyzstan (McCarthy et al., 2008); 49.4% in Baltistan (Anwar et al., 2011); 59% in south Gobi, Mongolia (Janečka et al., 2011); and 56.7% (115 of 203 scats) in Tost Uul, South Gobi (Shehzad et al., 2012a,b). Misidentified scats were shown to be mainly those of gray wolf (*Canis lupus*) or red fox (*V. vulpes*) but also included stone marten (*M. foina*) (Janečka et al., 2011; McCarthy et al., 2008; Shehzad et al., 2012a,b).

Studies elsewhere have shown that the problem of misidentified carnivore scats is not confined to the snow leopard: similar results have been reported for Arabian leopard (*Panthera pardus nimr*) in Israel (Perez et al., 2006); gray wolf in Kyrgyzstan (error rate 28%; Jumabay-Uulu et al., 2014); large and small carnivores in Venezuela (Farrell et al., 2000); and medium-sized carnivores in UK (Davison et al., 2002; Hansen and Jacobsen, 1999).

This recent research demonstrates that DNA analysis is required to confirm identification of snow leopard scats before they are analyzed. As a consequence, there may be doubt about the accuracy of some details of earlier dietary studies because it cannot be guaranteed that all scats in the sample analyzed were in fact those of snow leopard.

RECENT DIETARY STUDIES

Five recent studies have used molecular tools to confirm species identification prior to analysis of the scat contents, confirming that the scats were all those of snow leopard. DNA analysis and barcoding of scat contents is increasingly replacing microscopic identification to reconstruct dietary preferences (see Chapter 27.2).

Lovari et al. (2009) in Sagarmatha NP, Nepal, found that Himalayan tahr was the most frequent prey item (48% summer, 37% autumn); followed by domestic cattle (15%, 27%); musk deer (20%, 15%); and birds (3%; 14%) with occasional voles.

Anwar et al. (2011) reported the diet at four sites in northern Pakistan. They reported that by biomass 70.7% was livestock, 20% Siberian ibex, 7.8% markhor, and 2.2% birds (chukar and Himalayan snowcock). No marmots or pikas were recorded, but vegetation was present in 29 scats.

Wegge et al. (2012), Phu Valley, Manang, Nepal, recorded 42% livestock and 58% wild prey, of which 9.2% was blue sheep. Royle's pika and voles were frequently present in scats but made up only 1.4% of the estimated biomass. Birds and eggs were also detected occasionally. Plant material occurred in 62% of all samples, often dominating the scat content.

Shehzad et al. (2012a,b), Tost Uul of southern Mongolia, reported Siberian ibex 70.4%, argali 8.6% (so wild ungulates, 79%); livestock 19.7% (sheep and goats); and less than 2% small mammals and birds (chukar, 1.2%).

Jumabay-Uulu et al. (2014), Sarychat-Ertash State Reserve in the Tien Shan, Kyrgyzstan, reported that argali made up ca. > 50% of the diet, ibex 30%, and marmots 8%, with unidentified smaller mammals making up the rest.

In a separate type of study, Johansson et al. (2015) used cluster analysis of locations of 19 satellite-collared snow leopards to track kills in South Gobi, Mongolia. Out of 249 kills, 65% consisted of Siberian ibex, 8% argali (a total of 73% wild ungulates), 20% sheep and goats, 4% horse, and 2% camels (27% livestock). This methodology is not conducive to tracking kills of small prey, which as a result may be underrepresented.

These studies based on unambiguous snow leopard samples confirm that the snow leopard's main prey consist of wild ungulates and livestock. The principal difference from many earlier studies lies in the lower proportions of medium and small prey items, such as marmots, hares, and other small mammals. In some cases, this reflects features of the respective study sites; for example, hares and marmots are absent from the Phu Valley (Wegge et al., 2012) and marmots from Tost Uul (Shehzad et al., 2012a,b). The difference may, however, also derive in part from the inadvertent inclusion in samples of scats of medium-sized predators, such as red fox, species that would be expected to feed more on small and medium-sized prey. It is interesting to note that a DNA analysis of scats of large and small carnivores in Venezuela showed large carnivores consuming mainly large and medium prey items, with small and medium-sized carnivores preferring small and medium prey, a result that contrasted with previous dietary analyses based on scat size (i.e., visual identification) (Farrell et al., 2000).

During the last few years, the technique of using DNA sequencing of scats for verification of the target predator has been taken one step forward. From genetic barcoding of short DNA fragments in the scats, the content can be identified to species or taxonomic group, thus providing information on what the predator has recently consumed (Valentini et al., 2009; Shehzad et al. 2012a,b) (see Chapter 27.2). When further refined, this method is likely to replace the rather laborious and time-consuming microscopic identification method.

DIETARY REQUIREMENTS AND OFFTAKE RATES

Based on radio tracking, Jackson and Ahlborn (1984) estimated that snow leopards make a large kill (mainly blue sheep) every 10–15 days, implying an annual requirement of 24–36 blue sheep. Wegge et al. (2012) estimated that an adult snow leopard's food requirements are 1168 kg/year and that in the Phu Valley, Nepal, an adult killed two blue sheep and one head of livestock per month.

Johansson et al. (2015) found that 19 satellite tagged snow leopards in Tost Uul, Mongolia, killed an ungulate (ibex, argali, livestock) every 8.0 (+/− 0.53 days) and that kill rates were higher than estimates calculated from scat-based studies. These authors also reported that ibex and argali were preyed upon in proportion to their abundance, but the diet consisted of 73% wild species and 27% livestock, despite numbers of the latter being one order of magnitude higher. Adult male snow leopards killed larger prey, and 2–6 times more livestock, than females and younger animals (Johansson et al., 2015).

Wegge et al. (2012) estimated that snow leopards harvested 15.1% annually of the blue sheep population in the Phu Valley, which they considered close to the annual recruitment rate. The return of snow leopards to Sagarmatha NP in Nepal after an absence of 40 years was followed by a 66% fall in the numbers of Himalayan tahr from 2003 to 2010, and then a reduction in the number of snow leopards (Ferretti et al., 2014; Lovari et al., 2009).

A female snow leopard and two young ate a whole blue sheep in less than 48 h (Jackson and Ahlborn, 1988). Unless disturbed, a snow leopard may remain on its kill for up to a week (Fox and Chundawat, 1988).

COMPETITORS

The gray wolf (*C. lupus*) is roughly equal in size and co-occurs across snow leopard range, though it tends to prey in more open areas and is less adapted to hunt prey in the precipitous terrain favored by the snow leopard. Eurasian lynx (*Lynx lynx*) and wild dog (*Cuon alpinus*) are also found across snow leopard range – lynx quite widely and wild dog locally and more rarely. Common leopard (*P. pardus*) distribution adjoins that of the snow leopard along the Himalayas and in parts of China, but it is mainly a forest species. Lovari et al. (2013) estimated 10,000 km^2 of overlap between the two species. The tiger (*Panthera tigris*) has a limited overlap with snow leopard in Bhutan (see Chapter 35). Several species of sympatric small and medium-sized carnivores such as foxes, small felids, and mustelids are unlikely to represent serious competitors of the snow leopard.

Jumabay-Uulu et al. (2014) compared diets of snow leopard and gray wolf in Sarychat-Ertash Reserve in the Tien Shan of Kyrgyzstan and reported that snow leopards took argali and ibex in proportion to their availability but wolves preyed more heavily on argali and underselected ibex; wolves also consumed more marmots than snow leopards (20% vs. 8%).

Lovari et al. (2013) reviewed the diets of snow leopard and common leopard. They concluded that as the two species are roughly similar in size and food habits, competition is likely where they occur in sympatry, around the upper forest limit, and that as the treeline rises with the effects of climate change, habitat for the common leopard will increase and that of the snow leopard decrease.

CONCLUSIONS

The broad outlines of snow leopard diet are becoming clear, and advances in fecal DNA analysis have provided a reliable tool for investigating snow leopard diet, giving dependable identification of scats and therefore results. Only a few studies utilizing this methodology have taken place so far, and these are not representative of the full geographical distribution of the snow leopard. Complementary studies, including from other areas such as the Altai and mountains of Central Asia and covering regional and seasonal differences, are thus needed to provide a fuller picture of snow leopard diet. The relationship between snow leopard density and prey density and off-take rates are another necessary area of research. Robust estimates of global numbers are not available for most key prey species and in some cases even the population trend is unknown – although there are frequent anecdotal reports of population declines at local level. The logistical and statistical issues that complicate the generation of reliable population estimates of mountain ungulates are well-known (see e.g., Wingard et al., 2011), but overcoming these problems will be essential in assessing the size of the main prey base and trends in the most important species.

References

Anwar, M.B., Jackson, R., Nadeem, M.S., Janečka, J.E., Hussain, S., Beg, M.A., Muhammad, G., Qayyum, M., 2011. Food habits of the snow leopard *Panthera uncia* (Schreber, 1775) in Baltistan, Northern Pakistan. Eur. J. Wildl. Res. 57, 1077–1083.

Bagchi, S., Mishra, C., 2006. Living with large carnivores: predation on livestock by the snow leopard (*Uncia uncia*). J. Zool. 268, 217–224.

Bannikov, A.G., 1954. Mammals of the Mongolian People's Republic. Academy of Sciences, Moscow, (In Russian).

Bhatnagar, Y.V., Lovari, S., 2008. *Hemitragus jemlahicus*. The IUCN Red List of Threatened Species. Version 2014.3. www.iucnredlist.org

Bhatnagar, Y.V., Ahmad, R., Subba Kyarong, S., Ranjitsinh, M.K., Seth, C.M., Ahmed Lone, I., Easa, P.S., Kaul, R., Raghunath, R., 2009. Endangered markhor *Capra falconeri* in India: through war and insurgency. Oryx 43 (3), 407–411.

Bird Life International, 2012. *Tetraogallus altaicus*. The IUCN Red List of Threatened Species. Version 2015.2. www.iucnredlist.org.

Blanford, W.T., 1888–1891. The Fauna of British India: Mammalia. Taylor & Francis, London.

Bold, A., Dorjzunduy S., 1976. Report on snow leopards in the southern spurs of the Gobi Altai. Proceedings of the Institute of General and Experimental Biology – Ulaanbaatar 11, 27–43. (In Mongolian with Russian summary).

Chundawat R.S., Rawat, G.S., 1994. Food habits of the snow leopard in Ladakh. In: Fox, J.L., Jizeng, D. (Eds.). Proceedings of the Seventh International Snow Leopard Symposium. International Snow Leopard Trust and Northwest Plateau Institute of Biology. International Snow Leopard Trust, Seattle, Xining, Qinghai, China, pp. 127–132.

Dang, H., 1967. The snow leopard and its prey. The Cheetal 10, 72–84.

Dash, Y., Szaniawski, A., Child, G., Hunkeler, P., 1977. Observations on some large mammals of the Translatai, Djungarian, and Shargin Gobi, Mongolia. Terre et Vie 31, 587–596.

Davison, A., Birks, J.D.S., Brookes, R.C., Braithewaite, T.C., Messenger, J.E., 2002. On the origin of faeces: morphological versus molecular methods for surveying rare carnivores from their scats. J. Zool. (London) 257, 141–143.

Devkota, B.P., Thakur, S., Jaromir, K., 2013. Prey density and diet of Snow Leopard (*Uncia uncia*) in Shey Phoksundo National Park, Nepal. Appl. Ecol. Environ. Sci. 1 (4), 55–60.

Duckworth, J.W., Widmann P., Custodio, C., Gonzalez, J.C., Jennings, A., Veron, G., 2008. *Paradoxurus hermaphroditus*. The IUCN Red List of Threatened Species. Version 2015.2. www.iucnredlist.org

Farrell, L.E., Roman, J., Sunquist, M.E., 2000. Dietary separation of sympatric carnivores identified by molecular analysis of scats. Mol. Ecol. 9, 1583–1590.

Fedosenko, A.K., Blank, D.A., 2001. Capra sibirica. Mamm. Species 675, 1–13.

Ferretti, F., Lovari, S., Minder, I., Pellizzi, B., 2014. Recovery of the snow leopard in Sagarmatha (Mt. Everest) National Park: effects on main prey. Eur. J. Wildl. Res. 60, 559–562.

Fox, J.L., Chundawat, R.S., 1988. Observations of snow leopard stalking, killing, and feeding behavior. Mammalia 52 (1), 137–140.

Fox, J.L., Nurbu, C., Chundawat, R.S., 1991. The mountain ungulates of Ladakh, India. Biol. Conserv. 58, 167–190.

Fox, J.L., Sinha, S.P., Chundawat, R.S., 1992. Activity patterns and habitat use of ibex in the Himalaya Mountains of India. J. Mammal. 73, 527–534.

Gonzalez, T., Tsytsulina, K., 2008. *Capreolus pygargus*. The IUCN Red List of Threatened Species. Version 2015.2. www.iucnredlist.org.

Hansen, M.M., Jacobsen, L., 1999. Identification of mustelid species: otter (*Lutra lutra*), American mink (*Mustela vison*) and polecat (*Mustela putorius*), by analysis of DNA from faecal samples. J. Zool. 247, 177–181.

Harris, R.B. 2014. *Pseudois nayaur*. The IUCN Red List of Threatened Species. Version 2014.3. www.iucnredlist.org.

Harris, R.B., 2015. *Cervus albirostris*. The IUCN Red List of Threatened Species. Version 2015.2. www.iucnredlist.org.

Harris, R.B., Reading, R., 2008. *Ovis ammon*. The IUCN Red List of Threatened Species. Version 2014.3. www.iucnredlist.org.

Heptner, V.G., Sludskii, A.A., 1972. Mlekopitayushchiye Sovetskogo Soyuza. Vol. 2 Part 2. Khishchniye. [Mammals of the Soviet Union, Carnivora]. Vysshaya Shkola, Moscow. In Russian; English translation 1992, E.J. Brill, Leiden, The Netherlands.

Hoffmann, R.S., Smith, A.T., 2005. Order Lagomorpha. In: Wilson, D.E., Reeder, D.M. (Eds.), Mammal Species of the World. Johns Hopkins University Press, Baltimore, MD, USA, pp. 185–211.

IUCN, (2015). IUCN Red List of Threatened Species. Version 2015.2. www.iucnredlist.org.

Jackson, R., Ahlborn, G., 1984. A preliminary habitat suitability model for the snow leopard, *Panthera uncia*, in West Nepal. International Pedigree Book of Snow Leopards-vol. 4Helsinki Zoo, Helsinki, Finland, pp. 43–52.

Jackson, R., Ahlborn, G.G., 1988. Observation on the ecology of snow leopard (*Panthera uncia*) in west Nepal. In: Freeman, H. (Ed.), Proceedings of the Fifth International Snow Leopard Symposium. International Snow Leopard Trust and Wildlife Institute of India, Seattle, pp. 65–87.

Jackson, R.M., Mishra, C., McCarthy, T.M., Ale, S.B., 2010. Snow leopards: conflict and conservation. In: Macdonald, D.W., Loveridge, A.J. (Eds.), Biology and Conservation of Wild Felids. Oxford University Press, Oxford, pp. 417–430.

Janečka, J.E., Jackson, R.M., Zhang, Y., Li, D., Munkhtsog, B., Buckley-Beason, V., Murphy, W.J., 2008. Population monitoring of snow leopards using noninvasive collection of scat samples: a pilot study. Anim. Conserv. 11, 401–411.

Janečka, J.E., Munkhtsog, B., Jackson, R.M., Naranbaatar, G., Mallon, D.P., Murphy, W.J., 2011. Comparison of noninvasive genetic and camera-trapping techniques for surveying snow leopards. J. Mammal. 92 (4), 771–783.

Johansson, Ö., McCarthy, T.M., Samelius, G., Andrén, H., Tumursukh, L., Mishra, C., 2015. Snow leopard predation in a livestock dominated landscape in Mongolia. Biol. Conserv. 184, 251–258.

Jumabay-Uulu, K., Wegge, P., Mishra, C., Sharma, K., 2014. Large carnivores and low diversity of optimal prey: a comparison of the diets of snow leopards *Panthera uncia* and wolves *Canis lupus* in Sarychat-Ertash Reserve in Kyrgyzstan. Oryx 48, 529–535.

Koizumi, T., Ohtaishi, N., Kaji, K., Yu, Y., Tokida, K., 1993. Conservation of white-lipped deer in China. In: Ohtaishi, N., Sheng, H.I. (Eds.), Deer of China: Biology and Management. Proceedings of the International Symposium on Deer of China, Elsevier, Oxford, UK, pp. 309–318.

Lhagvasuren, B., Munkhtsog, B., 2000. The yak population in Mongolia and its relation with snow leopards as a prey species. In: Jianlin, H., Richard, C., Hanotte, O., McVeigh, C., Rege, J. (Eds.), Proceedings of the Third International Congress on Yak production in Central Asian Highlands. International Livestock Research Institute, Lhasa, China, pp. 69–75.

Lovari, S., Herrero. J., Conroy, J., Maran, T., Giannatos, G., Stubbe, M., Aulagnier, S., Jdeidi, T., Masseti, M., Nader, I., de Smet, K., Cuzin, F., 2008. *Cervus elaphus*. The IUCN Red List of Threatened Species. Version 2015.2. www.iucnredlist.org.

Lovari, S., Boesi, R., Minder, I., Mucci, N., Randi, E., Dematteis, A., Ale, S.B., 2009. Restoring a keystone predator may endanger a prey species in a human-altered ecosystem: the return of the snow leopard to Sagarmatha National Park. Anim. Conserv. 12, 559–570.

Lovari, S., Ventimiglia, M., Minder, I., 2013. Food habits of two leopard species, competition, climate change and upper treeline: a way to the decrease of an endangered species? Ethol. Ecol. Evol. 25 (4), 305–313.

Lyngdoh, s., Shrotriya, S., Goyal, S.P., Clements, H., Hayward, M.W., Habib, B., 2014. Prey preferences of the snow leopard (*Panthera uncia*): regional diet specificity holds global significance for conservation. PLoS One 9, 1–11.

Mallon, D., 1984. The snow leopard in Mongolia. International Pedigree Book of Snow Leopards 4, 3–9.

Mallon, D.P., 1991. Status and conservation of large mammals in Ladakh, India. Biol. Conserv. 56, 101–119.

Mallon, D., Singh, N., Röttger, C., 2014. International single species action plan for the conservation of the argali *Ovis ammon*. Bonn, Germany. CMS Secretariat. CMS Technical Series No. XX.

McCarthy, K.P., Fuller, T.K., Ming, M., McCarthy, T.M., Waits, L., Jumabaev, K., 2008. Assessing estimators of snow leopard abundance. J. Wildl. Manage. 72, 1826–1833.

McCarthy, T.M., 2000. Ecology and Conservation of Snow Leopards, Gobi Brown Bears and Wild Bactrian Camels in Mongolia. PhD Dissertation. University of Massachusetts, Amherst.

Michel, S., Rosen Michel, T., 2015. *Capra falconeri*. The IUCN Red List of Threatened Species. Version 2015.2. www.iucnredlist.org.

Michel, S., Rosen Michel, T., Saidov, A., Karimov, K., Alidodov, M., Kholmatov, I., 2015. Population status of Heptner's markhor *Capra falconeri heptneri* in Tajikistan: challenges for conservation. Oryx 49, 506–513.

Novikov, G.A., 1956. Carnivorous mammals of the fauna of the USSR. Zool. Inst. Acad. Sci. USSR (Eng. Trans. 1962. Israel Program for Scientific Translations, 284 p.).

Oli, M.K., 1993. A key for the identification of the hair of mammals of a snow leopard (*Panthera uncia*) habitat in Nepal. J. Zool. (London) 231, 71–93.

Oli, M.K., Taylor, I.R., Rogers, M.E., 1993. Diet of the snow leopard (*Panthera uncia*) in the Annapurna Conservation Area, Nepal. J. Zool. (London) 231, 365–370.

Perez, I., Geffen, E., Mokady, O., 2006. Critically endangered Arabian leopards *Panthera pardus nimr* in Israel: estimating population parameters using molecular scatology. Oryx 40, 295–301.

Reading, R., Shank, C., 2008. *Capra sibirica*. The IUCN Red List of Threatened Species. Version 2014.3. www.iucnredlist.org.

Roberts, T.J., 1977. The Mammals of Pakistan. Ernest Benn, Tonbridge, Kent, UK.

Schaller, G.B., 1977. Mountain Monarchs. Wild Sheep and Goats of the Himalaya. University of Chicago Press, Chicago, USA.

Schaller, G.B., 1998. Wildlife of the Tibetan Steppe. University of Chicago Press, Chicago, USA.

Schaller, G., Li, H., Talipu, Lu, H., Ren, J., Qu, M., Wang, H., 1987. Status of large mammals in the Taxkorgan Reserve, Xinjiang, China. Biol. Conserv. 42, 53–71.

Schaller, G.B., Li, H., Talipu, J.R., Mingjiang, Q., 1988. The snow leopard in Xinjiang. Oryx 22 (4), 197–204.

Shackleton, D.M. (Ed.), 1997. Wild Sheep and Goats and Their Relatives. Status Survey and Conservation Action Plan for Caprinae. IUCN/SSC Caprinae Specialist Group, Gland, Switzerland, and Cambridge, UK.

Shehzad, W., McCarthy, T.M., Pompanon, F., Purevjav, L., Coissac, E., 2012a. Prey preference of snow leopard (*Panthera uncia*) in South Gobi, Mongolia. PLoS ONE 7 (2), e32104.

Shehzad, W., Riaz, T., Nawaz, M.A., Miquel, C., Poillot, C., Shah, S.A., Pompanon, F., Coissac, E., Taberlet, P., 2012b. Carnivore diet analysis based on next-generation sequencing: application to the leopard cat (*Prionailurus bengalensis*) in Pakistan. Mol. Ecol. 21, 1951–1965.

Smith, A.T., Xie, Y., 2008. A Guide to the Mammals of China. Princeton University Press, Princeton.

Sunquist, M., Sunquist, F., 2002. Wild Cats of the World. Chicago University Press, Chicago, IL.

Tsytsulina, K., 2008. *Marmota menzbieri*. The IUCN Red List of Threatened Species. Version 2015.1. www.iucnredlist.org.

Tulgat, R., Schaller, G., 1992. Status and distribution of wild Bactrian camels *Camelus bactrianus ferus*. Biol. Conserv. 62, 11–19.

Valdez, R., 2008. Ovis orientalis. The IUCN Red List of Threatened Species. Version 2015.2. <www.iucnredlist.org>.

Valentini, A., Miquel, C., Nawaz, M.A., Bellemain, E., Coissac, E., Pompanon, F., Gielly, L., Nacetti, G., Wincker, P., Swenson, J.E., Taberlet, P., 2009. New perspectives in diet analysis based on DNA barcoding and parallel pyrosequencing: the *trn*L approach. Mol. Ecol. Resour. 9, 51–60.

Wang, X., Hoffmann, R.S., 1987. *Pseudois nayaur* and *Pseudois schaeferi*. Mammal. Species 278, 1–6.

Wegge, P., Shrestha, R., Flagstad, Ø., 2012. Snow leopard *Panthera uncia* predation on livestock and wild prey in a mountain valley in northern Nepal: implications for conservation management. Wildl. Biol. 18, 131–141.

Wingard, G.J., Harris, R.B., Amgalanbaatar, S., Reading, R.P., 2011. Estimating abundance of mountain ungulates incorporating imperfect detection: argali *Ovis ammon* in the Gobi Desert, Mongolia. Wildl. Biol. 17 (1), 93–101.

Zhiryakov, V.A., Baidavletov, R.Zh., 2002. Ecology and behaviour of the snow leopard in Kazakhstan. Selevinia 2002, 1–4, (in Russian).

SECTION II

Conservation Concerns

CHAPTER

5

Livestock Predation by Snow Leopards: Conflicts and the Search for Solutions

Charudutt Mishra, Stephen R. Redpath†, Kulbhushansingh R. Suryawanshi**

*Snow Leopard Trust and Nature Conservation Foundation, Mysore, Karnataka, India

†Institute of Biological & Environmental Sciences, Aberdeen University, Aberdeen, UK

INTRODUCTION

In the tree line ecotone habitats of Western and Central Himalaya that form the southern edge of the snow leopard's current global distribution, a significant impact of pastoralism is seen in the pollen record as far back as 5400–5700 years before present (BP) (Miehe et al., 2009). Farther north on the Tibet–Qinghai Plateau, the center of the snow leopard's range, seasonal human forays are recorded as early as 30,000 years ago, and more permanent pastoral habitation about 8,200 years BP (Brantingham et al., 2007). In the Altai Mountains that form the northernmost and easternmost parts of the snow leopard's range, mobile groups of livestock breeders existed 5,000 years BP (Yablonsky, 2003). Snow leopards appear in petroglyphs and in kurgan (nomad burial mounds) artifacts across the region including the westernmost parts of their range in the Tien Shan (Davis-Kimball, 2003; Hussain, 2002; Saveljev et al., 2014). Humans and snow leopards have interacted and coexisted for a considerably long period.

Instances of livestock predation by snow leopards must date back to the beginning of pastoral use of their habitats several thousand years ago. Together with wolves (*Canis lupus*) – and other sympatric carnivores to a much smaller extent – snow leopards continue to cause considerable livestock mortality; studies report annual losses ranging from 3 to 12% of local livestock holdings in some areas (Hussain, 2000; Jackson and Wangchuk, 2004; Mishra, 1997; Namgail et al., 2007). Between these two main predators, studies attribute 20–53% of total

Snow Leopards
http://dx.doi.org/10.1016/B978-0-12-802213-9.00005-5

unintended livestock mortality to snow leopards (Li et al., 2013; Namgail et al., 2007; Suryawanshi et al., 2013). Disease, the other important cause of livestock mortality, provides a useful reference. Depending on the level of veterinary care and livestock vaccination available, the extent of livestock losses to disease in snow leopard habitats is reported to be similar to the two predators (Li et al., 2013), fewer (14%; Suryawanshi et al., 2013), or several times more.

Annual livestock losses to snow leopards and wolves can translate to considerable economic loss for livestock-owning households, sometimes equivalent to as much as half of the regional per capita income (Mishra, 1997). Understandably, the killing of snow leopards in response to livestock predation is believed to be one of the important causes of the species' endangerment (Jackson et al., 2010). Human-snow leopard conflicts are widespread and intense across large parts of Central Asia. Or are they?

REVISITING "HUMAN-SNOW LEOPARD CONFLICTS"

At the outset, it is useful to examine the terminology in use, because language and framing, along with material experience, influence human worldviews and the nature of social action that follows (Peterson et al., 2010). Contemporary ecological and conservation literature on snow leopards is dominated by the term "human-snow leopard conflict" or its variants, derived from long-established and widely used terms "human–carnivore conflict" or "human-wildlife conflict." The lead author of this chapter, Charudutt Mishra himself, is one of those responsible for the widespread use of these terms. It is pertinent to take a step back and ask, "what is human-snow leopard conflict?" The issue is not relevant to snow leopards alone, and this is not the first time that such a question is being posed (e.g., Marshall et al., 2007; Peterson et al., 2010; Redpath et al., 2013, 2014).

Conflicts occur when the interests of two or more parties clash, and one party tries to assert its interests over the other (Marshall et al., 2007). The term "human-snow leopard conflict" therefore suggests that because humans value livestock and snow leopards kill them, humans and snow leopards are antagonists, in conflict with each other. It implies that wild animals are aware of their goals as well as those of humans, and that they seek to undermine human interests (Peterson et al., 2010). The inappropriateness of this framing is obvious, and as we shall now see, discontinuing its use is important for conservation.

Snow leopards are a landscape species, with home ranges of about 100 to even more than 1000 km^2 (McCarthy et al., 2005; Johansson et al., unpublished data). Snow leopard habitats are extensively used for livestock grazing. These realities dictate that snow leopards cannot be adequately conserved through protected areas alone, and a great need and potential for snow leopard conservation efforts lies beyond protected areas (e.g., Li et al., 2014). The continued survival of snow leopards into the future hinges heavily on the willingness of local people to tolerate them, and their ability to continue to coexist with them. Against this background, terminology that inappropriately projects snow leopards and humans as antagonists is unhelpful. It undermines the quest of conservationists to promote human-snow leopard coexistence. From a conservation perspective, the issue is therefore best viewed – and framed – as animal damage to something that humans value, which imposes economic and psychological costs on humans, and not as "human-snow leopard conflict."

Does this mean there is no conflict? Among pastoral peoples, hunting has often been a traditional practice. On the Tibet-Qinghai Plateau, for example, hunting of wild animals served as an additional source of protein and fat for the nomads; as a source of hides and other body parts for trade, medicine, ritual, and cultural items;

and for control of predators and perceived competitors (Huber, 2012). In large parts of the snow leopard's range, traditionally, people would kill predators. Even today, traditional hunting of snow leopards is reported to be prevalent in 8 of the 12 range countries (Snow Leopard Network, 2014). In modern times, however, snow leopards have been accorded the highest protection status in all range countries, rendering their killing a serious crime, and making it difficult to escape punishment and monetary losses when killings are discovered by the state. Even though enforcement is often weak, as a traditional measure to try to control predator populations is curtailed, it understandably creates frustration among the farmers and a sense of loss of their control over the situation (Mishra and Suryawanshi, 2014, 2015). On the other hand, livestock losses to predators continue, and with livestock becoming a global economic asset (Berger et al., 2013), affected people perhaps lose their tolerance for the predators even more.

This is where the conflict actually lies. It is not a conflict between humans and snow leopards, but a conflict between competing human interests, specifically those of livestock production and wildlife conservation. There are similar conservation conflicts (Redpath et al., 2013) or disagreements emerging between the objectives of snow leopard conservation and those of other human interests, such as mineral extraction and linear developments in snow leopard habitats (Snow Leopard Network, 2014), though our focus here remains on the specific issue of livestock predation.

UNDERSTANDING CONFLICTS OVER LIVESTOCK PREDATION

To manage conservation conflicts effectively requires that they be understood comprehensively. Understanding conflicts over livestock predation has two important dimensions, the actual patterns and causes of carnivore damage to livestock, and the perception and psyche of the affected people (Mishra and Suryawanshi, 2014). Therefore, understanding them requires a multidisciplinary approach that combines ecology, social sciences, and human psychology.

Ecological Underpinnings of Livestock Predation by Snow Leopards

Why do snow leopards kill livestock? Snow leopards are reported to prey on the entire diversity of livestock species, from small-bodied goats and sheep to large bodied yaks and Bactrian camels (Johansson et al., 2015; Mishra, 1997; Suryawanshi et al., 2013). Recent research suggests that snow leopards specialize in predation on ungulates, with the contribution of small-sized prey in their diets being rather low and perhaps overestimated in earlier studies (Johansson et al., 2015). Livestock therefore represent a potentially suitable prey type for snow leopards. They tend to occur at higher densities compared to wild ungulates (Berger et al., 2013). They also have degenerated antipredatory abilities such as a reduced ability to detect predators and escape from them, and a loss of camouflage coloration (Zohary et al., 1998). Thus, livestock may represent an attractive group of prey, although preying on livestock is risky because of the possibility of retaliatory persecution or carcass recovery and meat retrieval by people.

Given that livestock are a potentially attractive prey, it is pertinent to ask whether livestock killing by snow leopards is an opportunistic response, or whether it is a part of the active foraging strategy? Can trends in livestock predation by snow leopards be predicted as livestock or wild ungulate populations increase or decrease?

At a fundamental level, ecological theory predicts that the extent of livestock predation would depend on the interplay of functional (prey preference) and numerical (relation between prey abundance and snow leopard abundance) responses of snow leopards to livestock and wild ungulates. Available evidence suggests that snow leopards kill livestock opportunistically, and prefer wild ungulates despite their much lower abundance compared to livestock (Johansson et al., 2015). The abundance of snow leopards appears to be primarily determined by the abundance of wild ungulates, and not by the abundance of livestock (Suryawanshi, 2013). Although representing a potentially suitable prey, livestock at high densities may actually have a negative, indirect effect on snow leopard abundance by causing a reduction in wild ungulate abundance through forage competition (Mishra et al., 2004; Sharma et al., 2015). Model results of functional and numerical responses of snow leopards based on multisite data suggest that if livestock abundance in a habitat increases, the extent of livestock predation by snow leopards can be expected to increase. When wild ungulate abundance in a habitat increases, surprisingly, once again, the extent of livestock predation by snow leopards is predicted to increase, and subsequently stabilize (Suryawanshi et al., 2013). Thus, increasing wild ungulate abundance is important for snow leopard conservation but is likely to increase the extent of livestock predation.

A closer look reveals considerable spatiotemporal variation in the extent of livestock killing by snow leopards within and between landscapes, presumably reflecting the influence of local conditions and pastoral practices. For instance, in a 2-year period of monitoring over an area of 4000 km^2 in Spiti Valley (India), we recorded instances of livestock predation by snow leopards in only 14 of the 25 villages in the study area. Over 90% of these instances were recorded between spring and summer, the period when livestock is grazed extensively in the pastures (Suryawanshi et al., 2013). On the other hand, in Tost Mountains (Mongolia), predation on livestock and the relative contribution of livestock to snow leopard diet were lowest in summer and highest in winter. Here, a majority of livestock is moved to the adjoining steppe areas in summer, reducing their overlap with snow leopards (Johansson et al., 2015).

An even closer examination suggests that although snow leopards sometimes get inside poorly constructed corrals and can cause extensive livestock losses (Jackson et al. 2010), a large majority of instances of snow leopard attacks on livestock take place in the pastures, especially on stragglers (Johansson et al., 2015; Suryawanshi et al., 2013). Within the pastures, livestock are especially vulnerable to snow leopard predation in more rugged areas (Johansson et al., 2015) and areas with relatively higher abundance of wild ungulates (Suryawanshi et al., 2013).

Human Underpinnings of Livestock Predation by Snow Leopards

Although the act of livestock predation, as we have seen, is rooted in evolutionary and behavioral ecology of snow leopards and livestock, important human and societal aspects must be considered. Lax herding of livestock, for instance, leads to stragglers that become especially vulnerable to predation (Johansson et al., 2015). Similarly, poor construction and placement of corrals is the main cause of snow leopard attacks inside corrals. Although rare, such attacks are especially damaging because they usually result in surplus killing (Jackson et al., 2010) and instill more fear in people.

Understanding the extent and correlates of actual livestock damage by snow leopards is important, but is not sufficient for effective conflict management. For instance, livestock losses to carnivores reported by farmers tend to get exaggerated, often unconsciously (Mishra, 1997). Such perceptions can have considerable influence on people's attitudes and behaviors toward carnivores (Mishra and Suryawanshi, 2014). There is often a dichotomy between how affected farmers perceive the issue and the reality of carnivore-caused damage (Suryawanshi et al., 2013). Of the 14 villages in our study area in Spiti Valley where we recorded livestock killing by snow leopards, people had actually perceived snow leopards to be a threat to livestock in only 7 of them. We found that people's threat perception was better explained by the ownership of large-bodied, locally valuable livestock, rather than the actual level of livestock predation (Suryawanshi et al., 2013).

People's perceptions influence their attitudes and presumably their behavior toward snow leopards. Attitudes, once again, show complex and nonlinear patterns; the extent of livestock loss incurred or perceived is only one of the several factors that influence people's attitudes and their willingness to coexist with snow leopards. In a recent study, we found that at the level of the individual respondent, women tended to have more negative attitudes toward snow leopards compared to men, and attitudes generally became more positive with the respondent's level of education and the number and extent of additional income sources available (Suryawanshi et al., 2014). In this study, we also showed that as one scaled up from the individual to the community (village), attitudes toward snow leopards were influenced more by other factors such as village size and the abundance of large-bodied livestock in the entire village. Such scale dependence suggests that in areas of high livestock predation, individuals can develop a strong negative attitude toward snow leopards despite personally not having lost any livestock to them (Suryawanshi et al., 2014). As we shall see, these complexities are relevant for understanding the needs and current limitations of conflict management efforts.

MANAGING CONFLICTS OVER LIVESTOCK PREDATION

Two fundamental lessons for conflict management emerge from the preceding discussion. First, because the issue of livestock predation involves multiple dimensions, any conflict management program, to be effective, must try to address these dimensions, including the reality of livestock damage and the economic cost to the farmers and their perceptions and psyche (Fig. 5.1). Second, scale-dependence in people's attitudes – together with the fact that conservation of landscape species like the snow leopard needs local community support – suggests that the suite of conflict management initiatives should together aim to reach out to the entire community, as well as to most communities within a landscape, and not just those farmers who have suffered livestock predation losses.

Three-Pronged Strategy for Addressing Conflicts

We propose that effective conflict management requires a suite of initiatives that (i) reduce the extent of livestock losses to large carnivores, (ii) share or offset carnivore-caused livestock losses, and (iii) improve the social carrying capacity for carnivores (Mishra and Suryawanshi, 2014, 2015).

Reducing Livestock Losses

As previously mentioned, lax herding is an important cause of livestock losses to large carnivores. Generally, more responsible and vigilant herding and exploring the use of well-trained dogs where appropriate could help reduce livestock losses to predators. A system of rewarding herders for vigilant herding (least number of livestock lost to carnivores) built into one of our livestock insurance initiatives presumably helped in reducing the extent of livestock predation to about half of the baseline levels (Mishra and Suryawanshi, 2014). Predator attacks are not evenly distributed in the landscape, and avoiding pastures or exercising greater care while herding in areas that are relatively rugged can help cut down livestock losses to snow leopards by reducing the number of stragglers. The losses that take place inside corrals can be curtailed through collaborative predator proofing (see Chapter 14.1).

Traditionally, increasing the abundance of wild ungulate prey of snow leopards has been

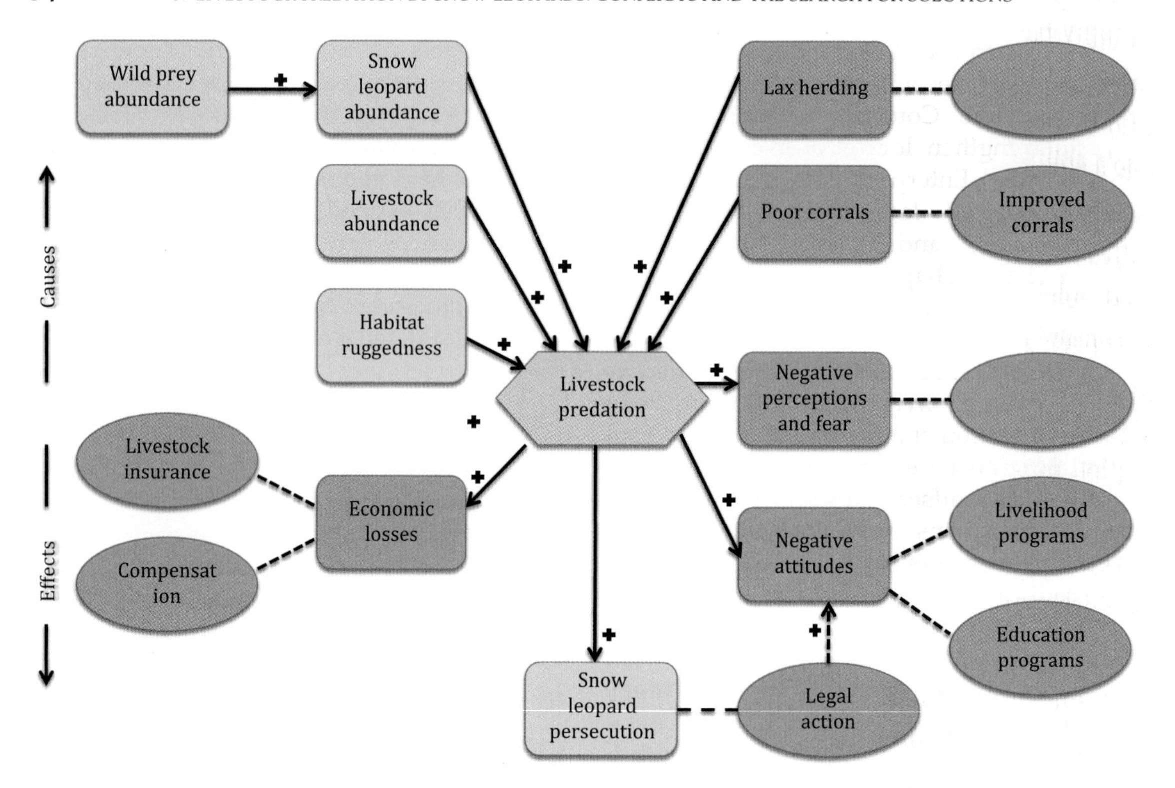

FIGURE 5.1 **A conceptual representation of the proximate ecological (green) and anthropogenic (brown) causes and effects (rectangles) of livestock predation by snow leopards and the commonly employed mitigation measures (ovals).** Blank ovals highlight aspects of this conservation conflict that have received relatively less attention from conservationists. The figure shows that conservation conflicts over livestock predation have multiple dimensions, and therefore their management requires multipronged initiatives. Most current conservation programs, however, tend to be single-initiative focused.

assumed to be useful in reducing livestock losses by deflecting predation pressure to wild ungulates (e.g., Mishra et al., 2004). As we have seen, however, more recent research suggests that increasing the abundance of wild ungulates (see Chapter 6), while being a desirable conservation outcome, can actually lead to an increase in the extent of livestock predation.

Offsetting Livestock Losses

In the extensive livestock production systems, characteristic of snow leopard landscapes, some amount of livestock losses to large carnivores will be inevitable, despite comprehensive measures to protect livestock. In some cases, such as the snow leopard habitats in India, governments have tried to help farmers through compensation programs, but these have largely been ineffective over the longer term in addressing the problem (Mishra, 1997). Smaller-scale community-managed livestock insurance programs have worked better and are currently in operation in several countries including China, India, Mongolia, Nepal, and Pakistan (see Chapter 13.3).

Improving Social Carrying Capacity

Snow leopard habitats are multiple-use landscapes, and snow leopard conservation will remain difficult unless the ability of local communities to coexist with them is strengthened, and they become more willing partners in conservation. Conservation linked initiatives to strengthen local livelihoods, such as Snow Leopard Enterprises (see Chapter 13.2), community-based low-impact tourism (see Chapter 13.1), and livestock vaccination programs (see Chapter 14.3), help improve people's ability and willingness to coexist with snow leopards. However, by themselves, single initiatives are inadequate and can come to be viewed solely as livelihood programs rather than conservation initiatives. Therefore, apart from simultaneous efforts to reduce livestock predation and offset the costs, conservation education programs are also essential to bring about greater awareness and address some of the psychological aspects of depredation related conflicts (see Chapter 17).

Improving the Current Approach to Livestock Predation Management

As we have seen, conservation conflicts over livestock predation by snow leopards and sympatric carnivores are complex and have multiple dimensions (Fig. 5.1), including the reality of damage and the perceptions and psyche of people. Most current community-based conservation programs, however, tend to be single-initiative focused, built around a livestock vaccination program, an insurance program, corral improvement, handicraft development, or other livelihood initiatives. Even though any collaborative conservation initiative with the affected communities can be helpful, such a single-initiative focus tends to be inadequate for conservation and conflict management. To take an example, an insurance program by itself can only help offset economic costs of livestock predation, but does little to reduce the extent of livestock predation in the first place, and perhaps doesn't address the issue of perception and fear either. Unless combined with other initiatives such as supporting and rewarding better herding, it may not even be able to offset the economic setbacks adequately or become economically sustainable. Further, typically, about half or a little more of the families in any community tend to get involved in insurance programs, which leaves the other half uninvolved in conservation or conflict management programs, and it is unreasonable to expect them to support snow leopard conservation. This is also a program where men tend to get more involved on behalf of the family, and there is often inadequate representation of women, who are much better represented in initiatives like the Snow Leopard Enterprises handicraft development program (see Chapter 13.2), again making the case for multipronged initiatives.

It may also be useful to point out that even though collaborative corral improvement is being employed by conservationists to help reduce the extent of livestock kills inside corrals, less effort has been made to help improve herding practices with the community. Systems that encourage more careful herding, especially in rugged areas, along with trials in the use of trained dogs and other approaches need to be explored more and can be valuable in reducing the extent of livestock predation.

Livestock predation by snow leopards becomes a conflict because local people are trying to make a living from livestock; the livestock gets killed by snow leopards; and conservationists try to protect snow leopards. Recognizing that this is a shared problem that requires information sharing, respectful dialogue, and a collaborative approach is the first step in effective conflict management. We propose that current and future community-based efforts to manage conflicts over livestock predation be made multipronged, such that they address the different aspects of management (reducing livestock predation, offsetting economic costs, and improving social carrying capacity) and also increase the coverage of families within a community, and of communities within the snow leopard landscape.

Acknowledgment

We are grateful to Foundation Segré–Whitley Fund for Nature for supporting our research and conservation programs.

References

Berger, J., Buuveibaatar, B., Mishra, C., 2013. Globalization of the cashmere market and the decline of large mammals in Central Asia. Conserv. Biol. 27 (4), 679–689.

Brantingham, P.J., Gao, X., Olsen, J.W., Ma, H., Rhode, D., Zhang, H., Madsen, D.B., 2007. A short chronology for the peopling of the Tibetan Plateau. Dev. Quat. Sci. 9, 129–150.

Davis-Kimball, J., 2003. Statuses of eastern early Iron Age nomads. Ancient West & East 1. 2, 332–356.

Huber, T. 2012. The changing role of hunting and wildlife in pastoral communities of Northern Tibet. In: Pastoral Practices in High Asia, Netherlands, Springer, 195–215.

Hussain, S., 2000. Protecting the snow leopard and enhancing farmers' livelihoods: a pilot insurance scheme in Baltistan. Mt. Res. Dev. 20 (3), 226–231.

Hussain, S. 2002. Nature and Human Nature: Conservation, Values and Snow Leopard. Unpublished report: Snow Leopard Survival Strategy Workshop, International Snow Leopard Trust, Seattle, USA.

Jackson, R.M., Wangchuk, R., 2004. A community-based approach to mitigating livestock depredation by snow leopards. Hum. Dimens. Wildl. 9 (4), 1–16.

Jackson, R.M., Mishra, C., McCarthy, T.M., Ale, S.B., 2010. Snow leopards: conflict and conservation. In: Macdonald, D.W., Loveridge, A.J. (Eds.), The Biology and Conservation of Wild Felids. Oxford University Press, Oxford, UK, pp. 417–430.

Johansson, Ö., McCarthy, T., Samelius, G., Andrén, H., Tumursukh, L., Mishra, C., 2015. Snow leopard predation in a livestock dominated landscape in Mongolia. Biol. Conserv. 184, 251–258.

Li, J., Yin, H., Wang, D., Jiagong, Z., Lu, Z., 2013. Human-snow leopard conflicts in the Sanjiangyuan region of the Tibetan Plateau. Biol. Conserv. 166, 118–123.

Li, J., Wang, D., Yin, H., Zhaxi, D., Jiagong, Z., Schaller, G.B., Mishra, C., McCarthy, T.M., Wang, H., Wu, L., Xiao, L., Basang, L., Zhang, Y., Zhou, Y., Lu, Z., 2014. Role of Tibetan Buddhist monasteries in snow leopard conservation. Conserv. Biol. 28, 87–94.

Marshall, K., White, R., Fischer, A., 2007. Conflicts between humans over wildlife management: on the diversity of stakeholder attitudes and implications for conflict management. Biodivers. Conserv. 16, 3129–3146.

McCarthy, T.M., Fuller, T.K., Munkhtsog, B., 2005. Movements and activities of snow leopards in Southwestern Mongolia. Biol. Conserv. 124, 527–537.

Miehe, G., Miehe, S., Schlütz, F., 2009. Early human impact in the forest ecotone of southern High Asia (Hindu Kush, Himalaya). Quaternary Res. 71, 255–265.

Mishra, C., 1997. Livestock depredation by large carnivores in the Indian trans-Himalaya: conflict perceptions and conservation prospects. Environ. Conserv. 24 (04), 338–343.

Mishra, C., Suryawanshi, K.R., 2014. Managing conflicts over livestock depredation by large carnivores. Successful Management Strategies and Practice. In: Human-Wildlife Conflict in the Mountains of SAARC Region. SAARC Forestry Centre, Thimphu, Bhutan, pp. 27–47.

Mishra, C., Suryawanshi, K.R., 2015. Conflicts over snow leopard conservation, livestock, production. In: Redpath, S., Young, J., Gutierrez, R., Wood, K. (Eds.), Conservation Conflicts. Cambridge University Press, Cambridge UK.

Mishra, C., Van Wieren, S.E., Ketner, P., Heitkonig, I.M.A., Prins, H.H.T., 2004. Competition between domestic livestock and wild bharal *Pseudois nayaur* in the Indian Trans-Himalaya. J. Appl. Ecol. 41, 344–354.

Namgail, T., Fox, J.L., Bhatnagar, Y.V., 2007. Carnivore-caused livestock mortality in Trans-Himalaya. Environ. Manage. 39 (4), 490–496.

Peterson, M.N., Birckhead, J.L., Leong, K., Peterson, M.J., Peterson, T.R., 2010. Rearticulating the myth of human–wildlife conflict. Conserv. Lett. 3, 74–82.

Redpath, S.M., Young, J., Evely, A., Adams, W.M., Sutherland, W.J., Whitehouse, A., Gutiérrez, R.J., 2013. Understanding and managing conservation conflicts. Trends Ecol. Evol. 28, 100–109.

Redpath, S.M., Bhatia, S., Young, J., 2014. Tilting at wildlife: reconsidering human–wildlife conflict. Oryx, 1–4.

Saveljev, A., Soloviev, V., Scopin, A., Shar, S., Otgonbaatar, M., 2014. Contemporary significance of hunting and game animals use in traditional folk medicine in northwest Mongolia and adjacent Tuva. Balkan J. Wildl. Res. 1, 76–81.

Sharma, R.K., Bhatnagar, Y.V., Mishra, C., 2015. Does livestock benefit or harm snow leopards? Biol. Conserv. 190, 8–13.

Snow Leopard Network, 2014. Snow Leopard Survival Strategy. Revised 2014 Version Snow Leopard Network, Seattle, Washington, USA.

Suryawanshi, K.R. 2013. Human carnivore conflicts: understanding predation ecology and livestock damage by snow leopards. PhD Thesis. Manipal University, India.

Suryawanshi, K.R., Bhatnagar, Y.V., Redpath, S., Mishra, C., 2013. People, predators and perceptions: patterns of livestock depredation by snow leopards and wolves. J. Appl. Ecol. 50, 550–560.

Suryawanshi, K.R., Bhatia, S., Bhatnagar, Y.V., Redpath, S., Mishra, C., 2014. Multiscale factors affecting human attitudes toward snow leopards and wolves. Conserv. Biol. 28, 1657–1666.

Yablonsky, L.T., 2003. The archaeology of Eurasian nomads. Archaeology. In: Hardesty, D.L. (Ed.), Encyclopedia of Life Support Systems (EOLSS). UNESCO EOLSS Publishers, Oxford.

Zohary, D., Tchernov, E., Horwitz, L.K., 1998. The role of unconscious selection in the domestication of sheep and goat. J. Zool. 245, 129–135.

CHAPTER

6

Living on the Edge: Depletion of Wild Prey and Survival of the Snow Leopard

Sandro Lovari, Charudutt Mishra***

*Ev.-K2-C.N.R., Bergamo, and Department of Life Sciences, University of Siena, Siena, Italy

**Snow Leopard Trust and Nature Conservation Foundation, Mysore, India

INTRODUCTION

Large carnivores require abundant prey, distributed over wide areas, for their long-term survival. Most large carnivores specialize on ungulate predation, and prey can include livestock that may be an ecologically acceptable surrogate (e.g., Gervasi et al., 2014; Valeix et al., 2012). Livestock is often a clumped, abundant, easy, and predictable prey, available over large areas, thus making a potential food source for most larger carnivores. Not surprisingly, local depletion of wild prey has been reported as one of the main determinants of the extent of predation on livestock by carnivores (e.g., Gusset et al., 2009; Kolowski and Holekamp, 2006; Meriggi and Lovari, 1996; Woodroffe et al., 2005a,b). However, preying on livestock can impose considerable costs on the carnivore, and the extent to which a species can benefit from livestock is variable. These costs can come in the form of retaliatory killing of the carnivore and removal of the kill by farmers (see Chapter 5). In the last two centuries, retaliatory killing over livestock predation is believed to have led to two known carnivore extinctions, the marsupial wolf (*Thylacinus cynocephalus*) in Australia and the Falkland Island wolf (*Dusicyon australis*) (Woodroffe et al., 2005b).

Some carnivores seem to adapt better than others to coexist with humans and to withstand retribution. Even though canids can make substantial use of alternative food sources (e.g., fruit, berries) in areas where wild prey is scarce (e.g., wolf *Canis lupus*: Meriggi et al., 1991; coyote *Canis latrans*: Cepek, 2004; red fox *Vulpes vulpes*: Cavallini and Lovari, 1991), felids have entirely carnivorous food habits (e.g., Macdonald, 1992). They may become more dependent than canids on livestock, which makes

Snow Leopards
http://dx.doi.org/10.1016/B978-0-12-802213-9.00006-7

them potentially more sensitive to human persecution. For example, in Asia, Lovari et al. (2013b) and Shehzad et al. (2015) have reported on viable populations of common leopards (*Panthera pardus*) occurring in areas (Deurali study area, Nepal; Ayubia National Park, Pakistan, respectively) where livestock was the staple diet, as wild ungulates were locally rare – or absent – because of overhunting. Furthermore, among canids, both sexes tend to disperse over large distances, with some exceptions (for a review: Macdonald and Sillero-Zubiri, 2004), whereas among cats female dispersal tends to be much shorter compared to that of males (e.g., Fattebert et al., 2015; Gour et al., 2013; Janecka et al., 2007). As a consequence, replacement of a killed female (by humans) may occur only after a long time in low-density populations of cats, slowing down a recolonization process (Lovari et al., 2009).

The snow leopard (*Panthera uncia*) is a specialized, cold-adapted cat species dwelling at relatively low densities (e.g., Sharma et al., 2014) on low-productivity, ecologically poor habitats compared to savannahs or forests (Mishra et al., 2010). Thus, reduction of wild ungulates or the local loss of even one main prey species may be expected to have a particularly heavy effect on its population density. Furthermore, the scarcity of wild prey is likely to increase predation on livestock, and the effects of retribution by humans can be especially important on a predator normally living at a low density, leading to slow local demise.

When wild prey get scarce, predators should either (i) adapt their population size to match that of their wild prey, if "specialists," or (ii) switch to alternative – domestic included – prey, if "generalists." Feeding specializations (e.g., anatomical, physiological, behavioral ones), body size, food requirements, prey preference, and ecological factors (e.g., availability and dispersion of prey; presence of competitors) determine the strategy used by a predator. The snow leopard is a specialist of mountain-dwelling ungulates of the higher elevations, with a narrow diet spectrum (Levin's standardized index, 0.2, Lovari et al., 2013b) compared to other large cats (e.g., common leopard: 0.5, Lovari et al., 2013b; tiger *Panthera tigris*: 0.3–0.6, Lovari et al., 2014). It specializes on ungulate predation, and, while occasionally killing livestock, prefers wild ungulates despite their abundance often being less than an order of magnitude compared to livestock (Johansson et al., 2015).

In this chapter, we will summarize the available information on survival of snow leopards in areas where its natural prey has been depleted: Sagarmatha (Mt. Everest) National Park, Nepal, where the snow leopard and its main prey have been continuously monitored for 8 years; and Spiti Valley, India where ungulates have been monitored in sample sites for nearly 15 years.

STUDY AREAS

Sagarmatha National Park (SNP), Nepal

The SNP (1148 km^2; 27° 20′ N; 86° 45′ E) lies in central Himalaya of northeastern Nepal. Snow leopards occur mainly between 3440 m and 4750 m above sea level (a.s.l.). Mixed *Betula-Rhododendron-Abies* forests (≤ 4500 m a.s.l.), alpine grassland, mosses, and lichens (≥ 4500 m a.s.l.) occur in the park. Musk deer (*Moschus chrysogaster*) dwell in forests and visit their ecotonal margins (≤ 4500 m a.s.l.) with an estimated density of about 0.3 ind/km^2 (Aryal et al., 2010), whereas Himalayan tahr (*Hemitragus jemlahicus*) has a present density of ca. 2–3 ind/km^2 (7–8 ind/km^2 in the 1990s, Lovari et al., 2005), inhabiting forests and grassland up to 4500 m a.s.l. Two more species of mountain ungulates (goral *Naemorhedus goral* and mainland serow *Capricornis sumatraensis*) are localized in the forest and are exceedingly rare (Lovari et al., 2005). Hillary and Doig (1962) mentioned the presence of bharal (*Pseudois nayaur*) in the park, but no sighting has been reported since then. Among predators, only the common leopard (in the forests) and the snow leopard (in the open habitats of the higher elevations) inhabit the park (Lovari et al., 2013a). The dhole (*Cuon alpinus*) and the golden jackal (*Canis aureus*) were present formerly, but were eliminated more than five decades ago because of human persecution (Lovari et al., 2005). The Tibetan wolf has become a rare winter visitor of the park, but it fails to establish a local population, most likely because of human persecution. Past overhunting and poaching before and partly after the park establishment in 1976 (Brower, 1991; Lovari et al., 2005) may have been responsible for the depauperate community of potential prey species for large carnivores.

The local spectrum of potential wild prey for the snow leopard is narrow compared to other areas (e.g., Jackson, 1996; Oli et al., 1993; Schaller, 1977; Schaller, 1998): in practice, just one species of a wild Artiodactyl (i.e., the Himalayan tahr) and several species of gallinaceous birds (Soldatini et al., 2012) are available to the cat in the open habitats it occupies. Domestic sheep and goats were removed from the park when it was established, but several thousands of domestic cattle (*Bos* sp.) are still present.

Spiti Valley, India

The Spiti Valley (32°00–32°42′ N; 77°37–78°30′ E; c. 12,000 km^2) is a high-altitude region (3300–6000 m a.s.l.) surrounded by the Greater Himalaya in the south, Ladakh in the north, and Tibet in the east. Agro-pastoral communities have inhabited the region for more than 2–3 millennia. Parts of Spiti are visited in summer by transhumant pastoralists from Ladakh for trade and from the Greater Himalaya for grazing (Mishra et al., 2003). The livestock assemblage in Spiti includes yak, cattle, cattle–yak hybrids, horse, donkey, sheep, and goat. Livestock graze in the pastures during most of the year except in the extreme winter. Meso-large carnivores include snow leopard, wolf, and red fox. The wild ungulates are the bharal and the ibex (*Capra sibirica*). Livestock outnumbers the wild ungulates by at least an order of magnitude (Mishra et al., 2001, 2004). Owing to the prevalence of Buddhism, initiation of conservation efforts more than 15 years back by the Nature Conservation Foundation, and the presence of the Forest Department, hunting of wildlife is now almost nonexistent.

SNOW LEOPARDS AND THEIR PREY IN SAGARMATHA NATIONAL PARK

Large predators, except the common leopard, were wiped out in the park before its establishment. Fleming (Undated) reports on one adult snow leopard male and one adult female with two cubs killed in the 1960s in the Namche area. The presence of the snow leopard was reported as accidental in the 1970s and in the 1980s (Ahlborn and Jackson, 1987; Brower, 1991). Most likely, vagrant individuals failed to establish breeding pairs up to the early 2000s (Ale and Boesi, 2005). Circumstantial evidence suggests that only in 2002–2003 a breeding pair established itself (Lovari et al., 2009). In absence of large predators (but for the common leopard in the forest, Brower, 1991), the numbers of Himalayan tahr had grown to at least 350 individuals by the end of the 1980s (Lovari, 1992), and remained around that figure until 2003 (Fig. 6.1). Since then, the number of Himalayan tahr decreased steadily for nearly a decade, with fewer than one-third left in 2010, because of heavy predation on their younger age classes by the snow leopard (Lovari et al., 2009; Ferretti et al., 2014). Himalayan tahr was the staple of the snow leopard diet (56% absolute frequency), followed by cattle *Bos* spp. (c. 25% absolute frequency; Fig. 6.2) (Lovari et al., 2013a). The frequency of occurrence of tahr in the diet of the snow leopard declined substantially between 2006 and 2010 (30% decrease), along with the decrease of

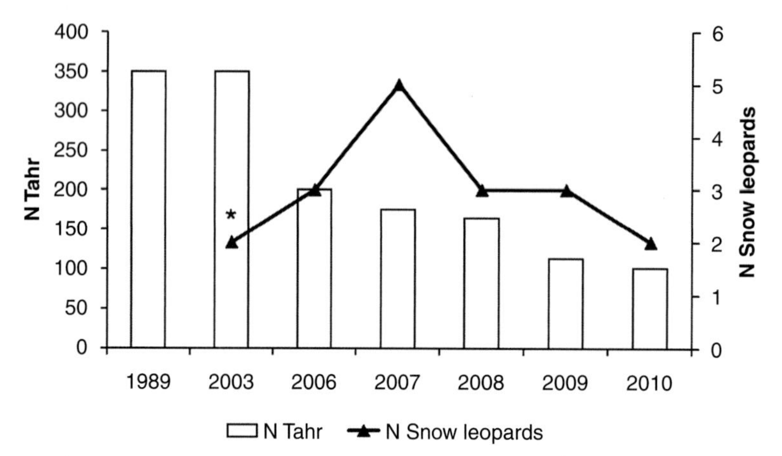

FIGURE 6.1 **Rise and fall of snow leopard numbers, following its return to SNP.** *Source: Data obtained through genotyped scats, and population dynamics of its staple prey. Himalayan tahr population from 1989 to 2010 estimated by monthly repeated counts (cf. methods, Lovari et al., 2013a). The asterisk marks the first formation of a breeding pair of snow leopards.*

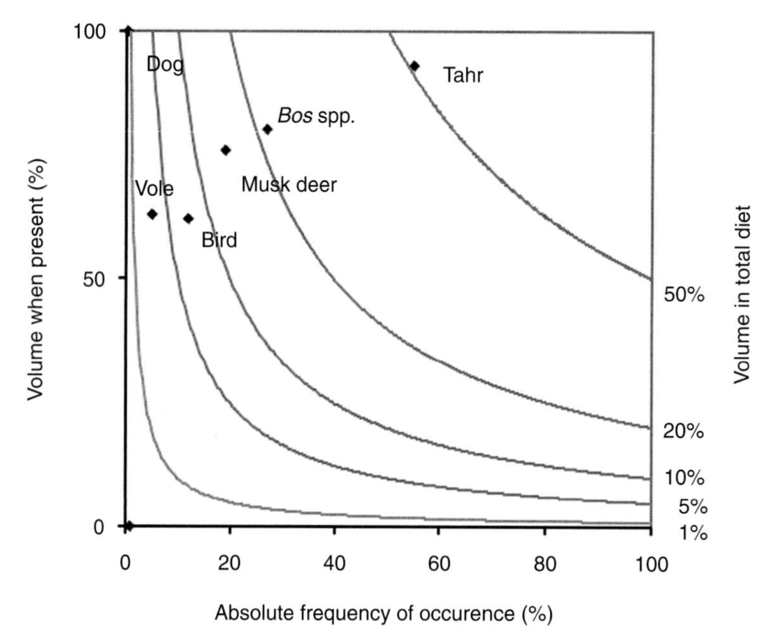

FIGURE 6.2 **Five-year food habits (pooled) of the snow leopard, in terms of estimated volume when present (%) versus frequency of occurrence (%), estimated through analyses of food remains in scats (*N* = 183, collected monthly along a fixed itinerary).** Isopleths connect points of equal relative volume. Himalayan tahr has been the staple prey. *Source: From Lovari et al. (2013a).*

the tahr population (Ferretti et al., 2014). Accordingly, there has been some indication of a partial reliance on alternative prey (Ferretti et al., 2014). After an initial quick growth of snow leopard numbers from 2002–2003 to 2007 (150% increase in 4 years), a fall of 60% occurred in the following 4 years, together with the decrease of tahr numbers (Ferretti et al., 2014; Fig. 6.1).

SNOW LEOPARDS AND THEIR PREY IN SPITI VALLEY

Spiti Valley, despite almost no hunting, has a relatively low density of wild ungulates. The overall density of ibex and bharal in representative sites of the valley was estimated at 1.26 individuals per km^2 (Suryawanshi et al., 2012). There is a high overlap in the diets of wild ungulates and livestock in the region (Bagchi et al., 2004; Mishra et al., 2004). The wild ungulates largely feed on graminoids, though they tend to expand their diets to include forbs and shrubs particularly during winter, a period of lean resource availability (Mishra et al., 2004; Suryawanshi et al., 2009).

Several lines of evidence suggest that the wild ungulate populations are resource limited. Their density is lower in areas with higher livestock density (Mishra et al., 2004). Bharal show reduced food intake in areas and periods of reduced forage availability (Kohli et al., 2014), which presumably results in reduced fecundity. Bharal have much lower kid-to-adult-female ratios in areas that are intensively grazed by livestock, which appears to ultimately lead to reduced abundance (Mishra et al., 2004). The wild ungulate density, ranging from 0.14 km^{-2} to 3.19 km^{-2}, is highest in those parts of Spiti that are least grazed by livestock (Suryawanshi et al., 2012).

How do snow leopards respond in such a livestock-dominated system? The contribution of livestock to the snow leopard diet in Spiti was reported to be rather high (40–58% of the diet; Bagchi and Mishra, 2006), although in hindsight this can partly be attributed to the fact that earlier studies were not able to confirm the identity of scats through fecal DNA. Recent research suggests a high possibility (31–57%) of misidentification of scats belonging to other species (foxes and dogs) as snow leopard feces (see Johansson et al., 2015). Our recent work (Suryawanshi et al., unpublished) suggests the contribution of livestock to snow leopard diet in various part of Spiti in the range from 0 to 40%, and from 20 to 40% in the study sites of Bagchi and Mishra (2006).

One area of Spiti belonging to the village Kibber, where livestock grazing was stopped with the support of the local people, recorded a fourfold recovery of bharal population and increased signs of use by snow leopards (see Chapter 14.2). Across Spiti, the relative use of areas by snow leopards, as recorded through camera trapping, appears to be largely determined by the local abundance of wild ungulates; areas with high abundance of wild ungulates recorded higher photographic capture rates as well as number of individual snow leopards (Sharma et al., 2015). The population of snow leopards in Spiti Valley is estimated at 18.6 (± 3.6 CI) individuals, and the density at 0.64 (± 0.18) individuals per 100 km^2 (Sharma et al., 2015).

IMPLICATIONS OF WILD PREY ABUNDANCE FOR CONSERVATION MANAGEMENT OF SNOW LEOPARDS

The distinction between specialist and generalist carnivores is a somewhat arbitrary one, because few carnivores show adaptations to use only one type (i.e., a limited variety) of wild prey. Natural selection has favored adaptable predators – who can survive on different types of prey when necessary – rather than *sensu stricto* specialists, who may suffer greater risks of extinction. The snow leopard tends to prey mainly on medium-sized herbivores occurring in open habitats, usually at high elevations, such as bharal, ibex, also marmots (*Marmota* spp.) (locally and seasonally), whereas other prey species are much less common in diet (Lovari et al., 2013b). Although it can occasionally prey on small mammals (<2 kg) or large ones (> 100 kg) (Lovari et al., 2013b), the snow leopard can be considered close to a "specialist." Lovari et al. (2013b) reported that occurrence of prey in the diet of the snow leopard in relation to main food categories, as well as to weight categories of prey, varied little between 16 study sites, which would support the view of a "specialist" carnivore.

Even though snow leopards kill livestock, their abundance seems to be determined by the abundance of wild ungulates, livestock killing being opportunistic (Johansson et al., 2015; Suryawanshi et al., unpublished). In fact, no statistically significant inverse correlation was found between occurrence of livestock and that of wild prey in diet, in a review of 16 studies (Lovari et al., 2013b).

In SNP, the near-absence of other wild and domesticated prey species of the preferred weight category, but for Himalayan tahr and partly musk deer, has concentrated predation by the snow leopard on tahr, determining a fall of tahr numbers and, in turn, its own decline, over a time span of 8 years. In Spiti Valley, on the other hand, which is a larger area, the larger wild ungulate populations are resource limited rather than top-down controlled. Depletion of wild ungulate populations due to high livestock density presumably causes a decline in abundance and habitat use of snow leopards locally. As livestock grazing was curtailed from one area of Spiti as part of a conservation program (see Chapter 14.2), the abundance of wild ungulates increased, and so did the use of the area by snow leopards (see above).

It has been suggested that the action of large carnivores can influence dynamics of prey populations (Hebblewhite et al., 2005; McLaren and Peterson, 1994; Ripple et al., 2014; Sinclair et al., 2003). In turn, cascade effects could result on other components of ecosystems, belonging to lower trophic levels (Hebblewhite et al., 2005; McLaren and Peterson, 1994; Ripple et al., 2014). Stable communities, holding rich and diverse assemblies of predators and prey, are influenced by both top-down control and bottom-up regulation of predator/prey numbers, mediated by body size of predators and prey (Sinclair et al., 2003). In turn, top-down control of prey dynamics can be expected to be heavier in simplified systems, where a large and diverse spectrum of wild prey is not available (e.g., Lovari et al., 2009; McLaren and Peterson, 1994). Predators cannot switch their action on alternative wild prey, if only one main wild prey is available, which would lead to a reciprocal dependence of their population dynamics over time or to a greater reliance on livestock.

Both case studies considered here point out the importance of wild ungulates in the diet and for the abundance of snow leopards, and the reciprocal impacts of snow leopards and wild prey on each other. As wild ungulate abundance increases in an area, it can be expected to increase habitat use and abundance of snow leopards. On the other hand, as snow leopard abundance increases, it can have a significant negative impact on the abundance of wild ungulates, especially if the ungulate populations are relatively small and alternate prey species are absent. Ultimately, this decline in wild ungulate abundance caused by snow leopards can lead to a reduction in snow leopard abundance itself.

The restoration of an assembly of wild prey species, through translocations (bharal and, perhaps, marmots) to areas such as SNP, and freeing up areas from excessive livestock grazing in others, would enhance the opportunities of persistence for the snow leopard and reduce the negative effects of concentrating predation on smaller populations of mountain ungulates such as the near-threatened Himalayan tahr (Aryal et al., 2013; Ferretti et al., 2014; Lovari et al., 2009).

Acknowledgments

We are grateful to G.B. Schaller, S.B. Ale, L. Corlatti, and in particular F. Ferretti for their suggestions and help. G.B. Schaller kindly revised our English. The research conducted in Sagarmatha National Park was funded by the Ev-K2-C.N.R. Committee (Italy) and backed by the Nepal Academy for Science and Technology. The Foundation Segré–Whitley Fund for Nature continues to support the work in Spiti Valley.

References

Ahlborn, G., Jackson R., 1987. A survey of Sagarmatha National Park for the endangered snow leopard. Final Report. Unpublished report: Department of National Parks and Wildlife Conservation, Kathmandu.

Ale, S.B., Boesi, R., 2005. Snow leopard sightings on the top of the world. CAT News 43, 19–20.

Aryal, A., Brunton, D., Raubenheimer, D., 2013. Habitat assessment for the translocation of blue sheep to maintain a viable snow leopard population in the Mt Everest region, Nepal. Zool. Ecol. 23, 66–82.

Aryal, A., Raubenheimer, D., Subedi, S., Kattel, B., 2010. Spatial habitat overlap and habitat preference of Himalayan musk deer *Moschus chrysogaster* in Sagarmatha (Mt. Everest) National Park, Nepal. Curr. Res. J. Biol. Sci. 2, 217–225.

Bagchi, S., Mishra, C., 2006. Living with large carnivores: predation on livestock by the snow leopard (*Uncia uncia*). J. Zool. 268, 217–224.

Bagchi, S., Mishra, C., Bhatnagar, Y.V., 2004. Conflicts between traditional pastoralism and conservation of Himalayan ibex *Capra sibirica* in the Trans-Himalayan mountains. Anim. Conserv. 7, 121–128.

Brower, B., 1991. Sherpa of Khumbu: People, Livestock and Landscape. Oxford University Press, New Delhi.

Cavallini, P., Lovari, S., 1991. Environmental factors influencing the use of habitat in the red fox *Vulpes vulpes* (L., 1758). J. Zool. 223, 323–339.

Cepek, J.D., 2004. Diet composition of coyotes in the Cuyahoga Valley National Park, Ohio. Ohio J. Sci. 104, 60–64.

Fattebert, J., Balme, G., Dickerson, T., Slotow, R., Hunter, L., 2015. Density-dependent natal dispersal patterns in a leopard population recovering from over-harvest. PLoS One 10, e0122355.

Ferretti, F., Lovari, S., Minder, I., Pellizzi, B., 2014. Recovery of the snow leopard in Sagarmatha (Mt. Everest) National Park: effects on main prey. Eur. J. Wildl. Res. 60, 559–562.

Fleming, R.L. Jr, (Undated). The natural history of the Everest area. A summary. Background paper for Heart of Himalayas Conservation programme, Kathmandu – Nepal.

Gervasi, V., Nilsen, E.B., Odden, J., Bouyer, Y., Linnell, J.D.C., 2014. The spatio-temporal distribution of wild and domestic ungulates modulates lynx kill rates in a multi-use landscape. J. Zool. 292, 175–183.

Gour, D.S., Bhagavatula, J., Bhavanishankar, M., Reddy, P.A., Gupta, J.A., Sarkar, M.S., Hussain, S.M., Harika, S., Gulia, R, Shivaji, S., 2013. Philopatry and dispersal patterns in tiger (*Panthera tigris*). PLoS One 8 (7), e66956.

Gusset, M., Swarner, M.J., Mponwane, L., Keletile, K., McNutt, J.W., 2009. Human wildlife conflict in northern Botswana: livestock predation by endangered African wild dog *Lycaon pictus* and other carnivores. Oryx 43, 67–72.

Hebblewhite, M., White, C., Nietvelt, C.G., McKenzie, J.A., Hurd, T.E., Fryxell, J.M., Bayley, S.E., Paquet, P.C., 2005. Human activity mediates a trophic cascade caused by wolves. Ecology 86, 2135–2144.

Hillary, E., Doig, D., 1962. High in the Thin Cold Air. Hodder and Stoughton, New Zealand, pp. 58–59.

Jackson, R., 1996. Home range, movements and habitat use of snow leopard *Uncia uncia* in Nepal. PhD thesis, University of London.

Janecka, J.E., Blankenship, T.L., Hirth, D.H., Kilpatrick, C.W., Tewes, M.E., Grassman, L.I., 2007. Evidence for male-biased dispersal in bobcats using relatedness analysis. Wildl. Biol. 13, 38–47.

Johansson, O., McCarthy, T., Samelius, G., Andren, H., Tumursukh, L., Mishra, C., 2015. Snow leopard predation in a livestock dominated landscape in Mongolia. Biol. Conserv. 184, 251–258.

Kohli, M., Sankaran, M., Suryawanshi, K.R., Mishra, C., 2014. A penny saved is a penny earned: lean season foraging strategy of an alpine ungulate. Anim. Behav. 92, 93–100.

Kolowski, J.M., Holekamp, K.E., 2006. Spatial, temporal, and physical characteristics of livestock depredations by large carnivores along a Kenyan reserve border. Biol. Conserv. 128, 529–541.

Lovari, S., 1992. Observations on the Himalayan tahr and other ungulates of the Sagarmatha National Park, Khumbu Himal Nepal. Oecol. Mont. 1, 51–52.

Lovari, S., Ale, S., Boesi, R., 2005. Notes on the large mammal community of Sagarmatha National Park. In: Proceedings of the First International Karakorum Conference. Islamabad. pp. 225–230.

Lovari, S., Boesi, R., Minder, I., Mucci, N., Randi, E., Dematteis, A., Ale, S.B., 2009. Restoring a keystone predator may endanger a prey species in a human-altered ecosystem: the return of the snow leopard to Sagarmatha National Park. Anim. Conserv. 12, 559–570.

Lovari, S., Minder, I., Ferretti, F., Mucci, N., Randi, E., Pellizzi, B., 2013a. Common and snow leopards share prey, but not habitats: competition avoidance by large predators? J. Zool. 291, 127–135.

Lovari, S., Ventimiglia, M., Minder, I., 2013b. Food habits of two leopard species, competition, climate change and upper treeline: a way to the decrease of an endangered species? Ethol. Ecol. Evol. 25, 305–318.

Lovari, S., Pokheral, C.P., Jnawali, S.R., Fusani, F., Ferretti, F., 2014. Coexistence of the tiger and the common leopard in a prey-rich area: the role of prey partitioning. J. Zool. 295, 122–131.

Macdonald, D.W., 1992. The Velvet Claw: A Natural History of Carnivores. BBC Books, London.

Macdonald, D.W., Sillero-Zubiri, C., 2004. Wild canids – an introduction and *dramatis personae*. In: Macdonald, D.W., Sillero-Zubiri, C. (Eds.), The Biology and Conservation of Wild Canids. Oxford University Press, Oxford, U.K, pp. 3–38.

McLaren, B.E., Peterson, R.O., 1994. Wolves, moose, and tree rings on Isle Royale. Science 266, 1555–1558.

Mishra, C., Bagchi, S., Namgail, T., Bhatnagar, Y.V., 2010. Multiple use of Trans-Himalayan rangelands: reconciling human livelihoods with wildlife conservation. In: Du Toit, J., Kock, R., Deutsch, J. (Eds.), Wild Rangelands: Conserving Wildlife While Maintaining Livestock in Semi-Arid Ecosystems. Blackwell Publishing, Oxford, pp. 291–311.

Mishra, C., Prins, H.H.T., Van Wieren, S.E., 2001. Overstocking in the Trans-Himalayan rangelands of India. Environ. Conserv. 28, 279–283.

Mishra, C., Van Wieren, S.E., Prins, H.H.T., 2003. Diversity, risk mediation, and change in a Trans-Himalayan agropastoral system. Hum. Ecol. 31, 595–609.

Mishra, C., Van Wieren, S.E., Ketner, P., Heitkonig, I.M.A., Prins, H.H.T., 2004. Competition between domestic livestock and wild bharal *Pseudois nayaur* in the Indian Trans-Himalaya. J. Appl. Ecol. 41, 344–354.

Meriggi, A., Rosa, P., Brangi, A., Matteucci, C., 1991. Habitat use and diet of the wolf in Northern Italy. Acta Theriol. 36, 141–151.

Meriggi, A., Lovari, S., 1996. A review of wolf predation in Southern Europe: does the wolf prefer wild prey to livestock? J. Appl. Ecol. 33, 1561–1571.

Oli, M.K., Taylor, I.R., Rogers, M.E., 1993. The diet of the snow leopard *Panthera uncia* in the Annapurna Conservation Area, Nepal. J. Zool. 231, 365–370.

Ripple, W.J., Estes, J.A., Beschta, R.L., Wilmers, C.C., Ritchie, E.G., Hebblewhite, M., Berger, J., Elmhagen, B., Letnic, M., Nelson, M.P., Schmitz, O.J., Smith, D.W., Wallach, A.D., Wirsing, A.J., 2014. Status and ecological effects of the World's largest carnivores. Science 343, 1241484.

Sharma, R.K., Bhatnagar, Y.V., Mishra, C., 2015. Does livestock benefit or harm snow leopards? Biol. Conserv. 190, 8–13.

Sharma, K., Bayrakcismith, R., Tumursukh, L., Johansson, O., Sevger, P., McCarthy, T., Mishra, C., 2014. Vigorous dynamics underlie a stable population of the endangered snow leopard *Panthera uncia* in Tost Mountains, South Gobi, Mongolia. PLoS One 9 (7), e101319.

Shehzad, W., Nawaz, M.A., Pompanon, F., Coissac, E., Riaz, T., Shah, S.A., Taberlet, P., 2015. Forest without prey: livestock sustain a leopard *Panthera pardus* population in Pakistan. Oryx 49 (2), 248–253.

Schaller, G.B., 1977. Mountain Monarchs: Wild Sheep and Goats of the Himalaya. University of Chicago Press, Chicago.

Schaller, G.B., 1998. Wildlife of the Tibetan Steppe. University of Chicago Press, Chicago & London.

Sinclair, A.R.E., Mduma, S., Brashares, J.S., 2003. Patterns of predation in a diverse predator-prey system. Nature 425, 288–290.

Soldatini, C., Pellizzi, B., Albores-Barajas, Y.V., 2012. The importance of integrating behavioural ecology into experimental design when censusing Himalayan Galliformes. Ital. J. Zool. 79, 120–127.

Suryawanshi, K., Bhatnagar, Y.V., Mishra, C., 2009. Why should a grazer browse? Livestock impact on winter resource use by bharal *Pseudois nayaur*. Oecologia 162, 453–462.

Suryawanshi, K.R., Bhatnagar, Y.V., Mishra, C., 2012. Standardizing the double-observer survey method for estimating mountain ungulate prey of the endangered snow leopard. Oecologia 169, 581–590.

Valeix, M., Hemson, G., Loveridge, A.J., Mills, G., Macdonald, D.W., 2012. Behavioural adjustments of a large carnivore to access secondary prey in a human-dominated landscape. J. Appl. Ecol. 49, 73–81.

Woodroffe, R., Lindsey, P., Romañach, S., Stein, A., Ranah, S.M.K., 2005a. Livestock predation by endangered African wild dogs. Biol. Conserv. 124, 225–234.

Woodroffe, R., Thirgood, S., Rabinowitz, A., 2005b. The impact of human-wildlife conflict on natural systems. In: Woodroffe, R., Thirgood, R., Rabinowitz, A. (Eds.), People and Wildlife: Conflict or Coexistence?. Cambridge University Press, New York, pp. 1–12.

CHAPTER

7

Monitoring Illegal Trade in Snow Leopards (2003–2012)

Aishwarya Maheshwari, Stephanie von Meibom***

***Independent Researcher, Saurashtra University and Wildlife Institute India, Ghaziabad, Uttar Pradesh, India**
****TRAFFIC – Europe, Frankfurt, Germany**

BACKGROUND

Poaching of snow leopards (*Panthera uncia*) for illegal trade has been identified as one of the major threats to the species across several parts of their natural range and has led to population declines (Snow Leopard Network, 2014; Snow Leopard Working Secretariat, 2013). Trade in snow leopards is not a new phenomenon: a review of snow leopard skin exports from Central Asia and Russia in the first two decades of the twentieth century puts the annual world trade in snow leopard skins at 1000 per year (Heptner and Sludskii, 1972). Following concerns about the high numbers of skins in trade, the species was included in Appendix I of the Convention on International Trade in Endangered Species of Wild Fauna and Flora (CITES) in 1975. However, despite this initiative, the killing of snow leopards in most parts of their range has persisted. In the late 1990s, reports of increased levels of snow leopard killing, especially in the Russian Federation and Central Asia following the break-up of the former Soviet Union, was a cause for concern (Anon, 2002; Dexel, 2002; Koshkarev, 1994; Koshkarev and Vyrypaev, 2000). Motives for such killings varied from mainly commercial interest to retaliatory killings to protect livestock, but whatever the reason, it is likely that the animal or its parts will ultimately enter trade (Theile, 2003).

Skins appear to be the main snow leopard product in demand, whereas other body parts found in trade include bones, claws, and meat (Cao, 2004; Nowell, 2000; Nowell and Xu, 2007; Theile, 2003). Recent evidence indicates that trade in skins is now moving toward rugs, luxury décor, and taxidermy (EIA, 2012). Demand for snow leopard products exists at both the national and international levels. Consumers are reported to include powerful and privileged individuals in the Central Asian range countries, Mongolia, Pakistan and Russia, while the Middle East and Europe have been cited as destinations for skins outside the snow leopard's range (Theile, 2003).

However, a lack of sufficient information prevents quantifying the extent of this threat.

Snow Leopards
http://dx.doi.org/10.1016/B978-0-12-802213-9.00007-9

Therefore in 2012, TRAFFIC, with financial support from USAID, began to collect and update information on snow leopard trade across the species' range and the data presented in this chapter is largely derived from this work (TRAFFIC, Ghost of the Himalayas: Snow Leopards in Illegal Trade, unpublished data).

LEGAL STATUS

CITES – Appendix I. All commercial international trade in snow leopards and their body parts and derivatives is prohibited. Resolution Conf. 12.5 (Rev CoP 5) of CITES covers all Asian big cats, including the snow leopard, and requires all CITES parties to report illegal trade and associated information to the CITES Secretariat.

Convention on the Conservation of Migratory Species of Wild Animals (CMS) – Appendix I (since 1985). For species listed in Appendix I, Parties to the Convention are requested to (i) conserve and restore the species' habitat; (ii) prevent, remove, compensate, or minimize adverse effects of activities or obstacles that seriously impede their migration; and (iii) prevent, reduce or control factors that are endangering or are likely to further endanger the species. With the accession of Kyrgyzstan to the Convention in 2014, seven of the 12 snow leopard range states are party to the CMS (Table 7.1).

National Level Legislation

Hunting and trade of snow leopards is prohibited by legislation in all 12 range states; however, the implementation and enforcement of these laws varies and is not always effective. Sometimes lack of awareness about the legislation in remote areas undermines the efficiency and implementation of national laws. Furthermore, lack of well-trained and well-equipped wildlife inspection and enforcement staff at the field level across most parts of the snow leopards' range hampers enforcement efforts. In addition, poor salaries of the frontline staff may also instigate their direct or indirect involvement in illegal activities related to poaching and trade in snow leopards (Dexel, 2002).

TABLE 7.1 Multilateral Environmental Agreements Applying to Snow Leopard Range States

Range State	Party to CITES	Date of Entry Into Force	Party to CMS	Date of Entry Into Force
Afghanistan	Yes	January 1986	No	–
Bhutan	Yes	November 2002	No	–
China	Yes	April 1981	No	–
India	Yes	October 1976	Yes	November 1983
Kazakhstan	Yes	January 2000	Yes	May 2006
Kyrgyzstan	Yes	September 2007	Yes	May 2014
Mongolia	Yes	April 1996	Yes	November 1999
Nepal	Yes	September 1975	No	–
Pakistan	Yes	July 1976	Yes	December 1987
Russia	Yes	January 1992	No	–
Tajikistan	No	–	Yes	February 2001
Uzbekistan	Yes	October 1997	Yes	September 1998

PREVIOUS TRAFFIC STUDY

In 2003, TRAFFIC published *Fading Footprints: The Killing and Trade of Snow Leopards,* a study that aimed to assess the magnitude of poaching and trade levels of snow leopard throughout its range (Theile, 2003). It covered the period 1993–2002 when a minimum of 260 snow leopards were recorded in trade. These were either seizures by enforcement authorities or observations by independent researchers of snow leopard body parts being sold. A large part of this report was based on published and unpublished literature, questionnaires, and market research.

CURRENT TRAFFIC STUDY

Methodology

For the current survey, a thorough desktop literature research was undertaken and country-specific questionnaires were developed to gather information. In addition, interviews were conducted with relevant experts and country representatives. Market surveys were also conducted in Kyrgyzstan and Afghanistan to collect up-to-date information on snow leopard trade in these two countries (see *Market surveys* on why these two were selected).

Literature Search

Information on snow leopard trade was extracted from published and unpublished sources such as reports on wildlife research, conservation surveys from the range states, newsletters, published scientific papers, and newspapers. A bibliography and news archive of the Snow Leopard Network was analyzed for information. Information on seizures by law enforcement agencies and on market observations by independent researchers (e.g., where snow leopard parts were offered for sale or researchers observed snow leopard parts for sale) was included in a database together with the date of the record and details of the geographical location. The number and type of item reported (e.g., skin, head/skull, teeth, bones, etc.) and whether it was converted into garments or other products (coats, hat, rugs, wall hanging) were recorded.

Questionnaires and Interviews

During the Global Snow Leopard Conservation Forum (GSLCF) held in Bishkek, Kyrgyzstan, October 21–24, 2013 (Snow Leopard Working Secretariat, 2013), interviews and a questionnaire survey were conducted with relevant experts and representatives from snow leopard range states. A questionnaire survey was also circulated within the Snow Leopard Network (SLN) – a worldwide network of more than 400 members working on or associated with snow leopard research and conservation.

Market Surveys

Three surveys were conducted in the 13 markets of Kabul, Mazar-E-Sharif ,and Herat in Afghanistan in September 2014, and in Bishkek in Kyrgyzstan in October 2013. These markets were chosen on the basis of prior reports of illegal trade from other organizations such as the Wildlife Conservation Society, and also because large sections of the markets specialize in fur and leather garments.

RESULTS

Illegal Trade in Snow Leopards

Between 2003 and 2012, a total of 57 records (seizure and observation) of a minimum total of 334 snow leopards in trade were identified through the literature research, while an additional 25 records of a minimum total of 78 snow leopards were recorded through questionnaires and market surveys. In total, 412 snow leopards (82 records) were reported in trade during 2003–2012.

Of these 82 records, 37 (46%) were reported as seizures by enforcement agencies: however, these seizures accounted for only 139 (30%) of snow leopards detected in trade. In contrast, individual observations accounted for 37 (46%) of records, which represented 276 (56%) of snow leopards in trade. China (54%) and Afghanistan (29%) together contributed 83% of the volume of trade in snow leopards recorded between 2003 and 2012.

Skins of snow leopard were the most frequently observed body part in trade. Out of the 82 records, 54 (66%) were of skins alone. Ten records included bones, skull, or skeletons with skins. There were no records of bones being traded on their own. Two records were of live snow leopards being smuggled: one from the Afghanistan–Pakistan border and another from the Tajikistan–Afghanistan border of the Pamir region (Gorno-Badakhshan). Three instances were of snow leopard skins that had been converted into products, that is, rugs or fur coats. Furthermore, two records were of whole bodies of snow leopards that were observed for sale in China, and almost 70 snow leopard claws were also observed in Xinjiang province, China.

The country-specific accounts of trade in snow leopard products that follow are based on data gathered during this study. A brief historical perspective for each country has also been compiled from published sources to illustrate existing trends in the overall trade in snow leopards.

Afghanistan

A few studies have documented trade in snow leopards in Afghanistan (Manati, 2009). Habibi (2003) noted: "*...hunting of fur animals is also being conducted at an unsustainable level. As a result those species whose furs have good market value are slowly disappearing.*" Wildlife trade in Afghanistan declined but did not disappear during the conflict between 1992 and 2001 (Mishra and Fitzherbert, 2004) and a resurgence due to the influx of foreigners and military personnel has recently been reported (Kretser et al., 2012). International military presence in conflict zones enhances the potentially negative impacts of war on wildlife through illicit trade and increased hunting pressure to supply that trade (Dudley et al., 2002; Formoli, 1995). Recently, Kretser et al. (2012) showed how the presence of military forces increased demand for wildlife products in Afghanistan. They reported the presence of snow leopard pelts at the base headquarters of the International Security Assistance Forces (ISAF), located in central Kabul. Furthermore, the Wildlife Conservation Society (WCS) also reported 13 snow leopards pelts for sale on Chicken Street, Kabul, during November 2011 (Kretser et al., 2012). During this study, reports of 119 snow leopards in trade were recorded during 2003–2012 in Afghanistan, some 25% of the total recorded across all range states.

Bhutan

Little information is available on snow leopard trade in Bhutan except for a report quoting the country's Nature Conservation Department (NCD) that states one snow leopard every year has been poached between 1992 and 2007 (Anon, 2009). However, the authenticity of this statement needs to be verified as it was rejected by the Ministry of Agriculture and Forests (MoAF), which stated it was a typographical error and referred to common leopard (*Panthera pardus*) rather than snow leopard. To date, the MoAF has made no official reports of snow leopards in trade in Bhutan.

China

A total of 291 snow leopards were reported in China, some 60% of the total recorded in trade during this study. The majority of these records were from the northwestern part of the country, mainly in the provinces of Gansu (a minimum of 121 snow leopards in trade), Xinjiang (98), and Qinghai (40). For the Tibet Autonomous Region, 5 cases involving 13 skins were reported (EIA, 2012; Li and Lu, 2014).

India

Between 2006 and 2012, six seizures and one observation of a snow leopard skin in trade were reported from India. Of the seizures, five were from northern India (two from Uttarakhand; two from Jammu and Kashmir; one from Delhi) and one from Assam in northeast India. The observation was made during a large mammal survey in the west of Arunachal Pradesh (Mishra et al., 2006). WWF-India also conducted snow leopard baseline surveys during 2011 and 2012 in the west of Arunachal Pradesh and found snow leopard killing persists in remote locations (WWF, unpublished). Interviews with local communities revealed that snow leopards were often killed in retribution for livestock depredation, which might have resulted in significant population declines, while recently some selling of snow leopard skins had taken place in Arunachal Pradesh.

Kazakhstan

Snow leopard trade has been identified as one of the major threats in Kazakhstan (Loginov, 2012), although little recent information is available, just old records such as those from Bo (2002), Chestin (1998), and Dexel (2002).

Kyrgyzstan

In Kyrgyzstan, apart from a single seizure of a snow leopard skin in 2009, no reports of trade were found during the literature search. In the questionnaire response, Nature and Biodiversity Conservation Union (NABU) estimated that up to 10 snow leopards had been traded between 2000 and 2003 in northern Kyrgyzstan. NABU (Anon, 2013) recently noted that fur trade is persistent and threatens remaining populations of snow leopards in Kyrgyzstan.

Mongolia

Mongolia once had an official trophy hunting program for snow leopards (the only range state to have had such a program), but it was terminated in the late 1980s (Dexel, 2002). With 17 snow leopard skins in one seizure in 2004, a total of 25 snow leopards in trade were documented from a total of 6 records between 2003 and 2012, accounting for 5% of the total recorded during the study.

Nepal

The literature search found no information on snow leopard trade from Nepal. However, the questionnaire surveys revealed seizures of two snow leopard skins from Api Nampa Conservation Area. Yonzon (2003) reported that Nepali shepherds exchange snow leopard pelts and bones for domestic livestock (sheep and goats) and also sometimes money from Tibet.

Pakistan

No information on snow leopard trade from Pakistan was found through the literature research for the period 2003–2012. A report by Khan (2002) on behalf of TRAFFIC on the availability of snow leopard pelts in Pakistan provides the only data available up to that year, which came from surveys conducted in four major cities, Peshawar, Islamabad, Lahore, and Karachi (Theile, 2003). WWF-Pakistan reported that six snow leopard skins were observed in the Hispar, Hopar, Bar, and Rakaposhi Valleys of Gilgit–Baltistan province in 2003–2012.

Russian Federation

Apart from targeted poaching, snow leopards face considerable risk from being unintended victims in snares set for musk deer

(*Moschus* sp.), another high-value species in trade (Paltsyn et al., 2012). Snaring of snow leopards is the main threat to their existence in the Argut River basin and in adjoining territories along the Katunsky, Southern Chuisky, and Northern Chuisky Ridges, as well as in Sayano-Shushensky Nature Reserve and its buffer zone, and on Sengelen Ridge. A total of 15 snow leopards (single record) were recorded in trade in the Altai region bordering Mongolia during 2003–2006 (Paltsyn et al., 2012). While Smelyansky and Nikolenko (2010) claimed 19 snow leopard pelts were for sale during 2003 and 2006 and experts estimated that approximately 10 snow leopards are killed every year in Russia (Paltsyn et al., 2012). Furthermore, Istomov (2008) reported a minimum of 7 snow leopards were caught in snares in Sayano-Shushensky Nature Reserve in the previous two decades and two were snared during the winter of 2004–2005. In addition, Paltsyn et al. (2012) also reported the capture of live snow leopard cubs.

Tajikistan

In Tajikistan, Panthera contributed to four seizures (two live snow leopards; a cub in 2009 and an adult female in 2014; one skeleton and skin in 2013; and one skeleton in 2014, all in the Pamir region of Gorno-Badakhshan) and gained knowledge of two trapped snow leopards taken alive to Dushanbe in 2014 that subsequently died (Tatjana Rosen, Panthera, Tajikistan, personal communication). Rosen also claimed that many houses in Dushanbe have snow leopard skins hanging on their walls.

Uzbekistan

Although there are old reports of snow leopards being killed for their fur for sale to foreigners in Uzbekistan (Dexel, 2002; Theile, 2003), no current information could be obtained for 2003–2012.

RECENT RECORDS OF SNOW LEOPARDS IN TRADE (2013–2014)

Some additional information on snow leopard trade for the years 2013 and 2014 was obtained through questionnaires and interviews: 16 snow leopards were seized in China in 2013 (9 fresh skins and 5 specimens) from Qinghai and Xinjiang provinces and 2 skins of snow leopards were seized in Gansu province in 2014. Three further skins were seized, one in Mongolia in 2014 and two in Tajikistan, one each in 2013 and 2014.

Records of Trade Outside the Geographical Range of Snow Leopard

Seven records of trade in snow leopard skins during 2003–2012 were found during the literature search (accounting for a total of eight snow leopards) in nonrange states. Australia (Sibylle Noras, Snow Leopard Network, personal communication), Georgia (Anon, 2014a), Netherlands (Anon, 2012), the United Arab Emirates (Anon, 2014b), the United Kingdom (Anon, 2006), and United States (Snow Leopard Network Blog, 2007) accounted for one record each, and two snow leopard skins were observed for sale in the infamous illegal wildlife market in Myanmar (Oswell, 2010). As of now, on the basis of available information, countries outside the snow leopard's natural range do not appear to be a significant market.

CONCLUSIONS

The present study has provided substantial information to certify that poaching of and trade in snow leopards is ongoing. Based on the current analysis, 481 individuals is the minimum number of snow leopards recorded in trade between 2003 and 2012, slightly under one snow leopard per week.

Recent years have seen an increase in international interest and attention to snow leopards and Asia's high mountain regions, while the 2013 Bishkek declaration clearly stated the need to "Take firm action to stop poaching and illegal trade of snow leopards and other wildlife by adopting comprehensive legislation, strengthening national law-enforcement systems, enhancing national, sub regional, regional and international collaboration and developing effective mechanisms to eliminate the illegal demand for snow leopard and other wildlife products."

The data obtained during this study are likely to present only a portion of the actual poaching and trade levels for various reasons, such as lack of a mechanism to monitor and record snow leopard trade at the country level, the secretive nature of illegal trade, and a lack of coordination in timely and active reporting of seizures of snow leopards in illegal trade.

One important focus for future studies could be to quantify the total loss of snow leopards including those killed for retaliatory reasons and not solely for illegal trade. That would certainly increase the estimates for the total number of snow leopards lost over a given time period. Improved information management on snow leopard trade by individual range states, and coordination between these countries, would help to compile and analyze effectively current dynamics. Until these existing gaps in information on snow leopard trade are filled, potential interventions to help curb poaching and illegal trade in this iconic species will not approach the necessary level of effectiveness.

References

Anon., 2002. Strategy for Conservation of the Snow Leopard in the Russian Federation, WWF Russia, Moscow, Russian Federation.

Anon., 2006. Tiger and leopard fur coats found in raid. http://www.thisislocallondon.co.uk/news/topstories/1002978.tiger_and_leopard_fur_coats_found_in_raid/%20-http:/news.bbc.co.uk/2/hi/uk_news/england/london/6112688 (accessed September 2012).

Anon., 2009. Bhutan: One Snow Leopard a Year Lost to Poaching. http://bigcatrescue.blogspot.in/2009/06/bhutan-one-snow-leopard-year-lost-to.html (accessed October 2014).

Anon., 2012. Dutch police find 40 boxes of rare animal bones. http://www.expatica.com/nl/news/country-news/Dutch-police-find-40-boxes-of-rare-animal-bones_313568.html (accessed October 2014).

Anon. 2013. The NABU's Snow leopard conservation programme in Kyrgyzstan. Background paper. http://www.nabu.de/imperia/md/content/background_paper_snow_leopard_final_eng.pdf. (accessed October 2014).

Anon., 2014a. TRAFFIC Bulletin Seizures and Prosecutions, V 16–26, March 1997–October 2014. Downloaded from http://www.traffic.org

Anon., 2014b. Customs seize skins of rare animals at Dubai airport. http://www.ipsnotizie.it/wam_en/news.php?idnews=1259 (accessed October 2014).

Bo, W., 2002. Illegal trade of snow leopards in China: an overview. Contributed papers to the Snow Leopard Survival Strategy Summit, Seattle, WA.

Cao, Y. 2004. Three men prosecuted for illegally killing and selling snow leopard skin and bone. Lanzhou Morning News. (in Chinese). http://www.gansudaily.com.cn/20040520/706/2004520A01342018.htm (accessed 20.05.2004).

Chestin, I., 1998. Wildlife Trade in Russia and Central Asia. TRAFFIC Europe, Brussels, Belgium.

Dexel, B., 2002. The Illegal Trade in Snow Leopards – A Global Perspective. Naturschutzbund Deutschland (NABU), Berlin, Germany.

Dudley, J.P., Ginsberg, J.R., Plumptre, A.J., Hart, J.A., Campos, L.C., 2002. Effects of war and civil strife on wildlife and wildlife habitats. Conserv. Biol. 16 (2), 319–329.

EIA, 2012. Briefing on snow leopards in illegal trade – Asia's forgotten cats. Environmental Investigation Agency. www.eia-international.org

Formoli, T.A., 1995. The impacts of the Afghan Soviet war on Afghanistan's environment. Environ. Conserv. 22, 66–69.

Habibi, K., 2003. Mammals of Afghanistan. Zoo Outreach Organization, Coimbatore, India.

Heptner, V.G., Sludskii, A.A., 1972. Mammals of the Soviet Union, III, Carnivores (Feloidea) (R.S. Hoffmann, Trans., 1992). Vysshaaya Shkola Publishers, Moscow, Russia. Smithsonian Institute and the National Science Foundation, Washington DC, USA

Istomov, S.V. 2008. Working group report on the results of winter fieldwork. Evaluating the snow leopard population in Sayano-Shushensky Nature Reserve and its buffer zone.

Khan, J., 2002. Availability of Snow Leopard Pelt in Pakistan. Unpublished project report to TRAFFIC.

Koshkarev, E.P., 1994. Snow leopard poaching in Central Asia. Cat News 21, 18.

Koshkarev, E.P., Vyrypaev, V., 2000. The snow leopard after the break-up of the Soviet Union. Cat News 32, 9–11.

Kretser, H.E., Johnson, M.F., Hickey, L.M., Zahler, P., Bennett, E.L., 2012. Wildlife trade products available to US military personnel serving abroad. Biodivers. Conserv. 21, 967–980.

Li, J., Lu, Z., 2014. Snow leopard poaching and trade in China 2000–2013. Biol. Conserv. 176, 207–211.

Loginov, O., 2012. Conservation strategy of the snow leopard in Kazakhstan. http://www.snowleopardnetwork.org/actionplans/Kazakhstan_strategy_English_Dec11.pdf.

Manati, A.R., 2009. The trade in leopard and snow leopard skins in Afghanistan. TRAFFIC Bull. 22 (2), 57–58.

Mishra, C., Fitzherbert, A., 2004. War and wildlife: a post conflict assessment of Afghanistan's Wakhan corridor. Oryx 38 (1), 102–105.

Mishra, C., Madhusada, M.D., Datta, A., 2006. Mammals of the high altitude of western Arunachal Pradesh, eastern Himalaya: an assessment of threats and conservation needs. Oryx 40 (1), 1–7.

Nowell, K., 2000. Far from a Cure. The Tiger Trade Revisited. TRAFFIC, Cambridge, UK.

Nowell, K., Xu, L., 2007. Taming the tiger trade: China's markets for wild and captive tiger products since the 1993 domestic trade ban. TRAFFIC East Asia. www.traffic.org.

Oswell, A.H., 2010. The big cat trade in Myanmar and Thailand. TRAFFIC Southeast Asia. www.traffic.org.

Paltsyn, M.Y., Spitsyn, S.V., Kuksin, A.N., Istomov, S.V., 2012. Snow Leopard Conservation in Russia. WWF-Russia, Moscow, Russian Federation. Available from: http://www.altaiproject.org/wp-content/uploads/2012/09/Russian-Snow-Leopard-Conservation-2012.pdf (accessed September 2014).

Smelyansky, I.E., Nikolenko, E.G., 2010. Анализ рынка диких животных и их дериватов в Алтае-Саянском экорегионе – 2005-2008 гг. [Analysis of the market for wild animals and their derivatives in the Altai-Sayan Ecoregion—2005–2008]. Krasnoyarsk, 150.

Snow Leopard Network, 2014. Snow Leopard Survival Strategy. Revised Version 2014. Snow Leopard Network, Seattle, Washington, USA.

Snow Leopard Network Blog, 2007. Father and son charged with selling illegal skins, including snow leopard. http://www.snowleopardnetwork.org/blog/?p=42.

Snow Leopard Working Secretariat, 2013. Global Snow Leopard and Ecosystem Protection Program, Bishkek, Kyrgyz Republic.

Theile, S., 2003. Fading Footprints: The Killing and Trade of Snow Leopards. TRAFFIC International, Cambridge, UK.

Yonzon, P., 2003. Draft of the snow leopard conservation action plan for the Kingdom of Nepal. http://www.resourceshimalaya.org/content_files/nepal_sAnUmAn4e-86754ab8f46.pdf (accessed October 2014).

CHAPTER

8

Climate Change Impacts on Snow Leopard Range

John D. Farrington, Juan Li***

*World Wildlife Fund, Thimphu, Bhutan

**Center for Nature and Society, College of Life Sciences, Peking University, Beijing, China

INTRODUCTION

Temperatures on the planet have been rising at an unprecedented rate since the second half of the twentieth century. According to the Intercontinental Panel on Climate Change Fifth Assessment Report (IPCC AR5), the global mean surface temperature increased about 0.72°C over the 1951–2012 period, and it is predicted that it will rise a further 0.3–4.8°C by 2081–2100 relative to the 1986–2005 period (IPCC, 2013). It is anticipated that climate change will be a major cause of global biodiversity loss by the end of twenty-first century due to rapid alteration of species' range, habitat, phenology, and life cycles, with mountaintop and polar species already showing severe range contractions (see Walther et al., 2002). Climatic warming is having a particularly large impact on snow leopard habitat in the highest mountains on earth. Here we examine current impacts of climate change in snow leopard range as well as possible future impacts.

CLIMATE CHANGE PHENOMENA IN SNOW LEOPARD RANGE

Temperatures

From 1951 to 2012, the Earth's global mean surface temperature increased at a rate of 0.12°C/decade (IPCC, 2013). However, a sampling of various sites from north to south across snow leopard range has revealed far higher warming rates during this period, ranging from 0.16°C to 0.90°C/decade, with the most rapid warming occurring since the 1970s (Table 8.1). Notably, many authors agree that these warming rates show a strong seasonal variation, with warming in winter being dramatically higher than the annual average. For example, Liu and Chen (2000) report that during the 1955–1996 period, temperatures on the Tibetan Plateau rose at an average annual rate of about 0.16°C/decade. However, during winter months the rate of warming was about 0.32°C/decade, twice the annual mean. This may be due in part to the loss

Snow Leopards
http://dx.doi.org/10.1016/B978-0-12-802213-9.00008-0

TABLE 8.1 A Selection of Climatic Warming Rates from Snow Leopard Range Areas, Listed North to South

Location	Period	Warming rate	Source
Global mean surface temperature	1951–2012	0.12°C/decade	IPCC 2013
Altai-Sayan Ecoregion, Russia (multiple station average)	1976–2008	0.58°C/decade	Kokorin (2011)
Naryn, Kyrgyzstan (2039 m)	1930–1989	0.32°C/decade	Marchenko et al. (2007)
Tibetan Plateau (average of 97 stations above 2000 m)	1955–1996	0.16°C/decade	Liu and Chen (2000)
Tibetan Plateau (average of 66 stations)	1961–2007	0.37°C/decade	Li et al. (2010)
Central Tibet Autonomous Region (average of 5 stations)	1981–2006	0.43°C/decade	Lei et al. (2013)
Dingri, Tibet, China (4032 m, Qomolangma Nature Preserve)	1959–2007	0.62°C/decade	Yang et al. (2011)
Trans-Himalayan Region, Nepal (average of 2 stations)	1977–1994	0.90°C/decade	Shrestha et al. (1999)
Himalayan Region, Nepal (average of 6 stations)	1977–1994	0.57°C/decade	Shrestha et al. (1999)

of snow and ice cover at higher elevations from climatic warming and a subsequent reduction in albedo, which in turn accelerates warming in areas formerly covered by snow and ice (Liu and Chen, 2000). The implications of rapid warming in snow leopard range areas are tremendous for glaciers, permafrost, precipitation, and weather phenomena, which in turn will have large consequences for ecosystems, wildlife, and human livelihoods in these areas.

Precipitation

The IPCC AR5 report states that, in general, time series reconstructions for virtually all land areas on the planet show little change in land-based precipitation since 1901, although the report also states that mid-latitude land areas of the Northern Hemisphere show a "likely" increase in precipitation, with high confidence after 1951 (IPCC, 2013). Recent precipitation trends for snow leopard range are less distinct than for temperatures, with significant variation from region to region. Nevertheless, current reports indicate a slight overall, though generally statistically insignificant, increase in precipitation of 0.54–1.35 mm/year since about the mid-twentieth century at selected snow leopard range areas north of the Himalaya. However, in the Nepal Himalaya no statistically significant precipitation trend is reported (Aizen et al., 1997; Kokorin, 2011; Shrestha et al., 2000; Wu et al., 2007; Yang et al., 2011).

Glaciers

The IPCC AR5 report states that over the last decade, the Asian mountains have been one of the earth's five regions making the largest contribution to total global glacier ice loss (IPCC, 2013). A sampling of glacier research across the snow leopard's range revealed that nearly all glaciers in the region examined experienced a net retreat since the mid-twentieth century. The reported declines in total glacier area in literature reviewed ranged from a 0.4% decrease for 278 glaciers in China's West Kunlun Mountains during 1970–2001 to a 31% decline in area for 212 glaciers in the Bhagirathi No. 2 Basin in India (Aizen et al., 2006; Bolch et al., 2012; Kokorin, 2011; Pan et al., 2012; Shangguan et al., 2007). Notably, continental-type glaciers in the interior of high altitude zones, such as in the central Tibetan Plateau and the Ak-Shyrak region of Kyrgyzstan, are retreating at significantly lower rates than temperate-type glaciers located in warmer climatic zones (e.g., see Aizen et al., 2006; Pan et al., 2012). This decrease in ice cover has various implications for snow leopard range, including loss of time-released water needed by both humans and wildlife, an acceleration of the feedback effect of increased warming at high altitude resulting from declining ice and snow cover, and an increase in glacial lake outburst flood (GLOF) hazards.

Permafrost

The IPCC AR5 report states that permafrost temperatures in most of the world's permafrost regions have increased since the early 1980s due to warming air temperatures and reduced snow cover (IPCC, 2013). Due to the generally high elevations and cold environments in which snow leopards live, the majority of snow leopard range is either partly or completely underlain by continuous or discontinuous permafrost (e.g., see Marchenko et al., 2007). Notably, at the heart of the snow leopard's range on the Tibetan Plateau, the areal extent of permafrost is 1,401,000 km^2, or about 54.3% of the plateau's total area (Zhao and Li, 2009). In snow leopard range, the lower elevational limit of permafrost is currently about 1800 m in the Russian Altai, increasing southward to a high of about 5400 m in the Nepal Himalaya (Fukui et al., 2007a,b; Zhao et al., 2010; Zhao and Li, 2009).

As elsewhere, climatic warming has resulted in both permafrost warming and degradation throughout the snow leopard's range, with measured thickness of reviewed seasonally thawed permafrost active layers increasing up to 1.6 m over various monitoring periods since the late 1960s (Sharkhuu et al., 2007; Zhao et al., 2010). This has been accompanied by actual warming of permafrost itself at measured rates of up to 0.07°C/year (Zhao et al., 2010). As a result, permafrost has declined in areal extent while the lower elevational limit of permafrost has shifted upward (Zhao and Li, 2009).

The impact of climate change-induced permafrost degradation can be severe on highly sensitive permafrost-controlled ecosystems in snow leopard range. A major impact is lowering of the water table as permafrost melts, resulting in conversion of alpine meadows to less productive steppe grassland-type ecosystems, which in turn can result in increased ground heat absorption and further permafrost degradation as vegetation cover declines. Permafrost melt also causes loss of shallow permafrost-controlled surface water features such as seeps, springs, streams, wet meadows, and ponds, which in turn contributes to desertification processes (Wang et al., 2006; Xue et al., 2009). For snow leopards, declines in grassland productivity and drying up of water holes resulting from permafrost degradation have the potential to greatly reduce the carrying capacity of wild prey species and livestock on affected grasslands, reducing the available prey base.

Wetlands

Extensive alpine wetlands exist in much of the snow leopard's range, including saline lakes, freshwater lakes, ponds, marshes, bogs, rivers, and extensive alpine wet meadows, most notably on the Tibetan Plateau. Climate change is having a profound impact on vast areas of these wetlands with various implications for snow leopard habitat. Wang et al. have reported that the area of alpine wetlands on the Tibetan Plateau has declined 37% in recent decades, with wet meadows alone declining in area by 28% between about 1990 and 2005, largely as a result of permafrost degradation (Wang et al., 2006; Qiu, 2012). At the same time, counterintuitively, many large closed basin saline lakes on the plateau are rising at a rapid rate due to increasing precipitation, increasing rates of glacial melt, and decreasing rates of evaporation resulting from increased cloud cover (Lei et al., 2013; Zhang et al., 2013).

One prominent example is that of Seling Co Lake (elevation 4544 m), located in the central Tibet Autonomous Region, which has an area of 2178 km^2. From 2003 to 2009, water surface level and lake area of Seling Co increased by 4.37 m and 186.9 km^2, respectively (Zhang et al., 2013).

The rise of Seling Co Lake has had profound impacts on local lakeshore herding communities, inundating thousands of hectares of highly productive lakeshore pastures. This has caused significant hardship for the herders involved, forcing them to reduce their herd sizes and move to less productive upland pastures which has resulted in significant income loss (Dawa Tsering, WWF, personal communication). This shift is also resulting in increased grazing competition on upland pastures between livestock and blue sheep populations, the primary wild prey of local snow leopards. Loss of highly productive valley bottom wet meadows elsewhere on the plateau, due to either permafrost degradation or inundation by rising lake levels, can be expected to have similar negative impacts on herding incomes. At the same time, this pasture loss is expected to increase grazing competition between livestock and blue sheep, and increase potential for human-snow leopard conflict.

Pasturelands

It is generally accepted that the fragile alpine grasslands that make up the vast majority of snow leopard habitat are being adversely affected by climate change. However, it is extremely difficult to differentiate and quantitatively assess the respective contributions of climate change and overgrazing to the widespread grassland degradation that is occurring across snow leopard range. In general, overgrazing damage either predates or has occurred concurrently with recent climatic warming trends over the last 3–4 decades. Nevertheless, it seems fairly safe to say that climate change has at the very minimum exacerbated overgrazing damage.

In the case of Kyrgyzstan, during the collective period from 1941–1989, livestock numbers increased from 3,491,700 head to 11,803,000 head, which greatly exceeded local pasture carrying capacity and resulted in widespread pasture degradation that is still visible today. However, with livestock privatization in the early 1990s, livestock numbers plummeted reaching a low of 4,877,800 head in 1996 at which level they remained relatively stable through 2003 (Farrington, 2005; Kulov, 2007). Anecdotally, livestock numbers in the northern Tibetan Plateau's Qinghai Province followed a similar downward trajectory following the beginning of livestock privatization in western China in the 1980s. Reasons for this decline in livestock numbers in Qinghai include both a decline in grassland productivity and the boom in the lucrative caterpillar fungus market, which now provides the entire annual cash income for many herding families (see Chapter 10.4). For example, in one typical interview a 49 year-old herder from Zadoi County said that in 1990 his family had 170 yaks, 200 sheep, and 22 horses, but by 2012 only had 70 yaks. He attributed this decline initially to a severe snowstorm in 1996 that killed off many animals and later to what he estimated to be a 70% decline in grassland productivity over the preceding two decades. In contrast, in Mongolia livestock numbers soared following privatization in the 1990s, increasing from 25,856,900 head in 1990 to 43,288,400 in 2008, resulting in widespread overgrazing damage (Shagdar, 2002; Trachtenberg 2009).

The combination of overgrazing and climate change has led to a variety of impacts, perhaps most visibly on the Tibetan Plateau (Wilkes, 2008). As discussed earlier, permafrost degradation can lead to conversion of alpine meadows to less productive alpine steppe-type grasslands. On the Tibetan Plateau, this drying up of alpine meadows is also leading to the widespread cracking of the underlying meadow turf layer, which in turn results in mass breaking up of the turf layer into desiccated, hydrologically disconnected blocks and eventually sheet erosion, known locally as "black beach" erosion, and desertification (Schaller, 2012). In addition, warming temperatures may be adversely affecting pastures by reducing productivity and pasture species diversity (Klein et al., 2007). Herder interviews in Qinghai in 2010–2012 revealed that herders also attributed declining plateau grassland productivity to recent increases in frequency of extreme weather events, possibly the result of climatic warming, such as spring drought and unseasonal snow falls in spring and summer. Furthermore, herders interviewed speculated that warming temperatures are contributing to widespread outbreaks of grass-consuming black caterpillars (*Gynaephora* spp.) and black-lipped pikas (*Ochotona curzoniae*), although others have speculated that pika outbreaks are a symptom of pasture degradation rather than a cause (Hong et al., 2014; Smith and Foggin, 1999).

In the Chang Tang region of Tibet, climate change may be contributing to grassland degradation in snow leopard habitat simply by warming the region to the point where pastures that were formerly used only seasonally in summer are now occupied year-round. When combined with increasing human and livestock populations as well as the increasing use of motor vehicles and fencing, year-round occupation of former summer lands will no doubt adversely impact snow leopards and their prey (Dawa Tsering, WWF, personal communication). Regardless of the cause of grassland degradation, be it related to climate change, increasing livestock populations, or other reasons such as fencing of pastures or construction of new roads, declines in pasture productivity will result in declines in herding and caterpillar fungus incomes, increased grazing competition between snow leopard prey species and livestock, as well as increased economic incentive for poaching of snow leopards and their prey.

Treeline Shift

The world's highest treeline is located at an elevation of 4900 m in Baxoi County in the southeastern Tibet Autonomous Region (Miehe et al., 2007). With continued global warming, treeline in the vicinity of snow leopard habitat is expected to shift upward significantly in coming decades, particularly in the greater Himalaya Region. Although snow leopards are known to occasionally descend to the forest edge in parts of southern Siberia, the Tian Shan, Mongolian Altai, Bhutan Himalaya, and Pakistan Hindu Kush, particularly in winter, given that their primary habitat is alpine grassland, upward shift of treeline with global warming could potentially result in a large-scale loss of snow leopard habitat (Forrest et al., 2012).

At present, upward rates of treeline shift resulting from climate change are still a matter of great conjecture because, as with grasslands, it is difficult to differentiate the impact of climatic warming on treeline elevation shift from anthropogenic impacts such as grazing, woodcutting, and pasture burning (Baker and Moseley, 2007). Nevertheless, based on historical photos, Baker and Moseley (2007) estimated that treeline at one site on Baima Snow Mountain in northwest Yunnan had shifted upward in elevation by 45 m between 1923 and 2003, which they attributed to a combination of climatic warming and a 1988 pasture burning ban. In Himachal Pradesh in the Western Himalaya, Dubey et al. (2003) estimate that treeline is presently increasing in elevation at a rate of 14–19 m/decade, depending on slope aspect. In the future, Forrest et al. (2012) estimate that snow leopard habitat in the Himalaya could be reduced by 30% due to an upward shift in treeline and subsequent loss of alpine grasslands used by both snow leopards and their prey.

Weather Phenomena

The IPCC AR5 report states that it is likely that the frequency and intensity of extreme weather events such as heat waves, heavy precipitation, drought, and flood have increased in various parts of the world in recent decades as a direct impact of climate change (IPCC, 2013). In snow leopard range, these increasingly frequent extreme weather events take on a variety of forms.

In Mongolia, Fernandez-Gimenez et al. (2012) report that severe winter snow disasters, known as "dzud," are increasing in frequency. Dzud generally involve deep snow or other conditions, such as partial melting and freezing of snow cover, that leave livestock unable to forage on grass, resulting in starvation and high mortality of livestock. Three dzud events occurred in Mongolia during 1951–1970, two during 1971–1990, while six occurred from 1991 to 2010, with the 2009–2010 dzud killing 20% of Mongolia's livestock (Fernandez-Gimenez et al., 2012). However, it is difficult to gauge how much more severe the impact of the 2009–2010 dzud may have been due to recent increases in livestock populations that have left far less forage on the ground prior to snowfalls than in years past.

On the Tibetan Plateau, notable snow disasters occurred in October 1985 and the winter of 1997–1998 in which millions of head of livestock died, with more localized snow disasters occurring at other times over the past three decades (Schaller, 2012; Klein et al., 2011). Numerous Tibetan livestock herders interviewed during 2008–2012 also noted that over the past two decades, the frequency of spring droughts and unseasonal spring and summer snowfalls at high altitude summer pastures has increased, while rainfall events were changing from protracted multiday bouts of drizzle and light rain to sudden short intense thunderstorms. Notably, these intense thunderstorms are no doubt increasing the rate of erosion on the plateau as well as resulting in faster runoff and less infiltration, which may be reducing the volume of groundwater flowing to springs and streams during the long dry plateau winter.

The impact of extreme weather events on snow leopards and their prey species has not been formally studied. However, Schaller (2012) presents observations of the large loss of Tibetan antelope, Tibetan gazelle, and Tibetan wild ass as well as livestock in southwest Qinghai province following the October 1985 blizzard. Although it could not be corroborated, a 31-year-old herder interviewed in Zadoi County, Qinghai, in 2010 stated that during the 2008 snow disaster there, 40 of his yaks died while some families lost up to 100 yaks. And it was the first time he had seen blue sheep killed by a snowfall, ranging from 100 to 400 blue sheep per herd, depending on the location. If even remotely true, such events have the potential to severely impact the food supply of snow leopards.

Spring droughts and spring snowfall are also extremely problematic for herders, and presumably wild snow leopard prey species as well. These events both delay or reduce the growth of grass precisely at the time when nutritious grass shoots are most needed for livestock that have recently given birth, which can be on the brink of starvation due to lack of forage at the end of the long, harsh plateau winter. Again such extreme weather events can have possible consequences for snow leopards if they lead to a long-term reduction in prey availability.

PREDICTING FUTURE IMPACTS OF CLIMATE CHANGE ON SNOW LEOPARD RANGE

Extensive modeling of future climate change impacts on temperature, precipitation, glacier melt, runoff, and vegetation has been conducted for snow leopard range areas in recent years (e.g., see Rees and Collins, 2006). However, modeling done specifically to gauge future climate change impacts on suitability of habitat for snow leopards has largely been limited to the work of Forrest et al. (2012), focusing on the Himalaya, and the Snow Leopard Trust (2010), focusing on western China.

As a first step in examining possible future changes to the distribution of suitable snow leopard habitat throughout the species range, the maximum entropy (MaxEnt) algorithm was employed to map current snow leopard habitat based on available snow leopard presence data and 10 related environmental variables (Phillips

TABLE 8.2 Area of Current (2014) and Predicted Snow Leopard Habitat and Coverage by Currently Existing Nature Reserves up to 2080 as Predicted Under IPCC Scenario A1B

	Current (2014) and predicted habitat area (km²)			**No. of patches >11 km² and >1667 km² (in parentheses)**			**Habitat protected by nature reserves (%)**	
Country	**2014**	**2080 (A1B)**	**% change**	**2014**	**2080 (A1B)**	**% change**	**2014**	**2080 (A1B)**
Afghanistan	77,615	127,057	64	54 (2)	48 (1)	–11 (–50)	8	6
Bhutan	10,316	4693	–55	9 (1)	11 (1)	22 (0)	46	53
China	1,026,708	1,048,352	2	1761 (20)	2237 (32)	27 (60)	21	21
India	99,409	93,043	–6	20 (2)	25 (2)	25 (0)	24	23
Kazakhstan	37,468	62,080	66	41 (3)	38 (4)	–7 (33)	29	21
Kyrgyzstan	128,684	140,805	9	25 (1)	16 (1)	–36 (0)	4	4
Mongolia	237,402	415,879	75	361 (8)	381 (16)	6 (100)	18	14
Myanmar	2024	627	–69	5 (1)	6 (0)	20 (–100)	93	89
Nepal	33,128	25,531	–23	12 (1)	10 (2)	–17 (100)	44	47
Pakistan	103,021	104,053	1	91 (1)	83 (1)	–9 (0)	10	11
Russia	54,125	156,118	188	153 (1)	236 (4)	54 (300)	35	23
Tajikistan	90,086	94,440	5	8 (1)	8 (1)	0 (0)	4	5
Uzbekistan	11,754	12,269	4	6 (1)	6 (1)	0 (0)	43	42
Total	**1,911,740**	**2,284,947**	**20**	**2546 (43)**	**3105 (66)**	**22 (53)**	**19**	**18**

et al., 2006; Li et al., 2014). This analysis revealed that the most pervasive features of snow leopard habitat are high elevation, rugged mountains, low precipitation in the warmest annual quarter, and low annual average temperature. Results of the analysis are summarized in Table 8.2, which shows that in 2014, suitable snow leopard habitat covered an area of 1,911,740 km² across the 12 known snow leopard ranges states and Myanmar. However, one limitation of this baseline habitat analysis was a lack of snow leopard presence points across large areas of potential snow leopard range, such as areas of the Altai, Kunlun, and Hengduan Ranges.

Next, based on the optimal environmental features of snow leopard habitat listed above, the MaxEnt model was used to predict potential range-wide snow leopard habitat in 2080 based on the A1B scenario in the IPCC Special Report on Emissions Scenarios. Specifically, the A1B scenario projects levels of greenhouse gas, aerosol, and other pollutant emissions in a future world of rapid economic growth with a global population that peaks in the mid-twenty-first century and declines thereafter that is powered by a combination of both fossil and nonfossil energy sources (Nakicenovic et al., 2000).

Under this scenario, both the elevation and latitude of snow leopard habitat increase in the future, which concurs with other predictions of future mammalian range shifts (e.g., see Walther et al., 2002). However, a closer examination of model results reveals distinct habitat distribution trends in the north and south of the snow leopard's range, the boundary between which is roughly demarcated by the Kunlun Mountains at approximately 35°N (Fig. 8.1). In the northern range area, alpine grasslands currently not known to have snow leopards are predicted to have increases in temperature and precipitation, potentially becoming more productive habitat, and more suitable for snow leopards and their

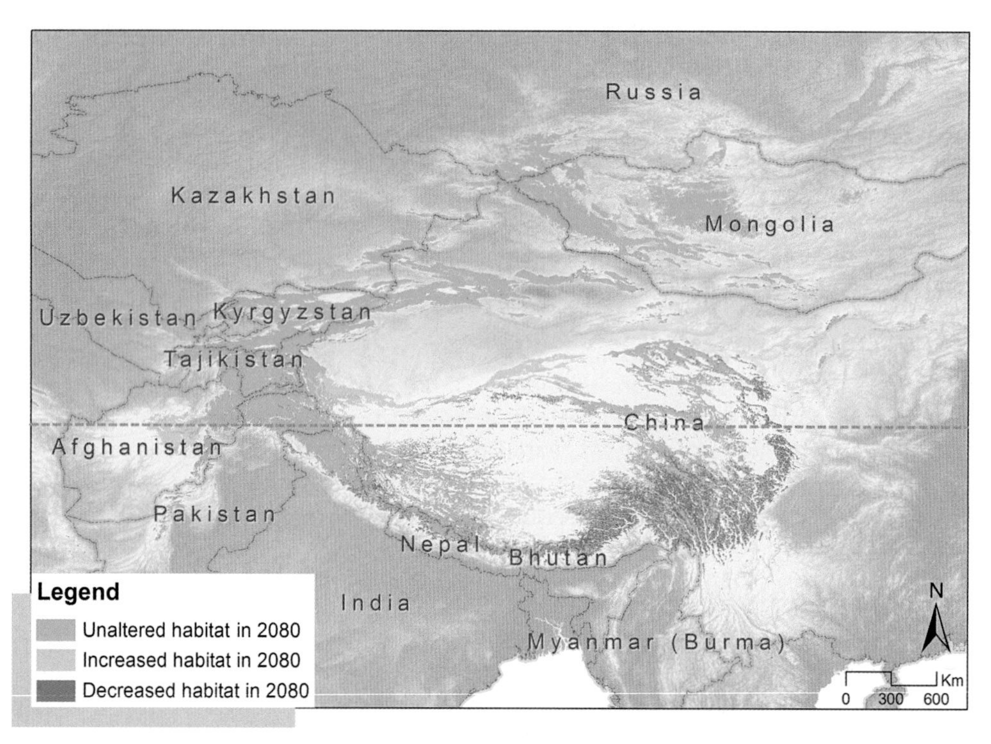

FIGURE 8.1 **Projected changes in distribution of snow leopard habitat by 2080 under the IPCC Scenario A1B.** Green, unaltered habitat areas; blue, new habitat areas; red, habitat loss; gray dashed line, boundary between the northern and southern snow leopard range areas (35°N).

prey. For the seven northernmost snow leopard range states (Afghanistan, Tajikistan, Uzbekistan, Kyrgyzstan, Kazakhstan, Russia, and Mongolia), the model predicts a total increase in suitable habitat of 58% by 2080 (Fig. 8.1, Table 8.2). In the southern range area, however, the model predicts that increased temperatures and precipitation will lead to an upward shift in treeline elevation and a loss of alpine and subalpine grasslands preferred by snow leopards, with most habitat loss occurring on the south slope of the Himalaya and the southeastern Tibetan Plateau. Notably, the model predicts a 23% decline in the area of snow leopard habitat in Nepal and a 55% decline in Bhutan by 2080 (Fig. 8.1, Table 8.2).

To further quantify predicted future changes in snow leopard habitat, the total number of habitat patches was calculated on both a range-wide and a north-south basis, where a habitat patch is defined as any continuous area of habitat at least 11 km^2 in size (Table 8.2). A separate count of habitat patches greater than 1667 km^2 in size was also made, as this is the smallest area required to support a snow leopard population of 50 individuals, the minimum number needed to prevent inbreeding in an area with a typical snow leopard density of 3 individuals/100 km^2 (Table 8.2) (Frankel, 1981; Snow Leopard Network, 2014).

Although the model predicts that the total area of suitable snow leopard habitat will increase 20% range wide by 2080, this area will be increasingly fragmented, with the total number of habitat patches increasing by 22% (Table 8.2). In the northern range area (north of 35°N latitude), total snow leopard habitat is predicted to increase in area by 45% by 2080, with the number of habitat patches remaining roughly the same as some patches expand and merge with others. In contrast, the total area of snow leopard habitat in the southern range area (south of 35°N) is predicted to decrease by 18% by 2080, but with the number of habitat patches increasing by 41%. This reflects a rapid rate of habitat fragmentation due to the creation of numerous mountaintop snow leopard habitat islands at higher elevations resulting from colonization of alpine grasslands by forest and shrublands. Thus, upward shift of the treeline is potentially a large threat to snow leopard populations in the southern range area and may one day prevent the dispersal of snow leopards, reducing population interactions needed to maintain the health of the species. From this north-south comparison of model results, it is clear that future conservation and climate adaptation strategies will need to be region specific.

One limitation of this model is that only the CSIRO Mk3 climate model was employed to predict future distribution of snow leopard habitat, when multimodel ensemble-averaged climate forecasts often generate better predictions than a single model alone (Fordham et al., 2011). Other limitations of the model include current gaps in snow leopard presence data, our limited ability to predict the future state of society and technology, and the numerous limitations of current climate models themselves, which necessarily simplify complex natural systems.

CONCLUSIONS

Temperatures in the Himalaya, Tibetan Plateau, and Central Asia are currently rising at a rate far faster than the global average. These regions constitute the home range of the snow leopard, and their alpine grasslands are suffering particularly adverse effects from climatic warming, in large part due to the outsized role permafrost, snow cover, and glaciers play in maintenance of these ecosystems. At the same time, these alpine grasslands are also rapidly degrading as a result of past and present overgrazing by livestock and other poor land use practices. Recent climate-based habitat models for snow leopard range predict a future loss of suitable habitat in the south of the species range, in large part due to upward treeline shift, but a possible expansion of suitable habitat in the north of the species range. Regardless of predicted future climatic conditions, the current rapid deterioration of snow leopard habitat from both climatic and direct human impacts can be expected to increase competition between livestock and snow leopard prey species for limited grazing resources as well as to increase human-snow leopard conflict in the form of livestock kills. One approach that will be necessary to mitigate climate impacts on both ecosystems and livelihoods in snow leopard range will be the development of appropriate climate adaptation strategies for improving natural resource management, increasing ecosystem resiliency, and reducing the dependence of local residents on their dwindling natural resource base. Although the sparse human populations of Asia's high mountain regions have contributed little to the buildup of greenhouse gas emissions in the atmosphere, these peoples and their ecosystems are suffering disproportionately from the consequences of global warming. Consequently, if snow leopards, their prey species, and habitat are to survive into the future, a great investment will need to be made in helping residents of snow leopard range adapt to their rapidly changing climate.

References

Aizen, V.B., Aizen, E.M., Melack, J.M., Dozier, J., 1997. Climatic and hydrologic changes in the Tien Shan, central Asia. J. Climate 10 (6), 1393–1404.

Aizen, V.B., Kuzmichenok, V.A., Surazakov, A.B., Aizen, E.M., 2006. Glacier changes in the central and northern Tien Shan during the last 140 years based on surface and remote-sensing data. Ann. Glaciol. 43 (1), 202–213.

Baker, B.B., Moseley, R.K., 2007. Advancing treeline and retreating glaciers: implications for conservation in Yunnan, P.R. China. Arct. Antarct. Alp. Res. 39 (2), 200–209.

Bolch, T., Kulkarni, A., Kääb, A., Huggel, C., Paul, F., Cogley, J.G., Frey, H., Kargel, J.S., Fujita, K., Scheel, M., Bajracharya, S., Stoffel, M., 2012. The state and fate of Himalayan glaciers. Science 336, 310–314, + supplementary text.

Dubey, B., Yadav, R.R., Singh, J., Chaturvedi, R., 2003. Upward shift of Himalayan pine in Western Himalaya, India. Curr. Sci. 85 (8), 1135–1136.

Farrington, J.D., 2005. De-development in eastern Kyrgyzstan and persistence of semi-nomadic livestock herding. Nomadic Peoples 9 (1–2), 171–197.

Fernandez-Gimenez, M.E., Batkhishig, B., Batbuyan, B., 2012. Cross-boundary and cross-level dynamics increase vulnerability to severe winter disasters (dzud) in Mongolia. Glob. Environ. Change 22, 836–851.

Fordham, D.A., Wigley, T.M., Brook, B.W., 2011. Multi-model climate projections for biodiversity risk assessments. Ecol. Appl. 21 (8), 3317–3331.

Forrest, J.L., Wikramanayake, E., Shrestha, R., Areendran, G., Gyeltshen, K., Maheshwari, A., Mazumdar, S., Naidoo, R., Thapa, G.J., Thapa, K., 2012. Conservation and climate change: assessing the vulnerability of snow leopard habitat to treeline shift in the Himalaya. Biol. Conserv. 150 (1), 129–135.

Frankel, O.H., 1981. Conservation and Evolution. Cambridge University Press, Cambridge.

Fukui, K., Fujii, Y., Ageta, Y., Asahi, K., 2007a. Changes in the lower limit of mountain permafrost between 1973 and 2004 in the Khumbu Himal, the Nepal Himalayas. Global Planet. Change 55 (4), 251–256.

Fukui, K., Fujii, Y., Mikhailov, N., Ostanin, O., Iwahana, G., 2007b. The lower limit of mountain permafrost in the Russian Altai Mountains. Permafrost Periglac. 18 (2), 129–136.

Hong, J., Ni, Y.F., Du, G.L., Yun, X.J., 2014. Current situation and cause analysis of grassland pests on grassland in China. Pratac. Sci. 31 (7), 1374–1379, (In Chinese with English abstract).

IPCC, 2013. Climate Change 2013: The Physical Science Basis. Contribution of Working Group I to the Fifth Assessment Report of the Intergovernmental Panel on Climate Change. Cambridge University Press, Cambridge and New York.

Klein, J.A., Harte, J., Zhao, X.Q., 2007. Experimental warming, not grazing, decreases rangeland quality on the Tibetan Plateau. Ecol. Appl. 17 (2), 541–557.

Klein, J.A., Yeh, E., Bump, J., Nyima, Y., Hopping, K., 2011. Coordinating environmental protection and climate change adaptation policy in resource-dependent communities: A case study from the Tibetan Plateau. In: Ford, J.D., Berrang-Ford, L. (Eds.), Advances in Global Change Research Volume 42: Climate Change Adaptation in Developed Nations. Springer, Netherlands, Dordrecht, pp. 423–438.

Kokorin, A.O. (Ed.), 2011. Assessment Report: Climate Change and its Impact on Ecosystems, Population and Economy of the Russian Portion of the Altai-Sayan Ecoregion. WWF, Russia, Moscow.

Kulov, S., 2007. Total economic valuation of Kyrgyzstan pastoralism. IUCN, Gland, Switzerland.

Lei, Y.B., Yao, T.D., Bird, B.W., Yang, K., Zhai, J.Q., Sheng, Y.W., 2013. Coherent lake growth on the central Tibetan Plateau since the 1970s: characterization and attribution. J. Hydrol. 483, 61–67.

Li, J., Wang, D., Yin, H., Zhaxi, D., Jiagong, Z., Schaller, G.B., Mishra, C., McCarthy, T.M., Wang, H., Wu, L., Xiao, L., Basang, L., Zhang, Y., Zhou, Y., Lu, Z., 2014. Role of Tibetan Buddhist monasteries in snow leopard conservation. Conserv. Biol. 28, 87–94.

Li, L., Yang, S., Wang, Z., Zhu, X., Tang, H., 2010. Evidence of warming and wetting climate over the Qinghai-Tibet Plateau. Arct. Antarct. Alp. Res. 42, 449–457.

Liu, X.D., Chen, B.D., 2000. Climatic warming in the Tibetan Plateau during recent decades. Int. J. Climatol. 20, 1729–1742.

Marchenko, S.S., Gorbunov, A.P., Romanovsky, V.E., 2007. Permafrost warming in the Tien Shan Mountains, Central Asia. Global Planet. Change 56, 311–327.

Miehe, G., Miehe, S., Vogel, J., Sonam Co, Duo, La., 2007. Highest treeline in the northern hemisphere found in southern Tibet. Mt. Res. Dev. 27, 169–173.

Nakicenovic, N., Alcamo, J., Davis, G., de Vries, B., Fenhann, J., Gaffin, S., Gregory, K., Grubler, A., Jung, T.Y., Kram, T., 2000. Special Report on Emissions Scenarios, Working Group III, Intergovernmental Panel on Climate Change (IPCC). Cambridge University Press, Cambridge.

Pan, B.T., Zhang, G.L., Wang, J., Cao, B., Geng, H.P., Wang, J., Zhang, C., Ji, Y.P., 2012. Glacier changes from 1966–2009 in the Gongga Mountains, on the south-eastern margin of the Qinghai-Tibetan Plateau and their climatic forcing. The Cryosphere 6, 1087–1101.

Phillips, S.J., Anderson, R.P., Schapire, R.E., 2006. Maximum entropy modeling of species geographic distributions. Ecol. Model. 190 (3-4), 231–259.

Qiu, J., 2012. Thawing permafrost reduces river runoff: China's Yangtze River is receiving less water as climate warms. Nature News (January 6, 2012). doi:10.1038/nature.2012.9749 http://www.nature.com/news/thawing-permafrost-reduces-river-runoff-1.9749 (accessed 01.01.2015).

Rees, H.G., Collins, D.N., 2006. Regional differences in response of flow in glacier-fed Himalayan rivers to climatic warming. Hydrol. Process. 20 (10), 2157–2169.

Schaller, G.B., 2012. Tibet Wild: A Naturalist's Journeys on the Roof of the World. Island Press, Washington, DC.

Shagdar, E., 2002. The Mongolian Livestock Sector: Vital for the Economy and People, but Vulnerable to Natural Phenomena. Economic Research Institute for Northeast Asia, Niigata City, Japan.

Shangguan, D.H., Liu, S.Y., Ding, Y.J., Li, J., Zhang, Y., Ding, L.F., Wang, X., Xie, C.W., Li, G., 2007. Glacier changes in the west Kunlun Shan from 1970 to 2001 derived from Landsat TM/ETM+ and Chinese glacier inventory data. Ann. Glaciol. 46, 204–208.

Sharkhuu, A., Sharkhuu, N., Etzelmüller, B., Heggem, E.S.F., Nelson, F.E., Shiklomanov, N.I., Goulden, C.E., Brown, J., 2007. Permafrost monitoring in the Hovsgol mountain region, Mongolia. J. Geophys. Res.-Ea. Surf. 112 (F2).

Shrestha, A.B., Wake, C.P., Mayewski, P.A., Dibb, J.E., 1999. Maximum temperature trends in the Himalaya and its vicinity: an analysis based on temperature records from Nepal for the period 1971-94. J. Climate 12 (9), 2775–2786.

Shrestha, A.B., Wake, C.P., Dibb, J.E., Mayewski, P.A., 2000. Precipitation fluctuations in the Nepal Himalaya and its vicinity and relationship with some large scale climatological parameters. Int. J. Climatol. 20 (3), 317–327.

Smith, A.T., Foggin, J.M., 1999. The plateau pika (*Ochotona curzoniae*) is a keystone species for biodiversity on the Tibetan plateau. Anim. Conserv. 2, 235–240.

Snow Leopard Network. 2014. Snow Leopard Survival Strategy, Revised Version 2014. Snow Leopard Network. www.snowleopardnetwork.org.

Snow Leopard Trust, 2010. Climate Change Impacts on Snow Leopard Range: Prioritizing Conservation Efforts to Mitigate Human-Wildlife Conflict. Snow Leopard Trust, Seattle, WA.

Trachtenberg, E., 2009. Mongolia Livestock Situation. USDA Foreign Agricultural Service-Agricultural Trade Office, Beijing.

Walther, G., Post, E., Convey, P., Menzel, A., Parmesan, C., Beebee, T.J., Fromentin, J., Hoegh-Guldberg, O., Bairlein, F., 2002. Ecological responses to recent climate change. Nature 416, 389–395.

Wang, G.X., Li, Y.S., Wu, Q.B., Wang, Y.B., 2006. Impacts of permafrost changes on alpine ecosystem in Qinghai-Tibet Plateau. Science in China, Series D: Earth Sciences 49, 1156–1169.

Wilkes, A., 2008. Towards mainstreaming climate change in grassland management policies and practices on the Tibetan Plateau. Working Paper no. 67. World Agroforestry Centre – ICRAF China, Beijing.

Wu, S.H., Yin, Y.H., Zheng, D., Yang, Q.Y., 2007. Climatic trends over the Tibetan Plateau during 1971-2000. J. Geogr. Sci. 17 (2), 141–151.

Xue, X., Guo, J., Han, B.S., Sun, Q.W., Liu, L.C., 2009. The effect of climate warming and permafrost thaw on desertification in the Qinghai–Tibetan Plateau. Geomorphology 108 (3–4), 182–190.

Yang, X.G., Zhang, T.J., Qin, D.H., Kang, S.C., Qin, X., 2011. Characteristics and changes in air temperature and glacier's response on the north slope of Mt. Qomolangma (Mt. Everest). Arct. Anarct. Alp. Res. 43, 147–160.

Zhang, G.Q., Xie, H.J., Yao, T.D., Kang, S.C., 2013. Water balance estimates of ten greatest lakes in China using ICESat and Landsat data. China Sci. Bull. 58 (31), 3815–3829.

Zhao, L., Li, R., 2009. Changes in permafrost along the Qinghai-Tibet highway in the Yangtze source region. In: Farrington, J.D. (Ed.), Impacts of Climate Change on the Yangtze Source Region and Adjacent Areas, Qinghai-Tibet Plateau, China. China Meteorological Press, Beijing, pp. 97–112.

Zhao, L., Wu, Q.B., Marchenko, S.S., Sharkhuu, N., 2010. Thermal state of permafrost and active layer in Central Asia during the International Polar Year. Permafrost Periglac. 21, 198–207.

CHAPTER

9

Diseases of Free-Ranging Snow Leopards and Primary Prey Species

Stéphane Ostrowski, Martin Gilbert
Wildlife Health & Health Policy Program, Wildlife Conservation Society, Bronx, NY, USA

INTRODUCTION

This review aims to present the major infectious diseases that may affect free-ranging snow leopards, and those that may impact the abundance of their natural ungulate prey. It is beyond the scope of this review to cite the numerous studies that have documented neoplasia, degenerative diseases, congenital malformations, and infectious diseases (occasionally lethal) in captive snow leopards. In addition it should not be considered as a comprehensive review of infectious diseases; it concentrates only on those with a perceived lethality in nature.

Snow leopards and their ungulate prey inhabit cold arid environments. Because microbial abundance in soil correlates negatively with precipitation (Blankinship et al., 2011), it is predicted that they encounter lower microbial abundance than their counterparts in more mesic, temperate or tropical environments, and may have evolved correspondingly lower immune indices. This circumstance is preoccupying from a conservation standpoint as it may render these species particularly vulnerable to the emergence of pathogens disseminated by fast-spreading populations of domestic species, and to changes in pathogen distribution resulting from climatic changes. The present review shows that at least for snow leopard prey species, disease is already a significant local threat, and may be following an increasing trend, whereas data deficiencies prevent a full evaluation of the disease threat to the snow leopards themselves.

DISEASES IN FREE-RANGING SNOW LEOPARDS

Causes of Mortality in Snow Leopards

No publications currently provide a comprehensive account of mortality of free-ranging snow leopards. Natural deaths due, for example, to starvation or natural accidents are rarely observed. Human-induced casualties due to poaching, traffic accidents, or poisoning are almost never reported quickly enough for forensic investigations to be performed efficiently. Surveillance of wild populations for infections, based on antemortem

Snow Leopards
http://dx.doi.org/10.1016/B978-0-12-802213-9.00009-2

testing of blood and feces has been limited. At the time of writing, laboratory investigations of infectious agents circulating in the snow leopard population of the Tost Mountains, in the South-Gobi Province of Mongolia are underway and should provide valuable information with significant sample sizes (Ö. Johansson, personal communication). Incidental reports, confirmed by comprehensive mortality studies carried out in other nondomestic cats (Schmidt-Posthaus et al., 2002) tend to suggest that noninfectious causes are probably responsible for a significant proportion of deaths in free-ranging snow leopards. Hussain Ali, who extensively surveyed the Khunjerab area in northern Pakistan, reports in his diary to have examined 14 dead snow leopards between 2000 and 2008. Two had been poached, four were found dead alongside uneaten carcasses of Siberian ibex (*Capra sibirica*) and fallen rocks, and presumably died accidentally in the course of chases in steep terrains, three had fallen over cliffs with no dead prey around and possibly as a result of avalanche or rock slide, one was found on the Karakoram Highway and was possibly a road kill or had fallen over a cliff, three (an adult female with her two subadult cubs) could have been poisoned, and one odd case was found dead on top of a juniper tree. Interestingly, of the seven animals that had fallen over cliffs, associated or not to a prey chase, five were young (<2-year-old) animals.

Snow leopards can also be victims of poisons, either intentionally as retaliation to livestock depredation or unintentionally during indiscriminate poisoning campaigns. However, the nature of poisons, extent of use, and impact on snow leopard populations remain largely unstudied. Poaching is usually underestimated because it is rarely reported or missed, such as in the case of alleged starved or "fallen-over-cliff" individuals (Fig. 9.1), but it has been reported

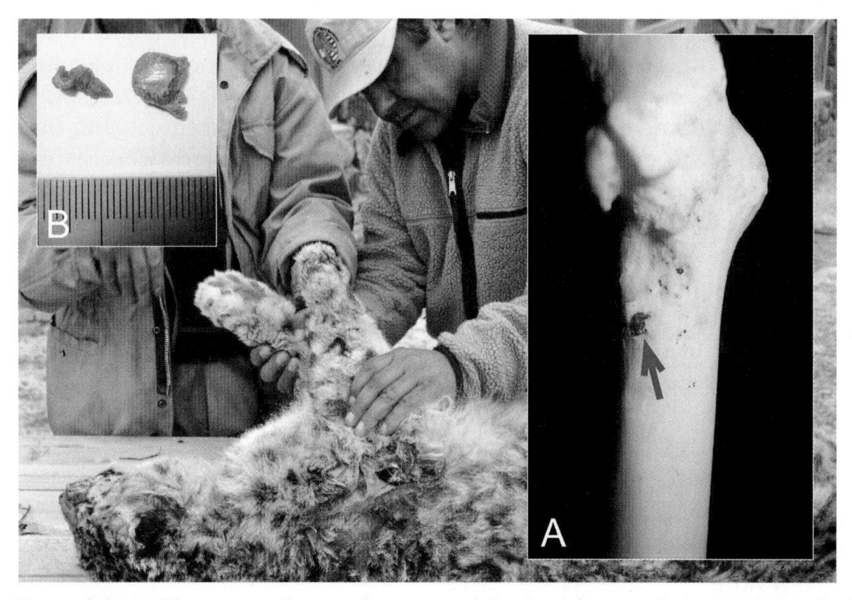

FIGURE 9.1 **Determining with accuracy the genuine cause of death of free-ranging snow leopards (*P. uncia*) could prove a daunting task.** This specimen was found dead at the base of a steep cliff over which local people alleged it had accidentally fallen. However, a thorough necropsy of the animal revealed the presence of bullet fragments in the radius bone (A) (arrow), and an hemorrhagic track through shoulder muscles with additional fragments of a broken up 0.22 caliber rimfire bullet (B), supporting that the fall was consecutive to a gunshot. Wakhan District, Afghanistan, December 2010. *Source: Photo courtesy of S. Ostrowski and Inayat Ali.*

FIGURE 9.2 **A veterinarian saws the skull of a dead radio-collared snow leopard (*P. uncia*) in an attempt to collect brain tissue for laboratory investigations.** Snow leopards are susceptible to a range of neurotropic infectious agents, such as the viruses responsible of rabies and canine distemper. Tost Mountains, South-Gobi Province, Mongolia, August 2011. *Source: Photo courtesy of T. Lhagvasumberel.*

to cause a significant proportion of the mortality of adult radio-collared felids (e.g., Amur tiger *Panthera tigris altaica* in Russia; Goodrich et al., 2008). Infectious diseases have almost never been reported or successfully verified by postmortem evaluations in free-ranging snow leopards (Fig. 9.2; Batzorig et al., 2011), but similarly to poaching this cause of mortality could easily be underestimated because of the difficulty in detecting and/or investigating cases in the inaccessible terrain they occupy. Infectious diseases may be a normal feature of an otherwise healthy snow leopard population, but effects may be exacerbated by increased stress and occur as a spillover from domestic carnivores or where populations are already compromised by declining numbers.

Infectious Diseases

Selected Viral Diseases

A wide variety of viral agents have been found in captive felids, including snow leopards, of which some have severe and sometimes fatal consequences to the host. Clinical disease has been recorded much less frequently in free-ranging wild felids, with a single poorly documented account of rabies infection from 1940 (Heptner and Sludskii, 1992), representing the only clinically significant report of viral disease in a free-ranging snow leopard. However, the true incidence of viral infections in wild snow leopards is hampered by the lack of surveillance in the remote locations they occupy, and the species is likely to be susceptible to a range of pathogens found in other free-ranging felids (Table 9.1). Viral infections will only have a negative impact on population viability if they reduce reproductive output either directly or by increasing host mortality that is additive to other causes of death. Contributory risk factors include increased pathogenicity of circulating viruses, the presence of a more abundant reservoir population (domestic or wild), and an increased susceptibility of the host such as in case of coinfection, chronic stress, or decreased genetic variability.

One viral pathogen of particular importance to populations of other *Panthera* species is canine

TABLE 9.1 Selected Microbial Diseases Potentially Responsible for Morbidity and Mortality in Free-Ranging Snow Leopards

Disease	Symptoms in felids	Mode of transmission	Perceived lethality	Reference (captive snow leopard)	Reference (free-ranging large nondomestic cats)
VIRUS					
Rabies	CNS disease; death	Bite injury; saliva	High	-	Pfukenyi et al., 2009[a]
Canine distemper	CNS disease, pneumonia; death	Ingestion; inhalation	High	Fix et al., 1989	Seimon et al., 2013
Feline immunodeficiency	Pneumonia; diarrhea; death	Bite injury	Possibly high	Barr et al., 1989[b]	Roelke et al., 2006[c]
Bluetongue	Lethargy; pneumonia; death	Ingestion; mosquito bite	Possibly high	-	Alexander et al., 1994[d]
Feline panleukopenia	Fever; diarrhea; vomiting; death	Ingestion; transplacentally	Moderate	Fix et al., 1989	Schmidt-Posthaus et al., 2002
Feline leukemia	Anemia; immunosuppression; death	Bite injury; body fluids	Moderate	-	Meli et al., 2009[e]
Feline papillomavirus	Oral warts; neoplasia	Damaged oral mucosa	Low	Sundberg et al., 2000[f]	-
Feline coronavirus	Enteritis; diarrhea; peritonitis	Ingestion	Low	Kennedy et al., 2002[g]	Heeney et al., 1990[h]
Feline calicivirus	Upper respiratory tract disease	Inhalation	Low	-	Hofmann-Lehmann et al., 1996[i]
BACTERIA					
Tuberculosis	Lower respiratory tract disease; emaciation; death	Inhalation	High	Helman et al., 1998	Michel et al., 2006
Plague	Pneumonia; death	Ingestion; flea bite	High	-	Wild et al., 2006[j]
Anthrax	Sudden death	Ingestion	Moderate	-	Jager et al., 1990
Pseudotuberculosis	Diarrhea; vomiting; lethargy; anorexia;	Ingestion	Moderate	-	Ryser-Degiorgis and Robert, 2006[k]
Tularemia	Lethargy; oral ulcers; enlarged lymph nodes	Ingestion; inhalation; insect bite	Low	-	Girard et al., 2012[l]

TABLE 9.1 Selected Microbial Diseases Potentially Responsible for Morbidity and Mortality in Free-Ranging Snow Leopards (*cont.*)

Disease	Symptoms in felids	Mode of transmission	Perceived lethality	Reference (captive snow leopard)	Reference (free-ranging large nondomestic cats)
PROTOZOA					
Babesiosis	Fever; anemia; jaundice	Tick bite	Low	-	Munson et al., 2008
Hepatozoonosis	Anemia; emaciation	Tick bite	Low	-	Khoshnegah J. et al., 2012[m]
Toxoplasmosis	CNS disease, pneumonia	Ingestion	Low	Ratcliffe and Worth 1951	Smith et al., 1995[n]
FUNGI					
Dermatophytosis	Focal/coalescing skin lesions; alopecia;	Skin contact	Low	-	Rotstein et al., 1999[o]

[a]*Pfukenyi, D.M., Pawandiwa, D., Makaya, P.V., Ushewokunze-Obatolu, U., 2009. A retrospective study of wildlife rabies in Zimbabwe between 1992 and 2003. Trop. Anim. Health Pro. 41, 565–572.*

[b]*Barr, M.C., Calle, P.P., Roelke, M.E., Scott, F.W., 1989. Feline immunodeficiency virus infection in nondomestic felids. J. Zoo Wildl. Med. 20, 265–272.*

[c]*Roelke, M.E., Pecon-Slattery, J., Taylor, S., Citino, S., Brown, E., Packer, C., VandeWoode, S., O'Brien, S.J., 2006. T-Lymphocyte profiles in FIV-infected wild lions and pumas reveal CD4 depletion. J. Wildl. Dis. 42, 234–248.*

[d]*Alexander, K.A., MacLachlan, N.J., Kat, P.W., House, C., O'Brien, S.J., Lerche, N.W., Sawyer, M., Frank, L.G., Holekamp, K., Smale, L., McNutt, J. W., Laurenson, M.K., Mills, M.G.L., Osburn, B.I., 1994. Evidence of natural bluetongue virus infection among African carnivores. Am. J. Trop. Med. Hyg. 51, 568–576.*

[e]*Meli, M.L., Cattori, V., Martínez, F., López, G., Vargas, A., Simón, M.A., Zorrilla, I., Muñoz, A., Palomares, F., Lópes-Bao, J.V., Pastor, J., Tandon, R., Willi, B., Hofmann-Lehmann, R., Lutz, H., 2009. Feline leukemia virus and other pathogens as important threats to the survival of the critically endangered Iberian Lynx (*Lynx pardinus*). PloS One. 4, e4744.*

[f]*Sundberg, J.P., Van Ranst, M., Montali, R., Homer, B.L., Miller, W.H., Rowland, P.H., Scott, D.W., England, J.J., Dunstan, R.W., Mikaelian, I., Jenson, A.B., 2000. Feline papillomas and papillomaviruses. Vet. Pathol. 37, 1–10.*

[g]*Kennedy, M., Citino, S., McNabb, A.H., Moffatt, A.S., Gertz, K., Kania, S., 2002. Detection of feline coronavirus in captive Felidae in the USA. J. Vet. Diag. Invest. 14, 20–522.*

[h]*Heeney, J.L., Evermann, J.F., McKierman, A.J., Marker-Kraus, L., Roelke, M.E., Bush, M., Wildt, D.E., Meltzer, D.G., Colly, L., Lukas, J., Manton, V.J., Caro, T., O'Brien, S.J., 1990. Prevalence and implications of feline coronavirus infections of captive and free-ranging cheetahs (*Acinonyx jubatus*). J. Virol. 64, 1964–1972.*

[i]*Hofmann-Lehmann, R., Fehr, D., Grob, M., Elgizoli, M., Packer, C., Martenson, J.S., O'Brien, S.J., Lutz, H., 1996. Prevalence of antibodies to feline parvovirus, calicivirus, herpesvirus, coronavirus and immunodeficiency virus, and feline leukemia antigen and the interrelationships of these infections in free-ranging lions in East Africa. Clin. Vacc. Immunol. 3, 554–562.*

[j]*Wild, M.A., Shenk, T.M., Spraker, T.R., 2006. Plague as a mortality factor in Canada lynx (*Lynx canadensis*) reintroduced to Colorado. J. Wildl. Dis. 42, 646–650.*

[k]*Ryser-Degiorgis, M.-P., Robert, N., 2006. Causes of mortality and diseases in free-ranging Eurasian lynx from Switzerland – an update. Proceedings of the Iberian Lynx.* Ex situ *conservation seminar series. Sevilla and Doñana, Spain, pp. 36–41.*

[l]*Girard, Y.A., Swift, P., Chomel, B.B., Kasten, R.W., Fleer, K., Foley, J.E., Torres, S.G., Johnson, C.K., 2012. Zoonotic vector-borne bacterial pathogens in California mountain lions (*Puma concolor*), 1987–2010. Vector-Borne Zoonotic Dis. 12, 913–921.*

[m]*Khoshnegah, J., Mohri, M., Mirshahi, A., Mousavi, S.J., 2012. Detection of* Hepatozoon *sp. in a Persian leopard (*P. pardus ciscaucasica*). J. Wildl. Dis. 48, 776–780.*

[n]*Smith, K.E., Fischer, J.R., Dubey, J.P., 1995. Toxoplasmosis in a bobcat (*Felis rufus*). J. Wildl. Dis. 31, 555–557.*

[o]*Rotstein, D.S., Thomas, R., Helmick, K., Citino, S.B., Taylor, S.K., Dunbar, M.R., 1999. Dermatophyte infections in free-ranging Florida panthers (*Felis concolor coryi*). J. Zoo Wildl. Med. 30, 281–284.*

distemper virus (CDV). In 1994, populations of lions (*P. leo*) in the Serengeti National Park declined by two-thirds during an outbreak of CDV, a loss of more than 1,000 animals (Roelke-Parker et al., 1996). More recently, CDV has been identified in Amur tigers in the Russian Far East, where it has contributed to local population declines (Seimon et al., 2013). Snow leopards are susceptible to CDV infection, with two cases recorded in captive animals (Fix et al., 1989; Silinski et al., 2003), although both of these were concurrent with other pathogens. In one case, CDV infection in two leopards was assumed to be a sequel to prior infection with feline panleukopenia virus, whereas in the other immunosuppression related to CDV was thought to have predisposed to an acute infection with *Toxoplasma gondii*. Coinfections have also been associated with outbreaks of CDV in free-ranging lions, with climatic conditions promoting high tick burdens and *Babesia* infection contributing to the high mortality recorded during CDV outbreaks in Serengeti in 1994 and Ngorongoro in 2001 (Munson et al., 2008). This may explain the occurrence of so-called "silent" outbreaks among lions in Eastern and Southern Africa, where infection is evident without apparent sickness or mortality (Munson et al., 2008). However, at least some captive and wild outbreaks appear to have been uncomplicated by coinfections (Seimon et al., 2013), and so other factors may contribute to clinical severity, such as strain virulence or additional external stressors.

Classical transmission of CDV is thought to require direct contact between an infected animal and a susceptible host, as it is inactivated by ultraviolet radiation, drying, and moderate temperatures (Green and Appel, 2006). Longer survival at low temperatures raises the possibility of indirect transmission in cold environments (such as viral contamination of carcasses attended by scavengers). However, the most likely source of infection in solitary *Panthera* species is assumed to be direct transmission when predating infected animals (Gilbert et al., 2014). During the early stages of infection the virus replicates in the respiratory epithelium, leading to respiratory distress signs and, often, purulent oculonasal discharge. The virus spreads systemically by infecting leucocytes, resulting in immunosuppression related to lymphopenia. Infected animals may die at this stage or improve in condition if the immune system is able to overcome the systemic infection. However, in a proportion of animals, the virus infects the central and peripheral nervous system, leading to degenerative neurological signs. For these animals, death is probably inevitable, either from the effects of the disease or as a result of sequelae such as inappropriate behavior. In wild tigers, CDV infections are most evident in these later stages, when neural deficits manifest as aberrant behavior, with a reduced aversion to people and observations of tigers in villages or along roadsides. Whether this is evident in all cases or is an effect of observation bias is unknown. It is unknown whether snow leopards would show similar behavioral signs, but CDV should be considered in any cases where snow leopards present with respiratory infection, ocular and/or nasal discharge, or aberrant behavior, fearlessness, muscle twitching, and/or convulsions.

Populations of large felids are too small and occur at densities that are too low to maintain CDV circulation in the long term. Therefore, infections in free-ranging large felids are the result of spillover from more abundant reservoir hosts and possibly short chains of infection among conspecific contacts. Most terrestrial carnivore species are thought to be susceptible to CDV, and so the exposure of snow leopard populations will depend on the presence of the virus in the wider carnivore community within their habitat. A more abundant susceptible host with high population turnover could act as a reservoir species, or several epidemiologically connected species could act as a reservoir community (Haydon et al., 2002; Viana et al., 2015).

In areas that are sparsely settled, the number of domestic dogs may be insufficient to maintain CDV on their own, but the virus could persist if it were to circulate among more abundant carnivores (such as wild canids and/or mustelids), either in concert with or independent of domestic dogs. Modeling has also shown that CDV circulation occurs over wide spatial scales (Almberg et al., 2010), and so the status of CDV in snow leopard habitat could be influenced by transmission in distant locations (such as urban centers) if they are epidemiologically connected to remote carnivore communities.

Selected Bacterial Diseases

Several bacterial agents have the potential to be lethal to free-ranging snow leopards (Table 9.1). However, as in many nondomestic cats, clinical diseases due to bacterial agents are probably most commonly due to ubiquitous bacteria associated with accidental injuries, gingival and dental lesions, and infected wounds (Schmidt-Posthaus et al., 2002). These bacterial infections self-resolve in most cases or remain benign in immunocompetent individuals.

In contrast, mycobacterial infections and particularly tuberculosis due to *Mycobacterium bovis* have caused significant morbidity and mortality in large free-ranging felids, including lions and leopards (*P. pardus*; Michel et al., 2006). In wild felids the disease seems to be primarily acquired from feeding on an infected carcass. Assessing the presence of *M. bovis* in natural prey species and livestock is therefore a crucial indicator of the risk that tuberculosis poses to free-ranging felids. In captivity, snow leopards infected by *M. bovis* have been found with symptoms of weight loss, persistent cough, and lesions of granulomatous inflammation of the lungs (Helman et al., 1998). This disease should therefore be considered in abnormally thin and emaciated free-ranging snow leopards with clinical signs or lesions of pulmonary disease.

Anthrax, caused by *Bacillus anthracis*, has been associated with deaths in free-ranging felids in Africa, infected after eating an infected carcass (Jager et al., 1990). The disease has been reported from most states of the snow leopard distribution range. A radio-collared snow leopard found dead in the Gobi desert in April 2011 with marked neck edema, a common sign of anthrax in felids (Jager et al,. 1990), and unclotted bloody discharge from the nostrils, was suspected of anthrax (K. Smimaul, personal communication), although no confirmatory test was carried out and the disease is not known to be endemic in this part of Mongolia (Odontsetseg et al., 2007).

Nondomestic felids are known to succumb to other bacterial infections, usually as incidental hosts (Table 9.1), yet the extent to which snow leopards are susceptible and the occurrence of responsible agents across their range are largely unknown. The exception to this is *Yersinia pestis*, the agent of plague, which is potentially dangerous to any carnivore it would infect, and occurs in enzootic or epizootic cycles in marmot populations across the snow leopard habitat (Gage and Kosoy, 2005).

Selected Parasitic Infections

Several reports of ectoparasite infestations in free-ranging felids include cases of highly debilitating mange caused by the mite *Sarcoptes scabiei* in the Eurasian lynx (*Lynx lynx*) in European countries (Ryser-Degiorgis et al., 2005) and cheetahs (*Acinonyx jubatus*) in Kenya (Mwanzia et al., 1995). The responsible mites are fairly host-specific, yet most will parasitize humans. Animals affected by these mites can have hair loss and various degrees of encrusting dermatitis affecting more prominently head, feet, and tail. Monitoring of Eurasian lynx populations has however shown that the disease does not constitute a threat to the long-term survival of this species and persists only in areas where it is endemic in coexisting red fox (*Vulpes vulpes*) populations (Ryser-Degiorgis et al., 2005). The snow leopard is susceptible to sarcoptic mange (Peters and Zwart, 1973), but to date the disease has not been documented with certitude in free-ranging animals, although debilitated individuals with hair loss suggestive of mange have been recorded locally (*Pamir Times*, 2011). Anecdotal cases of notoedric mange and facial demodicosis, caused by *Notoedres cati* and *Demodex cati*, respectively, have been recorded in captive snow leopards and manifested as localized hair loss (Fletcher, 1978, 1980).

Hemoparasites such as piroplasms, including *Babesia*, *Hepatozoon*, *Trypanosoma* and *Cytauxzoon* spp., appear to be common in free-ranging felids. However, except in a few rare instances (*Babesia* spp.), most cases of infections in wild felids are subclinical. Hemoparasites associated with clinical disease have yet to be documented in captive or free-ranging snow leopards.

Toxoplasma gondii is a protozoal parasite that infects most species of warm-blooded animals. Felids are the only known definitive hosts for this parasite, and therefore serve as the main reservoir. Cats rarely develop clinical toxoplasmosis, although captive Pallas' cats (*Otocolobus manul*) have been reported susceptible to the infection, resulting in high neonatal mortality (Kenny et al., 2002). A mortality case has also been reported in a captive snow leopard (Ratcliffe and Worth, 1951).

Metazoan parasites whether nematodes, cestodes, or trematodes are common in free-ranging felids and do not generally cause clinical disease. Infestations with ascarids such as *Toxascaris leonina* and *Toxocara cati* seem common in free-ranging snow leopards (Mozgovoi, 1953, cited in Ganzorig et al., 2003), and all fresh feces from a sample of 8–9 adult animals collected in Afghanistan (5), Pakistan (2), and China (1–2) between 2008 and 2013 had eggs of *Toxascaris* spp. (SO, personal observation). A case of mortality due to a ruptured aortic aneurysm caused by larvae of *Spirocerca lupi* was reported in a snow leopard 3 months after being brought into captivity, but authors suggested that the infection was acquired during captivity (Kelly and Penner, 1950).

DISEASES IN SNOW LEOPARD NATURAL UNGULATE PREY SPECIES

Sarcoptic Mange in Blue Sheep and Other Prey Species

The blue sheep (*Pseudois nayaur*) is an important prey species for snow leopards (Bagchi and Mishra, 2006). In 2007, an outbreak of sarcoptic mange was reported among blue sheep in extreme northern Pakistan (Fig. 9.3), and caused hundreds of fatalities (Dagleish et al., 2007). The disease, first reported by local herders in 1996, occurred throughout the year, affecting both sexes and all age groups and reducing the species' population over the ensuing decade. Infected animals were in poor condition, presented with severe and extensive skin lesions, especially on the forelegs and chest, and were reluctant to flee when approached (Fig. 9.4). Dagleish et al. (2007) suggested that the severity of lesions in blue sheep could have been the result of protein/energy malnutrition, and perhaps also weak immune response due to lack of previous exposure to the ectoparasite. Although the origin of the initial infection was not determined with certainty, the authors believed that the most likely source of infection were infected domestic livestock encroaching into the natural habitat of blue sheep. The gregarious social behavior of the blue sheep may have promoted the dissemination of this novel parasite, which may have been aided by malnutrition due to food competition with livestock (Mishra et al., 2004). The consequences of this mange outbreak appeared to be severe for blue sheep, but potentially also for snow leopards, which rely on this species as an important source of food. The prevalence of mange in blue sheep appeared to have decreased by 2010–2011 (Hussain Ali, personal communication), possibly as a result of host adaptation or selection of more tolerant animals. However, the corollary impacts on the snow

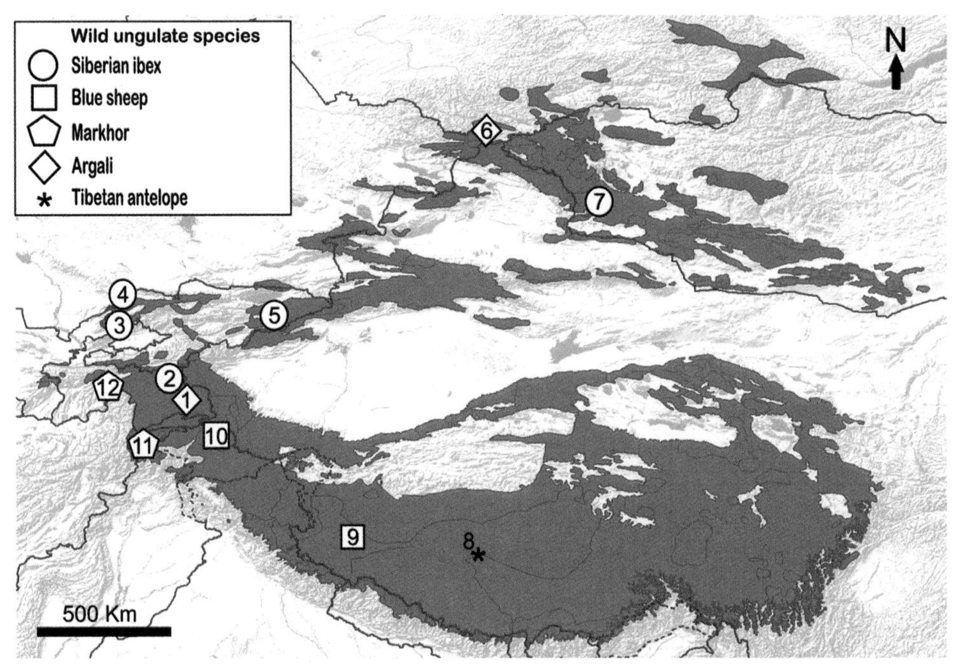

FIGURE 9.3 **Geographical locations of reported disease outbreaks in natural ungulate prey species plotted over snow leopard distribution range (in pink).** Details of outbreaks are provided hereinafter (disease name/date of outbreak/mountain range/country/reference). **1.** Rinderpest and anthrax/1895–1898/Pamir Mountains/Tajikistan/Meklenburtsev (1948). **2.** Goats pleuropneumonia disease/1940s/Pamir Mountains/Tajikistan/Heptner, V.G., Nasimovich, A.A., Bannikov, A.G., 1961. Mammals of the Soviet Union. Artiodactyla and Perissodactyla. Vyssahya Shkola, Moscow, USSR (in Russian). **3.** Sarcoptic mange/late 1960s/Pamir-Alai Mountains/Uzbekistan/Vyrypaev, V.A., 1973. The status of the Siberian ibex population in the west part of the Chatkal Range (Central Asia). In: Sokolov, V.E. (Ed.) The Rare Mammal Species in the USSR and Their Conservation. Nauka, Moscow, USSR (in Russian). **4.** Sarcoptic mange/1968–1971/Pamir-Alai Mountains/Kazakhstan/Fedosenko, A.K., Savinov, E.F., 1983. The Siberian ibex. In: Gvozdev, E.V., Kapitonov, V.I. (Eds.), Mammals of Kazakhstan. Nauka of Kazakh SSSR, Alma-Ata, pp. 92–143 (in Russian). **5.** Sarcoptic mange/late 1960s/Tien Shan Mountains/Kyrgyzstan/Yanushevich et al., 1972. cited in Fedosenko, A.K., Blank, D.A., 2001. **6.** Tuberculosis/1989/Altai Mountains/Russia/Fedosenko and Blank (2005). **7.** Sarcoptic mange/ongoing/Altai Mountains/Mongolia/E. Shiilegdamba personal communication, 2014. **8.** Contagious caprine pleuropneumonia/2012/Tibetan Plateau/China/Yu et al. (2013). **9.** Peste des petits ruminants/2007–2008/Tibetan Plateau/China/Bao et al. (2011). **10.** Sarcoptic mange/1997–2007/Karakoram Mountains/Pakistan/Dalgleish et al., 2007. **11.** Foot-and-mouth disease/2011/Hindu Kush Mountains/Pakistan/Shabbir Mir, 2011. Deadly disease kills 12 markhors in Chitral. Express Tribune, April 9, 2011. Available from: http://tribune.com.pk/story/146166/deadly-disease-kills-12-markhors-in-chitral [accessed 06.03.2015.]. **12.** Caprine mycoplasmosis/2010/Pamir-Alai Mountains/Tajikistan/Ostrowski et al., 2011.

leopard population size or dynamics have not been measured. This outbreak raised concerns on the risk posed by sarcoptic mange to other blue sheep populations. Fortunately no similar reports have emerged from elsewhere across the species' range since 2007.

Mange has a long history of affecting populations of Siberian ibex, possibly the main prey species for snow leopards range wide. Outbreaks can lead to considerable local mortality, such as in the Aksu-Zhabagly Reserve in Kazakhstan (where ca. 80% of the population was infected in 1968–1971) and in the Chatkal Range of Uzbekistan and in Kyrgyzstan. The disease also seems to be endemic in ibex in Khovd Province of the Mongolian Altai (Fig. 9.3). Clinically

FIGURE 9.4 **An adult male blue sheep (*Pseudois nayaur*) presenting severe and extensive skin lesions on the forelegs and chest due to *Sarcoptes scabiei*, the ectoparasite responsible of sarcoptic mange.** Debilitated and indifferent to humans, this animal was easily handled by yak herders. Shimshal area, Gilgit-Baltistan Province, Pakistan, July 2000. *Source: Photo courtesy of D. Butz.*

sarcoptic mange in ibex is similar to that described in blue sheep, with front legs and thorax being affected first, yet the neck and head also appear to be affected at a later stage. Mortality of infected ibex is particularly high during years of heavy snow (Fedosenko and Blank, 2001). Further studies are required to understand the impact of this infectious agent on intra- and interspecific population dynamics of snow leopards and associated prey.

Mycoplasmosis in Markhor and Other Prey Species

Throughout its range, the markhor (*C. falconeri*) has to forage in close proximity to domestic goats (Woodford et al., 2004), and is therefore prone to infections of contagious agents transmitted by these animals. In autumn 2010 an outbreak of *Mycoplasma capricolum* pneumonia killed at least 64 markhor in the southwest of the Hazratishoh Range in Tajikistan (Fig. 9.3; Ostrowski et al., 2011). Several live specimens were observed with clinical signs of labored breathing and the most relevant necropsy findings noted in the field were an abundant serous to mucopurulent nasal discharge; and internally, severe pneumonia associated with a variable level of yellow pleural fluid (Fig. 9.5). The clinicopathologic features of the disease resembled contagious caprine pleuropneumonia (CCPP) caused by *M. capricolum* subsp. *capripneumoniae* (Frey, 2002), a highly fatal disease-affecting goats in the Middle East, Africa, and Asia. However, a closely related species, *M. capricolum* subsp. *capricolum*, associated with respiratory diseases in domestic ruminants was identified, using sensitive molecular techniques, as the most probable causative agent of the fatal pneumonia outbreak in the markhor (Ostrowski et al., 2011). Although the origin of the infection remained unknown, domestic goats, which were occasionally coming in contact with markhor may have been the source of the outbreak. A serological survey carried out 8 months after the outbreak confirmed a CCPP prevalence of 10.1% (95% CI: 6.3–15.2%) in sympatric domestic goats (Peyraud et al., 2014). The susceptibility of ruminants to *Mycoplasma* infections may be exacerbated by environmental and nutritional factors. The disease appeared

FIGURE 9.5 **An adult male Heptner's markhor (*C. falconeri heptneri*) found dead in the southwest of the Hazratishoh range, Tajikistan.** *M. capricolum* subsp. *capricolum* associated with respiratory diseases in domestic ruminants was identified as the most probable causative agent of the fatal pneumonia that killed this markhor, along with at least 63 others, in September 2010. *Source: Photo courtesy of State Veterinary Department of Tajikistan.*

in autumn, when livestock and guard dogs force markhor to retreat to suboptimal pastures (Woodford et al., 2004). It was also the end of the dry season, when contact between markhor and livestock increases around dwindling water sources. The consequences of this pneumonia outbreak appeared to be locally severe for markhor but potentially also for the snow leopard, which relies on this species as a source of food. Community guards swiftly implemented control measures, including burning of carcasses and disinfecting contaminated grounds with slaked lime. Shepherds were also asked to avoid using water sources concomitantly to wild ungulates (Stefan Michel, personal communication). As a possible consequence, the markhor population recovered from the outbreak, increasing from an estimated 145 specimens in March 2011 to 236 animals in February 2012 (Michel et al., 2015).

The 2010 outbreak highlighted the general risk posed by mycoplasmas to ungulate prey species throughout the snow leopard's range. Although there have been no similar reports involving markhor populations since 2010, a relative lack of disease surveillance and limited access to sensitive molecular techniques for detection of mycoplasmas in nondomestic ruminants may mask any actual incidence. Primary outbreaks of diseases caused by various *Mycoplasma* species have had serious consequences for wild Caprinae in Europe. Mortality occasionally occurs due to the disease, and more frequently from associated starvation or falls due to disease-induced behavioral modifications (Giacometti et al., 2002). In Central Asia, the confirmation of CCPP presence in Tajikistan and China in 2009 (Office International des Epizooties, 2009; Chu et al., 2011), followed by a massive outbreak that claimed the lives of ca. 2400 endangered Tibetan antelopes (*Pantholops hodgsonii*) during September–December 2012 (Fig. 9.3; Yu et al., 2013), raise the specter of an increasing risk of *M. capricolum* outbreaks in susceptible prey species, which include blue sheep, Siberian ibex, argali (*Ovis ammon*), markhor, and Himalayan tahr (*Hemitragus jemlahicus*).

Peste des Petits Ruminants in Blue Sheep and Other Prey Species

Serologic and molecular evidence indicated that peste des petits ruminants (PPR) infection emerged between July and November 2007 in goats and sheep in southwestern Tibet, China (Wang et al., 2009). The disease likely existed for several years in Tibet without being recognized, possibly emerging via cross-border movements of infected livestock from India (Muniraju et al., 2014). The virus has now been reported in all countries bordering southwestern China (i.e., Afghanistan, India, Nepal, Pakistan, and Tajikistan), which encompass a large portion of the snow leopard range. Owing to the wide susceptibility of small ruminants (Furley et al., 1987), it was unsurprising that a fatal outbreak occurred in blue sheep in October 2007 in the same county as the domestic outbreak and then in an adjacent county in January 2008 (Fig. 9.3; Bao et al., 2011). Sick animals presented with ocular and nasal mucopurulent discharge, occasionally associated with diarrhea and lameness. The impact of PPR on this blue sheep population is unknown, but the fact that PPR morbillivirus could be responsible for high morbidity and mortality (> 50%) in wild ungulates (Elzein et al., 2004), and that 19 blue sheep as well as 6 Mongolian gazelles (*Procapra gutturosa*) were found dead in the course of the Tibetan outbreak suggest that it could have been significant. Several studies have shown that PPR infections are not self-sustaining in wildlife (Elzein et al., 2004). Similarly the closely related rinderpest *Morbillivirus* was reported in argali in the Pamirs prior to its eradication (Fig. 9.3) (Meklenburtsev, 1948), with infections attributed to spillover contamination while sharing pastures or water sources with infected livestock (Barrett et al., 2006).

CONCLUSIONS

The current lack of baseline information on the health of free-ranging snow leopards prohibits an assessment of the potential impact of infectious disease on their populations directly. The remote and inaccessible habitat occupied by the species reduces the chances of detecting disease-related mortality and complicates efforts to transfer diagnostic samples to laboratories for testing. To address these information deficits, researchers should be encouraged to include at least minimal sample collection protocols (appropriate to local circumstances) whenever opportunities arise to handle a snow leopard (e.g., through research projects, conflict situations, or when responding to a debilitated or dead individual). A description of minimal sampling protocols is beyond the scope of this review; therefore researchers are encouraged to seek the advice of a veterinarian with relevant wildlife experience in advance of these situations arising. The population impact of infectious disease should not be dismissed, despite the relatively low rate of contact among conspecifics, and comparatively low densities of other domestic and wild carnivore species occupying snow leopard habitat. Modeling of another wide-ranging solitary felid, the Amur tiger has shown profound impacts on population viability, even when opportunities for disease exposure are infrequent (Gilbert et al., 2014). These effects may be exacerbated by other physiological stressors, such as food availability, climate-related habitat changes, or by the genetic stresses of inbreeding depression. Pathogens with the potential to impact snow leopard populations are likely to be contracted from other carnivore species, therefore whenever possible, introduction of domestic dogs and cats into local carnivore communities is to be discouraged. Ultimately, a population's ability to withstand the pressures of disease will be maximized by maintaining snow leopards in numerically large subpopulations that are as interconnected as terrain and land use permit.

Snow leopards prey on whatever small and large mammals are available. However, their staple prey, accounting for more than 40% of their diet and without which they cannot survive, are large ungulates, including the blue sheep, Siberian ibex, argali, markhor, and the Himalayan tahr. Monitoring the health condition of these species appears therefore of crucial importance to support snow leopard conservation at local and global scales. In the Asian context of generalized increasing encroachment of livestock into wild habitats, domestic ungulates are the prime target for disease surveillance schemes as they are the most likely source of disease spillover to snow leopard prey. Moreover, livestock can be responsible for upslope range-shift of mountain ungulates into less suitable, stressful foraging habitat, even exceeding in magnitude the worst scenarios of climate-driven effects on their ecology (Mason et al., 2014). Therefore controlling the risk of disease outbreaks in snow leopard prey requires a complex and holistic approach that enforces prevention of disease spillover from livestock to wild ungulates and implements multifaceted controls over livestock numbers and their range use. Limiting other controllable stressors (such as human disturbance) and whenever possible maximizing genetic variability of small fragmented populations through enhanced subpopulation connectivity are also recommended to reduce disease susceptibility (Lafferty and Gerber, 2002). Vaccination of livestock is frequently not available (e.g., sarcoptic mange) or inefficacious (CCPP vaccination in Pakistan; Samiullah, 2013), and when efficiently implemented, may further enhance encroachment of livestock into wild ungulate habitat and nutritional competition as a consequence of increased livestock survival and productivity. Therefore unless implemented in combination, health managers should tend to prioritize measures that limit contact between livestock and wild ungulates rather than prophylactic actions on livestock.

Acknowledgments

We would like to thank the editors and P. Zahler for giving us the opportunity to write this chapter; E. Shiilegdamba, Ö. Johansson, and K. Smimaul for sharing their knowledge on diseases of snow leopard and prey in Mongolia; and Hussain Ali for allowing us to use the information he collected in Khunjerab area, Pakistan. We are also grateful for S. Michel for providing information on disease outbreaks in markhor, and for T. Lhagvasumberel, D. Butz, and the State Veterinary Department of Tajikistan for providing photographs. The writing time for this chapter was generously supported by the Wildlife Health & Health Policy Department of Wildlife Conservation Society.

References

Almberg, E.S., Cross, P.C., Smith, D.W., 2010. Persistence of canine distemper virus in the Greater Yellowstone ecosystem's carnivore community. Ecol. Appl. 20, 2058–2074.

Bagchi, S., Mishra, C., 2006. Living with large carnivores: predation on livestock by the snow leopard (*Uncia uncia*). J. Zool. 268, 217–224.

Bao, J., Wang, Z., Li, L., Wu, X., Sang, P., Wu, G., Ding, G., Suo, L., Liu, C., Wang, J., Zhao, W., Li, J., Qi, L., 2011. Detection and genetic characterization of peste des petits ruminants virus in free-living bharals (*Pseudois nayaur*) in Tibet, China. Res. Vet. Sci. 90, 238–240.

Batzorig, B., Tserenchimed, S., Sugir, S., Purevjat, L., Lhagvasumberel, T., 2011. A case of snow leopard mortality in Mongolia. Annual Proceedings of the State Central Veterinary Laboratory; Ulaanbaatar, Mongolia, p. 109–110 (in Mongol).

Barrett, T., Pastoret, P.-P., Taylor, W.P., 2006. Rinderpest and Peste des Petits Ruminants: Virus Plagues of Large and Small Ruminants. Academic Press, London.

Blankinship, J.C., Niklaus, P.A., Hungate, B.A., 2011. A meta-analysis of responses of soil biota to global change. Oecologia 165, 553–565.

Chu, Y., Yan, X., Gao, P., Zhao, P., He, Y., Liu, J.Z., Lu, Z., 2011. Molecular detection of a mixed infection of goatpox virus, Orf virus, and *Mycoplasma capricolum* subsp. *capripneumoniae* in goats. J. Vet. Diagn. Invest. 23, 786–789.

Dagleish, M.P., Qurban, A., Powell, R.K., Butz, D., Woodford, M.H., 2007. Fatal *Sarcoptes scabiei* infection of blue sheep (*Pseudois nayaur*) in Pakistan. J. Wildl. Dis. 43, 512–517.

Elzein, E.M.E., Housawi, F.M.T., Bashareek, Y., Gameel, A.A., Al-Afaleq, A.I., Anderson, E., 2004. Severe PPR infection in gazelles kept under semi-free range conditions. J. Vet. Med. B. Infect. Dis. Vet. Public Health. 51, 68–71.

Fedosenko, A.K., Blank, D.A., 2001. Capra sibirica. Mamm. Species 675, 1–13.

Fedosenko, A.K., Blank, D.A., 2005. Ovis ammon. Mamm. Species 773, 1–15.

Fix, A.S., Riordan, D.P., Hill, H.T., Gill, M.A., Evans, E.B., 1989. Feline panleukopenia virus and subsequent canine distemper virus infection in two snow leopards (*Panthera uncia*). J. Zoo Wildl. Med. 20, 273–281.

Fletcher, K.C., 1978. Notoedric mange in a litter of snow leopards. J. Am. Vet. Med. Assoc. 173, 1231–1232.

Fletcher, K.C., 1980. Demodicosis in a group of juvenile snow leopards. J. Am. Vet. Med. Assoc. 177, 896–898.

Frey, J., 2002. Mycoplasmas of animals. In: Razin, S., Hermann, R. (Eds.), Molecular Biology and Pathogenicity of Mycoplasmas. Kluwer Academic/Plenum Publishers, New York, pp. 73–90.

Furley, C.W., Taylor, W.P., Obi, T.U., 1987. An outbreak of peste des petits ruminants in a zoological collection. Vet. Rec. 121, 443–447.

Ganzorig, S., Oku, Y., Okamoto, M., Kamiya, M., 2003. Specific identification of a taeniid cestode from snow leopard, *Uncia uncia,* Schreber, 1776 (Felidae) in Mongolia. Mong. J. Biol. Sci. 1, 21–25.

Gage, K.L., Kosoy, M.Y., 2005. Natural history of plague: perspectives from more than a century of research. Annu. Rev. Entomol. 50, 505–528.

Giacometti, M., Janovsky, M., Jenny, H., Nicolet, J., Belloy, L., Goldschmidt-Clermont, E., Frey, J., 2002. *Mycoplasma conjunctivae* infection is not maintained in alpine chamois in eastern Switzerland. J. Wildl. Dis. 38, 297–304.

Gilbert, M., Miquelle, D.G., Goodrich, J.M., Reeve, R., Cleaveland, S., Matthews, L., Joly, D.O., 2014. Estimating the potential impact of canine distemper virus on the Amur tiger population (*Panthera tigris altaica*) in Russia. PLoS One 9 (10), e110811.

Goodrich, J.M., Kerley, L.L., Smirnov, E.N., Miquelle, D.G., McDonald, L., Quigley, H.B., Hornocker, M.G., McDonald, T., 2008. Survival rates and causes of mortality of Amur tigers on and near the Sikhote-Alin Biosphere Zapovednik. J. Zool. 276, 323–329.

Green, C.E., Appel, M.J., 2006. Canine Distemper. In: Green, C.E. (Ed.), Infectious diseases of the dog and cat, third ed. Elsevier, St Louis, MO, pp. 25–41.

Haydon, D.T., Cleaveland, S., Taylor, L.H., Laurenson, M.K., 2002. Identifying reservoirs of infection: a conceptual and practical challenge. Emerg. Infect. Dis. 8, 1468–1473.

Helman, R.G., Russel, W.C., Jenny, A., Miller, J., Payeur, J., 1998. Diagnosis of tuberculosis in two snow leopards using polymerase chain reaction. J. Vet. Diagn. Invest. 10, 89–92.

Heptner, V.G., Sludskii, A.A., 1992. Mammals of the Soviet Union, vol. 2. Part 2 (Carnivores: Hyaenas and Cats). Amerind Publishing, New Delhi, India.

Jager, H.G., Booker, H.H., Hubschle, O.J., 1990. Anthrax in cheetahs (*Acinonyx jubatus*) in Namibia. J. Wildl. Dis. 26, 423–424.

Kelly, A.L., Penner, L.R., 1950. *Spirocerca* from the snow leopard. J. Mammal. 31, 462.

Kenny, D.E., Lappin, M.R., Knightly, F., Baier, J., Brewer, M., Getzy, D.M., 2002. Toxoplasmosis in Pallas' cats (*Otocolobus felis manul*). J. Zoo Wildl. Med. 33, 131–138.

Lafferty, K.D., Gerber, L.R., 2002. Good medicine for conservation biology: the intersection of epidemiology and conservation biology. Conserv. Biol. 16, 593–604.

Mason, T.H.E., Stephens, P.A., Appolonio, M., Willis, S.G., 2014. Predicting potential responses to future climate in an alpine ungulate: interspecific interactions exceed climate effects. Glob. Change Biol. 20, 3872–3882.

Meklenburtsev, R.N., 1948. Pamir argali *Ovis polii polii* Blyth. Bulletin of the Moscow Society of Naturalists, Biology Department 53, 65–84 (in Russian).

Michel, A.L., Bengis, R.G., Keet, D.F., Hofmeyr, M., de Klerk, L.M., Cross, P.C., Jolles, A.E., Cooper, D., Whyte, I.J., Buss, P., Godfroid, J., 2006. Wildlife tuberculosis in South African conservation areas: implications and challenges. Vet. Microbiol. 112, 91–100.

Michel, S., Rosen Michel, T., Saidov, A., Karimov, K., Alidodov, M., Kholmatov, I., 2015. Population status of Heptner's markhor *Capra falconeri heptneri* in Tajikistan: challenges for conservation. Oryx 49(3), 506–513.

Mishra, C., Van Wieren, S.E., Ketner, P., Heitkönig, I.M.A., Prins, H.H.T., 2004. Competition between domestic livestock and wild bharal *Pseudois nayaur* in the Indian Trans-Himalaya. J. Appl. Ecol. 41, 344–354.

Muniraju, M., Munir, M., Parthiban, A.R., Banyard, A.C., Bao, J., Wang, Z., Ayebazibwe, C., Ayelet, G., El Harrak, M., Mahapatra, M., Libeau, G., Batten, C., Parida, S., 2014. Molecular evolution of peste des petits ruminants virus. Emerg. Infect. Dis. 20, 2023–2033.

Munson, L., Terio, K.A., Kock, R., Mlengeya, T., Roelke, M.E., Dubovi, E., Summers, B., Sinclair, A.R.E., Packer, C., 2008. Climate extremes promote fatal co-infections during canine distemper epidemics in African lions. PLoS One 3, e2545.

Mwanzia, J.M., Kock, R.A., Wambna, J.M., Kock, N.D., Jarrett, O., 1995. An outbreak of sarcoptic mange in free living cheetah (*Acinonyx jubatus*) in the Mara region of Kenya. Proceedings AAZV/WDA/AAWV Joint Conference, East Lansing, Michigan, 95–102.

Office International des Epizooties, 2009. Contagious caprine pleuropneumonia, Tajikistan. May 15. http://web.oie.int/wahis/public.php?page=event_summary&reportid=8610 (accessed 03.11.2014.).

Odontsetseg, N., Tserendorj, S., Adiyasuren, Z., Uuganbayar, D., Mweene, M.S., 2007. Anthrax in animals and humans in Mongolia. Rev. Sci. Tech. 26, 701–710.

Ostrowski, S., Thiaucourt, F., Amirbekov, M., Mahmadshoev, A., Manso-Silván, L., Dupuy, V., Vahobov, D., Ziyoev, O., Michel, S., 2011. Fatal outbreak of *Mycoplasma capricolum* pneumonia in endangered markhor. Emerg. Infect. Dis. 17, 2338–2341.

Pamir Times, 2011. Sick snow leopard captured by villagers in Skardu, dies. Available from http://pamirtimes.net/2011/02/27/sick-snow-leopard-captured-by-villagers-in-skardu-dies/.

Peyraud, A., Poumarat, F., Tardy, F., Manso-Silván, L., Hamroev, K., Tilloev, T., Amirbekov, M., Tounkara, K., Bodjo, C., Wesonga, H., Gacheri Nkando, I., Jenberie, S., Yami, M., Cardinale, E., Meenowa, D., Reshad Jaumally, M., Yaqub, T., Shabbir, M.Z., Mukhtar, N., Halimi, M., Ziay, G.M., Schauwers, W., Noori, H., Rajabi, A.M., Ostrowski, S., Thiaucourt, F., 2014. An international collaborative study to determine the prevalence of contagious caprine pleuropneumonia by monoclonal antibody-based cELISA. BMC Vet. Res. 10, 48.

Peters, J.C., Zwart, P., 1973. Sarcoptesräude bei Schneeleoparden. Erkrankungen der Zootiere. Verhandlungsbericht des XV Symposium uber die Erkrankungen der Zootiere 15, 333–334.

Ratcliffe, H.L., Worth, C.B., 1951. Toxoplasmosis of captive wild birds and mammals. Am. J. Pathol. 27, 655–667.

Roelke-Parker, M.E., Munson, L., Packer, C., Kock, R., Cleaveland, S., Carpenter, M., O'Brien, S.J., Pospischil, A., Hofmann-Lehmann, R., Lutz, H., Mwamengele, L.M., Mgasa, M.N., Machange, G.A., Summers, B.A., Appel, M.J.G., 1996. A canine distemper virus epidemic in Serengeti lions (*Panthera leo*). Nature 379, 441–445.

Ryser-Degiorgis, M.-P., Bröjer, C., Hård af Segerstad, C., Bornstein, S., Bignert, A., Lutz, H., Uggia, A., Gavier-Widén, D., Tryland, M., Mörner, T., 2005. Assessment of the health status of the free-ranging Eurasian lynx in Sweden. In: Proceedings of the First Workshop on Lynx Veterinary Aspects, Doñana, November 2005. Ministry of the Environment, Spain. .

Samiullah, S., 2013. Contagious caprine pleuropneumonia and its current picture in Pakistan: a review. Vet. Med. Czech. 58, 389–398.

Schmidt-Posthaus, H., Breintenmoser-Wursten, C., Posthaus, H., Bacciarini, L., Breitenmoser, U., 2002. Pathological investigation of mortality in reintroduced Eurasian lynx (*Lynx lynx*) populations in Switzerland. J. Wildl. Dis. 38, 84–92.

Seimon, T.A., Miquelle, D.G., Chang, T.Y., Newton, A.L., Korotkova, I., Ivanchuk, G., Lyubchenko, E., Tupikov, A., Slabe, E., McAloose, D., 2013. Canine distemper virus: an emerging disease in wild endangered Amur tigers (*Panthera tigris altaica*). mBio 4, e00410–e413.

Silinski, S., Robert, N., Walzer, C., 2003. Canine distemper and toxoplasmosis in a captive snow leopard (*Uncia uncia*) – a diagnostic dilemma. Verhandlungsbericht des Symposium uber die Erkrankungen der Zootiere 41, 107–111.

Viana, M., Cleaveland, S., Matthiopoulos, J., Halliday, J., Packer, C., Craft, M.E., Hampson, K., Czupryna, A., Dobson, A.P., Dubovi, E.J., Ernest, E., Fyumagwa, R., Hoare, R., Hopcraft, J.G.C., Horton, D.L., Kaare, M.D., Kanellos, T., Lankester, F., Mentzel, C., Mlengeya, T., Mzimbiri, I., Takahashi, E., Willett, B., Haydon, D.T., Lembo, T., 2015. Dynamics of a morbillivirus at the domestic-wildlife interface: canine distemper virus in domestic dogs and lions. Proc. Natl. Acad. Sci. 112, 1464–1469.

Wang, Z., Bao, J., Wu, X., Liu, Y., Li, L., Liu, C., Suo, L., Xie, Z., Zhao, W., Zhang, W., Yang, N., Li, J., Wang, S., Wang, J., 2009. Peste des petits ruminants virus in Tibet, China. Emerg. Infect. Dis. 15, 299–301.

Woodford, M.H., Frisina, M.R., Awan, G.A., 2004. The Torghar conservation project: management of the livestock, Suleiman markhor (*Capra falconeri*) and Afghan urial (*Ovis orientalis*) in the Torghar Hills, Pakistan. Game Wildl. Sci. 21, 177–187.

Yu, Z., Wang, T., Sun, H., Xia, Z., Zhang, K., Chu, D., Xu, Y., Cheng, K., Zheng, X., Huang, G., Zhao, Y., Yang, S., Gao, Y., Xia, X., 2013. Contagious caprine pleuropneumonia in endangered Tibetan antelope, China, 2012. Emerg. Infect. Dis. 19, 2051–2052.

CHAPTER

10

Resource Extraction

OUTLINE

Snow Leopards
http://dx.doi.org/10.1016/B978-0-12-802213-9.00010-9

SUBCHAPTER

10.1

Introduction

Peter Zahler

WCS Asia Program, Wildlife Conservation Society, New York, NY, USA

Humans have been pulling resources out of and off of the earth from the beginning of our time as a species. Hunting and gathering food items, cutting wood for heat, cooking, and building materials, digging minerals out of the earth to shape tools and weapons – and of course extracting rare metals and precious gems to trade for resources otherwise not immediately available.

As our population has swelled to more than 7 billion, our resource needs and footprint upon the earth's surface (and below it) have expanded concomitantly. These impacts have especially been felt in the tropics and in highly productive landscapes such as temperate grasslands, where villages have ballooned into cities, small farms have exploded into commercial agricultural enterprises, and wildlife habitat has been reduced to small islands of protected areas in a sea of humanity. Even in these protected areas resource extraction by humans has taken a terrible toll, leaving behind "empty forests" where little in the way of economically important wildlife or other natural resources remain.

The snow leopard, by dint of the characteristics of its habitat, has quite literally been able to stay above the fray. Asia's highest mountains are some of the most inhospitable terrains in the world (to humans), a vertical landscape of rock and ice with long, brutal winters, intense solar radiation, and low oxygen. Snow might fall in any month of the year, chilling winds blow throughout the year, and avalanches and landslides (often generated by regular tremors and earthquakes in this geologically active region) can strip a hillside or valley of even the hardiest vegetation.

The people of the high mountains of Asia have long scratched a living from this difficult environment, primarily living in small, scattered communities, keeping small herds of hardy livestock, and often moving up and down the mountains on a seasonal basis to avoid the worst of the winter's weather. The geologic isolation and low impact of the local population has long left Asia's great mountains much as they have been for thousands of years.

However, as an ancient saying goes, the only thing that is constant is change – and finally, inexorably, the snow leopard's high mountain landscape is changing. Local human communities are growing in size, as are their populations of livestock, and thus their grazing pressure. Competition with and disturbance of wild ungulates (snow leopard prey) is increasing, and hunting pressure – greatly amplified by radical improvements and increasing availability of firearms – has led to rapid decreases in wildlife populations. Meanwhile efforts to find new resources, from timber to minerals to electricity (via hydropower), have led to a push into this high-elevation landscape from below, as roads are constructed and industry gains a toehold in these mountains.

This chapter will look at a suite of growing threats to snow leopard habitat, and thus to the snow leopard itself. None of these threats are really new to Asia's high mountains, but whether they are direct efforts at extraction – such as mining, clear-cutting, and even caterpillar fungus collection, or indirect effects related to these incursions into the mountains such as the construction of roads, fences, and other linear infrastructure – all of them have been increasing over the past few years to the point where their impacts put snow leopard populations at risk.

SUBCHAPTER

10.2

Emerging Threats to Snow Leopards from Energy and Mineral Development

Michael Heiner, James Oakleaf*, Galbadrakh Davaa**, Bayarjargal Yunden**, Joseph Kiesecker**

***Conservation Lands, Fort Collins, The Nature Conservancy, CO, USA**
****Mongolia Country Program, The Nature Conservancy, Ulaanbaatar, Mongolia**

INTRODUCTION

Global population growth, projected to reach 9.6 billion by 2050 (Gerland et al., 2014), and increasing energy consumption, projected to increase 56% by 2040, are expected to drive increased investments in mining and energy development, particularly in developing countries (US EIA, 2013). Consumption of fossil fuels (oil, natural gas, and coal) grew more than threefold between 1965 and 2012 (Butt et al., 2013). By 2035, consumption of oil is projected to increase by more than 30%, natural gas by 53%, and coal by 50% (Institute for Energy Research, 2011). Technological advancements such as horizontal drilling in conjunction with hydraulic fracturing have spurred a rapid increase in unconventional gas production with most resources still untapped (Kerr, 2010). Climate change policy and concerns regarding future energy security continue to stimulate an unprecedented rise in the production of biofuels, as well as solar and wind development (Haberl et al., 2010). Mineral extraction is projected to increase 60% by 2050 (Kesler, 2007). The cumulative anthropogenic change resulting from this development could place 20% of the world's remaining natural lands, over 19 million km^2, under high threat of future conversion (Oakleaf et al., 2015). As a result, the viability of threatened and endangered species in these landscapes will face significant threats without proactive planning and conservation interventions.

Proactively identifying areas at risk of conversion and strategic planning to balance goals for development and conservation will be critical to maintain viability of species and ecological systems. Here we highlight areas of potential conflict between future development and snow leopard habitat using a novel assessment of global development risk. We also recommend steps to improve the implementation of impact mitigation to safeguard species such as the snow leopard. Finally, to illustrate this process, we provide an example of these principles in action drawing from recent mitigation planning work in Mongolia.

IMPACTS OF MINING AND ENERGY DEVELOPMENT

Although the literature examining the impacts of mining and energy development on wildlife is uneven, the greatest impact is typically habitat loss, either directly from land cover conversion, or though avoidance of noise and light pollution (Jones et al., 2015). Generally, the indirect impacts of roads and supporting infrastructure are greater and affect a larger area than direct disturbances. Petroleum, natural gas, wind, and solar energy development require a large network of roads and transmission lines, and require a significantly larger area per unit of energy produced than coal mining (McDonald et al., 2009). Though mining creates a significantly smaller direct footprint than energy development such as petroleum, solar photovoltaic, and natural gas developments, some wildlife can use habitat within natural gas and wind developments, while mine pits have a longer duration and effective reclamation is more costly and often does not restore all habitat values.

For the snow leopard, the most common anthropogenic threat is reduction in natural prey, either through competition – overgrazing and range degradation – or hunting, which results in predation on livestock and retribution killings (McCarthy and Chapron, 2003). The most significant threat from mining and energy development may be increases in human populations and the resulting cascade of indirect impacts. New large mines and energy developments may be followed by dramatic changes in land use, primarily livestock grazing, and increased killing following livestock predation and poaching for illegal wildlife trade (see Chapters 5, 6, and 7).

Another indirect threat from mining and energy development to snow leopards and their prey species, and other wide-ranging wildlife, is loss of habitat and genetic connectivity due to fragmentation from movement barriers created by linear infrastructure supporting mining and energy operations. Linear infrastructure (see Chapter 10.3) includes roads, railways, fences and transmission lines that form barriers either by directly blocking movement, such as fences, or by producing traffic, noise, and light pollution that wildlife avoid. Roads supporting mines often carry a high volume of truck traffic, and the related noise and light pollution often create a more distinct barrier than the direct surface disturbance.

DEVELOPMENT THREATS ACROSS SNOW LEOPARD RANGE

To assess future threats to snow leopards from energy development, we compared snow leopard range with data from a global, spatially explicit analysis of development threats. Oakleaf et al. (2015) developed spatial models of energy development potential for multiple sectors including coal, conventional and unconventional oil and gas, solar and wind (Fig. 10.2.2), as well as biofuels, mining, and agricultural and urban expansion (not shown). With this information, it's possible to evaluate the pattern of development threats at a regional scale, by sector, and as a cumulative index across sectors. To help facilitate future planning and guide conservation interventions for snow leopards, we examined development threats in regions defined by McCarthy and Chapron (2003): The Northern Range (NRANG), Commonwealth of Independent States and Western China (CISWC), Karakorum/Hindu Kush (KK/HK), and the Himalaya (HIMLY), shown in Fig. 10.2.1.

In the Northern Range, the energy sectors projected to grow fastest are coal, conventional and unconventional oil and gas, and wind. Across snow leopard range, coal deposits are most abundant in the Northern Range, particularly in Mongolia (Fig. 10.2.2A). Two large oil and gas fields lie in the basins north and south of the

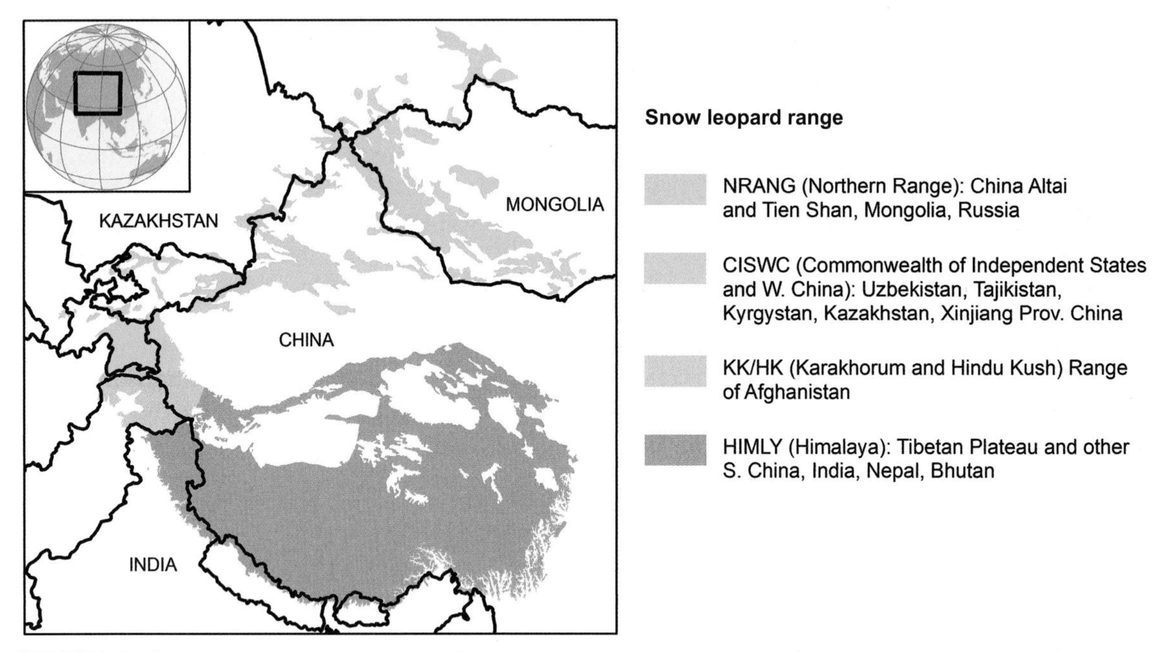

FIGURE 10.2.1 **Snow leopard range.** Global snow leopard range (see Chapter 3) divided into four regions defined by McCarthy and Chapron (2003).

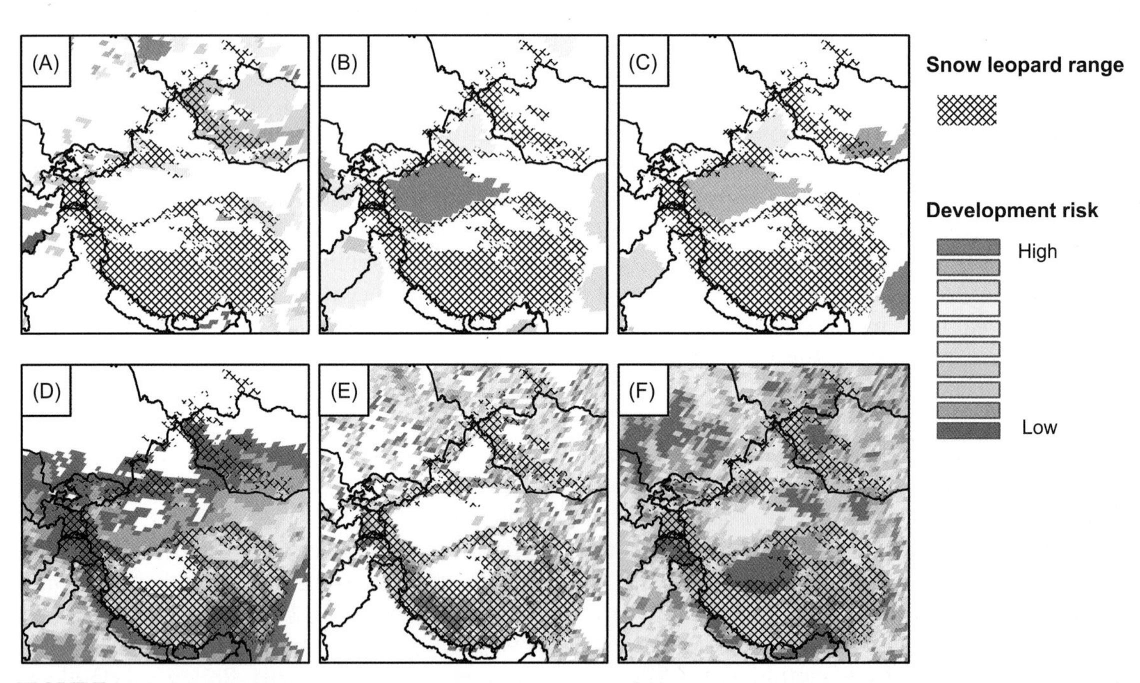

FIGURE 10.2.2 **Development risk across snow leopard range.** Snow leopard range and projected development threats by energy sector: (A) Coal; (B) conventional oil and gas; (C) unconventional oil and gas; (D) solar; (E) wind; and (F) cumulative: coal, conventional oil and gas; shale oil and gas, solar, wind, biofuels, agriculture expansion and urban expansion.

Tian Shan Mountains in Xinjiang, Western China (Figs 10.2.2B,C), which is thought to be important for metapopulation connectivity. The large spatial footprint of oil and gas development and supporting transportation infrastructure could lead to growth in human populations and related conflicts in areas adjacent to the Tian Shan, and also reduce movement and genetic connectivity across this metapopulation bottleneck.

In the CISWC and KK/HK, only wind is projected to grow significantly (Fig. 10.2.2E). However, coal deposits and oil and gas fields lie to the north, west, and east of snow leopard range in this region (Fig. 10.2.2A,B,C). As with the Tian Shan, increased human populations and transportation infrastructure could increase conflicts and reduce connectivity.

In the Himalayas, the projected fastest-growing energy sectors are solar (Fig. 10.2.2D), wind (Fig. 10.2.2E), and hydropower. Across snow leopard range, solar, and wind potential are greatest in the Himalayan Plateau. Though we did not evaluate hydropower, the Southern Himalayas are experiencing unprecedented hydropower construction, where China is the world leader (Zarfl et al., 2015), and the Indian Himalayas contain the world's highest density of planned dams (Grumbine and Pandit, 2013). There are also coal deposits at the northeastern edge of the plateau in Gansu, central China (Fig. 10.2.2A).

The areas of snow leopard range at greatest cumulative risk are the northern and southern edges of the Tian Shan Mountains and the southern and western edges of the Himalayan Plateau (Fig. 10.2.2F). The latter is thought to be one of the best habitats for the species in China (McCarthy and Chapron, 2003). As discussed earlier, the primary threats are indirect impacts of increased human populations and infrastructure that may impede movement between large areas of occupied habitat, particularly in the areas around the Tian Shan Mountains and across the string of occupied habitat patches in southern and western Tibet.

MITIGATION POLICY AND PRACTICE

Environmental licensing processes, such as Environmental Impact Assessment (EIA), play a critical role in controlling environmental damage from development projects (see Chapter 20). In most countries, developers obtain a permit before construction and operation can begin, and currently EIA has been legally adopted in almost all countries in the world (Morgan, 2012) and is one of the most effective policy mechanisms to incorporate environmental information into real-world decision making (McKenney and Kiesecker, 2010; Saenz et al., 2013). EIA and impact mitigation are a comprehensive process that examines the environmental impacts of planned developments with the goal of preventing environmental damage through the application of the mitigation hierarchy: avoid, minimize, restore, or offset. To avoid impacts on biodiversity, measures are taken to prevent impacts from the outset, such as careful spatial or temporal placement of elements of infrastructure. In minimization, measures are taken to reduce the duration, intensity, and/or extent of impacts that cannot be completely avoided. Restoration measures seek to rehabilitate degraded ecosystems or restore cleared ecosystems after impacts that cannot be completely avoided and/or minimized. To offset impacts, measures are taken to compensate for any residual adverse impacts that cannot be avoided, minimized, and/or restored. Offsets can take the form of management interventions such as restoration of degraded habitat, or protecting areas where biodiversity loss is projected (Kiesecker et al., 2009). Offsets have great potential as a source of funding to underwrite the costs of effective protection and management, given the appropriate sequence – avoid, minimize, and then offset – is followed.

Attempts to meet biodiversity goals through application of the mitigation hierarchy have gained wide traction globally with increased development of public policy (e.g., Saenz et al., 2014), lending standards (e.g., IFC, 2012), and corporate policy (e.g., Rio, 2004; Teck, 2013). Despite this increased interest, effective implementation must overcome several challenges: (1) to follow the appropriate sequence of the mitigation hierarchy, and (2) to evaluate cumulative impacts at the landscape level. Though most mitigation frameworks emphasize the sequence of avoid and minimize before offset, existing guidance is limited and allows broad interpretation (McKenney and Kiesecker, 2010). An assessment of critical habitat where development impacts should be avoided is often skipped or does not consider cumulative impacts of current and future development. A landscape- or watershed-level assessment is necessary to identify habitats that require special measures to minimize impacts or higher offset requirements. Finally, offset design must consider habitat distribution and cumulative impacts at the landscape level, both to ensure ecological equivalence of the impact and offset, and to achieve "additionality" (i.e., maximize conservation benefit by aligning with broader regional priorities and goals).

Mitigation of impacts to movement and connectivity of wide-ranging species is a difficult and urgent challenge, requiring improvements in both policy guidance and practical mitigation measures. Effective mitigation of barriers created by infrastructure requires a combination of siting analysis, wildlife passage structures, and traffic curfews, all of which require local studies to implement, and more engineering and ecology research. Wingard et al. (2014) provide guidance for mitigation of linear infrastructure, focused on wildlife in Central Asia, including snow leopards. Climate change will likely require further shifts in ranges and habitat use and compound the threat of habitat loss (see Chapter 8), fragmentation and movement barriers (Singh and Milner-Gulland, 2011), and Bull et al. (2013) propose offset design that accounts for these shifting ranges and habitat.

LANDSCAPE-LEVEL MITIGATION IN ACTION: MONGOLIAN GOBI CASE STUDY

The Mongolian Gobi region spans 510,000 km^2, or the southern third of Mongolia. This region is a mix of desert, desert grasslands, and mountains, and supports a large assemblage of native wildlife, including a northeastern extension of snow leopard range in the Altai Mountains of Southwestern Mongolia. Relative to other parts of snow leopard range, habitat here is more arid and the geographic pattern consists of smaller patches of low mountain ranges separated by desert valleys, and snow leopards may travel larger distances and cover larger home ranges as a result (McCarthy et al., 2005). The Tost Uul mountain range in Southern Mongolia supports one of the most stable and productive snow leopard populations in the global range (Jackson et al., 2009; Sharma et al., 2014). It also faces high development pressure, with 25 mining exploration leases, two large active mine leases, and the Oyu Tolgoi coal mine lying within 20 kilometers adjacent to the mountain range (MMRE, 2012).

Mineral resources exploration and exploitation in Mongolia is increasing dramatically, and mining development in the Gobi region is occurring faster than the national trend, with 24% of the Gobi Region currently leased for exploration and another 32% available for lease (MMRE, 2012). Since 2009, The Nature Conservancy (TNC) has been working in Mongolia to apply the Development by Design framework (Kiesecker et al., 2010) to balance development and conservation goals. The two focal areas are landscape-level conservation planning to guide mitigation decisions, specifically regarding avoidance and offsets, and policy guidance to strengthen licensing and mitigation.

In 2013, in collaboration with national and provincial governments, universities, and NGOs, TNC produced a landscape-level conservation plan for the Mongolian Gobi region

(Heiner et al., 2013). This plan identified a set of priority conservation areas, referred to as a "portfolio," that could maintain the biodiversity and ecological processes representative of the region, given adequate protection and management as high-quality core habitat within a larger landscape matrix that supports habitat use and movement. The portfolio includes the Tost Uul mountain range and was designed to (a) include national protected areas, (b) meet representation goals for the amount and distribution of focal biodiversity elements defined by a mapped ecosystem classification and modeled habitat distributions of 33 threatened or endangered species according to the National Red Lists, and (c) optimize for ecological condition based on an index of disturbance and cumulative anthropogenic impacts. Representation goals are based on the Mongolian government commitment to protect 30% of all natural habitats (Master Plan for Mongolia's Protected Areas, 1998).

The portfolio sites and supporting information provide critical information to guide mitigation, specifically:

1. Identification of areas where avoidance of development is recommended, through designation of protected areas to meet government commitments to protect 30% of natural habitats, based on a systematic landscape-level conservation plan.
2. One component of IFC PS 6: identify critical, natural and modified habitats based on a landscape-level conservation plan, with specific mitigation requirements for each category.
3. A basis for optimal offset design that ensures ecological equivalence, based on the regional ecosystem classification and directed to portfolio sites that maximize conservation benefits and align with regional conservation goals.

The conservation planning methods developed and applied in the Gobi region have been ratified by the Mongolian Academy of Sciences to establish credibility as a basis for mitigation policy (MAS, 2014). Following completion in 2010 of a similar regional conservation plan for the Eastern Mongolian Grasslands, 31 new protected areas covering 27,350 km^2 have been designated based on the conservation portfolio, including 4 national protected areas (3,310 km^2) and 27 provincial protected areas (24,040 km^2). In 2012, the national EIA law was amended to require offsets for all new active mineral and petroleum leases. The Mongolian Ministry of Environment is developing practical regulatory guidance to meet this new offset requirement, based on the conservation portfolio and supporting information provided by the regional conservation plans. In June 2015, the conservation planning process for the remaining Western and Central regions will finish, completing a national portfolio and information system to guide protection and mitigation across Mongolia.

CONCLUSIONS

Mining and energy development are projected to expand in snow leopard range, particularly in Southern and Western Tibet, north and south of the Tian Shan Mountains in Western China and in the Mongolian Altai Mountains. Though the immediate footprint and infrastructure will likely avoid most core snow leopard habitat in mountains, increased human populations in adjacent areas may lead to increased snow leopard conflicts and mortality, and infrastructure and land use change may reduce movement and metapopulation connectivity. Improvement in implementation of mitigation policy may be a mechanism to reduce the impacts. A significant obstacle to effective mitigation is that most mitigation assessments and planning operate at the level of individual projects and ignore cumulative impacts to habitat and connectivity at the landscape level. Effective conservation of wide-ranging species such as snow leopards will require mitigation policy and planning that function at the landscape level. Specifically, necessary improvements should (1) include effective identification and avoidance of impacts in critical habitat as the first step in the mitigation hierarchy, including areas for wildlife movement; (2) be based on a landscape-level assessment of cumulative impacts and habitat distributions; (3) effectively minimize barrier effects of linear infrastructure; and (4) incorporate effective application of biodiversity offsets to design and fund conservation measures. More general threats to snow leopards are lack of conservation policy, institutional capacity, enforcement, and community awareness (McCarthy and Chapron, 2003). Effective mitigation of mining and energy development may be a mechanism to strengthen and fund improvements in policy and institutional capacity.

SUBCHAPTER

10.3

Linear Infrastructure and Snow Leopard Conservation

Peter Zahler

WCS Asia Program, Wildlife Conservation Society, New York, NY, USA

The term "linear infrastructure" refers to roads and rail lines, fences, pipelines, power and communications lines, canals and irrigation ditches, and similar structures that are built over distances and in more or less a straight line. Linear infrastructure, by definition, can create barriers to wildlife movements. It is also the lifeblood of national and regional economies and can cross multiple jurisdictions (local, regional, and even international borders), both of which present multiple issues when attempting to minimize or mitigate the threat from these development projects.

Linear infrastructure poses the greatest threat to migratory or nomadic species, who have adapted to traveling long distances to find food, avoid inclement weather, or locate mates or birthing grounds (Wingard et al., 2014). Although the snow leopard is listed on Appendix I of the Convention on the Conservation of Migratory Species (CMS) of Wild Animals, it is not a migratory species *sensu stricto*. Its placement on this convention is based more on the fact that individuals sometimes perform long-distant movements (Karlstetter and Mallon, 2014), and significant populations exist near and cross over national borders between countries, with individuals known to have large home ranges. What is well known is that snow leopards are solitary, territorial habitat specialists that naturally live at relatively low densities; therefore, young snow leopards likely must move significant distances to find an appropriate home territory (dispersal), and both young and adult snow leopards may also have to travel long distances to find food or a mate.

Because of the snow leopard's tendency to move long distances, linear infrastructure poses a suite of threats to the species. Probably the primary threat is the building of border fences. It has been estimated that up to a third of the snow leopard's known or potential range is located less than 50–100 km from the international borders of the 12 range countries (Jackson, R., unpublished data). This is no surprise, as snow leopards live on high mountain ridges, which politicians and governments often view as easily defined physical definitions of political boundaries. Unfortunately, border fences serve as significant barriers to movement, both for snow leopards (bisecting territories or halting natal dispersal or searches for mates) and for their prey – especially argali, which are known to make considerable seasonal movements in parts of their range (Mallon et al., 2014).

A second threat from linear infrastructure comes from "lowland" infrastructure projects that may impede snow leopard movements from one mountain range to another. In the Mongolian Gobi and part of Tibet, snow leopard habitat consists of relatively small rocky massifs separated by open desert valleys, so the ability to move between massifs in search of prey or mates may be especially important. Snow leopards have been noted moving across large expanses of flatlands between small ranges, especially in Mongolia (McCarthy et al., 2005; Munkhtsog et al., 2012; Sharma et al., 2014). Construction of roads, railroads, and fences across these lowlands can be significant barriers to wildlife movement and result in the complete isolation of already partially fragmented populations (Batsaikhan et al., 2014). Even though fences tend to produce the most significant barrier effect to wildlife movement (Harrington and Conover, 2006), linear infrastructure does not have to be a complete physical barrier to be an impediment to movement – many wildlife species, including snow leopards, tend to avoid human activities including vehicle traffic (and thus roads) due to hunting pressure and harassment (Wolf and Ale, 2009). This may seem to be a minor threat, given that many range countries only have populations of 100–400 snow leopards, but population fragmentation or even loss of occasional individuals due to direct or indirect effects of linear infrastructure has the potential to be a significant issue. It may become an even more pervasive threat as climate change alters habitats, making contact between widely dispersed subpopulations of snow leopards – and their prey – more difficult and more important, as well as increasing the pervasiveness of conflict-related threats (Aryal et al., 2014; Forrest et al., 2012).

The third threat from linear infrastructure is the increase in human activity associated with development, especially from the creation of new roads. Roads provide easy access to previously uninhabited (or low human population) mountain regions, which can lead to disturbance (both to snow leopards and especially to wild ungulate prey species), further development (and associated negative impacts, including new or growing settlements), and poaching (Wingard and Zahler, 2006). Poaching may be the most pernicious effect on snow leopards and their prey species – historically wildlife in the mountains of Asia have been partly protected by their isolation and the difficulty of moving into and around the world's greatest mountains. As roads slowly but inexorably reach into these regions, however, the ability of poachers to easily access wildlife – including snow leopards – will only grow.

As mentioned, dealing with linear infrastructure is complicated by its importance to local and regional economies, by its placement in terms of national management and administration (often involving multiple ministries, ranging from development and infrastructure to environment), and by its tendency to cross jurisdictions, resulting in multiple local and national agencies – and sometimes multiple countries – involved in decision making.

However, tools are available to help improve the impact of linear infrastructure on wildlife, including snow leopards. They can be considered at three levels: international, national, and private sectors.

At the international level, conventions and agreements can be brought to bear on plans for linear infrastructure. The Convention on Biological Diversity sets out the following approach: avoid irreversible losses of biodiversity, seek alternative solutions that minimize biodiversity losses, use mitigation to reduce the severity of impacts, and compensate for unavoidable losses by providing substitutes of at least similar biodiversity value. All of these approaches can be applied to linear infrastructure. As mentioned, the CMS lists the snow leopard in Appendix I. Parties that are Range States to Appendix I species such as the snow leopard are obliged to afford them strict protection, including mitigating obstacles to migration and controlling other factors that might endanger them.

At the national level, environmental assessments are mandated for reviewing infrastructure projects, including linear infrastructure.

Strategic environmental assessments (SEAs) are higher-level policy reviews that create the legislative "enabling environment" for decision making related to infrastructure and other development projects. Environmental impact assessments (EIAs) are reviews of individual development projects, which determine whether the project is allowable under existing law and what, if any, mitigation efforts must be performed to limit the impact of the project. Unfortunately few countries have legislation that specifically mentions, much less describes, actions to be taken relative to linear infrastructure (Wingard et al., 2014).

Finally, the private sector – specifically global lenders such as the World Bank, International Finance Corporation (IFC), European Bank for Reconstruction and Development (EBRD), and the Asian Development Bank – have produced standards that stress avoidance and minimization of development impacts on the environment, including impacts from linear infrastructure. If a project is expected to receive loans from one of these institutions, the project must adhere to the standards and requirements of that institution, and these standards tend to focus on the mitigation hierarchy of (i) avoid, (ii) minimize, and (iii) compensate (Wingard et al., 2014). The IFC standards are considered the strictest and most "environmentally friendly," particularly the Performance Standard 1 (PS1) – Assessment and Management of Environmental and Social Risks and Impacts, and Performance Standard 6 (PS6) – Biodiversity Conservation and Sustainable Management of Living Natural Resources. However, the other lending institutions also have strong guidelines that can be referenced when discussing the potential or existing threats from linear infrastructure on the snow leopard (or its prey).

It is important to recognize that linear infrastructure does not have to have significantly adverse effects on snow leopards or key prey species if the tools mentioned already are used to minimize or mitigate effects, and that all stakeholders have a role in that process. Government personnel have a responsibility to adhere to national legislation and international obligations (and to work toward improving legislation related to linear infrastructure, including SEA and EIA legislation as described). Civil society has a right to hold government responsible for that adherence. Biologists and conservationists with particular expertise on snow leopards can and should play a role throughout the process; they are the ones who can sensibly and accurately inform government, industry, and lenders of the risk of a project to snow leopards. They must be involved in the planning and design of a development project, including options to avoid or mitigate impacts. They are also best positioned to provide advice (and in some cases actually perform) monitoring and evaluation of effects to determine whether adaptations need to be incorporated into the project to limit negative impacts.

As mentioned, fencing (especially border fencing) is the form of linear infrastructure that has the greatest impact on snow leopards and their prey species. Fences disrupt daily, seasonal, and dispersal movements, cause direct mortality from entanglement and indirect mortality from blocking animals from key resources or avoidance of other dangers such as inclement weather, and lead to habitat and population fragmentation (Ito et al., 2013; Lkhagvasuren et al., 2012). Serious consideration should always be given as to whether fences (which are expensive to construct and maintain) are worth the cost, given various issues including their impact as wildlife barriers. If a fence is deemed necessary for some reason, the least amount of fencing possible should be used to achieve its objectives, and serious consideration should be given to make the fence "wildlife friendly" (e.g., barbless) so that large mammals are able to move under, over, or through the fence (Olson, 2013). Any fences that are obsolete or no longer needed should be removed – removal costs are often lower than long-term maintenance.

Most roads in snow leopard habitat are low-traffic mountain tracks, often dirt, and are unlikely to have significant impact on snow leopard movements. However, some larger roads with fairly significant traffic bisect snow leopard range, such as the Karakoram Highway in northern Pakistan (Wang et al., 2014), and as mentioned snow leopards occasionally cross valleys and lowland areas for dispersal or other reasons, and thus may have to cross major roads (or even railways) to complete their movements. In both cases, consideration should be given to providing snow leopards and their ungulate prey species with options for crossing that minimize the roadway's interference with movements. Any efforts to mitigate impacts on wild ungulates will probably also benefit snow leopards. Options might include overpasses, underpasses, and in areas known to be travel routes for wildlife, warning signs and speed limit reductions (Wingard et al., 2014).

Linear infrastructure has been shown to be a significant threat to large mammals in Asia. Although the snow leopard is not often thought of as a species that can be negatively impacted by development initiatives, continued encroachment into snow leopard habitat by the construction of roads and fences and the existing (and growing) barrier provided by border fencing are issues that conservationists should both monitor and attempt to plan for, manage, or mitigate at the appropriate stages of planning and development.

SUBCHAPTER

10.4

Harvesting of Caterpillar Fungus and Wood by Local People

John D. Farrington
World Wildlife Fund, Thimphu, Bhutan

CATERPILLAR FUNGUS

Caterpillar fungus (*Ophiocordyceps sinensis*; Tibetan: *yartsa gunbo*) parasitizes ground-dwelling ghost moth larvae (*Thitarodes* spp.) to produce a needle-like mushroom with its root encased in the exoskeleton of a dead caterpillar (Fig. 10.4.1) (Winkler, 2008). This fungus is only found at altitudes of about 3000–5200 m in steep alpine meadows of the Tibet and Himalaya Region (Shrestha and Bawa, 2013; Winkler, 2008). Its range lies almost entirely within snow leopard range areas, occurring along the Himalaya from Uttarakhand in northwest India to northwest Yunnan, north along the eastern Tibetan Plateau to northeast Qinghai, and southwest across the eastern Tibetan Plateau to Lhoka Prefecture in southern Tibet (Winkler, 2008). Although its medicinal properties have been known for more than 500 years, since 1993 the Chinese market for caterpillar fungus has exploded. With this boom have come profound cultural and ecological impacts that have consequences for snow leopards and their habitat, particularly shifting patterns of natural resource use by local livestock herders.

A review by Wang and Yao (2011) found 56 potential insect hosts of *O. sinensis*, primarily "ghost moth" larvae of the genus *Thitarodes* (37 species). Ghost moth larvae are grass root-boring caterpillars that have a life cycle of up to 5 years, most of that spent underground. Adult ghost moths only emerge to live 2–5 days, at which time they mate and lays eggs (Winkler, 2005). Infection of ghost moth larvae by *O. sinensis* spores likely occurs by penetration of the still soft larva's exoskeleton shortly after hatching, after which a period of dormancy occurs before the fungus concentrates in the larva's lipid reserves, depriving it of nutrients leading to starvation. Upon death of the larva, the fungus absorbs the remaining insect tissue within the exoskeleton as it emerges aboveground (Cannon et al., 2009).

The earliest known mention of *O. sinensis* use in traditional medicine comes from a fifteenth-century Tibetan medical text, although not mentioned in a Chinese medical text until 1757 (Winkler, 2005). Historically, caterpillar fungus was used to treat a wide array of ailments, including lung, kidney, liver, and cardiovascular diseases as well as male sexual dysfunction, boosting the immune system, and general strengthening (Winkler, 2005; Zhu et al., 1998). However, widespread consumption of caterpillar fungus only began in 1993, when the coach of China's women's track team attributed a series of world-record-breaking performances by

FIGURE 10.4.1 ***Cordyceps sinensis* in alpine meadow.** *Source: Photo courtesy of Daniel Winkler, 2008.*

his runners to high altitude training and a tonic made from caterpillar fungus and turtle's blood (Hart, 2012; Steinkraus and Whitfield, 1994). Although six of the coach's athletes failed tests for performance enhancing drugs in 2000, the caterpillar fungus boom was already well underway. Demand for the fungus grew rapidly, and by late 2006 top-quality caterpillar fungus was retailing for about USD 32,000 per kg in China's eastern cities. And by August 2012, this price had soared to USD 110,000 per kg in Beijing, almost three times the current price of gold (Lo et al., 2013; Winkler, 2008). Given that *O. sinensis* occurs almost entirely within snow leopard range areas, the rising economic importance of this resource may be having a potentially large impact on snow leopards, their prey, and habitat.

Caterpillar fungus is harvested by family teams that slowly comb alpine meadows for thin well-hidden mushroom shoots. Once located, a small pickaxe is used to remove a patch of grass, together with roots and topsoil, to reveal a mushroom emerging from the exoskeleton of its deceased larval host. The entire mushroom and attached caterpillar remains are removed intact, carefully cleaned, and sold to itinerant caterpillar fungus traders. With an estimated 140,000 kg annual harvest in the Tibetan Plateau and Himalaya Region, caterpillar fungus is currently a multimillion-dollar industry and the primary source of cash income for many residents of the region (Winkler, 2009).

The economic riches to be earned send tens of thousands of people into mountains each spring to harvest caterpillar fungus, emptying out towns and schools (Yeh and Lama, 2013). Although no formal studies have been conducted on the impact of caterpillar fungus collection on snow leopards and their prey, disturbance to these species is no doubt large during the harvest period, which generally lasts from about mid-May to mid-July. In the Nepal Himalaya, thousands of people can enter a valley known to be a good collecting site (Som Ale, personal communication, 2014). Presumably such large human influxes cause blue sheep with newborn lambs to seek refuge away from prime, fungus-producing alpine meadows. On the eastern Tibetan Plateau, a similar influx occurs, with collectors traveling to collection sites by vehicles ranging from motorcycles to trucks, often bringing 2–6 loudly barking Tibetan mastiffs to guard their camp and harvest, causing further disturbance to wildlife. Direct degradation of snow leopard habitat caused by caterpillar fungus collectors includes the widespread digging up of alpine meadow turf and consequent grassland degradation, particularly if uprooted turf patches are not replaced; widespread cutting of alpine shrubs for fuel; dumping of large amounts of trash in alpine meadows; and open defecation by collectors, which contaminates water sources (Lu Zhi, personal communication, 2014; Thukten, 2014).

However, herder interviews indicate that the caterpillar fungus trade has had some benefits for alpine grassland habitat. The story of one yak herder interviewed in 2010 in Zadoi County, Qinghai, was typical. He stated that 10 years ago, his family had 120 yaks but now only kept 50, with all meat and dairy products produced used solely for family consumption. Meanwhile the family's entire cash income was now generated by three family members who annually collected 500–600 pieces of caterpillar fungus each. Another collector interviewed in Zadoi in 2011 stated that her family had completely stopped herding livestock 8 years earlier and now earned their entire annual income by collecting fungus. Many such successful collectors have left the pasturelands altogether and moved to nearby towns, where they now reside 9–10 months per year, with obvious benefits for wildlife being reduced disturbance, reduced grazing competition with domestic livestock, and reduced grazing pressure on grasslands in general (Winkler, 2011). Nevertheless, some former herders retain their pasture rights and lease their pastures to other herders.

With the value of caterpillar fungus having increased by about 1000% between 1997 and 2011, a fungal gold rush was triggered, sending legions of people to join the harvest (Winkler, 2011). Initially, there seemed to be no decline in the harvest from year to year. However in 2011, 17 collectors and traders interviewed in three different areas of Qinghai all reported declining production while many collectors reported that areas formerly rich in caterpillar fungus now had reduced amounts or none at all (Winkler, 2011). At the same time, with legalization of the caterpillar fungus trade in Nepal in 2001, the officially traded national harvest increased from 3.1 kg in 2002 to 2442.4 kg in 2009, before dropping precipitously to just 1170.8 kg in 2011, with one possible cause cited, being overharvesting (Shrestha and Bawa, 2013). In addition, a second large threat to caterpillar fungus is the widespread degradation of alpine meadows on the eastern Tibetan Plateau that have seen entire hillsides lose their turf layer and consequently the alpine meadow grasses that ghost moth larvae depend on for their existence (Schaller, 2012).

While potentially causing great disturbance to snow leopards, their habitat, and prey for two months each year, the caterpillar fungus harvest has brought relative prosperity to large numbers of herders, with somewhat unexpected benefits for alpine meadows. One benefit has been large voluntary reductions in livestock numbers kept on mountain pastures in fungus-collecting areas of the Tibetan Plateau, since fungus sales now provide the entire year's cash income for many herders. A second benefit is that many successful fungus collectors have given up livestock herding altogether and moved to nearby towns for ten months per year, reducing grazing pressure on fragile high-altitude pastures. However, with this prosperity has come increasingly intense competition for a fungal resource that may be in decline. The spiraling growth in the number of collectors has resulted in individual collectors harvesting less per person than in previous years, which at times has led to violent disputes over harvesting areas (Hansen, 2011; Shrestha and Bawa, 2013). With recent reports of steep declines in the amount of caterpillar fungus being harvested, the inherent sustainability of the range-wide caterpillar fungus harvest has come into question. It now appears increasingly possible that caterpillar fungus, if not properly managed, may be a one-time ecological windfall rather than the inexhaustible resource it once appeared to be. Should the profitability of caterpillar fungus collection decline, be it due to increased competition for a limited resource, drastic declines or disappearance of the fungus, or simply due to a drop-off in Chinese demand, a reverse flow of former herders from towns back to alpine pastures could ensue, with a subsequent increase in disturbance to snow leopard range areas.

FUEL WOOD HARVEST

Forests in snow leopard range areas are generally patchily distributed, with only the extreme southern, eastern, and northern edges of the species' range having extensive areas of continuous forest cover. The heart of the snow leopard's range in the central Tibetan Plateau is a treeless landscape of vast grasslands, although limited patchy shrub cover does exist. Regardless of the dominant vegetation cover type, wood is in high demand throughout the snow leopard's range.

In the Mongolian Altai, local timber is used for home construction, household fences, and fuel (Laurie et al., 2010). In the Kyrgyzstan Tian Shan, managers of some protected areas selectively cut trees within protected areas, the sale of which pays protected area salaries and expenses. On the Tibetan Plateau, timber is prized for roof beams for mud brick herder cabins. Although yak dung is the primary fuel of the plateau, low-growing juniper shrubs, often the only woody plant in the grassland landscape, are harvested by the cartload for kindling yak dung fires and for incense used in religious observance. Juniper shrubs are also a primary fuel for many migrants to Tibet who are culturally averse to cooking with yak dung (Tsering and Farrington, 2015). In the Tajikistan Pamir, teresken (*Krascheninnikovia ceratoides*) is the dominant shrub of high-altitude deserts and the primary fuel for local inhabitants who harvest it, root and all (Breckle and Wucherer, 2006). In the Himalaya, trees are cut extensively for firewood, which is often transported long distances to upland areas, and also used for constructing homes and tourist lodges (Stevens, 2003). In Buddhist areas, trees are used in large numbers for prayer flag masts.

Discontinuous patches of relict *Juniperus tibetica* and *J. convallium* forest can be found widely scattered over a 650 km stretch of grasslands in southern Tibet, with some *J. tibetica* trees being over 800 years old (Brauning and Grießinger, 2006; Miehe et al., 2008). Miehe et al. (2008) speculate that these forest patches were once connected and that their isolation is a result of both fuel wood gathering and intensive livestock grazing over the last 600 years that has prevented forest regeneration. In the Mongolian Altai, logging of Siberian larch (*Larix sibirica*) intensified after 1990 and the region's limited forest cover is now being exploited at an unsustainable rate, particularly at the forest edge where tree stumps outnumber living trees and regeneration is hindered by livestock grazing (Dulamsuren et al., 2014; Khishigjargal, 2013). In upland areas of Yunnan and Bhutan, encroaching shrublands have long been burned off to maintain alpine pasturelands, although burning is currently banned (Baker and Moseley, 2007; Wangchuk et al., 2013).

Conversion of forests to grasslands in snow leopard range areas may appear beneficial to snow leopards and their prey as it expands their habitat. However, loss of forest and shrub cover is generally the result of increased human pressure on mountain landscapes, and encroachment on snow leopard habitat. In addition, both woodcutting and grazing contribute to desertification of alpine grasslands, while deforestation of upland areas contributes to downstream flooding. Due to the important role that tree and shrub cover plays in preventing flooding and desertification, the Chinese government has implemented widespread woodcutting bans since the late 1990s. These include bans in the upper Yangtze and Yellow River basins on the Tibetan Plateau, as well as bans on logging of natural forest in the western provinces of Qinghai, Gansu, Sichuan, and Yunnan, which overlap portions of the Tibetan Plateau (Yang, 2001). In 2002, the Tibet Autonomous Region also protected most forests with woodcutting bans, including extensive areas of juniper shrublands (Miehe et al., 2008; Dawa Tsering, personal communication, 2015).

It is predicted that global warming will result in increased rates of tree growth and increased forest coverage at the expense of alpine grasslands in snow leopard range areas, particularly in the Himalaya (Dulamsuren et al., 2014; Forrest et al., 2012). Nevertheless, given the limited forest cover in most snow leopard range areas, and the current unsustainable exploitation of wood resources for fuel, construction, and religious observance, every effort should be made to improve management of these woodlands for the benefit of humans, biodiversity, and control of desertification and flooding.

SUBCHAPTER

10.5

Synthesis

Peter Zahler

WCS Asia Program, Wildlife Conservation Society, New York, NY, USA

SYNTHESIS

The rapid escalation of extractive industry and other development interests across much of the snow leopard's global range presents a new threat to this big cat – one that range states and the global conservation community are still trying to grasp and understand, even as the intensification of human use and extraction accelerates. In particular, the Central Asian states and Mongolia have entered a period of rapid economic growth, which is being accompanied by growing development in sectors such as mining and energy (estimates for Mongolia's mineral wealth range from USD 1.3 trillion to 2.75 trillion from its 10 biggest mines alone). These development projects require infrastructure (power, water, waste water, landfills, roads, or railroads), as well as the development of worker camps, residential areas and associated services, all of which put further stress on the environment from a variety of impacts. Unless effective regulations and planning procedures are introduced, and companies are provided with guidance on how best to mitigate and protect this threatened biodiversity, the region runs the risk of suffering irreparable environmental damage.

The term "gold rush" can cover more than a rush for precious minerals – this same scramble competition mentality has led to significant impacts on snow leopard habitat from caterpillar fungus collectors, as described by Farrington in this chapter. The "timber mafia," as large-scale forestry companies are dismissively called in parts of western Asia, move into regions and quickly (and often illegally) clear-cut entire valleys of pine, spruce, and deodar cedar, leaving devastated regions behind, with increased erosion and slumping soils, radically different microclimates, loss of understory and other nonwoody plant species, and even altered hydrological regimes.

Hydropower projects, while unlikely to directly affect snow leopards or their habitat (as most projects will occur at lower elevations), can still have significant knock-on effects. An example is the planned Diamer-Bhasha Dam in Gilgit-Baltistan, Pakistan, an estimated USD 8 billion project that would dam the middle Indus River in Diamer District. While the dam itself would only flood the arid, lower-elevation (~1,500 m) Indus gorge area, it would also displace an estimated 30,000 people. Many of these people would have little other choice but to move up the nearby side valleys, and the impact of their resource needs on higher elevation forests and rangelands (and wildlife) could be significant. Similarly, the influx of new people could significantly alter or disrupt existing community-based protection mechanisms, increase poaching and otherwise negatively affect snow leopards, their prey species, and their habitat.

Ecosystems in Asia's alpine mountain regions are particularly fragile due to limited water and nutrients, both of which limit primary productivity. This is coupled with the "sky island" structure of high mountains, where alpine habitat is often naturally separated by intervening valleys and lower slopes – even, in cases such as Mongolia's Gobi and China's Tibetan Plateau, by long stretches of intervening arid flatlands – means that this fragility has a cascading effect that results in often unpredictable, long-range movements characteristic of nomadic species. Snow leopards themselves are known to travel long distances for various reasons (see Chapter 19), as do some prey species such as argali. Due to this fragility and the behavioral responses associated with the harsh alpine environment, development in this ecosystem can have greater and more varied impacts than expected. Already, poorly designed development initiatives – roads, mining, gas and oil, and other natural resource extraction – threaten to fragment these mountain landscapes and negatively affect snow leopards and other wildlife that depend on large tracts of open land for their survival.

What is clear from each of these cases is that there is an obvious need – and a real opportunity – for landscape level planning to avoid or mitigate these threats. Protected areas cover only a small fraction of snow leopard habitat, and even those that exist in the high mountains often have little in the way of staffing, resources, and technical capacity for management. If these new extractive and infrastructure threats are to be dealt with, effective national-level planning and mitigation measures must be implemented to avoid damaging the high mountain regions that the snow leopard calls home.

Opportunity still exists to put into place strong legislation and effective monitoring and enforcement mechanisms to ensure that development does not severely negatively affect Asia's high mountains. Development is happening, and with it, the impacts that have already been felt across much of the rest of the world. What is different about the snow leopard's world is that its extraordinary geography and isolation has meant that this encroachment has been slow and late in coming. This gives snow leopard range states and the global community a chance to learn from mistakes made elsewhere while applying new methods of avoidance and mitigation to development projects and extraction activities. However, if this opportunity is not taken in the near future, the snow leopard's high mountain environment will undoubtedly face the slow but steady fragmentation and degradation that has already been experienced in many other once-wild landscapes around the globe.

References

Aryal, A., Brunton, D., Raubenheimer, D., 2014. Impact of climate change on human-wildlife-ecosystem interactions in the Trans-Himalaya region of Nepal. Theoret. Appl. Climatol. 115, 517–529.

Baker, B.B., Moseley, R.K., 2007. Advancing treeline and retreating glaciers: implications for conservation in Yunnan, P.R. China. Arct. Antarct. Alp. Res. 39 (2), 200–209.

Batsaikhan, N., Buuveibaatar, B., Chimed-Ochir, B., Galbrakh, D., Ganbaatar, O., Lkhagvasuren, B., Nandintsetseg, D., Berger, J., Calabrese, J.M., Edwards, A.E., Fagan, W.F., Fuller, T.K., Heiner, M., Ito, T.Y., Kaczensky, P., Leimgruber, P., Lushchekina, A., Milner-Gulland, E.J., Mueller, T., Murray, M.G., Olson, K.A., Reading, R., Schaller, G.B., Stubbe, A., Stubbe, M., Walzer, C., Von Wehrden, H., Whitten, T., 2014. Conserving the world's finest grassland amidst ambitious national development. Conserv. Biol. 28, 1736–1739.

Brauning, A., Grießinger, J., 2006. Late Holocene variations in monsoon intensity in the Tibetan-Himalayan region – evidence from tree rings. J. Geol. Soc. India 68 (3), 485–493.

Breckle, S.W., Wucherer, W., 2006. Vegetation of the Pamir (Tajikistan): land use and desertification problems. In: Spehn, E.M., Liberman, M., Korner, C. (Eds.), Land Use Change and Mountain Biodiversity. CRC Press, Boca Raton, FL, pp. 227–239.

Bull, J.W., Suttle, K.B., Singh, N.J., Milner-Gulland, E.J., 2013. Conservation when nothing stands still: moving targets and biodiversity offsets. Front. Ecol. Environ. 11 (4), 203–210.

Butt, N., Beyer, H.L., Bennett, J.R., Biggs, D., Maggini, R., Mills, M., Renwick, A.R., Seabrook, L.M., Possingham, H.P., 2013. Biodiversity risks from fossil fuel extraction. Science 342 (6157), 425–426.

Cannon, P.F., Hywel-Jones, N.L., Maczey, N., Norbu, L., Tshitila, Samdup, T., Lhendup, P., 2009. Steps towards sustainable harvest of *Ophiocordyceps sinensis* in Bhutan. Biodivers. Conserv. 2009 (18), 2263–2281.

Dulamsuren, C., Khishigjargal, M., Leuschner, C., Hauck, M., 2014. Response of tree-ring width to climate warming and selective logging in larch forests of the Mongolian Altai. J. Plant Ecol. 7 (1), 24–38.

Forrest, J.L., Wikramanayake, E., Shrestha, R., Areendran, G., Gyeltshen, K., Maheshwari, A., Mazumdar, S., Naidoo, R., Thapa, G.J., Thapa, K., 2012. Conservation and climate change: assessing the vulnerability of snow leopard habitat to treeline shift in the Himalaya. Biol. Conserv. 150 (1), 129–135.

Gerland, P., Raftery, A.E., Ševcíková, H., Li, N., Gu, D., Spoorenberg, T., Alkema, L., Fosdick, B.K., Chunn, J., Lalic, N., Bay, G., Buettner, T., Heilig, G.K., Wilmoth, J., 2014. World population stabilization unlikely this century. Science 346 (6206), 234–237.

Grumbine, R.E., Pandit, M.K., 2013. Threats from India's Himalaya dams. Science 339 (6115), 36–37.

Haberl, H., Beringer, T., Bhattacharya, S.C., Erb, K.H., Hoogwijk, M., 2010. The global technical potential of bio-energy in 2050 considering sustainability constraints. Curr. Opin. Environ. Sustain. 2, 394–403.

Hansen, E., 2011. The killing fields. Outside Magazine. Available from: http://www.outsideonline.com/1909346/killing-fields [accessed 29.02.2016].

Harrington, J.L., Conover, M.R., 2006. Characteristics of ungulate behavior and mortality associated with wire fences. Wildl. Soc. Bull. 34, 1295–1305.

Hart, S., 2012. Scandal as controversial Chinese athlete Wang Junxia enters IAAF Hall of Fame. The Telegraph.

Heiner, M., Bayarjargal, Y., Kiesecker, J.M., Galbadrakh, D., Batsaikhan, N., Ganbaatar, M., Odonchimeg, I., Enkhtuya, O., Enkhbat, D., von Wehrden, H., Reading, R., Olson, K., Jackson, R., Evans, J., McKenney, B., Oakleaf, J., Sochi, K., Oidov, E., 2013. Applying Conservation Priorities in the Face of Future Development: Applying Development by Design in the Mongolian Gobi. The Nature Conservancy, Ulaanbaatar, Mongolia.

Institute for Energy Research, 2011. Energy Information Association Forecast. www.instituteforenergyresearch.org/2011/09/22/eia-forecast-world-energy-led-by-chinato-grow-53-percent-by-2035/

International Finance Corporation (IFC), 2012. Performance Standards on Environmental and Social Sustainability. http://www.ifc.org/wps/wcm/connect/115482804a0255db96fbffd1a5d13d27/PS_English_2012_Full-Document.pdf?MOD=AJPERES (accessed August 5, 2014).

Ito, T.Y., Lhagvasuren, B., Tsunekawa, A., Shinoda, M., Takatsuki, S., Buuveibaatar, B., 2013. Fragmentation of the habitat of wild ungulates by anthropogenic barriers in Mongolia. PLoS One 8 (2), e56995.

Jackson, R., Munkhtsog, B., Mallon, D.P., Naranbaatar, G., Gerlemaa, K., 2009. Camera-trapping snow leopards in the Tost Uul region of Mongolia. Cat News 51, 18–21.

Jones, N.F., Pejchar, L., Kiesecker, J.M., 2015. The energy footprint: how oil, natural gas, and wind energy impact land for biodiversity and ecosystem services. BioScience 65 (3), 290–301.

Karlstetter, M., Mallon, D, 2014. Assessment of gaps and needs in migratory mammal conservation in Central Asia. Convention on Migratory Species Technical Report, Bonn, Germany.

Kerr, R.A., 2010. Natural gas from shale bursts onto the scene. Science 328, 1624–1626.

Kesler, S, 2007. Mineral supply and demand into the 21st century, Proceedings for a Workshop on Deposit Modelling, Mineral Resource Assessment, and Their Role in Sustainable Development, US Geological Survey Circular, 1294.

Kiesecker, J.M., Copeland, H., Pocewicz, A., McKenney, B., 2010. Development by design: blending landscape level planning with the mitigation hierarchy. Front. Ecol. Environ 8 (5), 261–266.

Kiesecker, J.M., Copeland, H., Pocewicz, A., Nibbelink, N., McKenney, B., Dahlke, J., Holloran, M., Stroud, D., 2009. A framework for implementing biodiversity offsets: selecting sites and determining scale. BioScience 59, 77–84.

Khishigjargal, M., 2013. Response of Tree-Ring Width and Regeneration in Conifer Forests of Mongolia to Climate Warming and Land Use. Ph.D. dissertation. Biodiversity and Sustainable Land Use Center – Biodiversity, Ecology, and Nature Conservation Section. University of Göttingen, Göttingen.

Laurie, A., Jamsranjav, J., van den Heuvel, O., Nyamjav, E., 2010. Biodiversity conservation and the ecological limits to development options in the Mongolian Altai: formulation of a strategy and discussion of priorities. Cent. Asian Surv. 29 (3), 321–343.

Lkhagvasuren, B., Chimeddorj, B., Sanjmyatav, D., 2012. Analyzing the effects of infrastructure on migratory terrestrial mammals in Mongolia. Saiga News 14, 13–14.

Lo, H.C., Hsieh, C.Y., Lin, F.Y., Hsu, T.H., 2013. A systematic review of the mysterious caterpillar fungus *Ophiocordyceps sinensis* in dongchongxiacao and related bioactive ingredients. J. Tradit. Complement. Med. 3 (1), 16–32.

Mallon, D., Singh, N., Röttger, C., 2014. International Single Species Action Plan for the Conservation of the Argali: *Ovis ammon*. CMS Technical Series, Bonn, Germany.

McCarthy, T.M., Chapron, G., 2003. Snow Leopard Survival Strategy. ISLT and SLN, Seattle, WA.

McCarthy, T.M., Fuller, T.K., Munkhtsog, B., 2005. Movements and activities of snow leopards in Southwestern Mongolia. Biol. Conserv. 124, 527–537.

McDonald, R.I., Fargione, J., Kiesecker, J., Miller, W.M., Powell, J., 2009. Energy sprawl or energy efficiency: climate policy on natural habitat for the United States of America. PLoS One 4, e6802.

McKenney, B., Kiesecker, J.M., 2010. Policy development for biodiversity offsets: a review of offset frameworks. Environ. Manage. 45, 165–176.

Miehe, G., Miehe, S., Will, M., Opgenoorth, L., Duo, L., Dorgeh, T., Liu, J., 2008. An inventory of forest relicts in the pastures of Southern Tibet (Xizang AR, China). Plant Ecol. 194 (2), 157–177.

Mongolian Academy of Sciences (MAS), Institute of Biology. Resolution No: 02, May 27, 2014.

Mongolian Ministry of Mineral Resources and Energy (MMRE), 2012 Mineral Leases GIS.

Master Plan for Mongolia's Protected Areas. 1998. Ulaanbaatar, Mongolia.

Morgan, R.K., 2012. Environmental impact assessment: the state of the art. Impact Assessment and Project Appraisal 30, 5–14.

Munkhtsog, B., Olonbaatar, G., Nansalmaa, A., Tuvd, L., Tseveenravdan, D. 2012. Dispersal of young snow leopards. Unpublished report: WWF-Mongolia.

Oakleaf, J.R., Kennedy, C.M., Baruch-Mordo, S., West, P.C., Gerber, J.S., Jarvis, L., Kiesecker, J.M, 2015. A world at risk: aggregating development trends to forecast global habitat conversion. PlosOne 10 (10), e0138334.

Olson, K.A., 2013. Saiga crossing options: guidelines and recommendations to mitigate barrier effects of border fencing and railroad corridors on saiga antelope in Kazakhstan. Report prepared for the Convention on Migratory Species, Bonn, Germany.

Rio Tinto, 2004. Rio Tinto's Biodiversity Strategy. Rio Tinto plc. London, UK and Melbourne, Australia.

Saenz, S., Walschburger, T., González, J.C., León, J., McKenney, B., Kiesecker, J.M., 2013. Development by design in Colombia: making mitigation decisions consistent with conservation outcomes. PLoS ONE 8 (12), e81831.

Schaller, G.B., 2012. Tibet Wild: A Naturalist's Journeys on the Roof of the World. Island Press, Washington, DC.

Sharma, K., Bayrakcismith, R., Tumursukh, L., Johansson, O., Sevger, P., McCarthy, T., Mishra, C., 2014. Vigorous dynamics underlie a stable population of the endangered snow leopard *Panthera uncia* in Tost Mountains, South Gobi, Mongolia. PLoS One 9 (7), e101319.

Shrestha, U.B., Bawa, K.S., 2013. Trade, harvest, and conservation of caterpillar fungus (*Ophiocordyceps sinensis*) in the Himalayas. Biol. Conserv. 159, 514–520.

Singh, N.J., Milner-Gulland, E.J., 2011. Conserving a moving target: planning protection for a migratory species as its distribution changes. J. Appl. Ecol. 48, 35–46.

Steinkraus, D.C., Whitfield, J.B., 1994. Chinese caterpillar fungus and world record runners. American Entomol. 40 (4), 235–239.

Stevens, S., 2003. Tourism and deforestation in the Mt. Everest region of Nepal. Geogr. J. 169 (3), 255–277.

Teck Resources Limited. 2013. Teck's Biodiversity Strategy. Vancouver, Canada. Available at http://www.tecksustainability.com/sites/base/pages/ourstrategy/biodiversity (accessed August 5, 2014).

Thukten, K., 2014. Habitat Impact Assessment of *Ophiocordyceps* Growing Areas Under Central Range. Wangchuck Centennial Park, Bumthang.

Tsering, D., Farrington, J., 2015. The Indus headwaters region. In: Abidi-Habib, M., Garstang, R., Saeed Khan, R., 2015. Water in the Wilderness: Life in the Coast, Deserts and Mountains of Pakistan. Oxford University Press Pakistan, Karachi (in press).

US Energy Information Administration. 2013. International Energy Outlook 2013, p. 312.

Wang, X.L., Yao, Y.J., 2011. Host insect species of *Ophiocordyceps sinensis*: a review. ZooKeys 127, 43–59.

Wang, Y., Wang, Y.D., Tao, S.C., Chen, X.P., Kong, Y.P., Shah, A., Ye, C.Y., Pang, M., 2014. Using infra-red camera trapping technology to monitor mammals along Karakorum Highway in Khunjerab National Park, Pakistan. Pak. J. Zool. 46, 725–731.

Wangchuk, K., Gyaltshen, T., Yonten, T., Nirola, H., Tshering, N., 2013. Shrubland or pasture? Restoration of degraded meadows in the mountains of Bhutan. Mt. Res. Dev. 33 (2), 161–169.

Wingard, J., Zahler, P., 2006. Silent steppe: the illegal wildlife trade crisis in Mongolia. Mongolia Discussion Papers, East Asia and Pacific Environment and Social Development DepartmentWorld Bank, Washington, DC.

Wingard, J., Zahler, P., Victurine, R., Bayasgalan, O., Buuveibaatar, B., 2014. Guidelines for addressing the impact of linear infrastructure on migratory large mammals in Central Asia. Convention on Migratory Species Technical Report, Bonn, Germany. http://www.cms.int/sites/default/files/document/COP11_Doc_23_3_2_Infrastructure_Guidelines_Mammals_in_Central_Asia_E.pdf

Winkler, D., 2005. Yartsa gunbu-*Cordyceps sinensis*, economy, ecology & ethno-mycology of a fungus endemic to the Tibetan Plateau. Memorie della Societa Italiana di Scienze Naturali e del Museo Civico di Storia Naturale di Milano 33, 69–85.

Winkler, D., 2008. Yartsa gunbu (*Cordyceps sinensis*) and the fungal commodification of Tibet's rural economy. Econ. Bot. 62 (3), 291–305.

Winkler, D., 2009. Caterpillar fungus (*Ophiocordyceps sinensis*) production and sustainability on the Tibetan Plateau and in the Himalayas. Asian Med. 5 (2), 291–316.

Winkler, D., 2011. Sustainable Resource Management of Caterpillar Fungus in Yushu Tibetan Autonomous

Prefecture, Qinghai Province, China. Eco-Montane Consulting, Kirkland.

Wolf, M., Ale, S., 2009. Signs at the top: habitat features influencing snow leopard *Uncia uncia* activity in Sagarmatha National Park, Nepal. J. Mammal. 90, 604–611.

Yang, Y.X., 2001. Impacts and effectiveness of logging bans in natural forests: People's Republic of China. In: Durst, P.B., Waggener, T.R., Enters, T., Cheng, T.L. (Eds.), Forests Out Of Bounds: Impacts and Effectiveness of Logging Bans in Natural Forests in Asia-Pacific. FAO Regional Office for Asia and the Pacific, Bangkok, Thailand, pp. 81–102.

Yeh, E.T., Lama, K.T., 2013. Following the caterpillar fungus: nature, commodity chains, and the place of Tibet in China's uneven geographies. Soc. Cult. Geogr. 14 (3), 318–340.

Zhu, J.S., Halpern, G.M., Jones, K., 1998. The scientific rediscovery of an ancient Chinese herbal medicine: *Cordyceps sinensis*, Part I. J. Alt. Complement. Med. 4 (3), 289–303.

Zarfl, C., Lumsdon, A.E., Berlekamp, J., Tydecks, L., Tockner, K., 2015. A global boom in hydropower dam construction. Aquat. Sci. 77 (1), 161–170.

SECTION III

Conservation Solutions *In situ*

CHAPTER

11

The Role of Mountain Communities in Snow Leopard Conservation

Rodney M. Jackson, Wendy Brewer Lama***

*Snow Leopard Conservancy, Sonoma, CA, USA

**KarmaQuest Ecotourism and Adventure Travel, Half Moon Bay, CA, USA

INTRODUCTION

Commencing in the late 1980s and continuing through the present time, conservationists have shifted their focus from the top-down approach centered on strict enforcement of wildlife laws and protected area (PA) sovereignty (the "guns and fences" protectionism paradigm) to one that increasingly incorporates local community concerns and interests (e.g., Western et al., 1994). These initiatives have assumed many forms, from the comanagement of a PA (MacKinnon et al., 1986) to people-centered, large-scale integrated conservation-development programs (ICDPs) funded by multilateral agencies like the United Nations, World Bank, and US Agency for International Development (USAID) (Wells and Brandon, 1992). Such programs have been articulated through numerous publications commissioned by institutions such as the World Commission on Protected Areas (WCPA) of International Union for Conservation of Nature (IUCN), many of which have highlighted biodiversity or PA comanagement (e.g., Borrini-Feyerabend et al., 2004). Then drawing on the threats-based approach proposed by Salafsky and Margoluis (1999) a series of manuals for practitioners emerged from the Biodiversity Conservation Network (BCN), a major collaborative program funded by international conservation organizations and USAID (Margoluis and Salafsky, 1998).

Over the past two decades, community-based conservation initiatives targeting the endangered snow leopards have proliferated rapidly. Conservation practitioners face many challenges, from how best to engage and motivate local people to protect a species often perceived as being a pest, to the logistics of operating in some of the world's most remote and forbidding high-elevation terrain. Actively involving stakeholders through participatory protocols remains

Snow Leopards
http://dx.doi.org/10.1016/B978-0-12-802213-9.00011-0

a vital ingredient in the design, implementation, and monitoring of robust, long-lasting, and locally adapted solutions to conservation dilemmas. Strategies that stress communities' collective and positive visions for change, and resolve underlying human–wildlife conflicts (HWCs), while also meeting the community's aspirations for a better future, lead to successful and more sustained outcomes (Jackson, 2015).

People who live in the snow leopard's habitat are characterized by a rich culture, a high level of ethnic diversity, and a marginalized, agropastoralism lifestyle. More than 40% of them live below national poverty levels with average annual parity incomes of USD 250–400 (Jackson et al., 2010). Livestock depredation from snow leopards and wolves (*Canis lupus*) is pervasive, along with competition for pasturage among domestic and wild ungulates (e.g., Anwar et al., 2011; Li et al., 2013; Namgail et al., 2007). Most communities in these remote areas where the topography is steep, soils are shallow, rocky, and often infertile, and natural disasters such as floods, landslides, and severe winters are frequent, rely on traditional animal husbandry. Few settlements in the Himalayan belt are accessible by road, and many lack infrastructure and basic services such as schools or health clinics. With a burgeoning tourism industry, especially in the Himalayan region, economic benefits from adventure tourism are spreading beyond the agropastoral sector, broadening opportunities for local people to earn supplemental income – especially within PAs that usually form core habitat for snow leopard (Jackson and Fox, 1997).

Local communities may also suffer from the presence of a national park without receiving any of benefits (Wells and Brandon, 1992). Unless people's legitimate concerns, especially those related to HWC and resource access are adequately addressed, conservation efforts are likely to fail. Resolving HWC dominates most efforts at community engagement, although it is in people's perceptions and participation that more effort is required (Karanth and Nepal, 2012; see Chapter 5). Clearly, remedial measures and holistic resolutions satisfactory to both parties are required – efforts that in the long run must also improve people's livelihood options. With more secure income, families are better able to tolerate the loss of livestock (in effect their "bank account") and thus are more willing to coexist with snow leopards and other predators (Jackson and Wangchuk, 2004; Lama et al., 2012).

A BRIEF OVERVIEW OF COMMUNITY INVOLVEMENT IN SNOW LEOPARD CONSERVATION

Nepal is widely acknowledged as the first range country to effectively involve mountain communities in snow leopard conservation. The Annapurna Conservation Area Project (ACAP) is widely touted as the world's first example of a community-driven PA (e.g., Müller et al., 2008). The establishment of the Makalu-Barun (MBCA) and Kangchenjunga Conservation (KCA) areas followed, while the Community Buffer Zone regulations are now the hallmark for enabling community participation in the management of Nepal's national parks and PAs. Snow Leopard Conservation Committees (SLCC) were established in ACAP in 1994 and in KCA in 2000 in an effort to help address local herders' concerns related to reports of increased livestock depredation presumed to be linked to recovery of the snow leopard population. ACAP established its first 12-member snow leopard committee in Manang, considered prime habitat, in order to address retributive killing of snow leopards, which are viewed as a keystone species (Ale, 1997; Jackson and Fox, 1997). Each SLCC was comprised of two persons from each of the six Village Development Committees (the lowest administrative unit in Nepal).

Two important considerations framed the involvement of local people in snow leopard and wildlife conservation: (1) the presence of a traditionally robust, highly cohesive society; and

(2) the predominant practice of Tibetan Buddhism, whose religious precepts do not readily sanction killing of wildlife. The ACAP managers were able to reduce poaching of musk deer (*Moschus* spp.) while discouraging further retributive killing of snow leopards in reaction to livestock depredation. ACAP deposited funds in several village conservation accounts and a local religious leader was honored with a plaque for helping stop snow leopard killing and wildlife poaching by local residents.

In Ladakh, India, Afghanistan, Tajikistan, and Tyva Republic, Russia, retaliatory killing linked to depredation involving multiple sheep and goat losses has been much reduced and even eliminated by encouraging communities to predator-proof their nighttime livestock corrals so that snow leopards and other predators can no longer gain entry (Jackson and Wangchuk, 2001). Kunkel and Hussain (see Chapter 13.3) describe lessons learned from community-managed livestock insurance programs in Pakistan and Nepal. Also in Nepal, the KCA developed a similar program that links improved livestock management with support for alternative livelihoods that generate supplemental income to help temper economic loss due to livestock depredation (Gurung et al., 2011). In Kyrgyzstan, Mongolia, and India, handicrafts production and livestock insurance programs are providing local people with tangible economic benefits and rewards for wildlife conservation (Mishra et al., 2003; see Chapter 13.2).

In Ladakh, a UNESCO sponsored ecotourism program enabled a team of conservationists, local tour operators, village committees, and the Jammu & Kashmir Wildlife Protection Department to develop homestay and wildlife tourism that generates substantial benefits for people living in small settlements along the region's most popular trekking route (Jackson et al., 2010; see Chapter 13.1). Snow leopard treks by one international company alone have contributed more than USD 30,000 to snow leopard conservation over a 7–8 year period. Tourists regularly observe snow leopards, suggesting that the villagers' commitment to stop killing the cats has a positive effect on either the snow leopard's survival rates or their freedom to descend with prey species to lower elevations during winter.

With these and other examples in mind, we take a look at some of the "best practices" for engaging local communities in furthering the conservation about snow leopards, highlighting successes as well as pitfalls, and where it is necessary to draw upon the literature involving other carnivore species. We conclude with a list of "best practices" for strengthening community-based snow leopard and biodiversity conservation.

RATIONALE FOR ADOPTING COMMUNITY-BASED BIODIVERSITY PROTECTION AND MANAGEMENT MODELS IN SNOW LEOPARD RANGE COUNTRIES

From the tropics to the Polar Regions, it is increasingly apparent that local people and communities *must* play a greater role in conservation if representative landscapes, habitats, and species, along with their functional and intact ecosystems, are to be preserved. Conservation initiatives that are launched and embraced by people whose livelihoods depend upon natural resources in the area, and that offer broad-reaching environmental, ecosystem, and socio-economic benefits are far more likely to succeed than prescribed top-down actions targeting a narrow set of beneficiaries.

In the case of the endangered snow leopard, there are a number of compelling reasons for expanding the role of local and often remote communities in protecting this large feline, its prey, and habitat:

- With a relatively small percentage of the snow leopard range falling within PAs combined with their far-ranging habits, it is likely that most snow leopards reside outside or overlap the boundaries where protective measures are weak (Jackson and Fox, 1997).

- As noted already, HWC is a major threat that unless addressed, will continue to jeopardize snow leopard populations and their long-term security by undermining efforts at both protection and conservation. Local herders hold the key to resolve such conflict by reducing risks to livestock through improved livestock guarding and animal husbandry practices in particular, by predator-proofing vulnerable nighttime enclosures (Jackson et al., 2010; Jackson and Wangchuk, 2001; see Chapters 14.1 and 33).
- The resources for monitoring this sparsely distributed species that largely inhabits remote and harsh terrain are exceedingly limited. Regulatory agencies and rangers themselves are ill disposed to enforce protective measures. Indigenous people who have lived in such areas for generations can serve as the "eyes and ears" for stemming illegal poaching and wildlife trade. While not empowered by law to make arrests, they are best positioned to provide timely reports to the PA authority, forest department, or the police for follow-up action (see Chapter 40).
- Local people, and especially youth, can be recruited and trained to spread environmental awareness and monitor animal activity, in effect serving as environmental stewards. For example, in Nepal, youths known as "Snow Leopard Scouts," working in tandem with local herders, are helping the Snow Leopard Conservancy (SLC) to monitor snow leopards in the Annapurna Conservation Area using remote camera traps. High-altitude herders in Bhutan regularly observe snow leopards during summer grazing season and sit on a snow leopard committee with national park staff that documents sightings.

IMPROVING SNOW LEOPARD CONSERVATION

Our base of knowledge on how to conserve such a rare and wide-ranging carnivore as the snow leopard has grown exponentially over the past few decades. As discussed already, there is strong evidence that regulation alone is not sufficient to protect a species that lives in such inhospitable terrain and for which humans are its greatest threat. But how exactly does the carrot-and-stick approach work? The following paragraphs briefly describe what are considered key "best practices" and the more important lessons learned:

Defining What Is Meant by Participation

For decades, local participation in conservation has been advocated by academics and practitioners without recognizing the broad range of participatory possibilities. As indicated in Table 11.1, community engagement in conservation projects may vary widely, from passive participation to self-mobilization and full empowerment, in which communities are vested with managerial authority and expected to deliver substantial conservation outputs (Horwich and Lyon, 2007). At least till now, the majority of snow leopard-related conservation initiatives have focused at the lower level of the chart, usually soliciting information from villagers but limiting their participation or role in decision making. In part, this may be due to the fact that many projects evolved from ones undertaken by academics whose agendas center around a research thesis. It also reflects the early forms of participatory research employing Participatory Rural Appraisal (PRA) (Pretty et al., 1995). Bringing this gradation of participatory practices to the attention of conservationists is a first step toward encouraging a higher level of local involvement, enablement, and self-reliance. Community-based conservation projects must be grounded in interactive

TABLE 11.1 A Typology of Participation

Level of participation		Description
MOST	Self-mobilization	People independently take initiatives to change systems. Develop contacts for external inputs, but retain control over how resources are managed – this should be the long-term goal for development initiatives!
	Interactive participation	People participate in information generation and subsequent analyses leading to action plans and their implementation. Involves different methodologies seeking various local perspectives, thus involving people in decision making regarding use and quality of information.
	Functional participation	Participation through forming groups with predetermined objectives. Such participation generally occurs only after major decisions have been taken.
	Participation for materials and incentives	Participation involves people taking incentives in cash or in kind for their services provided. Disadvantage is that there is no stake in being involved once the incentives end.
LEAST	Participation by consultation	People consulted and their views taken into account. However, they are not involved or empowered with decision making.
	Participation in information giving	People answer questions, but lack any opportunity to influence context of interviews; also findings often not shared with them.
	Passive participation	Responses of the participants not taken into consideration and outcome predetermined. Information shared belongs only to the external institutions.

Adapted from Pretty, J.N., et al., 1995. A Trainer's Guide for Participatory Learning and Action, IIED articipatory Methodology Series. International Institute for Environment and Development, London.

participation, with the goal being to encourage self-mobilization (ideally through learning by the example of others).

The sooner communities are involved in the planning process using the higher level of participatory approaches (e.g., PRA or a later adaptation, participatory learning and action – PLA) or other empowering processes, the better. Above all, snow leopard conservation programs need to recognize and incorporate traditional knowledge with scientifically derived information (Jackson, 2015). The common practice of limiting participation to household or key informant attitudinal interviews must be replaced by giving communities, through their elders, elected leaders, spokespersons, and minority voices, a *seat at the table* so that they can share their aspirations for improving and sustaining livelihoods that must ultimately lead to more collaborative land and natural resource stewardship. A system of shared responsibilities and benefits is more robust than one plagued by lack of ownership and driven by top-down handouts from government or NGOs.

However, narrowly focusing on integrating local people in PAs or endangered species management does not encompass the breadth of solutions needed. We should provide a suitable framework to enable a transformative shift that assures local communities far wider roles and greater responsibilities in helping restore critical ecosystem functions and enhancing habitat carrying capacity for the area's fauna, flora, and associated biodiversity (Müller et al., 2008). Land use and habitat management action must be implemented at the landscape level through clusters of communities working in harmony with government, NGOs and in some cases the private sector (e.g., where tourism is involved). One option for educational outreach is to focus on the snow leopard as a keystone or umbrella species representative of a healthy mountain ecosystem – and which in the case of the Tibetan-Qinghai Plateau and the Himalayan Mountains constitutes the watershed for more than 2 billion people living downstream (Snow Leopard Working Secretariat, 2013). Simultaneously we need to transform local perceptions of snow leopards to one of valued assets rather than despised pests that should be eliminated (Jackson and Wangchuk, 2004). Under the scenario whereby snow leopard populations are healthy, genetically diverse, and have access to connected habitats, humans will also obtain important ecological and economic benefits (Dinerstein et al., 2013).

Integrating Cultural Conservation with Snow Leopard Conservation

Extending across a large part of continental Asia, the snow leopard's range coincides with exceptionally rich cultural and ethnic diversity that helps define each region's human landscape. Like the snow leopard, these people, their ways of life, and their identities are becoming endangered, and should be seen as assets worthy of protection. Beyond the goal of cultural conservation *per se*, protection of these mountain peoples' traditional values, religious beliefs, and sustainable livelihoods are part and parcel of snow leopard conservation, for the reasons mentioned previously: without community participation, conservation efforts are incomplete.

Alleviating Human-Wildlife Conflicts

Although a significant proportion of the human population of these areas engage in agropastoralism, their husbandry practices, supplementary livelihoods, and the underlying patterns of HWC vary widely. Therefore, these issues must be addressed via relatively site-specific, case-by-case approaches because no simple magic bullet exists for alleviating HWC or for fostering coexistence between people and snow leopards. Providing mechanisms for learning from and valuing traditional knowledge more robustly builds trust between government, conservationists, and local people. The primary goal should be to engender multiple stakeholder contributions with each party having clearly defined roles and responsibilities, thus better ensuring that mutually acceptable and effective solutions are devised and implemented.

The levels of poverty existing in snow leopard range requires establishing effective linkages between conservation and development objectives, usually best achieved by fostering locally appropriate and suitably diversified livelihood strategies. These must be linked to the simultaneous alleviation of HWC and initiatives that curb poaching of the snow leopard's wild prey populations – using a multidimensional suite of incentives and disincentives aimed at ensuring local people perceive snow leopards (and other predators) more valuable alive than dead (Jackson, 2015). Since imbalances in rural populations can result from gender and/or social inequities, it is important to seek the support of all social and economic groups whose activities or livelihoods depend upon addressing HWC (Borrini-Feyerabend et al., 2004; Margoluis and Salafsky, 1998). The participation of women and youth in snow leopard conservation has proven particularly effective, but is not always easy: traditional social hierarchies often put older males at the top of the decision-making ladder, making it uncomfortable in some societies for women and youth to take on leadership roles.

Yet unless all segments of society are involved in and benefiting from intervention strategies (such as developing homestays or nature guiding for supplemental income, with a portion of revenues accruing to a community-managed conservation fund), the incentives to motivate conservation actions will not be widely enough shared to be effective.

Best Practices for Engaging Local Communities

When engaging stakeholders, both project sponsors and community members need to discuss and adopt a set of "best practices" for designing and implementing interventions. As a start, community participation should be underpinned by a project philosophy that emphasizes and values empowerment, equity, trust, and learning (Reed, 2008). Project planning must be facilitated by persons skilled in participatory methods and guided by persons knowledgeable in the key topics under consideration, ideally leading to collective decision making. We recommend that snow leopard practitioners adopt the Open Standards for the Practice of Conservation, which brings together common concepts, approaches, and terminology in conservation project design, management, and monitoring (www.conservationmeasures.org). It is also imperative that the team has a sound understanding of underlying threats to snow leopards, their prey, and the mountain ecosystem (i.e., the driving factors and results chain-derived intervention; see Margoluis and Salafsky, 1998, for details).

We have found the planning process known as Appreciative Planning and Participatory Action (APPA) initially developed by The Mountain Institute, the SLC, local NGOs, and community partners in Nepal, India (Sikkim and Ladakh), and China to be especially helpful in fulfilling these criteria. APPA's main tenet is to build upon positive values and recognize them as a foundation for envisioning an even better future. APPA has been used successfully to support snow leopard and biodiversity conservation in areas as culturally and geographically diverse as the Tibetan Plateau and the Himalaya of Nepal and India as well as Central Asia (Jackson, 1999). The APPA method is flexible and adaptable, and encompasses a set of protocols for engaging communities in designing highly participatory and self-driven solutions that have been used for reducing livestock depredation or other conflicts between livestock, wild ungulates, and the rangelands both depend upon. Equally, APPA enables stakeholders to identify realistic, cost-effective, livelihood- and income-generation activities that are compatible with the existing opportunities, environmental constraints or concerns, local economic realities, and stakeholder perceptions or skill sets. Readers are referred to the following publications for further information on the APPA process: The Mountain Institute (2000); Ashford and Patkar (2001); and SLC (2009).

Conservation practitioners are also urged to make full use of tools contained in the PRA toolbox (Pretty et al. 1995), with the aim of providing local people with the necessary skills to act more on their own rather than waiting for the government or NGOs to address their problems. Snow leopard conservationists are also referred to the handbook prepared by SNV, 2004 for field practitioners, development workers, and facilitators involved in community mobilization, organizing, and enterprise development.

Other key elements of effective community-implemented conservation initiatives:

- Community-driven and/or managed conservation actions targeting snow leopards, their prey and habitat must be environmentally and economically sustainable and socially responsible. Interventions need to be based upon sound science, incorporate adaptive management with clearly imbedded stakeholder responsibilities and transparent budgets, and be monitored by each party using verifiable indicators for measuring progress (Table 11.2). All of this can be developed using the APPA process.

- Cost-sharing by the community and its beneficiary members is imperative to achieve sustainable and successful outcomes. This applies equally to whether the intervention is infrastructure (e.g., improved corral, trail, mini-hydro, and nontimber product development), livelihood skills training, or governance and institutional development. Rather than cash contributions from local people (many of who do not have the resources), the facilitating organization should seek other forms of cost-sharing such as collective or individual agreements from beneficiaries to train other members of the community and/or to return a portion of their profits to a collectively managed community conservation or development fund. This has proven quite successful with the Homestay program in Ladakh (Jackson et al., 2010; see Chapter 13.1) as well as with the community-managed savings and credit program currently underway in the Sagarmatha (Mt. Everest) National Park (Snow Leopard Conservancy, USA, unpublished report).
- Rather than creating a new governance structure, local people and their communities are best engaged through their existing institutions, such as livestock herder groups, village development committees (VDCs), and existing livelihoods associations. Mother or women's groups have proven especially effective for developing snow leopard-linked handicrafts production (Mishra et al., 2003; see Chapter 13.2) and traditional homestays (Lama et al., 2012).
- Biodiversity conservation is best accomplished through implementing multidimensional activities that target major threats and their direct or indirect

TABLE 11.2 Conditionality and Best Practices for Community-Based Wildlife Conservation Interventions

Projects are most likely to succeed if their sponsors and beneficiary communities endorse and include the following conditionality within a project's implementation and operational framework:

- *Ensure tangible conservation results and benefits for people, rangelands, and wildlife.* Project activities should be implicitly linked with snow leopard and mountain biodiversity conservation (i.e., must have positive impact and not adversely impair species, habitat, or rangeland resources). Specific and clearly defined actions are needed to benefit snow leopards, their prey, and habitat as well as helping to improve local people's livelihoods and income-generating opportunities.
- *Require reciprocal investment by participating organizations or institutions.* Each stakeholder (whether villager, NGO, or government) must make a reciprocal (cofinancing) contribution, within their means, in support of the agreed-to project actions or activities. This may be in the form of cash or in-kind services like materials and labor, which are valued using existing market rates and prices.
- *Full participation by all involved parties.* There must be strong commitment to active and equitable participation from each involved stakeholder group throughout the life of the project (from planning to implementation, monitoring, and evaluation and reporting). In addition, project supported activities should benefit as many households as possible, and especially those who are more marginalized.
- *Responsibility for project facilities.* The beneficiary community must be willing to assume all or a significant responsibility for repairing and maintaining any infrastructural improvements (e.g., predator-proofed corral) that may be provided by the project.
- *Monitoring and compliance performance.* Stakeholders should be willing to employ their own simple but realistic indicators for measuring project performance and impact, according to an approved community monitoring and evaluation plan. Similarly, the project donors or implementing agency will monitor project activities, outputs, and results.

Adapted from Jackson, R., 1999. Snow leopards, local people and livestock losses. Cat News 31, 22–23.

socioenvironmental drivers. In addition, projects must be supported by well-designed monitoring plans employing transparent and verifiable indicators in order to ensure accountability and wise use of limited human resources and funds (Margoluis and Salafsky, 1998).

- Project managers and/or donors should disperse funds based on performance and in support of those project activities with transparent criteria and well-defined targets and that were developed with and endorsed by the key stakeholders. The development of annual or biannual work plans is another area where the community needs to be fully engaged and empowered from project onset and throughout the implementation, evaluation, and monitoring phases (including short and long term).
- Helping communities establish relationships and build partnerships with other stakeholders strengthens them institutionally, enhances their economic sustainability, and allows better shared learning. Peer-to-peer exchanges are an effective way for communities to learn by example. Study tour exchanges can be worthwhile if planned transparently. First, with community participation, identify the study tour participant selection criteria and then solicit nominations of individuals who meet the criteria (this can preclude the assumption that only male village leaders should go on the study tour). Secondly, find ways that both the host and visitor parties share and learn equally, focusing on each side's "assets" or "what's working well."

CONCLUSIONS

Coexistence with snow leopards can be best achieved by empowering rural communities and helping them forge more harmonious and ecocentric relationships with their environment, one in which snow leopards are perceived as valued assets rather than pests to be eliminated. Donors and practitioners alike need to accept the fact that building community capacity for sustained environmental planning, HWC alleviation, and enterprise development is usually time consuming and expensive. Sustainable conservation measures cannot be accomplished quickly: the current practice whereby community-based projects are funded and technically supported for 3–5 years is often unrealistic. The time frame for effective community engagement needs to be extended, because capacity building typically does not take root rapidly. It takes time for communities to convert to a market-driven economy from a traditional bartering one where greater external decision making is done at higher political levels, and communities need to have room to learn from their mistakes. Furthermore, careful guidance by technically and socially astute advisers and facilitators is a critical component for building sustainable and environmentally smart rural communities, especially in remote areas where the broader civil society structure and influence is weak.

Scaling up of incentivized conservation programs, especially those dependent upon broad-based community support, represents special challenges that require our urgent attention (Mishra et al., 2003; Jackson et al., 2010). We need to be careful not to rely upon "one-fits-all" solutions; even closely spaced valleys may have different ecosystems, economies, and ethnic groups whose values, livelihoods, and resource strategies evolved to address a specific set of environmental, social, and political factors (Jackson et al., 2010). However, this opens the door for indigenous knowledge to finally play an important supportive role to any external biological and social science expertise applied during project life. Community-driven initiatives must become mainstream though supported through national policy and conservation frameworks.

The real long-term challenge lies with moving communities beyond their harsh and insecure subsistence livelihood into more economically viable and environmentally friendly activities. Local people must perceive snow leopards and other large carnivores as being worth "more alive than dead." Toward this end, we need to examine options for remote mountain communities to provide and support ecosystem services by imbedding appropriate and attractive incentive packages (Dinerstein et al., 2013). The Global Snow Leopard and Ecosystem Protection Program (GSLEP) (see Chapter 45), endorsed in 2013 by all 12 snow leopard range countries, offers a possible blueprint for this transformational process to take place – but only if range country governments and multilateral institutions step forward with a commitment to large-scale funding, high-level community engagement and empowerment, investment in the required training of participatory process facilitators and conservation-linked product development, and the political will to diversify decision making among their traditionally marginalized constituents.

Acknowledgments

We extend our respect and appreciation to the communities with whom we have worked to develop and test new methods for engaging communities who share their environment with snow leopards. Mr Rinchen Wangchuk, Ms Nandita Jain, Ms Lamu Sherpa, and Mr Renzino Lepcha were key team members instrumental to developing the APPA protocols that evolved through field sessions with diverse rural communities in Nepal, India, and China. We dedicate this paper in honor of Nepal's leading conservationists, Dr Chandra Gurung and Mr Mingma Sherpa, who among others who lost their lives in September 2006 in a tragic helicopter accident shortly after handing over management of the Kangchenjunga Conservation Area to the local community. This paper is also dedicated to the memory of Rinchen Wangchuk, cofounder of Snow Leopard Conservancy–India Trust, who played a key role in developing our community-based approaches to snow leopard conservation, and who passed away all too early from amyotrophic lateral sclerosis (ALS) in March 2011.

References

Ale, S., 1997. The Annapurna Conservation Area Project: a case study of an integrated conservation and development project in Nepal. Jackson, R., Ahmad, A. (Eds.), Proceedings of the Eighth International Snow Leopard Symposium. International Snow Leopard Trust, Seattle and WWF-Pakistan, Lahore, pp. 155–169.

Anwar, M.B., Jackson, R.M., Sajid, N.M., Janečka, J.E., Hussain, S., Beg, M.A., Muhammad, G., Qayyum, M., 2011. Food habits of the snow leopard, *Panthera uncia* (Schreber, 1775) in Baltistan, Northern Pakistan. Eur. J. Wildl. Res. 57, 1077–1083.

Ashford, G., Patkar., S., 2001. The Positive Path: Using Appreciative Inquiry in Rural Indian Communities (International Institute for Sustainable Development/ Myrada, Winnipeg, Canada). http://myrada.org/myrada/docs/ai_the_postive_path.pdf

Borrini-Feyerabend, G., Kothari, A., Oviedo, G., 2004. Indigenous and local communities and protected areas: towards equity and enhanced conservation. Best Practice Protected Area Guidelines Series No. 11, Phillips, A. (Ed.), IUCN Programme on Protected Areas, Gland.

Dinerstein, E., Varma, K., Wikramanayake, E., Powell, G., Lumpkin, S., Naidoo, R., Korchinsky, M., Del Valle, C., Lohani, S., Seidensticker, J., Joldersma, D., Lovejoy, T., Kushlin, A., 2013. Enhancing conservation, ecosystem services, and local livelihoods through a wildlife premium mechanism. Conserv. Biol. 27, 14–23.

Gurung, G.S., Thapa, K., Kunkel, K., Thapa, G.J., Kollmar, M., Boeker, U., 2011. Enhancing herders' livelihood and conserving the snow leopard in Nepal. Cat News 55, 17–21.

Horwich, R.H., Lyon, J., 2007. Community conservation: practitioner's answer to critics. Oryx 41, 376–385.

Jackson, R., 1999. Snow leopards, local people and livestock losses. Cat News 31, 22–23.

Jackson, R.M., 2015. HWC ten years later—successes and shortcomings of approaches to global snow leopard conservation. Hum. Dimens. Wildl. 20 (4), , Available from: http://dx.doi.org/10.1080/10871209.2015.1005856

Jackson, R.M., Mishra, C., McCarthy, C., Ale, S.B., T.M., 2010. Snow leopards: conflict and conservation. In: Macdonald, D.W., Loveridge, A.J. (Eds.), Biology and Conservation of Wild Felids. Oxford University Press, Oxford, pp. 417–430.

Jackson, R.M., Wangchuk, R., 2004. A community-based approach to mitigating livestock depredation by snow leopards. Hum. Dimens. Wildl. 9, 307–315.

Jackson, R., Wangchuk, R., 2001. Linking snow leopard conservation and people-wildlife conflict resolution: grassroots measures to protect the endangered snow leopard from herder retribution. Endangered Species Update 18, 138–141.

Jackson, R., Fox, J.L., 1997. Snow leopard conservation: accomplishments and research priorities. In: Jackson, R., Ahmad, A. (Eds.), Proceedings of the Eighth International Snow Leopard Symposium. International Snow Leopard Trust, Seattle and WWF-Pakistan, Lahore, pp. 128–145.

Karanth, K., Nepal, S.K., 2012. Local residents' perception of benefits and losses from protected areas in India and Nepal. Environ. Manag. 49, 372–386.

Lama, W. B., Jackson, R., Wangchuk, R., 2012. Snow leopards and Himalayan homestays: catalysts for community-based conservation in Ladakh. In: Chettri, N., Sherchan, U., Chaudhary, S., Shakya, B., (Eds.), Mountain Biodiversity Conservation and Management. Working Paper 2012/2, International Centre for Integrated Mountain Development, Kathmandu, Nepal, 9–11.

Li, Juan, Hang, Yin, Dajun, Wang, Zhala, Jiagong, Zhi, Lu., 2013. Human-snow leopard conflicts in the Sanjiangyuan Region of the Tibetan Plateau. Biol. Conserv. 168, 118–123.

MacKinnon, J., MacKinnon, K., Child, G., Thorsell, J., 1986. Managing Protected Areas in the Tropics. International Union for the Conservation of Nature and natural Resources, Gland, Switzerland.

Margoluis, R., Salafsky, N., 1998. Measures of Success: Designing, Managing, and Monitoring Conservation and Development Projects. Island Press, Washington, DC.

Mishra, C., Allen, P., McCarthy, T., Madhusudan, M.D., Bayarjargal, A., Prins, H.H.T., 2003. The role of incentive programs in conserving the snow leopard. Conserv. Biol. 17, 1512–1520.

Müller, U., Gurung, G., Kollmair, M., Müller-Böker, U., 2008. "Because the project is helping us to improve our lives, we also help them with conservation" – Integrated conservation and development in the Kangchenjunga Conservation Area, Nepal. In: Haller, T., Galvin, M. (Eds.), People, Protected Areas and Global Change. Participatory Conservation in Latin America, Africa, Asia and Europe. NCCR North-South, University of Bern, Switzerland, pp. 363–399.

Namgail, T., Bhatnagar, Y.V., Mishra, C., Bagchi, S., 2007. Pasoral nomads of the Indian Changthang: pastoral production, land use and socioeconomic changes. Hum. Ecol. 35, 497–504.

Pretty, J.N., Guijt, I., Scoones, I., Thompson, J., 1995. A Trainer's Guide for Participatory Learning and Action, IIED Participatory Methodology Series. International Institute for Environment and Development, London.

Reed, M.S., 2008. Stakeholder participation for environmental management: a literature review. Biol. Conserv. 141, 2417–2431.

Salafsky, N., Margoluis, R., 1999. Threat reduction assessment: a practical and cost-effective approach to evaluating conservation and development projects. Conserv. Biol. 13, 830–841.

SLC, 2009. Community Engagement and Building Capacity for Lasting Conservation: Process and Tools. Snow Leopard Conservancy, Sonoma, CA. http://snowleopardconservancy.org/publications/.

Snow Leopard Working Secretariat, 2013. Global Snow Leopard and Ecosystem Protection Program (GSLEP). Snow Leopard Working Secretariat, Bishkek, Kyrgyz Republic.

SNV (Nepal) (with ICIMOD), 2004. Developing Sustainable Communities: a toolkit for development practitioners, Kathmandu, Nepal, http://www.icimod.org/publications/index.php/search/publication/51 (accessed January 5, 2015).

The Mountain Institute, 2000. Community-Based Tourism for Conservation and Development: A Resource Kit. The Mountain Institute, Washington, DC.

Wells, M., Brandon, K., 1992. People and parks: linking protected area management with local communities. The World Bank, Washington, DC.

Western, D., Wright, R.M., Strum, S.C., 1994. Natural Connections: Perspectives in Community-Based Conservation. Island Press, Washington, DC.

CHAPTER

12

Building Community Governance Structures and Institutions for Snow Leopard Conservation

*Peter Zahler**, *Richard Paley***

*WCS Asia Program, Wildlife Conservation Society, New York, NY, USA
**Afghanistan Program, Wildlife Conservation Society, Kabul, Afghanistan

THE CASE FOR GOVERNANCE AND SNOW LEOPARD CONSERVATION

Improving governance is not usually considered the province of conservation biologists. The topic – as one that involves the complex interactions of history, culture, politics, economics, sovereignty, and security – is deemed the responsibility of political scientists, policy think tanks, and international experts. However, in a growing case history of examples, conservation and governance do not just overlap but are integrally entwined and successfully support each other. Conservationists are interested in areas of high biodiversity and landscapes that are still relatively intact. These places also are usually the most rural, remote, and isolated locations on the planet. That does not mean that they are uninhabited by humans – but it usually means that those people are similarly rural, remote, and isolated.

Local communities in these landscapes often suffer from a near-complete lack of services from state government due to their geographic (and thus also political) isolation, which in turn leads to comparative greater levels of poverty than their national counterparts. They are often independent (not always by choice) from both national authority and support, and because of their poverty and lack of alternatives, they usually depend largely if not entirely on the local natural resource base for their survival and livelihoods.

Unfortunately, these communities are also facing unprecedented change. Slow but inexorable population increases are putting increasing strain on natural resources – forests, wildlife, and rangelands – while external commercial interests are pushing farther into the most remote landscapes in search of profit. The combination of pressures leaves these communities facing a crisis without the necessary tools to adapt to these events.

Snow Leopards
http://dx.doi.org/10.1016/B978-0-12-802213-9.00012-2

This is where conservation can come into play. Organizations dedicated to wildlife conservation are finding that partnerships with local communities are the surest way to achieve their goals and objectives. The reason that this works is that while the goals of conservation organizations and local communities may differ to some degree – a wildlife organization wants to save wildlife, while a community may simply want to improve their well-being – for isolated rural communities the fastest and simplest way to improve matters is often to improve sustainable natural resource management, which if implemented appropriately will simultaneously benefit wildlife.

The description of distant and isolated communities struggling to make ends meet is especially apt when discussing the area where snow leopards are found. Although Asia as a continent (ranging from the hyper-dense countries of China, India, and Bangladesh to Mongolia with the world's lowest human population density), is estimated to have roughly 64 people/km^2, the area where snow leopards are found contains approximately 8 people/km^2. To put it another way, if the geography defining snow leopard distribution across Asia was its own country, it would be in the global top five for lowest human population density.[1] This is no surprise, as snow leopards live mostly above the tree line in some of the world's most inaccessible mountain terrain. Cold temperatures and low productivity make this a difficult place to eke out a living, and the people who share the snow leopard's domain are mostly poor livestock herders, on the edge of survival, well "off the grid" of both national and international support or even awareness.

The "toolbox" for wildlife conservation has grown enormously in the past few decades, from the initial centralized and top-down model of classical, exclusive protected area management to including local communities in conservation initiatives that incorporate the way local people interact with their resource base. Options for framing these initiatives vary from straightforward community-based conservation, privately owned or managed conservation areas, and comanagement arrangements between communities and government (and in some cases communities, government, and industry) for both protected area and nonprotected area management.

A critical principle in achieving better management of natural resources is that of collaborative or comanagement. Comanagement is the acknowledgment that even if a resource falls under the jurisdiction of the government in law, it is still incumbent upon the government to engage other key stakeholders in the management of that resource. The stakeholders in question can be nongovernment organizations (NGOs), civil society organizations (CSOs), commercial interests, other government agencies, or even private individuals, but the principle of comanagement particularly applies to the communities who depend on a particular resource for their livelihoods. Comanagement is now incorporated into the conservation policies and legislation of many countries, but realizing it in practice is a challenge. It requires active and competent community organizations that are not only technically capable in management implementation but also speak with unity and legitimacy on behalf of constituents and actively participate in the partnership arrangements that are implicit in the concept of comanagement. This is where conservation and governance meet.

CONSERVATION AND GOOD GOVERNANCE: LAND TENURE AND REPRESENTATION

A significant amount of attention has been directed recently toward a specific aspect of conservation and governance – the political

[1]Population density data estimates use the global UN-adjusted 2.5′ data from: http://sedac.ciesin.columbia.edu/data/set/gpw-v3-population-density/data-download

empowerment of people and restitution of rights, especially land tenure. The focus has been on tropical forest systems (especially in Latin America and Southeast Asia) where local livelihoods (and ownership) often find themselves at odds with both protected area management and extractive/agricultural industry (Nelson, 2010; Painter and Castillo, 2014; Porter-Bolland et al., 2012; Reyes-Garcia et al., 2013; Sheil et al., 2006). The natural resource under consideration is usually tropical forests, which in Asia at least are often found near large human populations and provide multiple "goods," ranging from nontimber forest products (a local product; e.g., Ingram, 2012) to timber (a national product; e.g., Wiersum et al., 2013) to carbon sequestration (a global product; e.g., Biermann, 2010). This proximity and complexity have led to multiple stakeholders and polarized arguments related to local versus national management of resources, decentralization, and local empowerment versus commodities-based management (Bawa et al., 2011).

By comparison, snow leopard habitat is rarely thought of as a zone with multiple natural resource products of high value – instead it is a cold, high-elevation landscape with low biodiversity and low productivity. Although the natural resources of these areas are critical to the people who live in them (usually poor, transhumant pastoralists largely dependent on pastures for livestock production), they are often considered "wasteland" with little to offer the larger community, either nationally or internationally. (The perhaps significant if often overlooked exception is that snow leopard habitat contains the origins of many of the most important rivers in Asia, which provide critical ecosystem services to billions of people.) Because of the lack of focus and interest in these marginal mountain landscapes, land tenure rights are often less complicated and less contentious in high mountain environments than in the more heavily populated lowlands.

However, it should be stressed that any attempt to improve or build governance for conservation, regardless of location or existing state involvement, needs to begin with a full study and assessment of existing rights and responsibilities related to land tenure and resource use. The first focus needs to be on community-government relations and rights (whether inside or outside protected areas). For even though land tenure may be less contentious in mountain areas, it is no more likely to be clearly articulated in law or titles – and without a clear description of rights, community involvement in governance is difficult if not impossible (Knudsen, 1997).

A secondary but also critical aspect of any governance assessment is the status of different community groups within the landscape of interest. Disenfranchised groups are often found within a larger community, and truly representative governance needs to incorporate all members of society. This representation obviously includes the issue of gender, but extends to tribal or family groups that are, for various reasons, not normally fully included in the existing governance systems. Efforts that attempt to build or improve governance for conservation but do not include all members of society are likely to fail as internal struggles – or more simply, the lack of inclusion of all resource users, and thus all resource use – potentially lead to unintended consequences, including a lack of effective conservation outcomes.

If an assessment of rights and ownership uncovers issues, these have to be dealt with. Without clear rights and responsibilities and full representation, communities will not feel empowered to manage wildlife appropriately, and government will tend to ignore them as a potential management partner. Land tenure and representation are hugely complicated issues that are also highly context-dependent and frequently extremely contentious, but they are topics in which conservation organizations have shown leadership, providing clear examples that can be followed in this process (Painter, 2009).

BUILDING GOVERNANCE INSTITUTIONS

For a conservation organization to help in building better governance for natural resource management, probably the most important aspects for success are an on-ground presence and a long-term commitment to the area. Working successfully with communities can often boil down to building a sense of trust and partnership, which can only really happen over time and with a regular presence that provides a clear sign that the organization is not going to disappear after a three- or five-year funding cycle (Ming'ate et al., 2014). A prolonged, even decades-long commitment must be made, and that commitment must be conveyed convincingly to the local communities. It must also continue through periods of difficulty and even conflict; an external presence and support during times of strife can sometimes be the difference between success and the collapse of local civil society and governance in an area (Hart and Hart, 2003; Plumptre et al., 2001; Zahler, 2005).

Communication is another key consideration. It needs to be two-way – and initially it should be one-way, from communities to the support project personnel. Outsiders rarely have a complete sense of the real problems facing local people or the underlying complexities involved in those problems. Conservation organizations may have an in-depth scientific understanding of the threats faced by these communities (and their ecosystems), and the tools to help mitigate or solve these problems, but determining the best path forward involves understanding the local politics, culture, and structures. As well, existing traditional systems need to be understood and incorporated into any plans for applying conservation solutions. Only once those aspects are understood and fully integrated should there be an effort to provide ideas for solutions back to the communities.

Building that constituency for support then becomes the key next step. If aspects of local needs and considerations are properly incorporated, the process becomes one designed to solve the community's problems, which is much easier than attempting to apply an outside solution to a problem that local people have not yet articulated or even identified. As an example, providing ideas for improving rangeland practices after the local residents have expressed concern for degraded rangelands is much easier than introducing ideas to improve resilience to climate change, which may in fact be one of the ultimate causes of rangeland degradation in an area but which may not be perceived by local people as an underlying threat.

It is critical to have key leaders within the community who understand the issues and can bring the larger community with them. These may be traditional leaders such as religious leaders (Dudley et al., 2009) or respected and active individuals within the communities. The traditional leaders are particularly crucial because they may see new and improved governance as a challenge to their established powers. Without their buy-in, any efforts to create new governance institutions are liable to fail. Moreover, community members are likely to have some respect for existing leaders even if they acknowledge that they are not providing adequate governance in a rapidly changing world.

The actual design of governance institutions must be suited to the national and local context, but there are some guiding principles to consider. If possible, it is best to base the design of new governance institutions on existing systems or at least ensure that new structures are entirely compatible with existing legal and policy frameworks. Too often parallel systems create a plethora of institutions each with slightly different but overlapping mandates, leading to confusion, competition, and conflict.

Another advantage of building on established systems is that in most cases, they are already endorsed by government, which will give them legitimacy and assist in their official acceptance. There must be some procedure to creation that is as transparent and democratic as possible in order to promote a sense that the institution is truly representative of all local people (rather than, for example, only the established local elites). Any community institution brought into being must have some central guiding document, a constitution or bylaws, which outlines in clear terms the purpose of the organization and some simple rules governing its structure and *modus operandi*. This document is normally a requirement of national laws related to the formation and existence of community social organizations, but it also helps provide a framework and focus to any community governance institution, and a clear mandate together with roles and responsibilities for its members. Consultations with communities and local government should be as broad as possible to establish the need for the institutions that are envisaged. Initial elections to whatever body will oversee the organization should be open to the public regardless of the selection procedure. Once agreed upon, the bylaws should be disseminated as widely as possible so all community members are clear on the purpose of the organization and can monitor that it is brought into being in accordance with the agreed process.

EARLY SUPPORT FOR NEW GOVERNANCE INSTITUTIONS

In many instances where new governance institutions are created there is little historical or cultural precedent for this type of organization. Communities will either be used to defer decisions onto established elites (or in some cases centralized control) or to unstructured and ad hoc public gatherings. Frequently, communities will have people of the required capabilities, but not always the necessary knowledge or experience to make informed and sometimes technical decisions. Nor will the members of the new institution necessarily be well versed in dealing with a range of outside agencies such as government officials, representatives of commercial interests, or the donor community. Moreover, some fundamental concepts such as acting on behalf of the community rather than the individual or their family will not necessarily be practiced immediately by the new institution's decision makers. From the outset therefore, the institution will need mentoring from both technical and philosophical perspectives. This mentoring can be provided by an established NGO that has expertise in the field, is known to the community, and enjoys its confidence (Fraser et al., 2010; Zahler et al., in press). The institution also will often need a degree of financial support, at least until it can generate its own income from membership fees or similar mechanisms. This support will be necessary both to run its day-to-day operations (as many of its officers will be rural farmers or small business people unable to afford transport or subsistence costs while undertaking their duties) and to fund its interventions. The mentor organization can play a critical role in identifying and securing external donor assistance while building the community organization's capacity to generate funds for itself.

As in all institutions and communities, a wide range of agendas and perceptions can be found among the constituents, and conflicts are inevitable. Often these conflicts are resolved using traditional mechanisms, whether privately or at broader community meetings. It can also be helpful for third parties to be in involved, most obviously the NGO that is mentoring the fledgling institution. Alternatively, sensitive and respected local government officials can provide useful facilitation in conflict resolution. Regardless, it is important that the deeper social conflicts that are often entangled in such community disputes be considered and addressed, and ultimately resolved. Otherwise these long-simmering disputes, sometimes having nothing to do with the issue at hand, can derail efforts at governance building and conservation of any kind (Madden and McQuinn, 2014).

Establishing robust and effective community governance institutions is a long-term process that usually entails multiple setbacks, some due to internal tensions and others to external pressures. The longer the institution can rely on a degree of support from a mentor institution, the more likely it is to survive these vicissitudes and emerge as a fully independent and self-sustaining entity. It is important throughout this potentially protracted period that the mentor organization intervenes only when requested and as necessary. In that way, a space will be maintained for the nascent institution to gradually assume full ownership and responsibility for itself and its activities, and avoid the dependence on external support, whether real or perceived, that can undermine its legitimacy and power and lead to criticism and backlash against the entire process (Zaidi, 1999).

As with any initiative, it is critical to monitor and evaluate progress. To assess governance activities and report on the impact of associated conservation outcomes, Wilkie and Cowles (2012) suggest looking at three principal attributes that are essential for the long-term functioning of any governance institution. They are (i) legitimacy – this is when stakeholders perceive that the governance institution is governing in their interests and does so with formal or informal authority, accountability, transparency, fairness, and participation; (ii) capacity – when the institution has the skills and knowledge to plan, implement, and monitor conservation actions, the staff, and financial resources; and (iii) power – when the institution has the political power to exert its authority and not have its decisions and actions undermined by others. A number of tools have been created to do this (see, for example, Stephanson and Mascia, 2014; Wilkie and Cowles, 2012).

COMPLETING THE CIRCLE: BUILDING LINKAGES AND COMANAGEMENT PROCESSES WITH GOVERNMENT

Even though community governance institutions are often created to fill a vacuum in the reach of effective government in rural areas, it does not mean that they can or should operate entirely independently of, or as an alternative to, formal government structures. In order to achieve full legitimacy they must build functioning relations with local government so that they are eventually perceived as a valued partner, facilitating rather than opposing government at the local level.

Building effective partnerships with government is not always easy owing to mutual distrust between communities and government officials. Often the reason communities are keen to create their own governance institutions in the first place is the government's poor record in delivering basic services and a lack of confidence among communities that the government has the ability or desire to act in their interests. Government officials, for their part, often view community organizations with a degree of suspicion as potential challengers to their authority, or with contempt as ill-educated or unenlightened and unworthy of receiving devolved responsibility for decision making. These barriers are difficult to overcome, but can be gradually dissipated by displays of unity and competence on the part of the community and its representative organization that generates respect and recognition from the government. Similarly, if government officials are seen to be supportive and cooperative toward the community's efforts to achieve effective governance over its natural resources, they engender confidence that the government recognizes the community's interests as important.

Therefore, any mentoring organization needs to be interacting with government stakeholders at the same time that it is helping to build community-level governance institutions. The ability of a national or international organization to simultaneously build these (initially often disparate) relations and act as an objective third party to bring these two groups together cannot be understated. This process can be facilitated through direct meetings and/or combined training workshops on basic conservation issues and concepts, field methodologies, and so on. These can help to build relationships and trust, which are critical steps to building actual comanagement processes. Once communities come to know and trust government staff (especially those whose mandate overlaps with the community interests, such as wildlife and forestry department specialists), and once government officials realize that communities are both interested in seeing government mandates achieved and are able to help government achieve them (ranging from collecting field data in distant sites for national reporting to actual implementation of conservation actions to monitor or improve conditions), developing comanagement systems becomes a natural objective that is shared by both parties. Creating platforms for this process – meetings, workshops, local study tours, field training – is where the external, mentoring organization can play a key role.

CONCLUSIONS

A profound and important shift has occurred in snow leopard conservation work, which began with focused field research on snow leopards and their prey, moved to analyzing human-snow leopard conflict with local people (Bagchi and Mishra, 2006), and is now implementing community conservation initiatives (Baral and Stern, 2009; Jackson and Wangchuck, 2004). A number of these initiatives are also attempting to build new or improve existing governance institutions based on recognition by all parties that current local systems are not always capable of dealing with new and growing pressures from a range of internal and external sources (Simms et al., 2011).

Several key lessons on local governance have been learned in recent years from these initiatives. For new or improved governance to be sustainable, it is important that each of these lessons be incorporated into any governance-building process. They include:

- Assess and resolve issues related to land tenure and ownership.
- Ensure that all stakeholders have a voice, including women and other possible disenfranchised community members.
- Commit to long-term support to governance building.
- Understand existing power structures and incorporate existing traditional systems into any attempts to improve governance.
- Ensure that local needs and concerns are incorporated into governance design.
- Find ways to include existing leaders as advocates in the process.
- Base new governance institutions on existing systems or at least ensure that new structures are entirely compatible with existing legal and policy frameworks.
- Be sure that there is a central guiding document that defines the purpose of the organization and some simple rules governing its structure and activities, and that this document is disseminated widely to all stakeholders.
- Have the new governance institution be recognized and endorsed by government.
- Monitor and evaluate progress, including the three principal attributes: legitimacy, capacity, and power.
- Build functioning relations and linkages with local government so that the new institution is perceived as a valued partner for comanagement.

The snow leopard's remote, high-mountain world is slowly but inexorably changing. These changes are also happening to local communities, and often the changes are happening in part because of these people. Providing these communities with the governance tools to properly manage their fragile, alpine landscape will be critical to ensure that snow leopards remain one of the great wild predators of Asia.

PAKISTAN: BUILDING LOCAL CONSTITUENCIES FOR CONSERVATION

In Diamer District of Gilgit-Baltistan (previously the Northern Areas), initial wildlife research activities in the early 1990s uncovered the fact that a significant ecological crisis was underway. Diamer District is the only official "tribal area" in Gilgit-Baltistan; from an environmental perspective this means that local communities own the rights to the natural resources of the region, including both forests and wildlife. Diamer also contains a significant amount of the limited natural conifer forest remaining in the province (and a significant percentage of the forest remaining for the country). Unfortunately, these forests were being cut down rapidly as outside interests (colloquially termed the "timber mafia") would buy rights for clear-cutting forests at well under market value from economically naïve communities. The results were severe and dramatic, as communities lost a critical resource that provided timber for construction, fuel wood for heat and cooking, and nontimber forest products (such as pine nuts, morel mushrooms, and medicinal plants). There was also some evidence that erosion from the loss of forest cover was affecting grazing areas and even water sources.

At the same time significant overhunting was negatively affecting wildlife populations. An influx and subsequent widespread availability of high-powered weapons from regional conflicts, coupled with a lack of local rules or government enforcement efforts, meant that markhor (*Capra falconeri*) and urial (*Ovis orientalis*), the two species in the district that are key prey for snow leopard, were disappearing rapidly, with markhor appearing to exist only in small and fragmented populations and urial believed to have been entirely extirpated in the region.

Extensive, multiyear consultations with local communities made it clear that they were very concerned about these changes, but they felt powerless to cope with the new paradigms that these changes had thrust upon them. Over the next few years, efforts were made to encourage the creation of a new governance system based initially on natural resource management, one that would provide the communities with a new method of decision making that could directly address these concerns and determine steps they could take to alter and mitigate the environmental destruction occurring in their landscapes.

These governance structures were initially developed in 23 communities in Diamer and southern Gilgit Districts. Coupled with extensive conservation education and outreach efforts, these organizations developed their own bylaws aimed initially at natural resource rules (e.g., use of trees for construction and other purposes, and hunting of certain species of wildlife such as markhor) but that grew to include other social rules related to development, construction, and other activities. Each of the Wildlife Conservation Social Development Organizations (WCSDO) helped identify (usually two) community members to serve as volunteer wildlife rangers, who were mandated with monitoring wildlife on a monthly basis and patrolling to identify and halt possible violations based on local bylaws.

This governance initiative now has expanded to include 65 communities over 5 districts in Gilgit-Baltistan, impacting over 400,000 villagers. It has also led to the creation of new multicommunity conservancies, based on providing revenues from markhor trophy hunting across multiple communities (four permits are allowed in Gilgit-Baltistan, and communities have joined together into multivalley conservancies to ensure more equitable sharing of the 80% of revenues that are legally allowed to be returned locally). Finally, a platform has been created to bring these community-led organizations together with local government departments – the Mountain Conservation and Development Programme (MCDP).

The results of these efforts have been striking. The new governance structures led to communities refusing the advances of the timber mafia and maintaining significant forest cover. Markhor have shown clear signs of recovery, with the population appearing to have increased from fewer than 1000 animals to approximately 1500 in only about a decade, a 50% increase. A small herd of urial, previously thought extirpated in the region, was discovered by community rangers in the Indus Valley region of northeast Diamer and southern Gilgit, and the population has been closely monitored and protected and now numbers more than 100 animals (Zahler and Khan, 2004; Wildlife Conservation Society (WCS), unpublished data).

AFGHANISTAN: TOP-DOWN MEETS BOTTOM-UP

Wakhan District in Badakhshan Province is one of most geographically remote areas of Afghanistan. Its craggy mountains and high-elevation alpine meadows also make up approximately 70% of the confirmed range of the snow leopard in Afghanistan. The 15,000 inhabitants of the district are comprised mainly of ethnic Wakhi and Kyrgyz. In recent years monitoring in parts of the Wakhan has revealed a trend toward dramatic increases in livestock numbers and a corresponding degradation of the rangeland (Ostrowski and Rajabi, 2014). Among other consequences, this expansion has brought livestock into greater competition with snow leopard prey species such as urial (*Ovis orientalis*), ibex (*Capra sibirica*), and Marco Polo sheep (*Ovis ammon*) and increased the risk of disease transmission (Ostrowski et al., 2012). It has also brought wild predators such as snow leopards into closer proximity to people and livestock, with the result that killing of snow leopards in retaliation for livestock predation continues (Ranger patrol reports, Afghanistan, unpublished data).

A program was started to build community capacity for conservation, but the program simultaneously worked with its partners in the National Environmental Protection Agency (NEPA) and the Ministry of Agriculture, Irrigation and Livestock (MAIL) to ensure that community participation in natural resource management was firmly enshrined in Afghan legislation and government policy (including assistance in drafting the Environment Law of 2007, the Interim Protected Area Tarzulamal of 2009, and the National Protected Area System Plan in 2010). The country's first national park at Band-e-Amir provided a testing ground and the work there has provided a model for establishing collaborative management institutions elsewhere in Afghanistan (Zahler, 2010; Zahler and Lawson, 2012).

After consultations over a period of almost 4 years, the program assisted the Wakhi and Kyrgyz people of Wakhan District to establish their own community institution, the Wakhan-Pamir Association (WPA), for managing natural resources across the landscape. During the process of establishing the WPA, those involved were assiduous in following the processes laid down in the Law on Social Organizations

AFGHANISTAN: TOP-DOWN MEETS BOTTOM-UP *(cont'd)*

(2002). A chairperson and board were elected, who with external guidance drew up a set of by-laws outlining the purpose of the organization, its structure, and rules for governance. Thereafter the organization was formally registered with the Ministry of Justice. Unlike in Pakistan, a governance institution already existed at local level in Afghanistan, the Community Development Council (CDC). In 2003, the government of Afghanistan began establishing CDCs as part of its National Solidarity Program. However, these institutions have tended to focus on conventional development with little attention on conserving the natural resource base. Furthermore, the limited geographical reach of each CDC inevitably prevents them from making decisions at landscape level, which is essential for effective conservation. The WPA was conceived therefore as an umbrella organization for the CDCs of the Wakhan. Its board members are elected from incumbent CDC representatives and thus make collective decisions relating to resource management across the entire district. An additional rationale for using CDCs as a foundation for the WPA's structure was that it would confer upon the WPA a degree of immediate legitimacy, as CDCs are an established part of local governance and are accepted by communities and government alike. It also avoided the pitfall, characteristic of many attempts at institution building, of creating parallel organizations that either duplicate or frustrate each other's activities.

Ultimately however, community institutions are judged by their constituents according to whether they are achieving their objectives. Since its formation the WPA, with ongoing mentoring and support from WCS, has grown steadily in confidence and institutional capacity, to the point where in 2014 it secured its first direct grants (as opposed to those channeled through third-party organizations) from international donors to implement resource management projects. WPA engages in a wide array of activities ranging from afforestation to promoting nature-based tourism, and including interventions aimed at conserving the snow leopard. The WPA has been involved in the construction of more than 30 predator-proof corrals across Wakhan District designed to reduce human-snow leopard conflict. It is increasingly taking management responsibility for 55 Wakhi and Kyrgyz community rangers, initially equipped and trained by WCS, who play a prominent role in monitoring wildlife and illegal hunting. These local rangers have gained expertise in camera trapping and snow leopard collaring techniques.[2] The WPA is also playing a pivotal role in the establishment of several protected areas vital for the conservation of snow leopards, working in partnership with NEPA, MAIL, local government, and WCS to develop management plans, demarcate boundaries, and secure the approval and local communities. These protected areas include the provisionally declared Wakhan National Park and two more strict protection zones within its boundaries (Big Pamir and Teggermansu Wildlife Reserves), all of which will be managed in accordance the principles of comanagement.

[2]Since 2011, the Wakhan community rangers have captured more than 5000 camera trap photographs of snow leopards.

References

Bagchi, S., Mishra, C., 2006. Living with large carnivores: predation on livestock by the snow leopard (*Uncia uncia*). J. Zool. 268 (3), 217–224.

Baral, N., Stern, M.J., 2009. Looking back and looking ahead: local empowerment and governance in the Annapurna Conservation Area, Nepal. Environ. Conserv. 37 (1), 54–63.

Bawa, K.S., Rai, N.D., Sodh, I.D.S., 2011. Rights, governance, and conservation of biological diversity. Conserv. Biol. 25 (3), 639–641.

Biermann, F., 2010. Beyond the intergovernmental regime: recent trends in global carbon governance. Curr. Opin. Environ. Sustain. 2, 284–288.

Dudley, N., Higgins-Zogib, L., Mansourian, S., 2009. The links between protected areas, faiths, and sacred natural sites. Conserv. Biol. 23 (3), 568–577.

Fraser, J., Wilkie, D., Wallace, R., Coppolillo, P., McNab, R.B., Painter, R.L.E., Zahler, P., Buechsel, L., 2010. The emergence of conservation NGOs as catalysts for local democracy. In: Manfredo, M.J., Vaske, J.J., Brown, P., Decker, D.J., Duke, E.A. (Eds.), Wildlife and Society: the Science of Human Dimensions. Island Press, Washington, DC, pp. 44–56.

Hart, J.A., Hart, T.B., 2003. Rules of engagement for conservation. Conserv. Practice 4 (1), 14–22.

Ingram, V., 2012. Governance of non-timber forest products in the Congo Basin. ETFRN News 53, 36–45.

Jackson, R.M., Wangchuck, R., 2004. Community-based approach to mitigating livestock depredation by snow leopards. Hum. Dimens. Wildl. 9 (4), 1–16.

Knudsen, A.J., 1997. Mountain protected areas in Northern Pakistan: the case of Khunjerab National Park. CMI working paper. Chr. Michelsen Institute, Bergen, Norway.

Madden, F., McQuinn, B., 2014. Conservation's blind spot: the case for conflict transformation in wildlife conservation. Biol. Conserv. 178, 97–106.

Ming'ate, F.L.M., Rennie, H.G., Memon, A., 2014. NGOs come and go but business continues: lessons from co-management institutional arrangements for governance of the Arabuko-Sokoke Forest Reserve in Kenya. Int. J. Sust. Dev. World 21 (6), 526–531.

Nelson, F. (Ed.), 2010. Community Rights, Conservation and Contested Land: The Politics of Natural Resource Governance in Africa. Earthscan, London, UK.

Ostrowski, S., Yacub, T., Zahler, P., 2012. Transboundary Ecosystem Health in the Pamirs. University of Veterinary and Animal Science, Lahore, Pakistan.

Ostrowski, S., Rajabi, A.M., 2014. Update on Wakhi Livestock Numbers in Big Pamir (2006–2014). Unpublished report: Wildlife Conservation Society, Afghanistan.

Painter, M., Castillo, O., 2014. The impacts of large-scale energy development: Indigenous people and the Bolivia-Brazil Gas Pipeline. Hum. Organ. 73 (2), 116–127.

Painter, M., 2009. Rights-based conservation and the quality of life of indigenous people in the Bolivian Chaco. In: Rights Based Approaches: Exploring Issues and Opportunities for Conservation. J. Campese, T. Sunderland, T. Greiber and G. Oviedo (Eds.), pp. 163-184. Center for International Forestry Research (CIFOR), International Union for Conservation of Nature (IUCN), and Commission on Environmental, Economic and Social Policy (CEESP), Bogor, Indonesia.

Porter-Bolland, L., Ellis, E., Guariguata, M., Ruiz-Mallen, I., NegreteYankelevich, S., Reyes-Garcia, V., 2012. Community managed forests and forest protected areas: an assessment of their conservation effectiveness across the tropics. Forest Ecol. Manag. 268, 6–17.

Plumptre, A.J., Masozera, M., Vedder, A., 2001. The Impact of Civil War on the Conservation of Protected Areas in Rwanda. Biodiversity Support Program, Washington, DC.

Reyes-Garcia, V., Ruiz-Mallen, I., Porter-Bolland, L., Garcia-Frapolli, E., Ellis, E.A., Mendez, M.E., Pritchard, D.J., Sanchez-Gonzalez, M.C., 2013. Local understandings of conservation in southeastern Mexico and their implications for community-based conservation as an alternative paradigm. Conserv. Biol. 27 (4), 856–865.

Sheil, D., Puri, R., Wan, M., Basuki, I., van Heist, M., Liswanti, N., Rukmiyati, Rachmatika, I., Samsoedin, I., 2006. Local people's priorities for biodiversity: examples from the forests of Indonesian Borneo. Ambio 15, 17–24.

Simms, A., Moheb, Z., Salahudin, Ali, H., Ali, I., Wood, T., 2011. Saving threatened species in Afghanistan: snow leopards in the Wakhan Corridor. Int. J. Environ. Stud. 68 (3), 299–312.

Stephanson, S., Mascia, M.B., 2014. Putting people on the map through an approach that integrates social data in conservation planning. Conserv. Biol. 28 (5), 1236–1248.

Wiersum, K.F., Lescuyer, G., Nketiah, K.S., Wit, M., 2013. International forest governance regimes: reconciling concerns on timber legality and forest-based livelihoods. Forest Pol. Econ. 32, 1–5.

Wilkie, D., Cowles, P., 2012. Guidelines for Assessing the Strengths and Weaknesses of Natural Resource Governance in Landscapes and Seascapes. United States Agency for International Development, Washington, DC.

Zahler, P., Wilkie, D., Painter, M., Carter J.C., In press. The role of conservation in promoting stability, security in at-risk communities. In: Bruch, C., Muffett, C., Nichols, S. (Eds.), Strengthening Post-Conflict Peacebuilding Through Natural Resource Management: Vol. 6: Governance and Institutions. United Nations Environment Programme, Geneva.

Zahler, P., Lawson D., 2012. Improving livelihoods and governance through resource management in Afghanistan. In: Chettri, N., Sherchan, U., Chaudhary, S., Shakya, B. (Eds.), Mountain Biodiversity Conservation and Management: Selected Examples of Good Practices and Lessons Learned from the Hindu Kush-Himalayan Region. International Centre for Integrated Mountain Development (ICIMOD) Working Paper 2/2012, Kathmandu, Nepal, pp. 36–38.

Zahler, P., 2010. Conservation and governance: Lessons from the reconstruction effort in Afghanistan. State of the Wild III: A Global Portrait of Wildlife, Wildlands, and Oceans. Island Press, Washington, DC, pp. 72–80.

Zahler, P., 2005. Conservation and conflict: the importance of continuing conservation work during political upheaval and armed conflict. In: Guynup, S. (Ed.), State of the Wild: A Global Portrait of Wildlife, Wildlands, and Oceans. Island Press, Washington, DC, pp. 243–249.

Zahler, P., Khan, M., 2004. Status and new records of Ladakh urial (*Ovis orientalis vignei*) in Northern Pakistan. Caprinae: Newsletter of the IUCN/SSC Caprinae Specialist Group, 1–3.

Zaidi, S.A., 1999. NGO failure and the need to bring back the state. J. Int. Dev. 11, 259–271.

CHAPTER

13

Incentive and Reward Programs in Snow Leopard Conservation

OUTLINE

Snow Leopards
http://dx.doi.org/10.1016/B978-0-12-802213-9.00013-4

SUBCHAPTER

13.1

Himalayan Homestays: Fostering Human-Snow Leopard Coexistence

Tsewang Namgail, Bipasha Majumder**, Jigmet Dadul**

***Snow Leopard Conservancy India Trust, Leh, Ladakh, Jammu & Kashmir, India**
****Independent, Mumbai, Maharashtra, India**

INTRODUCTION

Human-snow leopard conflict is a serious conservation issue through most of the cat's range in Asia (Hussain, 2003; Ikeda, 2004; Li et al., 2013; Snow Leopard Network, 2014). Snow leopards are notorious for killing multiple livestock in corrals at night. Such multiple killings inflict serious damage to the economy of local people who rely on livestock production as the mainstay for their existence in harsh and remote mountains. In the absence of any appropriate incentive and/or compensation, the ire of the affected communities ultimately gets directed back to these predators, which are often killed in retaliation (Bagchi and Mishra, 2006; Namgail et al., 2007). Several incentive programs have been devised to mitigate this conflict and to foster coexistence between humans and snow leopards throughout their range (Allen and Macray, 2002; Hussain, 2000; Mishra et al., 2003).

The Ladakh region of the Indian Trans-Himalaya harbors the maximum number of snow leopards within India (Fox et al., 1991; Snow Leopard Network, 2014). The region supports a diverse assemblage of mammalian herbivores (Namgail, 2009; Namgail et al., 2013), which in turn support a good number of snow leopards (Jackson et al., 2006). As in other parts of its range, snow leopards kill a large number of livestock in Ladakh every year, which brings the cat in direct conflict with livestock farmers (Namgail et al., 2007). Although some level of conflict between pastoralists and large carnivores seems inevitable in areas where livestock production is the mainstay of people's economy, the level of conflict can be reduced by providing incentives to the affected communities.

It was against this background that the Snow Leopard Conservancy with support from The Mountain Institute and UNESCO started the Himalayan Homestay Program in 2002 in the Hemis National Park in Ladakh, where human-wildlife conflict had reached a new high. While looking for solutions, it was observed that although about 5000 tourists visited the park every year, the local communities who suffered due to wildlife presence, received minimal or no benefit from tourism (Wangchuk and Jackson, 2002). Recognizing the high rate of livestock depredation by the snow leopard, inequality in the distribution of income from tourism and undue pressure on pastures from pack animals accompanying tourists, we began a dialog with the local communities to initiate the Himalayan Homestay Program.

The homestay concept came about from a series of workshops in Ladakh, in which various governmental and nongovernmental organizations, travel agents, and local people played a pivotal role that proved vital to its ultimate design, growth, and success. The first idea of traditional homestays came from a village woman who, having observed the growth in trekking and proliferation of infrastructure, including numerous guesthouses in Leh, visualized a different approach for rural areas. Thus, under a pilot program, a dozen or so households interested in pioneering the homestays were given initial investment support such as blankets, mattresses, buckets, bed sheets, and other fundamental items required to run a homestay. Initially, potential villager entrepreneurs were trained in hospitality, hygiene, housekeeping, and other management skills. For those who were not able to start a homestay, other income-generating opportunities, such as nature guiding, solar showers, ecocafes, and so on, were offered as alternatives. A mechanism was also put in place whereby 10% of the proceeds from the homestays were to be set aside for environmental protection in and around the villages. In return, the villagers assured abstinence from any retaliatory killing of snow leopards and other large predators.

Currently, Snow Leopard Conservancy–India Trust (SLC-IT) is running the program in 27 villages across Ladakh covering more than 130 households. More than 200 nature guides have been trained. Ten ecocafes have been established, and five solar shower facilities have been provided. The households in a village follow a rotation system whereby they take turns at hosting tourists to ensure equitable distribution of income from the program. Over the years, the program has helped the communities get directly involved in the conservation process in their respective areas. In fact, the program has been so successful in offsetting the livestock loss to snow leopards, that the local government adopted it in its effort to conserve wildlife in protected areas of Ladakh. It is also emulated by several organizations across the Himalayas. In this chapter, we assess the efficacy of this program in mitigating human-snow leopard conflict, and more importantly in influencing people's attitude toward the snow leopard.

SURVEY METHODS

We carried out a questionnaire survey in the months of September and October 2014. Data were collected largely through semistructured interviews with local communities in the Sham Valley, one of our homestay areas in Ladakh. We interviewed one individual, preferably the head, from each homestay households and randomly selected nonhomestay households in seven villages: Ulley, Yangthang, Tarutse, Saspotsey, Hemis Shukpachan, Ang, and Tia. Although we strove to obtain accurate information during the household surveys through cross-checking, some people, concerned with tax-related issues, may have provided incorrect figures. Information on modern developmental activities such as road and bridge construction was collected by interviewing the village headmen, and also by opportunistically interviewing knowledgeable people in the villages. We compared the responses of homestay and nonhomestay households to see potential impact of the homestay program on the economy, ecology, and socioculture of the villages. We used χ^2 to see if the differences are statistically significant.

RESULTS

We interviewed 28 persons from the homestay households, and 33 persons from nonhomestay households. Among the homestay respondents,

54% ($n = 28$) were females, while 46% were males. Similarly, there were more female respondents (67%, $n = 33$) and fewer male respondents (33%) among the nonhomestay households. Farmers dominated the respondents among both homestay (89%) and nonhomestay households (88%). Among the homestay households, tourism emerged as the third most important source of income after agriculture and livestock grazing, whereas among the nonhomestay households, military service was the third most important source of income after the aforementioned traditional sources of income.

Economic Impact

Sixty-four percent of the homestay respondents agreed that the Himalayan Homestay Program promoted ecotourism, and that the program facilitated greater influx of tourists in the region. During the interviews, it became apparent that 99% of the homestay owners had started the homestay enterprise for supplementary cash income. When asked how homestays are benefiting their villages, the majority of respondents mentioned higher cash income and cultural preservation. The homestay households had significantly higher cash income than the nonhomestay households ($\chi^2 = 53.77$, $p < 0.001$). For instance, 14% of the homestay households have an annual cash income of over Indian Rs. 50,000, while only 3% of the nonhomestay households are in this income category. Among the homestay households, 53% had a meager annual income of less than Indian Rs. 10,000 prior to the program, but this percentage plunged to 3% after starting the program. The homestay households use the increased income for children's education, home maintenance, better health, clothes, and food. Seventy-nine percent of the respondents strongly believe that ecotourism enhances cash income, and they want to remain part of the Himalayan Homestays.

Ecological Impact

All the respondents knew most of the wild mammals and birds found in their area. Sixty-four percent of the homestay owners liked the presence of snow leopards and wildlife as they brought in tourists and thus revenue, while only 54% of the nonhomestay respondents liked the presence of snow leopards and other wildlife, though they were wary of them killing their livestock. However, there was no significant difference between homestay and nonhomestay households in the level of their liking of snow leopard ($\chi^2 = 1.12$, $p = 0.289$). Both homestay (100%) and nonhomestay respondents (88%) agreed that the Himalayan Homestays reduced pressure on the pastureland in the area, as trekkers do not need horses to carry their camping gear.

The Himalayan Homestays also helped in creating awareness about environmental cleanliness, which helps in sustaining the environment and the growth of the region's flora and fauna. The environment-conscious trekkers seem to have an impact on the homestay owners, 85% of whom now feel that it's important to keep their houses and villages clean. One is likely to think that an increase in the number of tourists also increases the garbage problem, but this seems to be unfounded in the Sham valley; the majority of respondents (86% homestay and 85% nonhomestay households) said that the increase in tourists did not increase garbage in the village as most of the tourists dispose the garbage appropriately, and some even carry it back out.

Sociocultural Impact

Homestay and nonhomestay households differed significantly in their views on the influence of tourists on their culture and traditions ($\chi^2 = 13.52$, $p < 0.01$). When asked if the tourists affected the local culture and traditions, 68% ($n = 28$) of the homestay respondents gave an affirmative answer. Some of the most visible changes are cleaner houses and appreciation of local food and costumes. However, the majority of nonhomestay respondents (51%) see no influence on their culture and traditions. Furthermore, 82% of the homestay owners felt that they have learned a lot from tourists, especially things like maintaining schedules and cleanliness, change in eating habits and learning to cook new dishes, hospitality, and last but not the least, new languages. Furthermore, as 68% of respondents agreed, the homestay program has made them more social and confident than before, as tourists foster pride in local cultures and traditions.

Tourists also seem to have influenced local people's view on livelihood options. Overall 62% of the respondents felt that it is important to educate their children so that they can get good jobs. It also became apparent that the number of livestock, especially sheep and goats are decreasing. The younger generation, unlike their parents and grandparents, do not want to be involved in agriculture and livestock rearing for their future livelihood, and many of them are migrating to Leh, the main town in the area. Despite these social dynamics, 77% of all homestay respondents want to remain part of the Himalayan Homestay Program.

Discussion

Ladakh is fast becoming an important tourist destination in India. The number of tourists in this Trans-Himalayan region increased from approximately 500 in 1974, when the region was first opened to tourists, to more than 150,000 in 2011 (Rajashekariah and Chandan, 2013). At this rate of increase, the transient population of tourists will soon surpass the local population (300,000). Ladakh, however, witnessed a boom in tourism and tourism-related activities in the last decade. Tourist season was earlier confined to a few summer months, but with the growing number of wildlife tourists visiting to see the snow leopard, the season now extends into winter.

The Himalayan Homestay Program was developed taking into cognizance the growing animosity of people against snow leopards and the booming tourism industry. People in the past viewed the snow leopard as a pest and a threat to their livelihoods, but now they view it as a tourism asset, bringing revenue. Our survey revealed that the annual income of homestay households is significantly higher than the nonhomestay households. Survey results also showed that people of homestay households like the presence of snow leopards in the mountains surrounding their villages more than those of nonhomestay households, although statistically insignificant. It is important to note, however, that the nonhomestay households also benefit from facilities like livestock insurance, environmental education, nature guiding, solar showers, and ecocafes, which are allied to the homestay program.

A large proportion of the interviewees mentioned that the homestay program facilitated a greater influx of tourists in the region. SLC-IT achieved this through aggressive marketing and value addition. For instance, we have forged strong links with various travel agents, who offer tourists packages including homestays. In our marketing efforts, we also highlight the conservation-linkage of the program such as the homestay-operators' abstinence of retaliatory killing of snow leopards, and the flow of 10% of the proceeds toward a village conservation fund. We have also been opening new trails to connect remote villages with the network of trekking routes in different parts of Ladakh, and developing skills of homestay operators regularly.

The survey results also showed that the homestay operators learn various things like the importance of hygiene, sanitation, and gastronomy from the tourists visiting their homes. They also enjoy meeting people from different parts of the world and learning about their culture and traditions. "*I was born in this valley and lived here all my life as a farmer. I hated it. Now that visitors come from distant places and appreciate our mountains and culture, it makes me proud to be a Yangthang pa*," said Skarma Lungstar pa from Yangthang, one of the survey villages.

CHALLENGES AND THE WAY FORWARD

Urbanization and the consequent change in the social fabric has resulted in the younger generation aspiring for "jobs" outside their villages rather than getting involved in traditional labor-intensive livelihood activities such as agriculture and livestock production. This is likely to make the younger generation lose connection with their way of life, which might affect the quality of the homestays. Secondly, with a boom in the tourism industry, many households have opened guest houses/homestays, either of their own accord or through other agencies, resulting in competition with the Himalayan Homestays. The owners of these guest houses/homestays do not contribute toward environmental protection, and they often invite the tourists to their homes, pretending that they are part of the Himalayan Homestay Program. These factors may dilute the uniqueness of the Himalayan Homestay brand as well as deal a massive blow to conservation in the region as people start catering only to tourism for business without a linkage to conservation. Hence it is imperative to educate and involve the younger generation and other agencies at the grassroots level to help them understand the critical linkage between sustainable tourism and conservation.

The rotation system to distribute the income equitably among all homestay operators in a village needs strengthening, as households that are either influential or are close to the road heads get a larger share of the tourists. Almost all the homestay owners are farmers, and when they are busy in agricultural fields, some hosts save time by packaging simple meals such as flat bread and jam for lunch, which are disdained by some guests. Therefore, we are currently standardizing the menu. Finally, frequent training of homestay owners and community members can help them take more ownership of the program, and hence can promote more holistic nature conservation in their villages.

CONCLUSIONS

The Himalayan Homestay initiative has been able to achieve what it set out to do: reduce human-wildlife conflict in areas with high livestock depredation, and involve communities in the conservation process. The homestay owners have not only been able to improve their economic conditions, but also been able to contribute to the positive ecological and sociocultural aspects from tourism. The changing socioeconomic fabric of Ladakh will dictate the future of wildlife conservation and how we continually improve the initiative to keep the message of conservation at the fore. Even though imminent challenges lay ahead, the Himalayan Homestay Program has provided a distinct advantage of tolerance and understanding toward snow leopards and other wildlife, and the villagers want to remain part of the program.

Acknowledgments

We thank the The Mountain Institute, UNESCO, Panthera, Royal Bank of Scotland, and Snow Leopard Conservancy USA for supporting our homestay program in Ladakh. We thank the late Mr Rinchen Wangchuk, Rodney Jackson, Nandita Jain, and Wendy Lama for conceptualizing and initiating the program. We thank Stanzin Khakhyab for his assistance in analyzing the survey data. We thank Tsewang Dolma for her support and encouragement. We thank Tsering Angmo, Rigzen Chorol, Mohd. Hasnane, Thupstan Dolker, Tsering Lazes, and Nawang Gyalson for their help during the questionnaire survey. Last but not least, we thank all the villagers for participating in the program and this survey.

SUBCHAPTER

13.2

Handicrafts: Snow Leopard Enterprises in Mongolia

Bayarjargal Agvaantseren, Priscilla Allen**, Unurzul Dashzeveg*, Tserenadmid Mijiddorj*, Jennifer Snell Rullman*[†]

*Mongolian Snow Leopard Conservation Foundation, Ulaanbaatar, Mongolia

**Independent, Seattle, WA, USA

[†]Snow Leopard Trust, Seattle, WA, USA

VISION

A 1997 survey among nomadic herders in snow leopard habitat in Mongolia indicated ambivalent attitudes toward snow leopards (Allen et al., 2002). Despite high levels of awareness of laws protecting snow leopards, herders reported that retaliatory killing did occur when snow leopards preyed on livestock. Of the 116 interviewees, 49% had experienced depredation of livestock by snow leopards. When asked what actions could ameliorate the challenges of sharing habitat with an endangered predator, herders suggested support to increase income generation from their livestock products, specifically wool. In response to this request, independent conservationists P. Allen and A. Bayarjargal created Snow Leopard Enterprises (SLE) with funding from the David Shepherd Conservation Foundation (UK), WWF-Mongolia, International Snow Leopard Trust (USA), British Embassy (Mongolia), and support from the Mongolian Union for the Conservation of Nature and the Environment and the Great Gobi Biodiversity Project. The goal was to enable herders to add value to their livestock products by creating finished items instead of selling raw wool at wholesale rates. In exchange for the opportunity to increase household income by adding value to livestock products, herders made a commitment through a conservation contract to specific conservation actions that benefit snow leopards.

HOW SLE WORKS

SLE now operates under the umbrella of the Snow Leopard Trust (SLT) and the Mongolian Snow Leopard Conservation Foundation (SLCF). The program works with herder communities whose pastures overlap with snow leopard habitat in the seven provinces where snow leopards are found. Communities establish norms and select a representative. The representative communicates with the central office to coordinate training in product manufacture and design, order distribution and collection, organization of environmental education and workshops, and to support other logistical requirements. Twice a year, SLE staff visits the approximately 30 participating communities to place orders and to purchase products. In

September, contracts are renewed after the conservation criteria have been reviewed and compliance surveys completed.

Products are then labeled, packaged, and prepared for market. Some are sold on the Mongolian tourist market; the majority is shipped to the SLT headquarters in USA. From there products are distributed to consumers through web sales, events, and niche markets. Marketing efforts have been especially successful in US zoos. Sales revenue covers payment to the artisans, running costs, all in-country salaries, as well as some monitoring and compliance costs. The program also funds a small grant program for community-based conservation initiatives. In 2014, SLE involved more than 450 herder households from 30 nomadic herder communities and had generated over USD 1 million in cumulative sales since 2000.

CONSERVATION CONTRACT, COMPLIANCE, AND CONSEQUENCES

Participation in SLE is directly linked to the conservation of snow leopards through a multipartied conservation contract. Stakeholders include local protected area administrations and/or environmental agencies, SLE administration, community coordinators, and individual herder participants. The contract requires participants to:

- Prevent poaching of snow leopards or their prey species within their area of responsibility
- Follow protected area regulations
- Support conservation awareness

Communities that comply receive a 20% conservation bonus in addition to the income earned from product sales. A single violation within the area that the community is responsible for results in the loss of the bonus for the entire community. The bonus is often the difference between breaking even and making a clear profit so this model creates substantial peer pressure against poaching.

Data from protected area administrations, environmental agencies, herder communities, anti-poaching units, and SLT researchers are compiled and analyzed to determine levels of compliance with the conservation contract. If poaching has occurred in a SLE area, the community loses its conservation bonus and the violation is investigated, often resulting in legal action by local authorities. To date, no cases of snow leopard poaching have occurred in SLE program areas. Several ibex and argali—primary snow leopard prey—were poached; bonuses were withheld.

Besides the punitive impact of withholding the bonus as a result of contract violations, SLE offers positive incentives to support conservation initiatives. First, SLE provides handicraft skills training to ensure high-quality products that follow consistent, marketable designs. Second, training in wildlife monitoring techniques is available for community members interested in becoming rangers. Qualified rangers support snow leopard sign surveys and monitoring of prey species such as ibex and argali. Finally, workshops are offered on a variety of topics depending on local needs: land rights, community advocacy, and reducing herd sizes while increasing yield with healthier animals.

ECONOMIC AND SOCIAL IMPACT

Participants earn an average of USD 150 from participating in the manufacture and sale of handicrafts for SLE, raising annual incomes by 4% compared to the national average for rural households (National Statistical Office, 2011). Due to their remoteness, many SLE households' annual income is far below the national average. Therefore the economic impact of SLE is much greater; in many cases incomes are almost doubled. According to a 2004 survey, the additional cash was spent on basic household goods and family needs, such as food, clothing, school tuition, and school books. Most SLE participants are women (98%) who enjoy financial empowerment and elevated status as decision makers within households and the community. This was noted as one of the most positive outcomes of the program (Mallon, 2006). Other perceived benefits include collaboration and networking with organizations like Snow Leopard Conservation Foundation (SLCF), national parks, and local government institutions. Men in the community cited the economic benefits and greater freedom granted by additional cash within a predominantly barter economy. Women emphasized the changing attitudes about women's roles, increased self-esteem, and pride in being part of an international market (Bayarjargal, 2004; Mallon, 2006).

Site-specific educational workshops have also yielded positive results. Posters and follow-up workshops have been found to be effective methods to share best practices on avoiding predator attacks (e.g., burying or burning carcasses, using lights and sounds and unpredictable distractions for several days following an attack). By having access to locally relevant information and training, communities have become curious to learn about snow leopard ecology and actively engaged in their conservation.

CONSERVATION IMPACT

Since the program was initiated in 1998, poaching of snow leopards and their prey has decreased in areas where SLE is active according to provincial wildlife management reports. Incidences of prey species poaching have occurred and resulted in the loss of several hundred dollars of bonus moneys for the affected communities. In some cases SLE community members helped identify and locate the poacher in order to demand restitution as well as impose a fine. In at least one case, members of the poacher's family joined SLE – making a commitment to snow leopard conservation in exchange for the opportunity to generate income. Local government and protected area officials have a positive view of the program, and consider the environmental component of the program effective, especially in reducing poaching of snow leopards (Mallon, 2006). Additionally, the small grant program, funded from 10% of the total purchase funds, has resulted in several locally relevant conservation projects initiated by SLE community members.

In contrast to attitudes expressed when the program began, most herders in SLE areas no longer regard snow leopards as a significant threat to their livestock. Attitude surveys indicate that tolerance for livestock depredation is higher in SLE communities than in nonparticipating communities. Herders in the SLE community of Tost, South Gobi, said they would tolerate up to 30 livestock losses to predators, whereas members of the non-SLE community of Baysah, South Gobi, would tolerate a maximum of 5 losses per community per year (Tserennadmid, 2011). These tolerance levels are significant considering the economic value of livestock to herders. Total livestock losses throughout SLE program areas in 2011 represented approximately USD 38,000, or about USD 170 per household on average. As mentioned previously, the additional average income per household generated by SLE provides an average of USD 150, almost entirely offsetting this economic loss. In reality, predation losses are uneven whereby some households have devastating losses while others have none. Support programs and mechanisms to provide community-based insurance (Mishra et al., 2003) are beginning to smooth such discrepancies.

CHALLENGES AND OPPORTUNITIES

Governmentally delineated borders initially determined the area that a community was responsible for in terms of the conservation contract. These areas were often much too large for a community to be able to manage. Also, some areas did not overlap with key snow leopard habitat. In order to strengthen the conservation link and to create true stewardship, SLE helped communities to determine "community responsible areas" (CRA) by mapping land use patterns, sacred sites, water sources, and other important landmarks and natural resources. Hand-drawn maps were converted to geo-referenced GIS maps, designating land use patterns, landscape features, and CRA boundaries. Beyond being a tool for SLE contract compliance, these maps have enabled communities to develop locally informed and empowered management plans and even to register with the Mongolian government and establish legal rights over resource use.

Another challenge is tracking the impact of SLE on the conservation of snow leopards and prey populations. The challenges of monitoring snow leopard populations make it difficult to say whether there is a response to a particular conservation intervention. Our initial monitoring tool, snow leopard sign surveys, was not a reliable indicator of conservation success. In 2011 we adopted the Threat Reduction Assessment (TRA) (Margoluis and Salafsky, 2001). We determined priority threats to snow leopards, created targets to reduce threats, and developed specific interventions. Measurable indicators and a "threat reduction index" allow us to determine our level of impact. Results will be monitored over the long term.

Emerging challenges and threats require diligence and flexibility. Policy changes might impact the price of raw wool, creating pressure on the pricing structure of the SLE products; mining exploration might cause SLE communities to lose pasturelands; fluctuating tourist numbers may shift sales opportunities. By being mindful of emerging challenges and open to evolving opportunities, SLE remains a resilient and successful chapter in snow leopard conservation efforts.

SUBCHAPTER

13.3

A Review of Lessons, Successes, and Pitfalls of Livestock Insurance Schemes

*Kyran Kunkel**, *Shafqat Hussain***, *Ambika Khatiwada*[†]

*Conservation Science Collaborative and Wildlife Biology Program, University of Montana, Missoula, MT, USA

**Department of Anthropology, Trinity College, Hartford, CT, USA

[†]National Trust for Nature Conservation, Sauraha, Chitwan, Nepal

PROBLEMS AND SOLUTIONS

More than 50% of the human population in snow leopard range is engaged in agropastoralism, and more than 40% are below the poverty level. In most of snow leopard range, livestock biomass is an order of magnitude above natural prey. Studies throughout the range indicate that snow leopard predation on livestock can incur a high cost to herders and communities in many places. For example, in Bhutan, villagers on average lose more than two thirds of their annual cash income to snow leopard depredations (Wang and Macdonald, 2006).

A primary limiting factor for snow leopard in much of its range is retaliatory killing in response to predation on livestock (Snow Leopard Network, 2014). Thus, a high priority to improve snow leopard survival and populations, and to improve human livelihoods, is to address that factor. One of the most pragmatic, efficient, and arguably successful approaches to reducing snow leopard mortality has been to provide compensation or insurance for losses to reduce hostility and retaliatory killing (Gurung et al., 2011; Mishra et al., 2003; Rosen et al., 2012).

Direct compensation programs have a long history but are often not successful (Nyhus et al., 2003). Livestock insurance schemes were developed largely in response to failures and shortcomings in compensation programs. Insurance programs are usually more comprehensive and locally driven than compensation schemes, and these factors likely account for their greater success (Gurung et al., 2011; Rosen et al., 2012).

Snow leopard conservation practitioners, along with local communities and herders, have been pioneers in developing innovative and context-specific interventions, such as livestock micro-insurance schemes against snow leopard predation. Most insurance programs are developed and run by local communities and there is direct involvement of herders. This results in greater buy-in and incentives to improve husbandry, reduce reporting fraud, and obtain quicker responses. They are more economically and logistically sustainable as a result of buy-in and local funding and investment. Importantly, these programs often also include locally designed innovations. Finally, the local community and herders are often part of the snow leopard monitoring programs and that improves local conservation education, capacity, and interest.

HISTORY AND DESIGN

One of the longest running Community Managed Livestock Insurance Schemes (CMLIS) is managed by Project Snow Leopard, PSL (later became Baltistan Wildlife Conservation and Development Organization, BWCDO), which was started in one village in the Baltistan region of northern Pakistan in 1999. The scheme is based on a simple premise that snow leopard predation is a stochastic shock to farmers and they will be better off by spreading this risk amongst them. Thus it operates on a simple logic upon which most insurance is based. Under the insurance schemes villagers pay a premium amount per head of livestock, both small animals such as goats and sheep and large animals such as dzo (yak and cattle hybrid) and cows, as determined by the historical loss rate of domestic livestock to snow leopard predation, while the project subsidizes the premium up to 50%. The project raises the money from private donors to subsidize insurance premium payments. The scheme is designed to keep fraudulent claims to a minimum, and by making local herders active partners, keep transaction and monitoring costs to a minimum. Under the project a Village Insurance Committee is established that makes decisions on all compensation claims that are filed by the participating villagers.

After running for seven years, the scheme in Baltistan was expanded to ten villages in 2006, and currently it serves 26 villages, insuring about 18,000 animals. Since 2007, PSL has paid out compensation of approximately US$ 24,000. The scheme is successful in halting the decline in snow leopard population in the project area and may have even contributed to increase in cat numbers according to population estimates in Hussain (2003) and Anwar et al. (2011). In addition to compensating farmers, PSL also gives additional incentives to participating villagers, such as small infrastructure and educational schemes. The major challenges that remain are finding continued sources of funding.

IMPORTANT FACTORS FOR DESIGN, IMPLEMENTATION, AND SUCCESS

Morrison et al. (2009) concluded that effectiveness of insurance programs is contingent upon strong community coherence, sustained dialog, and strong partnership between the community and conservationists or managers, and people's economic ability to participate in them.

Economic Sustainability and Scaling Up

Long-term economic sustainability of CMLIS appears to have been reached in some programs via the creation of endowments. These endowments are generally established from outside funding. They are relatively small investments for big returns and address the problem of a general global market failure in carnivore conservation; developed nations highly value carnivore conservation in developing nations where resources are limited and carnivores are costly to livelihoods (Dickman et al., 2011). Addressing this failure is key (see economic incentives section that follows), and outside funding of these endowments is one way to do that.

CMLIS programs need to be scaled up to secure not just individual snow leopards, but whole populations. Several communities over a large landscape need to be involved to achieve this. In Kangchenjunga Conservation Area (KCA) in Nepal, Gurung et al. (2011) were successful in establishing CMLIS in several villages and this program is expanding and could link a globally important transnational conservation region (India, Bhutan, China, and Nepal). An important part of scaling up is a regional assessment of where the priority populations of snow leopard are and focusing resources there. To date, though, we are not aware of a program that has been able to assess and monitor and thus ensure that a viable snow leopard population has been secured. Such is, however, true also for any snow leopard conservation programs we are aware of.

Additional Factors

In order to make these programs sustainable, funding proactive and preventative husbandry measures such as increased numbers of herders, scare devices, guard dogs, and secure corrals is needed. Livestock grazing levels should be assessed to determine impacts on overall biodiversity and if too high, incentives can be developed to reduce grazing intensity (Mishra et al., 2003).

Rosen et al. (2012) indicated that the programs should also be focused on multiple species, as wolves (*Canis lupus*), dhole (*Cuon alpinus*), common leopards (*Panthera pardus*), bears (*Ursus spp.*), and other large carnivores also kill livestock in snow leopard range and there is a need to address these species to make programs ecologically and economically sustainable. The program of Gurung et al. (2011) in Nepal is now addressing snow leopard and dholes.

All of these efforts would greatly benefit by developing best management practices and then an idea-sharing platform for lessons learned, perhaps including practitioner exchanges among sites. In an assessment to determine the most effective tiger (*Panthera tigris*) conservation strategies, Gratwicke et al. (2007) found that one of the best performing suites of activities was mitigating human-tiger conflict. But, given the diversity of approaches and potential outcomes, best practices were difficult to ascertain. They indicated that there was a need for better communication between the different groups working on human-tiger conflict issues so that the experts themselves could share lessons learned and come up with a set of best practices that would be applicable in each landscape.

One of the most important factors is having a system for building local capacity so that technical support and monitoring is always improving. Incorporating local students into these programs is a proven way to ensure that.

Finally, the programs often do not cover the full cost of losses and we believe there is need for additional incentive programs to be developed to offset that and further incentivize conservation. The ultimate goal is to have carnivores viewed as worth more alive than dead. These incentive programs include investments in education (Rosen et al., 2012), alternative livelihoods (Gurung et al., 2011; Mishra et al., 2003; Rosen et al., 2012), and direct conservation payments (Dickman et al., 2011).

SUCCESSES OF CMLIS FOR SNOW LEOPARDS AND COMMUNITIES

Gurung et al. (2011) reported that since initiation of the CMLIS in KCA, Nepal, the number of participating herders has increased. The rate of snow leopard killing by locals went from 1–3/year to 0/year during the first 4 years. Based on sign surveys, snow leopard abundance increased over that period. The correlation between sign density and abundance is, however, questionable (McCarthy et al., 2008). In 2014, farmers and herders participating in a CMLIS program in the villages of Tapethok and Yamphudin were asked about instances of snow leopard and dhole retaliatory killings after implementation of CMLIS. Of 120 households and 40 key informants interviewed during the study, 74% respondents reported that CMLIS has contributed to the conservation of snow leopard and dholes, 23% reported no change, and 3% reported that the program did not have any contribution toward snow leopard/dhole conservation. The endowment funds (established for all

KCA CMLIS) appears to be self-sustaining and has yielded USD 2199 in excess funds that have been used in the promotion of their livestock herding profession, livestock-based enterprises, and other income generation activities such as tourism-based enterprises, kitchen gardening, cooking and baking training, communication development (e.g., purchasing of telephone and operation of Internet services), and handicrafts. Further, CMLIS has been expanded to more communities for dhole and snow leopard, and neighboring communities are asking for additional CMLIS sites. The number of yaks lost per year, however, remained similar in the first few years, indicating that proactive husbandry measures need work.

Experiences from KCA suggest a need for improving local capacity for record keeping, effective administration, and communication. Distant herders sometimes become reluctant to insure their livestock, as they are unable to communicate the depredation cases to CMLIS authorities in time to claim the relief fund. Assignment of one permanent member to look after one CMLIS can probably solve the administration and record keeping-related issues for effective management.

Rosen et al. (2012) reported that in 13 years since the start of project PSL, the affected communities of Gilgit–Baltistan have understood the value the international community places on snow leopard; they have adapted by largely accepting the presence of snow leopard and are now participating in a mutually respectful partnership that merges local and global interest in conservation and more harmonious coexistence with carnivores. Since 2006, about USD 13,000 has been spent on corral improvement and small-scale infrastructure projects, with another USD 4,200 spent on community general education programs. The total area of snow leopard habitat in the PSL project area is about 5000 km^2, and includes 19 snow leopards and is thus nearing protection of a viable snow leopard population (Anwar et al., 2011).

No known carnivore persecution has taken place since the CMLIS program site in Kibber, India, was initiated in 2002 (Mishra and Suryawanshi, 2014). Insurance programs have recently been started in Afghanistan, Bhutan, China, and Tajikistan. The program started in the Wakhan of Afghanistan in 2010 however has not been able to continue due to funding shortages (Anthony Simms, WCS, personal communication).

Despite these apparently positive outcomes for snow leopard, we recommend much more rigorous monitoring of these efforts to ensure they are priorities for snow leopard conservation and that they are working. Good data on snow leopard numbers and mortality prior to initiation are needed to assess success. Comparing sites with and without CMLIS programs would be a good way to assess the programs overall. A few sites with intensive demographic monitoring would also be valuable to assess cause-specific mortality and population density before and after CMLIS implementation. The size of the area that needs to be impacted to secure a population would also be a valuable course of investigation.

DIRECT CONSERVATION PAYMENTS

Overall, insurance schemes appear to reduce economic losses and thus animosity to snow leopards, but some do not provide adequate incentives for local people to actually deliver conservation. Incentives may be a way to produce substantial benefits for long-term conservation and poverty alleviation. Recently, the idea of direct conservation payments has attracted much attention (Dickman et al., 2011). Such payments are linked specifically to the production of the desired environmental output (e.g., maintenance of carnivores on private land) rather than to indirect inputs assumed to affect the production of that service.

Conservation payments have several benefits for people and predators: they are likely to provide additionality, as they create a direct incentive for maintaining carnivores, whereas communities are less constrained and able to act in the manner optimal to their specific conditions to reach the desired endpoint, often resulting in greater cost effectiveness (Ferraro and Simpson, 2002; Zabel and Holm-Müller, 2008). Payments are usually independent of levels of depredation, thereby avoiding moral hazard, and entail low transaction costs for livestock-keepers, as they do not have to search for depredated livestock or submit claims for compensation. Furthermore, unlike schemes linked to protected areas, which can impose substantial opportunity costs, these payments actually reduce the costs of maintaining traditional lifestyles in areas where humans and carnivores coexist, helping people maintain their cultural integrity and avoid traditional pastoral poverty traps.

Based on successes to date, we believe CMLIS is one of the most effective and cost efficient strategies for enhancing snow leopard populations while simultaneously enhancing local livelihoods and therefore should be a priority in rangewide stewardship and funding of snow leopard and large carnivore conservation in Asia.

References

Allen, P., Macray, D., 2002. Snow leopard enterprises description and summarized business plan. Contributed Papers to the Snow Leopard Survival Strategy Summit. International Snow Leopard Trust, pp. 15–24.

Allen, P., McCarthy, T., Bayarjargal, A., 2002. Conservation de la panthère des neiges (*Uncia uncia*) avec les éleveurs de Mongolie. In: Chapron, G., Mountou, F. (Eds.), L'Etude et la Conservation des Carnivores. Societé Française pour l'Etude et la Protection des Mammifères, Paris, pp. 47–53, (in French).

Anwar, B., Jackson, R., Nadeem, M.S., Janecka, J.E., Hussain, S., Beg, M.A., Mohammad, G., Qayyum, M., 2011. Food habits of the snow leopard *Panthera uncia* (Schreber, 1775) in Baltistan, Northern Pakistan. Int. Eur. J. Wildl. Res. 57 (5), 1077–1083.

Bagchi, S., Mishra, C., 2006. Living with large carnivores: predation on livestock by the snow leopard (*Uncia uncia*). J. Zool. 268, 217–224.

Bayarjargal, A., 2004. Women's development through Nature Conservation: can protected areas improve women's livelihood? Case Study in Altai Tavan Bogd National Park, Western Mongolia. MA thesis, Development Studies Center, Kimmage, Dublin, Ireland.

Dickman, A.J., Macdonald, E.A., Macdonald, D.W., 2011. A review of financial instruments to pay for predator conservation and encourage human–carnivore coexistence. Proc. Natl. Acad. Sci. 108, 13937–13944.

Ferraro, P., Simpson, R., 2002. The cost-effectiveness of conservation payments. Land Econ. 78, 339–353.

Fox, J.L., Sinha, S.P., Chundawat, R.S., Das, P.K., 1991. Status of snow leopard *Panthera uncia* in northwest India. Biol. Conserv. 55, 282–298.

Gratwicke, B., Seidensticker, J., Shrestha, M., Vermilye, K., Birnbaum, M., 2007. Evaluating the performance of a decade of Save The Tiger Fund's investments to save the world's last wild tigers. Environ. Conserv. 34, 255–265.

Gurung, G.S., Thapa, K., Kunkel, K., Thapa, G.J., Kollmair, M., Mueller Boeker, M., 2011. Enhancing herders' livelihood and conserving the endangered snow leopard in Kangchenjunga Conservation Area of Nepal Himalaya. Cat News 55, 17–21.

Hussain, S., 2000. Protecting the snow leopard and enhancing farmers' livelihoods: a pilot insurance scheme in Baltistan. Mt. Res. Dev. 20, 226–231.

Hussain, S., 2003. The status of the snow leopard in Pakistan and its conflict with local farmers. Oryx 37, 26–33.

Ikeda, N., 2004. Economic impacts of livestock depredation by snow leopard *Uncia uncia* in the Kanchenjunga Conservation Area, Nepal Himalaya. Environ. Conserv. 31, 322–330.

Jackson, R.M., Roe, J.D., Wangchuk, R., Hunter, D.O., 2006. Estimating snow leopard population abundance using photography and capture-recapture techniques. Wildl. Soc. Bull. 34, 772–781.

Li, J., Yin, H., Wang, D., Jiagong, Z., Lu, Z., 2013. Human-snow leopard conflicts in the Sanjiangyuan Region of the Tibetan Plateau. Biol. Conserv. 166, 118–123.

Mallon, D., 2006. SLE Evaluation: Final report. Unpublished report: Snow Leopard Trust, Seattle WA, USA.

Margoluis, R., Salafsky, N., 2001. Threat reduction assessment: a practical and cost-effective approach to evaluating conservation and development projects. Conserv. Biol. 13, 830–841.

McCarthy, K.P., Fuller, T.K., Ming, M., McCarthy, T.M., Waits, L., Jumabaev, K., 2008. Assessing estimators of snow leopard abundance. J. Wildl. Manag. 72 (8), 1826–1833.

Mishra, C., Allen, P., McCarthy, T., Madhusudan, M.D., Bayarjargal, A., Prins, H.H.T., 2003. The role of incentive programs in conserving the snow leopard. Conserv. Biol. 17, 1512–1520.

Mishra, C., Suryawanshi, K. 2014., Managing conflicts over livestock depredation by large carnivores Pages 27–47 in South Asian Association for Regional Cooperation—Human-Wildlife Conflict in the Mountains of SAARC Region - Compilation of Successful Management Strategies and Practices.

Morrison, K., Victurine, R., Mishra, C., 2009. Lessons learned, opportunities and innovations in human wildlife conflict compensation and insurance schemes. Report prepared for Wildlife Conservation Society TransLinks Program.

Namgail, T., 2009. Mountain ungulates of the Trans-Himalayan region of Ladakh, India. Int. J. Wilderness 15, 35–40.

Namgail, T., Fox, J.L., Bhatnagar, Y.V., 2007. Carnivore-caused livestock mortality in Trans-Himalaya. Environ. Manag. 39, 490–496.

Namgail, T., van Wieren, S.E., Prins, H.H.T., 2013. Distributional congruence of mammalian herbivores in the Trans-Himalayan Mountains. Curr. Zool. 59, 116–124.

National Statistical Office of Mongolia, 2011. Census and Survey. National Statistical Office of Mongolia, Ulaanbaatar.

Nyhus, P., Fischer, F., Madden, F., Osofsky, S., 2003. Taking the bite out of wildlife damage: the challenge of wildlife compensation schemes. Conserv. Prac. 4, 37–40.

Rajashekariah, K., Chandan, P., 2013. Value Chain Mapping of Tourism in Ladakh. WWF-India, Lodhi Estate, New Delhi.

Rosen, T., Hussain, S., Mohammad, G., Jackson, R., Janecka, J., Michel, S., 2012. Reconciling sustainable development of mountain communities with large carnivore conservation. Mt. Res. Dev. 32, 286–293.

Snow Leopard Network, 2014. Snow Leopard Survival Strategy. Revised 2014 Version. Snow Leopard Network, Seattle, WA, USA. http://www.snowleopardnetwork.org/docs/Snow_Leopard_Survival_Strategy_2014.1.pdf.

Tserennadmid, M., 2011. Pastoral practice and herders' attitude towards wildlife in South Gobi, Mongolia. MSc Thesis, Wildlife Institute of India, Saurashtra University.

Wang, S., Macdonald, D., 2006. Livestock predation by carnivores in Jigme Singye Wangchuck National Park, Bhutan. Biol. Conserv. 129, 558–565.

Wangchuk, R., Jackson, R.M., 2002. A community-based approach to mitigating livestock-wildlife conflict in Ladakh, India. Snow Leopard Conservancy–India Trust, Ladakh, India.

Zabel, A., Holm-Müller, K., 2008. Conservation performance payments for carnivore conservation in Sweden. Conserv. Biol. 22, 247–251.

CHAPTER

14

Livestock Husbandry and Snow Leopard Conservation

OUTLINE

Snow Leopards
http://dx.doi.org/10.1016/B978-0-12-802213-9.00014-6

SUBCHAPTER

14.1

Corral Improvements

*Ghulam Mohammad**, Sayed Naqibullah Mostafawi***, Jigmet Dadul*†, Tatjana Rosen*‡

***Baltistan Wildlife and Conservation Development Organization (BWCDO), Skardu, Gilgit-Baltistan, Pakistan**

****Wildlife Conservation Society, Kabul, Afghanistan**

†Snow Leopard Conservancy India Trust, Leh, Ladakh, Jammu & Kashmir, India

‡Tajikistan and Kyrgyzstan Snow Leopard Programs, Panthera, Khorog, Gorno-Badakhshan, Tajikistan

INTRODUCTION

Livestock depredation is one of the key sources of snow leopard mortality across much of its range. Depredation rates due to snow leopards and other carnivores, such as wolves, vary widely from under 1% in parts of Mongolia or China (Schaller, 1998; Schaller et al., 1994) to more than 12% of livestock holdings in hotspots in Nepal (Jackson et al., 1996) or India (Bhatnagar et al., 1999; Mishra, 1997), but they generally average 1–3% (Mallon, 1991; Oli et al., 1994; 2000; Namgail et al., 2007; Maheshwari et al., 2010, 2013; Wegge et al., 2012; Li et al., 2013; Snow Leopard Network, 2014). Snow leopards break into livestock corrals, often as a result of poor maintenance and design, killing many domestic goats and sheep and inflicting substantial economic damage to the lives of the local people affected. Local people will then often respond by killing the snow leopard and selling its parts. Predator proofing of livestock corrals has improved livestock protection (Bhatnagar et al., 1999; Jackson and Wangchuk, 2001, 2004), by significantly reducing depredation on livestock by snow leopards.

Many corrals commonly seen in snow leopard habitat do not have proper roofs and have gaps in the walls allowing a snow leopard to easily climb in. Other times, poor maintenance and local people's lack of material and financial resources to make improvements have led to collapse of some of the roof structures and walls, leaving the livestock vulnerable to predation.

Here is where predator proofing of corrals has emerged as an important conflict mitigation tool across many of the snow leopard range countries, including Afghanistan, India, Nepal, Pakistan, and Tajikistan. Predator proofing of a corral typically involves securing the corral so that no snow leopard can enter it. One style of structure that has proved effective has at least 2.4 m high stonewalls, and an open roof covered by 8×8 cm wire mesh, supported with wooden poles every 50 cm. The structure will have no windows, or if it does they will be covered in mesh wire or have tight fitting bars. There will be a single closely fitting wooden door that can be securely locked at night. Corrals can come in different sizes, depending on how big the herd is and the needs of the families. Communal corrals may be larger and accommodate as many as 700 sheep and goats.

DESIGN OF CORRALS ACROSS DIFFERENT COUNTRIES: EXAMPLES FROM TAJIKISTAN, AFGHANISTAN, PAKISTAN, AND INDIA

In Tajikistan, the international cat conservation NGO Panthera has supported the predator proofing of more than 100 corrals to date. Most of them are small and meet the needs of single families. However, 10 of them are communal and one in particular is large enough for as many as 700 sheep and goats. The gaps in the wire mesh used to cover the roof are wide enough to let the snowfall through, preventing accumulation and possible collapse of the roof. The first of the program's corrals were built in 2013, and since then no livestock depredation events have been recorded. The corrals are also too new to have suffered structural problems. As in other snow leopard range countries where predator-proof corrals are built, local communities in Tajikistan are under obligation to maintain the predator-proof corrals as well ensure that all livestock is safely secured in the corrals at night.

In the Wakhan corridor of Afghanistan, snow leopards and wolves are known to frequently predate on livestock, causing significant loss of income and threatening food security. In retaliation for these incidents, snow leopards and wolves are killed. The predation problem stems largely from the livestock corrals used in the Wakhan. The "traditional corral" used communally by families offers little protection against predators. These low-walled structures have no roof, allowing the predators easy access. When inside and surrounded by terrified livestock, snow leopards often adopt a frenzied behavior and kill multiple animals (Simms et al., 2011). The Wildlife Conservation Society (WCS) and the Wakhan-Pamir Association (WPA) have been addressing this problem through the construction of "predator-proof corrals" (Simms et al., 2011).

Since 2010, 36 of these corrals have been built in snow leopard trouble spots. The corrals have had success: there has not been a single domestic animal predated while housed in a predator-proof corral and retaliation killing has ceased in these areas. Demand for more corrals is high among the communities in Wakhan. The corrals have thick and high stonewalls and a wire mesh roof, which prevents snow leopards and wolves gaining access. The corrals are approximately 15 m long and 8 m wide, with 2 m high walls. They are built from local materials – stone, timber, and wire – and they are able to house more than 500 sheep and goats.

In India, more than 100 predator-proof livestock corrals have been built in the Ladakh region, benefiting more those communities. Families report finding snow leopard pug marks outside the improved corrals but no depredation. They report that in the past, herders and their guard dogs used to sleep at night outside the corrals to fend off potential snow leopard attacks, but this is no longer necessary because snow leopard attacks do not occur where the corrals have been predator proofed.

In northeast Pakistan, the Baltistan Wildlife Conservation and Development Organization (BWCDO) has constructed more than 30 predator-proof corrals. The average size of one corral is 15 × 6 m, and it accommodates 300 small animals. Most of the corrals are constructed by the local communities on a cost-sharing basis. Like in Tajikistan, India, and Afghanistan, there have been no complaints of snow leopard attacks on livestock in predator-proof corrals. Prior to the construction of predator-proof corrals, many attacks on domestic livestock were reported. In one instance in 2004, in the village of Hushe in the Ganche valley, a snow leopard attacked an unprotected corral and killed 18 goats and sheep. Despite the loss, the community released the snow leopard. In 2011, in Manthal, near Skardu, a snow leopard that had attacked livestock was heavily beaten by the community and later died.

MEASURING THE SUCCESS OF CORRAL IMPROVEMENTS AND DOCUMENTING PROBLEMS

Currently, only limited empirical evidence demonstrates the impact of systematic measures to improve livestock husbandry, such as through predator proofing of corrals, on depredation rates and the retaliatory killing of snow leopards. To begin with, formalized records of improved corrals are currently still lacking, despite the fact that many structures have already been predator proofed (Snow Leopard Network, 2014). As described already, none of the predator-proofed corrals has experienced a depredation – while in some cases, in the same villages, in Tajikistan and in Pakistan, the nonpredator-proofed corrals have experienced livestock losses. Although the design is different, there are some interesting parallels between predator proofing of corrals and boma improvement efforts in east Africa to deter lion predation on livestock (a *boma* is the name in Swahili for a livestock enclosure that normally is not predator-proof). Lichtenfeld et al. (2015) provide an evidence-based account of the significant impact predator-proof bomas have on reducing livestock depredation and the retaliatory killing of lions and leopards. Over a period of 10 years, there were only two recorded depredations, in both cases because the gate of the fortified boma enclosure was not properly constructed, and a leopard was able to enter through the gate and kill sheep and goats.

In other instances, in Hushe, Baltistan in Pakistan, for example, a door to a regular corral was not properly secured and the sheep and goats ran out to be killed by snow leopards and wolves. Therefore, poor oversight or maintenance could likely be the only reason that could lead to compromising the effectiveness of predator-proof corrals.

HOW TO IMPROVE THE SUSTAINABILITY OF BUILDING PREDATOR-PROOF CORRALS TO ENABLE THEIR MORE WIDESPREAD USE IN THE FUTURE

Significant funds and human resources are necessary to improve livestock corrals across the greater part of the snow leopard range (Snow Leopard Network, 2014). The key is to ensure that corral improvements benefit high-risk depredation sites and high-density snow leopard areas. There are situations when conservationists respond to the request to build a predator-proof corral without concrete proof that a snow leopard is truly responsible for the alleged depredation. This determination can be achieved through a combination of interviews of livestock owners, depredation records, and diet assessments as well as camera trap surveys of snow leopards. At times it may also be more cost efficient to predator-proof community corrals versus single-family corrals.

Although local labor can be provided for free, and materials such as rocks and wood can be sourced locally at little or no cost, the expense of purchasing and transporting wire mesh, often from China or Iran, remains high and beyond the budgets of many small mountain communities.

The other challenge is the maintenance of the corrals over time. As predator-proof corrals age and parts, such as doors, require replacement, the risk is that local people may neglect to perform repairs, eventually causing the corrals to cease being predator-proof and exposing their livestock to predation. The key is that local people understand that it's up to them and their financial resources at that point to ensure that the predator-proof corrals don't break down. It is also critical that an agreement is reached between conservation organizations and the beneficiaries of the corrals that clearly states the responsibilities of each party. The maintenance of the corrals should be the responsibility of the beneficiaries of the corrals. With these considerations in mind, predator proofing of corrals remains an effective measure for reducing conflicts and is thus worthy of investment.

SUBCHAPTER

14.2

The Role of Village Reserves in Revitalizing the Natural Prey Base of the Snow Leopard

Charudutt Mishra, Yash Veer Bhatnagar*, Pranav Trivedi*, Radhika Timbadia*, Ajay Bijoor*, Ranjini Murali*, Karma Sonam*, Tanzin Thinley*, Tsewang Namgail**, Herbert H.T. Prins*[†]

***Snow Leopard Trust and Nature Conservation Foundation, Mysore, Karnataka, India**
****Snow Leopard Conservancy India Trust, Leh, Ladakh, Jammu & Kashmir, India**
[†]Environmental Sciences Group – Resource Ecology, Wageningen University, Wageningen, The Netherlands

INTRODUCTION

Ungulates play important roles as drivers of ecosystem functions and as prey of large carnivores (Mishra et al., in press). Wild mountain ungulates such as the bharal or blue sheep (*Pseudois nayaur*), Siberian ibex (*Capra sibirica*), argali (*Ovis ammon*), and Himalayan tahr (*Hemitragus jemlahicus*) form the main prey of the endangered snow leopard, which specializes in feeding on ungulates (Jackson et al., 2010; Johansson et al., 2015). The abundance of wild ungulates in any area is the key determinant of snow leopard abundance (Suryawanshi, 2013).

The primary productivity of the dry and cold landscapes in which snow leopards occur is in general low compared to tropical and temperate rangeland systems (Jackson et al., 2010; Mishra et al., 2010; Namgail et al., 2012), and they support relatively low densities of wild ungulate prey, typically ranging from < 1 to about 5 per km^2 (Mishra et al., 2004; Suryawanshi et al., 2012; Tumursukh et al., 2015). Many mountain ungulate species occurring here are wild relatives of livestock, representing an important genetic resource. Most of their populations are in decline, and the conservation status of species such as markhor (*Capra falconeri*) (Near Threatened; IUCN Red List), tahr, and argali (Near Threatened) is a matter of immediate concern (see Chapter 4). Thus, conserving wild ungulates and enhancing their abundance is of conservation value by itself, and is also an important aspect of snow leopard conservation.

Snow leopards occur in multiple-use landscapes that are extensively and often pervasively used for livestock grazing (Jackson et al., 2010). Overstocking of rangelands with livestock can lead to competition for forage and for space with wild ungulate prey of the snow leopard (Mishra et al., 2001, 2004). Decline in populations of wild ungulates due to competition from livestock is a matter of concern in all snow leopard range countries. It is considered to be a serious threat to snow leopards in 6 of the 12 range countries, and a medium intensity threat in the other 6 (Snow Leopard Network, 2014).

In areas where wild ungulate populations are resource-limited due to excessive livestock grazing in their habitats, reducing livestock grazing pressures can potentially assist in their population recovery. However, the strong livelihood dependence of local human communities on snow leopard habitats for grazing makes it difficult to reduce livestock populations. How does one manage multiple-use snow leopard landscapes to improve the abundance of wild ungulate prey? One proven method is village reserves, which are grazing set-asides created in partnership with local communities with the idea of facilitating the recovery of wild ungulate populations.

VILLAGE RESERVES IN OPERATION

Our first village reserve was established in Spiti Valley in 1998, and it continues to be protected by the local community to this day. Research had shown that wild ungulate abundance in Spiti Valley was reduced due to overstocking of rangelands with livestock (Mishra et al., 2001, 2004). In 1998, we started working with the local community to establish a grazing set-aside in the rangelands of the village Kibber (Mishra et al., 2003). After several rounds of discussions and trust building, we conducted joint surveys with the community to select the land where a village reserve could be established. An agreement was signed with the village council whereby they agreed to stop grazing livestock in the designated area (a valley of around 500 ha at about 4500 m altitude) in exchange for compensation for lost grazing. The local communities in the region have a history of leasing out parts of their grazing land to migratory herders from other parts of the Himalayas, and we used those rates as objective standards to negotiate the extent of compensation (agreed upon at c. USD 0.425 annually for Kibber according to current exchange rate of USD 1 equaling approximately 60 INR). The compensation amount goes to the village council, and it is used for community work. Additionally, two guards were appointed locally to ensure that free-ranging livestock did not enter the reserve area. The agreement was for a period of 5 years, and it has been renewed each time it expired. The initial size of the village reserve included about 500 ha of grazing land, which was subsequently increased in 2004 to about 2000 ha, with the compensation being increased threefold.

Protection from grazing by livestock and other forms of resource extraction (grass, dung, etc.) was accompanied by a significant increase in bharal abundance in the grazing set-aside and surrounding pastures of Kibber region, perhaps due to an initial aggregational response, and what appears to have been a subsequent numerical response (Fig. 14.2.1). The response of bharal became evident by 2002, and showed a monotonic increase till 2008 after which the population has fluctuated at the higher end (Fig. 14.2.1). The abundance of bharal in the village reserve and surrounding pastures of Kibber is about five times greater today compared to the period before and up to 2 years after the establishment of the village reserve. We have continued to also monitor the livestock population of the surrounding pastures, which has remained stable since 1998 (when the village reserve was established).

How would such a local increase in wild ungulate abundance impact snow leopards? The village reserve is too small in area to meaningfully estimate snow leopard abundance. We also do not have information on relative use of the area by snow leopards over the years, given that reliable techniques and equipment (such as camera traps) for field monitoring were not yet available when we started this program. However, it is reasonable to speculate that the Kibber village reserve may have elicited a response

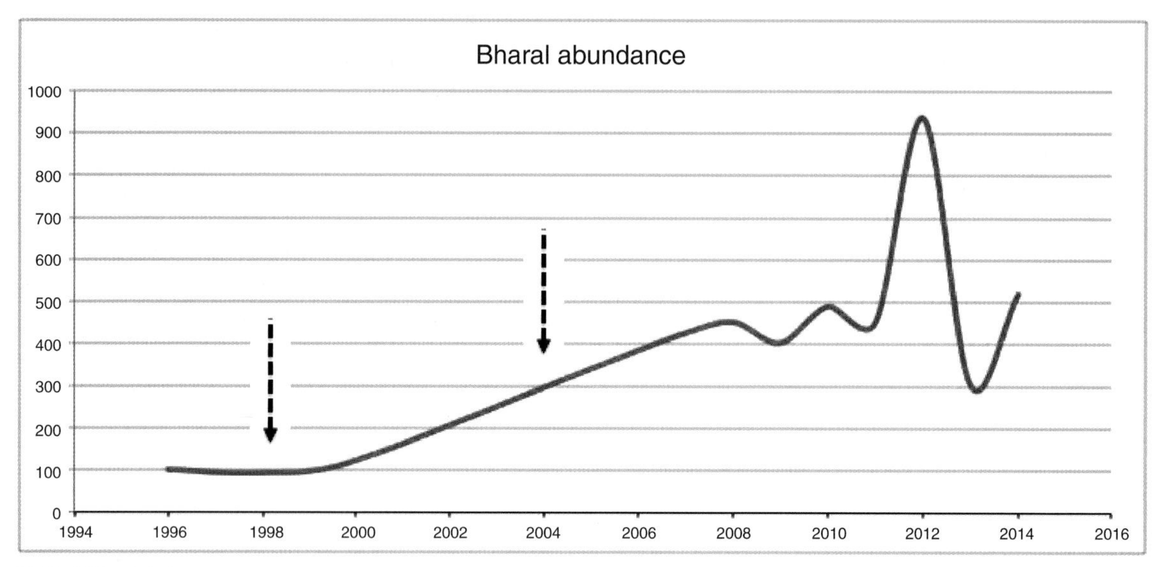

FIGURE 14.2.1 **Trends in abundance of bharal *Pseudois nayaur* in the Kibber region, Spiti Valley, India.** The population was monitored one to several times each year during spring and autumn, periods when herds are congregated. Figures presented are maximum counts for a given year. Dashed arrow on the left indicates the establishment of a 500 ha village reserve or grazing set-aside in the area, which was then expanded to 2000 ha in 2004 (arrow on the right) to cover about 20% of the total grazing land of the village. The spike in bharal population seen in 2012 was possibly caused by the temporary movement of a few herds into the area from an adjoining region (separated from the Kibber region by a deep gorge and stream).

from snow leopards. In a recent study, we assessed relative habitat use of snow leopards using camera traps in Spiti Valley (Sharma et al., 2015). We sampled snow leopards photographically in 10 study sites within Spiti (average area 70 km^2), placing 10 cameras in each site for a period of 60 days. We found that snow leopard habitat use was most influenced (positively) by the local abundance of wild ungulates. Among all 10 study sites, the site that included the Kibber village reserve (and another smaller reserve of c. 4.3 km^2 established in 2004 in the adjoining village of Chichim) recorded the highest intensity of habitat use as well as the number of snow leopards (adults) captured in camera traps.

Today, seven more villages in Spiti Valley and Ladakh are running village reserves in partnership with us. They are aimed at facilitating the recovery of various ungulate species, including bharal, urial (*O. vignei*), argali, ibex, and Tibetan gazelle (*Procapra picticaudata*). The reserves range in size from 1 km^2 to 400 km^2. The selection of any reserve area is usually based on a combination of its ecological potential (quality habitat for wild ungulates) and its topographical (for determining reserve boundaries) and administrative characteristics. The terms of agreement with the communities are variable; some involve similar compensation for lost grazing as in Kibber, others have involved provisioning of desired amenities (e.g., solar lanterns) or employment in the form of village reserve guards. In the case of the wide-ranging argali, we are experimenting with a "floating reserve" for the past 3 years, where livestock grazing is not curtailed, but the herders refrain from taking livestock to pastures where argali have been sighted recently. So far, we have not recorded an increase in ungulate abundance in any of the aforesaid reserves, though none of them have showed a decline either. Most of these village reserves are relatively recently established. In the case of the Tibetan gazelle, the reserve was set up in 2005, but recovery may have been prevented either due to the small size (1 km^2) of the reserve, due to other forms of disturbance, or due to demographic or Allee effects because this small population is on the brink of local extinction (Bhatnagar et al., 2007).

Snow leopards require large areas, with individual home ranges typically varying from 100 km^2 to more than 1000 km^2 (McCarthy et al., 2005; O. Johansson et al., unpublished data). It is often not feasible to create protected areas or desirable to exclude humans and livestock from areas that are large enough to support even a number as small as 20 adult snow leopards. Village reserves provide a potential alternative; they represent voluntarily created "core" areas with local community support in landscapes that are otherwise used by people. If planned at a landscape level (Appendix 1), a series of village reserves could actually help facilitate an increase in the abundance of snow leopards through recovery of their wild prey. Although we have developed the village reserve model to reduce competitive pressures on wild ungulates, the model could also be adapted for enabling ungulate recovery in areas where their populations are limited by hunting rather than livestock. We suggest that village reserves should be a part of a repertoire of conservation initiatives with local communities that are necessary for snow leopard conservation.

Acknowledgment

We are grateful to Van Tienhoven Foundation and Fondation Segré-Whitley Fund for Nature for supporting our work.

SUBCHAPTER

14.3

The Ecosystem Health Program: A Tool to Promote the Coexistence of Livestock Owners and Snow Leopards

*Muhammad Ali Nawaz**,**, *Jaffar Ud Din***, *Hafeez Buzdar***

*Department of Animal Sciences, Quaid-i-Azam University, Islamabad, Pakistan

**Snow Leopard Foundation, Islamabad, Pakistan

INTRODUCTION

Livelihood systems throughout the snow leopard's (*Panthera uncia*) range are predominantly agropastoral where livestock plays a central role. Indeed, the average livestock holding per household is 34 animals in northern Pakistan (goats = 47%; sheep = 31%; cattle = 19%; others = 3%). Livestock serves as the main source of cash income in such communities besides providing milk, milk products, and meat. Families sell an average of 4.4 animals each year, which constitutes some 13% of their holdings. Animals sold are usually replaced by births. However, disease and predation, which often operate in parallel, cause considerable livestock loss, affecting the local economy. The annual loss to disease in northern Pakistan ranges between 3% and 14% across different valleys, averaging 7.8% of livestock holdings. This translates into a 60% drop in disposable income from livestock sales.

Predation by snow leopards and other large carnivores is another major threat to livestock and a key cause of retaliatory killings. However, the disease death toll is estimated to be 1.5–5 times greater than that by predation. In fact, it was found that herders were more inclined to tolerate occasional losses to predation if losses to disease were reduced or controlled. It is this set of facts that motivated a conservation-based incentive program for local communities, the Snow Ecosystem Health Program (EHP), which is administrated by the Snow Leopard Foundation (SLF), an independent Pakistan conservation NGO.

The EHP promotes the peaceful coexistence of livestock owners and snow leopards through indirect compensation for livestock predation and by improving overall ecosystem health. The program is a conflict mitigation and management tool that increases both incomes and tolerance towards snow leopards by reducing disease-based livestock mortality.

The EHP is named as such because the risks associated with livestock health cannot be isolated from wildlife or human beings. Thus, an ecosystem approach to health issues is being increasingly recognized and touted as they examine animal and human health issues holistically. According to Osofsky et al. (2005), "the state of health of an ecosystem can be judged by criteria very similar to those used for evaluating the health of a person or animal, namely, homeostasis (having a balance between system components), absence of disease, diversity and complexity, stability and resiliency, and vigor and scope for growth."

PROGRAM IMPLEMENTATION MECHANISM

The EHP makes vaccines available to rural communities, empowering them economically. In return, the communities agree to limit their herd sizes – accounting for reproduction – to ensure that there is no increase in competition with wild prey. Communities are also required to refrain from poaching snow leopards and their primary wild prey. The program is implemented in seven steps:

Site Selection

Site selection is based on the presence of snow leopards and prey species, depredation pressure, and livestock movement in the area as determined by snow leopard and prey surveys and herder interviews.

Social Mobilization

Program orientation is provided through meetings with concerned communities. The community, upon entering the program, signs a *Conservation Agreement* that specifies the responsibilities of the parties (Table 14.3.1).

Training

Trainees are selected from communities in consultation with local community organizations. Centrally located and conducted by relevant experts, the trainings enhance understanding of livestock diseases, while teaching trainees how to vaccinate animals according to vaccination calendars. Trained vaccinators are referred to as *Ecosystem Health Workers (EHWs).*

Vaccine Delivery

The EHW's role is an important one. The vaccine delivery system involves determining vaccine quantity and type in consultation with vaccination calendars and local officials of the Livestock Department. Vaccines are then purchased from the production source by SLF to ensure quality and validity and given to the EHWs along with registration forms. The completed forms are later returned to the implementing agency. Considered self-employed, the vaccinators receive a small payment from each household for their services.

Vaccination Fund

A fund is created in each community and serves two purposes. First, it encourages farmers to pay for vaccines and formalizes the system of payment, procurement, and vaccination. Second, it provides a small financial buffer against minor emergencies such as unexpected disease outbreaks or increases in vaccine costs.

TABLE 14.3.1 Key Points in Conservation Agreements with Local Communities

Local community	Implementing agency
• Select persons for vaccine-administration training • Pay the community share of vaccine costs • Maintain constant herd sizes by selling animals • Record and report snow leopard and wild ungulate sightings, and report predation • Protect snow leopards and wild ungulates	• Arrange for community members to receive training in administering vaccines • Develop vaccination calendars • Provide vaccines at subsidized rates • Monitor the program and its environmental impacts

Cost Sharing

The concerned conservation agency, in this case SLF, bears the full cost of vaccines in the first year while participants begin contributing to the vaccination fund. In the second year, the conservation agency covers 75% of the cost of vaccines and provides the remaining 25% to strengthen the vaccination fund. The unpaid portion of the year's vaccination cost is borne by the participants. In the third year, half the cost of vaccines is provided by the conservation agency and the remaining 50% goes into the vaccination fund. Again, the unpaid portion of the year's vaccination is covered by the participants. Finally, in the fourth year, the participants pay for their vaccinations themselves while the agency's contribution goes into the vaccination fund. Cost sharing stops at this point.

Monitoring

The program is monitored biannually and involves implementing agency staff, the Wildlife Department, and the community organization. Monitoring entails reviewing vaccine administration, examining its impacts on livestock health and community well-being, and looking for signs of snow leopard poaching. In addition, the impact of the Conservation Agreements is assessed through periodic specialized snow leopard surveys that assess the occurrence and abundance of snow leopards.

PROGRAM SUCCESS IN RESOLVING CONFLICTS

The EHP was initiated in Kuju village in district Chitral, Pakistan, in 2003. It has been replicated in 10 other district villages and six valleys in Gilgit-Baltistan (GB) in the past 9 years (Fig. 14.3.1).

The program has trained 66 persons over the last 12 years, resulting in 2–4 active EHWs in each community who successfully vaccinated 11,000 head of livestock in Chitral and 96,000 in GB in 2014. The EHP was assessed through household structured interviews in 2008 and 2013 and focused on program effectiveness and its impact on livestock health and productivity, herders' incomes, and changes in attitudes toward snow leopard conservation. Both reviews concluded that the EHP had been a positive experience for the communities and reduced livestock mortality and improved livelihoods. These measures contributed significantly in reducing retaliatory killings of snow leopards, as evidenced by zero poaching in program communities.

REDUCTION IN DISEASE-CAUSED MORTALITY AND IMPACTS ON COMMUNITY WELL-BEING

Livestock production and productivity had been affected by the widespread occurrence of vectors and diseases prior to the EHP. The program improved both productivity and household incomes. Review of results showed that livestock losses to disease varied significantly across the program area. Overall, livestock mortality fell from 2.2 to 1.1 animals per household – a reduction of 50% through vaccination. Goat mortality constituted 82% of the loss to disease, compared to 12% for cattle and 6% for other stock.

Disease-related mortality was responsible for an average household loss of PKR 13,999 (USD 140). Comparing this figure to the average annual household cash income of PKR 96,000 (USD 960) in program villages, the loss from disease-related mortality was equivalent to 1.75 months of household income. Mortality reductions can be regarded as savings. The financial gain per household from animals sold was estimated at PKR 11,604 (USD 116). Income gained through vaccination programs therefore enables a general improvement in standards of living.

Healthy livestock are a principal source of animal protein, a basic requirement for body growth and maintenance (FAO, 1996). A major direct effect of animal diseases on human beings

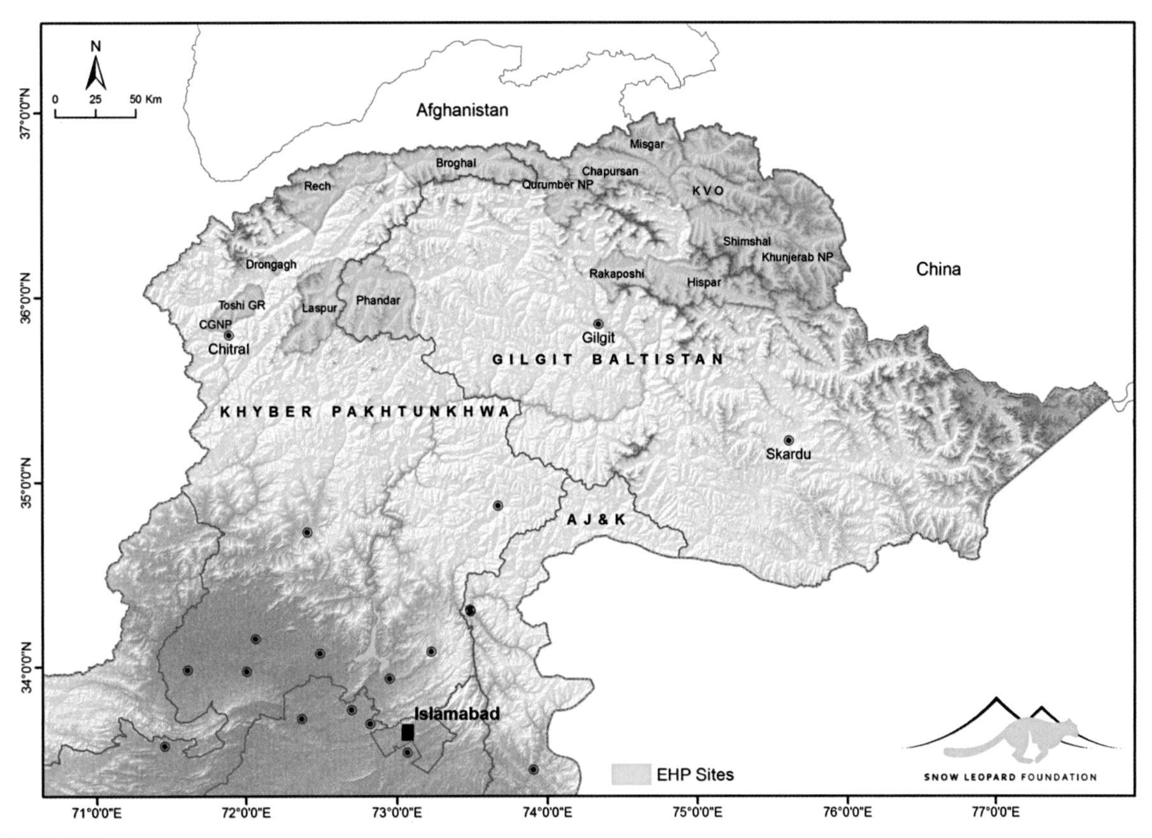

FIGURE 14.3.1 **Ecosystem Health Program operations in 2014.**

is the loss of protein and milk. The latter is particularly important for children. Animal products also supply other nutrients, minerals, and vitamins (Hush-Ashmore and Curry, 1992). The review found increases in both livestock and local consumption, an overall positive impact of vaccination services on livestock health and productivity and on human well-being.

STABILIZING HERD SIZE AND AVOIDING PRESSURE ON THE ENVIRONMENT

Herd size is largely dependent on range resources that vary considerably with altitude, aspect, season, labor availability, and cropping patterns. Households within snow leopard range traditionally sell relatively few cattle (1.8 per household). Household consumption is also low at 1.5 animals. The largest share is considered a long-term investment, especially for families possessing no other sources of income. Community Conservation Agreements stipulated that participating communities would *not* increase their livestock holdings. Compliance was evaluated through surveys where respondents were questioned about their herds. Some 72% of herders reported stable herd sizes, and another 14% felt that herd sizes had decreased. However, the remaining 13% reported increased herd sizes. Though the livestock numbers slightly increased in few household, the average herd size for the participating communities did not increase, hence the agreement was considered to be in compliance.

Livestock census data also support community perceptions about herd sizes. Average household herd size in program communities in 2013 was calculated as 26 (15 goats, 6 sheep, 5 cattle). The corresponding figure for 2008 was 36 (24 goats, 3 sheep, 6 cattle). Livestock holdings appear to be declining in many areas due to labor shortages.

In addition, communities have not reported any expansion of grazing lands in the past 5 years – herders were reportedly using existing pastures that also support the snow leopard's natural prey species.

ENHANCED TOLERANCE TOWARD SNOW LEOPARDS

Mountain communities are generally hostile toward carnivores. For example, a recent survey in Musk Deer National Park in Azad Jammu and Kashmir (AJK), where there are currently no ongoing carnivore conservation efforts, showed that 72–86% of people wanted to reduce or eliminate populations of snow leopards, brown bears, and wolves (Ahmad, 2015). Predation results in a loss of 0.3–1.1 livestock per household in EHP sites, with snow leopards reportedly responsible for the majority of livestock kills. Despite the substantial economic loss associated with predation, 83% of EHP site survey respondents reported a desire to see increased or maintained snow leopard populations in their valleys. Just 6% preferred a population decrease and 10% wanted complete elimination. These figures probably indicate higher tolerance for snow leopards in our program sites, over ecologically similar nonprogram sites. However, we have no baseline attitudinal data to document a change in our sites over time. Still, there have been no reports of snow leopard or prey species being poached in the area since the program began.

CONCLUSIONS AND RECOMMENDED PRACTICES

The EHP has been effective in helping reduce livestock mortality and improving incomes. EHW training has strengthened local capacity and helped poor herders adopt technical management options for improved livestock systems. Herd sizes are stable or declining, despite reductions in mortality, and there have been no expansions of natural pastures. Overall acceptance of snow leopards is apparent, despite the associated economic loss.

The program is cost effective and lean in terms of staff time and overall budget (approximately USD 5000 per site for the 4-year cost-sharing period). Its low input, low maintenance, and high impact clearly distinguish it from other conservation programs.

The following measures are considered good practices that can strengthen program implementation and increase the chances of success:

Strengthening Community Organizations

Community organizations must be strong enough to deliver programs. Success in this regard requires effective social mobilization, organization, and engagement. Social mobilization helps idea acceptance and adoption, paving the way for collective action. People can be brought together to form community organizations that create and spread awareness about the social and economic benefits of organized action. This allows resource pooling and collective planning and management. When required, program interventions can even be revisited.

An ongoing systematic approach is required for effective engagement with community organizations (e.g., monthly meetings). Rights and responsibilities must be clearly defined and in line with overall conservation and development goals. The focus of such exercises is to strengthen community organizations through maximum representation and create an environment of empowerment.

Establishing Vaccination Funds

Community-level vaccination funds can help program communities attain financial self-sufficiency. Such funds encourage people to pay for their own vaccines and provide savings for minor emergencies. Even though resource mobilization and savings are not new concepts, regular contact and community meetings are necessary to motivate community members to deposit their monetary shares regularly. The fund also sustains the program after donor support is terminated.

Enhancing EHW Capacity

EHW capacity and commitment are important factors in program delivery and sustainability. Selection through community organizations is based on education, experience with livestock, and staying potential. Candidates possessing a high school education – or less – are unable to assimilate trainings effectively. Conversely, candidates possessing higher education levels were found to have low staying potential because they were likely to leave the community in search of better jobs. It was found that mid-career persons possessing a matriculation qualification performed better and stayed in the community longer. Proper training means that EHWs become committed, cost-effective workers, who make efforts to convince livestock owners of the merits of vaccination. Naturally, EHWs require remuneration to maintain their motivation. This compensation is generally based on the number of animals vaccinated. The EHW role can also be expanded to provide first aid services, manage reproductive disorders, and treat injuries and common diseases. Given suitable professional training, EHWs can become breadwinners for their families, and remote communities can receive cost-effective veterinary services at their doorsteps.

Program Monitoring

The entire process demands an effective monitoring system and regular field team visits to community organizations. The following measures are recommended:

- A participatory data monitoring system involving community organizations should be developed. Variables could include livestock population, disease-caused mortality, livestock depredation, and sightings of snow leopards and prey species. Such information could be collected by the field offices on prescribed formats. Random data checks at the household level are also recommended.
- Key performance indicators to meet program requirements need to be developed in collaboration with community organizations. These would make both communities and field teams part of the process, and therefore, accountable.

References

Ahmad, S., 2015. Carnivore's diversity and conflicts with human in Musk Deer National Park, Azad Jammu and Kashmir. M. Phil Thesis. Quaid-i-Azam University, Islamabad.

Bhatnagar, Y.V., Seth, C.M., Takpa, J., Haq, S.U., Namgail, T., Bagchi, S., Mishra, C., 2007. A strategy for conservation of Tibetan gazelle *Procapra picticaudata* in Ladakh. Conserv. Soc. 5, 262–276.

Bhatnagar, Y.V., Wangchuk, R., Jackson, R., 1999. A Survey of Depredation and Related Wildlife-Human Conflicts in Hemis National Park, Ladakh, Jammu and Kashmir, India. Unpublished Report: International Snow Leopard Trust.

FAO., 1996. The State of Food and Culture. FAO, Rome, Italy. http://www.fao.org/docrep/003/w1358e/w1358e00.HTM (accessed June 15, 2015.).

Hush-Ashmore, R., Curry, J.J., 1992. Nutritional Impacts of Livestock Disease Control, ILRAD Occasional Publication. International Laboratory for Research on Animal Diseases, Nairobi, Kenya.

Jackson, R., Wangchuk, R., 2004. A community-based approach to mitigating livestock depredation by snow leopards. Hum. Dimens. Wildl. 9, 307–315.

Jackson, R., Wangchuk, R., 2001. Linking snow leopard conservation and people-wildlife conflict resolution: grassroots measures to protect the endangered snow leopard from herder retribution. Endangered Species Update 18, 138–141.

Jackson, R., Ahlborn, G., Gurung, M., Ale, S.B., 1996. Reducing livestock depredation losses in the Nepalese Himalaya. In: Timm, R.M., Crabb, A.C. (Eds.), Proceedings of the Seventeenth Vertebrate Pest Conference. University of California, Davis, CA, USA.

Jackson, R., Mishra, C., McCarthy, T.M., Ale, S., 2010. Snow leopards, conflict and conservation. In: Macdonald, D., Loveridge, A. (Eds.), Wild Felids. Oxford University Press, pp. 417–430.

Johansson, Ö., McCarthy, T., Samelius, G., Andrén, H., Tumursukh, L., Mishra, C., 2015. Snow leopard predation in a livestock dominated landscape in Mongolia. Biol. Conserv. 184, 251–258.

Li, J., Yin, H., Wang, D., Jiagong, Z., Lu, Z., 2013. Human-snow leopard conflicts in the Sanjiangyuan Region of the Tibetan Plateau. Conserv. Biol. 166, 118–123.

Lichtenfeld, L., Trout, C., Kisimir, E., 2014. Evidence-based conservation: predator-proof bomas protect livestock and lions. Biodivers. Conserv. 24, 483–491.

Maheshwari, A., Sharma, D., Sathyakumar, S., 2013. Snow leopard (*Panthera uncia*) surveys in the Western Himalayas, India. J. Ecol. Nat. Environ. 5 (10), 303–309.

Maheshwari, A., Takpa, J., Kujur, S., Shawl, T., 2010. An investigation of carnivore-human conflicts in Kargil and Drass areas of Jammu and Kashmir, India. Report submitted to Rufford Small Grant, WWF India, 30 p.

Mallon, D.P., 1991. Status and conservation of large mammals in Ladakh. Biol. Conserv. 56, 101–119.

McCarthy, T.M., Fuller, T.K., Munkhtsog, B., 2005. Movements and activities of snow leopards in Southwestern Mongolia. Biol. Conserv. 124, 527–537.

Mishra, C., 1997. Livestock depredation by large carnivores in the Indian Trans-Himalaya: conflict perceptions and conservation prospects. Environ. Conserv. 24, 338–343.

Mishra, C., Bhatnagar, Y.V., Suryawanshi, K.R., in press. Species richness and size distribution of large herbivores in the Himalaya. In: Ahrestani, F., Sankaran, M. (Eds.), The Ecology of Large Herbivores in South and Southeast Asia. Springer, The Netherlands.

Mishra, C., Prins, H.H.T., Van Wieren, S.E., 2001. Overstocking in the Trans-Himalayan rangelands of India. Environ. Conserv. 28, 279–283.

Mishra, C., Allen, P., McCarthy, T., Madhusudan, M.D., Bayarjargal, A., Prins, H.H.T., 2003. The role of incentive programs in conserving the snow leopard. Conserv. Biol. 17 (6), 1512–1520.

Mishra, C., Van Wieren, S.E., Ketner, P., Heitkonig, I.M.A., Prins, H.H.T., 2004. Competition between domestic livestock and wild bharal *Pseudois nayaur* in the Indian Trans-Himalaya. J. Appl. Ecol. 41, 344–354.

Mishra, C., Bagchi, S., Namgail, T., Bhatnagar, Y.V., 2010. Multiple use of Trans-Himalayan rangelands: reconciling human livelihoods with wildlife conservation. In: Du Toit, J., Kock, R., Deutsch, J. (Eds.), Wild Rangelands: Conserving Wildlife While Maintaining Livestock in Semi-Arid Ecosystems. John Wiley & Sons Ltd, Chichester, UK.

Namgail, T., Fox, J.L., Bhatnagar, Y.V., 2007. Carnivore-caused livestock mortality in Trans-Himalaya. Environ. Manage. 39 (4), 490–496.

Namgail, T., Rawat, G.S., Mishra, C., Van Wieren, S.E., Prins, H.H.T., 2012. Biomass and diversity of dry alpine plant communities along altitudinal gradients in the Himalayas. J. Plant Res. 125, 93–101.

Oli, M.K., 1994. Snow leopards and blue sheep in Nepal: densities and predator:prey ratio. J. Mammal. 75, 998–1004.

Osofsky, S.A., Kock, R.A., Kock, M.D., Kalema-Zikusoka, G., Grahn, R., Leyland, T., Karesh, W.B., 2005. Building support for protected areas using a 'One Health' perspective. In: McNeely, J.A. (Ed.), Friends for Life: New Partners in Support of Protected Areas. IUCN, Gland, Switzerland, and Cambridge, United Kingdom, pp. 65–79.

Schaller, G.B., 1998. Wildlife of the Tibetan Steppe. University Chicago Press, Chicago.

Schaller, G.B., Tserendeleg, J., Amarsanaa, G., 1994. In: Fox, J.L., Jizeng, Du (Eds.), In: Proceedings of the Seventh International Snow Leopard Symposium. International Snow Leopard Trust, Seattle, pp. 33–42.

Sharma, R.K., Bhatnagar, Y.V., Mishra, C., 2015. Does livestock benefit or harm snow leopards? Biol. Conserv. 190, 8–13.

Simms, A., Moheb, Z., Salahudin, Ali, H., Ali, I., Wood, T., 2011. Saving threatened species in Afghanistan: snow leopards in the Wakhan Corridor. Int. J. Environ. Stud. 68 (3), 299–312.

Snow Leopard Network (SLN). 2014. Snow leopard survival strategy. Revised version 2014.1. www.snowleopardnetwork.org.

Suryawanshi, K.R., 2013. Human carnivore conflicts: understanding predation ecology and livestock damage by snow leopards. Ph.D. Thesis. Manipal University, India.

Suryawanshi, K.R., Bhatnagar, Y.V., Mishra, C., 2012. Standardizing the double-observer survey method for estimating mountain ungulate prey of the endangered snow leopard. Oecologia 169, 581–590.

Tumursukh, S., Suryawanshi, K.R., Mishra, C., McCarthy, T.M., Boldgiv, B., 2015. Status of the mountain ungulate prey of the endangered snow leopard in the Tost Local Protected Area, South Gobi, Mongolia. *Oryx*, 1–6.

Wegge, P., Shrestha, R., Flagstad, O., 2012. Snow leopard *Panthera uncia* predation on livestock and wild prey in a mountain valley in northern Nepal: implications for conservation management. Wildl. Biol. 18 (2), 131–141.

A. APPENDIX 1 SUGGESTED PRINCIPLE FOR LANDSCAPE-LEVEL PLANNING OF VILLAGE RESERVES (MISHRA ET AL., 2010)

Within a multiple use landscape matrix, it is desirable to have as many village reserves as possible. The guiding principle for recovery and management of wild ungulate population can be as follows:

1. In village reserves, the aim is to facilitate and maintain wild ungulate populations (N_v) at highest potential abundance (K), and enable conditions where birth rates (b_v) exceed mortality rates (m_v), and rates of emigration (e_v) are considerably higher than immigration rates (i_v) to enable spillover effects, that is (Eq. 1):

$$N_v \approx K, b_v > m_v, \quad \text{and} \quad e_v \gg i_v$$

2. For the intervening landscape units between any two village reserves, it is conceptually useful to estimate the desirable wild ungulate population size (N_m) – which will be a function of the trade-off between conservation and rangeland use objectives – and ensure that populations are maintained around that level (Eq. 2):

$$N_m = K - f(A), \quad \text{and} \quad b_m + i_m \geq m_m + e_m$$

where $f(A)$ is a function by which the wild ungulate population size is reduced below carrying capacity as a result of an acceptable level of anthropogenic pressure.

The size and number of village reserves should be large and adequately interspersed within a matrix of multiple-use landscape units to enable the conservation of viable wildlife populations. At a minimum, the coupled landscape-level guiding principle for village reserves and multiple-use landscape units should be to aim for the total spillover from village reserves to at least offset the net individuals lost from multiple-use units due to mortality and emigration, that is (Eq. 3):

$$\sum N_v (e_v - i_v) \geq \sum N_m (b_m - m_m - e_m)$$

This assumes that as the livestock grazing intensity in a multiple-use landscape unit increases, one can expect a decline in the density of wild ungulates. It will need to be counterbalanced by establishing a suitable village reserve in the proximity such that the inequality condition above continues to hold.

CHAPTER

15

Religion and Cultural Impacts on Snow Leopard Conservation

OUTLINE

Snow Leopards
http://dx.doi.org/10.1016/B978-0-12-802213-9.00015-8

SUBCHAPTER

15.1

Introduction

Betsy Gaines Quammen
Yellowstone Theological Institute, Bozeman, MT, USA

Snow leopards live in vast, rugged landscapes that also support some of the most remote human cultures also in the world. And, like the biodiversity that resides in these most difficult of climates and landscapes, the human communities have adapted to their environments culturally in amazing and unique ways. People have learned to interpret the divine in their geography, wildlife, and other natural elements. Their religions offer a special understanding of landscape–mountains, lakes, and animals are often characters in sacred stories. Whether communities adhere to Buddhism, Hinduism, Islam, shamanistic practices or a syncretistic blend of traditions, many of the people inhabiting the realm of the snow leopard see the world through their religion and religion through their world.

In order for us to protect the last members of the Earth's threatened species, it is imperative for us to understand the worldviews of people who live with vulnerable wild populations. From a practical point of view, it is of course imperative to determine conflicts between wildlife, humans (or livestock), and work to remedy those struggles. But gaining a further understanding of the local religion, rituals, and stories, we learn something even more effective, a clue that culture reinforces the importance of that species, as many religions view various species as sacred.

Of course religion can be practiced in many ways–ways that might harm species and ways that might encourage species protection. For centuries, Taoism encouraged the use of medicines that contain the parts of rare animals; however in the last few years, Taoist priests in China have begun to urge the discontinuation of threatened species in traditional medicine. By working with these religious leaders, the British organization Alliance of Religions and Conservation, is helping to slow poaching through relationship building, education, and reform.

When I discussed my personal interest in the power of religious belief with Dr George Schaller several years ago, he told me a story of his own experience connecting wildlife protection with local traditions. When he visited homes in his study area on the Tibetan Plateau, he gave his hosts a picture of Milarepa, the Buddhist ascetic who abhorred hunting and the eating of meat. This sometimes led to a discussion on the tenets of Buddhism and the implications of killing of snow leopards and other wildlife.

But as I've said, religions inspire multiple interpretations. During another conversation I had with Dr Alan Rabinowitz about Buddhism and the idea of the sacred nature of sentient beings, Alan expressed his frustrations with Southeast Asian Buddhists' insistence that it is karmically appropriate for a leopard that kills prey to be poached as a consequence. When I shared Alan's frustration with my friend Dekila Chungyalpa, the former director of the World Wildlife Fund Sacred Earth Initiative, she told me about a meeting with His Holiness the Dalai Lama during which he jokingly asked her if it was possible to teach tigers to eat a vegetarian diet.

Religions are stories. They are mirrors of culture. They are allegorical, not quantitative. When considering science and religion, Dr Stephen Jay Gould described these two ways of interpreting the world as "nonoverlapping magisteria." They can reside together. And when 85% of the people on this planet practice a faith of some sort, it is important to reach out and understand those traditions (ARC http://arcworld.org). In understanding spiritual practices, we find cultural frameworks that lead to unique methods in species protection.

Beliefs can be supportive of protecting nature depending on the way sacred text is read, interpreted, and practiced. This is where education and relationship building are imperative. During my work with religious leaders on conservation projects, I have not heard anyone say they aren't interested in helping. When priests, rabbis, and lamas are asked for their help and leadership, I have always found them to be willing to collaborate, and often deeply grateful for being asked. A religious leader who has information on species protection and the delicate balance of ecology often becomes a great partner in protecting wildlife.

Scientists often approach the world in a measured and analytical way. They have data to offer and numbers to back their positions. But in reaching the people who live with the snow leopard, religions have something much more forceful and persuasive, as readers will surely come to understand as they consider the four essays in this chapter. Religions have stories, dances, art, songs, and passion. If a scientist explains the parameters of a viable population of snow leopards to a herder in Tibet, it will mean nothing. But if a monk reminds the herder that Milarepa turned himself into a snow leopard during a snowstorm, next time the herder sees a leopard, he will see Milarepa. And that snow leopard may live to see another day.

SUBCHAPTER

15.2

Tibetan Buddhist Monastery-Based Snow Leopard Conservation

Juan Li, Hang Yin**, Zhi Lu*,***

***Center for Nature and Society, College of Life Sciences, Peking University, Beijing, China**

****Shan Shui Conservation Center, Beijing, China**

INTRODUCTION

Nature worship is pervasive in traditional religions and cultures that originated in Central and South Asia, including Buddhism, Daoism, Hinduism, Jainism, Shinto, Sikhism, and Zoroastrianism. These religions regard nature as a critical aspect of divinity that should be treated with reverence (Nasr, 1996), thus they play an important role in biodiversity conservation, which comes mainly in two forms. First, it is a tradition to protect sacred species and sacred natural sites around religious structures. Second, the religions indirectly protect the biodiversity by shaping their followers' view and behavior toward nature (Dudley et al., 2009).

As a major branch of Buddhism – one of the primary religions in Asia, Tibetan Buddhism teaches that all life is connected and interdependent, and respect and compassion should be shown for all living beings (Dorje, 2011). It has been practiced in the Tibetan Autonomous Region in China; Mongolia; Kalmykia, Buryatia, and Tuva in Russia; Bhutan, northern Nepal, and northern India for hundreds to thousands of years (Harvey, 2012) (Fig. 15.2.1). Tibetan sacred sites include sacred mountains, lakes, relics, forbidden areas, and pilgrim routes, and it has been shown that they may contribute significantly to biodiversity conservation (Anderson et al., 2005; Salick et al., 2007). Among them, Tibetan sacred mountains have the largest land cover, they are administered by monasteries and local communities, and Tibetan Buddhist monasteries are usually located near the sacred mountains and in the center of traditional Tibetan communities (Shen et al., 2012). They are not only the centers of culture and religion, but also important participants in environmental protection. They organize patrolling around the sacred mountains and educate the communities on environmental issues.

We find Tibetan Buddhism is crucially involved in snow leopard conservation due to its substantial overlapping range with snow leopards, particularly in the portion of the species' range where mainstream conservation strategies face many difficulties (Li et al., 2014).

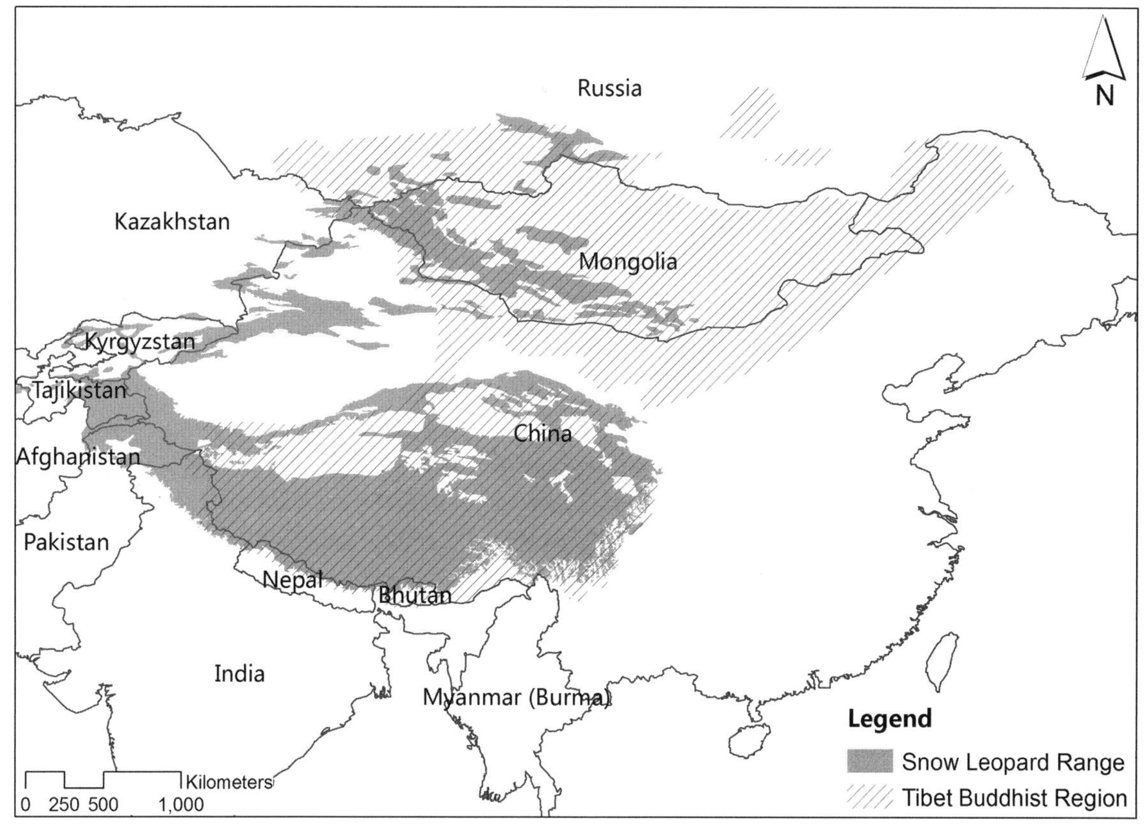

FIGURE 15.2.1 **Global range of snow leopards and Tibetan Buddhism.** The regions under the influence of Tibetan Buddhism could cover about 80% of global snow leopard range (see Chapter 3). *(Adapted from Encyclopedia Britannica, 2003, Atlas of Faiths).*

To be specific, the nature reserves only cover 0.3–27% of snow leopard habitat in the snow leopard-range countries except Bhutan (57%), and most of them are highly understaffed (McCarthy and Chapron, 2003). Incentive programs have been successful in addressing human-snow leopard conflict in several sites, but are difficult to replicate at a large scale for lack of baseline ecological and socioeconomic data (Mishra et al., 2003). In areas where such mainstream strategies are not practical, Tibetan Buddhism may be providing a crucial complementary strategy for environmental protection.

CONNECTIONS BETWEEN TIBETAN BUDDHISM AND SNOW LEOPARDS

Snow leopard global range overlaps substantially with the area of Tibetan Buddhism influence, including the whole Tibetan Plateau, part of Mongolia and Russia, Bhutan, Nepal, and northern India (Fig. 15.2.1). Snow leopards might have originated in the hinterland of Qinghai–Tibetan Plateau 3 million years ago, where they evolved to adapt to the cold and high-elevation environment, and then gradually spread to the surrounding mountains (Deng et al., 2011). Tibetan Buddhism originated five centuries BCE in northern India and later spread to the whole Tibetan Plateau and surrounding area. Although separated by millions of years, snow leopards and Tibetan Buddhism followed similar paths of range expansion and ultimately reached similar areas. More interestingly, snow leopards and Tibetan Buddhist monasteries selected similar sites on a smaller scale. Both of them usually select sites or habitats backed with high rugged mountains and containing rivers (Gyatsho, 1979). The snow leopard is adapted to its particular ecological niche in the rugged mountains with shorter limbs and strong chest muscles, which are the result of long-term evolution. Whereas, Buddhism stresses that monasteries should look like a lotus when viewed from the sky, so they prefer to build on the rugged mountains with large white rocks as a background.

In the process of long-term coexisting, Tibetan Buddhism also developed cultural connections with snow leopards. It is recorded in the Buddhist scriptures that snow leopards own the Rocky mountains, they are the leader of all carnivores and one of the protectors of the sacred mountains. These legends demonstrate that snow leopards have a sacred place in Tibetan Buddhism.

SCIENTIFIC STUDY OF MONASTERIES' ROLE IN SNOW LEOPARD CONSERVATION

To further explore the role of Tibetan Buddhist monasteries in snow leopard conservation, we selected the heart of global snow leopard habitat, the Sanjiangyuan Region in Qinghai Province, China, as our study area. The Sanjiangyuan Region extends over 360,000 km^2 and includes the 150,000 km^2 Sanjiangyuan National Nature Reserve. Tibetans account for 90% of the population living here, and they follow Tibetan Buddhism. At least 336 monasteries are formally recorded by the government in this region (Pu, 1990).

Using snow leopard presence points and relevant environmental variables (elevation, ruggedness, land cover, etc.), we built a snow leopard distribution model in the Sanjiangyuan Region. We found that 90% of monasteries were located within 5 km of snow leopard habitat, and 46% were located in the snow leopard habitat (Fig. 15.2.2). The distance of monasteries to snow leopard habitat is significantly lower than random points ($p = 0.000$). This fit well with the tight natural connections between Buddhist monasteries and snow leopards we described earlier. We also found that the sacred mountains around monasteries are estimated to protect 8342 km^2 of snow leopard habitat, while the core zones of Sanjiangyuan Nature Reserve only cover 7674 km^2. Furthermore, the Sanjiangyuan Nature Reserve contains only 21 conservation stations, with only 2–3 employees per station. In contrast, the 336 monasteries routinely organize active patrols around their sacred mountains. Hence monastery-based conservation might provide effective protection within snow leopard habitat in the Sanjiangyuan Region, especially in the most remote parts. Monastery-based conservation has other advantages for snow leopards. The rugged mountains that snow leopards inhabit usually serve as administrative boundaries. It is difficult for governments to manage these areas effectively due to boundary issues. In contrast, the influence of monasteries can cross-administrative boundaries.

Besides the organized patrols, monasteries also provide environmental education to the local communities. To understand the influence of Tibetan Buddhism on local people's hunting behavior, we did 144 semistructured household interviews across the Sanjiangyuan Region during 2009–2011. Most of them reported that they did not kill wildlife. Out of 144 interviewees, 34 people claimed that they did not kill wildlife because it is prohibited by the government; 25 people said that they did not kill wildlife

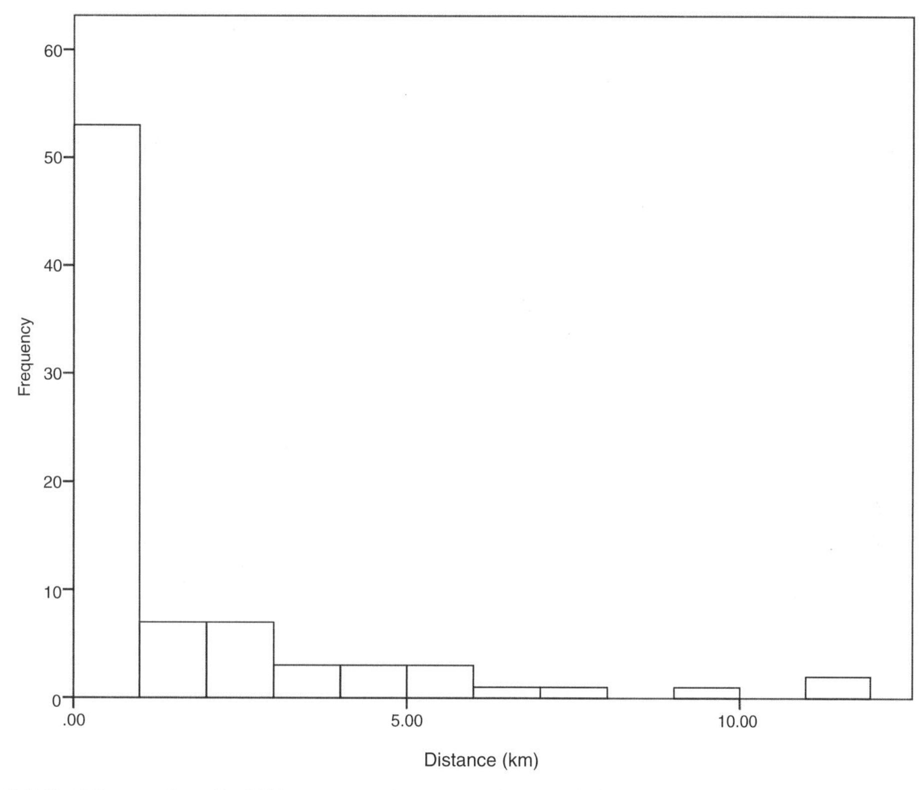

FIGURE 15.2.2 **Distance from Buddhist monasteries to snow leopard habitat in Sanjiangyuan Region.** Ninety percent of monasteries were located within 5 km of snow leopard habitat, 65% were within 1 km, and 46% were in snow leopard habitat.

because it was a sin in Buddhism; 35 people mentioned both reasons. Altogether 60 people attributed their nonkilling behavior to Buddhism, accounting for 42% of those interviewed. This seemed to indicate a strong religious influence on local people's behavior, which should play an important role in protecting snow leopards and other sympatric wildlife.

PILOT CONSERVATION PROJECTS COOPERATING WITH MONASTERIES

Since 2011, Shan Shui Conservation Center and Peking University, in collaboration with Panthera and the Snow Leopard Trust (SLT), have cooperated with four monasteries in the Sanjiangyuan Region. All four monasteries had previously done a lot of work on environmental conservation, including garbage collection, tree planting, patrolling sacred mountains, and so on. Monastery leaders readily accepted our offer to cooperate on snow leopard conservation. We provided them with funding to buy hiking boots, binoculars, cameras, and GPS units for patrollers. We also provided funds to compensate those households who lost livestock to snow leopards and other carnivores, and to make promotional materials related to snow leopard conservation. We trained monks in basic snow leopard survey methods, including identification of scrapes and scats in the field, and how to monitor blue sheep scientifically. As part of the collaboration, we requested the senior khenpo highlight the importance of snow leopards in the yearly ceremonies. We also distributed snow leopard posters and publicized the law of the People's Republic of China on the protection of wildlife at the same time.

In the process of this pilot conservation project, one problem we encountered was that monasteries and local communities do not have legal rights to evict illegal miners or poachers from their sacred sites, especially those from outside the community. We suggested that local government and nature reserves could confer some management rights to monasteries and communities to help them stop the illegal activities. Another suggestion we made was for nongovernmental organizations and local governments to train the monks and community members in wildlife monitoring techniques and ecology, to supplement their traditional conservation practices.

FUTURE PROSPECTS

Although this monastery-based conservation strategy has many advantages, a concern is that young people are losing faith in religion under the impact of modernization. In the process of modernization, the traditional religious beliefs are challenged and might be gradually substituted by reason and science. But one study indicated that Tibetan young people did not differ from their elders in use and appreciation of sacred sites (Allendorf et al., 2014). This question is complicated and still needs further investigation.

Another worry is that monasteries also shelter other animals like stray or feral dogs in addition to snow leopards. We recorded that there were more than 500 stray dogs in one monastery in Nangqian County in 2011. Although these dogs were fed by monks, they would still form packs to prey on blue sheep, the main food source of snow leopards. Feral or stray dogs may also indirectly harm snow leopards and other wildlife as disease reservoirs (Young et al., 2011). Further investigation is needed to understand the potential influence of the stray or feral dogs.

In summary, Tibetan Buddhist monasteries could have already contributed substantially to the snow leopard conservation given their tight connections. We have also shown the feasibility of cooperating with monasteries on snow leopard conservation in Qinghai province in China. So we suggest monastery-based conservation as a snow leopard conservation strategy could be extended to other areas under the influence of Tibetan Buddhism, which may cover as much as 80% of snow leopard range, and the whole mountain ecosystem will also be protected.

SUBCHAPTER

15.3

Shamanism in Central Asian Snow Leopard Cultures

Apela Colorado, Nargiza Ryskulova***

***Worldwide Indigenous Science Network, Lahaina, HI, USA**

****Worldwide Indigenous Science Central Asian Consultant, Bishkek, Kyrgyzstan**

"Out of sense of responsibility before our ancestors and future generations, we indigenous cultural practitioners stress the central role of snow leopard in the survival of humanity facing a civilization crisis, which threatens our mountains with the cold breath of the death." ***-Japarkul Raimbekov, Guardian of the sacred snow leopard site of Arashan Mountain.***

Science is increasingly looking to indigenous wisdom to extend baseline data, provide personal witness, raise new questions, and answer critical environmental issues for which science cannot find answers. "The snow leopard is a protector of sacred mountains, a unifying force and a source of spiritual power and wisdom. A link between the spirit and natural world, the snow leopard draws the attention of shamans in preserving its habitat." (Phalnikar, 2014).

What is a shaman and/or Indigenous Cultural Practitioner (ICP) and what is the relevance of indigenous wisdom to conservation? The term "shaman" derives from anthropology and religious studies and was meant as a metaconcept, an overarching term, and refers to a holy person with religious, magical powers that are able to interface the spiritual and material world. But the term is problematic. Although these terms have opened a place for indigenous wisdom with the Western mind, it has simultaneously reframed the indigenous way of knowing to suit the Western paradigm. Indigenous people became viewed as fodder for research. The ancient, earth-based way of knowing and being was deprecated, often reduced to overworked ideas of New Age Shamanism, thus losing all specificity – its strength.

The deepening of our global crises has brought conservationists and indigenous cultural practitioners (ICPs) an unexpected and paradoxical possibility, linking Western science with its direct opposite – indigenous science, a partnership that may save the snow leopard and its habitat. Until recent times, science and conservation have dismissed the "science"[a] of people indigenous to snow leopard habitat and looked at ancient knowledge and practices as anecdotal,

[a]Terms such as *traditional knowledge, local knowledge,* and even *traditional ecological knowledge,* if used to define the entire indigenous way of knowing and being is inaccurate and maintains western scientific colonialism. Knowledge can be extracted and exploited whereas pluralism in science necessitates communication, sharing of resources and relationship. The Science of Indigenous people addresses multiple dimensions while simultaneously secures data based on observation.

irrelevant, or even superstitious. At the Rio Earth Summit in 1992, "everybody was talking about the ecological knowledge of indigenous people, but certainly no one was talking about the spiritual origin of it as claimed by indigenous people themselves. ... We, scientists, were not talking about it because we were afraid we would not be taken seriously" (Narby, 1998). The result is a hidden exploitation. Data are taken from the community, empirically confirmed and used, but their origin cannot be discussed because it contradicts the axioms of science.

This perspective is a luxury we can no longer afford. As early as 1987, the United Nation's Brundtland Report, and in 1992, Article 21 of the Rio Earth Summit specifically called for researchers to take indigenous knowledge into account. Informally, this is happening. Conservationists have included local knowledge in their research, and increasingly enlist the support of communities to halt predation and increase protection. Unfortunately, the relationship has been built on a foundation of separation and usurpation of indigenous wisdom. Indigenous people are seen and used as a source of data but are not included as coequal partners. Also, indigenous wisdom has been vetted through Western lenses without any parameters or checks for reliability, vigor, or accountability. The monocultural scientific paradigm is failing to protect the snow leopard, yet without meaningful participation in conservation projects and policy formation, indigenous science is likewise limited.

Indigenous science is embedded wisdom. In *The Lost World of the Kalahari*, Van der Post (1958) proposes the term "Great Memory" for the capacity of indigenous peoples to remember events in the parahistory or great cycles of time. "Learning by analogy from the narrated oral history or story is more than mere historical records: it is equally the deep-rooted awareness of the natural laws of life itself. ... On the one hand, 'memory' means a process of learning, by rote, tribal history preserved through the centuries. On the other hand, together with this highly specialized memorizing art, there ... is another form of 'memory' – something approximating cross cultural or 'archetypal' memory, and the ability to bring the past into the present, along with the future (Deloria, 1999). The Great Memory involves more than the oral tradition of storytelling, which is a cultural outgrowth of it. It is a faculty, 'synonymous with a heightened, or deepened level of consciousness.'

In a 2003 interview with National Public Radio and the National Geographic Society, anthropologist and explorer Wade Davis elaborated, "Just as there is a biological web of life, there is also a cultural and spiritual web of life the 'ethnosphere.' It's really the sum total of all the thoughts, beliefs, myths, and institutions brought into being by the human imagination. It is humanity's greatest legacy, embodying everything we have produced as a curious and amazingly adaptive species. The ethnosphere is as vital to our collective well-being as the biosphere. And just as the biosphere is being eroded, so is the ethnosphere – if anything, at a far greater rate."

Indigenous science is a repository of data gleaned across the millennia and accrued in intimate association with the land, but its primary power lies in the propensity or ceremony, to access the Great Memory or ethnosphere. Chagat Almashev, PhD, Altaian, observes: "According to the concept of Altai people, information, knowledge, and wisdom are existing all around. When people, elders, deal with information, they use some verbs or notions to note these sacred objects. For instance, 'belem' means knowledge or information is existing around, everywhere. In order to catch this information, they use the verb 'ailanderar' ... make things go around and find the proper one, to look for any solution or problems or anything. It's a silence sitting around fire, talking – it's like virtual travel somewhere. It's a typical ceremony for ICPs. To travel and see things even the future they say 'belge.' It's all 'belim.'

What is it that enables the multidimensional functionality of indigenous science? How is it that a way of knowing can heal, communicate across species and "remember" knowledge in vast cycles of time? In China, the clarity that comes from the ability to decipher underlying meanings of symbols, including natural phenomena, is the "light of the gods" and brings the individual soul into harmony with the cosmos. "The principle behind divination known as synchronicity recognizes an essential link between inner and outer reality. ... The dynamic interaction between complementary pairs of opposites is believed to create images that mirror the structure of the human psyche and of life itself. The dialectic operates by exploiting contradictions against a common ground of similarity rather than by, linear logic. A dramatic illustration of indigenous science nexus is Prasena, a Tibetan practice that allows the 'unconscious mind to project visionary images onto the surface of a mirror' (Hope, 1997).

Altaian shaman Danil Mamyev puts it this way: "Indigenous knowledge never speaks in direct language, but it is important to know the meaning underneath." Shamans use metaphor–ways of thinking about one thing in terms of another (Bishkek Notes, p. 1, 9/14/13). Such practices permit the use of sensory, emotional, and cognitive information. The "image-based right hemisphere of the brain comes to dominate the left where most language processing takes place" (Hope, 1997). The combination of various mental states and the movement between them generates creativity, imagination, and access to other dimensions of knowing.

Japarkul Raimbekov, Kyrgyz indigenous cultural practitioner (ICP) and Guardian of the sacred snow leopard site of Arashan Mountain, elaborates on indigenous research protocols: "I started my spiritual journey 18 years ago. My vision and mission became to tell people about Altyn Dor, the Golden Age, the renewal of humanity. About five or so years ago, I met Apela [Colorado]. This fateful meeting happened on the sacred site 'Arashan' where my vision and Apela's became one. It was about Uluu Ot, the ancient fire of love that united all living beings. As I was speaking of my vision, the Kasiet – or aura – I perceived around her transformed into light. The spirit of the snow leopard manifested. This was exactly as it had happened in ancient times and so connected this modern time with the past. First came the fire, Uluu Ot, then the spirits of all animals and then humans. In this way we brought back the connection of humanity to the fire and to the spirits of the animals.

"Kyrgyz people don't use the term "shamans"; we have other terms, such as jaachy, kuuchu, bakshy. Each of these people has different abilities. Fire was the main tool for connecting to spirituality. They lit the fire, they called out to the ancestors and to the spirits of animals. Each animal also has a spirit animal. The highest of the animals for us was a snow leopard that lived on the top of the mountain and looked upon people and connected us to higher worlds. This link between humans, animals, and nature was very strong and progressive thanks to bakshys who renewed and held that connection. Their rituals and traditions instilled in people the order of the things. So we don't have to invent anything. We just have to revive things as they were, remember we are one with Nature and with all living beings, and that all things are creations of love and spirit, humans as well (interview and translation by Nargiza Ryskulova, November 2014).

SNOW LEOPARD WORK BRINGS THE SCIENCES TOGETHER

"This is what we've been looking for, a place where science and spirituality come together."
-Rodney Jackson, snow leopard scientist & founder, Snow Leopard Conservancy

There is an increasing recognition that cultural and biological diversity are deeply linked and that conservation programs should take into account the ethical, cultural, and spiritual values of nature.

Kyrgyz oral history records a time and sacred fire that unified people. In 2010, in furtherance of the vision at the sacred snow leopard site of Arashan, Kyrgyz elders reignited Uluu Ot[b] to call upon the spirit of the snow leopard, *"protector of sacred mountains, a unifying force and a source of spiritual power and wisdom"* (Tedlock, 2005). More than 50 ICPs from around the world lit ceremonial fires in their own lands at the same time. We did so to reawaken and renew our relationships, our science, and to create an ongoing forum to achieve these ends.

Three years after the sacred fires were lit, snow leopard ICPs and scientists met in Ashu, Kyrgyzstan, to draft a policy statement to present to a global snow leopard meeting. Given the infrastructure needs of such a high-profile meeting, one would imagine Russia or China would be a more likely venue. Instead, the Global Snow Leopard Conservation Forum met in Kyrgyzstan and there endorsed the Bishkek Declaration on Snow Leopard Conservation and the Global Snow Leopard and Ecosystem Protection Program (GSLEP). Prior to this meeting, the Worldwide Indigenous Science Network partnered with the Snow Leopard Conservancy to assemble an alliance of snow leopard cultural practitioners drawn from several Central Asian countries. Guardians made use of the convening to formulate their own global conservation strategy calling for a deep networking of sacred sites, shamans, and sacred species first within the cultural frame, and secondly working with scientists and conservationists to formulate new biocultural approaches in nature conservation and ecosystem management. The outcome of their meeting was the development of "The Indigenous Cultural Practitioners Statement to the Global Snow Leopard Conservation Forum."

> "The rock carvings transfer consciousness of what survives disaster. We must get into that mind set, to comprehend the ancient messages with our minds." ***-Ilarion Merculief, Aleut***

On October 23, 2013, history was made. Japarkul Raimbekov delivered the ICP statement to the global assemblage. Indigenous wisdom contributed to the critical plan for the survival of snow leopards in the wild. At the conclusion of the Ashu meeting, Japarkul led our group of ICPs and conservationists to Chopan Ata, a 42-ha, boulder-strewn, petroglyph site on the shores of Lake Issyk Kul. Conventional wisdom holds that ceremonial ways of the snow leopard are lost; as Japarkul was about to show us, this view is false. The midday August sun beat down upon us as we trekked from our bus to a massive boulder, more than a meter across (Fig. 15.3.1). Facing the stone, Japarkul raised his serpent staff ornamented with tiny bells and shook it as he called out to the ancestors, asking permission to enter the sacred knowledge and to be at the site.

Then, satisfied, he gently gestured to the incised images of snow leopards, ibex, and a human being and began to interpret. "The sheep are returning to the earth; the snow leopards are ascending. Both will become extinct. It is up to the human being to create balance to prevent this from happening." He looked at us expectantly with his traditional white kulpak hat and robe; it was as if the snow-peaked mountains behind us had just stepped forward. The silence was intense. He turned again and this time focused on the etching of the man. Two things were striking. The man held a bow and a line connecting the human with a snow leopard. Western science has taken this image to mean that early Central Asians used the snow leopard to hunt for them. This is a superficial interpretation. Japarkul explained, the connection is a metaphor for the need to align ourselves with the wisdom of the snow leopard and illustrates that in so doing we mediate the destructive shadow of humanity and create unity and harmony with nature.

We looked at the rock. It was as if our week's deliberations and the codification of the

[b]Unthinkable in linear time, Kygyz elders maintain that this particular type of ceremonial fire had last been ignited when humanity dispersed out of Central Asia.

FIGURE 15.3.1 **Petroglyph with snow leopard, ibex, and human images on the shore of lake Issykul, Kyrgyzstan.**

shamans' statement for global snow leopard protection had been anticipated and was now mirrored to us in this ancient message. This was the world's first gathering of snow leopard shamans and the shamans with conservationists, but it felt as if we had known each other and the path ahead through all time.

GOING FORWARD

> "We have come to crisis, we are at edge of time when we need to reconnect to natural world, animal world … connect our ways of knowing, and realize our internal potential to transform." *-Danil Mamyev, Altai elder*

If the knowledge of the primordial mind, its holism, sustainability, underlying physics, and outlook is to survive, elders must be given a voice in the world and supported to express it; their knowledge must be translated and organized so that future generations can access the life-sustaining wisdom. Ethical scientists must consider the impact of their research on indigenous people and our "master scientists," the ICPs, and change behavior to provide for real partnership and collaboration, including sharing resources. This collaboration can serve as a catalyst to strengthen culture, safeguard the earth, and foster the emergence of a new type of science – one that consciously links analytical and holistic thought. This is not meant to supplant Western science, but to accord it its proper place among diverse ways of knowing. Nor will a new, linked science displace the indigenous way of knowing.

As Slava, Altaian snow leopard ICP, puts it: "Conservationists need to recognize spiritual sites … the rituals and traditions and everything they bring. This is not about allowing us to work; it's about not hindering our work – and not bringing the scientific framework into our world."

Other ICPs are committed to working with conservationists. Altai ICP Danil Mamyev states, "The Uluu Ot has given us a new start. … We recognize the value of linking our traditional knowledge with conservation science and are ready for cooperation" (IUCN, Jeju South Korea, interview, September 2012).

True collaboration is like the Cholpon Ata cord that connects humanity with the snow leopard. To join with integrity we must allow indigenous science to function on its own terms and to permit scientists to stay in a dual consciousness, fully embodied by indigenous science while being entirely aware that indigenous science is the other, not the self. This is the way to honor all phenomena (Bosnak, 2008). The combined wisdom with the right technology and the energy of local communities can deliver change where the snow leopard and the planet need it most.

SUBCHAPTER

15.4

Snow Leopards in Art and Legend of the Pamir

John Mock

Secretary and Trustee-at-Large, American Institute of Afghanistan Studies, Boston University, Boston, MA, USA

Throughout the Pamir–Hindu Kush region that spans the mountains of present day Tajikistan, Afghanistan, and Pakistan, parallels in complex symbolic categories suggest the mountain communities share knowledge that transcends linguistic and social boundaries (Mock, 2011). This indigenous knowledge is contextually grounded in people's interactions with land, weather, and biodiversity over considerable time. Hence this indigenous knowledge can be said to have an ecological foundation (Kassam et al., 2010) and is recognized as essential to addressing complex environmental issues (Lynch and Hammer, 2013).

The symbolic linkages between biological and cultural diversity find representation in stories and art produced by indigenous people. This brief article presents examples of such representations from Afghanistan and Pakistan and discusses the role of indigenous knowledge in addressing environmental and conservation issues. Rock art uses rock surfaces as a "canvas" to tell stories grounded in indigenous knowledge that is specific and vital to the culture in which it was produced (Safinov, 2009). In the Pamir–Hindu Kush region, we find numerous depictions of wild ungulates, and hunters with spears and bows on foot and on horseback (Mock, 2013). The rock art was typically composed on several rock panels with multiple images on each panel. This clustering of rock art has led scholars to assume that such places were "sanctuaries" where symbolically significant compositions were made on rock. The context in which the person(s) who produced the rock art and the motivations for the production of particular images are often obscured by time. The "semantics" of the story of such rock art are also difficult to interpret. However, the significance of the rock art is not in doubt; it was produced to record important events or features of the environment that resonated significantly with the people who took the time to produce it. In Wakhan District of northeast Afghanistan, I observed a depiction of what appears to be a long-tailed felid, most probably a snow leopard (Fig. 15.4.1; Mock, 2013). This panel is located in the settled area of Wakhan, home to indigenous Wakhi people. Snow leopards (*Panthera uncia*) inhabit this area (Simms et al., 2011). Depictions of snow leopards in higher elevation areas of Wakhan have (so far) not been observed, which agrees with the observed presence of snow leopards in Wakhan; they seem to be more abundant in the side valleys of the settled area of Wakhan. It is not possible to say who made this depiction

FIGURE 15.4.1 **Rock art depicting what appears to be a snow leopard in Wakhan District of northeast Afghanistan.** *(Photograph by John Mock).*

of a snow leopard, when they made it, or why they made it. However, when viewed through the lens of indigenous knowledge of snow leopards contained in stories told today, we can understand some of the significance that snow leopards have to Pamir–Hindu Kush people and perhaps glimpse a motive for this rock art depiction.

One aspect of the region's indigenous knowledge is the concept of alpine places as pure and separate from the lower, less pure human domain (Steinberg, J. The Horns of the Ibex: Preserving and Protecting Mountain Cultural Landscapes in Central Asia. Sacred Mountains Program of The Mountain Institute, Washington, DC, unpublished report). Such high places are the realm of spiritual beings, and the plants and animals of the environment share an association with the cultural concept of purity and the presence of spiritual beings (Dodykhudoeva, 2004; Kassam, 2009; Mock, 1998). Plants such as primrose (*Primula macrophylla*) and wild rue (*Peganum harmala*) carry spiritual and ritual connotations and usages for mountain people (Mock and O'Neil, 2002). Wildlife such as ibex (*Capra sibirica*), urial (*Ovis orientalis*), argali (*Ovis ammon polii*), and snow leopard also carry connotations of purity and spirituality.

The mountain people of the region, and indeed throughout Central and South Asia, are familiar with *pari*, who are female supernatural beings of the high mountains. Wakhi people who live in Pakistan, China, Tajikistan, and Afghanistan have their own word for *pari*: *mergichan*. The *mergichan* inhabit the *mergich* realm, which is the realm of the mountains and high pastures. It is a pure, even sacred realm, where the supernatural mountain spirits tend their wild flocks of mountain sheep and ibex. Humans only enter the *mergich* realm during summer, and only after ceremonially announcing to the *mergichan* that the people will displace them for the summer and asking them for a favorable influence on the livestock and dairy production.

The *mergichan* are not malevolent beings, but their displeasure can be provoked. They are angered by "impure" actions that pollute the *mergich* realm. As powerful beings, they need to be propitiated to ensure the success of summer herding and dairy production by the community as a whole, and for success at hunting by individual men. The *mergichan* can assume the guise of a *mergich* animal, and in that form they can both harm and help men. Hunters cultivate a positive relationship with the *mergichan* and the *mergich* realm in order that the *mergichan* should reveal the location of the wild game to them through signs or through a dream. The knowledge of how to gain the favor of the *mergichan* includes knowledge of the habits of the wild game, the landscape within which they dwell, the vegetation they prefer for food, and a general respect and reverence for all that is *mergich*. From a Western perspective, this seems much like an environmental ethic.

Wakhi people regard the snow leopard as a *mergich* animal that lives in *mergich* areas. It rarely interacts with people, and it is hard to see, powerful, beautiful, and potentially dangerous. As such, it exemplifies many of the qualities of the *mergichan*, and so is an appropriate animal shape for them to assume. Currently, in the Wakhi community of Shimshal in Pakistan, villagers are engaged in a process of integrating their concept of *mergich* with a modern conservation ethic through the Shimshal Nature Trust (http://www.snt.org.pk). The younger generation sees this as a way to make the old concepts again relevant. A similar process is underway in the Wakhi villages that manage the buffer zone area of the Khunjerab National Park through the Khunjerav Villagers Organization (http://www.kvo.org.pk). These community-based organizations, which both maintain websites, are important examples of cooperative efforts to integrate old and new knowledge into a framework that can be shared between people inside and outside the Wakhi community to develop a new significance for the mountain landscape (Abidi-Habib and Lawrence, 2007; Ali and Butz, 2003).

The following Wakhi story from Shimshal village of how *mergichan* assumed the guise of a snow leopard and became the protective spirit partner of a Wakhi man exemplifies "old" knowledge that can serve as a scaffold for a new integrated significance. I recorded and translated the story as part of my research in northern Pakistan (Mock, 1998).

> It is like a miracle. Someone sees something and then it vanishes. My own father, a miracle happened to him. What sort of miracle? My father went with the people to Lemarz Keshk, below Furzin. In the evening my father went to the spring for water. When he went for water, he saw a woman with a white scarf, a pitek. He lay down into a low area so he was hidden, looking.
>
> "What sort of thing was this?" he thought. People never came there, so a woman would never go there either. Then his uncle came, and he said to him, "O Uncle, a woman came here, a woman with a white scarf stood here, and now she has vanished down here." He said, "Ya Maula, what can it be?"
>
> Well, it became dark. Night fell, and they ate dinner and slept. They had nothing but an old blanket. They both covered themselves with that, and slept. In the night, my father dreamt that two horses came with two riders. He dreamt one horse came and passed over him, and one came and sunk its teeth into his leg. He awoke suddenly and something heavy was on his body. He tried to sit up, but he couldn't. It was very heavy. He was still sort of asleep, and he moved a little, then shook his blanket and saw a snow leopard, with eyes like that. And it sat on top of him, like this. And it moved off of him and slowly went outside. It went out, and he felt a lot of pain. He said, "O Uncle, wake up! My leg hurts, something came on top of me. Go out and shine a light."
>
> He got up and they made a fire, and saw a lot of blood. Blood, he was bleeding. And then he was very scared. He said, "What thing was this, what happened?" They sat a while, but it didn't come. Then they closed the door with a stone and slept. While they slept, it grabbed the door and tossed the stone aside. It came and yanked their blanket and took it down to the trees. Then they got up and made a fire.
>
> "O God, what is this thing?" they said. They saw it seemed like a snow leopard. It came toward the door of the hut and his uncle was going to shoot at it when he fainted. He went unconscious and the snow leopard became a horse and went away. They sat and sat and it grew light.
>
> Their companions were at the settled area down river, where they cultivated barley. People were there for harvesting. My father came there, and said, "Someone come with us, and give us a dog, too, we were so frightened." They refused to come. He took a dog and tied it at the door that night. It came again and it grabbed that dog and tossed it far away. It came again and it wouldn't let them sleep all night long till dawn. It took the shape of a snow leopard. It put itself into a snow leopard skin.
>
> They returned to Shimshal. There, the *khalifa* (spiritual leader) said that it was a *pari*. The *pari* itself took on the guise of a snow leopard. It took on the skin of a snow leopard and then attacked them like that.
>
> Then what happened to my father? It happened like this, that this *pari* was with him continuously,

with my father. It came itself as a snow leopard. I myself and my brother Shifa, we both saw it. Our father was with us, at Arbob Purien. We were there together when it came. It came and I saw it first. I said, "Ya Ali, what thing has come? A snow leopard." It came and stood on the far side. It stood there and didn't come near us. It turned and left. My father was there, too. And until his dying day, that never was a danger to anyone, but to his dying day that *pari* was with him. A *pari* in the shape of a snow leopard. It would assume the shape of a snow leopard and come. That *pari* was ready to make friends with him. Whenever he was preparing to hunt somewhere, that *pari* said to him at night in a dream, "In such and such place go and hunt. To such and such place don't go and hunt, no game is there." Whether ibex, or whether small game, he would go and it would be there. Such miraculous things happened with him. You can ask other Shimshalis if such things were or not. They will tell you. Up until his death, this was with him.

Then it ended when he went from this world. Now it is no longer like this with us. *Pari* are not close to us. We live more at ease. My father was different. A snow leopard came to him. This event occurred.

SUBCHAPTER

15.5

The Snow Leopard in Symbolism, Heraldry, and Numismatics: The Order "Barys" and Title "Snow Leopard"

Irina Loginova

Snow Leopard Fund, Ust Kamenogorsk, Kazakhstan

SYMBOLISM, HERALDRY, AND NUMISMATICS

The rare and mysterious snow leopard has since ancient times been a totem and symbol in many nations of Central Asia. In those days, snow leopards were worshipped as sacred animals. People respectfully referred to the snow leopard as "the lord of the celestial mountains" or "the owner of the snowy peaks." From time immemorial, the tradition has survived of depicting this beautiful and graceful animal in rock drawings, on plates made of precious metal, to decorate clothing and weapons, and on the arms, coins, and standards of entire countries and individual cities. The image of the snow leopard even today is widely represented in art, literature, and in symbolism, numismatics, and heraldry.

Unfortunately, from the Middle Ages to the mid-twentieth century, attitudes about wildlife and nature in general took on an exploitative and unthinking character. The slogans of conquest and transformation prevailed and nature seemed inexhaustible.

Large predators were particularly affected by persecution. In Central Asia and Kazakhstan, the tiger (*Panthera tigris*) and cheetah (*Acinonyx jubatus*) were completely extirpated, but high and inaccessible mountains have kept the snow leopard from extinction. In the 1950s–1960s, it was declared a harmful predator, like the wolf, and was destroyed even in nature reserves. Despite negligible damage to livestock and no danger at all in relation to people, snow leopards were persecuted widely across their range, except where Buddhism was widespread – a religion that prohibits killing animals.

Only in 1948, when the International Union for Conservation of Nature (IUCN) was founded, and then later with other international and local environmental organizations, did the situation begin to improve. A variety of "Red Books" were published and relevant laws prohibited the killing of rare and endangered species of animals and plants. The theme of "Red Book species" became popular in philately, calendars, badges, and coins.

After the collapse of the Soviet Union and formation of independent states in Central Asia, the snow leopard appeared on the emblems of the cities of Almaty, Astana (Kazakhstan), Bishkek (Kyrgyzstan), and Samarkand (Uzbekistan).

The snow leopard has become a symbol of states in whose territory it resides – Kyrgyzstan and Kazakhstan. A winged snow leopard is depicted on the emblem of the Russian republic of Khakassia. And even Tatarstan – located in the middle reaches of the Volga, where not only the snow leopard, but also the common leopard (*Panthera pardus*), never occurred, has now adopted the "white leopard" as its symbol. Over the centuries, from the era of the great conquests of Genghis Khan, there have been great migrations of peoples, mixing of cultures and beliefs, and transformation of languages, symbols, and traditions. The snow leopard became the official symbol of Kazakhstan, proposed by President Nursultan Nazarbayev in his "Address to the Nation – Strategy 2030."

Archaeological finds in the Altai, Semirechye, in the Issyk *kurgan* (burial mound), and other sites in the vast extent of the former Golden Horde confirm the widespread symbolism of the snow leopard, common leopard, and tiger. From these "big three" large cats in Central Asia, the snow leopard survives the best, dwelling in the inaccessible mountains. For the ancient Saks and Oirats and other peoples, it was also sacred, as were the "Celestial Mountains" (Tien Shan), "Golden Mountains" (Altai), and "Roof of the World" (Pamir). Gold decoration on the pointed headdress of the Saka king – "winged snow leopard in the mountains" recovered in 1969 from the Issyk burial mound near Almaty – demonstrated the high standard of jewelry making of the Saka craftsmen of the Empire of the Great Kushans.

Snow leopards are depicted on gold and silver coins and postage stamps around the world, such as in Afghanistan, Bhutan, China, Mongolia, and Nepal. Many countries, even those where snow leopards can be seen only in zoos, regularly issue stamps with the image of this cat. The image of a snow leopard also appears on state awards and currency of Kazakhstan and Russia. Russia has issued a large number of collector's gold and silver coins of the highest quality showing the image of snow leopard, common leopard, tiger, and other rare animals. This series of coins is called "Save Our World."

On each note of the Kazakhstan *tenge* can be seen a watermark in the form of a snow leopard, replacing Lenin as depicted in the currency of the USSR. There are collectable silver and gold *tenge* with "barys" – snow leopard – on the face. One of the main orders (honorary awards) of Kazakhstan is also called "Barys." This order was established in 1999 and is awarded for outstanding achievements in strengthening statehood and sovereignty of the Republic of Kazakhstan; in securing peace, consolidation of society and the unity of the people of Kazakhstan; in the state, industrial, scientific, social, cultural, and social activities; in strengthening cooperation among peoples, mutual enrichment of national cultures, and friendly relations between states.

Sculptures depicting the snow leopard are not uncommon in cities and by highways. In almost every valley near most major roads in Kyrgyzstan and Kazakhstan, one can see a concrete sculpture of a snow leopard. As a rule, there is also a spring of pure water, car parking, and a view of the mountains. In the center of the capital of Kazakhstan, Astana, and in the city of Ust-Kamenogorsk, there are several sculptures of the snow leopard, and in Almaty there is a bronze sculpture of this rare cat in the middle of Gorky Park. Snow leopards are also loved by artists.

Two ice hockey teams in the top division of the Kazakhstan Hockey League have snow leopard–related names "Barys" of Astana and "Ak Bars" (meaning white leopard) from Kazan. The Astana Youth Team is proudly referred to as the "Snow Leopards." A snow leopard cub became the symbol of the ice hockey team "Kaztsink-Torpedo" from Ust-Kamenogorsk. It was devised by 12-year-old Georgiy Gaikov from Zyryanovsk and called "Barsik." This beautiful clay figurine won a competition in 2003 and became the mascot of the hockey club.

In January and February 2011, the Asian Olympic Games – Asiad 2011 – took place. They were held simultaneously in the two capitals of Kazakhstan, in Astana (the official capital) and Almaty (the southern capital). The mascot was a snow leopard cub named Irby. His image adorned billboards and posters in the streets of Almaty and Astana for a year and the image of Irby on souvenirs and as a puppet is still sold in the cities of Kazakhstan. In 2013, the student Olympics were held in Kazan, and the mascot selected was the snow leopard named Yuni – from the English pronunciation of the word *university*.

For mountaineers in Russia and Central Asia, the snow leopard is not only a symbol, but also a title. Climbers who manage to climb all five 7000 m peaks in the former Soviet Union (Pobeda Peak and Khan Tengri in the Tien Shan, Peak Somoni and Korzhenevskaya in the Pamirs, and Lenin Peak in the Pamir-Alai) are assigned the highest title of "Snow Leopard," which is equivalent to the rank of Master of Sport, international class. The award is "Snow Leopard Russia" with a picture of the snow leopard and the mountains.

CONCLUSIONS

The snow leopard is also widely used as the brand name of various products, not only in the region but also in the world. The snow leopard is equally popular as the name of shops, hotels, and campsites in the region. The symbolism of the snow leopard can play a significant role in shaping and strengthening its positive image as a living symbol of national pride and an object of the peoples of Central Asia, and therefore contribute to its conservation.

References

Abidi-Habib, M., Lawrence, A., 2007. Revolt and remember: how the Shimshal Nature Trust develops and sustains social-ecological resilience in northern Pakistan. Ecol. Soc. 12 (2), 35http://www.ecologyandsociety.org/vol12/iss2/art35/.

Ali, I., Butz, D., 2003. The Shimshal governance model – a Community Conserved Area, a sense of cultural identity, a way of life. Policy Matters 12, 111–120.

Allendorf, T.D., Brandt, J.S., Yang, J.M., 2014. Local perceptions of Tibetan village sacred forests in northwest Yunnan. Biol. Conserv. 169, 303–310.

Anderson, D.M., Salick, J., Moseley, R.K., Xiaokun, O., 2005. Conserving the sacred medicine mountains: a vegetation analysis of Tibetan sacred sites in Northwest Yunnan. Biodivers. Conserv. 14, 3065–3091.

Bosnak, R., 2008. Embodiment: Creative Imagination in Medicine, Art and Travel. Routledge, London.

Colorado, A., 2014. Scientific pluralism. In: Peat, F.D. (Ed.), The Pari Dialogues Volume 2: Essays in Indigenous Knowledge, Western Science. Pari Publishing, Pari, Tuscany, Italy.

Deloria, Jr., V., 1999. If You Think about It You Will See That It Is True. In: Deloria, B., Foehlner, K., Scinta, S. (Eds.), Spirit & Reason. Fulcrum Publishing, Boulder, Colorado, USA, pp. 40–60.

Deng, T., Wang, X., Fortelius, M., Li, Q., Wang, Y., Tseng, Z.J., Takeuchi, G.T., Saylor, J.E., Säilä, L.K., Xie, G., 2011. Out of Tibet: Pliocene woolly rhino suggests high-plateau origin of Ice Age megaherbivores. Science 333, 1285–1288.

Dodykhudoeva, L.R., 2004. Ethno-cultural heritage of the peoples of West Pamir. Collegium Antropol. 28 (Suppl. 1), 147–159.

Dorje, O.T., 2011. Walking the path of environmental Buddhism through compassion and emptiness. Conserv. Biol. 25, 1094–1097.

Dudley, N., Higgins-Zogib, L., Mansourian, S., 2009. The links between protected areas, faiths, and sacred natural sites. Conserv. Biol. 23, 568–577.

Encyclopedia Britannica, 2003. The Atlas of Faiths. Encyclopedia Britannica, Chicago.

Gyatsho, T.L., 1979. Gateway to the temple: manual of Tibetan monastic customs, art, building and celebrations. Ratna Pustak Bhandar, Katmandu, Nepal.

Harvey, P., 2012. An Introduction to Buddhism: Teachings, History and Practices. Cambridge University Press, Cambridge, UK.

Hope, J., 1997. The Secret Language of the Soul. Chronicle Books, San Francisco, USA.

Kassam, K.-A., 2009. Viewing change through the prism of indigenous human ecology: findings from the Afghan and Tajik Pamirs. Hum. Ecol. 37 (6), 677–690.

Kassam, K.-A., Karamkhudoeva, M., Ruelle, M., Baumflek, M., 2010. Medicinal plant use and health sovereignty: findings from the Tajik and Afghan Pamirs. Hum. Ecol. 38, 817–829.

Khunjerav Villagers Organization. http://www.kvo.org.pk (accessed March 3, 2015).

Li, J., Wang, D., Yin, H., Zhaxi, D., Jiagong, Z., Schaller, G.B., Mishra, C., McCarthy, T.M., Wang, H., Wu, L., Xiao, L., Basang, L., Zhang, Y., Zhou, Y., Lu, Z., 2014. Role of Tibetan Buddhist monasteries in snow leopard conservation. Conserv. Biol. 28, 87–94.

Lynch, A., Hammer, C., 2013. Editorial: protecting and sustaining indigenous people's traditional environmental knowledge and cultural practice. Policy Sci. 46, 105–108.

McCarthy, T.M., Chapron, G., 2003. Snow Leopard Survival Strategy. International Snow Leopard Trust.

Mishra, C., Allen, P., McCarthy, T., Madhusudan, M.D., Bayarjargal, A., 2003. The role of incentive programs in conserving the snow leopard. Conserv. Biol. 17, 1512–1520.

Mock, J., 1998. The Discursive Construction of Reality in the Wakhi Community of Northern Pakistan. Doctoral dissertation. University of California, Berkeley.

Mock, J., 2011. Shrine traditions of Wakhan, Afghanistan. J. Persianate Stud. 4, 117–145.

Mock, J., 2013. New discoveries of rock art in Afghanistan's Wakhan Corridor and Pamir: a preliminary study. The Silk Road 11, 36–53, Plates III–IV.

Mock, J., O'Neil, K., 2002. Trekking in the Karakoram & Hindukush, second ed. Lonely Planet Publications, Footscray, Australia.

Narby, J., 1998. The Cosmic Serpent: DNA and the Origins of Knowledge. Tarcher/Putnam Books, New York, USA.

Nasr, S.H., 1996. Religion and the Order of Nature. Oxford University Press.

Phalnikar, S., 2014. "Tying conservation with faith to protect a big cat," Deutsche Welle, http://dw.com/p/1DDFb.

Pu, W., 1990. Tibetan Buddhist Monasteries of Gansu and Qinghai. Qinghai People's Publishing House, Xining.

Safinov, D.G., 2009. On the interpretation of Central Asian and South Siberian rock art. Archaeology Ethnology & Anthropology of Eurasia 37 (2), 92–103.

Salick, J., Amend, A., Anderson, D., Hoffmeister, K., Gunn, B., Zhendong, F., 2007. Tibetan sacred sites conserve old growth trees and cover in the eastern Himalayas. Biodivers. Conserv. 16, 693–706.

Shen, X., Lu, Z., Li, S., Chen, N., 2012. Tibetan sacred sites: understanding the traditional management system and its role in modern conservation. Ecol. Soc. 17, 13.

Shimshal Nature Trust. http://www.snt.org.pk (accessed on March 3, 2015).

Simms, A., Moheb, Z., Salahudin, Ali, H., Ali, I., Wood, T., 2011. Saving threatened species in Afghanistan: snow leopards in the Wakhan Corridor. Int. J. Environ. Stud. 68 (3), 299–312.

Tedlock, B., 2005. The Woman in the Shaman's Body. Bantam Books, New York, USA.

Van der Post, L., 1958. Lost World of the Kalahari. William Morrow and Company, New York, USA.

Young, J.K., Olson, K.A., Reading, R.P., Amgalanbaatar, S., Berger, J., 2011. Is wildlife going to the dogs? Impacts of feral and free-roaming dogs on wildlife populations. Bioscience 61, 125–132.

CHAPTER

16

Trophy Hunting as a Conservation Tool for Snow Leopards

OUTLINE

Snow Leopards
http://dx.doi.org/10.1016/B978-0-12-802213-9.00016-X

SUBCHAPTER

16.1

The Trophy Hunting Program: Enhancing Snow Leopard Prey Populations Through Community Participation

Muhammad Ali Nawaz[*,**], *Jaffar Ud Din*[**], *Safdar Ali Shah*[†], *Ashiq Ahmad Khan*[**]

*Department of Animal Sciences, Quaid-i-Azam University, Islamabad, Pakistan
**Snow Leopard Foundation, Islamabad, Pakistan
†Wildlife Department, Khyber Pakhtunkhwa, Peshawar

INTRODUCTION

Mountain-dwelling communities in the snow leopard (*Panthera uncia*) range are growing in population and increasing their dependence on natural resources. However, the natural resource base, particularly pastures and wild habitats, are shrinking and gradually losing their productivity. This often spawns human-wildlife conflict where human livelihood needs and conservation become competing interests. For example, Pakistan's Gilgit-Baltistan (GB) province's human population has quadrupled in the past four decades and livestock numbers have grown at an annual rate of 3.5% since 1976 (Khan, 2003). Rangelands in Pakistan occupy about 20% of snow leopard habitat and provide critical grazing areas for wild and domestic ungulates. Rangelands assessed to be burdened and overgrazed 20 years ago (Khan, 2003) are experiencing even higher degradation due to progressively increasing livestock numbers.

In economic terms, livestock is a source of cash income and contributes more than 11% to the country's gross domestic product (GDP) [Government of Pakistan (GoP, 2010)]. On the other hand, the lack of economic gain associated with wild ungulates motivates communities to hunt them, which elimates competition on pastures and crop damage, besides providing meat.

Economic incentives are considered helpful in changing human attitudes toward wildlife (Mishra et al., 2003) and enable locally supported conservation actions. Incentives may include monetary compensation, arable land, grazing rights, employment, education, and access to amenities (Hotte and Bereznuk, 2001; Karanth, 2002). The sustainable-use approach of conservation states that the authority to regulate natural resources rests with local communities, who traditionally use and appreciate their utilitarian value, and are negatively affected by their degradation (IUCN, 1991). This approach promotes local participation and support for conservation by allowing extractive use of resources (Mishra et al., 2003). The trophy-hunting program in Pakistan operates along the same lines. It assumes that a combination of recreational hunting and other forms of wildlife uses such as ecotourism can generate sufficient benefits and incentivize local communities to conserve wildlife and habitats in the long term.

Pakistan is actively promoting community-based wild resources management as a conservation tool to ensure that the financial benefits derived from limited trophy hunting go directly to local communities. The idea is that the communities use an equitable share of these financial benefits to sustain the management program.

TROPHY ANIMALS IN NORTHERN PAKISTAN

Pakistan hosts 7 Caprinae species with 11 subspecies occupying habitats from the hills in the southern deserts to the high-alpine areas of the Himalayas (Hess et al., 1997). Five of them share habitat with the snow leopard, including the Himalayan ibex (*Capra sibirica*), blue sheep (*Pseudois nayaur*), markhor (*Capra falconeri*), Ladakh urial (*Ovis vignei*), and Marco Polo sheep (*Ovis ammon polii*). Trophy hunting is restricted to the first three as the populations of the Ladakh urial and Marco Polo sheep are too small and restricted.

The markhor is highly prized as a trophy animal for its long, twisted horns. There are two broad categories, namely the flare-horned markhor *(Capra falconeri falconeri)* and straight-horned markhor *(Capra falconeri megaceros)*. The former includes two subspecies; the Pir Panjal or Kashmir markhor found in Khyber Pakhtunkhwa (KP) province in areas like Chitral, Dir, and Swat, and the Astore markhor mainly found in GB. The straight-horned markhor is also subdivided into the Kabul markhor and Suleiman markhor.

HISTORY OF TROPHY HUNTING IN PAKISTAN

The idea of using trophy hunting as a conservation tool was floated in the mid-1970s by the late Major (retired) Amanullah Khan who witnessed declining populations of large ungulates in habitats where he used to hunt. The practice began in the late 1970s under Abdur Rehman Khan (late), the then Conservator of Wildlife, KP. He allowed limited markhor hunting in Chitral Gol National Park (CGNP), turning 50% of the revenue over to local communities. Similarly in the 1980s, Sardar Naseer Tareen, a tribal chief of Balochistan in Torghar managed trophy hunting as a commercial enterprise, successfully marketing the trophy hunting of straight-horned markhor. There were also other instances of trophy hunting in Durreji, Lasbella district in Balochistan, and Kirthar National Park in Sindh.

Trophy hunting was interrupted in 1991 by the then National Council of Wildlife (NCCW), which believed the practice was damaging ungulate populations. Trophy hunting was therefore restricted as part of the overall ban on the hunting of mammals by the Cabinet Division, GoP.

The first organized efforts to promote community-managed trophy hunting were initiated in GB, the former Northern Areas. The idea was put forward by Agha Syed Yehya Shah, a political and religious leader of Nagar Subdivision, Gilgit, and Ghulam Rasool, a retired divisional forest officer (Wildlife), Northern Areas. A proposal was submitted to the Agha Khan Rural Support Programme (AKRSP) with the idea that people could be compensated for the cost of meat in return for wildlife protection and sustaining this as a livelihood option through ibex trophy hunting, once the already-depleted population was recovered.

AKRSP sent this proposal to the Northern Areas Forest and Wildlife Department, the International Union for Conservation of Nature (IUCN), and the World Wide Fund for Nature (WWF)–Pakistan. Ashiq Ahmad Khan was asked to assess the feasibility and validity of the proposal in 1989–1990.

This was being done at a time when the Cabinet Division had already banned hunting countrywide – including in GB – and when hunting wild ungulates was considered more as a cultural practice and source of food, rather than anything else. In addition, there was no precedent for the government sharing hunting revenues with communities. Surveys were conducted to assess ibex populations and study the general environment and attitudes to see if a program like this could work. This was done recognizing the importance of the task and its demonstration value.

This occasion was regarded as a good opportunity to establish a replicable model for the conservation of both ungulates and predators. A report was prepared and submitted to the government of the then Northern Areas, recommending the trophy hunting program for the Bar Valley. It was suggested that the local community be granted a loan equivalent to the cost of meat for one year. The formula worked such that the money would stay as a loan and be returned by the community when the first trophy hunting took place. WWF–Pakistan agreed to pay the money and signed an agreement with the community. The money was generated through trophy hunting in 1994 and the loan was returned. However, it was donated back for the construction of a health unit in the Bar Valley.

Communities in other valleys followed the program with interest. There were calls for program replication once it had been established that an actual monetary benefit was available to those who protected wild animals in their areas. Several valleys were immediately closed for illicit hunting and social organizations were formed, each trying to initiate similar programs.

PROGRAM IMPLEMENTATION MECHANISM

Community organization and the availability of harvestable trophy animal populations are the main prerequisites for such programs. Nongovernment organizations (NGOs), including the Wildlife Conservation Society (WCS), Snow Leopard Foundation (SLF), IUCN, and WWF–Pakistan, play a vital role in organizing communities through social mobilization and assessing trophy animal populations. The procedure requires that communities form a social organization with wide participation through membership and elect office bearers. They can then approach the provincial government to initiate trophy hunting and have their area declared as a community-managed hunting area. The provincial governments evaluate each case and then approach the federal government for permit allocation if the case is deemed feasible. The major steps involved in implementing a trophy hunting program are permit allocation, marketing, fees, and trophy hunting revenue distribution.

Permit Allocation

Provincial governments are authorized to allocate hunting quotas for community-based trophy hunting programs (CTHPs) for species not covered by the Convention on International Trade in Endangered Species (CITES). There is a complete ban on big-game hunting everywhere else in Pakistan. The Ministry of Climate Change (MCC) allocates trophy hunting quotas for CITES species to the federating units for trophy hunting within community-managed areas. This is done through the Office of the Inspector General of Forests.

Marketing

Provincial governments issue permits to foreign hunters or outfitters through open bidding once quotas have been allocated to the provinces. Foreign hunters participate directly in the bidding process or purchase permits from successful outfitters. Outfitters also handle import/export permit applications for hunters' firearms.

Fees

Permit fees for hunting markhor have increased over time through competitive bidding in KP. It started from USD 35,000 in 1999 and reached the highest bid ever of USD 105,000 in 2014.

Permit fees in GB for markhor cost USD 62,000 in 2014–2015 for both foreign and local hunters. Recent ibex permit fees in GB amounted to USD 3,100–3,700 for foreigners, USD 1,100–1,500 for Pakistanis, and USD 500–600 for GB residents. The fee for blue sheep in 2014–2015 was USD 8,600 and USD 5,000 for foreigners and Pakistanis, respectively.

Distribution of Trophy Hunting Revenues

Fees were divided until 2000 with 25% going to the government as a license fee and 75% to the community as a trophy fee. The National Council for the Conservation of Wildlife (NCCW) changed this to 20/80 in the summer of 2000 to conform to the Conference of Parties (COP 11) of the Convention on Biological Diversity (CBD), in which Pakistan's 1998 annual report stated, "Village communities participating in markhor conservation will receive 80% of the revenue from hunting" (CITES, 2000).

Markhor conservancies in Chitral are spread over a large area encompassing numerous villages that protect and conserve animals during most of the year. Hunt locations are a matter of chance and hunters' personal preferences. This means that custodian communities may lose interest in markhor conservation if revenue is handed over to a single village. To this end, the following strategy has been adopted in KP to fully engage communities in conservation efforts:

- Half of the community share in Chitral's two markhor conservancies goes to the community where the hunt actually took place. The remaining 50% is divided equally among the remaining collaborative communities.
- The Kaigha conservation area of Kohistan operates differently; the total share goes to the Kaigha Conservation Committee, which represents the entire valley.

Communities in GB are well organized. Valley conservation committees (VCCs) have elected representation from all communities of the community-controlled hunting area (CCHA) and receive and mange community trophy hunting revenues. The governments of KP and GB distribute trophy fees to communities at annual ceremonies in Peshawar and Gilgit, respectively.

THE CURRENT STATUS OF TROPHY HUNTING PROGRAMS IN THE SNOW LEOPARD RANGE

Gilgit-Baltistan

The community-based trophy-hunting program in GB was first initiated in the Bar Valley in 1989 (Khan, 2011) and later replicated in seven other sites by IUCN during the period 1995–1999 (Mir, 2006). As of 2011, both the government and communities sought to expand the program's portfolio by establishing new community-managed conservation areas (CMCAs) or community-controlled hunting areas (CCHAs). This was a result of having evaluated program paybacks in the form of increased species populations and communities' socioeconomic status. Twenty-seven CMCAs, six game reserves (GRs), two wildlife sanctuaries, and five national parks have been notified since 2014. They cover a total area of 33,999 km^2 and constitute 47% of GB's total area of 74,496 km^2 (Parks and Wildlife Department, 2014). The GRs, wildlife sanctuaries, and national parks feed populations in the adjoining CMCAs. The ibex was initially designated as a trophy hunting species due to its wide distribution and high density in the region. However, the markhor was added to the trophy hunting list in 2001 and the blue sheep in 2004–2005. The period 2000–2014 saw 262 ibex, 30 markhor, and 19 blue sheep—311 animals—hunted by 198 foreigners and 111 Pakistanis, generating USD 346,958 (Table 16.1.1). On average 19 ibex, 2 markhor and 1 blue sheep were hunted per year, generating revenue of USD 24,783 (Parks and Wildlife Department, 2014).

The replication and expansion of the trophy hunting program continues; the management plans of about 20 more valleys in the districts of Diamer, Astore, and Hunza-Nagar covering roughly 5000 km^2 have been approved through the respective district conservation committees (DCCs) and have been presented to the Department for formal notifications as CMCAs (unpublished DCC meeting minutes). New quotas were proposed in 2014–2015. These were 4 markhor (USD 60,000 per trophy), 8 blue sheep (USD 8,000 per trophy), and 60 ibex (USD 3,000 per trophy). Most of the allocations were based on the results of the pretrophy quota allocation survey of ungulates conducted by SLF and the GB Parks and Wildlife Department in 2014 (Ali et al., 2014). Snow leopards have been reported in all GB protected areas and some of them, especially those in the Karakoram–Pamir area, are considered strongholds of the large cat (GoP, 2013).

Karakoram–Pamir

The wildlife wing of the KP Forest Department began the Chitral Conservation Hunting Program, a trophy hunting program for markhor, in 1983. This was not a community-based conservation program because all proceeds went to the government. Developed by Dr Mohammad Mumtaz Malik, Conservator of Wildlife, this program operated in cooperation with a hunting organization called the Shikar Safari Club. It lasted 8 years until the GoP banned the export of trophies along with all big game hunting throughout Pakistan. Members of the Shikar

TABLE 16.1.1 Revenue Generated Through Trophy Hunting of Wild Ungulates in GB, 2000–2014

Year	Ibex	Markhor	Blue sheep	Total revenue (USD)	Community share (USD)	Wildlife Department share (USD)
2000–01	15	1	0	15,000	12,000	3,000
2001–02	9	1	0	12,500	10,000	2,500
2002–03	11	2	0	62,000	49,600	12,400
2003–04	23	2	0	70,000	56,000	14,000
2004–05	23	3	2	109,000	87,200	21,800
2005–06	31	2	0	94,000	75,200	18,800
2006–07	20	3	4	177,700	142,160	35,540
2007–08	25	3	2	252,500	202,000	50,500
2008–09	24	1	2	105,520	84,416	21,104
2009–10	19	3	0	178,920	143,136	35,784
2010–11	17	1	2	103,500	82,800	20,700
2011–12	17	1	1	94,800	75,840	18,960
2012–13	12	3	6	196,800	157,440	39,360
2013–14	16	4	0	262,550	210,040	52,510
Total	262	30	19	1,734,790	1,387,832	346,958

Safari Club hunted 16 markhor during two licensed hunts per year in and around CGNP during the period 1983–1991 (Johnson, 1997).

In 1997, a COP meeting in Zimbabwe approved six annual export permits for markhor trophy hunting in Pakistan. Hunting was allowed in community-protected areas only.

The world's largest markhor population is found in CGNP and the surrounding valleys of Chitral district. The KP Wildlife Department organized communities residing in Chitral's markhor habitats to actively participate in the animal's conservation and enjoy program benefits. The Gehret and Toshi-Shasha conservancies were consequently established in Chitral district. The Gehret Conservancy is located on the northeastern side of Chitral town, has an area of about 600 km^2 and includes the valleys of Gehret, Kessu, Jughor, Nerdet, Kaghuzi, and Goleen Gol. The Toshi-Shasha Conservancy is located on the northwestern side of Chitral town, has an area of about 500 km^2, and encompasses several villages, including Seen-Lasht, Boriogh, Bothtoli, and Karimabad.

Some 57 markhor hunts have taken place in the Chitral and Kohistan district of KP since the inception of the trophy hunting program in 1999 (Table 16.1.2). The KP program has earned USD 3,306,700 so far. Eighty percent of this (USD 2,645,360) has been distributed among the custodian communities of the Chitral and Kohistan districts (Table 16.1.2). The program has contributed significantly to socioeconomic uplift and helped revive local species, as evidenced by the markhor's population growth in both districts.

TABLE 16.1.2 Revenue Generated Through Markhor Trophy Hunting in KP, 1999–2015

Year	Markhor hunted	Total revenue (USD)	80% share paid to community	20% government share remitted to treasury
1998–99	3	45,000	36,000	9,000
1999–00	2	44,300	35,440	8,860
2000–01	3	81,000	64,800	16,200
2001–02	1	28,000	22,400	5,600
2002–03	3	91,500	73,200	18,300
2003–04	4	132,000	105,600	26,400
2004–05	4	180,000	144,000	36,000
2005–06	4	212,500	170,000	42,500
2006–07	4	227,000	181,600	45,400
2007–08	4	284,000	227,200	56,800
2008–09	4	291,300	233,040	58,260
2009–10	3	220,000	176,000	44,000
2010–11	3	216,000	172,800	43,200
2011–12	4	313,500	250,800	62,700
2012–13	4	317,000	253,600	63,400
2013–14	4	357,100	285,680	71,420
2014–15	3	266,500	213,200	53,300
Total	57	3,306,700	2,645,360	661,340

KP currently allows markhor trophy hunts to foreigners only; there have been no legal markhor hunts by domestic hunters. Similarly, the trophy hunting of the Himalayan ibex by foreign hunters is quite limited. They are usually offered as part of flare-horned markhor hunting packages. It should be noted, however, that ibex hunts by local hunters have been increasing over time.

ACHIEVEMENTS, OPPORTUNITIES, AND LESSONS LEARNED

Achievements

The program's biggest achievement is the discernable recovery of ibex, markhor, and blue sheep populations that had previously reached alarming levels due to illicit and unsustainable hunting by local communities. The program has instilled a sense of ownership of wild ungulates among local communities; it is now much more difficult for outsiders to openly hunt wild ungulates.

Both government sources and local residents report increased ibex, markhor, and blue sheep populations – a result of reduced poaching. In addition, the KP Wildlife Department's annual markhor census in CGNP, which has been conducted every year since 1985, provides empirical evidence of population growth: 343 in 1985 to 1722 in 2014. Furthermore, dividing the data into the pre- (1985–1999) and post-trophy hunting periods (1999–2014) shows just how well the program has performed (Fig. 16.1.1).

After the data split, we estimated the finite rate of increase (λ) from annual markhor census with λ as the ratio of numbers in two successive years (Caughley, 1977). λ was calculated by the exponential rate of increase, β, which was estimated by regressing year against population size (Ln of N). The analysis shows that the markhor population was declining at an annual rate of 3.3% ($\beta = -0.03351, \lambda = 0.967$) due to poaching in the pre-trophy hunting period (Fig. 16.1.1a). Populations showed an impressive recovery rate of 14.3% ($\beta = 0.1337, \lambda = 1.143$) *after* the inception of the program in 1999 (Fig. 16.1.1b).

The gradual recovery of prey populations also benefits the snow leopard whose numbers have also grown, according to reports, and camera trapping studies indicate higher snow leopard photo captures in trophy hunting areas (SLF, 2014, unpublished data).

The program has been consistent with its rationale, and trophy hunting has generated substantial monetary amounts over the past 17 years. Local communities benefit directly from their share of hunting fees, which have amounted to USD 4.04 million so far (USD 2.65 in KP and USD 1.4 in GB). In addition, there has been a multiplier effect that has not yet been qualified. This can be considered a significant cash injection into the local economy with measurable long-term wildlife conservation and habitat benefits.

Other communities are still waiting for various government departments or private enterprises to initiate social development programs whereas the trophy-hunting program has already allowed considerable social development in wildlife-rich valleys. Wildlife departments, which are generally underfunded, have utilized their share of hunting fees (about USD 1 million) to enhance law enforcement and improve protected areas management and strengthen relationships with local communities.

Opportunities

Protected areas with potential trophy hunting sites are more likely to be accepted and protected by custodian communities. Khunjerab National Park has been accepted because of the government's commitment to start a trophy-hunting program in the park's buffer zone, and this remains an effective tool, even today. This has made the creation of larger protected areas much easier.

(a)

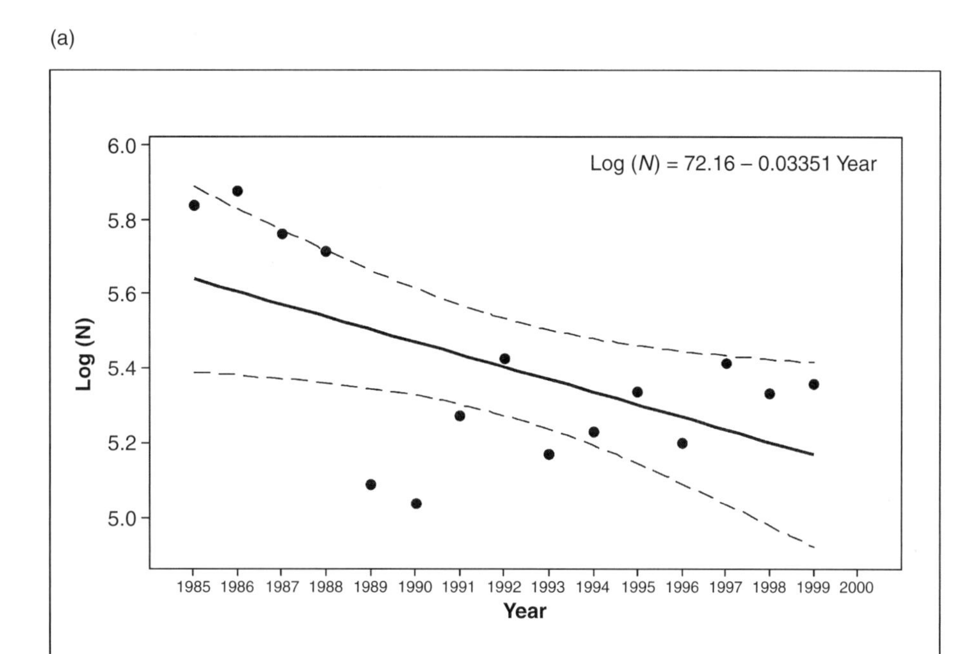

(b)

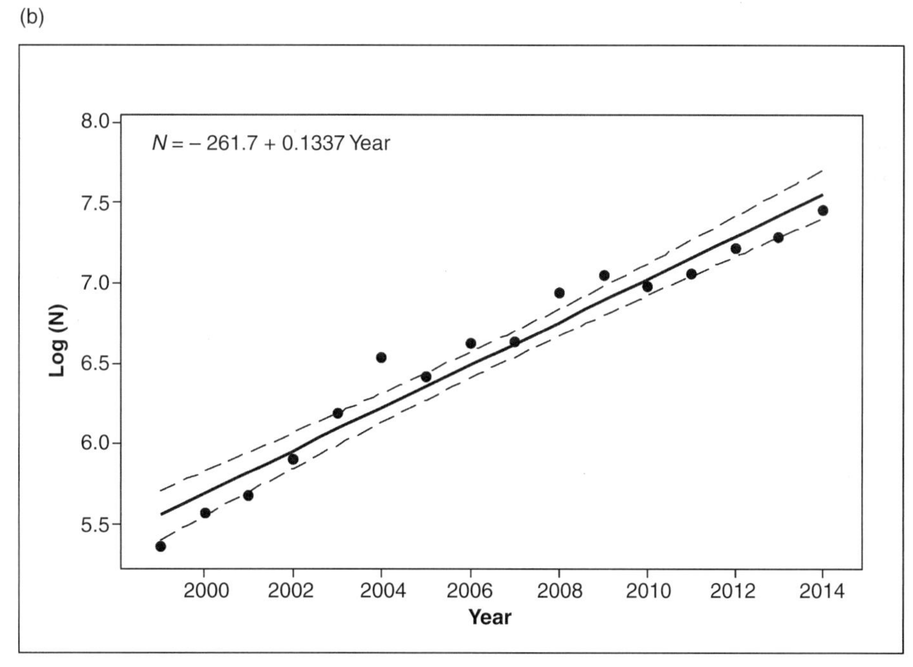

FIGURE 16.1.1 **Markhor population growth in CGNP, Pakistan.** (a) Before the trophy hunting program (1985–1999); (b) After the inception of the trophy hunting program (1999–2014). Growth is shown on a natural log scale. The solid line shows the regression line and the dashed line is the 95% confidence interval.

Social organizations are a basic requirement for ensuring species and ecosystem conservation, and creating social organizations is difficult where people see no incentives. The trophy-hunting program has provided people opportunities to come together and undertake social activities.

Snow leopard protection is part of the trophy-hunting program, but can only be achieved if taken seriously by plan supervisors. The program has made it possible to extract community commitments even for species that hold no attraction or those that people even actively dislike. Agreement flaws or improper monitoring are to blame where this is not the case.

Lessons Learned

Incentives for earning money from a natural resource are popular with communities. It is important to note that such programs can be wasteful if not planned properly. Such programs must insist on community commitment in the early stages.

Trust is an essential component of trophy hunting programs. However, communities need time to fully accept and own the concept. Launching such programs without constant monitoring may well result in failure.

Endangered species require more attention. This is often in contrast to communities' thoughts. To this end, trophy-hunting programs must be marketed as complete packages where animals of little or no interest to communities are accorded the same respect and status as animals of interest. This is especially important as certain species are protected by international conventions and obligations due to their endangered status.

Habitat protection is a basic component of species sustainability and is often ignored by communities when it comes to grazing animals in areas that are sources of income through trophy hunting programs. This is because livestock is individually owned while trophy-hunting benefits are shared by the community at large. Habitat protection must be stated as a priority in initial planning stages; communities are not otherwise likely to reduce livestock numbers, which cause overgrazing and disease.

Conservation decision-making processes are influenced by local, national, and international socioeconomic factors associated with the contexts in which they take place. Conservation can also significantly affect socioeconomic development and lead to improvements in people's lives.

Hunting was traditionally a sport of rulers and wealthy individuals. Large tracts of land were set aside as royal hunting grounds where access and entry were limited. Violations were dealt with severely, although people in far-flung areas *did* hunt on a limited scale for survival. The trophy-hunting program provides an excellent opportunity to engage communities in decision-making processes that help erode the frustration they experienced in the past, turning it into a positive attitude toward conservation.

The conservation of human heritage may revitalize intangible aspects of cultural traditions. The practice of conservation can promote economic prosperity, support disaster recovery, and foster social cohesion among groups. However, conservation can also be used to shape political and economic development, following agendas that may not correspond to the needs or desires of communities.

Desirable Future

Aspects of any program that are needed to ensure success include the following:

- More effective agreements with local communities with essential components such as the protection of associated species (such as snow leopards) and their habitats, formalized on paper
- The formulation of local committees and trainings for regular system monitoring

- Frequent monitoring visits by responsible staff from the Wildlife Department to ensure agreement adherence
- Formal monitoring committees comprising relevant stakeholders and private sector and concerned organization representatives who conduct periodic site visits
- Developing conservation and social development plans that are vetted by the aforementioned monitoring committee
- Measures to prevent communities from inviting local hunters on their own and hunting immature animals
- Independent population assessments and improved scientific census techniques

SUBCHAPTER

16.2

Argali Sheep (*Ovis ammon*) and Siberian Ibex (*Capra sibirica*) Trophy Hunting in Mongolia

*Richard P. Reading**, *Sukh Amgalanbaatar***

***Department of Biological Sciences & Graduate School of Social Work, University of Denver, Denver, CO, USA**

****President, Argali Wildlife Research Center, Ulaanbaatar, Mongolia**

INTRODUCTION

Mongolia supports relatively large populations of argali sheep (*Ovis ammon*) and Siberian ibex (*Capra sibirica*), the preferred prey of snow leopards (*Panthera uncia*) in that country. Although the Mongolian Law on Fauna of 2000 prohibits general hunting of both species, labeling argali as Very Rare (i.e., endangered) and ibex as Rare (i.e., vulnerable or threatened), the Mongolian Hunting Law of 2000 permits limited trophy hunting using special use permits issued by the Mongolian Ministry for Nature, Tourism, and Green Development (hereafter Ministry) and its precursors (Amgalanbaatar et al., 2002; Wingard and Odgerel, 2001). Foreign hunters pay high fees for both species, but especially argali, because of their impressive size and large horns. The Altai argali subspecies (*O. a. ammon*) grow to be the largest sheep in the world; some males weigh more than 200 kg with lengths greater than 165 cm (Amgalanbaatar and Reading, 2000; Schaller, 1998).

Sustainable hunting programs require well-managed populations and local support, ideally with money generated from the hunting going back to conservation management and to benefit local communities (Amgalanbaatar et al., 2002; Harris, 1995; Harris and Pletscher, 2002; Shackleton, 2001; Wegge, 1997). We, and others, believe that the best approach entails developing community-based programs with external review and a high level of transparency (Amgalanbaatar et al., 2002).

Here, we evaluate trophy hunting for argali and ibex in Mongolia, briefly describing the social and ecological context, the challenges to and controversies surrounding trophy hunting, and providing recommendations for improvement.

CONTEXT

The Mongolian Red Book lists argali as Endangered and ibex as Near Threatened (Clark et al., 2006). Argali and ibex occur in scattered and fragmented populations across much of northern, western, central, and southern Mongolia, inhabiting mountains, valleys, canyons, plateaus, and areas with rocky outcrops (Amgalanbaatar and Reading, 2000; Mallon et al., 1997). Argali prefer rolling hills, plateaus, and gentle slopes, while ibex generally occur in more rugged, steep, and mountainous terrain (Schaller, 1998).

Data from the Mongolian Academy of Sciences suggest that populations of argali and ibex have remained relatively stable over the past 15–20 years, following decades of decline (Amgalanbaatar et al., 2012; Bold et al., 1975; General and Experimental Biology Institute, 1986; Harris et al., 2010; Institute of Biology, 2001; Lhagvasuren et al., 2010; Lushchekina, 1994) (Fig. 16.2.1). Unfortunately, survey methods preclude rigorous population estimates, but since methods remained comparable, the trends are likely correct. Harris et al. (2010) provide the most rigorous estimates of mountain ungulate densities for the country, with point estimates of 19,701 argali (95% confidence limits (CL) of 9,193–43,135) and 24,371 ibex (95% CL of 13,840–43,873) in 2009. Major threats to argali and ibex in Mongolia include poaching, competition with livestock, and rapidly increasing natural resources extraction (Amgalanbaatar et al., 2012; des Clers, 1985; Mallon et al., 1997; Reading et al., 2010, 2015).

Argali compete with livestock, particularly domestic sheep and goats, for limited forage and water (Amgalanbaatar and Reading, 2000; des Clers, 1985; Mallon et al., 1997; Schuerholz, 2001). Livestock numbers in Mongolia increased dramatically after the fall of communism, but especially after 1993 when most herds were privatized (Fig. 16.2.2) (Reading et al., 2010, 2015). Total livestock numbers in Mongolia increased from 24.7 million animals in 1989 to 45.15 million in 2013, or an increase of 82.8% (Mongolian Statistical Office, 1999, 2013). Cashmere goats, in particular, have increased almost 5 times from 4.96 million in 1989 to 19.23 million in 2013 (Mongolian Statistical Office, 1999, 2013; Fig. 16.2.2) due to their highly marketable wool.

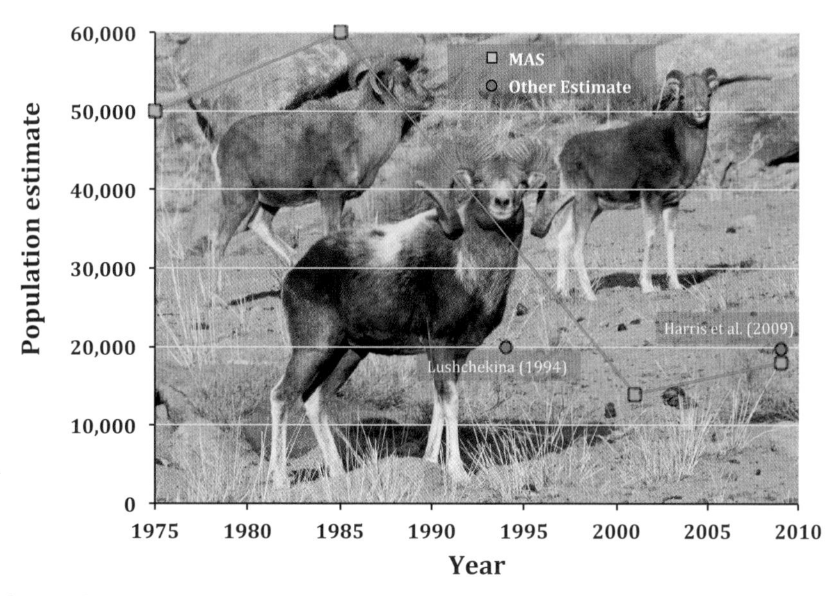

FIGURE 16.2.1 **Argali *Ovis ammon* population estimates for Mongolia, 1975–2009.** *(Source: Data from MAS, Mongolian Academy of Sciences' Institute of Biology; Lushchekina, A., 1994. The status of argali in Kirgizstan, Tadjikistan, and Mongolia. Report to the USFish and Wildlife Service, Office of Scientific Authority, Washington, DC; Harris, R.B., Wingard, G., Lhagvasuren, B. 2010. 2009 National assessment of mountain ungulates in Mongolia. Unpublished Report: Institute of Biology, Mongolian Academy of Sciences, Mongolian Ministry of Nature, Environment, and Tourism, and Worldwide Fund for Nature–Mongolia, Ulaanbaatar, Mongolia)*

Goats heavily impact environments they inhabit, resulting in substantial degradation of pasturelands (Berger et al., 2013; Schuerholz, 2001). In addition, as livestock numbers increase, herders dominate water sources and move their animals into more marginal lands that were traditionally little grazed, often displacing wild ungulates (Amgalanbaatar and Reading, 2000; Lushchekina, 1994; Mallon et al., 1997; Schuerholz 2001).

Poachers in Mongolia kill argali and ibex for meat (subsistence poaching) and to sell parts, particularly horns, in Asian markets (commercial poaching) (Amgalanbaatar and Reading, 2000; Amgalanbaatar et al., 2002; des Clers, 1985; Mallon et al., 1997; Wingard and Zahler, 2006). Limited law enforcement, especially outside of protected areas, and the increasing ease of international commerce have led to large increases in poaching (Wingard and Zahler, 2006; Zahler et al., 2004). Even formerly abundant and widespread species like red deer (*Cervus elaphus*) and Siberian marmots (*Marmota sibirica*) are now critically endangered due to poaching (Clark et al., 2006).

Mining represents the third greatest threat to argali and ibex in Mongolia. Mining companies have leased more than 45% of Mongolia for exploration or extraction (Asia Foundation, 2007; Reading et al., 2010, 2015; see Chapter 10.2). In addition, tens of thousands of illegal wildcat miners (i.e., the so-called "ninja" miners of Mongolia), have destroyed millions more hectares (Farrington, 2005; World Bank, 2006). Many of these areas represent prime argali and ibex habitat. Habitat loss degradation due to global warming and off-road vehicle use represent additional threats (Amgalanbaatar et al., 2002).

Trophy Hunting

A well-managed, community-based trophy-hunting program in Mongolia could offer a sustainable approach to management that would generate resources for effective conservation

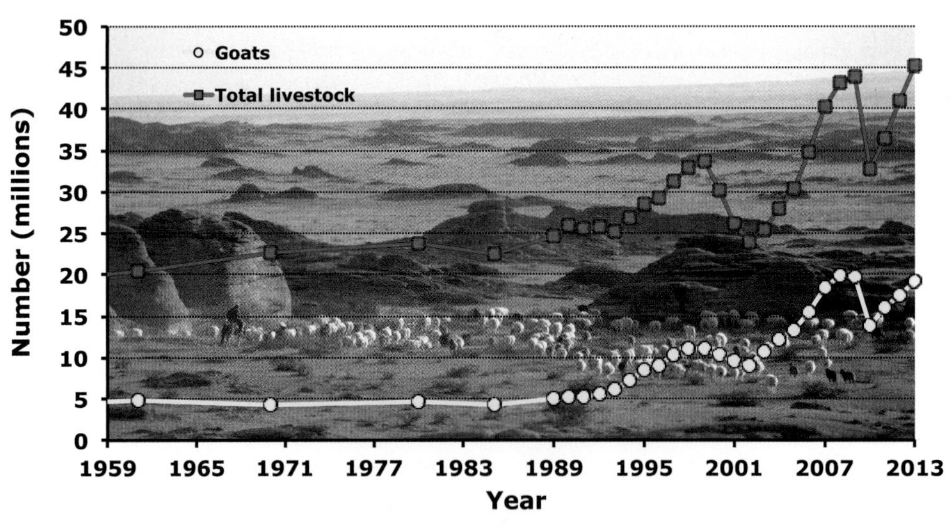

FIGURE 16.2.2 **Number of livestock and cashmere goats in Mongolia, 1959–2013.** *(Source: Mongolian Statistical Office, 1999. Mongolian Statistical Yearbook 1998. Mongolian Statistical Office, Ulaanbaatar, Mongolia; and Mongolian Statistical Office, 2013. Mongolian Statistical Yearbook 2013. Mongolian Statistical Office, Ulaanbaatar, Mongolia.)*

actions and to build and maintain support among local populations. Unfortunately, thus far, developing such an approach has proved elusive.

The Mongolian Scientific Authority, currently the Institute of Biology of the Mongolian Academy of Sciences, issues recommendations to the Ministry for the number of argali and ibex trophy licenses to issue each year based on recent research (when available) or expert opinion (more common). The Mongolian Cabinet makes the final decision on number of licenses. The number of licenses issued for argali always fell well below the recommended numbers until 2002, when licenses exceeded the recommendation by a factor of two (Amgalanbaatar et al., 2002). Under Mongolian law, the Ministry must actively manage and conduct population surveys for hunted populations every 4 years and local governments must survey populations every year following a hunt using funds generated by the hunting (Wingard and Odgerel, 2001). Unfortunately, to our knowledge, none of these surveys have occurred to date. Mongolian law specifies that some of the proceeds from hunting licenses go to federal and local governments (Amgalanbaatar et al., 2002; Wingard and Odgerel, 2001). However, we have been unable to clarify how much, if any, of the money that trophy hunting generates actually goes to where Mongolian law states it should (Amgalanbaatar et al., 2002).

CHALLENGES AND CONTROVERSIES

Trophy hunting for mountain ungulates in Mongolia faces several challenges and controversies. One the greatest challenges to managing mountain ungulates, especially argali, stems from the large fees trophy hunters are willing to pay to harvest an animal, creating a strong enticement for corruption. For example, companies typically charge > USD 50,000 to hunt an Altai argali (Amgalanbaatar et al., 2002). Lack of transparency in how the government issues licenses to hunting companies and spends the income trophy hunting generates have led to allegations of corruption and calls for community-based approaches (Amgalanbaatar et al., 2002 and citations therein; Schuerholz, 2001).

Mongolia lacks a management plan for argali and ibex and other than a heavily studied population of each species in Ikh Nart Nature Reserve (Wingard et al., 2011), most populations receive little attention. The Institute of Biology conducts only nationwide surveys of mountain ungulates at a gross level on an irregular basis (i.e., in 1975, 1985, 2001, and 2009; Amgalanbaatar et al., 2012). Thus, even hunted populations of argali and ibex remain unsurveyed and unstudied, and the Ministry has yet to share data on animals harvested (Amgalanbaatar et al., 2002; Schuerholz, 2001). The Mongolian media has argued that corruption characterizes the Ministry's management of trophy hunting (summarized in Amgalanbaatar, 2002).

Increased local opposition to trophy hunting represents a major challenge to developing community-based approaches. Opposition stems not only from the lack of tangible benefits from trophy hunting, but from actual financial disincentives and resentment that foreigners can legally hunt argali and ibex while local people cannot (Amgalanbaatar et al., 2002). Because trophy hunting companies and most of their staff are based in the capital, Ulaanbaatar, little money from trophy hunting reaches the local people (Schuerholz, 2001). Worse, the federal government reduces aid to local governments based on the number of hunting licenses the region receives, so at best local governments can break even from license revenues and if they do not sell all of their licenses, they can actually lose money (Amgalanbaatar et al., 2002). As a result, many local governments have created local protected areas or petitioned the federal government to create federal protected areas in former hunting areas for argali and ibex, because protected areas in Mongolia do not allow hunting (Amgalanbaatar et al., 2002). In addition, protected areas often attract international conservation projects that bring benefits to local communities and most protected areas allow at least some grazing, and so remain popular with local people.

RECOMMENDATIONS

Trophy hunting could generate income for sustainable conservation management of argali and ibex, while also providing benefits to local people (Amgalanbaatar et al., 2002; des Clers, 1985; Harris and Pletscher, 2002; Johnson, 1997; Liu et al., 2000; Shackleton, 2001; Wegge, 1997). We argue that doing so will require community-based approaches characterized by (1) transparency and accountability in all aspects of the program, (2) external review and oversight, (3) a mix of top-down and bottom-up authority that enjoys local community support, (4) strong and active local involvement, and (5) effective, adaptive argali conservation management using funds generated by hunters.

Transparency includes providing the public with information about where money generated from trophy hunting goes and the results of conservation management actions, especially surveys of exploited populations (Amgalanbaatar et al., 2002; Harris and Pletscher, 2002; Wegge, 1997). In addition, governments, individuals, and organizations involved in the program must be held accountable for their actions. Trophy hunting generates large sums of money, so transparency and accountability can help stem the potential for corruption.

External review and oversight would also help stop corruption, direct funds to conservation and local people, and ensure program efficacy. We support the proposal from Amgalanbaatar et al. (2002) to create an external committee comprised of people with pertinent expertise, who do not stand to gain or lose from trophy hunting (see also Schuerholz, 2001). Such a committee could oversee a community-based approach to trophy hunting that relied on a mix of top-down and bottom-up authority. Incorporating a bottom-up approach would ensure strong and active local involvement. Mongolian law requires the involvement of the Ministry and other government officials, but combining that with community-operated trophy hunting organizations, in which benefits also accrued to the local community, would help develop and maintain strong local support (Harris and Pletscher, 2002; Johnson, 1997; Wegge 1997). See Schuerholz (2001) and Amgalanbaatar et al. (2002) for more details.

Finally, we recommend approaching argali and ibex trophy hunting from an adaptive management perspective (Holling, 1978). This requires gathering substantial data on the consequences of different management actions (especially target population dynamics), frequent and frank evaluation, and altering approaches based on those evaluations (Kleiman et al., 2000; Shackleton, 2001; Wegge, 1997). Money generated by trophy hunting should go to support such community-based trophy hunting programs, including paying for monitoring, hiring, and training rangers to stop poaching, improving livestock management in hunting areas, and covering the costs of program oversight and management.

CONCLUSIONS

Trophy hunters seek to harvest argali and ibex in Mongolia due to the animals' large body size and horns. Yet, little, if any, money generated by trophy hunting of mountain ungulates in Mongolia goes to conservation management or to benefit local people. We support calls by others (Amgalanbaatar et al., 2002; Schuerholz, 2001) for dramatic changes to trophy hunting in Mongolia. We suggest that community-based trophy hunting programs could provide money and strong incentives for sustainable conservation of these charismatic species, as well as engender enduring local support. Furthermore, we believe transparency, external oversight, mix of top-down and bottom-up authority, substantial local involvement, and adaptive management should characterize such a program. Although the challenges to changing trophy hunting in Mongolia are formidable, we argue that the benefits to hunters, local people, and, most importantly, argali, ibex, and their ecosystem are worth the hard work it will require.

ACKNOWLEDGMENTS

Funding for this work was provided by the Denver Zoological Foundation. We thank C.-O. Bazarsad, S. Dulamtseren, T. Galbaatar, R. Harris, H. Mix, Z. Namshir, O. Shagdarsuren, G. Wingard, and J. Wingard.

SUBCHAPTER

16.3

Hunting of Prey Species: A Review of Lessons, Successes, and Pitfalls – Experiences from Kyrgyzstan and Tajikistan

Stefan Michel, Tatjana Rosen***

***Denver Zoological Foundation, Khorog, GBAO, Tajikistan**
****Tajikistan and Kyrgyzstan Snow Leopard Programs, Panthera, Khorog, Gorno-Badakhshan, Tajikistan**

DEVELOPMENT OF HUNTING MANAGEMENT OF MOUNTAIN UNGULATES IN THE POST-SOVIET ERA

During the market-oriented reforms of the later 1980s, private business became involved in the emerging trophy hunting by foreigners, providing local services such as permits, transportation, accommodations, and guides for hunters who bought their tours via western outfitters. The system of hunting management areas, as it evolved during the Soviet period, where game was harvested by managers to fulfill state plans, or by domestic sport-hunters, allowed for development of area-based private game management. Now, hunting management areas are assigned to legal entities providing them with rights and responsibilities concerning game on the assigned territory, but not with land-use rights (e.g., livestock grazing). Contracts on hunting concessions are concluded for 10 (Tajikistan) or 15 years (Kyrgyzstan).

KYRGYZSTAN

In Kyrgyzstan, there were nearly 100 hunting concessions offering hunts on argali (*Ovis ammon*) and Asiatic ibex (*Capra sibirica*) by 2010. Many concessions were too small for the conservation of ungulate populations sufficient for a biologically sustainable and economically viable operation. In this situation, rangers were not employed, guards of hunting camps poached for subsistence and trade in meat, quotas were exceeded and hunters guided into areas outside of the actual concessions. The new law "On hunting and game management" (Kyrgyz Republic, 2014) defines minimum area sizes (70,000 ha for argali; 20,000 ha for ibex), and fewer but larger blocks are assigned. The government expects from these and other reforms a strengthened motivation to sustainably manage wildlife.

National conservation NGOs claim that hunting quotas for foreigners present a threat to the snow leopard (*Panthera uncia*) population due to reduction of prey. Anecdotal information collected by the authors from hunters and outfitters suggests a decline in the trophy quality, attributed to overhunting. The annual trophy hunting quota of 80 argali out of estimated 15,000 (0.5%) and 350 Asiatic ibex out of estimated 40,000 (1%) should not by itself lead to a population decline. Because trophy hunters are primarily after mature males, and the target species have a polygynous mating system, the impact on the reproduction is likely negligible. However, long-term data on age and size of harvested trophy males have not been assessed for trends (Almaz Musaev, Hunting Department, personal communication, 2015). If trophy sizes declined, it could be related to the take of younger males (in absence of older ones) or represent a genetic change (i.e., smaller trophies of same age males than decades ago) due to selective hunting for large males, as found by Coltmann et al. (2003) in a population of bighorn sheep.

Hunting by locals (quota 1200 ibex in 2012) and poaching have a higher impact on the ungulate populations than trophy hunting of 350 old male ibex. Recently, the government reduced the domestic ibex quota (2013: 800; 2014: 400) and announced a 2-year moratoria in specific local areas. Until 2014, domestic hunting permits were not area-bound, leading to open access and lack of sustainable management of game populations. The new hunting law allows hunting only in assigned areas, and hunters have to obtain permits from the area managers (Almaz Musaev, personal communication, 2015).

This new regulation, and facilitation by NGOs, has encouraged the establishment of the first community-based organizations aimed at rehabilitating ungulate populations while providing hunting opportunities for members and outsiders, thus creating income for wildlife management and improved livelihoods.

TAJIKISTAN

In Tajikistan the types of hunting management areas include private commercial concessions, family-based management areas, and the more recently formed "community-based conservancies." Examples of each follow.

"Murgab" Concession

This commercial trophy hunting concession in the southeastern Pamirs, covering roughly 2000 km^2 and managed by the same family since the early 1990s, has the reputation of doing the most successful anti-poaching work of all concessions in Tajikistan by controlling the area through its own staff, informal networks, and collaboration with police, national security service, and border guards. Their annual quota for Marco Polo argali (*O. a. polii*) is about 40 rams, which are all taken; a slightly lower number of ibex is taken by trophy hunters.

Several researchers have surveyed areas of the concession, applying different methods. The data somewhat uniformly indicate a growth in argali numbers until recently. Estimated by Fedosenko, A.K., Lushchekina, A.A., 2005. Population status of the Pamir argali (Marco Polo sheep) in Tajikistan. Unpublished manuscript. pp. 23 in November 1995 about 1500 argali in the concession area, and Schaller and Kang (2008) in February and March 2005 recorded 2200 argali in an area covering about 37% of the concession. Surveys during the winters of 2009/10, 2010/11, and 2011/12 by Valdez et al. (2016) yielded counts of 8649, 8392, and 7663 argali, respectively. Although the latter multiyear data set may indicate a slight decrease, the results might be influenced by migration patterns.

Panthera (unpublished data) in September 2013 conducted point surveys from 40 random observation points within 1600 km^2 of the "Murgab" concession and of the Madiyan and Pshart valleys about 70 km away that are not managed as hunting concession but are otherwise ecologically comparable. In the concession, they estimated densities of 5.3 argali and 1.7 ibex per km^2, while in the unmanaged area they found densities of only 0.1 argali and 0.9 ibex per km^2. Livestock densities were also higher in the concession area. The differences in numbers of individual snow leopards detected on camera traps in summer 2012: 16 in "Murgab" and 6 in the unmanaged area matched these differences in prey densities (Kachel, 2014). The relative difference (~2x) in snow leopard density estimates between the two sites (0.87 and 0.46 per km^2) was approximately the same as the relative difference in ibex densities (Kachel, personal communication, 2014).

Not all commercial concessions have the reputation the Murgab one does, and none we are aware of controls poaching to the same extent. To ensure sustainability of harvest levels, the quota should be determined by the government specifically for each concession, rather than issuing a countrywide quota as currently practiced.

M-Sayod: A Family-Based Markhor Management Area

The conservancy M-Sayod, established in 2004 by a family of traditional hunters, covers an area of 120 km^2 of the southwestern edge of the Darvaz Range. The area has become a stronghold of Tajik markhor (*Capra falconeri heptneri*) as well as snow leopards, which have been observed hunting markhor. Surveys in 2012 and 2014 yielded counts of 320 and 388 markhor, respectively (Alidodov et al., 2012, 2014). A camera trap survey covering ~40 km^2 of the management area in 2013 yielded six individually recognizable snow leopards (Panthera, unpublished data). The conservancy, previously financed through hunts of wild boar and ibex as well as tourism, in 2013 and 2014 hosted the first legal trophy hunts for markhor with a quota of three per season. The conservancy agreed to allocate 30% of the revenue to such projects as rehabilitation of irrigation canals to motivate local support for conservation activities. Further, the budget of the local district of Darvaz received 60% of the permit fees (about USD 40,000 per animal in total), while the central government kept the remaining portion, ostensibly for conservation activities.

Community-Based Conservancies

During and after Tajikistan's civil war (1992–1997), poaching was rampant, especially in the Pamirs where food was insufficient, arms were easily accessible, and enforcement of hunting regulations was virtually absent. Ibex and argali populations suffered from this intensive pressure.

Since 2008, with support from foreign organizations, the authors of this subchapter started a facilitation and empowerment process in selected valleys of the Pamirs aimed at traditional hunters and other community members interested in the sustainable use of wildlife. We chose model sites that could be controlled by the communities and are large enough to host at least a few hundred ibex. During the participatory analysis and planning processes, local hunters understood that past declines of ibex and argali were a direct effect of unregulated and intensive hunting, and that such declines in prey had also resulted in lowering of snow leopard numbers. Even though poaching was considered much less intense than in the early 2000s, continuous pressure prevented a recovery of ungulate populations, reducing hunting opportunities as well as prey availability for snow leopards. Local hunters agreed to establish legally recognized control over the areas used by them, to prevent community members as well as outsiders from poaching, and, after recovery of the ibex and argali populations, to start a regulated use, based on surveys and agreed quotas.

In Tajikistan the administrative levels below the district (*jamoats* and villages) have no formal authority to manage natural resources. Thus it was not possible for communities to take over hunting management. One alternative was the village organizations and their associations at the *jamoat* level, and vesting management in these community-based nongovernmental organizations, which had previously been established with assistance from the Aga Khan Development Network as local institutional structures for rural development. These organizations, however, did not see conservation, wildlife management, and hunting as part of their mandate, and traditional hunters had no interest in integrating "their resource" in a broader institutional context. Hence, community members wishing to manage local wildlife and hunting established local NGOs and included in their bylaws the conservation and sustainable use of game animals in designated areas, ecotourism, and support of the well-being and development of their communities.

The first of these NGOs, "Parcham" in the Ravmeddara Gorge of Bartang Valley, was registered in November 2008 and acquired land-use rights over 470 km^2 assigned by the authorities of Rushan District. Soon after, the State Committee on Environmental Protection of GBAO Region recognized the 12 active members of Parcham as volunteer game wardens. The State Forestry Agency in September 2011 assigned to Parcham the rights and responsibilities on game management in this area. None of the members of Parcham were paid by external donors, but they did receive assistance in institutional development and equipment funded by the German government. In late autumn of 2012, the first hunting tourist visited the area and took an ibex (Fig. 16.3.1): for the first time members of Parcham and the community earned legal income from wildlife use, as well as meat and a contribution to a microcredit scheme.

Following this example, other communities established similar organizations and applied for the assignment of hunting management

FIGURE 16.3.1 **International hunter and local villagers celebrate first successful ibex hunt in Parcham, GBAO, Tajikistan.**

areas. Some attempts were unsuccessful where communities lacked sufficiently energetic organizers, private concessionaires had already been assigned the rights, or where areas were not suitable. To date three additional conservancies have been established in the Pamirs: "Darshaydara," managed by the NGO Yoquti Darshay (2010; 413 km^2); "Zong," managed by the NGO Yuz Palang (2013; 415 km^2) and "Alichur," managed by the NGO Burgut (2013; 927 km^2). The area of these community-based conservancies at the end of 2014 covered 2248 km^2, protected and managed by 40 volunteer rangers who received initial equipment support through the same program as Parcham and from the international NGO Panthera.

Supporting NGO scientists, together with the traditional hunters, surveyed populations of mountain ungulates through direct counts. Populations initially appeared to increase, although real trends in population sizes are difficult to determine due to the short time frame since conservancy establishment, variations in survey effort, and detectability. Still, these surveys show minimum population numbers observed on the territory of each conservancy. Ungulate density per km^2 conservancy area ranges from 0.5 to 1.13 in Ravmeddara, Darshay, and Zong (ibex) and 0.36–1.33 in Alichur (ibex and argali together). The total numbers recorded for the first three sites in December 2014 were 1104 ibex, and for Alichur 728 ibex and 508 argali in December 2015. Ungulates are now less shy and easier to observe, possibly a response to reduced poaching pressure. Survey data are used for internal quota setting for trophy hunting, allowing for 1 male ibex or argali to be taken per 100 animals recorded, if at least 5 males are estimated being older than 8 years. Camera trapping yielded records of 15 individual snow leopards (Panthera, unpublished data): Ravmeddara (Winter 2011/12: 6 snow leopards); Darshaydara (Summer 2013: 5); Zong (Summer 2013: 1), Alichur: Spring 2013: none; Summer 2014: at least 3).

CHALLENGES

In Kyrgyzstan, management plans and regular surveys are required of each concession holder and these requirements are enforced. The Hunting Department encourages concessionaires to involve scientists from the National Academy of Sciences in these activities and supports the establishment of conservancies through the allocation of areas to community-based organizations. However, the performance of private concessions is still poorly controlled, few have effective anti-poaching and sometimes hunts occur outside the assigned areas.

In Tajikistan, state institutions such as the Committee for Environmental Protection, the State Forest Agency, and the Academy of Sciences collaborate with the community-based conservancies and the private hunting concessions, and support their work to a limited extent. Management plans and surveys are required, but these requirements are not enforced, and wildlife surveys are conducted almost exclusively in the context of externally funded projects or lack any reliability.

In Kyrgyzstan, as the reallocation under the new hunting law led to a reduction of hunting blocks, political resistance emerged, which may challenge not only this process but sustainable hunting in general. Quota allocation is formally based on survey results, but the practice is to distribute the centrally established quota with little consideration of actual local population numbers and trends.

In Tajikistan allocation of hunting grounds is on a first-come, first-served basis, and in the case of competition for blocks, political influence and financial power decide. The situation is complicated by legal and institutional uncertainty. The State Forest Agency is formally authorized to assign hunting grounds, but district authorities often assign them as well without proper documentation and without informing the State Forest Agency. Quotas are set arbitrarily for ibex and argali and are not area-specific. The timing of quota setting and hunting seasons do not take into account market demand.

In community-based conservancies in Tajikistan, during the three hunting seasons of 2012/13 through 2014/15, 12 foreign hunters legally harvested 11 Asiatic ibex in three conservancies. While nature tourism provides some income for conservancies and community members, hunting tourism provides much more substantial income per client.

Market access for the community-based conservancies is difficult as most foreign outfitters have already established relations with concessions, and the international demand for hunts on Asiatic ibex is limited. The more lucrative argali hunts have thus far not been allocated to community-based conservancies. In Kyrgyzstan, private concessions currently control all areas with huntable argali populations. In Tajikistan an association of private concessions has the right to distribute the countrywide argali quota, so far preventing the allocation of a quota to community-based conservancies. Thus, these conservancies have only limited cash income and rely on the motivation of their members stimulated by the occasional income from tourism and hunting, meat obtained in hunts by foreign hunters, and the option of some subsistence hunting. Achieving sustainability of the latter is a challenge as wildlife populations in the conservancies are too small to sustain higher harvest, and subsistence hunting brings a risk of reducing trophy hunting and ecotourism opportunities.

Allocation of revenues from permit fees includes shares for the local administrative levels, which is of some significance in the case of argali and markhor. Although in Kyrgyzstan these revenues have encouraged some interest in the conservation of argali at local level (Musaev, A., personal communication, 2012), in Tajikistan so far the lack of local budget authority and transparency impede the use of these revenues for community development and thus the creation of incentives at the local level.

In both countries, investment by concessionaires in the management of ungulates is insufficient. In Kyrgyzstan, concessionaires may get reimbursed a percentage of the permit fee if they show evidence of investment in wildlife management. In the past these activities were often of questionable value (predator control or purchase of forage for "emergency" feeding of wildlife). The noncommercial community-based conservancies, while having much less funding, are more effective, because their members carry out their activities either voluntarily or in the context of other activities such as herding of livestock.

Some private concessions have been accused of illegal activities, for example, guiding hunters into protected areas, changing a hunter's trophy for a poached trophy of larger size, selling hunts exceeding the allocated quota, and even illegal trophy hunts on endangered species such as the snow leopard and Tien Shan brown bear (*Ursus arctos isabellinus*). In some community-based conservancies internal control, peer pressure and support from some community members are not yet sufficient to ensure full compliance, and the work is further hampered by outsiders, including police and other officials, poaching or hunting without authorization by the conservancies.

CONCLUSIONS AND PROSPECTS

Hunting management of its prey is far from a panacea for the conservation of the snow leopard. However, in societies where hunting bans cannot be realistically enforced, and where protected areas are limited in space and insufficiently guarded, incentive-based management of mountain ungulates is an important and effective approach. Some private concessions have shown impressive conservation results, while others obviously fail to sustain the resource they use. Hardly any of them provide incentives to the local communities. The assignment of rights to manage and use wildlife to local community institutions is a new development in the countries of the former Soviet Union. First experiences in Tajikistan and Kyrgyzstan show that this approach can reestablish a sense of ownership, and that even limited revenues from hunting and nature tourism create incentives for local people to refrain from poaching and to protect mountain ungulates.

References

Ali, H., Din, J.U., Younus, M., 2014. Pre-Trophy Hunting Quota Allocation Survey of Major Ungulate Species in Selected CCHAs of District Hunza-Nagar, Gilgit-Baltistan. Parks and Wildlife Department, Gilgit-Baltistan.

Alidodov, M., Kholmatov, I., Karimov, K., Michel, S., 2012. Uchet vintorogogo kozla na khrebte Hazratishoh i na Darvazskom khrebte, Respublika Tajikistan. Committee for Environmental Protection under the Government of the Republic of Tajikistan, Dushanbe.

Alidodov, M., Amirov, Z., Oshurmamadov, N., Saidov, K., Holmatov, I., 2014. Uchet vintorogogo kozla na khrebte Hazratishoh i na Darvazskom khrebte, Tajikistan. Forestry Agency under the Government of the Republic of Tajikistan, Dushanbe.

Amgalanbaatar, S., Reading, R.P., 2000. Altai argali. In: Reading, R.P., Miller, B. (Eds.), Endangered Animals: Conflicting Issues. Greenwood Press, Westport, CT, USA, pp. 5–9.

Amgalanbaatar, S., Reading, R.P., Lhagvasuren, B., Batsukh, N., 2002. Argali sheep (*Ovis ammon*) trophy hunting in Mongolia. Pirineos 157, 129–150.

Amgalanbaatar, S., Reading, R.P., Dorzhiev, Ts.Z., 2012. Argali sheep (*Ovis ammon*) current population level and the issues of trophy hunting in Mongolia. Vyestnik Spyevipusk V 2012, 290–293, In Russian.

Asia Foundation, 2007. Mine Licensing: A Mongolian Citizens Guide. The Asia Foundation, Ulaanbaatar, Mongolia.

Berger, J., Buuveibaatar, B., Mishra, C., 2013. Globalization of the cashmere market and the decline of large mammals in Central Asia. Conserv. Biol. 27 (4), 679–689.

Bold, A., Dulamtseren, S., Avirmed, D., Tsendjav, D., Khotolkhuu, N., 1975. Mongolian Game Mammal and Bird Population Sizes and Recommended National Quotas by Year. Biological Institute of the Scientific Commission, Mongolian Academy of Sciences, Ulaanbaatar, Mongolia, in Mongolian.

Caughley, G., 1977. Analysis of Vertebrate Populations. Wiley, London.

CITES, 2000. Convention on International Trade in Endangered Species of Wild Fauna and Flora, Eleventh meeting of the Conference of the Parties, Gigiri (Kenya), April 10–20, 2000. Interpretation and implementation of the Convention Quotas for species in Appendix I: Markhor. Available from: http://www.cites.org/eng/cop/11/doc/28_02.pdf.

Clark, E.L., Munkhbat, J., Dulamtseren, S., Baillie, J.E.M., Batsaikhan, N., Samiya, R., Stubbe, M. (compilers and editors), 2006. Mongolian Red List of Mammals. Regional Red List Series, vol. 1. Zoological Society of London, London. (in English and Mongolian).

Coltmann, D.W., O'Donoghue, P., Jorgenson, J.T., Hogg, J.T., Strobeck, C., Festa-Bianchet, M., 2003. Undesirable evolutionary consequences of trophy hunting. Nature 426, 655–658.

des Clers, B., 1985. Conservation and utilization of the Mongolian argali (*Ovis ammon*): a socio-economic success. In: Hoefs, M. (Ed.), Wild Sheep: Distribution, Management, Conservation of the Sheep of the World, Closely Related Mountain Ungulates. Northern Wild Sheep and Goat Council, Whitehorse, Yukon, Canada, pp. 188–197.

Farrington, J.D., 2005. The impact of mining activities on Mongolia's protected areas: a status report with policy recommendations. Integr. Environ. Assess. Manage. 1 (3), 283–289.

General and Experimental Biology Institute, 1986. Mongolian Game Mammal Population Size and Recommended Quotas. Ecological Laboratory, General and Experimental Biology Institute, Mongolian Academy of Sciences, Ulaanbaatar, Mongolia, In Mongolian.

GoP, 2010. Draft National Rangelands Policy. Ministry of Environment, Islamabad.

GoP, 2013. Pakistan National Snow Leopard Ecosystem Protection Priorities. Climate Change Division, Islamabad.

Harris, R.B., 1995. Ecotourism versus trophy-hunting: Incentives toward conservation in Yeniugoui, Tibetan Plateau, China. In: Bissonette, J.A., Krausman, P.R. (Eds.), Integrating People, Wildlife for a Sustainable Future. The Wildlife Society, Bethesda, MD, pp. 228–234.

Harris, R.B., Pletscher, D.H., 2002. Incentives toward conservation of argali *Ovis ammon*: a case study of trophy hunting in Western China. Oryx 36 (4), 1–9.

Harris, R.B., Wingard, G., Lhagvasuren, B. 2010. 2009 National assessment of mountain ungulates in Mongolia. Unpublished Report: Institute of Biology, Mongolian Academy of Sciences, Mongolian Ministry of Nature, Environment, and Tourism, and Worldwide Fund for Nature-Mongolia, Ulaanbaatar, Mongolia.

Hess, R., Bollman, K., Rasool, G., Chaudhry, A.A., Virk, A.T., Ahmad, A., 1997. Pakistan. In: Shackleton, D.M (Ed.), Wild Sheep and Goats and Their Relatives: Status Survey and Conservation Action Plan for Caprinae. IUCN, Gland.

Holling, C.S., 1978. Adaptive Environmental Assessment and Management. John Wiley and Sons, New York.

Hotte, M., Bereznuk, S., 2001. Compensation for livestock kills by tigers and leopards in Russia. Carnivore Damage Prevention News 3, 6–7.

Institute of Biology, 2001. Population Assessment of Argali in Mongolia, 2001. Institute of Biology, Mongolian Academy of Sciences, Ulaanbaatar, Mongolia, (In Mongolian).

IUCN, UNEP [United Nations Environment Programme], WWF, 1991. Caring for the Earth: A Strategy for Sustainable Living. IUCN, Gland.

Johnson, K.A., 1997. Trophy hunting as a conservation tool for Caprinae in Pakistan. In: Freese, C.H. (Ed.), Harvesting Wild Species: Implications for Biodiversity. John Hopkins University Press, Baltimore.

Kachel, Sh., 2014. Evaluating the Efficacy of Wild Ungulate Trophy Hunting as a Tool for Snow Leopard Conservation in the Pamir Mountains of Tajikistan. MS. thesis in Wildlife Ecology, University of Delaware, 87 pp.

Karanth, K.U., 2002. Limits and opportunities in wildlife conservation. In: Terborgh, J., Schaik, C.V., Davenport, L., Rao, M. (Eds.), Making Parks Work: Strategies for Preserving Tropical Nature. Island Press, Washington, DC.

Khan, M.I., 2003. NASSD Background Paper: Communication for Sustainable Development. IUCN Pakistan, Northern Areas Programme, Gilgit.

Khan, A.A., 2011. An Assessment of the Community-Based Trophy Hunting Program in Gilgit-Baltistan. Parks and Wildlife Department, Gilgit-Baltistan, Pakistan.

Kleiman, D.G., Reading, R.P., Miller, B.J., Clark, T.W., Scott, J.M., Robinson, J., Wallace, R., Cabin, R., Fellman, F., 2000. The importance of improving evaluation in conservation. Conserv. Biol. 14 (2), 1–11.

Law of the Kyrgyz Republic, On hunting and game management, 2014.

Liu, C.G., Lu, J., Yu, Y.Q., Wang, W., Ji, M.Z., Guo, S.T., 2000. A comprehensive evaluation on management of three international hunting grounds for argali in Gansu. Chinese Biodivers. 8, 441–448, in Chinese.

Lkhagvasuren, B., Amgalanbaatar, S. Harris, R.B., 2010. Population estimates of mountain ungulates in Mongolia. Unpublished report: Institute of Biology, Mongolian Academy of Sciences, Ulaanbaatar, Mongolia.

Lushchekina, A., 1994. The status of argali in Kirgizstan, Tadjikistan, and Mongolia. Report to the US Fish and Wildlife Service, Office of Scientific Authority, Washington, DC.

Mallon, D.P., Dulamtseren, S., Bold, A., Reading, R.P., Amgalanbaatar, S., 1997. Mongolia. In: Shackleton, D.M., IUCN/SSC Caprinae Specialist Group (Eds.), Wild Sheep and Goats and Their Relatives: Status Survey and Conservation Action Plan for Caprinae. IUCN/SSC Caprinae Specialist Group, Gland, Switzerland and Cambridge, UK, pp. 193–201.

Mir, A., 2006. Impact Assessment of Community-Based Trophy Hunting in MACP Areas of NWFP and Northern Areas. IUCN Pakistan, Islamabad.

Mishra, C., Allen, P., McCarthy, T., Madhusudan, M.D., Bayarjargal, A., Prins, H.H., 2003. The role of incentive programs in conserving the snow leopard. Conserv. Biol. 17, 1512–1520.

Mongolian Statistical Office, 1999. Mongolian Statistical Yearbook 1998. Mongolian Statistical Office, Ulaanbaatar, Mongolia.

Mongolian Statistical Office, 2013. Mongolian Statistical Yearbook 2013. Mongolian Statistical Office, Ulaanbaatar, Mongolia.

Parks and Wildlife Department, Gilgit-Baltistan, 2014. Brief Note on Trophy Hunting Program for Gilgit-Baltistan, 2014–15. Parks and Wildlife Department, Gilgit-Baltistan.

Reading, R.P., Bedunah, D.J., Amgalanbaatar, S., 2010. Conserving Mongolia's grasslands with challenges, opportunities, and lessons for North America's Great Plains. Great Plains Res. 20 (1), 85–108.

Reading, R.P., Wingard, G., Selenge, T., Amgalanbaatar, S., 2015. The crucial importance of protected areas to conserving Mongolia's natural heritage. In: Wuerthner, G., Crist, E., Butler, T. (Eds.), Protecting the Wild: Parks and Wilderness, the Foundation for Conservation. Island Press, Washington, DC, pp. 257–265.

Law of the Republic of Tajikistan, On hunting and game management, 2014.

Schaller, G.B., 1998. Wildlife of the Tibetan Steppe. University of Chicago Press, Chicago, IL, USA, 373 pp.

Schaller, G.B., Kang, A., 2008. Status of Marco Polo sheep *Ovis ammon polii* in China and adjacent countries: conservation of a vulnerable species. Oryx 42, 100–106.

Schuerholz, G., 2001. Community based wildlife management (CBWM) in the Altai Sayan Ecoregion of Mongolia feasibility assessment: opportunities for and barriers to CBWM. Report to WWF-Mongolia, Ulaanbaatar, Mongolia.

Shackleton, D.M., 2001. A Review of Community-Based Trophy Hunting Programs in Pakistan. Mountain Areas Conservancy Project, Pakistan, 59 p.

Valdez, R., Michel, S., Subbotin, A., Klich, D., 2016. Status and population structure of a hunted population of Marco Polo argali *Ovis ammon polii* (Cetartiodactyla, Bovidae) in Southeastern Tajikistan. Mammalia 80 (1), 49–57.

Wegge, P., 1997. Preliminary guidelines for sustainable use of wild caprins. Wild Sheep, Goats, their Relatives: Status Survey, Action Plan for Caprinae, Shackleton, D., the IUCN, SSC., Caprinae Specialist Group, (Eds.), IUCN, Gland, Switzerland, pp. 365–372.

Wingard, J.R., Odgerel, P., 2001. Compendium of Environmental Law and Practice in Mongolia. GTZ Commercial and Civil Law Reform Project and GTZ Nature and Conservation and Buffer Zone Development Project, Ulaanbaatar, Mongolia, 409 pp.

Wingard, J.R., Zahler, P., 2006. Silent steppe: the illegal wildlife trade crisis in Mongolia. Mongolia Discussion Papers, East Asia and Pacific Environment and Social Development Department, World Bank, Washington, DC.

Wingard, G.J., Harris, R.B., Amgalanbaatar, S., Reading, R.P., 2011. Estimating abundance of mountain ungulates incorporating imperfect detection: Argali *Ovis ammon* in the Gobi Desert, Mongolia. Wildl. Res. 17 (1), 93–101.

World Bank, 2006. Mongolia: A review of environmental and social impacts in the mining sector. Mongolia Discussion Papers, East Asia and Pacific Environment and Social Development Department, World Bank, Washington, DC.

Zahler, P., Lhagasuren, B., Reading, R.P., Wingard, J.R., Amgalanbaatar, S., Gombobaatar, S., Barton, N.W.H., Onon, Y., 2004. Illegal and unsustainable wildlife hunting and trade in Mongolia. Mongolian J. Biol. Sci. 2, 23–32.

CHAPTER

17

Environmental Education for Snow Leopard Conservation

Darla Hillard, Mike Weddle**, Sujatha Padmanabhan[†], Som Ale*[,‡], Tungalagtuya Khuukhenduu[§], Chagat Almashev[¶]*

*Snow Leopard Conservancy, Sonoma, CA, USA
**Jane Goodall Environmental Middle School, Salem, OR, USA
[†]Kalpavriksh Environment Action Group, Pune, Maharashtra, India
[‡]Biological Sciences, University of Illinois, Chicago, IL, USA
[§]Nomadic Nature Conservation, Ulaanbaatar, Mongolia
[¶]Foundation for Sustainable Development of the Altai, Gorno-Altaisk, Altai Republic, Russia

INTRODUCTION

Across the 12 snow leopard (*Panthera uncia*) range countries, the need for environmental education (EE) has become more critical as poaching of the cats and their prey and conflicts between livestock herders and the cats, intensify. Efforts to date have been largely limited to identifying key objectives and target audiences, and the implementation of relatively small-scale EE activities with a focus on snow leopards and their habitat. School-based EE has introduced teachers to experiential learning rather than the traditional rote system. Few range country EE programs have started with pilot projects aimed at testing and evaluating how EE is best presented, whether the audience is school children, youth, adults, or a combination of both. We found none that attempted to assess, qualitatively or quantitatively, how (or if) EE led to any changes in conservation behavior over the long term. It has become clear that changing behaviors that negatively affect the local or global environment or an endangered species such as the snow leopard is complex, time-consuming, and demanding in terms of resources (Jacobson et al., 2006). Although a discussion of the many components involved in designing,

Snow Leopards
http://dx.doi.org/10.1016/B978-0-12-802213-9.00017-1

implementing, and evaluating EE programs is beyond the scope of this paper, we focus on different approaches to EE for school-age children, and the need for moving from awareness to action.

WHAT IS EE?

EE came into practical existence in 1975 during a conference in Belgrade (then Yugoslavia), where delegates ratified the Belgrade Charter outlining the basic structure of EE. Two years later, UNESCO and the UN Environment Program held the Intergovernmental Conference on Environmental Education in Tbilisi, Republic of Georgia, where the goals, objectives, and guiding principles of EE were established (McCrea, 2006).

Most individuals develop their interest, understanding, and attitudes toward animals and conservation through an accumulation of experiences from different sources at different times (i.e., helping to care for a family pet, learning about biology in school, watching nature shows, reading about animals, and visiting parks, farms and zoos). None of these experiences alone can be said to "cause" someone to know or care about animals (Falk, 2014).

Good EE teaches students a sense of place, of interconnectedness and stewardship. All environmental education can be reduced to three elements: knowledge, compassion, and action.

These elements need not be addressed in any prescribed order. Students may start by acquiring knowledge about an environmental issue: habitat loss for an endangered species, for example. After learning about the species, they may develop compassion and then they want to take action. Alternatively, the students might start by taking action. They might want to learn why they are taking this particular action, which brings them knowledge. From that knowledge they develop compassion. Perhaps the student hears or sees a particular situation that engenders compassion. The student wants to gain knowledge of the situation and then wants to take action.

Many techniques are available for creating education and outreach for conservation. School curricula often include environmental topics, but too few offer comprehensive programs or focus on achieving conservation goals. Yet the need for conservation education continues to increase as problems become more complex (Jacobson et al., 2006).

CHALLENGES IN TEACHING SNOW LEOPARD-FOCUSED EE

Throughout the snow leopard range countries, public education is free. Annual incomes in rural regions range from about USD 400 to USD 2000. Generally, urban centers have excellent capacity in terms of facilities and access to materials, to deliver EE. But rural primary schools, particularly in remote villages, often lack such basic amenities as electricity, classroom furniture, or even pencils and paper. Teachers are not always recruited from the local community, and may not be viewed as – or feel themselves to be – fitting into the culture. Thus absenteeism and turnover are high. Where there is a functional school, teachers may already have a full classroom schedule and be reluctant to add EE. Even where EE is part of the official curriculum, teachers seldom have access to relevant EE curricula or teaching guides. Another challenge is the lack of trained personnel.

EE employs the principles of experiential education (learning by direct experience), which has been a widely recognized teaching model since the 1970s. Many of the range country school systems still use rote teaching. Even though this approach may have its place in certain subjects such as mathematics, range country teachers and students are invariably excited by experiential, interactive EE, where the classroom is often a mountain meadow or streamside.

The launching of the Global Snow Leopard and Ecosystem Protection Plan (see Chapter 45) is promising in terms of greater individual government support for EE. But not all range country governments are stable, and politics can also be a constraint. A case in point is Nepal, where in the early 2000s, the Snow Leopard Conservancy's (SLC) EE programs were cut short by the Maoist revolution and the flight into exile of three key individual partners. In Russia, the current regime is supportive of snow leopard conservation and education, highlighting EE efforts on the official government web page. Conversely, government officials have been caught poaching by helicopter, which flies in the face of conservation efforts. In addition, strained relations between Russia and the United States have led to extreme scrutiny and tightened restrictions on all Russian nonprofits with links to western NGOs, such as requiring their staff to register as foreign agents.

DIFFERENT APPROACHES TO SNOW LEOPARD EE

The goals of EE are the same across the snow leopard range countries: to instill in children a sense of place and interconnectedness, knowledge and appreciation for the biodiversity of their homelands, and to encourage them to understand the importance of harmonious coexistence between humans and wildlife and issues of wildlife conservation so they can become future stewards of their natural environment.

Since 1986, professionals involved in captive management of snow leopards, field-based research, conservation, and education have met every few years to exchange information and updates. Since the first meeting, EE has had a presence in snow leopard conservation planning (Ale, 1995; Cecil, 1986; Dexel, 2003; Hunter et al., 1992; Jafri and Shah, 1992; Mallon and Nurbu, 1986). We are not attempting in this space to present a comprehensive overview of the numerous EE programs targeted specifically for snow leopards and their habitat – sponsored by the SLC, the Snow Leopard Trust (SLT), WWF and others – in snow leopard range countries over the past three decades. Instead, we illustrate several different "approaches" to EE focused on snow leopards and their high mountain habitat, as representative of current programs rangewide, aimed at primary and secondary school students.

RI GYANCHA, INDIA – A SCHOOL-BASED APPROACH

Ladakh lies in the trans-Himalayan region of Jammu and Kashmir, the northernmost state in India. These high mountains are home to a unique assemblage of wildlife, including the endangered snow leopard and Tibetan wolf (*Canis lupus*) as well as native herbivores. Ladakh's population of less than 0.3 million exerts significantly less pressure on the region than in other parts of India. This, coupled with the strong value of nonviolence that encourages a harmonious coexistence with wildlife, has meant that the area's native fauna has survived over the years. However, a wide network of roads has brought change along with modern development to the people, and tourism has seen a phenomenal increase. These changes have led to rapid changes in people's aspirations and lifestyles, which have had a negative impact on the region's fragile environment.

Literacy is currently over 70%. Path-breaking innovations in education include the creation of locally relevant primary textbooks for Leh district. However, while access to education has become easier even for children who live in remote villages, there is a need to improve the quality of education. In a place as unique as Ladakh, education must help children strengthen their relationships with their rich natural and cultural heritage.

In 2006, the SLC–India Trust (SLC-IT) initiated a collaborative EE program with the Pune-based NGO Kalpavriksh. The program, funded by the SLC–United States (SLC–US) and Association for India's Development, was prompted in part by the dearth of localized educational resources. The fun and activity-filled program focused mainly on Ladakh's wild biodiversity, threats faced by snow leopards and other fauna and flora, and conservation actions to tackle them. The program was field-tested in ten schools in the Trans-Himalayan region of Ladakh during the first 3 years. The EE content was put together from many sources: articles, research studies, interviews with local persons, and information gathered from the Wildlife Department and local NGOs. Content included posters, nature activity cards, worksheets, and a board game. There is often a wealth of information that exists that does not percolate down to the school curriculum, information that is crucial for the health of our planet and the survival of species. A successful EE initiative should distil relevant information from various sources and creatively bring it into the program.

The EE materials were published in 2010 in the form of a resource kit titled "Ri Gyancha" (Jewels of the Mountains) for educators in Ladakh. Besides ready-to-use educational material, the kit also includes a handbook with valuable information on Ladakh's wildlife as well as a description of more than 80 activities that could be part of an EE program. Ri Gyancha was released by His Holiness the Dalai Lama and Shri Jairam Ramesh, the then Minister of State for Environment and Forests. Government and private school teachers were trained in the use of the kit, and collaborations were developed with other local NGOs so that the program could reach more schools. A revised edition of Ri Gyancha was published in 2014, affirming the great need for localized information and resources.

The program has been implemented in 45 schools in communities that have reported conflict with snow leopards. In each school, the program is implemented over two years, and often culminates with a small project that tackles a village-level issue initiated with help from the community and a special event in the village where children take the lead and share what they have learned with their families. Ishey Dolma, a Class 7 student reported, "There have been many changes in me... I got a lot of knowledge about plants, animals and birds. I didn't know anything before and through these games I learnt many things."

NOMADIC NATURE TRUNKS, MONGOLIA – THINKING "OUTSIDE THE BOX"

Mongolia's human population is less than 3 million. They are almost equally divided between the capital, Ulaanbaatar, and the mountains and remote steppes of a country half the size of India. The people have retained their traditional nomadic culture, adding modern conveniences such as solar power and motorized vehicles. Even though many rural children are sent to boarding schools, herder communities and families living in small towns need access to quality education.

Nomadic Nature Conservation (NNC) launched its first *Nomadic Nature Trunk* program in 2007 in the Eastern Steppe region, funded by Wildlife Conservation Society in partnership with Conservation Ink and Denver Zoo. NNC was founded by two women who saw the critical need for natural science and conservation education for remote communities in and around protected areas throughout Mongolia. "Nomadic Nature Trunks" are traveling classrooms that provide a 3-week curriculum, with interactive lesson plans and hands-on projects. Lessons are designed to promote positive perceptions of nature and the environment, increase scientific and cultural knowledge, and encourage environmental stewardship. Each trunk includes activities and materials such as puppets, posters, maps, animal tracks, books, and games focused on region-specific biodiversity and conservation concerns.

In 2012 the SLC partnered with NNC to produce a set of trunks with activities focused on EE in the Altai Mountain region. NNC staff held trainings to teach the proper use of the materials in "train-the-trainer" workshops. Teachers and administrators have expressed interest in the program because it is compatible with the Education for Sustainable Development (ESD) curriculum, recently revised to incorporate EE into other subjects. However, teachers report that they do not have sufficient access to the trunks, wish that they could keep them for longer, or that each school could have its own trunk. The trunks are also a resource for the wider community. Staff of Gobi Gurvansaikhan National Park used one as part of a cross-border summer camp with Mongolian, Chinese, and Russian participants. Javzansuren-she, a public awareness specialist, and Byamba, a herder in the Yamaat Mountains, both said the lessons are important and appreciated, they are appropriate for nomadic people, and provide knowledge about the wildlife of the mountains, causes of habitat loss, and importance of mountain ungulates.

SNOW LEOPARD SCOUTS – ENVIRONMENTAL CAMPS IN TWO REGIONS OF NEPAL

Home to some 5000 Sherpa people, Sagarmatha (Mt. Everest) National Park was officially established in 1976 to protect unique Himalayan ecosystems in eastern Nepal. Conflict between livestock herders and predators is intense. The region's rugged terrain and Sherpa culture attract over 35,000 visitors per year. Snow leopards are returning to the Everest ecosystem after an absence of two decades, as documented by Ale and Brown (2009). Although a significant portion of the economy revolves around tourism, most Sherpas are still herders and farmers.

Annapurna Conservation Area (ACAP), Nepal's largest protected area, comprises 7629 km^2 of mountainous landscape in the central part of the country. ACAP is managed by the National Trust for Nature Conservation, and grassroots conservation area management committees. More than 100,000 international tourists visited Annapurna in 2013 (Baral and Dhungana, 2014), making tourism a major source of revenue for park management. But less than 20% of ACAP's 90,000+ local inhabitants are engaged in trade and tourism (Bajracharya, 2003); most are subsistence farmers and herders. Snow leopards raid livestock, and crops are stolen by animals such as blue sheep and monkeys. In both these remote, virtually roadless areas, the poor quality of schools and rudimentary infrastructure are hurdles to snow leopard conservation.

The Snow Leopard Scouts EE program was initiated in 2011 in both Sagarmatha National Park and ACAP, with funding from SLC. The program operates as an extracurricular ecoclub. Middle school students who are technology-wise are paired with elder livestock herders, who know through long experience the snow leopard's habits and movements. At the annual snow leopard environmental awareness camp, the student-herder teams are taught techniques in snow leopard research and monitoring including remote-camera trapping, collection of scats, prey vantage-point counts and classification, and habitat characterization. Participants gain an understanding of the root causes of human wildlife conflict and brainstorm on how to resolve these using local means and resources. At the end of the camp they appreciate the significance of snow leopards as the apex predator for the high-altitude food web, and develop the skills needed for promoting their own awareness campaigns.

The vision is that these citizen scientists help to create a network of predator-friendly communities across the Nepal Himalayas, to enable snow leopards to travel and disperse along an essentially threat-free landscape. This, at its core, is also the policy of the Master Plan for Forestry Sector and National Biodiversity Conservation Strategy.

In the first year, ACAP teams captured 11snow leopard images on 13 trail cameras. That same year their camera held a surprise: the perfectly framed image of a forest leopard (*Panthera pardus*). It was suspected that the two species were sympatric in the lower parts of ACAP, but these student-herder teams delivered the proof and were rewarded by wide media coverage. Since then they have been annually monitoring local snow leopard populations. In 2012 one group identified 3 individuals and 2013 all 3 plus 2 more were identified and monitored through 7 strategically placed remote cameras.

As of 2014, four Scout groups were active in Sagarmatha NP and ACAP. Students helped spread snow leopard conservation messages through arts, village meetings, and street dramas. Two dozen teachers and more than 100 students and herders have participated in the program. It is worth noting that the Environmental Camp model is employed by the SLT, WWF, and other organizations involved in snow leopard conservation.

SNOW LEOPARD DAY FESTIVAL, ALTAI REPUBLIC, RUSSIA

The Altai Republic covers some 93,240 km^2 and supports just over 200,000 people, whose main occupations are herding and farming. The Altai Mountains run roughly along the republic's western border and south into Mongolia. In 1998, UNESCO recognized the Golden Mountains of Altai, as a World Heritage Site. National and international tourism has grown, and predictions are that visitation will reach 3 million per year by 2020 (Gordon et al., 2009).

To the indigenous Altaians, for whom cultural and spiritual practices are deeply intertwined and rooted in reverence for the natural world, snow leopards are sacred totem animals. Since the dissolution of the Soviet Union in 1991, the people are bringing back the ancient ceremonies and rituals that reaffirmed the community's spiritual connections to their sacred lands and animals.

However, also since 1991, poaching in the Altai and trade in wildlife parts has grown, partly in response to the sudden loss of state-sponsored jobs. Conservation and education efforts are particularly critical here, as both snow leopards and their prey were severely diminished by 2010. Collaborative research with Mongolian biologists indicated the possibility that snow leopards from Mongolia could – if protected – repopulate the Golden Mountains of the Altai.

In 2010, the local NGO Foundation for Sustainable Development of Altai (FSDA) and the Ukok Nature Park initiated the Altai Republic's first Snow Leopard Day festival. It was held in the small town of Kosh-Agach, gateway to Quiet Zone Ukok Nature Park, one of three areas within the World Heritage Site. The park's then deputy-director and cultural expert worked with middle school teachers to design art contests, traditional dances, plays, and other activities, embracing the sacred nature of the snow leopard. WWF-Russia, FSDA/Altai Assistance Project, the Republic's Education Ministry, and SLC provided funding and technical support. The program grew to include schools in four remote districts as well as a day of celebrations in the main square of Gorno-Altaisk, the republic's capital.

CROSS-BORDER EE EXCHANGES

International partnerships between western and range country schools give students the opportunity to learn about life and cultures in other parts of the world and to collaborate on meaningful real life environmental issues.

Salem is Oregon's state capital. With a population of around 150,000 it is a blend of urban and rural cultures. An hour's drive from the Pacific Ocean and the Cascade Mountains, Salem is ideally situated for student field trips. The temperate rainforests of Western Oregon are renowned for their complex ecosystems, fascinating wildlife, and rich cultural history.

The Jane Goodall Environmental Middle School (JGEMS) in Salem is named for Dr Goodall but has no connection to her organization. JGEMS started in 1995 as a Roots & Shoots Club, which evolved into a class and then an entire school focused on environmental science and community service. JGEMS students are selected via a lottery open to all in the school district. The only requirement is the willingness to work outside, get muddy, and work toward the JGEMS mission – to improve their community both locally and globally. Early in the program, JGEMS established partnerships with international wildlife conservation organizations. One JGEMS teacher spent a sabbatical year travelling around the world making connections with such organizations, including the SLC. Starting in the sixth grade, students learn in the classroom and field about old growth forest ecology, river and wetland ecology, taxonomy, and endangered species.

Students working with the SLC become experts on the cats, develop their own recovery plan, including what they would do for the animal where it lives as well as what they could do in Salem, Oregon, to help. One group exchanged art work with a school in northern India and raised money for a corral improvement project there. Besides raising money for their partner organization, the group also visits a local elementary school to talk to younger students about snow leopards and wildlife conservation in general. They present their recovery plan, including wetland ecology, taxonomy, and endangered species. Administrators and other staff attend as guests. The panelists ask questions of the group, and they must defend their plan.

ZOOS AND SNOW LEOPARD EE

Zoos that keep snow leopards maintain myriad programs to educate children generally about the cats, threats to their survival, and conservation challenges and opportunities. These zoos may also have *in situ* EE programs overseen by their own staff or they may help fund an outside partner such as the SLC or SLT.

MONITORING AND EVALUATION

In environmental education as in conservation management itself, there is a lack of a widespread culture of evaluation. Also, there is significant uncertainty about the differences between outputs, outcomes, and impacts, between the short, medium, and long-term effects of EE. Further, EE programs face the problem of measuring developments that are essentially long-term but having to do this under the eyes of stakeholders anxious for progress, in the short to medium term (Fein et al., 2001). "The real things, the ways in which environmental education can change someone's life, are much more subtle and difficult to measure" (Thomson and Hoffman, 2010). Despite our dedication to developing effective EE it is difficult to measure success by actual behavior changes. Even the term "behavior change" means different things to different people (Jacobson et al., 2006).

We found no broad-based qualitative and/or quantitative assessments of snow leopard range country EE. Although many programs have carried out immediate assessments (i.e., before-and-after tests or questionnaires), we found no long-term tracking of students, from their school years into young adulthood, when they would begin making life choices and when the impact, if any, of their EE experiences would begin to show. In the absence of such assessments, we might gain some insights from the experiences of the zoo EE community. Falk (2014) cites the Multi-Institutional Research Project, funded by the US National Science Foundation. In this project, a random sample of 1862 zoo and aquarium visitors agreed to fill out questionnaires, then participate in follow-up interviews a year later. The initial visit prompted many individuals to reconsider their role in environmental problems and conservation action, and to see themselves as part of the solution. In the follow-up interview, about half mentioned a particular animal or species as the highlight of their visit, and over half gave detailed descriptions of what they learned, describing how their visit either reinforced or added to their prior understandings. Most believed that zoos play an important role in species preservation and in increasing their visitors' awareness of conservation issues.

We can perhaps infer from the results of this study that where our snow leopard EE students have a direct or indirect experience of the cats in their natural habitat, a similar proportion will carry the impact into their future lives. A few examples:

In Ladakh, there have been two episodes of snow leopards either overtly threatening or actually raiding livestock pens. The latter incident resulted in severe economic loss to the herder, but in both cases, local educators from SLC-IT convinced the local communities to take nonlethal measures to save the lives of snow leopards.

Nepalese Snow Leopard Scout Ramesh Sunar grew up considering animals more as targets for his slingshot than as part of a healthy ecosystem. When his cameras captured the images of six wild snow leopards, Ramesh became the first "Scout Teacher," passing on his knowledge, skills, and passion to the younger Scouts. Ramesh proudly shares the trail camera photos with his classmates and the adults in his community and fully believes they are having an impact.

Cassidy Huun attended JGEMS from 2004 to 2006, and graduated in 2014 with a degree in biology from the University of Oregon. In a recent communication, she wrote: "My time at JGEMS provided me with the skills necessary to think of all situations as a whole. I learned about ecosystems and how so many different species have interactions that keep the system working. This way of thinking can be applied to so many situations and has greatly shaped the way I have lived my life. Everything I do has an impact on others and the environment. Environmental education has instilled in me a responsibility for all of my actions, but it is a responsibility that I am happy to take on."

FROM AWARENESS TO ACTION

Just as monitoring and tracking of EE is challenging, Jacobson et al. (2006) are among many EE specialists who realize that moving people from awareness to action is not a simple task, yet it is a critical one. "Environmental education has failed because it's not keeping pace with environmental degradation, with human impacts on the environment" (Nijhuis, 2011).

On August 20, 2013, WWF declared that the planet had reached Earth Overshoot Day – the point when humanity had used as much of our renewable natural resources as earth can regenerate in one year (WWF European Policy Office, 2013). The results of a survey about climate change, conducted at 10 zoos and 5 aquariums across the United States, illustrate the need to find ways in which the general public can feel empowered to turn what they learn about EE into action for conservation. The vast majority of survey respondents acknowledged the human role in climate change. Nearly two-thirds agreed that it is mostly human-caused, and 69% also wished to personally do more to address this issue, but many perceived significant obstacles, including perception of low personal impact on addressing climate change, pessimism that people in general are not willing to change their behavior or do what is needed to address the issues, and lack of knowledge about effective and affordable actions (Luebke et al., 2014). While these hard facts may be the foundations for a sense of hopelessness, the snow leopard EE community can effect positive change for the future of these cats.

CONCLUSIONS

Our tendency has been to forge ahead under our gut instinct that EE is a critical component of any conservation effort, and that exposure to EE at an early age will lead to adult conservationists. Surely, as we have illustrated, that is true to some extent, but we must also realize it's not enough to do EE programs and trust that they will create the passion and commitment in young adults to change the world. We need to move beyond the passive to the active, and both time and collective power are of the essence.

We should replace the term "environmental education" with "responsible citizenship education" (RCE) (Nijhuis, 2011). We believe it's safe to say it's not just in western countries that the environmental movement has become marginalized, and that environmentalists are often seen as "tree huggers" – antidevelopment or free-market protesters. Responsible citizenship is our individual and collective obligation, whether we acknowledge it, or like it, or not. We need to nurture more RCE specialists, train them in monitoring and evaluation, and see that these techniques are embedded into every project.

If ever there was a symbol of wild nature, and thus a unifying force for conservation action, snow leopards are it. They are beautiful, mysterious, rare, and sacred. And as virtually all the snow leopard EE programs attest, artists, poets, writers, and actors can be powerful motivators for responsible citizenship.

Joe Rohde, executive designer at Disney Imagineering, made a painting expedition to Mongolia to raise money for snow leopard conservation. In his blog, he wrote "I sat in a snow leopard's cave, looking out at its dwindling remnant of the world, and thought, 'If we can't save this creature... what can we save?' I think it is our time now, the artists, storytellers, spiritual leaders, the poets. The rational scientific arguments for wildlife conservation and sustainable living are well-made and well-distributed... and it is not enough. People are not rational beings. They are the stuff that dreams are made on, creatures of poetry, spirit, and emotion. If they are to change, as they must... then it will be story that changes them."

The Snow Leopard Network (SLN) was established in 2002 to facilitate the exchange of information and promote sound, scientifically based conservation of snow leopards through networking and collaboration between individuals, organizations, and governments (see Chapter 43). We believe this network should be mobilized to turn awareness to action:

- We should encourage artists, writers, and poets to join the SLN, and help motivate action.
- SLN should have a student division where young people with an interest in Central Asia's high mountain diversity can interact and seek international support, if necessary, for their ideas and action plans. Students need to know – as part of every EE program – how their political process works and what the opportunities are for action at

the local, state, and national levels to turn their beliefs into policy. We need the next generation of leaders in each of the range countries to be grounded in appreciation for snow leopards, their fragile mountain habitat, and the often-ancient cultures that share the land. Via the SLN, this student division could track their progress and support each other.

- Zoos that keep snow leopards should also join the SLN. With 175 million people walking through the gates of AZA-accredited zoos each year, these institutions have the power to be incredibly effective agents for education and conservation action. The SLN member organizations need their help in moving from awareness to action.
- SLN should house an online library of resources for teachers, NGOs, and club organizers. Everyone wants to reinvent the wheel, producing their own materials, but there are excellent resources available now that can be easily adapted for different conditions across the snow leopard's range countries.
- Philanthropy is in its nascent stages in the range countries, and few programs are self-sustaining financially, instead relying on funding from the Europe and the United States. The collective power of the SLN could be brought to bear on changing the culture of giving in the range countries.
- We need to ensure that donors understand why they must invest in EE for the long term, and that funding for evaluation and long-term monitoring are part of the investment.
- Social media and other online communications can help facilitate EE, and can motivate the kinds of changes we advocate here. Electronic communications should be encouraged and facilitated where possible, but at the same time we recognize that there are limitations. Remote villages in the snow leopard's habitat are generally without access to the Internet, even though the herder communities are especially important players. Where countries block access to social media for political reasons, it may be dangerous for individuals to make their voices heard.
- Finally, we need to continue to celebrate our successes, whether they are long or short term, while also learning from our failures.

References

Ale, S.B., 1995. The Annapurna Conservation Area Project: A Case Study of an Integrated Conservation and Development Project in Nepal, 155–169.

Ale, S.B., Brown, J.S., 2009. Prey behavior leads to predator: a case study of the Himalayan tahr and the snow leopard in Sagarmatha (Mt. Everest) National Park, Nepal. Israel J. Ecol. Evol. 55, 315–327.

Bajracharya, S.B., 2003. Community involvement in conservation: an assessment of impacts and implications in the Annapurna Conservation Area, Nepal. Masters Dissertation. Institute of Geography, University of Edinburgh, Edinburgh, UK.

Baral, N., Dhungana, A., 2014. Diversifying finance mechanisms for protected areas capitalizing on untapped revenues. Forest Policy Econ. 41, 60–67.

Cecil, R., 1986. Educational programming for snow leopard conservation. Proceedings of the Fifth International Snow Leopard Symposium, 247–248.

Dexel, B., 2003. Snow leopard conservation in Kyrgyzstan: enforcement, education and research activities by the German Society for Nature Conservation (NABU). International Pedigree Book for Snow Leopards (*Uncia uncia*) 8, 18–20.

Falk, J.H., 2014. Evidence for the Educational Value of Zoos and Aquariums. World Association of Zoos & Aquariums (WAZA) Magazine, Vol. 15.

Fein, J., Scott, W., Tilbury, D., 2001. Education and conservation: lessons from an evaluation. Environ. Educ. Res. 7 (4), 379–395.

Gordon, D., Zimmerman, L., Martin, M., 2009. The Golden Mountains of Altai: a treasure of biodiversity and culture. Altai Alliance, Pacific Environment. Available from: www.altaiproject.org/wp-content/uploads/2013/04/Altai-White-Paper_final.pdf (accessed 21.01.2016).

Hunter, D.O., Jackson, R., Freeman, H., Hillard, D., 1992. Project Snow Leopard: a model for conserving Central Asian biodiversity. Proceedings of the Seventh International Snow Leopard Symposium, 247–252.

Jacobson, S.K., McDuff, M.D., Monroe, M.C., 2006. Conservation Education and Outreach Techniques. Oxford University Press, Oxford, UK.

Jafri, R.H., Shah, F., 1992. The role of education and research in the conservation of snow leopard and its habitat in Northern Pakistan. Proceedings of the Seventh International Snow Leopard Symposium, 271–277.

Luebke, J.F., DeGregoria Kelly, L., Grajal, A., 2014. Beyond Facts: The Role of Zoos and Aquariums in Environmental Solutions. World Association of Zoos & Aquariums (WAZA) Magazine, vol. 15.

Mallon, D.P., Nurbu, C., 1986. A conservation program for the snow leopard in Kashmir. Proceedings of the Fifth International Snow Leopard Symposium, 207–214.

McCrea, E., 2006, The Roots of Environmental Education: How the Past Supports the Future. Environmental Education and Training Partnership (EETAP).

Nijhuis, M., 2011. Interview with Charles Saylan: Green Failure: What's Wrong With Environmental Education? Yale Environment 360.

Thomson, G., Hoffman, J., 2010. Measuring the Success of Environmental Education Programs. Sierra Club of Canada, BC Chapter.

WWF European Policy Office, 2013. In the red: humanity's demands exceed Earth's available resources. WWF European Policy Office. Available from: http://www.wwf.eu/?209773/In-the-red (accessed 21.01.2016).

CHAPTER

18

Law Enforcement in Snow Leopard Conservation

Nick Beale, Ioana Botezatu***

*Panthera, London, UK

**INTERPOL, Lyon, France

SNOW LEOPARDS – ILLEGAL KILLING AND TRADE

Conflict Cycles and Killings

Big cats, including snow leopards (*Panthera uncia*), face increasing conflict with people. Large areas of snow leopard range are in regions of human poverty where loss of livestock frequently leads to retaliatory killing of cats and there is a rising threat from poaching.

In the 1990s, snow leopard numbers in Kyrgyzstan and Tajikistan declined three- or fourfold, with poachers taking as many as 120 animals in a single year. The fur trade in Afghanistan reemerged after the fall of the Taliban and fresh demand from international aid workers and military – until conservationists launched a targeted awareness campaign (Snow Leopard Conservancy, 2015).

A lack of awareness, funding, staff, and training means that snow leopards – even in many protected areas – are effectively unprotected by any meaningful law enforcement (Snow Leopard Network, 2014). Broadly speaking there are three categories of poacher:

- Professional (those whose primary income is generated through hunting) – targeting specific species, including snow leopard
- Amateur – generalists or opportunists (including retaliatory killing)
- Trophy/sport hunters – involved in the trade and killing of a number of Asian big cat species

The most common poaching method is trapping snow leopards by leg traps or snares. The involvement of local residents is significant in that they may contribute in various ways to all types of poaching, whether as guides, hunters, or just a source of information.

Sport hunting is a real threat in some areas, with a rising number of documented cases of snow leopard prey species being illegally hunted by professional hunting guides and high-paying clients. Many international borders in the region are "porous" and a lack of meaningful enforcement allows for unchallenged movement of hunters, eager to obtain trophies

Snow Leopards
http://dx.doi.org/10.1016/B978-0-12-802213-9.00018-3

of prey species and potentially snow leopards. Local guides are generally willing to provide information for money, so developing a strong community-based informant network is crucial to countering this emerging threat.

Laws aimed at protecting snow leopards and other wildlife are often inadequate and underfunded, and tackling threats to wildlife and snow leopards tends to be a low priority. Success will only be possible through a concerted and coordinated effort by high-level agencies, including INTERPOL, the South Asian Wildlife Enforcement Network (SAWEN), the World Customs Organization (WCO), and the Convention on International Trade in Endangered Species (CITES).

High-Level Trade, Its Expansion and Criminal Linkages

Wildlife and other environmental crimes have the potential to undermine the rule of law and good governance. Despite cross-border trade in snow leopards being illegal, sections of the global community are neglecting their obligation to enforce this legislation. Criminals exploit weaknesses in law enforcement systems, and profits generated by illicit trade in wildlife may be channeled into other criminal activities, including drug trafficking, counterfeit products and other illicit supply chains, or weapons smuggling. Links with terrorists should be further investigated by police agencies.

Prices are high enough to act as a strong incentive for killing and trade in pelts, derivatives for Asian medicine, or live animals. The existence of wealthy buyers, coupled with factors such as local poverty and lack of education, stimulate the trade in wildlife. The demand for wildlife is facilitated by networks, the ease of transport, and rapidly evolving technologies that signal profound changes in global markets. The responsibility of police agencies or border forces has become increasingly relevant in the apprehension of individuals engaged in this activity. Wildlife can be transported from the source to sites thousands of kilometers away in just few hours. Poachers, transporters, facilitators, and individuals masterminding the trade from urban centers and actual end-buyers represent the actors in a corrosive black market. The key is to identify the criminal linkages that perpetuate this trade and share information among national agencies and between countries.

David Higgins, INTERPOL Environmental Security Head, in "The Dull Anger of Environmental Crime" (WWF, 2012) states, "… we need to broaden our thinking of who the criminals are and what a victim is. … Environmental crime is individuals within our community that are stealing from us as people. They are taking from our future and from our future generations." Such ideas should guide us all, but mainly politicians who can generate change and redefine mandates and priorities for law enforcement and good governance.

THE TWO ENDS OF THE ILLEGAL TRADE MARKET: WHAT MUST BE DONE?

Improving Protection and Counteracting Killings

Recent evidence showing that organized criminals are involved in the trade of Asian big cats and their body parts includes:

- The increasingly organized nature of poaching. Teams are being deployed and supported in distant and remote sites harboring big cats (often nations foreign to the poacher).
- Multiple shipments are being confiscated. Extrapolating from this, the true scale of shipments is far higher than the few seizures identified. Criminals are increasingly being represented by high-quality lawyers.
- Persecuted offenders are often reoffending. (EIA, 2014)

Clearly, counteractions are required to stem illegal killing and trade.

First, these crimes take place in remote areas often isolated from any meaningful law enforcement. Secondly, habitats are being increasingly disturbed and are under rising pressure from poor herding families for whom even small amounts of money make a huge difference to their quality of life. And finally, many law enforcement staff working in wildlife protection are inadequately paid, undervalued, and receive very little, if any, professional training. Low levels of motivation mean these individuals are not incentivized to risk their personal safety for a wild cat.

Detection, arrest, evidential procedure, and prosecution must improve. An overhaul of current systems is crucial, and governments must acknowledge that organized criminals are operating freely, and that there is a correlation between smuggling wildlife products across borders and trade in people, arms, drugs, and terrorism. Governments must commit to increasing the number of personnel and the provision of professional training, funding, and equipment. Community engagement is also vital in remote areas where employing members of local communities in site-security is a way of harnessing the valuable knowledge and skills of those who may currently be engaged in illegal poaching.

The Upstream Trade in Illegal Products and Counteractions

Individuals engaged in wildlife crime need a certain level of organization and the ability to operate in networks, as well as being able to benefit from technology advances. A clear visualization of such networks can help law enforcement identify which individuals are the most connected, the most influential, the most entrepreneurial, and the most elusive. The most powerful counteractions law enforcement agencies have against snow leopard crime include employing sensitive information, sharing it transnationally in a professional manner, and ensuring appropriate levels of integrity and security. Even when such actions result in one or more players being removed, a criminal network's resilience should not be underestimated.

Proactive law enforcement through intelligence presents an opportunity to act in advance and to get ahead of the offenders and the crimes they are likely to commit. Even though high profile wildlife criminals often escape prosecution, successful investigations have resulted prosecution, such as the case of Anson Wong, convicted and sentenced to almost 6 years in jail and given a considerable fine, after a US Fish and Wildlife Service investigation (Christy, 2010).

Beyond national governmental agencies, several international actors have the ability to contribute in a meaningful way, such as the International System of Notices, which provides law enforcement alerts, allowing law enforcement agencies to share critical crime-related information, access global databases, and recommend a course of action. In 2013, INTERPOL and UNEP coorganized the International Environmental Compliance and Enforcement Conference where global leaders recognized that each country's biodiversity and natural resources are an important part of its economic well-being and security, yet are continuously threatened by overexploitation from criminal elements. The decision makers agreed that in order to maintain environmental security, governments need to invest in current activities, while developing innovative strategies and ensuring their efficient implementation within a global strategy.

PROTECTION: A LANDSCAPE SOLUTION AND LEGAL STRATEGY

Protection Through Patrolling

Patrolling by rangers, police, military, other governmental agencies, or NGOs is a proven way of improving overall protection for landscapes, habitats, and species. Patrolling, raises awareness, provides a deterrent, and ultimately promotes prosecution of illegal activity.

Snow leopard habitat is rugged and difficult to navigate. On this basis, a security team could implement a patrol strategy focusing on a specified "core area." A core area is a manageable area of good-quality habitat incorporated within a larger landscape in which a number of snow leopard home ranges could be located. Normal home ranges vary between 12 km^2 and 500 km^2 (Snow Leopard Network, 2014), hence, an area of 1000 km^2 would be a reasonable area in which to start a patrol routine. Patrolling a finite area enables a directed and concerted approach, with the aim of "locking down" a core area. This method is favored over stretching underfunded, badly equipped teams more thinly. As team efficiency increases and the number of team members grows, this core area can be replicated or expanded to increase the total area patrolled.

Reliable data on such things as threats and abundance of prey species are important to the establishment of long-term operational goals. The data will also allow effective planning for patrolling and for concentrating resources on areas of importance for wildlife or that have high threats. SMART (Spatial Monitoring and Reporting Tool) is a tool for planning, implementing, monitoring, and reporting law enforcement efforts that can assist patrol teams and protected area management in collecting and assessing data critical to the given protected area. This site security assessment should be conducted as an in-depth reconnaissance of the given area with particular attention to areas of known past incidence of illegal activity, local communities within or near to the core area, as well as entry and exit points (roads/paths and tracks). Included should be areas where human use is high, such as pastures. These "hot spots" should be visited regularly to ensure no illegal activity is occurring. As the data model grows, so can the complexity of patrol routes. Patrols need to be seen in order to have a deterrent effect, but having a team able to blend into their area of operations acting covertly can also bring great gains.

How a team conducts patrolling is partially determined by terrain limitations, but also by funding. Foot patrolling is the most common form of anti-poaching law enforcement, but in the right environment horseback and vehicles allow for significantly larger distance to be covered, albeit less covertly. Motorbikes or quad ATVs might be effective in some moderately rugged terrain occupied by snow leopards, but much of the cat's range is accessible only by foot. Patrol teams need to be trained and equipped and have the skills needed to be effective as wildlife law enforcement officers. Best practice training for law enforcement teams is covered in more detail next.

Ground-Level Investigations and Intelligence Capture

Alongside site-level ground force protection, overt and covert investigations are also important. Led by local and national law enforcement agencies, and frequently NGOs, these investigations aim to disrupt illegal wildlife killing and trade, ultimately leading to arrests. Wildlife crime has proven links to other serious and organized crime including terrorism (Neme et al., 2013). Comparable investigative systems, and advice on conducting investigations such as those used in drug and terror cases, should be made available to wildlife crime units.

Ground-level law enforcement actions and patrolling can harness information on localized and regional snow leopard crime. Information gathered through carefully managed informant networks often provides sensitive information on people, places, and organizations. As an example, Nepal has 26 community-based anti-poaching operations (CBAPO). Their role is to gather information on wildlife crime, trade networks, and trade routes specifically related to snow leopards.

The vast majority of this work is conducted by community members in their local areas.

Forensic and supporting evidence gathered by law enforcement officers at crime scenes or during executive actions provides the structure for prosecuting cases and leads to further investigations. Mishandled information can lead to failed prosecutions and can endanger lives. Good police work, and following evidential procedure, can lead to successful prosecution. Postarrest questioning is an invaluable source of information. In the vast majority of cases, an individual in a remote mountain community who is caught poaching snow leopards will have further useful information. If managed carefully, individuals who are currently or have previously been involved in wildlife crime can be the most useful informants.

Snares and traps all have their own unique "signature." Although not yet a common approach, snare wires and other traps could be a source of human touch DNA evidence. Collecting DNA from animal carcasses has also proven successful (Tobe et al., 2011). Obviously, good forensics can link individuals to a crime scene, and conversely, DNA of a poached animal can be found on the suspect, providing useful evidence for prosecution (Mascarenhas and Deshpande, 2014).

Often overlooked, but a powerful tool in prosecution, is photographic, video, and audio evidence. Patrolling staff have a unique opportunity to record what they find. Identifying individuals involved in crime and the use of such evidence in gathering further information through informant networks is critical to the progress of an investigation. Filming and recording a search, crime scene, or incident is vital to demonstrate evidence at the scene and prove such evidence was not mishandled. Questioning a suspect alongside audio and visual recording can also be effective.

Footage gathered from a covert position that allows a view of a snare or trap can be all that is needed for a successful prosecution if the criminal in question returns to service the trap. Working against traders and intermediaries is clearly a difficult and potentially dangerous role. But with modern covert photographic means, gathering strong evidence against a subject is possible.

Higher-Level, Cross-Border LE Investigations

Wildlife crime can only be effectively disrupted through complementarity, camaraderie, and the convergence of national law enforcement agencies and countries. Sensitive information on known criminals and ongoing cases should be shared and managed under a strict regime. Agencies need to share best practices and network using modern means of communication.

In some countries, wildlife crime is not yet part of mainstream law enforcement, and environmental crime is undervalued in their criminal justice systems, resulting in a lack of investigative follow-up when significant incidents occur. This is often true when organized criminals are linked to cases or suspected of managing the trafficking in wildlife items. This lack of follow-up often happens when key arrests are made, with the offenders escaping prosecution and conviction thus generating frustration both nationally and internationally.

Wildlife crime's recent upsurge, as well as its complexity, calls for a united government approach, linking law enforcement, environmental management authorities, and veterinary institutes.

In the case of snow leopards, the seizure of pelts in a nonrange country may constitute a serious international incident. Depending on the specifics of the incident, a team of law enforcement officers, specialists, and translators may be deployed to assist the investigation's lead authority. INTERPOL has an investigative support team or an incident response team that can be deployed at the request of a member country.

LE Technologies for Conservation

Wildlife law enforcement lacks capacity in all areas, notably technical surveillance capacity. While technology is extensively used as a capacity multiplier across all disciplines of law enforcement, even basic night vision equipment is largely unavailable to most wildlife teams. Well-funded poachers or well-equipped professional hunters are likely to have access to modern technologies, including thermal imaging scopes and light-intensifying binoculars and range finders. Yet anti-poaching teams are often not deployable at night due to a lack of basic security equipment.

Covert and overt CCTV, drones, and remote trigger systems are used as a matter of course across the law enforcement world, and these items would make a significant difference to teams with limited numbers and capacity. The issues of large area coverage and remote rugged terrain still persist in the case of snow leopard range, however, and modern technologies can assist. For example, an unmanned aerial vehicle (UAV) with a thermal imaging camera could be an exceptional tool in a cool/cold mountain habitat, enabling easy detection of people, livestock, and wildlife, including snow leopards. However, it would take a trained operator to make a distinction between individuals surveyed by a UAV.

Technological aids are useful to teams operating at ground level. But intelligence units also require rapid modernization to deal with information flow and intelligence case developments. Several systems are now available; for example, Palantir, i2, and others are exceptional software tools capable of bringing together information from all available sources. These systems help map links between individuals, financial transactions, communications, and areas of suspected trade and poaching (Fig. 18.1).

Good communication is a significant issue for teams operating on the ground in remote environments. Communications are also important in ensuring staff health and safety. Whether as basic as mobile phones or HF/digital radios, working communications should be a priority.

BREAKING THE CHAIN

LE Agencies, Intelligence Needs

Snow leopard and other wildlife crime calls for a national multiagency coordinated response involving police, customs, compliance, and enforcement agencies for environment, wildlife, and forestry management authorities. It also relates to finance and tax agencies, prosecutors and the judiciary, transportation and health departments, and potentially labor authorities. Each agency has its own capabilities, experience, and ability to complement each other's activities. The flow of information between these national entities should be enhanced. Working in partnership with other national agencies may require both formal and informal partnerships, with the most common method used to establish a formal partnership being a Memorandum of Understanding (MOU). Examples of such legal documents are given in the INTERPOL National Environmental Security Task Force (NEST) manual, such as the New Zealand Memorandum of Understanding between three distinct national agencies on the Illegal Trade in Wildlife and Endangered Species. What is valid at the national level can also be expanded at the international level. This occurs where there is growing expectations for international organizations and stakeholders to form an intergovernmental task force for environmental security.

There is broad agreement that intelligence-led enforcement can effectively disrupt and

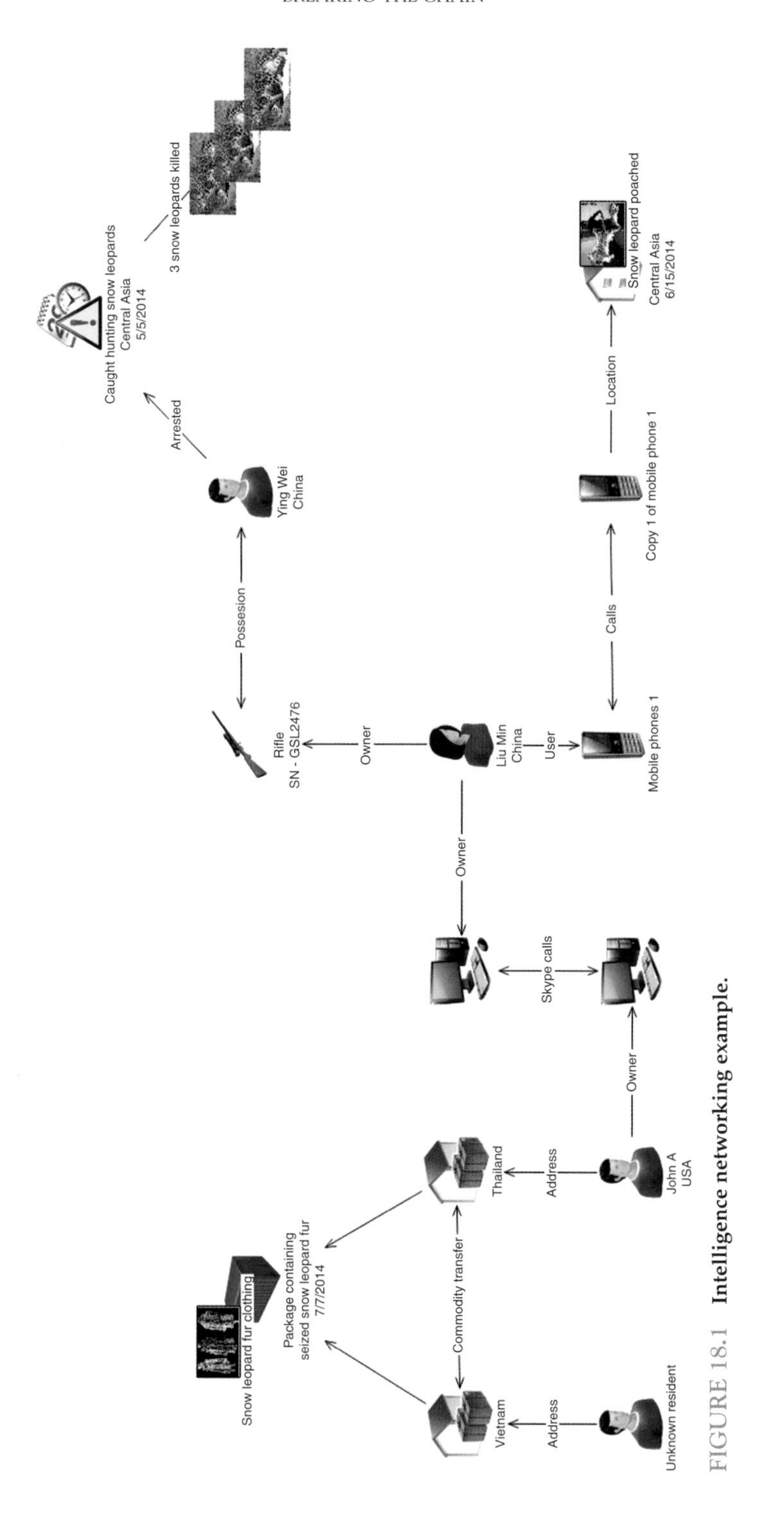

FIGURE 18.1 **Intelligence networking example.**

dismantle major transnational criminal syndicates engaged in the illegal trade of Asian big cats. However, continued improvement in the understanding on how to task intelligence is needed. And this perspective was also supported by the chair of the INTERPOL Wildlife Crime Working Group, Sheldon Jordan, Director of Environment for Canada's Wildlife Enforcement Division in Quebec City. He said, "Managers should ask relevant questions for the intelligence analysts to answer. They should also have an understating of how to integrate both strategic and tactical intelligence into the appropriate levels of investigative planning."

International discussions tend to focus on the best way to manage information related to wildlife crime. They also focus on central information management systems that could be developed in dedicated environmental agencies or incorporated into already existing national mechanisms and with international organizations. Snow leopard range countries have at least a central information management system located with the police. Such systems allow officers to plan, collect, collate, enhance, and analyze information and circulate it on a need-to-know basis. These systems create a central database of known criminals and incidents with a dedicated unit that is able to analyze the information as well as secure means of entering and circulating relevant information. When sharing information both nationally and across borders, several data quality principles must be considered, such as accuracy, adequacy, as well as relevance and timelines. It also should provide a minimum amount of data such as:

- A brief description of the offense (What)
- Place where the offense occurred (Where)
- Method of discovery (How)
- Contraband products (Scientific and common names of the species involved, quantity)
- Identity of persons involved (Name, surname, date of birth, and ID)

International Collaborative and Coordinated LE Approach

Global organizations such as INTERPOL and WCO, as well as regional agencies such as ASEANAPOL and EUROPOL, have a duty to connect agencies, implement decisions, facilitate action, present innovative solutions, drive positive change, and ultimately assist in the identification of crimes and criminals. Work through these agencies does not preclude bilateral law enforcement work, as countries can collaborate directly through their embassies as well as judiciary channels.

A number of United Nations agencies and international conventions are also engaged in the international wildlife crime dialog by supporting countries through capacity development initiatives. These include – but are not limited to – United Nations Environmental Program (UNEP), United Nations Development Program (UNDP), United Nations of Office of Drugs and Crime (UNODC), the World Bank through their Global Tiger Initiative, as well as conventions such as the Convention of International Trade in Endangered Species of Wild Fauna and Flora (CITES) and the Convention on the Conservation of Migratory Species of Wild Animals (CMS), which have provisions either on poaching or illegal trade in snow leopards.

Regionally, one network solely focused on snow leopard conservation dialogue is the Global Snow Leopard and Ecosystem Protection Program (GSLEP) with a secretariat located in Bishkek, Kyrgyzstan (see Chapter 45). The South Asian Wildlife Enforcement Network (SAWEN) is another platform for cooperation. Many of these initiatives have been largely supported either through the UN Global Environment Facility (GEF) mechanisms or funding from US Agency for International Development (USAID).

Another platform where Asian big cat conservation has been discussed and activities initiated is the International Consortium for Combatting Wildlife Crime (ICCWC). By a letter of understanding, five international partners have agreed to develop a joint work program to combat wildlife crime. This ambitious collaborative effort was signed during the Global Tiger Summit in 2010.

The Green Customs Initiative (GCI) is another partnership comprised of 11 secretariats of multilateral environmental agreements (MEA) and intergovernmental organizations cooperating to prevent the illegal trade in environmentally sensitive commodities, including endangered species.

International organizations such as INTERPOL bring decision makers and other high-level governmental representatives to annual and biannual meetings, where they demonstrate continued commitment toward maintaining environmental security. INTERPOL's General Assembly in consultation with the INTERPOL Environmental Compliance and Enforcement Committee and the INTERPOL Working Groups on Wildlife, Fisheries, and Pollution Crime design annual strategies and recommend law enforcement action. These strategic discussions also include civil society, academia, and the private sector.

NEXT STEPS

Training and Best Practice for Anti-poaching Site Security Teams

Any professional team relies on high skill levels. Anti-poaching, site security, and law enforcement teams are no different. Wildlife law enforcement is complicated, with issues spanning policing, investigations, physical area security, and armed conflict. The strategies and tactics needed to be effective are learned through training and operational experience, which both take time.

Best practices, standard operating procedures, and a team's actions are critical to safe and effective operations, and working within a legal structure and expectations requires good management. Training team members to their fullest potential is an ongoing cycle, and time must be allowed for selection, basic training, and more specialized training. Team structure should allow for experienced members to mentor new recruits. Training should be conducted by professionals and trainers who are experienced in core skills and training. Military, police, intelligence specialists, customs professionals, and scientists can all have relevant input to training.

Skills that can and should be taught to a team include, but are not limited to:

- Patrol and/or operational tasking, planning, briefing, and debriefing
- Field craft (camouflage and concealment, tactical movement, obstacle crossing)
- Navigation (map and compass)
- Advanced navigation and GPS
- Wildlife patrolling and patrol types and strategies
- Expedition skills
- Wilderness first aid
- Wildlife law enforcement roles and responsibilities
- Hard stops, suspect capture, and question resolution
- Retasking while on operations
- Observation posts and ambush
- Operational/situational awareness
- Basic surveillance
- Close target reconnaissance
- Human tracking
- Evidence handling and crime scene management
- Managing an agent network
- Operational planners course
- Contact drills (firearms training)

Best Practices in Law Enforcement Collaboration

In order to create strong law enforcement collaboration against wildlife crime, genuine long-term political will and continued departmental support are needed. Governments have an increasing responsibility to invest in their criminal justice systems and expand their mandates to include the environment.

A simple way to present best practices is to break law enforcement activities into four broad categories: information and intelligence management, capacity development, investigations and operations, and raising awareness.

We have shown in previous sections that information and intelligence provide investigators with opportunities to prevent crime from occurring. They also provides an in-depth understanding at the strategic level, as well as at the operational level, with a clear picture of existing links between criminals and networks, routes, or modus operandi. In order to avoid fragmented management of information, appropriate channels for information sharing should be explored and utilized. Nongovernmental organizations actively engaging in grassroots-level conservation efforts often hold information that may be especially valuable to law enforcement officials.

Capacity development can mean a range of activities from providing necessary infrastructure to technical support and training, and may also require reallocation of resources. One immediate action recommended to governments is to organize national consultations with all relevant agencies to discuss increased interagency cooperation, as well as preparedness for environmental compliance and enforcement.

Dialogue through intergovernmental platforms should be consistent for law enforcement to identify common snow leopard-related operational and investigative response needs. The use of standardized templates for planning international operations, as well as clear objectives and shared responsibilities between agencies, leads to operational success. Availability of modern investigative tools also helpful.

Raising awareness of wildlife crime and widely communicating best practices in law enforcement will be strong deterrents in stemming the killing and trade in snow leopards. There needs to be a balanced investment toward raising awareness and strengthening law enforcement responses.

References

Christy, B., 2010. The kingpin: An exposé of the world's most notorious wildlife dealer, his special government friend, and his ambitious new plan. Natl. Geogr. 217, 78–107.

EIA, 2014. In cold blood: combating the illegal wildlife trade. Environmental Investigation Agency (EIA). London, UK. Available from: http://www.eia-international.org/wp-content/../EIA-In-Cold-Blood-FINAL.pdf (accessed 15.05.2015).

Mascarenhas, A., Deshpande, V., 2014. Tiger poaching: 3 get 5-year jail in 'fastest' conviction. Indian Express online. Available from: http://indianexpress.com/article/india/india-others/tiger-poaching-3-get-5-year-jail-in-fastest-conviction/#sthash.nxylS1qN.dpuf (accessed 15.05.2015).

Neme, L., Crosta, A., Kalron, N., 2013. Terrorism and the ivory trade. LA Times, October 14, 2013. Available from: http://www.latimes.com (accessed 15.05.2015).

Snow Leopard Conservancy, 2015. Threats to Snow Leopard Survival. Available from: http://snowleopardconservancy.org/threats-to-snow-leopard-survival (accessed 15.05.2015).

Snow Leopard Network, 2014. Snow Leopard Survival Strategy - Revised Version. Available from: http://www.snowleopardnetwork.org/docs/Snow_Leopard_Survival_Strategy_2014.1.pdf (accessed 15.05.2015).

Tobe, S.S., Govan, J., Welch, L.A., 2011. Recovery of human DNA profiles from poached deer remains: a feasibility study. Sci. Jus. 51, 190–195.

WWF, 2012. The dull anger of environmental crime. Available from: http://wwf.panda.org/?206939/david-higgins-manager-interpol-environmental-crime-programme (accessed 10.06.2015).

CHAPTER

19

Transboundary Initiatives and Snow Leopard Conservation

*Tatjana Rosen**, *Peter Zahler***

*Tajikistan and Kyrgyzstan Snow Leopard Programs, Panthera, Khorog, Gorno-Badakhshan, Tajikistan

**WCS Asia Program, Wildlife Conservation Society, New York, NY, USA

TRANSBOUNDARY CONSERVATION AND SNOW LEOPARDS

Snow leopards (*Panthera uncia*) are top predators that live in environments with relatively low productivity, and they are well-known for having large home range sizes, up to 500–800 km^2 (Jackson et al., 2008). Most protected areas are too small to harbor self-sustaining populations of snow leopards. It is therefore essential to design and implement conservation strategies at the scale of landscapes to ensure the long-term persistence of viable populations of snow leopards and their prey (Jackson et al., 2010). Larger populations of snow leopards are inherently more likely to persist, retain greater genetic variation, and are less vulnerable to the stochastic factors that can negatively influence population size and dynamics. Landscape-scale planning for intact metapopulations will safeguard dispersal corridors between core snow leopard populations, maintain genetic variation, and potentially improve resilience to climate change (SLN, 2014).

Political borders rarely coincide with ecological landscapes (López-Hoffman et al., 2010), and this is particularly true of mountain regions where national boundaries commonly follow ridgelines, and where snow leopards and mountain ungulates range on both sides. It has been estimated that up to a third of the snow leopard's known or potential range is located less than 50–100 km from the international borders of the 12 range countries (Jackson, R., SLC, unpublished data). Table 19.1 shows protected areas for all range countries that are located on or within approximately 10–30 km of an international boundary. These areas, as well as all other documented nontransboundary protected areas, are also shown in Fig. 19.1.

The need for transboundary cooperation has always been clear. Turning this clear need into reality has been more difficult, given the political sensitivities related to national sovereignty and international relations. Despite these

Snow Leopards
http://dx.doi.org/10.1016/B978-0-12-802213-9.00019-5

TABLE 19.1 Snow Leopard Conservation Landscapes

Countries	Landscape	Area (km^2)
Afghanistan	Wakhan National Park	10,951
Bhutan	Snow Leopard Habitat	12,110
China	Qilian Mountains	13,600
	Tianshan Mountains	2,376
	Pamirs	15,000
India	Hemis-Spiti	29,000
	Nanda Devi–Gangotri	12,000
	Kanchendzonga–Tawang	5,630
Kazakhstan	Zhetysu Alatau (Jungar Alatau)	16,008
	Northern Tien Shan	23,426
Kyrgyzstan	Sarychat	13,201
Mongolia	Altai	56,000
	South Gobi	82,000
	North Altai	72,000
Nepal	Eastern	9,674
	Central Complex	9,258
	Western	10,436
Pakistan	Hindu Kush	10,541
	Pamir	25,498
	Himalaya	4,659
Russia	Altai	48,000
Tajikistan	Pamir	92,000
Kyrgyzstan–Tajikistan	Alai – Gissar	30,000

Note: This information was compiled from the World Conservation Monitoring Centre's Protected Areas Data Unit (PADU) GIS dataset, supplemented by listings published by country protected area agencies, NGOs, and INGOs. Experts were contacted where information was known to be contradictory, out of date, or lacked recently proposed or established protected areas (e.g., Afghanistan, Kazakhstan, and Russia). There is still an urgent need to validate and update the database on protected areas within snow leopard range on a country-by-country basis.

inherent difficulties, surprising developments have occurred in transboundary cooperation in recent decades, and multiple initiatives are poised to significantly enhance transnational conservation of snow leopards and their associated biodiversity if allowed to proceed.

The most obvious argument that could be put forward as motivation for the establishment of transboundary conservation initiatives is that political boundaries and the processes that put them in place are infamous for ignoring the natural boundaries of, and processes within, ecosystems (WWF ICIMOD, 2001). As a result, ecosystems of various scales throughout the world are divided by international boundaries, with the result being that the various portions of these systems within the respective countries are subjected to different management regimes with different policy and legal frameworks and socioeconomic contexts. The outcome of this political division of ecosystems is often that their ability to function optimally and retain their natural species assemblages is highly compromised. The ability of both government agencies and NGOs to achieve biodiversity conservation targets under these circumstances is thus also compromised. This holds especially true for target species such as the snow leopard that move across political boundaries and whose populations often encompass multiple state jurisdictions.

RATIONALE FOR TRANSBOUNDARY COLLABORATION

There are multiple reasons for encouraging transboundary cooperation in snow leopard conservation. As mentioned, snow leopards cross international boundaries, which make protecting local subpopulations impossible without cooperation between neighboring range states. However, multiple aspects of landscape-level conservation are significantly enhanced by transboundary cooperation.

Political boundaries create barriers to the effective understanding of the biology of a species such as the snow leopard and its prey. Transboundary cooperation facilitates scientific

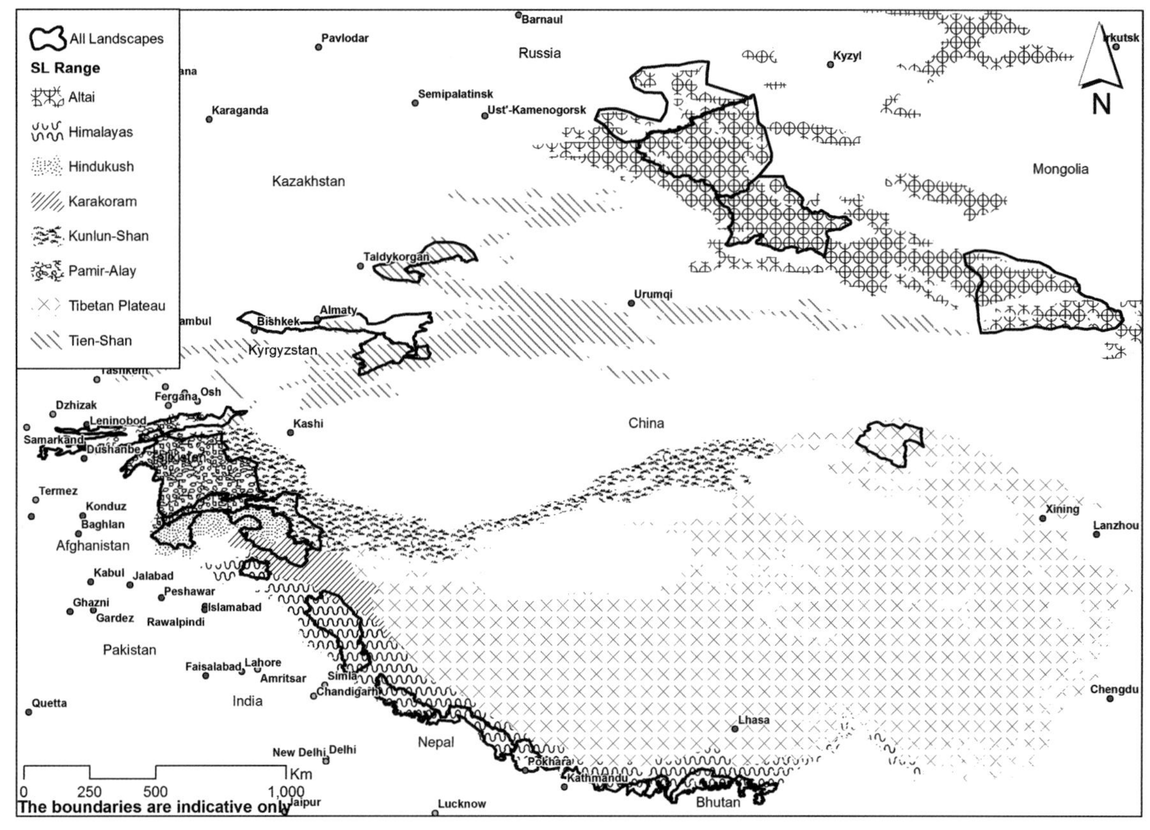

FIGURE 19.1 **PAs located within approximately 10–30 km of an international boundary.** These areas, as well as all other documented nontransboundary PAs are shown. GIS dataset, supplemented by listings published by country PA agencies, NGOs and INGOs. *(Map prepared by the Snow Leopard Conservancy.)*

research and monitoring of populations. Joint research programs can eliminate duplication of activities, enlarge perspectives, pool skills, standardize research methods, and lead to the sharing of equipment and data. Neighboring states, which often have different levels of technical expertise, knowledge, capacity, and financial resources, can benefit by combining their respective strengths through transboundary cooperation.

These same benefits accrue from transboundary cooperation aimed at improving the management of existing protected areas near border areas. The sharing of training materials (and even the training workshops themselves), along with the sharing of practices, experiences, and data, can help to improve management. Within snow leopard range, this transborder cooperation can also raise morale in what are some of the most isolated protected areas in the world, whose staffs are otherwise largely cut off from support and input even from their own national management agencies. Often the closest protected area support can be found across the border, and encouraging the building of transnational relations can positively influence and improve management.

The sharing of management techniques and systems extends beyond protected areas. Community-led conservation initiatives within snow leopard landscapes can also benefit from the sharing of practices and experiences. This can include sharing conservation education information and materials, sharing lessons related to improvements in rangeland management, livestock protection initiatives such as predator-proof corral construction methods, and even training (and deployment) of community rangers for wildlife monitoring and enforcement. Communities on either side of an international border are often closely linked, sharing language, customs, and even direct familial ties, which can lead to natural collaborations that can greatly enhance conservation.

It is not just species of concern such as snow leopards and their ungulate prey that can cross-political boundaries – certain threats such as disease can also be transboundary. Although an understanding of the threat of disease for snow leopards is still in its infancy (see Chapter 9), disease is well-known to affect snow leopard prey species. Infectious diseases, including brucellosis, Q fever, toxoplasmosis, contagious caprine pleuropneumonia, and peste des petits ruminants are known to occur among livestock in snow leopard habitat, and these diseases can pass to wild ungulates – fairly recent examples include a fatal outbreak of sarcoptic scabies among blue sheep (*Pseudois nayaur*) in Pakistan and a devastating pleuropneumonia outbreak in Tajik markhor (*Capra falconeri*) (Ostrowski et al., 2011). Disease pathogens are more easily monitored and outbreaks controlled if joint management is in place to monitor and prevent infection sources from crossing international borders.

Other threats to snow leopards and their prey are also transboundary in nature. Poaching and illegal trade across boundaries is perhaps the single greatest threat to snow leopards, and this threat is more easily controlled by transboundary cooperation. Transport of snow leopard pelts is known to cross international borders, and the fur trade is known to target international tourists, aid or development staff, and military personnel who then transport them across borders as souvenirs (Kretser et al., 2012; Mishra and Fitzherbert, 2004; Wingard and Zahler, 2006). This trade is best controlled through transboundary cooperation, including the sharing of information among enforcement agencies and personnel, but this can also be expanded to include bilateral training activities (e.g., among border guards) and coordinated data collection and management systems such as the use of SMART software (www.smartconservationtools.org) to collect and share law enforcement monitoring data (Stokes, 2010).

A significant threat to both snow leopards and their ungulate prey species are border fences that demarcate national boundaries (Wingard et al., 2014). These fences are often impenetrable barriers and stop critical movements – for snow leopards in their efforts to find new territories, prey, and mates, and for ungulates to find new grazing pastures and to avoid deep snows and other potentially fatal winter weather. Finding solutions to these threats may necessitate bilateral negotiations – especially in cases where both countries have constructed fences on their sides of the border, creating a double barrier.

Transboundary cooperation can also lead to the harmonization of laws and regulations related to wildlife management that can benefit snow leopards. An example of this is trophy hunting – a number of snow leopard range countries allow trophy hunting for wild caprids (e.g., argali or *Ovis ammon*, ibex, markhor), but rarely are these initiatives coordinated in any fashion, despite the fact that some populations of targeted species move across international borders and are thus impacted by trophy hunting on one or both sides. More coordinated sharing of monitoring data can improve each nation's adaptive planning and management for trophy hunting to ensure that offtake remains sustainable, which has obvious benefits to snow leopards that depend upon those same species for food (Rosen, 2012).

Finally, transboundary coordination can provide clear benefits outside of wildlife conservation such as increasing financial opportunities and improving livelihoods for local people. For example, the development of a high-profile transboundary protected area can provide jobs for local people in management of the park, and it can also attract high-end tourism into the region (trekking and climbing), which also creates jobs and income related to guiding, portering, cooking, and housing. Even without the creation of a transboundary protected area, transnational cooperation in tourism outreach related to neighboring national protected areas can significantly enhance international tourism and thus job opportunities, as many adventure tourists appreciate the opportunity to visit multiple countries during their vacations.

A last topic related to transboundary cooperation in wildlife conservation is the potential effect it can have on building relationships and thus peace and security in regions where relations are otherwise sensitive or even at odds. This use of wildlife conservation initiatives for what is sometimes called "track-two diplomacy" (informal dialog to open channels of communication between nations using noncontroversial topics such as wildlife conservation) is a subject unto itself and outside the reach of this chapter, but a great deal has been written on the subject and readers are encouraged to investigate the topic (e.g., see Sandwith et al., 2001; Schoon, 2012; Verma, 2011; Westing, 1998).

THE LEGAL FRAMEWORK FOR TRANSBOUNDARY CONSERVATION

The growing interest in transboundary conservation has led to the somewhat ad hoc emergence of international laws and policies providing policy support for the legal application of this concept. Many types of legal instruments can play a role in promoting transboundary cooperation (Vasilijević et al., 2015). These vary in their level of formality and include multilateral treaties such as the Convention on International Trade in Endangered Species of Wild Fauna and Flora (CITES) and the Convention on the Conservation of Migratory Species of Wild Animals (CMS), bilateral agreements, and Memoranda of Understanding (MoU).

Perhaps one of the most important transboundary frameworks for the conservation of snow leopard is provided by CMS, which entered into force in 1983 and aims to conserve terrestrial, aquatic, and avian migratory species throughout their ranges. By definition it is a convention that requires transboundary cooperation as a tool for conserving migratory species across their borders. One of the important outcomes of the eleventh meeting of the Conference of the Parties to the Convention on the Conservation of Migratory Species of Wild Animals (CMS COP11) was the adoption of the Central Asian Mammals Initiative (CAMI), an innovative and comprehensive framework for the conservation of 15 species of Central Asian mammals, including the snow leopard and one of its key prey species, the argali (Rosen Michel and Roettger, 2014). The CAMI is an opportunity to bring attention and resources to the plight and conservation of these species. As mentioned, the snow leopard is listed as Appendix I of CMS and is one of the 15 species of interest for CAMI.

One important element of transboundary conservation and collaboration is enforcement. The significant increase in illegal wildlife poaching and trafficking across the borders has resulted in the need to develop frameworks to address and respond to this threat. CITES plays a pivotal role in ensuring that all trade in endangered species is sustainable and legal. It is also a key partner in an innovative partnership called the International Consortium on Combating Wildlife Crime (ICCWC). In addition to CITES, the ICCWC brings together INTERPOL, the United Nations Office on Drugs and Crime, the World Bank, and the World Customs Organization. Each of the international organizations offers specialized expertise that supports national enforcement agencies and subregional and regional networks involved in the fight against illegal trade (Vasilijević et al., 2015).

In 2013, representatives from the 12 snow leopard range countries reviewed and endorsed the Global Snow Leopard and Ecosystem Protection Program (GSLEP) (see Chapter 45). The GSLEP seeks to address high-mountain development issues using the conservation of the charismatic and endangered snow leopard as a flagship. A significant focus of the GSLEP is on transboundary cooperation and coordination between the range states, and while each of the states has endorsed this program it remains to be seen whether significant transboundary progress will ensue from this initiative.

Even though formal agreements provide the strongest legal basis for long-term transboundary cooperation, informal agreements can also promote cooperative, friendly relations where the situation is not favorable to more formal arrangements. Informal approaches supplement, complement, and often enhance the more formal processes of governance. Informal arrangements can take the shape of collaborations between researchers, NGOs, or community-based conservancies from bordering countries. Such collaborations help ensure more effective implementation of wildlife conservation goals, and they normally do not require many resources or complex bureaucratic procedures.

CHALLENGES IN IMPLEMENTING TRANSBOUNDARY CONSERVATION

All transboundary initiatives are subject to certain limitations that include the inherent weakness of international or bilateral agreements, changes in political relationships that can affect implementation and progress, and changes in personnel and resources that decrease the effectiveness of efforts. Success will depend on a host of variables, including specificities of local circumstances, the degree to which participants share a vision, the capacity (often embodied in one person or a small staff) to coordinate and convene activities among the independent players, and generally, the ability of actors to artfully frame and implement the process.

Political indifference and lack of commitment toward common and regional issues shared by countries can impede the establishment of a transboundary conservation initiative. Often this lack of commitment results in an absence of implementation of agreements ratified between countries. Regional instability and insecurity can also negatively influence the effectiveness of existing transboundary agreements and collaborative initiatives (Vasilijević et al., 2015).

Additional management challenges specifically relate to the involvement of local communities (Jones, 2005). They range from the lack of government will to engage with local communities, both in terms of comanagement arrangements and ensuring their representation in decision-making processes (Dressler and Büscher, 2008), lack of local capacity, inadequate internal communication and transparency to facilitate local engagement with protected area authorities (Büscher and Schoon, 2009), and a failure to ensure that local communities benefit from initiatives (and thus buy into the effort and work toward its successful implementation).

Other challenges to transboundary projects include protracted processes associated with amendments of legal and policy instruments; different interpretations of and institutional responses to legal and policy implementation requirements; limitations and disparities in ecosystem and species management capacities, as well as in the capacities required to implement systematic conservation planning; and external social, economic, and/or political dynamics, both immediately adjacent to and far removed from the area, which add layers of complexity that can frustrate approaches unless they are fully understood and integrated into management plans. Significant challenges include avoiding unrealistic expectations that are easily created (stakeholder engagement processes need to be handled carefully to guard against this); the ability to ensure that benefits are equitably distributed to beneficiaries, particularly where the necessary structures and processes are either not in place or are unclear or not transparent; language barriers; cultural, historical, and political differences; development disparities, particularly as they relate to the access to resources and capacity for implementation; and a lack of leadership at (multiple) appropriate levels of governance. Other complexities are related to sharing governance responsibilities and/or appointing an objective nonpartisan representative to coordinate implementation, significant differences in terms of land uses and plans for adjacent areas, and conflicting resource management policies, such as adjacent areas that may or may not allow trophy hunting. Finally, the concern about loss of sovereignty often becomes an issue, despite the fact that international agreements related to cooperation and coordination do not supersede a nation's own laws and regulations.

TRANSBOUNDARY CONSERVATION INITIATIVES AND CURRENT STATUS OF TRANSBOUNDARY PROTECTED AREAS

In 1997, more than 31% of the protected areas within snow leopard range, totaling 276,123 km^2, were classified as existing or potential transboundary protected areas, or TPAs (Green and Zhimbiev, 1997). Participants at international conferences held over the past few decades have regularly advocated for transboundary collaboration for the conservation of snow leopards and associated biodiversity, including the establishment of transboundary protected areas. For example, at the 2008 conference on Range-wide Conservation Planning for Snow Leopards held in Beijing, participants called upon range countries to "develop mechanisms (e.g., Memoranda of Understanding) for promoting transboundary cooperation on matters such as trade, research and management relevant to snow leopard conservation that include, *inter alia*, the impacts of climate change on distribution and long-term survival of snow leopards, and where possible incorporating positive actions within conservation programs (e.g., carbon neutral projects)" (SLN, 2014).

Several transboundary, ecosystem-level projects within snow leopard range have been initiated. The GEF West Tien Shan project aimed to improve and increase cooperation among several protected areas, all of which hold snow leopards: Chatkalskiy Nature Reserve (Uzbekistan), Sary-Chelek and Besh-Aral Nature Reserves (Kyrgyzstan), and Aksu-Zhabagly (Kazakhstan). Objectives also included strengthening institutional capacity and national policies, supporting regional cooperation, and enhancing income generation within the protected areas.

The Tien Shan Ecosystem Development Project, also funded by GEF, was launched in 2009 to support management of protected areas and sustainable development in Kazakhstan and Kyrgyzstan. The Pamir-Alai Transboundary Conservation Area (PATCA) project was funded by the EU and examined the option of creating a transboundary protected area across the border between Kyrgyzstan and Tajikistan. A biological database was assembled but no further action was taken, although proposals to establish a protected area still exist.

The Altai-Sayan Ecoregion Project, which began in 2007, aimed to enhance cooperation on biodiversity conservation between Mongolia and Russia in the Altai-Sayan region, and the snow leopard was one of the project's focal species. Subsequently, the governments of Russia, Mongolia, and Kazakhstan prepared and signed agreements to establish the Uvs-Nuur and Altai Transboundary Nature Reserves in 2011–2012, with the UNDP-GEF Project "Biodiversity Conservation in Altai-Sayan Ecoregion" providing a coordinating role. The Altai Transboundary complex consists of the Katunskiy Biosphere Reserve (Zapovednik) (1516.4 km^2) in the Altai Republic, Russia, and the Katon-Karagaysky National Park (6435 km^2) of the Eastern Kazakhstan Region, Kazakhstan. The Uvs-Nuur complex includes the Ubsunurskaya Kotlovina Biosphere Reserve (Zapovednik) (3232 km^2) of the Tuva Republic of Russia, and eight protected areas in Mongolia (Tsagaan Shuvuut Uul Strict Protected Area, Uvs Nuur Strict Protected Area, Tesiin Gol Nature Reserve, Altan Els Strict Protected Area, Khankhokhii National Park, Khyargas Nuur National Park, and Turgen Uul Strictly Protected Area, totaling some 14,000 km^2). A threats assessment was completed in 2012, along with the drafting of the Altai-Sayan Ecoregion Conservation Strategy (WWF, 2012).

The Pamir International Protected Area has been proposed in the eastern Pamirs where the borders of Afghanistan, Pakistan, Tajikistan, and China meet (Schaller, 2007; Xie et al., 2007). This would encompass multiple reserves, including one in China, two in Pakistan, two in Tajikistan, and three (in development) in Afghanistan, totaling 35,870 km^2. The most significant PAs containing snow leopards are Zorkul NR (870 km^2) in Tajikistan, Wakhan National Park (over 10,000 km^2) in Afghanistan, Taxkorgan NR (15,863 km^2) in China, and Khunjerab National Park (6150 km^2) in Pakistan. More recent activities include a meeting on transboundary conservation in Dushanbe, Tajikistan (WCS, 2011), and a transboundary health initiative looking at livestock diseases that may be transmitted to wild snow leopard ungulate prey in the Pamirs of Tajikistan, Pakistan, and Afghanistan (Ostrowski et al., 2012).

Nepal has signed agreements with China and India to facilitate biodiversity and forest management, encompassing six border protected areas under the initiative known as the Sacred Himalayan Landscape. This effort covers about 39,021 km^2 in the eastern and central Himalaya, with 74% located in Nepal, 24% in Sikkim and Darjeeling areas of India, and the remaining 2% in Bhutan (SLN, 2014). The large Qomolangma Nature Reserve (34,000 km^2) is located on the Chinese side.

The Kailash Sacred Landscape (KSL) Conservation Initiative is a collaborative effort of ICIMOD, UNEP, and regional partners from China, India, and Nepal. It represents a sacred landscape significant to hundreds of millions of people in Asia and around the globe, as well as the source of four large rivers (Indus, Brahmaputra, Karnali, and the Sutlej), which serve as lifelines for large parts of Asia and the Indian subcontinent.

Bilateral initiatives exist in the Khangchendzonga landscape between Bhutan and Nepal and India and Nepal. One example of cross-border cooperation on the ground in this region is represented by a joint survey of the Kyrgyz range on the border between Kazakhstan and Kyrgyzstan by scientists from both counties (FFI, 2007).

The "Mountains of Northern Tien Shan" project has been developed for the period 2013–2016 with the assistance of German International Cooperation (GIZ) and the Nature and Biodiversity Conservation Union (NABU). Within this project, a transboundary protected area is planned encompassing three existing protected areas: Chon-Kemin Reserve (Kyrgyz Republic), Chu-Or National Park, and Almaty Reserve (Republic of Kazakhstan).

A new project to strengthen conservation in the Central and Inner Tien Shan of Kyrgyzstan is supported by UNDP and the State Agency on Environmental Protection and Forestry. Among its aims are the establishment of Khan Tengri Natural Park (more than 1870 km^2) in the east of the country. This protected area will border the Republic of Kazakhstan and link Sarychat-Ertash Reserve in Kyrgyzstan with Tomur Reserve in Xinjiang, China.

In 2010, an MoU was signed between Xinjiang Uygur Autonomous Regional Forestry Department (XUARFD) and the Gilgit-Baltistan Forest, Wildlife Parks and Environment Department,

Pakistan. The agreement was for the conservation of wildlife species along the Pakistan–China border area with regards to generating and sharing knowledge about wildlife species and their habitats and developing a joint management plan addressing the issues of wildlife species and their habitats together with suggestive measures for minimizing negative anthropogenic influences on the environment and helping socioeconomic development of the local communities. Following that, in 2011, China participated in a consultation aimed at providing a platform to share the progress made toward the conservation of the ecologically contiguous landscape between China and Pakistan and to develop a common strategic framework of action for the landscape (ICIMOD, 2012).

CONCLUSIONS

Despite the challenges described, the snow leopard's enormous global distribution and the number of range countries make transboundary conservation and collaboration critical in order for snow leopard conservation efforts to succeed. At the very least this will need to include the sharing of knowledge, intelligence on illegal poaching and trade in snow leopards and their parts, and best practices for conservation interventions. A concerted effort must be undertaken to overcome obstacles and facilitate processes to make snow leopard transboundary conservation and collaboration possible.

References

Büscher, B., Schoon, M., 2009. Competition over conservation: collective action and negotiating transfrontier conservation in Southern Africa. J. Int. Wildl. Law Policy 12 (1–2), 33–59.

Dressler, W., Büscher, B., 2008. Market triumphalism and the CBNRM 'crises' at the South African section of the Great Limpopo Transfrontier Park. Geoforum 39 (1), 452–465.

FFI, 2007. Central Asia snow leopard workshop, Bishkek. June 19–21, 2006. Meeting Report. Fauna & Flora International, Cambridge, UK.

Green, M., Zhimbiev, B., 1997. Transboundary protected areas and snow leopard conservation, 194–201. In: R. Jackson and A. Ahmad (Eds.). Proceedings of the Eight International Snow Leopard Symposium, Islamabad, November 1995. International Snow Leopard Trust, Seattle and WWF-Pakistan, Lahore.

ICIMOD, 2012. Towards developing the Karakoram-Pamir landscape: report of the regional consultation to develop future strategic programme for biodiversity management and climate change adaptation. Working Paper 2012/3.

Jackson, R., Mallon, D., McCarthy, T., Chundaway, R.A., Habib, B., 2008. *Panthera uncia*. The IUCN Red List of Threatened Species. Version 2014.3. www.iucnredlist.org (accessed December 20, 2014).

Jackson, R., Mishra, C., McCarthy, T.M., Ale, S.B., 2010. Snow leopards: conflict and conservation. In: Macdonald, D.W., Loveridge, A.J. (Eds.), Biology and Conservation of Wild Felids. Oxford University Press, UK, pp. 417–430.

Jones, J.L., 2005. Transboundary conservation: development implications for communities in KwaZulu-Natal, South Africa. Int. J. Sust. Dev. World 12 (3), 266–278.

Kretser, H., Johnson, M., Hickey, L.M., Zahler, P., Bennett, E., 2012. Supply and demand for wildlife trade products available to US military personnel serving abroad. Biodivers. Conserv. 21, 967–980.

López-Hoffman, L., Varady, R.G., Flessa, K.W., Balvanera, P., 2010. Ecosystem services across borders: a framework for transboundary conservation policy. Front. Ecol. Environ. 8, 84–91.

Mishra, C., Fitzherbert, A., 2004. War and wildlife: a post-conflict assessment of Afghanistan's Wakhan Corridor. Oryx 38, 102–105.

Ostrowski, S., Thiaucourt, F., Amirbekov, M., Mahmadshoev, A., Manso-Silvá, L., Dupuy, V., Vohobov, D., Ziyoev, O., Michel, S., 2011. Fatal outbreak of Mycoplasma capricolum pneumonia in endangered markhors. Emerg. Infect. Dis. 17, 2338–2341.

Ostrowski, S., Yacub, T., Zahler, P., 2012. Transboundary Ecosystem Health in the Pamirs. Wildlife Conservation Society, American Association for the Advancement of Science, University of Veterinary and Animal Science, Lahore, Pakistan, 34 p.

Rosen, T., 2012. Analyzing gaps and options for enhancing argali conservation in Central Asia within the context of the Convention on the Conservation of Migratory Species of Wild Animals. Report prepared for The Convention on the Conservation of Migratory Species of Wild Animals (CMS), Bonn, Germany, and the GIZ Regional Program on Sustainable Use of Natural Resources in Central Asia.

Rosen Michel, T., Roettger, C., 2014. The Central Asian Mammals Initiative: Saving the Last Migrations. UNEP/CMS Secretariat.

Sandwith, T., Shine, C., Hamilton, L., Sheppard, D., 2001. Transboundary protected areas for peace and co-operation. In: Phillips, A. (Ed.), Best Practice Protected Area Guidelines Series No. 7. World Commission on Protected Areas (WCPA), World Conservation Union (IUCN), Gland, Switzerland.

Schaller, G., 2007. A proposal for a Pamir International Peace Park. USDA Forest Service Proc RMRS 49, 227–231.

Schoon, M.L., 2012. Governance in Southern African transboundary protected areas. In: Quinn, M., Broberg, L., Freimund, W. (Eds.), Parks, Peace, and Partnerships. University of Calgary Press, Calgary.

SLN, 2014. Snow Leopard Survival Strategy. Snow Leopard Network, Seattle, Washington, USA, Revised 2014 Version.

Stokes, E.J., 2010. Improving effectiveness of protection efforts in tiger source sites: developing a framework for law enforcement monitoring using MIST. J. Integr. Zool. 5, 363–377.

Vasilijević, M., Zunckel, K., McKinney, M., Erg, B., Schoon, M., Rosen, T., 2015. Transboundary Conservation: A systematic and integrated approach. Best Practice Protected Area Guidelines Series No. 23, Gland, Switzerland: IUCN. xii + 107 pp.

Verma, K., 2011. Siachen – From battlefield to 'peace park'? The South-Asian Life & Times, October–December, 50–59.

WCS, 2011. The Tajik Pamirs: Transboundary Conservation and Management. Wildlife Conservation Society, United States Forest Service, Committee for Environmental Protection under the Republic of Tajikistan, 29 p.

Westing, A.H., 1998. Establishment and management of transfrontier reserves for conflict prevention and confidence building. Environ. Conserv. 25 (2), 91–94.

Wingard, J., Zahler, P., 2006. Silent Steppe: the illegal wildlife trade crisis in Mongolia. Mongolia Discussion Papers, East Asia and Pacific Environment and Social Development Department, World Bank, Washington DC.

Wingard, J., Zahler, P., Victurine, R., Bayasgalan, O., Buuveibaatar, B., 2014. Guidelines for addressing the impact of linear infrastructure guidelines on migratory large mammals in Central Asia. Convention on Migratory Species (CMS) Technical Report, Bonn, Germany.

WWF, 2012. Altai-Sayan ecoregional conservation strategy. World Wide Fund for Nature. Available from: http://wwf.ru/resources/publ/book/eng/843.

WWF, ICIMOD, 2001. Ecoregion-based conservation in the eastern Himalaya: Identifying important areas for biodiversity conservation. WWF Nepal, Kathmandu.

Xie, Y., Kang, A., Wingard, J., Zahler P. 2007. The Pamirs Transboundary Protected Area: A report on the 2006 International Workshop on Wildlife and Habitat Conservation in the Pamirs. Unpublished report: Wildlife Conservation Society, USAID, and the Governments of Tajikistan, Pakistan, Afghanistan, and China.

CHAPTER

20

Corporate Business and the Conservation of the Snow Leopard: Worlds That Need Not Collide

Paul Hotham, Pippa Howard, Helen Nyul, Tony Whitten

Fauna & Flora International, Cambridge, UK

INTRODUCTION

The snow leopard (*Panthera uncia*) lives in some of the most beautiful but sensitive and remote landscapes in the world. These landscapes are increasingly targeted by corporate business for mineral extraction, for example, precious metals and coal, and other forms of large-scale development including alpine tourism, road building, dams, pipelines, and other infrastructure.

Chapter 8 of *Snow Leopard Survival Strategy – Revised Version 2014.1* (Snow Leopard Network, 2014) highlights the potential future threats from the development sector – "Major infrastructural developments are either planned or under construction in different parts of the snow leopard's range, particularly in those countries undergoing rapid economic growth like India, China, Russia, and Kazakhstan. These include mineral exploration and extraction, new gas and oil pipelines, new road and rail transportation networks, and hydro-electric power facilities associated with large or medium-sized dams ..., [elsewhere] upstream water-storage facilities are expected to grow significantly." The chapter goes on to say that "... it becomes increasingly important for range countries to put into place, or act upon, existing regulations in order to minimize negative environmental impacts through careful planning, appropriate mitigation measures and related 'Best Practices'." (Snow Leopard Network, 2014). The strategy effectively makes an urgent call for action and for engagement in processes to ensure that the best possible practices are applied in relation to the planning and mitigation of potential impacts on the snow leopard and its habitats.

Snow Leopards
http://dx.doi.org/10.1016/B978-0-12-802213-9.00020-1

The range and scale of developments and corporate actors implied here is enormous. In order to more effectively review the issues with corporate business, the following chapter primarily focuses on the extractive industry.[1] However, the processes and principles described apply equally well to other large developments. Extractive industry often brings to mind large-scale operations; however, the scale of activity may extend from intensive high-impact but localized operations such as that found at the Kumtor Mine in the Tien Shan mountains of Kyrgyzstan, to low-impact but extensive activities such as that typically found with artisanal mining, which takes place in some areas of Mongolia, China, and the Himalayan countries.

The activities of extractive industry are always cited in literature as a threat, and while those impacts are often obvious at a local scale, there appears to be little scientific evidence compiled concerning their direct and indirect impacts on the snow leopard and its prey across its range. Similarly, while the impacts of other large-scale development can be projected, for example, the potential impacts of building a large scale dam in a mountain habitat can be predicted, their impacts on snow leopards are difficult to determine until site-specific proposals have been put forward.

Although we lack breadth of evidence for the snow leopard, the impacts of large-scale developments and the extractive industry on other Central Asian species, for example, the critically endangered saiga antelope, make it safe to assume that activities of an industrial nature are likely to both directly and indirectly affect the snow leopard: first through both acute or chronic damage to habitat and ecosystem function, and; second from the "ripple effect" of ancillary activities such as the development of road and rail networks and villages, which fragment and open up remote landscapes to hunting, poaching, and livestock grazing. On the other hand, the presence of more engaged and enlightened companies may provide opportunities for mobilizing resources for conservation and research activities, provide safe havens for snow leopard prey species, and create barriers to more unscrupulous companies exploiting the area.

Significantly more research and case work needs to take place into the large-scale development and extractive industry aspects of snow leopard conservation, especially a thorough quantitative assessment of their impacts on snow leopards, their prey species, and habitats.

In the meantime it would be wise to follow the precautionary principle "where there is a threat of significant reduction or loss of biological diversity, lack of full scientific certainty should not be used as a reason for postponing measures to avoid or minimize such a threat" (Convention on Biological Diversity and Rio Declaration 1992) at UNEP.

Today, many businesses are adopting approaches to mitigate their impacts on ecosystems, habitats, and species and are utilizing charismatic species such as the snow leopard as a focus for channeling efforts to meet corporate environmental and social responsibility (CSR) requirements. However, it is recognized that the application of CSR and other approaches is often nonexistent within companies based in countries that are not signatories to global agreements and protocols and may be inadequately implemented by companies that are based in countries carrying such obligations. First, it is useful to consider why a company should manage its impacts on biodiversity at all.

BUSINESS CASE FOR CONSERVATION

A range of business drivers encourage companies to engage in biodiversity conservation and impact mitigation measures. These drivers

[1]The *extractive industry* is made up of the mining, quarrying, dredging, and oil and gas extraction industries.

relate to the development of (i) operational efficiencies; (ii) competitive advantage, social license, and market positioning, and (iii) complying with legislation and lender bank requirements. Essentially, the business that understands how biodiversity supports or presents risks to its operation is able to make better informed decisions and improve its performance in relation to managing business risk, achieving social and environmental goals, and improving the financial bottom line. Sound management of biodiversity and ecosystem services is also an effective way of building trust and confidence with stakeholders and a means by which to engage them to secure the long-term effectiveness of mitigation and conservation activities.

Operational Efficiencies

Companies are increasingly identifying benefits to their operations through adopting environmental practices that incorporate sustainable use measures. For example, recycling water within the operation not only reduces the costs associated with extracting water and controlling effluent emissions but reduces the amount of water taken from habitats that snow leopards or their prey rely on.

CASE STUDY

Anglo American considers water to be a material issue, recognizing that there is the potential for water shortages, cost escalations, and growing legislative complexities related to increasing demand and competition for water resources, compounded by the potential effects of climate change. Given that 70% of Anglo's operations are in water-scarce regions, the business, social, and environmental case for minimizing the amount of water it uses, reusing as much as possible, and discharging as little as it can is indisputable. In 2013, Anglo American reports having saved 22% of its projected water usage for 2013 through developing closed-circuit systems at its operations and managing water use more effectively, for example, incorporating new technologies that help to reduce evaporation from the systems and finding new ways to suppress dust on haul roads rather than using water. Some operations are obtaining water from different sources rather than ground water. For example, Barro Alto mine in Brazil draws 20% of its water use from rainwater. These efficiencies helped Anglo American realize a savings of USD 85 million in water-related costs in 2013 (Anglo American Sustainability Reports, various).

Understanding social and environmental risks at a landscape or ecosystem scale benefits an operation by enabling it to manage risks and take advantage of opportunities. It also ensures timely interventions through the early application of a mitigation hierarchy to avoid, reduce, mitigate, and, where residual impacts may remain, offset impacts and dependencies on biodiversity and ecosystem services. The benefits of this approach include better operational risk management, which results in positive social and environmental outcomes. Importantly, the early application of the mitigation hierarchy within a landscape scale analysis of the mine at the concept stage provides opportunities for the effective analysis of alternatives whereby the operation can seek to maximize avoidance of habitat, for example, that is important to snow leopards.

Competitive Advantage, Social License, and Market Positioning

Companies have also realized competitive advantages when they start to employ better environmental practices. Competitive advantages within the extractives sector include obtaining the legal and social license to operate, which ultimately result in access to land. Businesses that implement policies and strategies to manage impacts on biodiversity and ecosystem services are more likely to be able to demonstrate that they can manage projects sensitively. This will become even more important as areas of extraction become increasingly remote and perhaps found within sensitive areas. Gunningham et al. (2004) stated that "good citizen measures are justified on the grounds that enhancing the firm's reputation for good environmental citizenship (and avoiding a reputation for bad environmental citizenship) will in the short or long run be good business" and further that companies with improved environmental performance have better reputations and "it is argued, will gain readiest access to the means by which to make future profit: development approvals, preferred access to prospective areas and products, the ear of government, the trust of regulators, the tolerance of local communities, and the least risk of being targeted by Environmental NGOs."

CASE STUDY

Anglo American has also formed an alliance with the IRBIS Mongolian Centre, a snow leopard conservation NGO affiliated with the Institute of Biology of the Mongolian Academy of Sciences, to support snow leopard conservation. They have reached an agreement to support a number of activities including (i) the preparation and publication of a book on snow leopard conservation for general public consumption in Mongolian; (ii) support for camera trap and GPS tracking surveys; and (iii) support for genetic and pathogen studies (G. Hancock, personal communication).

Market positioning requires the companies to set themselves high standards that might set the bar for other companies. For example, the middle-sized mining company Eramet wanted global recognition for its highly specific products and activities from its nickel mine in Halmahera, Indonesia. It is committed to adhering to IFC PS6 (see later) and will be one of the first pilots for the Business and Biodiversity Offset Program (BBOP) standard (BBOP, 2012).

Complying With Legislation and Lender Bank Requirements

Awareness of the importance of biodiversity and ecosystem services has increased, as has recognition of the value of these to local communities and the contribution of these services to national economies. This has fueled the growth in policies that refer to biodiversity and ecosystem offsets as a potential or required tool to meet government targets to balance development with environmental stewardship. There has been a significant rise in government policies, guidance, and legislation that require or enable biodiversity offsets since 1965, with acceleration during the past 5 years. Forty-five biodiversity market mechanisms (such as conservation banks) have been developed globally and 27 more are in development; a marked increase from the 39 documented the previous year (Madsen et al., 2011).

Financial institutions have also been incorporating no net loss (NNL) and/or net gain (NG) objectives into their environmental safeguard systems (the term "net gain" is used here as a catch all for all commonly used terms referring to an intention to provide a net gain for biodiversity, including net positive impact and net positive outcome). International Finance Corporation (IFC) Performance Standard 6 (PS6) is the best-known financial lending requirement. PS6 requires a net (biodiversity) gain for impacts on critical habitat and NNL (of biodiversity) where feasible for impacts on natural habitat (IFC, 2012). Habitat that supports snow leopards could be considered critical habitat and all appropriate steps should be taken to avoid affecting this habitat, if it is to receive IFC funding.

CASE STUDY

IFC's eight Environmental and Social Performance Standards (PS) define IFC clients' responsibilities for managing their environmental and social risks. Where projects trigger a PS, they must show in a detailed report whether mitigation measures can be designed and implemented in a satisfactory way to adhere to the requirements set out in the relevant PS. The implementation of actions necessary to meet the requirements of the PSs is managed though the company's Social Environmental Management System, in the case of PS6 this is the Biodiversity Action Plan. PS6 recognizes that protecting and conserving biodiversity is fundamental to sustainable development. The requirements set out in PS6 have been guided by the Convention on Biological Diversity (CBD) and specifically address how companies can avoid or mitigate threats to biodiversity arising from their operations as well as sustainably manage natural resources. For example, paragraph 16 of IFC (2012) identifies critical habitat as areas of significant importance to (i) critically endangered and/or endangered species; (ii) endemic and/or restricted range species; (iii) migratory and/or congregatory species; (iv) highly threatened and/or unique ecosystems; and/or (v) areas associated with key evolutionary processes.

Many development/multilateral banks follow IFC PS6 guidelines or have developed similar approaches themselves, for example, European Bank of Reconstruction and Development (2008), Inter-American Developmental Bank (2011), and European Investment Bank (2012). In addition, financial institutions that abide by the Equator Principles (Equator Principles, 2013) have agreed to follow PS6 in their loan agreements. PS6 is thus becoming a major driver of biodiversity offsets within industry, even for companies that do not normally use multilateral finance, for the following reasons:

- PS6 is viewed as leading practice by many stakeholders. Therefore, corporations are increasingly using PS6 as a global best-practice benchmark.
- The Equator Principles Financial Institutions (more than 75 institutions) have committed to follow PS6 for all relatively large projects in developing countries.
- Nations (especially non-OECD) that own a percentage of extractives projects often obtain their financing from development/multilateral banks, which increasingly follow PS6 or similar.
- In joint-venture or multipartner projects, one partner may have PS6-related financing, which can impact schedules and costs for all partners.

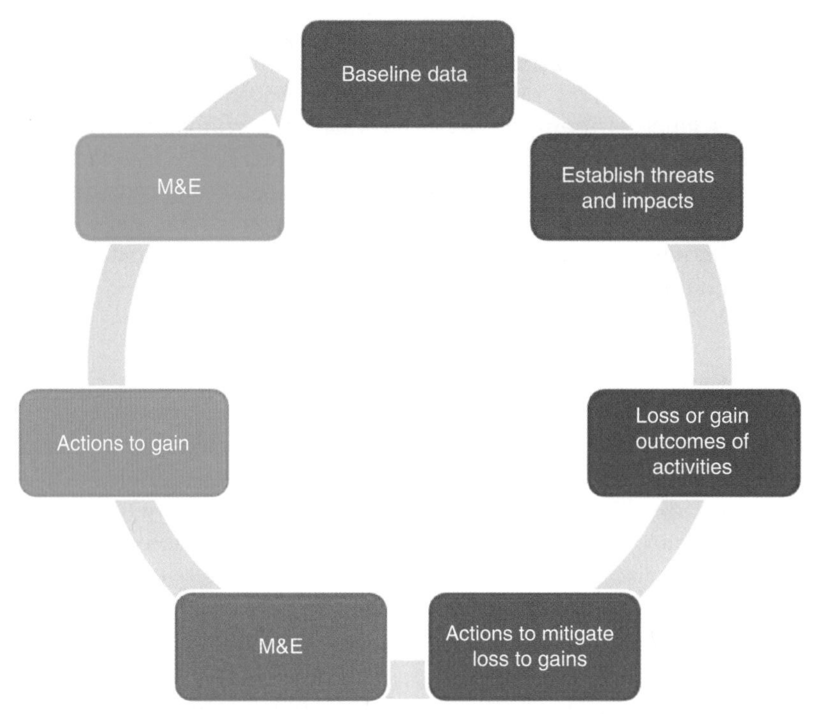

FIGURE 20.1 **Applying best practices as a company in snow leopard territory.**

- Purchase of small or medium-sized companies or projects that were started with bank finance results in inheritance of loan conditions.

Having explored the business case for why a company should manage its impacts on biodiversity, we now explain the five key steps in applying the mitigation hierarchy and best practices (Fig. 20.1).

Stage 1: Baseline Data – Identifying Priority Sites for Snow Leopard

The Environmental and Social Impact Assessment (ESIA) process (stage 2) relies on baseline information to predict and evaluate project impacts, to assess alternative actions, and to support mitigation and monitoring plans. Baseline data need to be specific and relevant, up to date, reflect the seasonality of the site, and collected using standardized methodologies. If there is an expected time lag before implementation, for example, dams or offshore oil and gas developments, it is important to understand baseline trends to estimate what the baseline will look like when the project is implemented, taking into account other projects that are likely to occur in the interim (Abaza et al., 2004).

Detailed information on snow leopard ranges is becoming increasingly available as the number of field surveys and use of camera-trapping and genetic analysis increases. The Global Snow Leopard and Ecosystem Protection Program (GSLEP) identified 23 landscapes to be secured by the year 2020 (Chapter 45). The value of planning across large spatial scales is widely advocated (McCarthy and Chapron 2003; Snow Leopard Network, 2014).

By overlaying data on mining developments, infrastructure, and habitat with snow leopard ranges, potential zones of conflict between snow leopard conservation and current or potential developments can be identified. These then translate into conservation priorities to be addressed.

At a regional and site level, further studies of extractive concession areas must be undertaken to identify habitats that support snow leopards and thus better understand the potential level of risk posed to populations of snow leopards. Ideally this work would be undertaken at the project concept stage so that appropriate mitigation plans can be developed to avoid impacts, then minimize, and then restore them.

Project planners should use the biodiversity baseline to design, construct, and plan the operational phases of the development ensuring maximum avoidance of impact. Aspects to be aware of include:

- Areas of high biodiversity value or endemism and critical threatened ecosystems
- Sensitive sites such as rivers, wetlands, and ridges
- Graves and other culturally significant sites
- The latest published versions of conservation plans

Practices to prevent and limit the biodiversity impacts caused by developments include:

- Avoiding activities in sensitive environments
- Disturbing as few sites as possible, minimizing the footprint
- Using existing access roads and disturbed areas in preference to disturbing new areas
- Using lighter or more effective equipment
- Managing chemicals, hydrocarbons, and waste to prevent pollution
- Timing activities to avoid disturbance of seasonal parameters
- Rehabilitating disturbances as soon as possible

Stage 2: Environmental and Social Impact Assessment (ESIA) – Establish Threats and Impacts

ESIA is arguably one of the most important and widely used tools to guide sustainable development. ESIA uses relevant studies to ensure that environmental concerns are integrated into decision making. Specifically, it aims to identify the environmental impacts of development proposals early on in project planning, and investigate how to avoid or minimize them (Abaza et al., 2004; International Association for Impact Assessment, 2003; Retief et al., 2011). When implemented correctly, ESIA can integrate environmental issues into decision making in a way that considers conservation needs alongside those of developers, governments, and societies alike. ESIAs need to be undertaken early in a project to result in proper application of the mitigation hierarchy (specifically avoidance through the alternatives analysis); incorporate meaningful stakeholder participation; integrate broad environmental, social, and economic impacts; and coordinate the implementation of all activities under strong governance and legal frameworks. The ESIA process should assess all direct and indirect environmental impacts from project conception and exploration through to closure.

ESIA SYSTEM

ESIA systems differ from country to country, but generally involve the following stages:

1. *Screening*: Assess development proposals against regulations in order to determine which proposals require a full or partial ESIA.
2. *Scoping*: Determine the key issues and impacts (direct and indirect) that are relevant for decision making and require further study and may include creation of a Terms of Reference (ToR).
3. *Baseline assessment*: Undertake in-depth studies on key issues such as water resources, terrestrial ecology, aquatic ecology, and social aspects, including natural resource use and cultural heritage.
4. *Identification and evaluation of impacts*: Identify potential environmental impacts and develop actions to avoid, mitigate, or offset them.

5. *ESIA report*: Describe the business case for proposed development, legislative context, existing environment, and predicted impacts; document the significance of residual impacts; and outline mitigation, management, and offset measures.
6. *Review and decision making*: Determine whether the ESIA report satisfies its ToRs and whether the development should be approved or rejected.
7. *Follow-up*: Monitor development impacts against the predetermined baseline and assess the effectiveness of mitigation measures; ensure that developers comply with conditions of approval; and evaluate the effectiveness of the ESIA as a whole to ensure that learning can be fed back into the ESIA process to improve future practice (e.g., Morrison-Saunders et al., 2007; Abaza et al., 2004).

Stage 3: Apply the Mitigation Hierarchy to Develop Management Actions to Mitigate Impacts

Proper use of the mitigation hierarchy should seek to avoid impacts, then minimize, then restore, and finally only use offsets to compensate for the residual impacts after all other options have been exercised. The hierarchy is used to establish an order of preference for mitigation measures in order to achieve NNL or a NG of biodiversity, starting with avoidance and working through the other stages as necessary.

1. *Avoidance*: measures taken to avoid operational impacts on biodiversity, such as careful placement of project infrastructure or temporal planning of activities. The biggest opportunity for avoidance is during options analysis in the ESIA phase and project development. Avoidance measures significantly reduce impacts on biodiversity, thereby reducing future costs of restoration, offsets, and closure.
2. *Minimization/reduction*: measures taken to reduce the duration, intensity, and/or extent of impacts (including primary, secondary, and cumulative impacts, as appropriate) that cannot be completely avoided. It can sometimes be difficult to distinguish between avoidance and minimization because some actions have aspects of both.
3. *Restoration*: the re-establishment of ecosystem structure (diversity), composition (species), and function (processes) to bring it back to its predisturbance state or to a healthy state close to the original. Restoration differs from rehabilitation in that restoration is a long-term process that accounts toward NNL or NG of biodiversity.
4. *Rehabilitation*: the preparation of safe and stable landforms on sites that have been disturbed, followed by revegetation with the aim of establishing a specific habitat type. Rehabilitation is important for improving basic ecosystem functions such as erosion control and water quality regulation.
5. *Offset*: measures taken to environmentally compensate for any residual adverse impacts that cannot be avoided, reduced or restored, in order to achieve NNL or a NG of biodiversity. Offsets include positive management interventions, such as restoration of degraded habitat, arrested degradation or averted risk, protecting areas important for biodiversity conservation (BBOP, 2009a,b).
6. *Additional conservation actions*: a broad range of activities that benefit biodiversity or the ecosystem, but where effects or outcomes are difficult to quantify in terms of biodiversity and ecosystem service gains. Examples include scientific research, environmental education, and building capacity and expertise in conservation organizations. These form an essential part of a company's contribution to biodiversity conservation, often underpinning the success of other mitigation actions and are highly valued by interested stakeholders. However, they are NOT offsets (Fig. 20.2).

CASE STUDY

The presence of the Kumtor Mine has provided benefits to snow leopard conservation in the Central Tien Shan region of Kyrgyzstan. In the late 1990s, multilateral lenders (EBRD and IFC) engaged an NGO to review the biodiversity aspects of Kumtor's Environmental Impact Assessment. The review made recommendation on mitigation measures for the mine, ranging from no-hunting policies to wildlife monitoring, which Kumtor adopted and continues to report results against in their annual environmental reports. The lenders also encouraged the Kyrgyz government to establish the Sarychat-Ertash Reserve (SCER) adjacent the mine site. Once established, the lenders and Kumtor mobilized funds and resources to support NGO-led initiatives that have built the capacity of the reserve, developed a reserve management plan and supported biodiversity-focused research and community development efforts. A strict no-hunting policy is still being applied on the mine site that acts as a barrier to poachers. The number of argali on the reserve has increased from 750 to 2500 head, making it the largest population in Kyrgyzstan. The population of ibex has stabilized at 750–850 head. Research also confirms that the number of snow leopards present in and around the SCER has increased significantly to 18 individuals (Prizma, 2014).

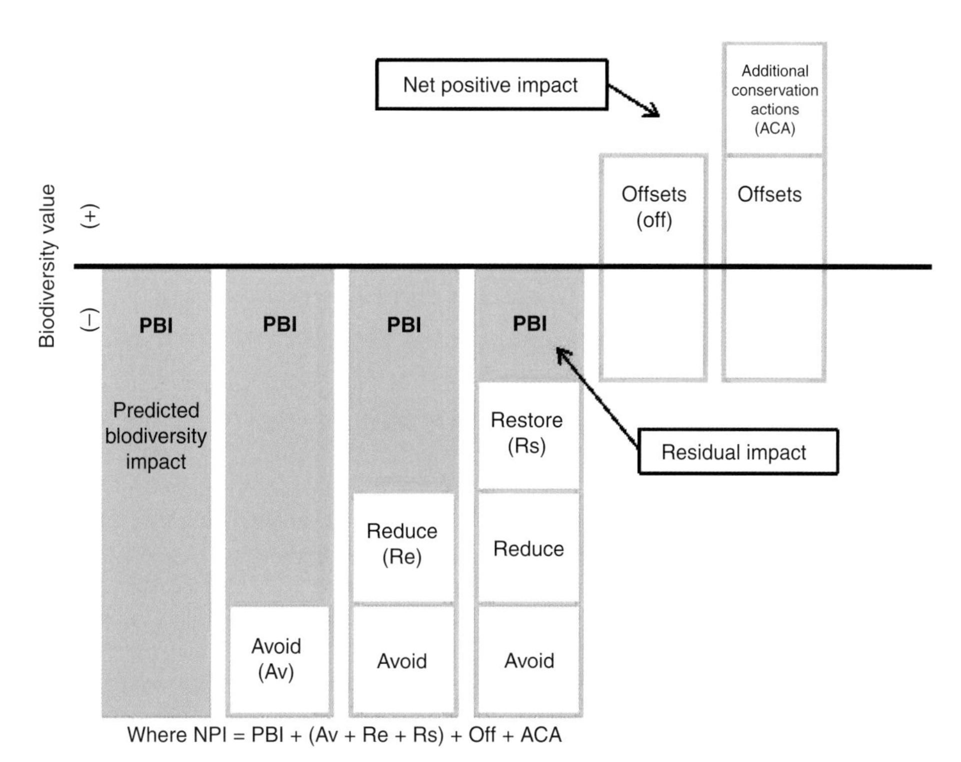

FIGURE 20.2 **Mitigation hierarchy effect on predicted biodiversity impacts (after BBOP, Rio Tinto, 2004).**

ENVIRONMENTAL MANAGEMENT PLANS

Following the ESIA, the company must include measures to prevent or mitigate biodiversity impacts in an Environmental Management Plan (EMP). Site exploration should not commence before legal permission is granted and an approved EMP is prepared. EMPs should take into account the identified direct and indirect impacts on biodiversity and ensure that activities avoid, minimize or restore impacts in areas with high biodiversity value, for example, known snow leopard habitat.

Adequate management, mitigation, and rehabilitation measures should be identified for each phase of the project within the EMP and the full cost of rehabilitation and the long-term management of impacts before the project commences. If a company already operates in habitats that support snow leopards and an ESIA was not required previously, managers should mitigate the impact of the mine through minimization, restoration, and offsetting, aiming for NNL or a NG to that habitat. All companies have an excellent opportunity for conserving threatened ecosystems on those parts of a property unaffected by extractive activities.

BIODIVERSITY OFFSETTING

Most initiatives aiming toward a NNL or net positive outcome (NPO) for biodiversity are framed around the mitigation hierarchy as outlined earlier, and in particular utilize the concept of biodiversity offsets. Biodiversity offsets operate through various mechanisms which include: securing or setting aside land, management actions, and defined conservation and livelihoods activities. They can be used to expand or buffer existing protected areas, create new protected areas, enhance or restore habitats, and protect or manage species. In each case, a conservation gain should be achieved that is relative and related to the impacts of development on a particular species, habitat, or ecosystem. If positive outcomes are achieved through the timely application of the mitigation hierarchy, biodiversity offsets may not be required

A wide range of organizations are working on offsets toward a NNL or NG of biodiversity including companies with a NNL or NPO commitment, such as De Beers, BC Hydro, Norsk Hydro, and Rio Tinto; companies such as Eni and Anglo American that have a commitment to the application of the mitigation hierarchy; financial institutions such as the EBRD and IFC; intergovernmental organizations, for instance the CBD and the IUCN; and a variety of nongovernmental organizations collaborating directly with the private sector in the field including Birdlife International, Fauna & Flora International, and The Nature Conservancy.

CASE STUDY

The United Nations Development Programme (UNDP), in partnership with the Mongolian Ministry of Environment and Green Development, Ministry of Mines and Energy, and Ministry of Industry and Agriculture, with funding from the Global Environment Facility (GEF), is collaborating with the private sector to develop the Land Degradation Mitigation and Offsets in Western Mongolia project. Starting in 2015, the project aims to reduce the negative impacts of mining on rangelands in snow leopard habitat of the western mountain and steppe region by incorporating the mitigation hierarchy and offset into landscape level planning and management. The project will then work with selected mining companies, in close cooperation with local government and communities, to pilot best practice approaches. The aim is to conserve biodiversity and enhance ecosystem services while maintaining ecological function, including pastureland and water quality and quantity. The project will also capture lessons, to be fed into other initiatives seeking to address large scale development impacts in sensitive landscapes. Key criteria provide a decision support framework for selecting sites where the approaches will be implemented. These include the ecology, cultural and historical heritage, socioeconomics, and institutional context. The project sites are highly biodiverse, being remote and in excellent ecological condition. In addition, Sutai Khairkhan, a nationally significant sacred mountain on the border of the Hovd and Gobi Altai Aimags, lies within this snow leopard landscape; the project will contribute to the protection of this important area. The project will also promote collaboration between the different land users to their mutual benefit while protecting vital resources through sustainable development initiatives. Existing activities and sustainable pastoralism projects will contribute to the suite of potential activities.

BIODIVERSITY ACTION PLANS

Biodiversity action plans (BAPs) are a common framework used by companies to help them manage activities specifically related to biodiversity and ecosystem services. BAPs are developed in a variety of ways. The overarching goal of a BAP however should be to serve as a repository of information related to biodiversity and ecosystem services that highlights their importance within the company's area of influence (direct and indirect impacts) as well as the management of those impacts to meet specific objectives, such as NNL or NG. The process of identifying risks and setting targets through the application of the mitigation hierarchy should be clearly communicated in the BAP, with all activities once identified being integrated into the company's environmental management system (EMS) should they have one.

Another important aspect of BAPs is that they should also align with national biodiversity strategy and action plans (NBSAPs). NBSAPs articulate the targets countries are committing to deliver to achieve the CBD. Alignment with the NBSAP and, for example, snow leopard conservation objectives, will ensure that the company's objectives are appropriate and help meet overarching national targets and commitments.

Stage 4: Monitoring and Evaluating the Effectiveness of Actions

Responsible companies must be able to manage risks, ensure that operations deliver conservation objectives as planned, and identify and respond to unexpected problems in a timely manner. Through monitoring changes to natural systems over time, companies can evaluate their impacts and respond appropriately to meet agreed environmental objectives and targets. Monitoring and evaluation of biodiversity through a plan-do-check-act cycle is a key component of project implementation and adaptive management that helps businesses:

- *Manage risks* by analyzing the effects of actual or predicted impacts and identifying new issues as they appear.
- *Meet targets* by reviewing environmental performance to measure success against company objectives and improve approaches.
- *Increase environmental benefits* by ensuring the greatest possible environmental outcomes at each project stage.
- *Create business benefits* by communicating transparently with stakeholders and being able to demonstrate performance.
- *Learn and adapt* by using analysis to evaluate the success of current methods, adapt management and potentially improve company policies.

Stage 5: Modify and Update Actions

Monitoring approaches vary greatly depending on the landscape and biodiversity in question, but all use indicators to identify and measure change. Indicators help identify issues and understand trends by presenting information about complex and changing ecosystems in a way that businesses and their stakeholders can understand.

It is useful to recognize that many companies are familiar with international conservation framework conventions and are "signed" up to their implementation. This includes the Convention on the Conservation of Migratory Species of Wild Animals (CMS) and the CBD, both of which make explicit reference to how the private sector can focus on maintaining habitat and species viability and support conservation efforts in locations. CMS has also recently produced guidelines on mitigating the impact of linear infrastructure and related disturbance on mammals in Central Asia (Conservation of Migratory Species of Wild Animals, 2014).

OPPORTUNITIES

While researching for this chapter the authors received a comment from the extractives sector regarding the reluctance of some groups to engage with the sector for reasons including fear of reputational impact, negative reaction from memberships, and the belief that extractive companies are inherently evil. As we bring this chapter to a close, it is useful to consider the opportunities that, given the chance, corporate business could provide in support of conservation.

It is undeniable that the corporate world has a significant impact on biodiversity and communities but where there are impacts there are also opportunities for change. The land area under corporate management is vast. Often community livelihoods are overwhelmingly dependent on successful businesses. The extractive industries, in particular, frequently operate in areas of highest biodiversity value that often have low livelihood opportunities for local people. The activities and attitudes of business therefore exert a significant influence on the long-term viability and sustainability of an environment on which we all depend. Some companies are making efforts to support worthwhile conservation activities, but they struggle with many groups being unwilling to see them as a genuine partner.

The corporate sector increasingly views partnership and collaboration with conservation groups as the best way to tackle the demanding and complex issues associated with operating in biologically sensitive environments. The corporate sector needs an active and engaged conservation sector in order to find ways to improve biodiversity performance. Positive and transparent engagement between the sectors can result in reduced environmental impacts, improved business biodiversity performance, and business practices toward conservation, while providing tangible benefits to conservation. Benefits include enhanced funding for conservation, increased number and area of designated sites and enhanced conservation capacity.

Although challenging, NGO-corporate partnerships have the potential to elevate biodiversity conservation to new levels, provided that they are well designed and benefit from clear objectives and equal commitment on both sides. While not compromising their roles as defenders of communities and biodiversity, NGOs and other actors could benefit from understanding the position of corporate business and using processes such as ESIAs, IFC PS6, and the mitigation hierarchy as leverage points to hold business to account and as a means to positively engage to ensure that the best possible effort is put into the conservation of the snow leopard and its habitat. The key to a good working partnership is the early identification of common goals and objectives, the development of a partnership work plan, and good communication.

CONCLUSIONS

Although we lack significant quantifiable evidence for the snow leopard, the impacts of large-scale developments and the extractives industry present a current and significant future threat to biodiversity. However, the corporate and conservation worlds need not always collide.

The presence of progressive and engaged companies provides opportunities for mobilizing resources for wider biodiversity conservation; provide safe havens for snow leopard and their prey species and creates barriers to more unscrupulous companies exploiting snow leopard range areas. They also present an opportunity for positive engagement by NGOs and other actors in key planning and decision-making processes.

For reasons of operational efficiency, competitive and market advantage, legislative compliance, and good citizenship, companies are increasingly looking to mitigate their impacts through the application of good practices such as ESIA and the mitigation hierarchy. Companies should always conduct credible participatory ESIAs and commit to avoid, minimize, or mitigate any environmentally damaging effects. Where there is doubt, companies should adhere to the CBD's precautionary principle, especially when operating in the most important snow leopard landscapes. It is also imperative that companies communicate what they do, not only to demonstrate that they are behaving responsibly but to establish widely recognized best practice norms that encourage recalcitrant companies to adopt similar practices. Some companies are trying to support worthwhile conservation activities and these provide opportunities to ensure that the best possible effort is put into the conservation of the snow leopard and its habitat.

References

Abaza, H., Bisset, R., Sadler, B., 2004. Environmental impact assessment and strategic environmental assessment: towards an integrated approach. United Nations Environment Programme, Geneva. Available from: http://www.unep.ch/etu/publications/textONUBr.pdf

BBOP, 2009a. Business, Biodiversity Offsets and BBOP: An Overview. Forest Trends, Washington, DC, USA.

BBOP, 2009b. Compensatory Conservation Case Studies. Forest Trends, Washington, DC, USA.

BBOP, 2012. Biodiversity Offsets: Principles, Criteria and Indicators. Forest Trends, Washington, DC, USA.

Conservation of Migratory Species of Wild Animals, 2014. Convention on Migratory Species. Available from: <http://www.cms.int/sites/default/files/document/COP11_Doc_23_3_2_Infrastructure_Guidelines_Mammals_in_Central_Asia_E.pdf.

Convention on Biological Diversity, 1992. Precautionary Principle in the Convention on Biological Diversity. Available from: <https://www.cbd.int/convention/articles.shtml?a=cbd-00.

Equator Principles, 2013. Equator Principles III – 2013. Available from: www.equator-principles.com

European Bank of Reconstruction and Development, 2008. Performance Requirements. Available from: www.ebrd.com/downloads/about/sustainability/ESP_PR01_Eng.pdf

Gunningham, N., Kagan, R.A., Thornton, D., 2004. Social license and environmental protection: why businesses go beyond compliance. Law Social Inquiry 29, 307.

IFC, 2012. Performance Standard 6 Biodiversity Conservation and Sustainable Management of Living Natural Resources. International Finance Corporation, Washington, DC, USA, www.ifc.org.

International Association for Impact Assessment, 2003. Impact Assessment and Project Appraisal, vol. 21, no. 1. Beech Tree Publishing, Surrey, pp. 5–11. Available from: https://www.iaia.org/../sections/sia/IAIA-SIA-International-Principles.pdf

Madsen, B., Carroll, N., Kandy, D., Bennett, G., 2011. State of Biodiversity Markets Report: Offset and Compensation Programs Worldwide. Forest Trends, Washington, DC, USA.

McCarthy, T.M., Chapron, G. (Eds.), 2003. Snow Leopard Survival Strategy. Snow Leopard Trust and Snow Leopard Network, Seattle, WA.

Morrison-Saunders, A., Marshall, R., Arts, J., 2007. EIA Follow-Up International. Best Practice Principles. Special Publication Series No. 6. Fargo, USA.

Prizma, 2014. Creating paper parks or biodiversity value in Kyrgyzstan? Available from: http://prizmablog.com/2012/05/15/creating-paper-parks-or-biodiversity-value-in-kyrgyzstan/

Retiefa, F., Welmana, C.N.J., Sandhama, L., 2011. Performance of environmental impact assessment (EIA) screening in South Africa: a comparative analysis between the 1997 and 2006 EIA regimes. S. Afr. Geogr. J. 93 (2), 154–171.

Rio Tinto, 2004. Rio Tinto Biodiversity Strategy. Rio Tinto, www.riotinto.com/documents/ReportsPublications/RTBidoversitystrategyfinal.pdf

Snow Leopard Network, 2014. Snow Leopard Survival Strategy – Revised Version 2014. Snow Leopard Trust and Snow Leopard Network, Seattle, WA, USA. Available from: http://www.snowleopardnetwork.org/

SECTION IV

CONSERVATION SOLUTIONS *EX SITU*

CHAPTER

21

Role of Zoos in Snow Leopard Conservation: Management of Captive Snow Leopards in the EAZA Region

Leif Blomqvist, Alexander Sliwa***

*Nordens Ark Åby Säteri, Hunnebostrand, Sweden

**Cologne Zoo, Cologne, Germany

INTRODUCTION

The snow leopard (*Panthera uncia*) has a long history in European zoos with the taxon exhibited for the first time in Antwerp Zoo in 1851 (Rieger, 1980). Although the first litter was born in Wroclaw Zoo in 1910, it is safe to say that prior to the 1960s, most snow leopards, not only in Europe but also in North America, imported from the wild did not survive long and only a handful reproduced. Before 1960, when collection from the wild was the usual means of acquiring snow leopards for zoos, only four wild-caught pairs in European zoo collections bred. Of these pairs, descendants of the successful breeding pair in Copenhagen from the late 1950s, "Hassan" (studbook #85) and "Muddi" (studbook #86) are still represented in the current population, while the gene lines of the three other pairs have gradually disappeared.

Snow Leopards in Focus in the 1970s

Snow leopard reproduction thus remained sporadic in Europe, and it was not until the late 1970s that births were to become more common. Due to poor breeding results and high neonatal mortality, the species became a subject of intense focus and the desire to establish a self-sustaining population. This was encouraged in particular by three European zoo directors, Ilkka Koivisto (Helsinki), Peter Weilenmann (Zurich), and Walter Encke (Krefeld) by initiating a global Snow Leopard Conference hosted by Helsinki Zoo in 1978. The aim of the symposium was to establish standard management

Snow Leopards
http://dx.doi.org/10.1016/B978-0-12-802213-9.00021-3

protocols to improve breeding and to devise methods by which essential information could be compiled and exchanged between zoo professionals (Blomqvist, 1978a).

At the end of the 1970s, the future of the snow leopard population still seemed bleak in Europe, and in 1979 the 29 pairs produced 6 litters, meaning only 21% reproduced. The cub mortality rate was high, and 38% of the population consisted of wild-born individuals while the age structure of the stock indicated an unstable population heading for extinction.

The need for closer cooperation between zoos had become even more evident after the mid-1970s when the Convention on International Trade in Endangered Species (CITES), which strictly regulated the trade in wild animals, came into force. Zoos had to build up self-sustaining populations of animals in order not to have to rely on the import of wild individuals. Captive propagation now became the prime mean for replenishing zoo collections. As a result the European Endangered Species Program (EEP) was established in 1985, with breeding programs for 19 threatened species kept in European zoo collections. The number of programs expanded quickly, and in 1987, the snow leopard was integrated into the EEP program.

Global Studbook 1976

All isolated populations, whether in the wild or in captivity, lose part of their gene diversity with each generation. It may therefore seem surprising that a number of founders that bred several decades ago are still represented in the current stock in Europe and North America. The reason for this is simple: captive snow leopards have been recorded in an international studbook since 1976 (Blomqvist, 1978b), first maintained by the author at Helsinki Zoo and since 2010 at Nordens Ark in Sweden.

The international studbook allows one to follow each individual in detail by ancestry, location, and dates of birth and death. Instead of being merely a bookkeeping tool, the studbook has aspired to be a proactive tool for population management. Thanks to the studbook, a number of animal exchanges between the main breeders in Europe and North America took place in the 1970s and 1980s, and many of the old founders that bred during these years are therefore still represented on both continents.

Several papers describe the types of analyses of studbook data that help to characterize captive populations under intensive management (Mace, 1986; Wharton and Freeman, 1988). Once completed these analyses are translated into facts and figures, which, along with information on the species' basic biology, form a picture of the population status and of the biological constraints faced by the studbook keeper. These facts and their synthesis provide the basis for understanding the importance of each individual in the population and the type of recommendations most appropriate for that particular animal (Leus, 2011).

Today the vast majority of snow leopards worldwide are part of regional, cooperative breeding programs functioning under regional zoo associations. Most of the jointly managed snow leopards live in Europe and North America, with smaller populations maintained in Australia, India, and Japan (Gillespie and Hibbard, 2013; Jha, 2013; Jha and Rai, 2013; Taniguchi and Namaizawa, 2013).

BREAKTHROUGHS IN THE 1980s

The breakthrough came in Europe in the 1980s, when the number of cubs finally exceeded the number of deaths. This improvement can be attributed to the higher status the taxon attained through being managed in an international studbook in addition to its incorporation in the pan-European breeding program. As illustrated in Fig. 21.1, the captive population continued to prosper with high growth rates until the mid-1990s, when it finally peaked in 1996 with 218 animals in 75 institutions. Due to lack

FIGURE 21.1 **Development of the snow leopard EEP, 1987–2013.**

FIGURE 21.2 **Births and deaths in snow leopard EEP, 2005–2013.**

of holding capacity, breeding restrictions were then imposed with the result that the population declined to 183 animals by 2006. During the past years the number of births again exceeded the number of animals lost (Fig. 21.2), and at the beginning of 2014, the EEP population stood at 216 (107 males, 109 females) animals in 81 institutions, a net increase of three individuals compared with the previous year (Blomqvist, 2013).

As felid exhibits increase in size to meet the modern animal welfare and exhibit needs, the number of different species that can be maintained at sustainable population levels decreases. Due to space limitations and competition with other endangered cat species of the same size, the breeding program for snow leopards is currently being managed for a zero population growth of about 220 animals.

GOAL OF THE EEP: TO MAINTAIN A GENETICALLY INTACT POPULATION WITH HIGH GENE DIVERSITY

Gene diversity (GD) is considered vital to all organisms for their future evolution. A sufficient level of genetic variability must therefore be retained to allow the captive populations to adapt to potential environmental changes. The primary goal for the breeding program has consequently been to keep the population demographically robust and genetically representative of its wild counterparts and to sustain these characteristics for a foreseeable future. Preserving the gene pool maximizes the chances that the snow leopard stock will adapt to a variety of environmental conditions in the future.

It is well known that all isolated populations, whether they exist in the wild or in captivity, inevitably lose a part of their GD with each successive generation. The speed of this process depends partly on the size of the population, but also on the time that has elapsed (Gilpin and Soulé, 1986). Small and isolated populations as in the case of the snow leopard are known to lose their GD more rapidly than larger ones. In order to stabilize the population at the current level, necessary breeding restrictions can quickly destabilize the age structure of the population.

Founder Representation

Although 143 wild-caught snow leopards have arrived in Europe and the former Soviet Union, only 36% of them (25 males and 26 females) have bred in the region. Among these 51 individuals, breeding success has been variable. Several founders that have reproduced in the past and had only a few offspring, no longer have a visible representation in the current gene pool, while successful breeders with several offspring that have also bred are strongly represented. Due to the disparities in the founder representation and to losses in bottlenecks, some of the GD that can be found in wild populations will therefore always be lost regardless of how successfully the population is managed.

Because of the uneven breeding success, the founder contribution is highly skewed among the 214 living descendants, which do not contain the genomes of all the original founders. Thanks to the early establishment of the studbook and the numerous animal exchanges between North America, Japan, and Europe, gene lines from other continental founders have also been added into the captive stock. The current population therefore shows traces of 56 founders, which significantly exceeds the minimum of 20 unrelated wild individuals generally recommended to capture 97.5% of the GD of the wild population (Leus et al., 2011). One of the founders is still alive in Zurich, where she delivered her sixth litter in 2014. A potential founder is also alive in Kazan in Russia, which also needs to be incorporated in the gene pool.

Effective Population Size

Many individuals in the population have not bred due to a variety of factors such as age, poor health, and sterility. As a result of these factors, the "effective population size" (N_e) of potentially breeding animals is substantially smaller than the actual population size (N). The ratio of N_e to N is therefore an important indicator of the demographic and genetic health status of the population, informing the rate at which GD is lost. The smaller the N_e, the more GD has been lost. In wild populations the ratio of N_e/N is usually about 0.1 (Frankham, 1995), whereas in captive

TABLE 21.1 Genetic and Demographic Summary of EEP Population, January 1, 2014

	Current	Potential
Population size (N)	107.109	
Number of descendants	106.108	
Effective population size (N_e)	48.46	
N_e/N	0.4432	
Percentage of pedigree known	100	
Gene diversity (GD)	0.9555	0.9799
Mean kinship (MK)	0.0445	
Founders (F)	56	1
Founder genome equivalents (F_{ge})	11.23	24.84
Mean inbreeding (F)	0.0193	

conditions it is to some extent possible to decide how many and which individuals will breed and with whom. Captive populations therefore have an N_e/N ratio that is larger than that of wild populations, usually ranging from 0.2 to 0.4 (Frankham et al., 2002; Mace, 1986). In the snow leopard EEP the N_e/N ratio exceeds these figures being 0.44 (Table 21.1). In other words, 44% or 48 males and 46 females of the population consist of successful breeders.

Founder Genome Equivalents

An even better measure of the health status of the captive stock takes into consideration not only the unequal founder contribution, but also the loss of alleles due to pedigree effects. This is the basis for the concept of "Founder genome equivalents" (F_{ge}) or the number of founders required to achieve the observed levels of GD in the population if all wild-caught animals were equally represented and would have retained all their alleles.

Analyses undertaken by the software package PM2000, show that the F_{ge} value for the snow leopard EEP is 11.23, but would be 24.84 if all the wild-born animals had bred in an optimal manner (Table 21.1). The loss of alleles due to pedigree effects has consequently eroded the levels of diversity. As presented in Table 21.1, the European stock has lost 5% of its GD and is currently equivalent to the same amount of diversity that can be found in only 11 animals randomly caught from the wild. Again, with optimal management, which is impossible to achieve, the GD could be increased to that found in about 25 unrelated animals.

SUGGESTIONS FOR IMPROVEMENT

The EEP population of snow leopards with its 216 individuals is the largest subpopulation worldwide and contains 95.5% of the GD existing in the other subpopulations. The mean inbreeding coefficient (F) for snow leopards is the lowest among the existing breeding programs and has decreased from 0.022 in 2001 (Blomqvist, 2003) to 0.019 in 2013. The population therefore seems to be healthy and not at any imminent risk of extinction. One has to remember, however, that the smallest population size for which genetic drift is balanced by mutation is estimated to have an effective population size of 500 animals (Frankham et al., 2002).

The health status of the population cannot be ascertained only in terms of numbers. If a population is not constantly managed, its status can change quickly and as stated earlier, small populations are more prone to extinction than large ones. With well-planned management where pairings are based on mean kinship values and where breeding with underrepresented gene lines are prioritized and closely related individuals are not mated, the GD can be further improved.

Genetic drift and inbreeding depression can also be alleviated by introducing new genes through exchanges with other continental programs. The EEP has, since its establishment, been proactive in exchanging animals with other regional programs, thus improving the GD not only in its own population but also on a global scale. Possibilities to recruit additional wild-caught animals are highly limited and should not be recommended although semen collection from males that are captured as part of the long-term ecological study in Mongolia (Johansson et al., 2013; McCarthy et al., 2010) could open new opportunities to improve the founder base in the captive stock. A sizeable population of founders not linked with the rest of the captive population is, however, alive in Chinese zoos (Tan and Liao, 1988). If descendants of these founders could be incorporated with the jointly managed population, the EEP would have excellent prognosis for the future. An incorporation of snow leopards in Chinese zoos into the globally managed population is therefore a top priority for the years to come.

TOWARD GLOBAL MANAGEMENT

Recent analyses show that a number of zoo populations are both too small and based on too few founders to fill the conditions for future sustainability (Lees and Wilcken, 2009). The contribution of such small populations to species conservation is therefore of less importance than commonly advertised. The World Zoo Conservation Strategy has consequently urged all conservation-orientated networks to intensify their efforts toward more intensive species conservation.

Because of the growing concern of long-term sustainability of wild animal populations in human care, Global Species Management Programs (GSMP) have already been established for *inter alia* Amur tigers (*Panthera tigris altaica*; Cook and Arzhanova, 2013) and red pandas (*Ailurus fulgens*; Glatston, 2013). By linking ongoing snow leopard breeding programs into a universal GSMP, the global census as well as the effective population size would increase with the positive effects this would have on the GD. The establishment of a snow leopard GSMP would also facilitate animal transfers between regional breeding programs and encourage mutual cooperation between zoos. By working more closely together, husbandry and management practices would be easier to share than they are today, resulting in a global approach for species management. A higher profile of the snow leopard would without doubt promote its conservation *in situ* and *ex situ*.

WHY KEEP SNOW LEOPARDS IN CAPTIVITY?

The actual number of wild snow leopards is difficult to estimate due to its secretive nature and the remote, inaccessible areas it inhabits. The current knowledge is therefore inadequate to generate reliable figures of the free-ranging population, and it is well documented that severe declines took place in the former Soviet Central Asian republics after the breakup of the Soviet Union (Jackson et al., 2008; Koshkarev and Vyrypaev, 2000; McCarthy and Chapron, 2003; Theile, 2003). The reasons for the decline include habitat loss and fragmentation, illegal poaching, and persecution by herders for killing their livestock. As local human populations continue to move deeper into the mountain areas, these threats are not likely to disappear and the future of the wild population is therefore far from safe.

Enlightened zoo visitors therefore expect captive-bred snow leopards to be released back into the wild. Due to political and biological aspects, this is currently not a realistic option for the snow leopard. In a scenario of a further decline in the wild it is, however, important to keep in mind that a self-sustaining backup population in captivity must be maintained if we are to safeguard the species from extinction.

It is also fair to say that the *ex situ* population has played an important role in providing useful knowledge for ongoing field research (Johansson et al., 2013). Because of the snow leopard's secretive nature and its inaccessible habitats, it is difficult and expensive to study snow leopards in the wild. Field workers must often rely on indirect methods such as GPS telemetry to gather basic data on species ecology, habitat use, home range size, and seasonal movements (see Chapter 26). The use of telemetry for radio-collaring and tracking wild snow leopards would not have been possible if exact data on safe anesthesia from zoological collections had not been available. Zoos also offer many advantages to behavioral researchers as well as for collection of precise life history information for the secretive life of the snow leopard.

The snow leopard is a "flagship" species that generates funding support for broader conservation efforts, thus protecting many other species in its range that have decreased in numbers. With more than 700 million worldwide visitors per year, the global zoo community has a huge potential to play an important role in environmental education and wildlife conservation. A number of European zoos display graphics oriented toward field conservation, and many of them are among the main providers of conservation funding. The global zoo community is spending more than USD 350 million every year on wildlife conservation (Gusset et al., 2014). The snow leopard can be seen as an icon for biodiversity conservation in the alpine ecosystem of Central Asia, and it is without doubt easier to attract funding for such charismatic species than for a number of other species. Fundraising activities for *in situ* conservation have during recent years developed rapidly among zoos and offer excellent possibilities to finance not only projects run by the zoos themselves, but more commonly, also to donate funds to other conservation agencies working with snow leopards in the wild.

References

Blomqvist, L., 1978a. First international snow leopard conference in Helsinki, 7th – 8th March 1978. Int. Zoo News 25, 5–6.

Blomqvist, L., 1978b. First report of the snow leopard studbook, *Panthera uncia*, and the 1976 World register. Int. Zoo Yearb. 18, 227–231.

Blomqvist, L., 2003. Captive status of the snow leopard in Europe 2001. International Pedigree Book of Snow Leopards (*Uncia uncia*), vol. 8. Helsinki Zoo, Helsinki, Finland, pp. 27–30.

Blomqvist, L., 2013. Snow leopard EEP 2012. International Pedigree Book of Snow Leopards, *Uncia uncia*, vol. 10. Nordens Ark Foundation, Sweden, pp. 24–32.

Cook, J., Arzhanova, T., 2013. Amur Tiger (*Panthera tigris altaica*). EEP Status and Recommendations. Zoological Society of London, London, UK.

Frankham, R., 1995. Effective population size/adult population size ratios in wildlife: a review. Genet. Res. 66, 95–107.

Frankham, R., Ballou, J.D., Briscoe, D.A., 2002. Introduction to Conservation Genetics. Cambridge University Press, Cambridge, UK.

Gillespie, J., Hibbard, C., 2013. ASMP snow leopard report 2012. International Pedigree Book of Snow leopards, *Uncia uncia*, vol. 10. Nordens Ark Foundation, Sweden, pp. 21–23.

Gilpin, M.E., Soulé, M.E., 1986. Minimum viable populations: processes of species extinction. In: Soulé, M.E. (Ed.), Conservation Biology: The Science of Scarcity and Diversity. Sinauer Associates, Sunderland, MA, USA, pp. 19–34.

Glatston, A., 2013. Red pandas go global – officially. WAZA Zoo News 1, 21–22.

Gusset, M., Fa, J.E., Sutherland, W.J., 2014. A horizon scan for species conservation by zoos and aquariums. Zoo Biol. 33, 375–380.

Jackson, R., Mallon, D., McCarthy, T., Chundaway, R.A., Habib, B., 2008. *Panthera uncia*. The IUCN red list of threatened species. Version 2014.2. Available from: <http://www.iucnredlist.org> (01.01.2015).

Jha, A.K., 2013. Conservation breeding of snow leopard at Padmaja Naidu Himalayan Zoological Park. International Pedigree Book of Snow Leopards, *Uncia uncia*, vol. 10. Nordens Ark Foundation, Sweden, pp. 14–16.

Jha, A.K., Rai, U., 2013. Conservation breeding programme of snow leopard. Ex situ updates 2/1: 2–4. Central Zoo Authority, New Delhi, India.

Johansson, Ö., Malmsten, J., Mishra, C., Lkhagvajav, P., McCarthy, T., 2013. Reversible immobilization of free-ranging snow leopards (*Panthera uncia*) with a combination of medetomidine and tiletamine-zolazepam. J. Wildl. Dis. 49, 338–346.

Koshkarev, E.P., Vyrypaev, V., 2000. The snow leopard after the break-up of the Soviet Union. Cat News 32, 9–11.

Lees, C.M., Wilcken, J., 2009. Sustaining the Ark: the challenges faced by zoos in maintaining viable populations. Int. Zoo Yearb. 43, 6–18.

Leus, K., 2011. The global captive population of the red panda – possibilities for the future. In: Glatston, A.R. (Ed.), Red Panda. Biology and Conservation of the First Panda. Academic Press, London, UK, pp. 335–356.

Leus, K., Bingaman Lackey, L., van Lint, W., de Man, D., Riewald, S., Veldkam, A., Wijmans, J., 2011. Sustainability of European Association of Zoos and Aquaria Bird and Mammal Populations. WAZA Magazine 12, 11–14.

Mace, G.M., 1986. Genetic management of small populations. Int. Zoo Yearb. 24/25, 167–174.

McCarthy, T.M., Chapron, G., 2003. Snow Leopard Survival Strategy. ISLT and SLN, Seattle, USA.

McCarthy, T., Murray, K., Sharma, K., Johansson, Ö., 2010. Preliminary results of a long-term study of snow leopards in South Gobi, Mongolia. Cat News 53, 15–19.

Rieger, I., 1980. Some aspects of the history of ounce knowledge. International Pedigree Book of Snow Leopards, *Panthera uncia*, vol. 2. Helsinki Zoo, Helsinki, Finland, pp. 1–36.

Tan, B., Liao, Y., 1988. The status of captive snow leopards in China. In: Freeman, H. (Ed.), In: Proceedings of the Fifth International Snow Leopard Symposium. International Snow Leopard Trust and Wildlife Institute of India, Bombay, India, pp. 151–166.

Taniguchi, A., Namaizawa, H., 2013. JAZA snow leopard annual report. International Pedigree Book of Snow Leopards, *Uncia uncia*, vol. 10. Nordens Ark Foundation, Sweden, pp. 17–20.

Theile, S., 2003. Fading footsteps: the killing and trade of snow leopards. TRAFFIC International.

Wharton, D., Freeman, H., 1988. The snow leopard, *Panthera uncia*, a captive population under the Species Survival Plan. Int. Zoo Yearb. 27, 85–98.

CHAPTER

22

Role of Zoos in Snow Leopard Conservation: The Species Survival Plan in North America

Jay Tetzloff

Miller Park Zoo, Bloomington, IL, USA

Snow leopards (*Panthera uncia*) were first exhibited in the United States in 1903. The first captive birth occurred in 1945 although births were not common until the 1960s and were responsible for much of the growth that occurred from the early 1970s to the early 1990s.

In the late 1970s, it became evident that wildlife populations were declining in the wild and access to collection animals was becoming increasingly more difficult. This realization inspired a group of zoo professionals to create the Species Survival Plan® (SSP) concept as a cooperative breeding and conservation program administered by the Association of Zoos and Aquariums (AZA). AZA's first SSPs were created in 1981. The Snow Leopard SSP began in 1982 (Figs 22.1 and 22.2).

AZA and its member institutions recognize that cooperative management is critical to the long-term survival of professionally managed Animal Programs and are fully committed to the goals and cooperative spirit of the SSP Program partnerships.

Each SSP Program coordinates the individual activities of participating member institutions through a variety of species conservation, research, husbandry, management, and educational initiatives. The Snow Leopard SSP works under the supervision of the Felid Taxon Advisory Group (TAG), which manages all feline AZA Animal Programs.

The mission of the Felid TAG is to bring together animal managers and scientists to:

- further conservation of felids in the wild;
- effectively manage felids in AZA zoos throughout North America;
- support scientific research concerning felid species.

Expert advisors who cooperatively work together to maximize genetic diversity and appropriately manage the demographic distribution and long-term sustainability of TAG recommended Animal Programs within AZA member institutions lead SSP Programs. Each SSP Program manages the breeding of a select species

Snow Leopards
http://dx.doi.org/10.1016/B978-0-12-802213-9.00022-5

Population Analysis &
Breeding and Transfer Plan

Snow Leopard (*Uncia uncia*)
AZA Species Survival Plan®
Yellow Program

AZA Species Survival Plan® Coordinator
Jay Tetzloff, Miller Park Zoo (jtetzloff@cityblm.org)

AZA Studbook Keeper
Lynn Tupa, Albuquerque Biological Park (ltupa@cabq.gov)

AZA Population Advisor
Colleen Lynch, Population Management Center (clynch@riverbanks.org)

5 September 2014

FIGURE 22.1

FIGURE 22.2

through a Breeding and Transfer Plan. Breeding and Transfer Plans summarize the current demographic and genetic status of the population and recommend breeding pairs and transfers. Breeding and Transfer Plans are designed to maintain a healthy, genetically diverse, and demographically stable population for the long term.

Advisors, often members of corresponding Scientific Advisory Groups, play a critical role in advising, designing, and executing conservation management decisions within AZA Animal Programs. Current advisors in the fields of education, field conservation, husbandry, nutrition, pathology, reproduction, and veterinary aid the Snow Leopard SSP in making science-based decisions.

The SSP Coordinator works with Institutional Representatives (IR), the AZA Regional Studbook Keeper, the TAG, the Wildlife Conservation Management Committee (WCMC), and the AZA Conservation & Science Department, as well any associated governmental agencies, to develop, oversee, promote, and support the cooperative animal management, conservation, and research initiatives of the SSP Program. The primary responsibility of the SSP Coordinator is to regularly complete and distribute an SSP Breeding and Transfer Plan for the managed population. Additional responsibilities include leadership and organization of the SSP Program in building and appropriately managing a sustainable population, and communication of recommendations and guidelines to the appropriate stakeholders. The SSP Coordinator serves as the primary contact and AZA expert for a particular species and abides by the duties and responsibilities set forth by the AZA, WCMC, and the TAG. Specifically, SSP Coordinators:

- Actively advocate and develop sustained interest on the part of member institutions to participate in the SSP Program and build a sustainable population.
- Provide routine SSP Program updates to IRs.
- Serve on, or as an Advisor to, the appropriate TAG and attend relevant meetings.
- Maintain contact with counterparts in other regional associations to facilitate interregional cooperation, if applicable.

The SSP maintains a management group composed of the Coordinator, Vice Coordinator (AZA Regional Studbook Keeper), and seven members elected by SSP IRs. The steering committee serves as the voting body for SSP Program business, and all members are integrally involved in the SSP Program appointments, publications, and meetings.

POPULATION MANAGEMENT AND SUSTAINABILITY

The breeding plan provides breeding (or "do-not-breed") recommendations to maintain demographically stable populations with the greatest possible genetic diversity for a healthy and sustainable population over the long term. Sustainability of the population is related to many factors including its gene diversity, demographic stability, husbandry expertise, and so on.

A combination of variables has contributed to the reduced long-term sustainability of many of AZA's managed Animal Program populations.

Because there is a variety of causes, there is no single answer, direction, or solution. These variables include insufficient:

- Knowledge of current SSP population sustainability duration and genetic diversity
- Number of holding and breeding spaces needed to increase the sustainability of the SSP
- SSP planning capacity
- Institutional awareness surrounding the topic of sustainability
- Institutional commitment to provide additional holding or breeding spaces
- Permitting and/or regulatory availability to move animals
- Advanced breeding expertise

Current gene diversity for the managed Snow Leopard SSP population is 95% and is equivalent to the genetic variation that would be present in 9–10 founders. Most AZA-managed populations have set a genetic goal of maintaining 90% gene diversity for 100 years. When gene diversity falls below 90%, it is expected that reproduction will be increasingly compromised by, among other factors, lower birth weights, smaller litter sizes, and greater neonatal mortality. Gene diversity inevitably decreases over time due to random genetic processes, as offspring are produced and as previous generations pass away. With current population parameters, gene diversity can be maintained at or above 90% for 40 years and would decline to 83% at the end of 100 years, given current projected growth rate.

With good genetic management, including prioritizing breeding of low mean kinship animals, the time to 90% gene diversity can be extended. Pairing animals with the lowest mean kinships helps to equalize founder representation. Effective population size and population growth rate, also important to the long-term maintenance of gene diversity, have historically been high in this population and should be maintained at their current levels.

A strategy to increase cubs being produced has been implemented at some institutions with the philosophy of giving a female more opportunities to produce cubs. If a female was successful in the summer, the cubs would be moved to another part of the zoo or another institution altogether. By removing the cubs, the female may cycle again that next winter. In some ways, one might consider this similar to a double-clutching strategy in bird species. The goal is to maximize the amount of offspring from a given pair. Some institutions have been successful with this concept while some have not. Further long-term research will be needed to fully understand whether this strategy should be considered and/or continued.

HUSBANDRY

The snow leopard is known as a seasonal breeder. This biological fact causes a husbandry and management dilemma. Should a breeding pair be kept together in the nonbreeding season? Research has begun to determine whether a particular management style increases the chances of cubs being born. Some institutions leave their pair together year-round while others only introduce a pair at the beginning of the breeding season. If separated, it means that either the male or female would not be on exhibit and visible to the public. Separating could mean less of an exhibit in the eyes of the public and an increase in the amount of time staff must spend on the species due to multiple "groups" and shifting that is required. The personality of the individual cats also plays into the strategy of the zoo. Not all pairs can be housed together during the nonbreeding season even if the institution wants to keep them together. Aggression to the point of a fatality is rare in snow leopards but constant aggression during the nonbreeding season can be detrimental for the pair during the breeding season. On the opposite side of aggression, some pairs may be more stressed if separated. The holding institution must understand their snow leopards and devise a strategy based on that and the general philosophy of the staff in how they manage their felines.

Animal Care Manuals (ACMs) provide a compilation of animal care and management knowledge that has been gained from recognized species experts based on the current science, practice, and technology of animal management. These manuals compile and organize our understanding of basic requirements, best practices, and animal care recommendations to advance the capacity for excellence in animal care and welfare. These dynamic manuals are considered works in progress as practices continue to evolve through scientific learning. Once completed, the use of information within each manual should always be in accordance with all local, state, and federal laws and regulations concerning the care of the species specified.

Recommendations included in the manuals are not exclusive management approaches, diets, medical treatments, or procedures, and may require adaptation to the specific needs of individual animals and particular circumstances in each institution.

ACMs provide up-to-date information gained from a large body of expertise including biologists, veterinarians, nutritionists, reproduction physiologists, behaviorists, and researchers. Each relevant area should be as comprehensive as existing knowledge allows and could include:

Taxonomic information	Ambient environment	Habitat design and containment
Records	Transport	Social environment
Nutrition	Veterinary care	Behavior management

NUTRITION

A formal nutrition program is recommended to meet the behavioral and nutritional needs of all snow leopards. Diets should be developed using the recommendations of veterinarians as well as nutrition advisors of the Snow Leopard SSP. No detailed studies have been conducted to determine the specific nutrient requirements of snow leopards. Consequently, until these data become available, the domestic cat can serve as a model for most nutrient parameters. In the wild, snow leopards typically eat most of the prey they capture and kill, including some bones, fat, and viscera. The addition of bones, appropriate carcass parts, or whole prey is behaviorally enriching and may have a positive effect on dental and digestive health. Therefore, the SSP makes this a requirement for the keeping of snow leopards.

DISEASE RECOGNITION AND MANAGEMENT

Disease issues in snow leopards are similar to other large felid species. A retrospective study of snow leopards in North American zoos taken from records from 2000 to 2008 revealed the following disease issues in order of most common to least common:

1. Parasites: roundworms most common
2. Wounds and trauma
3. Arthritis and hip dysplasia
4. Vomiting
5. Diarrhea
6. Hematuria/hematochezia
7. Renal failure
8. Cystitis
9. Upper respiratory infection
10. Lingual papilloma, fibropapilloma
11. Septicemia
12. Ear infections
13. Coloboma, eyelid agenesis, microphthalmia

Most of these issues are common in all felids and many are related to renal disease and old age.

Currently the Snow Leopard SSP is seeking to unravel the issues of congenital coloboma and eyelid deformities and the prevalence and significance of papillomavirus. Colobomas or eyelid deformities in which all or a portion of the lid margin does not form have continued to be found in several cubs each year. It is usually seen in cubs at birth and is a congenital defect. It is currently found in about 6–10% of the North American population. This incidence could be higher, and current research is ongoing to determine the prevalence and significance of this defect. It is still unknown whether it is related to a genetic component and is probably multifactorial. An investigation by Gripenburg et al. (1985) was unable to prove whether colobomas in snow leopards were genetically related, but did rule out a chromosomal abnormality. Several factors from vitamin A deficiencies or overdoses, to viral infection during eye development in the womb, have been suggested but none proven. Cases have been seen in Europe as well. The defects are usually mild and range from small defects to eyelid agenesis and some deeper into the fundus. Symptoms range from none to epiphora and blepharospasm from facial hair similar to entropion. It can be corrected by entropion surgery, pedicle flap, or reconstructive surgery. There has been one case of a litter with two cubs with more severe deformities including anophthalmia, blindness, and microphthalmia, while a third cub in the same litter had severe congenital defects of the heart, bilateral colobomas, and one underdeveloped eye (Helmick, K., Woodland Park Zoo, USA, personal communication). Affected and normal cubs may occur in the same litter, and the same parents may produce normal litters and abnormal litters. It continues to be a complex issue to resolve.

Papillomavirus is known to be present in the captive snow leopard population, but what is unknown is its prevalence, mode of transmission, significance to reproduction, and overall impact on the SSP. Research is being done to assess prevalence in the North American population. In felids, it appears to be species specific and to be eight distinct different viruses (Sundberg et al., 2000). In domestic cats it is more commonly found in immune deficient animals. (i.e., FIV, Chediak–Higashi syndrome) (August, 2005) The virus is a member of *lambda papillomavirus* genus and tentatively named in snow leopards UuPV-1 for oral lesions and UuPV-2 for cutaneous lesions (Sundberg et al., 2000). Individuals infected with UuPV-1 may have whitish flat plaques sublingually. UuPV-2 papillomas appear as small black masses in the skin on the head, neck, and limbs. Transformation of the papillomas to cutaneous squamous cell carcinoma, with an effective mortality rate of 100%, has been reported in two captive snow leopards (Joslin et al., 2000). It is rare to see both clinical lesions in the same animal. The SSP is recommending complete surgical removal of papillomas when found. Other modalities have been used such as cryotherapy and possible beta radiation therapy (August, 2005). Currently, the SSP is not limiting movement because of the virus or clinical disease. It is probable that most of the snow leopards have been exposed or are positive but only some show clinical symptoms. Current research is trying to answer these questions.

REPRODUCTION

Snow leopards reach sexual maturity between the ages of 2 and 3 years (Wharton and Freeman, 1988), but exceptions do occur and care should be taken when male offspring are kept with females. Snow leopards are highly seasonal breeders in both captivity and the wild. In the captive North American population, breeding occurs from January through June. Parturition occurs in late spring or early summer, with 89% of births in April, May, or June (Wharton and Mainka, 1997; Sunquist and Sunquist, 2002). The estrous cycle of the snow leopard has been reported to be 25–38 days, (Schmidt et al., 1993; Graham et al., 1995), but a recent study suggests that the estrous cycle is much shorter (12.7 ± 0.6 days; Reichert-Stewart et al., 2014).

During estrus (4–8 days), pairs will copulate 12–36 times per day, with intromission lasting 15–45 s (Wharton and Mainka, 1997; Reichert-Stewart et al., 2014). Spontaneous ovulation has not been observed in snow leopards (Reichert-Stewart et al., 2014). If the female ovulates but does not conceive after mating, she will experience a pseudopregnancy, or nonpregnant luteal phase, and return to estrus in 45.7 ± 5.7 days with a range of 11–72 days (Schmidt et al., 1993; Brown et al., 1994; Reichert-Stewart et al., 2014). If pregnancy is achieved, gestation is 93–103 days (Wharton and Mainka, 1997; Reichert-Stewart et al., 2014). Litter sizes range from 1 cub to 5 cubs, averaging about 2.3 cubs per litter (Wharton and Mainka, 1997; Sunquist and Sunquist, 2002).

Although male snow leopards produce spermatozoa throughout the year, some reports indicate a decline in both the quantity and quality of sperm produced during the summer and fall nonbreeding season (Johnston et al., 1994; Roth et al., 1997). During the breeding season, 2.5 ± 0.5 mL of semen containing 172.8 ± 37.3 million sperm (37.0% ± 4.8% normal morphology) was collected from proven males ($n = 5$) using electroejaculation (J. Herrick, unpublished data).

Assisted reproductive technologies (ART) could play an important role in the management of captive populations, but these techniques remain experimental and results are sporadic and are not yet efficient for use in snow leopards. To date, only a single cub has been produced from artificial insemination (AI) in snow leopards, and no cubs have been produced following in vitro fertilization/embryo transfer (IVF/ET), but it should be noted that few attempts have been made to produce cubs using these techniques in the past 20 years (Roth et al., 1997). Research to improve the efficiency of AI and IVF/ET should be done in cooperation with SSPs who could identify suitable candidates. Population managers should not rely on AI or IVF for routine offspring production, but these techniques may be applicable for individuals with behavioral issues or physical conditions that limit/prevent copulation. In addition, these techniques provide the only opportunity to introduce new genetics into the captive population without removing animals from the wild. Semen collection and sperm cryopreservation can be performed under field conditions, allowing valuable genetic material to be collected whenever a wild male is anesthetized. Although hypothetical at this time, work with other felid species indicates this is a feasible approach for snow leopards (Swanson, 2006). Given the potential of ART for management of captive populations, continued development of these techniques for use in snow leopards is important.

Some challenges have come from the advances in veterinary care. Snow leopards are living longer in captivity, which can create management issues. Female snow leopards are no longer fertile after the age of 15 years, so while they may still remain a great exhibit animal to educate and engage guests, they will not be reproducing. The aged female could be housed with a male that is still fertile, which creates a management issue for the institution that may not have the space to bring in another fertile female to pair with their breeding male. The strategy for the SSP in regards to new or renovated exhibits is to encourage enough spaces to successfully manage a breeding pair, their offspring for 2 years, and an older animal (usually female).

Although contraception is currently not recommended by the Snow Leopard SSP due to the desire to increase the breeding possibilities and health concerns with the implants, the AZA Wildlife Contraception Center (WCC) at the St. Louis Zoo was created in 2000 to assess contraception efficacy, reversibility, and safety for animals not recommended for breeding. The mission of the AZA WCC is to provide information and recommendations to the AZA community about contraceptive products that are safe, effective, and reversible. These recommendations are used by zoo professionals to make informed decisions on how to sustainably manage their animal collections. The WCC includes scientists, veterinarians, and animal managers with research and management expertise in wildlife contraception.

EXHIBIT DESIGN

Zoos have come a long way since the 1970s, when the goal for an exhibit was to be able to keep it clean and see the animal. Showing off the animal in sterile cages that had bars for containment was the norm.

Snow leopard exhibits today focus on being naturalistic. Exhibits are creative, functional, and safe and may include rocks (or artificial rockwork), logs, and ample space for the snow leopards to utilize. Containment is glass and/or mesh that enhance guest viewing. The guest can be "entertained" while experiencing the snow leopard. Natural sounds pumped in and interpretive and interactive graphics are part of the guest engagement and education. Story lines also can be intertwined in multiple exhibit areas.

Modern exhibits include a management area that works well for the staff. Quality and quantity of space are important for successful management. SSP institutions are recommended to provide enough space so they may hold a postreproductive animal (usually a female) and a breeding pair. The institution should have the capacity to hold any cubs born for 2 years. The behavior of the snow leopard is incorporated into any quality design. Animal enrichment and training incorporated into an exhibit design is a newer concept. Training walls, where the public can observe animal-staff interactions are becoming common. Designated feeding and training times provide zoo staff opportunities to "show off" individual snow leopards. Another new enrichment concept is a trail system where different route choices are offered to snow leopards daily, mimicking to at least to a small degree choices made in the wild. Complexity of exhibits of today is key for a successful exhibit.

EDUCATION

Even though a good deal of Snow Leopard SSP activities focus on managing the population, one cannot ignore the important role zoo audiences can play in snow leopard survival. Educating guests and facilitating an emotional connection with this animal is at the heart of saving the species. These connections may be brief, but can create a lifelong memory, which may, in turn, spur action to help the wild snow leopard population.

Many zoo educators take the audience age into account when developing education programs for snow leopards. Although helping to preserve the species is the goal, younger audiences (younger than 11–12 years old) first need to develop appreciation and empathy for these animals if we want them to take action on their behalf in the future. This approach is described in *Beyond Ecophobia* (Sobel, 1996).

Educators often engage these younger audiences with questions. Developmentally appropriate questions may include those focused on the snow leopard's adaptations, habitat, and its role in the ecosystem or care at the zoo. Questions may include asking the child to find the snow leopard in the exhibit/habitat (which works well with a naturalistic exhibit) or having them provide explanations for why they think the snow leopard has such a long tail or why the exhibit contains a box of hay (enrichment). Institutions that have the capacity may want to provide additional experiential interactions with the younger audience. A keeper-led training session in which the snow leopard's trained behaviors are demonstrated can be an unforgettable experience. Tangible items, such as skulls or pelts, provide an additional opportunity for these younger audiences to experience snow leopards in a concrete way. Through these guest-focused interactions, zoos can provide emotional and cognitive connections that will lead to long-lasting memories and, eventually, conservation action.

With audiences older than 12 years old, engaging in discussions related to conservation issues is developmentally appropriate and important as these youth are starting to wonder and care more about the world beyond their backyard (Sobel, 1996). Many AZA institutions tell stories related to human-snow leopard conflict in the wild; stories that can continue to deepen an individual's emotional connection and interest in learning more.

Successful snow leopard conservation depends on the local people living in snow leopard regions valuing these animals enough to want to save them. Some AZA institutions have partnerships with organizations such as Snow Leopard Trust, Snow Leopard Conservancy, Panthera, or the Wildlife Conservation Society, among others, with successful snow leopard conservation projects. Zoos may provide educational materials via these partnerships that help local people develop a better understanding of the ecological role of snow leopards and instill a will to coexist with them. In turn, sharing these conservation stories with zoo guests can inspire them to take action to help. Zoos can explain that these programs or partnerships in the local communities are funded by the guests' admission to the institution. Opportunities can be provided to make direct contributions to snow leopard conservation organizations.

The zoo guest is important in the conservation of this species. Animals in AZA institutions are animal ambassadors for their wild relatives. Animal ambassadors help create those meaningful connections, and in turn give zoos the opportunity to create conservationists who want to help to their wild relatives.

COLLABORATION AND CHALLENGES

Collaboration is the key to any conservation breeding program. No program can be successful without the individual institutions working together. The issue is that the institutions are

FIGURE 22.3

just that, individual institutions with their own agenda, philosophy, and mission. The institutions in the Snow Leopard SSP have worked hard to think about what is in the best interest of the captive North American population. The nearly 70 institutions in the Snow Leopard SSP work as well together as any program in AZA. Zoos are constantly making decisions that benefit the entire population even if it may not benefit their particular institution.

The collaboration is not only on the regional level but reaches out into the global scale through multiple programs across the world with the largest being in Europe (European Endangered Species Programs). Periodically, snow leopards are exchanged across programs. The exchanges are beneficial for both programs and for the international global captive population. By trading within the two programs, new genetics are passed along to bolster the sustainability of both programs. Exchanges must be balanced against the fact the import/export process for an institution in the United States takes about a year for completion, which can be a drain of time and resources.

The Snow Leopard Species Survival Plan strives to educate and engage zoo guests about this amazing animal. The sustainability of the captive snow leopard population could be crucial in saving this endangered species in the wild (Fig. 22.3).

References

August, J.R., 2005. Consultations in Feline Internal Medicinevol. 5 Elsevier Health Sciences, 800 pages.

Brown, J.L., Wasser, S.K., Wildt, D.E., Graham, L.H., 1994. Comparative aspects of steroid hormone metabolism and ovarian activity in Felids, measured noninvasively in feces. Biol. Reprod. 51, 776–786.

Graham, L.H., Goodrowe, K.L., Raeside, J.I., Liptrap, R.M., 1995. Non-invasive monitoring of ovarian function in several felid species by measurement of fecal estradiol-17β and progestins. Zoo Biol. 14, 223–237.

Gripenburg, U., Blomqvist, L., Pamilo, P., Soderlund, V., Tarkkanean, A., Wahlberg, C., Varvio-Aho, S.L., Virtaranta-Knowles, K., 1985. Multiple ocular coloboma (MOC) in snow leopards (*Panthera uncia*). Clinical report, pedigree analysis, chromosomal investigations and serum protein studies. Hereditas 102, 221–229.

Johnston, L.A., Armstrong, D.L., Brown, J.L., 1994. Seasonal effects on seminal and endocrine traits in the captive snow leopard (*Panthera uncia*). J. Reprod. Fert. 102, 229–236.

Joslin, J.O., Garner, M., Collins, D., Kamaka, E., Sinabaldi, K., Meleo, K., Montali, R., Sundberg, J.P., Jenson, A.B., Ghim, S.-J., Davidow, B., Hargis, M., Clark, T., Haines, D., 2000. Viral papilloma and squamous cell carcinomas in snow leopards. Proc. AAZV/IAAAM, 155–158.

Reichert-Stewart, J.L., Santymire, R.M., Armstrong, D., Harrison, T.M., Herrick, J.R., 2014. Fecal endocrine monitoring of reproduction in female snow leopards (*Uncia uncia*). Theriogenology 82, 17–26.

Roth, T.L., Armstrong, D.L., Barrie, M.T., Wildt, D.E., 1997. Seasonal effects on ovarian responsiveness to exogenous gonadotrophins and successful artificial insemination in the snow leopard (*Uncia uncia*). Reprod. Fert. Dev. 9, 285–295.

Schmidt, A.M., Hess, D.L., Schmidt, M.J., Lewis, C.R., 1993. Serum concentrations of oestradiol and progesterone and frequency of sexual behavior during the normal oestrous cycle in the snow leopard (*Panthera uncia*). J. Reprod. Fert. 98, 91–95.

Sobel, David, 1996. Beyond Ecophobia. Orion Society, Great Barrington, MA, pp. 2–6, 9–12.

Sundberg, J.P., Van Ranst, M., Montali, R., Homer, B.L., Miller, W.H., Rowland, P.H., 2000. Feline papillomas and papillomaviruses. Vet Pathol. 37 (1), 1–10.

Sunquist, M., Sunquist, F., 2002. Wild Cats of the World. University of Chicago Press, Chicago, IL.

Swanson, W.F., 2006. Application of assisted reproduction for population management in felids: the potential and reality for conservation of small cats. Theriogenology 66, 49–58.

Wharton, D., Freeman, H., 1988. The snow leopard (*Panthera uncia*): a captive population under the Species Survival Plan. Int. Zoo Yearb. 27, 85–98.

Wharton, D., Mainka, S.A., 1997. Management and husbandry of the snow leopard (*Uncia uncia*). Int. Zoo Yearb. 35, 139–147.

CHAPTER

23

The Role of Zoos in Snow Leopard Conservation: Captive Snow Leopards as Ambassadors of Wild Kin

OUTLINE

http://dx.doi.org/10.1016/B978-0-12-802213-9.00023-7

SUBCHAPTER

23.1

Kolmården Wildlife Park: Supporting Snow Leopards in the Wild, Sharing the Message at Home

Thomas Lind

Kolmården Wildlife Park, Kolmården, Sweden

At the beginning of twenty-first century, Kolmården Wildlife Park made the strategic decision to remove the polar bears from the park, and in the context of our collection planning we would acquire a mountain living, climate-appropriate endangered species that would be attractive to the public and provide educational opportunities. With those criteria in mind and the polar bear exhibit and its large rock formation about to be vacant, snow leopards were an obvious fit. Our first pair of snow leopards (*Panthera uncia*) arrived in 2006 from Nordens Ark Zoo in Sweden and Marwell Zoo in the United Kingdom.

Kolmården Wildlife Park, through the Kolmården Fundraising Foundation (KIS), has a long history of *in situ* conservation and development, through collaboration and financial support to carefully selected organizations and projects around the world. Projects that Kolmården supports include Amur tigers (*Panthera tigris altaica*), African painted dogs (*Lycaon pictus*), and Cross River gorillas (*Gorilla gorilla diehli*), to name just a few. In Spring 2008, Kolmården was contacted by the Snow Leopard Trust (SLT) and researchers at Grimsö Wildlife Research Station (part of the Swedish University of Agricultural Sciences) to see if we would be interested in collaboration on a comprehensive long-term study of snow leopards in South Gobi, Mongolia. The proposed snow leopard project fitted well into our conservation priorities, especially with our new cats on exhibit. And the fact that the leader of the field study was a Swedish PhD student brought national pride, so it was not difficult for us to decide to support it.

The science director of the Snow Leopard Trust was invited to Kolmården later that year and gave a presentation for the Foundation and the staff, which convinced us even more that it was a project that Kolmården would be proud to be a part of.

With snow leopards on exhibit, and the Foundation supporting an exciting new project in the field, we decided that the snow leopard would be Kolmården's *Animal of the Year* for 2009. As part of that process and prior to the opening of the park in May 2009, we substantially added to our interpretive displays and signage in and around our snow leopard exhibit, and the Mongolian field project was prominently depicted (Fig. 23.1.1). Fund-raising activities for the project were already in full swing by time the zoo opened on May 1, in the form of collection

FIGURE 23.1.1 (A) and (B) Prominent signage explains KIS's support for both snow leopard research and conservation projects.

boxes, SMS donations, and the selling of handicrafts from Snow Leopard Enterprises (another SLT program in Mongolia) in our souvenir shops. The Year of the Snow Leopard was a huge success for the zoo, for our fund-raising, and for snow leopards in the wild through our support for the Mongolian field study.

Our funding of that program was to continue for the next 4 years, which would continue to allow us to be even more involved, including collaboration on a film for Swedish television about the study and the role of Swedish zoos in supporting it, and a visit to the project site by a Foundation Board member. So, in May 2011 we visited the research camp in the Tost Mountains of South Gobi, Mongolia. The film crew had already been there a few weeks and researchers managed to catch a two leopards during that time providing a lot of good footage for the film, so we also had high hopes of seeing a snow leopard during our stay there.

However, like Peter Matthiessen in his famous book *The Snow Leopard*, we knew the animal was there, but never saw one during our brief stay. Still, to be in that place – and see how the research in the field works and to see the tracks of a snow leopard – was very exciting. To visit *in situ* projects that we are funding is important, and a firsthand experience allows us to share an even more convincing conservation message with our visitors at the Park and get them more excited about helping save endangered wildlife half a world away. One positive outcome was having Kolmården name one of the study animals in the wild. For the eighth female snow leopard to be caught and fitted with a GPS collar, we decided that she would be named "Dagina," which means Beautiful Princess. For the years that she was part of the study, information sent from the field to our zoo allowed Dagina to be "tracked" by our visitors. She was a long-distance ambassador telling her story, while the public watched our zoo cats and better understood their wild cousins.

So in what other ways can captive snow leopards help to save snow leopards in the wild? Thanks to its beauty and agility the snow leopard inspires people and we can then make the public aware of the vulnerability of the species in the wild. Snow leopards in Kolmården are important for fund-raising efforts to support field conservation, but they can also provide useful information to field researchers, such as data on types of anesthesia best used to capture wild snow leopards during collaring. Zoos have allowed researchers to test camera traps in their exhibits, to better plan snow leopard surveys. And zoo cats have been fitted with some of the first GPS collars to see how snow leopards adapted to them. Beyond just funding field research, zoos can provide information that improves field methods and makes them as safe as possible for wild cats.

Sharing information with our visitors about snow leopards is a key function for our staff. It's quite common to stand at the exhibit and hear visitors say they do not see any snow leopards, and when you point out a cat lying on the rocks they are amazed by how well camouflaged they are. Such occasions give us the opportunity to talk with people and get them fascinated by more than just their ability to hide on a bare rock, but also to be touched by their beauty and concerned with their plight in the wild. Our commitment to *in situ* conservation, such as the Mongolia project, can then be explained. There are many who have never seen a snow leopard before, or for that matter even heard of them. This moment, when they first become acquainted with the majestic cat, will hopefully stay in their minds forever.

Newborn animals are always one of the most popular sights for zoo visitors, and few draw more attention than snow leopard cubs. So we were excited when just 2 years after our snow leopards arrived, our female, Binu, gave birth to two cubs. Unfortunately she was not interested in the cubs and failed to care for them, requiring us to remove them for hand rearing.

Despite the effort one cub died after 9 days. We learned that another Swedish zoo, Nordens Ark, had a litter born just 1 day prior to ours, so with the advice of the EEP coordinator, a plan was made to introduce our female cub, now named Irma, to the female with two cubs in Nordens Ark. It was a bold move, which had not been previously attempted. A camera was installed in the den at Nordens Ark to follow the events when Irma was introduced. One of Irma's soft toys was sent ahead and placed in the den to allow the female to learn Irma's smell. At 24 days old she was transported to the new zoo. With the female locked out, Irma was rubbed with straw from the natal den, and the other two cubs were rubbed with Irma's blanket. When the female was allowed back in she went straight to the cubs and began to lick all three. But having never suckled since birth, Irma was unable to do so with her surrogate mother and began to lose weight and had to be removed back to Kolmården for continued hand rearing. But given how well the introduction had gone, once on solid food Irma was reintroduced to her adoptive family. This attempt was successful and by autumn she was fully integrated and showing the same behavior and growth as the other two cubs.

Irma was moved to Twycross Zoo in the United Kingdom in 2010, and a year later she gave birth to two healthy cubs that she reared successfully. Management of snow leopard breedings and exchanges falls under guidance of the European Endangered Species Program (EEP, see Chapter 21). Success stories like Irma's allow zoos to continue to breed, care for, and exhibit healthy snow leopards who serve as ambassadors for the cats in the wild.

In the few short years that Kolmården has had snow leopards in its animal collection, they have become one of our most popular attractions. An example of that comes from our "Endangered Species Ice Cream Day," when we hold a race and every child who crosses the finish line gets an ice cream, a certificate, and SEK 20 (about USD 2.40) to donate to any conservation project being supported by the KIS. The snow leopard gets the most support. Our goal is not just to raise funds, but to inspire children and adults to a lifelong commitment to animals and nature. With Ice Cream Day, we want to playfully highlight that everyone can contribute to the conservation of endangered species. With the support and commitment of our visitors, as well as all employees at Kolmården, we are able to support *in situ* conservation and take great pride in doing so.

More information about KIS and the various conservation projects it supports can be found at www.stiftelsenkolmarden.se

SUBCHAPTER

23.2

Woodland Park Zoo: From a Zoo Came a True Snow Leopard Champion

Fred W. Koontz

Vice President of Field Conservation, Woodland Park Zoo, Seattle, WA, USA

The role that zoo animals play in developing concern in zoo visitors for animal welfare and environmental protection is a topic of active investigation (Clayton et al., 2011; Kelly et al., 2014; Luebke and Matiasek, 2013). However, the inspirational affect that zoo animals have on zoo volunteers and employees, and consequently, the impact that these persons have on wildlife conservation remains unstudied. One example of the latter is Helen Freeman (Fig. 23.2.1), a volunteer-turned-employee of Seattle's Woodland Park Zoo, who by her own words had "a love affair with the snow leopards" for 35 years (Freeman, 2005).

Helen Freeman was a volunteer docent in 1972 when Woodland Park Zoo acquired their first two snow leopards (*Panthera uncia*), Nicholas and Alexandra, captured in the mountains of Kyrgyzstan, then known as Kirghizia, a republic of the Soviet Union. Little published information was available to guide the zoo's curator and keepers in caring for the cats. For example, Lee Crandall's classic reference *The Management of Wild Mammals in Captivity* (1964) devoted less than two pages to *Panthera uncia* and presented only a few bits of husbandry information. Upon the snow leopards' arrival, Freeman immediately took up observations in the zoo's Feline House. Nicholas and Alexandra showed signs of stress in acclimating to captivity, which sparked Freeman's empathy and desire to improve their situation (Freeman, 2005). This heartfelt reaction toward these two individual animals would lead to Freeman's career-long endeavor to understand and care for all snow leopards serving as ambassadors in zoos – and to protect their wild counterparts in Central Asia.

Helen Elaine Maniotas, born in 1932, graduated from Washington State University in 1954 with a degree in business administration. After marrying Stanley Freeman and raising two sons to school age in the early 1970s, she volunteered at Woodland Park Zoo and studied animal behavior at the University of Washington, receiving her BS degree in 1973. This second degree obtained at midlife secured her employment at Woodland Park, serving as Behavioral Research Coordinator and Curator of Education.

Freeman's interest in animal behavior, environmental education, and wildlife conservation came at an important time in zoo history. The 1970s to mid-1990s were a period of rapid change in North American and European zoos, most significantly (i) old-styled barred cages were replaced with naturalistic enclosures; (ii) wildlife

FIGURE 23.2.1 Helen Freeman with her beloved snow leopards.

conservation became the highest priority for modern zoos dictating new animal acquisition ethics (Koontz, 1995), development of multizoo breeding programs (Hutchins and Wiese, 1991), and support for field conservation (Conway, 2003); and (iii) animal behavior studies became recognized by zoo leaders as essential to improve animal care (Kleiman, 1992). Freeman was one of a small group of zoo biologists working in the 1970s and 1980s whose work and enthusiasm served as catalysts for building today's animal conservation breeding programs for endangered species.

Freeman's research on Nicholas and Alexandra was one of the first zoo studies that quantified animal behavior in a systematic way by using a behavioral catalog of snow leopard activities (Freeman, 1975), and by combining breeding records of zoos she demonstrated that a zoo's geographic location had little effect on the timing of parturition (Freeman and Braden, 1977). When Freeman sought to help zoos breed more snow leopard cubs by recognizing signals of mating compatibility and recognizing the limitations of small sample size at any one zoo, she convinced Bronx, Brookfield, Calgary, Portland, and Woodland Park zoos to use trained volunteers to collect standardized behavioral observations together (Freeman, 1983). This multinational effort revealed key correlates to snow leopard breeding success, which allows curators to make informed decisions on whether to change pairs after a certain period of time (Freeman, 1983). Historically, this project served as an early example of the utility of multizoo observational studies and the legitimacy of using trained volunteers.

Aided by Freeman's applied research, Nicholas and Alexandra went on to parent 29 cubs. Freeman and her coworkers took advantage of these births to document snow leopard maternal behavior, cub development, and management procedures (Freeman and Hutchins, 1978; O'Connor and Freeman, 1982). Ironically, after 75 years of mostly failing to breed snow leopards, by 1982 improved husbandry of the species in North American zoos was at the point of causing a snow leopard population explosion (Wharton and Freeman, 1988). As a result, Freeman's focus changed from producing more cubs to managing captive populations.

The Association of Zoos and Aquariums (AZA) in 1980 launched their Species Survival Plan (SSP) Program to cooperatively manage specific, and typically endangered or threatened, species populations within AZA-accredited zoos and aquariums and their partners (see Chapter 22). In 1984, with Helen Freeman as its first Species Coordinator, AZA's Snow Leopard SSP was initiated, at a time when fewer than 40 SSP programs existed (Wharton and Freeman, 1988). Under Freeman's leadership, the Snow Leopard SSP program excelled and became a model for many subsequent conservation breeding programs. Today, one of 460 AZA SSPs, the Snow Leopard SSP maintains a population of about 150 animals from a founder base of 38 wild snow leopards in 60 managed locations within North America; and SSP planned breedings manage loss of genetic heterozygosity and produce demographic stability, averaging 14 cubs born each year (AZA, 2014).

While helping many zoos throughout the 1970s to successfully manage captive snow leopards, Freeman realized the cats were struggling in the wild. In 1981, Freeman founded the International Snow Leopard Trust (now known as Snow Leopard Trust or SLT) with a mission to protect snow leopards and their habitat. She served as executive director until 1996, and thereafter assisted the SLT Board until her death in 2007. SLT, among many successes, became known for its leadership in developing a global snow leopard conservation strategy (McCarthy and Chapron, 2003) and for its community-based partnership approaches to wildlife conservation (SLT, 2014). At SLT, Freeman traveled widely to Asia, Europe, and throughout North America on behalf of snow leopards, ultimately becoming one of the world's foremost experts on the conservation of the species.

During her career, Freeman rightfully received many awards. Largely unrecognized, however, it is the broader role she played in the modernization of North American and European zoos. Her pioneering work in applied animal behavior, conservation breeding programs, species-focused field studies, and community-based conservation undoubtedly inspired many others volunteering and working in zoos.

Today, SLT and Woodland Park Zoo continue Freeman's legacy through their conservation, education, and research efforts. Woodland Park's snow leopards inspire our children and adult guests with their playful behavior and naturalistic setting, and zoo employees and volunteers still speak of Freeman often, celebrating how one person can make a difference. Her legacy also lives on at all snow leopard SSP zoos, where the cats each year inform millions of guests about the plight of wild snow leopards and efforts to save them. Since 2013, SLT and Woodland Park have worked together to support the Snow Leopard Foundation of Kyrgyzstan – the homeland of Nicholas and Alexandra, the two ambassador cats at Woodland Park Zoo that Helen Freeman fell in love with, starting an avalanche of concern for snow leopards living in zoos and in the wild.

SUBCHAPTER

23.3

Bronx Zoo: Ambassadors from the Roof of the World

Patrick R. Thomas

Vice President and General Curator, Wildlife Conservation Society and Associate Director, Bronx Zoo, New York, NY, USA

The Wildlife Conservation Society (WCS) has had a long and successful history with snow leopards (*Panthera uncia*) at its New York-based zoos. The Bronx Zoo was the first zoo in the Western Hemisphere to exhibit the species when it acquired a male in 1903, and snow leopards have been on display nearly continuously since then. For most of that time, the snow leopards were exhibited alongside other big cats in the zoo's Lion House. In 1986, the snow leopards were moved to the Himalayan Highlands, a series of naturalistic habitats that received the Association of Zoos and Aquariums (AZA) Exhibit Award for exhibit excellence. The Himalayan Highlands was one of the first exhibits to incorporate conservation as one of the main messaging themes, highlighting the pioneering work of Rodney Jackson and George Schaller. In 2009, WCS opened a second snow leopard facility, the Allison Maher Stern Snow Leopard Exhibit at the Central Park Zoo, and WCS manages the two populations cooperatively through its integrated animal collection plan.

Beginning in the 1960s, the Bronx Zoo began an extremely successful snow leopard breeding program that to date has produced 75 surviving cubs, more than any other zoo in North America. Snow leopards born at the zoo have been sent to zoos in seven countries and more than 30 US states.

The Bronx Zoo has also been an active and integral member of the AZA's Snow Leopard Species Survival Plan (SSP) since its inception in 1984, and Dan Wharton, a WCS staff member who worked at both the Bronx and Central Park Zoos, served as Species Coordinator for the program for 20 years (1987–2007). Because WCS has conservation programs and projects in more than half of the 12 snow leopard range countries, it is uniquely qualified to meet the USFWS's *in situ* enhancement requirement when applying for permits to import or export zoo-born snow leopards from other regional programs, including the European Association of Zoos and Aquaria (EAZA) and Southeast Asian Association of Zoos and Aquariums (SEAZA), to enhance the genetics of the regional populations.

The Bronx Zoo has always been known for its exhibits and display of animals in naturalistic habitats, and the snow leopard exhibit is no exception. The Himalayan Highlands has three exhibits for snow leopards as well as exhibits for red pandas (*Ailurus fulgens*), alpine pheasants (tragopans, *Tragopan* spp. or monals, *Lophophorus* spp.) and white-naped cranes (*Grus vipio*). Each of the snow leopard exhibits depict a different habitat frequented by the cats: a mountain meadow, a scree slope with running stream, and a birch-shaded alpine hillside with large rocky outcroppings. The Central Park Zoo has two snow leopard exhibits: a meadow and a granite hillside.

The exhibits and corresponding enrichment program for snow leopards at both zoos provide the animals with mental and physical stimulation, offer the cats some choice and control over their environment, and encourage a wide spectrum of species-typical behaviors. This allows visitors to observe active, engaged snow leopards, and provides an opportunity for zoo guests to see animals as they would appear in nature. The animals serve as ambassadors, inspiring the two zoos' approximately 3 million annual visitors to care about snow leopards and their conservation. The interpretive panels at the exhibits highlight the threats facing snow leopards (loss of prey species, retaliatory killings for losses of livestock, and poaching for their pelts) and what WCS field staff are doing to conserve them. Attached to the exhibits at both the Bronx and Central Park Zoos, but not on public display, are a series of holding enclosures. The snow leopards are brought off exhibit into these enclosures each evening, and all breeding, parturition, and rearing of young cubs is done in the off-exhibit areas where the animals can be closely managed.

The Bronx Zoo engages in a wide range of research. Much of it involves applied studies, with the goal of producing results that will not only benefit snow leopards in zoos, but also enable conservationists to better protect animals in nature. The Bronx Zoo historically (and currently) has had one of the largest snow leopard populations of any zoo in the world and utilizes it, whenever possible, to advance knowledge about the species and their *in situ* and *ex situ* management. One study evaluated the effectiveness of a proprietary scent developed by a WCS biologist to attract and stimulate snow leopards to rub up against hair traps. This scent had been used previously to assess the genetics of local Canada lynx (*Lynx canadensis*) and ocelot (*Leopardus pardalis*) populations by extracting DNA from hair follicles that were collected on the traps. Another study involved using digital facial images of the zoo's snow leopards from different angles to assess the ability of field researchers to use camera traps to identify individual snow leopards by their facial markings and/or vibrissae patterns. An offshoot of this study was to test various digital camera traps in the snow leopards' exhibits to see which models were most effective at "capturing" snow leopards in different environmental conditions. The Bronx Zoo also participated in a multizoo project that involved sending fecal samples from all the snow leopards to Working Dogs for Conservation to help train dogs to not only identify snow leopard feces from the scat of other carnivores, but also to discern the stool samples of individual snow leopards.

The zoo has also spearheaded academic studies involving snow leopards. One masters thesis from Columbia University, New York, focused on the variables that influence reproductive success in the SSP population; the results of this study have been sent to the SSP to help inform breeding recommendations. A current PhD dissertation project from Fordham University, New York, is conducting a noninvasive genetic assessment of snow leopard population structuring across its range, assessing what region(s) SSP founder animals originated from for population management, and comparing the genetic versus computer model-based relatedness of the SSP animals for *ex situ* management.

Even though the beauty, grace, and athleticism of all snow leopards are capable of inspiring awe and can create a desire among zoo visitors to help save the species, possibly no other snow leopard has served as a better ambassador than Leo, a wild-born animal that was orphaned when his mother was illegally killed. Leo was born in the Naltar Valley in northern Pakistan in 2005, and was still a young cub at the time of his mother's death. He would not have been able to survive on his own in the wild, but was successfully hand-reared by a Pakistan wildlife official.

Because no facilities in Pakistan were equipped with the housing, resources, and expertise to properly care for an adult snow leopard, the Pakistani government, in a unique collaboration with the US State Department, IUCN-Pakistan, World Wildlife Fund-Pakistan, and WCS, made the decision to send Leo to the Bronx Zoo until a suitable facility could be designed and constructed in Pakistan. Leo was just over a year old when he left Pakistan.

As a growing cub in Pakistan, Leo had gained tremendous notoriety – he was even visited by then-President Pervez Musharraf, who flew to northern Pakistan to see him. The unselfish decision to send Leo to a world-class snow leopard facility in the United States where he could best be cared for fostered tremendous good will between people of two nations with different cultures during a time of great political instability. There were two "turning over" ceremonies in Pakistan that officially loaned Leo to the Bronx Zoo. Both were heavily covered by the Pakistani and international press, and the ceremony in Islamabad resulted in front page news coverage in *The Dawn*, Pakistan's oldest and most widely read English language newspaper.

Leo arrived at the Bronx Zoo in August 2006, and was moved to his new home at the Himalayan Highlands where he is on exhibit each day (Fig. 23.3.1). An interpretive panel at his exhibit tells his story and helps visitors understand the threats wildlife face from human–animal conflict. Leo has been nearly as popular in the Bronx as he was in Pakistan. The former first lady of Pakistan, Sehba Musharraf, has visited him in New York, as have numerous Pakistani government ministers. He was also visited by Claudia McMurray, then the US Assistant Secretary of State for Oceans, Environment, and Science. Ms. McMurray was instrumental in facilitating Leo's transfer to the United States and the Bronx Zoo. In addition to widespread news coverage, a children's book chronicling Leo's early life, *Leo the Snow Leopard: The True*

FIGURE 23.3.1 Leo, a wild-born orphan snow leopard from Pakistan, in the exhibit at Bronx Zoo. *Source: Photo courtesy of Julie Larsen Maher, Wildlife Conservation Society.*

Story of an Amazing Rescue, was written by Jukiana, Isabella, and Craig Hatkoff and published by Scholastic Press. This book serves to inspire our next generation to care about conserving this emblematic species, and a portion of the proceeds of the book's sales go to snow leopard conservation.

After Leo's arrival at the Bronx Zoo and once he had settled into his new home, he was introduced to a female of similar age in order to socialize him. Even though the two animals were compatible, they never reproduced. The fact that both animals were sexually naïve when they were first paired possibly contributed to their inability to successfully breed. Leo finally produced his first offspring, a male cub, in 2013 with a female who had prior breeding experience. Now that he has experience, it is hoped that Leo will produce many more offspring that will help improve the gene diversity in the SSP population.

Each year, approximately 2 million Bronx Zoo visitors see Leo and learn about his story. And while he will never roam the high alpine slopes in northern Pakistan, his story will hopefully ensure that those mountains will be a safe haven for many generations of snow leopards.

References

AZA. 2014. Snow leopard. http://felid-tag.org/snow-leopard/ (accessed 31.12.2014).

Clayton, S., Fraser, J., Burgess, C., 2011. The role of zoos in fostering environmental identity. Ecopsychology 3 (2), 87–96.

Conway, W., 2003. The role of zoos in the 21st century. Int. Zoo Yearb. 38, 7–13.

Crandall, L.S., 1964. The Management of Wild Mammals in Captivity. University of Chicago Press, Chicago.

Freeman, H., 1975. A preliminary study of the behaviour of captive snow leopards. Int. Zoo Yearb. 15 (1), 217–222.

Freeman, H., 1983. Behavior in adult pairs of captive snow leopards (*Panthera unica*). Zoo Biol. 2, 1–22.

Freeman, H., 2005. Life, Laugher and the Pursuit of Snow Leopards. Snow Leopard Trust, Seattle, WA, USA.

Freeman, H., Braden, K., 1977. Zoo location as a factor in the reproductive behavior of captive snow leopards, *Uncia uncia*. Zool. Garden JF (Jena) 47, 280–288.

Freeman, H., Hutchins, M., 1978. Captive management of snow leopard cubs. Der Zoologischer, 49–62.

Hutchins, M., Wiese, R.J., 1991. Beyond genetic and demographic management: the future of the Species Survival Plan and related AAZPA conservation efforts. Zoo Biol. 10, 285–292.

Kelly, L.D., Luebke, J.F., Clayton, S., Saunders, C.D., Matiasek, J., Grajal, A., 2014. Climate change attitudes of zoo and aquarium visitors: implications for climate literacy education. J. Geosci. Ed. 62, 502–510.

Kleiman, D.G., 1992. Behavioral research in zoos: past, present, and future. Zoo Biol. 11, 301–312.

Koontz, F., 1995. Wild animal acquisition ethics for zoo biologists. In: Norton, B., Hutchins, M., Stevens, E., Maple, T. (Eds.), Ethics on the Ark. Smithsonian Institution Press, Washington and London, pp. 127–145.

Luebke, J.F., Matiasek, J., 2013. An exploratory study of zoo visitors' exhibit experiences and reactions. Zoo Biol. 32, 407–416.

McCarthy, T.M., Chapron, G. (Eds.), 2003. Snow Leopard Survival Strategy. Snow Leopard Trust and Snow Leopard Network, Seattle, WA, USA.

O'Connor, T., Freeman, H., 1982. Maternal behavior and behavioral development in captive snow leopard (*Panthera uncia*). International Pedigree Book of Snow Leopards 3, 103–110.

SLT. 2014. Snow leopard enterprises. http://www.snowleopard.org/learn/community-based-conservation/snow-leopard-enterprises

Wharton, D., Freeman, H., 1988. The snow leopard *Panthera uncia*: a captive population under the Species Survival Plan. Int. Zoo Yearb. 27, 85–98.

CHAPTER

24

Rescue and Rehabilitation Centers and Reintroductions to the Wild

OUTLINE

Snow Leopards
http://dx.doi.org/10.1016/B978-0-12-802213-9.00024-9

SUBCHAPTER

24.1

Rescue, Rehabilitation, Translocation, Reintroduction, and Captive Rearing: Lessons From Handling the Other Big Cats

Dale G. Miquelle, Ignacio Ignacio Jiménez-Peréz**, Guillermo López*[†], *Dave Onorato*[‡], *Viatcheslav V. Rozhnov*[§], *Rafael Arenas**, *Ekaterina Yu. Blidchenko*[§], *Jordi Boixader*[††], *Marc Criffield*[‡], *Leonardo Fernández**, Germán Garrote**, Jose A. Hernandez-Blanco*[§], *Sergey V. Naidenko*[§], *Marcos López-Parra**, Teresa del Rey**, Gema Ruiz**, Miguel A. Simón*[¶], *Pavel A. Sorokin*[§], *Maribel García-Tardío**, Anna A. Yachmennikova*[§]

***Wildlife Conservation Society, New York, NY, USA**

****The Conservation Land Trust, Scalabrini Ortiz, Argentina**

[‡]Florida Fish and Wildlife Conservation Commission, Naples, FL, USA

[§]A.N. Severtsov Institute of Ecology and Evolution, Russian Academy of Sciences, Moscow, Russia

[†]LIFE+ Iberlince Project: Recovery of the Historic Distribution Range of the Iberian Lynx, Andalusia, Spain

[††]Iberian lynx *Ex Situ* Conservation Program, Andalusia, Spain

[¶]Consejería de Medio Ambiente de la Junta de Andalucía, Jaén, Spain

INTRODUCTION

The capacity to rescue, rehabilitate, and exploit captive-reared individuals and/or translocate or reintroduce wild individuals back into native habitat can be useful components of the conservation "toolkit" for ensuring long-term survival/recovery of large felid populations. What these processes all have in common is that they require "hands-on" management; that is, animals are either captured in the wild or managed in captivity with the intent of release back into the wild. Individuals in distress (wounded, diseased, starving) may be captured in the wild (rescued), held in captivity (for rehabilitation), and released back into the wild (reintroduction). Additionally, individuals may be raised in captivity (from wild or captive sources) for restoration or supplementation of a wild population. In nearly all cases, this type of management of wild felids is highly controversial. When the primary goal is long-term persistence of a wild population, "rescue" attempts of lone individuals may often seem to have little to do with conservation objectives, and often require excessive investments of time, labor, and funding (Jackson and Ale, 2009). For this and other reasons, such management protocols need to be critically examined before they are applied to a large felid population in the wild.

With social media instantaneously spreading information of conflicts, interactions, and appearances of animals in distress, the global demand for effective and professional responses will only grow in the foreseeable future. This new form of "instantaneous spotlighting" represents both a problem and an opportunity. The global public will expect responses, and the absence of an appropriate response (because it is not a conservation priority) will often be deemed unacceptable. However, the value of such rescue efforts as a media tool to influence local, national, and international opinions should not be underestimated, and may be an important component of an education/awareness program to foster support for conservation goals. Therefore, while in the short term rescue and rehabilitation efforts may require significant financial and logistical investments, payoffs in terms of public and political support for conservation initiatives may make such efforts worthwhile.

Captive rearing is a labor intensive, major financial investment that should only be considered when effective alternatives do not exist. Some of the same skills needed for rescue and rehabilitation can be applied to captive breeding programs intended to supplement or restore wild populations. Both types of activities require the ability to care for animals that will eventually be released into the wild and therefore these animals must not only retain a healthy fear of humans but must be capable of successfully hunting wild prey. While some projects have invested millions of dollars in rewilding efforts for large felids, others appear to have had some success on a "shoestring" budget. Therefore, determining the key components of a captive breeding or rehabilitation program will assist in managing the financial costs of such programs and improve chances of success.

Among an increasing number of incidents where handling wounded or stray snow leopards (*P. uncia*) has been necessary, some such events have garnered significant attention in the press, with expectations of an appropriate response. Given the growing threats of human encroachment and climatic changes on snow leopard habitat, the need for rescue and rehabilitation of individuals will no doubt only grow. Additionally, given the already patchy nature of snow leopard habitat and continuing fragmentation due to human activities, the need for genetic restoration may arise more quickly than most are willing to admit. At the same time, with changing environmental conditions, it may not be long before consideration of supplementation, restoration, or even assisted colonization into newly suitable habitat will be potential or necessary options for snow leopards. To prepare such management regimes in the future, it is useful to consider examples and lessons learned from similar work with other large felids. In the chapter, we review four case studies from around the world to understand what potential opportunities and pitfalls lay ahead for their implementation with snow leopards.

CASE STUDY 1. PLANNING A JAGUAR REINTRODUCTION IN ARGENTINA: COMBINING SCIENCE, PUBLICITY, AND PUBLIC POLICY

The jaguar (*Panthera onca*) is the largest terrestrial predator in the Neotropics, where it has been extirpated from 54% of its original range, and is globally classified as Near Threatened (Caso et al., 2012). In Argentina, the species is classified as Critically Endangered after a 95% decline of its historical range (Aprile et al., 2012), and it is presently found in three disjunct populations that total approximately 200 individuals within the Yungas, Chaco, and Atlantic Forest ecoregions in the northern part of the country (Di Bitetti et al., in press).

During the twentieth century, the jaguar has disappeared from the Iberá region in Corrientes province, Northeastern Argentina (Parera, 2004). In 1983, the provincial government of Corrientes established the 1.3 million ha Iberá Nature Reserve (INR) to protect an entire river basin covered by wetlands, grasslands, and patches of forest. In 1999, the Conservation Land Trust (CLT) purchased 150,000 ha inside INR with the intention of creating a 700,000 ha national park. As negotiations for creation of the park proceeded, CLT began exploring the potential for restoring viable populations of extirpated mammals, including the jaguar. CLT led a two-staged process, first assessing feasibility of a jaguar reintroduction, then planning how best to conduct it.

Assessing the Feasibility of a Jaguar Reintroduction into INR

Kelly and Silver (2009) recommended that any such initiative should "be contingent on the proper combination of suitable habitat, socio-cultural tolerance by local and national communities and the resources in time, money and expertise to carry out such a project responsibly," largely following recommendations from other experts (Kleiman et al., 1994; Macdonald, 2009). Accordingly, the feasibility study addressed three main issues.

Habitat Suitability

To assess whether there is sufficient suitable habitat within INR for a viable jaguar population, De Angelo (2011) conducted a GIS-based habitat suitability assessment using existing information on habitat types, presence of cattle and human populations, roads and human access, watercourses, and distribution/abundance of potential prey. De Angelo (2011) estimated that INR contained 250,000 ha of quality habitat and 400,000 ha of suboptimal habitat that is currently protected and almost devoid of humans or livestock. Within this potential jaguar core area, based on densities obtained from similar landscapes, "70 to 90 jaguars could create a significant population with high chances of long-term survival, even though it would need proper genetic and demographic management" (De Angelo, 2011).

Public Support

To determine whether there was sufficient support for a large carnivore reintroduction Caruso and Jiménez Pérez (2013) assessed the attitudes of both urban and rural residents of Corrientes Province. They discovered that 95% of the people support the return of jaguars with results independent of the respondents' gender, age, or location. These results were obtained prior to any media campaigns to promote jaguar reintroduction, suggesting that jaguars were already viewed positively by the public. Caruso and Jiménez Pérez (2013) suggested that for Corrientes Province the jaguar could act as a bridge between a proud provincial heritage and a hopeful future in which ecotourism could provide a means of economic development. With such strong local support, the only other groups that could thwart jaguar reintroduction would be scientists, conservationists, or government officials. Engagement with these groups is discussed later.

Capacity and Commitment

The final component of a preliminary assessment was identifying an organization willing to commit long-term resources and expertise to lead such a complex endeavor. In this case, CLT would clearly fill this role, as it had already invested more than USD 5 million/year for conservation activities in INR for the previous

16 years and was committed to maintaining this level of investment until the lands are donated to the government and all extirpated species are reestablished. Also, since 2006 CLT had established and trained a team that had experience in successful reintroduction programs for giant anteaters and pampas deer (Jiménez Pérez, 2013; Jiménez Pérez et al., 2013).

Planning and Negotiating a Jaguar Reintroduction Plan

Once it was determined that INR had sufficient habitat, public support, organizational capacity, and commitment, the best technical strategy had to be designed to mesh with the best political strategy before negotiating with agencies responsible for authorization. To derive a technical reintroduction plan with high probability of success, representatives of CLT first visited other jaguar conservation projects, such as in the Brazilian Pantanal, and other felid reintroduction programs, such as the rewilding initiative at Pilanesberg National Park and the South China Tiger Project (both in South Africa), the Iberian Lynx *ex situ* conservation program in Spain (see later), and two tiger reintroduction projects in India. Second, over a period of 4 years CLT held meetings, workshops, and consultations with jaguar experts from Argentina, Brazil, Europe, and the United States, plus other international experts in large felid reintroductions. These activities included national and regional experts as advisors and supporters in an adaptive planning process that produced a strategic vision document (CLT, 2012. Visión estratégica para la reintroducción del yaguareté en la Reserva Natural Iberá (Corrientes, Argentina), unpublished report.), which has been updated regularly.

This strategic vision was based on the following premises: (i) with a total national population of only 200 individuals, it was not ethically or biologically justifiable nor politically feasible to translocate wild jaguars from other regions of Argentina to INR; (ii) at this phase of the program it was also not politically or administratively feasible to obtain permits for translocation of jaguars from a healthy population in neighboring countries such as Brazil; however, it was believed that such permits might be possible in the future if initial successes could be demonstrated; (iii) due to these constraints, the first animals to be released should be born from captive jaguars in a facility within INR and provided the opportunity to learn to hunt native prey while avoiding any positive stimulus from or affiliation to humans; (iv) the facility and first release site should be located within remote high-quality habitat inside INR to minimize dispersal to less secure areas; and (v) with an initial captive-born nucleus established, wild jaguars potentially translocated in the future would be less likely to show homing instincts or disperse away from the area.

Based on these principles, a three-step jaguar reintroduction plan for INR was derived. Phase One would include construction of a breeding facility for captive jaguars from zoos that would produce viable offspring born in large pens situated on the best habitat inside the reserve. Also, media reports regarding individuals born at the facility would assist in promoting the jaguar reintroduction program across Corrientes. In Phase Two, jaguars born and raised on-site (demonstrating the ability to successfully hunt and avoid human contact) would be released inside INR to establish a core population. Ideally, the establishment of such a nucleus would help to initiate Phase Three, which would include capturing, translocating, and releasing wild jaguars from viable populations in neighboring countries to boost demographic growth and genetic viability of the re-established population. Homing behavior of wild translocated animals would be minimized through soft-releases from acclimatization pens.

After receiving input from national and international experts, the program had to be reviewed and approved by the provincial and national wildlife authorities. Initial meetings with provincial authorities soon made clear that they were only willing to approve the first phase – an on-site breeding facility – before considering possible authorizations of jaguar releases. Approval of the first phase by provincial authorities required a year and a half of meetings and adjustments to the strategic document. Approval by national authorities required another 2 years of a protracted negotiations involving more meetings and formal presentations. In December 2014, an on-site jaguar breeding facility was approved by both regional and federal authorities. Construction had already began in a remote area of INR in September 2013 and the Experimental Jaguar Breeding Center was ready to hold two captive breeding pairs and their offspring by the time the plan was approved. Technical details are beyond the scope of this text, but are described in the management plan (Solís et al., 2014).

Negotiating and achieving final governmental authorization was probably the most taxing and, at the same time, most critical part of the entire process. With permits in hand, the first female jaguar was transferred to the breeding center in May 2015 after going through a quarantine phase. Along the route to the center, the jaguar was met with unexpected enthusiasm from local villagers. Around 70 people (including government officials, religious leaders, national park authorities, and local representatives) witnessed the first jaguar step into Iberá after a 60-year absence. By the end of 2015, two pairs of jaguars should be living at the Experimental Jaguar Breeding Center inside INR.

Conclusions on Developing a Reintroduction Plan

Following Seddon et al. (2014), the jaguar reintroduction project in INR is both a *rewilding* initiative to restore the ecological role of the jaguar as the top predator in this vast wilderness, and a *species conservation* initiative to recover a critically endangered species at the national level. Nevertheless, despite the existence of extensive high-quality habitat, strong support by local residents and the scientific community for the reintroduction effort, and the existence of a capable and committed organization to see the process to fruition, there were still extensive bureaucratic and political obstacles to overcome before approval and initiation of the project. Perhaps the most useful lesson is that delineation and approval of a feasible reintroduction plan requires wisdom, patience, persistence, communication skills, empathy with other points of view, and sufficient flexibility to adapt an "ideal" vision to political and bureaucratic constraints so that implementation becomes a reality.

CASE STUDY 2. THE IBERIAN LYNX: RESTORING A POPULATION ON THE VERGE OF EXTINCTION

The Iberian lynx (*Lynx pardinus*) is a medium-sized felid endemic to the Iberian Peninsula (Spain and Portugal). Formerly widespread, only about 100 individuals were found in 2002 in two isolated populations in Andalusia Province of southern Spain (Andújar-Cardeña and Doñana) (Guzmán et al., 2004). Given the small size of this remnant population, the Iberian lynx was the only felid species listed as "Critically Endangered" (IUCN, 2003). Starting in 2002, three consecutive EU-funded conservation projects were developed by the Regional Government for Environment of Andalusia to halt the decline of the remaining populations (mainly by increasing carrying capacity and decreasing mortality rates) and restore extinct populations through reintroductions (Simón et al., 2013). A captive-breeding program was initiated in 2004 with the main goal of providing individuals for release (Vargas et al., 2008). Here we summarize methods and results of the reintroduction program that began in 2006, and continues to the present, when the IUCN recently downlisted Iberian lynx to "Endangered" (IUCN, 2015).

Identification of Suitable Habitat

Optimal areas for reintroduction were selected in a two-stage process: (i) preselection based on GIS habitat suitability analyses; and (ii) definitive selection based on fine-scale field studies of habitat suitability, prey density, public support and potential threats. In the first stage a suitability model using presence/absence data (generated both by radio-tracking and phototrapping studies conducted between 2002 and 2006) was generated (Gill-Sánchez et al., 2011). In 2006, four areas were preselected in Andalusia as optimal for Iberian lynx reintroduction based on the habitat suitability analyses, size (a minimum of 10,000 ha), and the potential for integration into a metapopulation (Gaona et al., 1998; Palomares, 2001). Between 2007 and 2009, fine-scale studies collected data on habitat quality, rabbit (*Oryctolagus cuniculus*) density (the Iberian lynx's primary prey), public support, and threats analyses. All values derived by these methods were compared with the same variables obtained within the current range of Iberian lynx (Simón et al., 2013). By December 2009, two areas of Andalusia were selected for the first reintroductions: Guadalmellato and Guarrizas, approximately 50 km to the east and west, respectively, of the remnant Andújar-Cardeña population. First releases began in Guadalmellato in 2009 and Guarrizas in 2010.

Origin of Released Individuals

Both wild-caught and captive-raised individuals have been used in the reintroduction process. Both males and females were removed from a wild population for translocation when it was determined that removal would have no significant impact on the existent population (Palomares et al., 2002). Wild individuals were selected based on their age (young sexually mature), social status (only nonresident individuals), and genetic origin (closely related individuals were not released in the same area). Once captured, all individuals were given an extensive health assessment and placed in quarantine for 2–6 weeks to avoid introduction of any diseases into the recovering populations. Captive individuals were selected to maximize genetic diversity and were kept in large natural enclosures with a complex environment similar to that found in the wild. Contact with keepers was minimized. Although promoting natural behaviors is considered essential for the survival in the wild (Griffin et al., 2000; Hartmann-Furter, 2009), candidates for release were only fed live prey, mainly wild rabbits (80–100% of their diet in the phases prior to release). Mothers generally hunt and provide food for young cubs, but cubs are able to kill prey on their own as early as 103 days after birth (Yerga et al., 2012). A network of tunnels was built for rabbits in the training enclosures to mimic the complicated process of both locating and capturing prey in the wild. Moreover, unpredictability of food availability and the occasional use of hunger were used to promote exploration and foraging behavior, as well as to mimic natural conditions. To prevent association of humans with food, we developed a system of automatic feeders connected to a timer. To promote avoidance behavior toward humans, lynx-human contacts were kept to a minimum and negative stimuli (shouting or throwing water on them) discouraged contact. Human handling was avoided while social interactions with other lynx were encouraged.

A health assessment was conducted on all captive Iberian lynx to ensure only healthy individuals in good condition were released. When a limiting health problem or disease was detected during checkups, the individual was removed from the program. The sex ratio of released individuals was kept near to 1:1 for each release year.

Release and Monitoring

Released individuals were radio-tagged to allow postrelease monitoring. A soft-release approach was initially used, with enclosures of 1–8 ha built in areas of optimal habitat and prey density (Simón et al., 2013). Three enclosures were built in Guadalmellato and two in Guarrizas. Two to four individuals were soft-released at any given time, and time spent in the enclosure ranged between 7 and 180 days. Once a nucleus population (at least two pairs) of wild individuals had settled into a given area, captive-reared individuals were released directly into a site (hard releases). In cases where soft releases were performed within a stable adult home range, fights between incoming individuals and resident territorial individuals were recorded.

Results and Conclusions

From 2009 through 2014, 36 (17 males and 19 females) Iberian lynx were released in Guadalmellato and 37 (17 males and 20 females) in Guarrizas. After release, distances explored by individuals ranged from 0 km to 75 km. Newly released lynx usually settled into unoccupied areas with high prey abundance. Distance between the release site and the center of a home range ranged from 0 km to 32 km with no variation in dispersal distance between the sexes. Although data are still limited, there appear to be no differences in dispersal distances between soft- and hard-released individuals. Mean survival rate in the first year after release (73.9%), was nearly identical for captive-raised (72.7%) and wild caught individuals (73.2%). Reproduction has been recorded every year since the first release. The range of the Iberian lynx in AndújarCardeña has increased from 11,900 ha and 59 individuals in 2002 to 26,000 ha and 179 individuals in 2010 (Simón et al., 2012). Reintroductions into former Iberian lynx range will continue to reach a recovery goal of five populations totaling 300 individuals within Sierra Morena (Simón et al., 2013).

These results suggest that we have been fairly successful in both translocating wild Iberian lynx and developing protocols for raising young Iberian lynx for the reintroduction program. Similar survival rates of captive-reared and wild translocated lynx are a strong indicator of the success of the rearing program, and suggests that similar approaches may be successful with other medium and large felids. Although some human-Iberian lynx conflicts have been documented (Garrote et al., 2013), mitigation programs are in place. Well-designed programs with appropriate training to develop hunting skills and negative conditioning toward humans may be some of the key components to ensure success.

CASE STUDY 3. GENETIC RESTORATION AS A MANAGEMENT TOOL FOR ENDANGERED FELIDS: LESSONS LEARNED FROM THE FLORIDA PANTHER

Many large carnivore populations have become isolated and dwindled in size due to a combination of unregulated harvest and habitat loss associated with human development (Woodroffe, 2001). Inbreeding is inevitable in small populations and can have a major impact on long-term population viability and risk of extinction (Frankham et al., 2002). That scenario has transpired for the Florida panther (*Puma concolor coryi*), a subspecies of puma that once ranged across the southeastern United States, which is now restricted to <5% of its historical range (USFWS, 2008) and is afforded protection under the US Endangered Species Act (Public Law 93-205; Federal Register, 1967).

Identification of a Problem

In 1981, the Florida Fish and Wildlife Conservation Commission (FWC) initiated a research project on panthers that has been ongoing for >30 years (Onorato et al., 2010). Early research revealed that many of the remaining panthers exhibited congenital anomalies including high incidences of defects such as kinked tails, thoracic "cowlicks" of fur, cryptorchidism, atrial septal defects, depressed immune systems, and detrimental sperm characteristics (Roelke et al., 1993). These anomalies were presumed to be a consequence of low levels of genetic variation associated with inbreeding depression (Roelke et al., 1993). Minimum counts of panthers remaining in the wild in the early 1990s indicated the population likely consisted of as few as 20–30 individuals (McBride et al., 2008). The combination of the extremely small population size and observed impacts of inbreeding depression led wildlife managers in 1991 to establish a captive breeding program with kittens removed from the wild. The captive breeding program was discontinued in 1992 for two reasons: (i) heightened concerns that the genetic health of the wild population had reached a critical point where the continued survival of the panther population was in question, and (ii) logistical constraints regarding the necessary space for captive breeding facilities and the length of time needed before captive breeding might feasibly contribute to recovery (potentially several panther generations). A different, more expedient approach was necessary to reverse what appeared to be the panther's inevitable rendezvous with extinction.

Developing a Plan for Genetic Restoration

In 1994, the FWC, National Park Service, US Fish and Wildlife Service, and nongovernment organizations, along with experts in the field of carnivore biology and conservation genetics convened to develop an alternate approach to recover the Florida panther. The result was a plan to release eight female pumas from western Texas (*Puma concolor stanleyana*) into the wilds of South Florida (Seal, 1994). Pairing these females with wild male Florida panthers was predicted to improve levels of heterozygosity in the population (Seal, 1994). Pumas from Texas were chosen because Florida panthers historically experienced a level of gene flow with that subspecies when panthers were distributed across their historical range throughout the Southeastern United States. This periodic exchange of breeding animals ceased when panthers became isolated in South Florida. Genetic restoration would therefore mimic this gene flow that used to occur naturally. In theory, genetic restoration had the potential to improve the long-term outlook for the panther through revitalized genetic variation and in turn result in a population comprised of individuals that may genetically more closely resemble those that were historically distributed across the Southeastern United States. Females rather than males were selected for release because they would allow managers to more accurately document the level of genetic introgression by sampling kittens from litters of radio collared Texas females. Documenting the litters of uncollared female panthers sired by a male Texas puma would be nearly impossible.

The plan for genetic restoration was not without critics. For instance, Maehr and Caddick (1995) proposed that outbreeding depression could result and lead to the loss of localized adaptations. While such a scenario merited contemplation, the overall consensus was that genetic restoration held the most promising, expedient, and perhaps last chance to avert the extinction of Florida panthers.

Implementation and Results

Between March and July 1995, eight female pumas from west Texas were released at five different locations in South Florida (Onorato et al., 2010). Upon release, the Texas females adapted quickly to the vastly different landscape. By October 1995, less than 6 months after release, a Texas puma produced the first documented admixed litter. Accounting for the 90-day gestation period for pumas (Currier, 1983), fertilization by a male Florida panther occurred less than 3 months after release. Eventually, a minimum of 20 kittens born to Texas pumas were documented (Onorato et al., 2010). In subsequent years, breeding by the F1 generation of admixed panthers was verified, along with several backcrosses to the Texas pumas. An extensive assessment of the effects of genetic restoration on the panther highlighted the benefits that ensued (Johnson et al., 2010). Admixed panthers exhibited a lower prevalence of kinks, cowlicks, cryptorchidism, atrial septal defects, and sperm anomalies while showing marked improvements in genetic heterozygosity and survival rates of both kittens and adults (Benson et al., 2011; Hostetler et al., 2010; Johnson et al., 2010). Furthermore, population viability analyses demonstrated that genetic restoration had a significant impact improving the outlook for the long-term persistence of the panther population (Hostetler et al., 2013). The benefits associated with genetic restoration are perhaps best exemplified by the most recent minimum count data that shows an increase from 26 in 1995 to 133 panthers in 2013 (McBride and Sensor, 2013).

Conclusions

While other factors invariably played a role in the dramatic turnaround for the Florida panther (including acquisition and protection of >120,000 ha of land, wildlife highway underpasses, legal protection under the Endangered Species Act), genetic restoration can most certainly be deemed a success (Johnson et al., 2010; Onorato et al., 2010). In retrospect, there is little doubt that managers in 1994 assumed some level of risk when deciding to implement genetic restoration of a large carnivore *in situ*. Only limited control over the progression and outcome of genetic restoration was possible. While panthers remain endangered, the outlook for recovery today is much improved compared to two decades ago. The lessons learned from the experience with genetic restoration in Florida should be contemplated for use in other regions of the world where options of releasing conspecifics into the wild to reinvigorate dwindling wildlife populations have the promise of averting future extinctions.

CASE STUDY 4: RESCUE, REHABILITATION, AND REINTRODUCTION OF AMUR TIGERS INTO HISTORIC RANGE IN THE RUSSIAN FAR EAST

Historically the best habitat for Amur, or Siberian tigers (*Panthera tigris altaica*) probably existed in northeast China and the Korean peninsula, but today at least 95% of the remaining wild population resides in the Russian Far East. Tigers originally occurred along both sides of the Amur River – in Amur Oblast and the Jewish Autonomous Region (JAR) in Russia, and in the Lesser Khingan Mountains of Heilongjiang Province China (Heptner and Sludskii, 1992; Ma, 2000), but animals in these areas largely disappeared in the 1970s and 1980s. Today, although suitable habitat remains in this historic part of its range, nearly all Amur tigers reside in the Sikhote-Alin Mountains of Primorye and Khabarovsk Provinces, Russia and in southwest Primorye and the neighboring Changbaishan Mountains of Jilin Province, China (Hebblewhite et al., 2012; Miquelle et al., 2007).

Orphaned tiger cubs are not uncommon in the Russian Far East, perhaps because females with cubs are more likely to stand their ground and protect cubs, making them more susceptible to poachers with firearms. In the past most orphaned cubs went into captivity since there were no facilities for rehabilitation, although a few attempts were made to retain cubs in the wild (Goodrich and Miquelle, 2005). However, because these few individuals had little impact on dynamics of the Sikhote-Alin population, estimated in 2005 at 430–500 individuals (Miquelle et al., 2007), these rescues, rehabilitations, and releases back into the wild were little more than animal welfare cases.

In 2012 a tiger rehabilitation center in Alekseevka, Primorye, was completed in time to receive a female cub orphaned at approximately 4 months of age. In 2013, six more tiger cubs arrived at the facility, all 3–5 months of age. Instead of the standard practice of sending these cubs to zoos, or releasing them back into the remaining tiger population in the Sikhote-Alin, it was decided to attempt to use these cubs to recolonize lost habitat in the JAR and Amur Oblast, where tigers had been absent for nearly 40 years.

Rescue and Rehabilitation of Orphaned Cubs

In most cases local villagers or hunters contacted authorities when abandoned cubs appeared. In every instance attempts were made to confirm that cubs were indeed abandoned. Techniques for capturing cubs varied with their size and condition: some were so weak they could simply be wrapped in a jacket and carried out, while others in better condition required immobilization before handling. Cubs were often held in temporary facilities near the capture site for treatment (administering fluids, first feeding, medical needs) and physical assessments before being transferred to the rehabilitation center. Cubs were kept in quarantine for one month before release into one of the six enclosures at the center. Some cubs were kept together (a litter of three and a pair of females) while one male and one female were held separately. The female from the litter of three died while still in quarantine, apparently from a viral infection. No other tiger cubs showed any symptoms of illness. Blood was collected from all cubs upon arrival and prior to release, and screened for 17 pathogens. Tigers had antibodies to feline calicivirus (three of six), *Toxoplasma gondii* (three tigers), feline panleukopenia virus (five tigers), *Trichinella* sp. (one tiger), and were negative for all other 13 pathogens tested. Because these four pathogens are common in wild tigers of the Russian Far-East (Goodrich et al., 2012; Naidenko et al., 2012), all cubs were still considered suitable candidates for release.

Chain-linked fencing 4.5 m high (with an inward overhang of 1 m at the top) enclosed pens ranging in size from 0.3 ha to 0.7 ha. Natural vegetation was retained in the enclosures, but degree of cover varied in each, from largely forested with brushy undercover to mostly open tall grass fields. Water was available ad libitum. Human contact with animals throughout the rehabilitation period was minimized by placing sheeted material on enclosure fences to block visual contact, by installing multiple video cameras to monitor activities remotely, and by providing food through boxed enclosures that could be opened remotely. Tigers were fed almost exclusively wild game, although the first cub was fed some beef. At 7–8 months of age, small live prey (domestic rabbits and pheasants) were presented to cubs, who actively hunted these prey. When cubs reached 11 months of age, live young wild boar (*Sus scrofa*) and young sika deer (*Cervus nippon*) were released into the pens. Larger prey (subadult/adult wild boar and sika deer) were presented to tigers older than 15 months of age after their permanent teeth were fully developed. For the 6 months prior to release, tigers were provided only live natural prey items, and thus, were wholly dependent on their own abilities to capture prey. Intervals between presentation of live prey ranged from 7 to 12 days. Each cub had successfully killed at least eight wild boars and/or sika deer before it was considered ready for release.

Before a final decision was made to release individuals, they were tested for their reaction to human presence. A person walked along the perimeter of a tiger enclosure while observers (via the remote video system) scored the reactions of the tiger(s). None of the tigers attempted to approach the encroaching human or showed any signs of aggression.

Reintroduction of Cubs into Historic Range

Suitability assessments were conducted at potential release sites in the JAR and Amur Oblast (Aramilev, 2013). Yearly surveys on ungulate densities conducted in protected areas and hunting leases allowed assessments of whether prey densities were adequate to support tigers. The three sites selected for release were all in remote locations. Extensive discussions and agreements with government agencies and governors of these regions occurred prior to release, and special funds were allocated to ensure local wildlife agencies would be able to monitor tigers after release. Teams with experience in resolving human-tiger conflicts were also prepared to travel to the sites if necessary. Educational and outreach programs were conducted in villages close to release sites prior to release. No extensive surveys were conducted to assess public opinion of the reintroduction program, although informal assessments suggested many local people were opposed to the idea of returning tigers to the area (Aramilev, 2013).

Six tigers were released at the age of 18–20 months, mimicking the time when subadult tigers normally disperse from their natal home range (Goodrich et al., 2010). In May 2013, one tigress was released in Bastak Reserve (JAR); in late May 2014 two males and one female were released together in Zhelundinsky Wildlife Refuge (Amur Oblast); and in early June 2014, one female and one male were released into Zhuravliny Wildlife Refuge (JAR). All were affixed with GPS collars to allow monitoring after release.

The first female released in 2013 has survived 2 years in the wild, and four of the five released in May 2014 survived their first year in the wild. The first female remained in Bastak Reserve (her release site) after a small excursion to the north in her first summer (Rozhnov et al., 2014). She has commonly been in association with a male that apparently dispersed into the region prior to her arrival. Camera traps have regularly captured both her and the male in the reserve. The female released into Amur Oblast moved approximately 35 km south of the release site, and remained there for the summer before shifting another 50 km further south-east to Khinganskii Reserve (where wild boar are abundant) for the winter. The female released into Zhuravliny Refuge also made one excursion approximately 50 km to the west in summer, before returning and remaining in the area close to her release site.

The three males have traveled much more extensively. One male covered more than 1200 km between May and December 2013, including an extensive circle through the Lesser Khingan Mountains of Heilongjiang Province China, before returning to Russia after the Amur River froze over. GPS locations suggest he may have crossed paths with the female in Zhuravliny Refuge, possibly explaining his extended stay in this region prior to failure of his collar in February 2015 (camera traps set in this region may yet confirm his status and location). A second male (released in JAO) moved more than 600 km in an easterly direction following the Amur River, and also entered China, where he fed extensively on domestic animals. When he returned to Russia in December 2014, he continued to rely on domestic animals (dogs) and did not show adequate fear of humans, and was therefore captured and removed from the wild. The other five animals all appear to be surviving almost exclusively on wild prey. Of 82 kills located by examining clusters of GPS locations and snow-tracking (Miller et al., 2013), only 3 (2 young cattle and 1 dog) were domestic: 87% were wild ungulates (including 67% wild boar).

Conclusions

Goodrich and Miquelle (2005) suggested three indicators of success for translocating "problem tigers" in Russia that are applicable here: (i) Survival through the first winter with evidence of predation on wild prey; (ii) lack of conflict with people or domestic animals; and (iii) successful reproduction. One tiger was removed due to predation on domestic animals, but all others have concentrated mainly on wild prey. All five tigers remaining in the wild survived their first winter, and the one female with 2 years in the wild is consorting with a male, but has just approached breeding age (earliest age of first reproduction reported for wild Amur tigers is 3.5 years [Kerley et al., 2003]). Thus, while it is too early to claim complete success, it appears that all six cubs successfully learned how to hunt in the wild, and with one exception, have avoided conflicts with people. It is yet to be determined whether these tigers will successfully breed and begin the process of restoring a self-sustaining population into this former tiger range.

LESSONS LEARNED

Find Common Ground with Key Constituents

Rescue, rehabilitation, and recovery operations require direct handling of animals, and therefore will generally be controversial and highly scrutinized. Therefore, success will often be dictated by the preliminary work done in developing a defensible plan and garnering support from scientists, the appropriate agencies, political entities, and the public. Each of the case studies to varying extents was successful because they were able to identify key constituencies and found means of agreeing on programmatic goals and methods. The case study on jaguar reintroductions perhaps most clearly underscores the need for compromise and patience in dealing with a variety of interest groups. To conduct genetic restoration of the Florida panther, strong support from the scientific community was a critical component of garnering political support. The captive rearing program for Iberian lynx required years of debate among local, national, and international political and scientific organizations to reach a consensus plan that is now probably the best example of using captive-raised individuals to restore a wild felid population.

Develop Effective Public Outreach

In none of the studies did public opinion negatively impact the process, but given the strong sentiments that large carnivores generally evoke, these may be more the exception than the rule. Working to restore large felids into their former range (in three of the case studies) is likely to be more acceptable to the public than expanding historical range, but working with the public is likely to be an important component in nearly all situations (Garrote et al., 2013; Goodrich, 2010). Rescue work with tigers receives lots of publicity both locally and globally, so while these activities may be largely viewed as animal welfare issues, they help build public and financial support for the larger goal of tiger conservation. The importance of this aspect of animal rescue work should not be underestimated as a rationale for having the capacity to do such work professionally.

It is important to have a well-defined message and mechanism for getting information to the public. For example, a common misconception in Florida is that admixed panthers are much more aggressive than the "original" Florida panthers, and that they are not afraid of people. There is no scientific basis for these claims, but they are pervasive in segments of the public that are not in favor of Florida panther recovery. Be prepared to address such concerns with convincing evidence, and ideally, envisage potential criticisms beforehand, and address concerns before they work their way into the public consciousness.

Release of rehabilitated, captive-reared, or translocated animals can be effective media opportunities that are intensively covered and provide an opportunity to convey important messages to the public. In organizing such events, film/photo opportunities must be provided to media, but in a way that does not compromise the purpose of the release (or other event). Just as importantly, there should be a clear message that is defined beforehand to be conveyed to the media, remembering that they are your means of outreach to the public. Preparation of printed materials prior to the event is a good way to ensure that the proper information gets into the hands of journalists. However, it must be remembered that all events must be planned for maximum benefit to the animals, not to accommodating spectators.

Exploit the Knowledge and Experience of the Global Scientific, Zoo, and Conservation Communities to Develop a Defensible Plan

Most of the case studies relied on well-known experts, or visited multiple sites to assess and learn from others. Knowledge of how to construct facilities, plan releases, and handle animals (as examples) exists, but within a relatively small circle of specialists. These people need to review existing plans and advise on how best to proceed. Learning by trial and error is not recommended.

Ensure Suitable Habitat Exists, and Reasons for Extirpation are Known and Mitigated

If restoration into historical range is the goal, the existence of a sufficiently large tract of suitable habitat should be identified and sufficient prey densities confirmed before any plans are initiated. Identification of the release site must be a precursor to any captive-breeding programs intended to restore wild populations of felids. Just as importantly, it must be clear why the former population went extinct, and whether those threats have been sufficiently mitigated.

Disease Screening

For translocations or restorations, ensure that individuals selected are thoroughly screened for feline-specific diseases and other health problems. The occurrence of diseases may not present a problem if those same diseases are common in the wild population (as was the case for Amur tigers), but screening of both the captive and wild populations are necessary to make this determination. Screenings to assess the presence of recessive genetic disorders, if feasible, should be contemplated. A quarantine period should be adhered to.

Release of Captive-Reared Versus Wild Individuals

For restoration/supplementation of a population, it is almost always preferable to use wild individuals as a source. Wild individuals have demonstrated their ability to survive in the wild, and will be innately wary of humans, while some level of acclimatization to humans is almost unavoidable in captive conditions. However, wild individuals are likely to disperse further (Belden and McCown, 1996), although use of soft releases (see later) may temper that tendency. Wild Iberian lynx were released only into areas where a population was already established, thus tempering the likelihood of dispersal. When wild individuals are not available, captive-rearing is an expensive, time-consuming option, but one that has demonstrated success, as in the case with Iberian lynx.

With Captive-Reared Individuals, it is Essential to Minimize Human Contact, and Probably Preferable to Provide Negative Reinforcements

Holding pens should be designed to minimize human contact (visual, auditory, olfactory). It is especially important to avoid associating the presence of humans with food (e.g., a keeper bringing food to the enclosure). The larger the felid, the more important this issue becomes. The one tiger that was removed from the Russian Far East was kept in an enclosure with little cover and closest to the facilities where keepers worked, and was therefore forced into closer association with humans. Assessing how large felids respond to negative reinforcements (as was done with Iberian lynx) would be a useful line of investigation assisting future release programs.

Better to Release Females (First)

In most cases females are likely preferred over males as the source for genetic restoration into a population that is endangered. Use of females permits researchers to more effectively document progress related to genetic introgression as reproduction of females (if radiotagged) can be closely monitored, whereas pairings between translocated males and uncollared wild females already in the population are nearly impossible to monitor without extensive genetic monitoring of nearly the entire population.

Experience with tigers in the Russian Far East demonstrated that two of three rehabilitated male tigers dispersed long distances from the release site, while all three tigresses remained relatively close to the release site. Similar long-distance dispersals of male felids relative to females in translocation projects are documented (Belden and McCown, 1996). Therefore, it may be wise to establish a nucleus population by first releasing females, and then releasing males into the home ranges already established by females in hopes males will encounter females (through visual or olfactory cues) and be less likely to disperse.

Hard Versus Soft Releases

Soft releases were used initially for Iberian lynx to develop a nucleus population, but once populations were established, hard releases appeared to be effective. In Florida, though the sample size was small (8 animals), no observable difference was found in movements of Texas pumas after hard versus soft releases. Only hard releases were used for tigers in Russia, but only males traveled extensively.

Genetic Restoration

For genetic restoration: (i) determine the level of genetic introgression necessary to positively impact the population that is in peril; (ii) select conspecific stock from an area that previously may have interbred with the focal population (e.g., Texas pumas and Florida panthers) so that genetic restoration is not altering the population that is endangered, but mimicking what historically happened naturally; and (iii) be prepared to explain (with data, historic distribution patterns, genetics) why admixed animals are not significantly different from the original stock for which you are initiating genetic restoration.

Closely Monitor Released Individuals

The process of translocation and restoration of felids is still in its infancy, and much is still unknown. Therefore, clearly documenting the process and monitoring released individuals will greatly inform future efforts. Additionally, because skepticism and criticism are likely when releasing either wild or captive-reared/rehabilitated animals, scientifically defensible documentation of successes and failures are important. Basic information on kill rates (to document ability to hunt in the wild), encounters with humans, habitat selection, and reproduction are vital to determining success. With the increased use of GPS collars, it is now possible to collect multiple data points per day, which allow fine-scale analysis of movement patterns after release. Such collars are especially useful in remote areas because they have the ability to transmit data to researchers via cellular or satellite phone networks (see Chapter 26). However, the deficiencies of GPS collars (short battery life and high failure rate) need to be considered and compared to "traditional" VHF collars. To document reproduction and mortality, having radio-collars that last 3+ years is preferred and minimizes the need to recapture animals on an annual basis (something that is impossible in many situations). VHF collars can function for more than 4 years, but, for large felids, will often require the use of aerial telemetry to locate them, which is expensive, poses risks to staff, and may not be feasible in some cases.

Be Prepared for Conflicts and the Need to Remove Individuals

A response plan should exist prior to translocating animals so that agencies, the public, private landowners, and NGOs are aware how felid-human conflicts will be dealt with. Having this protocol developed and approved by all stakeholders prior to releasing a large predator is highly recommended.

SUBCHAPTER

24.2

NABU Snow Leopard Rehabilitation Center in Kyrgyzstan

Boris Tichomirow

Middle Asia Program, NABU, Berlin, Germany

The Nature and Biodiversity Conservation Union (NABU) was one of the first international environmental organizations to become active in the conservation of wildlife in Kyrgyzstan and especially of one of its most iconic species, the snow leopard (*Panthera uncial*). NABU started work in 1994, when it organized the first Issyk-Kul Conference at which it was decided to create a biosphere reserve around Lake Issyk-Kul in Kyrgyzstan. In 1998, at the second conference, the creation of the Issyk-Kul Biosphere Reserve, with a total area of 43,000 km^2, was announced. In 1999, NABU started its project "Snow Leopard," the first one for the Issyk-Kul Biosphere Reserve.

Since 2001, the anti-poaching-unit "Gruppa Bars" (Group Snow Leopard) works to protect snow leopards by patrolling their habitat and persecuting poachers. The four members of this group have experience in environmental and security related work (including weapons handling), and have higher scientific and legal education. The objectives of the "Gruppa Bars" are to reduce poaching of snow leopards and other endangered wildlife, *inter alia* through:

- assisting relevant government agencies in their operational work for nature protection, especially in combatting poaching
- awareness raising and outreach to the public
- monitoring snow leopards using camera traps

The "Gruppa Bars," together with employees of state agencies, regularly conducts visits to local communities, primarily in the habitat of the snow leopard, to raise awareness and reduce poaching; it has already detained more than 200 poachers and seized dozens of skins and hundreds of weapons and traps. Thanks to this work, poaching of snow leopards in Kyrgyzstan has reduced significantly. But not only snow leopards benefit: the group also protects other endangered wildlife, especially those listed in the Red Book of Kyrgyzstan. One of its greatest successes was the confiscation of 127 saker falcons (*Falco cherrug*), indicating an organized trade network, as well as 1022 steppe tortoises (*Agrionemys horsfieldi*) that were intended for sale in Moscow. The steppe tortoise is listed in the Red Book of Kyrgyzstan and is considered an endangered species in Central Asia. Since 2000, NABU has confiscated seven snow leopards from poachers or illegal keeping.

In 2002, NABU opened a Snow Leopard Rehabilitation Center in Sasyk-Bulak Valley of Issyk-Kul District, the first in Central Asia and the world's largest enclosure for snow leopards. The focus of NABU's work is the protection of natural habitats of wild flora and fauna. The rehabilitation center was built in order to keep confiscated animals in their natural habitat and release them back into the wild after their recovery. Unfortunately, most snow leopards could not be released, either because they were badly injured (were missing a paw after being caught in a steel trap), or because they were taken at a young age and never learned to hunt wild prey on their own. Therefore, the rehabilitation center was set up in order to enable those animals to live in close-to-natural conditions and close to their natural habitat, as well as for environmental education purposes.

The snow leopard enclosure covers 7000 m^2 and has held up to seven snow leopards. At present, three snow leopards are still living in the rehabilitation center. In addition, there are currently one lynx (*Lynx lynx*) and several wild birds such as black kite (*Milvus migrans*), kestrel (*Falco tinnunculus*), steppe eagle (*Aquila nipalensis*), and golden eagle (*A. chrysaetos*), all of which will be released back into the wild after they have recovered.

The main goals and objectives of the Rehabilitation Center are:

- saving wild animals listed in the Red Book
- promoting a positive attitude toward them among the local population
- treatment, care, and rehabilitation of injured wildlife
- release into the wild after rehabilitation.

In 2010, NABU rescued several starving animals, including snow leopards, several Tien Shan bears (*Ursus arctos issabellinus*), and rare species of eagles from the private zoo of ousted President Kurmanbek Bakiyev.

On July 20, 2011, a new phase began for the conservation of snow leopards: during the visit of Kyrgyz president R.I. Otunbayeva to the NABU Snow Leopard Rehabilitation Center, the Head of State expressed her support to Boris Tikhomirov, NABU director of the Central Asia Programs, for the organization of a World Forum on the Conservation of Snow Leopards. With the support of the Global Tiger Initiative (GTI) of the World Bank, the process started to develop a global protection program for the species. In October 2013, the Kyrgyz Government hosted the World Forum on snow leopard conservation in Bishkek, which was attended by delegations from all 12 countries of the snow leopard range, and resulted in the adoption of the high-level political Bishkek Declaration and Global Program on Snow Leopard and Ecosystem Protection Program (Chapter 45).

Hence, as ambassadors of their wild kin, the snow leopards in the NABU Rehabilitation Center led the President of a nation to take strong and positive steps to save the species across its vast range in Asia.

References

Aprile, G., Cuyckens, E., De Angelo, C., Di Bitetti, M.S., Lucherini, M., Muzzachiodi, N., Palacios, R., Paviolo, A., Quiroga, V., Soler, L., 2012. Familia felidae. In: Ojeda, R.A.V., Chillo, V., Díaz, G.B. (Eds.), Libro Rojo de los Mamíferos Amenazados de la Argentina,. Sociedad Argentina para el Estudio de los Mamíferos. Buenos Aires, Argentina, pp. 92–101.

Aramilev, V.V., 2013. Scientific basis for reintroduction of Amur tigers into Amur Oblast. Report for the governmental contract no. 0123200000313002189-Kot 30.09.2013. Institute of Geography, Far Eastern Branch of the Russian Academy of Sciences. pp. 44.

Belden, R.C., McCown, J.W., 1996. Florida panther reintroduction feasibility study. Final report 7507. Florida Game and Freshwater Fish Commission, Tallahassee, Florida USA, pp. 70.

Benson, J.F., Hostetler, J.A., Onorato, D.P., Johnson, W.E., Roelke, M.E., O'Brien, S.J., Jansen, D., Oli, M.K., 2011. Intentional genetic introgression influences survival of adults and subadults in a small, inbred felid population. J. Anim. Ecol. 80, 958–967.

Caruso, F., Jiménez Pérez, I., 2013. Tourism, local pride and attitudes towards the reintroduction of a large predator: the case of the Jaguar in Corrientes, Argentina. Endang. Spec. Res. 21, 263–272.

Caso, A., Lopez-Gonzalez, C., Payan, E., Eizirik, E., de Oliveira, T., Leite-Pitman, R., Kelly, M., Valderrama. C., 2012. Panthera onca, In: IUCN 2011. IUCN Red List of Threatened Species. Version 2011.2. www.iucnredlist.org (accessed 03.02.2012).

Currier, M.J.P., 1983. Felis concolor. Mammalian Species No. 200:1–7.

De Angelo, C., 2011 Evaluación de la aptitud del hábitat para la reintroducción del yaguareté en la cuenca del Iberá.

Unpublished report. http://www.proyectoibera.org/download/yaguarete/habitat_para_el_yaguarete_en_ibera.pdf

Di Bitetti MS, De Angelo CD, Quiroga V, Altrichter M, Paviolo A, Cuyckens E, Perovic P. In press., Estado de conservación del jaguar en la Argentina, In : Medellín, R.A., Chávez, C., de la Torre, A., Zarza, H., Ceballos, G., (eds.) El jaguar en el Siglo XXI: La Perspectiva Continental. Fondo de Cultura Económica, México, D.F., México.

Federal Register, 1967. Native fish and wildlife: endangered species. Federal Register, Department of the Interior, Fish and Wildlife Service, Washington DC, USA, pp. 4001.

Frankham, R., Ballou, J.D., Briscoe, D.A., 2002. Introduction to Conservation Genetics. Cambridge University Press, Cambridge, UK.

Gaona, P., Ferreras, P., Delibes, M., 1998. Dynamics and viability of a metapopulation of the endangered Iberian lynx (*Lynx pardinus*). Ecological Monographs 68, 349–370.

Garrote, G., López, G., Gil-Sánchez, J.M., Rojas, E., Ruiz, M., Bueno, J.F., de Lillo, S., Rodriguez-Siles, J., Martín, J.M., Pérez, J., García-Tardío, M., Valenzuela, G., Simón, M.A., 2013. Human–felid conflict as a further handicap to the conservation of the critically endangered Iberian lynx. Eur. J. Wildl. Res. 59, 287–290.

Gill-Sánchez, J.M., Moral, M., Bueno, J., Rodríguez-Siles, J., Lillo, S., Pérez, J., Martín, J.M., Valenzuela, G., Garrote, G., Torralba, B., Simón-Mata, M.Á., 2011. The use of camera trapping for estimating Iberian lynx (*Lynx pardinus*) home ranges. Eur. J. Wildl. Res. 57 (6), 1203–1211.

Goodrich, J.M., 2010. Human–tiger conflict: a review and call for comprehensive plans. Integr. Zool. 5, 300–312.

Goodrich, J.M., Miquelle, D.G., 2005. Translocation of problem Amur tigers *Panthera tigris altaica* to alleviate tiger-human conflicts. Oryx 39, 454–457.

Goodrich, J.M., Miquelle, D.G., Smirnov, E.N., Kerley, L.L., Quigley, H.B., Hornocker, M.G., 2010. Spatial structure of Amur (Siberian) tigers (*Panthera tigris altaica*) on Sikhote-Alin Biosphere Zapovednik, Russia. J. Mammal. 91, 737–748.

Goodrich, J.M., Quigley, K.S., Lewis, J.C.M., Astafiev, A.A., Slabi, E.V., Miquelle, D.G., Smirnov, E.N., Kerley, L.L., Armstrong, D.L., Quigley, H.B., Hornocker, M.G., 2012. Serosurvey of free-ranging Amur tigers in the Russian Far East. J. Wildl. Dis. 48, 186–189.

Griffin, A.S., Blumstein, D.T., Evans, C.S., 2000. Training captive-bred or translocated animals to avoid predators. Conserv. Biol. 14, 1317–1326.

Guzmán, N., García, F.J., Garrote, G., Pérez de Ayala, R., Iglesias, C., 2004. El lince ibérico (*Lynx pardinus*) en España y Portugal. Censo-diagnóstico de sus poblaciones. Dirección General para la Biodiversidad, Madrid.

Hartmann-Furter, M., 2009. Breeding European wildcats (Felis Silvestris Silvestris, Schreber 1777) in species-specific enclosures for reintroduction in Germany. In: Vargas, A., Breitenmoser, C., Breitenmoser, U., Conservación *Ex situ* del Lince Ibérico: un enfoque multidisciplinar. Funcacion Biodiversidad. pp. 453–461.

Hebblewhite, M., Zimmermann, F., Li, Z., Miquelle, D.G., Zhang, M., Sun, H., Mörschel, F., Wu, Z., Sheng, L., Purekhovsky, A., Chunquan, Z., 2012. Is there a future for Amur tigers in a restored tiger conservation landscape in Northeast China? Anim. Conserv. 15, 579–592.

Heptner, V.G., Sludskii, A.A., 1992. Mammals of the Soviet Union, Volume II, Part 2, Carnivora (Hyaenas and Cats). Smithsonian Institution Libraries and The National Science Foundation, Washington, DC.

Hostetler, J.A., Onorato, D.P., Nichols, J.D., Johnson, W.E., Roelke, M.E., O'Brien, S.J., Jansen, D., Oli, M.K., 2010. Genetic introgression and the survival of Florida panther kittens. Biol. Conserv. 143, 2789–2796.

Hostetler, J.A., Onorato, D.P., Jansen, D., Oli, M.K., 2013. A cat's tale: the impact of genetic restoration on Florida panther population dynamics and persistence. J. Anim. Ecol. 82, 608–620.

IUCN, 2003. IUCN Red List of Threatened Species. http://www.redlist.org

IUCN, 2015. IUCN Red List of Threatened Species. http://www.redlist.org

Jackson, R.M., Ale, S.B., 2009. Snow leopards: is reintroduction the best option? In: Hayward, M.W., Somers, M.J. (Eds.), Reintroduction of Top-Order Predators. Blackwell Publishing Ltd, West Sussex, UK, pp. 165–186.

Jiménez Pérez, I. (Ed.), 2013. Giant Anteater: A Homecoming to Corrientes. The Conservation Land Trust, Buenos Aires, Argentina.

Jiménez Pérez, I., Abuín, R., Antúnez, B., Pereda, I., Delgado, A., Cirignoli, S., Galetto, E., Jorge Peña, J., 2013. Proyecto de recuperación del venado de las pampas en la Reserva Natural Iberá y los bañados de Aguapey: informe de resultados y actividades (año 2013). Unpublished report. http://www.proyectoibera.org/download/venado/informe_proyecto_venados_corrientes_2013.pdf

Johnson, W.E., Onorato, D.P., Roelke, M.E., Land, E.D., Cunningham, M., Belden, R.C., McBride, R., Jansen, D., Lotz, M., Shindle, D., Howard, J., Wildt, D.E., Penfold, L.M., Hostetler, J.A., Oli, M.K., O'Brien, S.J., 2010. Genetic restoration of the Florida panther. Science 329, 1641–1645.

Kelly, M.J., Silver, S., 2009. The suitability of the jaguar (*Panthera onca*) for reintroduction. In: Hayward, M.W., Somers, M.J. (Eds.), Reintroduction of Top-Order Predators. Blackwell Publishing Ltd, West Sussex, UK.

Kerley, L.L., Goodrich, J.M., Miquelle, D.G., Smirnov, E.N., Nikolaev, I.G., Quigley, H.B., Hornocker, M.G., 2003. Reproductive parameters of wild female Amur (Siberian) tigers (*Panthera tigris altaica*). J. Mammal. 84, 288–298.

Kleiman, D.G., Stanley Price, M.R., Beck, B.B., 1994. Criteria for reintroductions. In: Olney, P.J.S., Mace, G.M., Feistner, A.T.C. (Eds.), Creative Conservation: Interactive

Management of Wild and Captive Animals. Chapman and Hall, London, pp. 287–303.

Ma, Y., 2000. Changes in numbers and distribution of the Amur tiger in northeast China in the past 100 years – a summary report. In: Miquelle, D.G., Zhang, E., Jones, M., Jin, T., (Eds.), Proceedings of the Workshop to Develop a Recovery Plan for the Wild North China Tiger Population. Harbin. pp. 12–14

Macdonald, D.W., 2009. Lessons learnt and plans laid: seven awkward questions for the future of reintroductions. In: Hayward, M.W., Somers, M.J. (Eds.), Reintroduction of Top-Order Predators. Blackwell Publishing Ltd, West Sussex, UK, pp. 411–448.

Maehr, D.S., Caddick, G.B., 1995. Demographics and genetic introgression in the Florida panther. Conserv. Biol. 9, 1295–1298.

McBride, R.T., Sensor, R., 2013. Florida Panther Annual Count. Rancher's Supply Inc, Ochopee, FL, USA, 166 pp.

McBride, R.T., McBride, R.T., McBride, R.M., McBride, C.E., 2008. Counting pumas by categorizing physical evidence. Southeastern Naturalist 7, 381–400.

Miller, C.S., Mark Hebblewhite, M., Petrunenko, Y.K., Seryodkin, I.V., DeCesare, N.J., Goodrich, J.M., Miquelle, D.G., 2013. Estimating Amur tiger (*Panthera tigris altaica*) kill rates and potential consumption rates using global positioning system collars. J. Mammal. 94, 845–855.

Miquelle, D.G., Pikunov, D.G., Dunishenko, Y.M., Aramilev, V.V., Nikolaev, I.G., Abramov, V.K., Smirnov, E.N., Salkina, G.P., Gaponov, V.V., Fomenko, P.V., Litvinov, M.N., Kostyria, A.V., Yudin, V.G., Korkisko, V.G., 2007. 2005 Amur tiger census. Cat News 46, 12–14.

Naidenko, S.V., Esaulova, N.V., Lukarevsky, V.S., Hernandez-Blanco, J.A., Sorokin, P.A., Litvinov, M.N., Kotlyar, A.K., Rozhnov, V.V., 2012. Occurence of infection diseases in Amur tigers in the south of their range. In: Seryodkin, I.V., Miquelle, D.G. (Eds.), Diseases and Parasites of Wildlife in Siberia and the Russian Far East. Dalnauka, Vladivostok, pp. 32–35.

Onorato, D., Belden, C., Cunningham, M., Land, D., McBride, R., Roelke, M., 2010. Long-term research on the Florida panther (*Puma concolor coryi*): historical findings and future obstacles to population persistence. In: Macdonald, D., Loveridge, A. (Eds.), Biology and Conservation of Wild Felids. Oxford University Press, Oxford, UK, pp. 453–469.

Palomares, F., 2001. Vegetation structure and prey abundance requirements of the Iberian lynx: implications for the design of reserves and corridors. J. Appl. Ecol. 38, 9–18.

Palomares, F., 2002. Efecto de la extracción de linces Ibéricos en las poblaciones donantes de Doñana y la Sierra de Andújar para posibles campañas de reintroducción. Technical report. Consejería de Medio Ambiente de la Junta de Andalucía, Sevilla.

Parera, A. (Ed.), 2004. Fauna de Iberá: composición, estado de conservación y propuestas de manejo. Fundación Biodiversidad Argentina. Unpublished report.

Roelke, M.E., Martenson, J.S., O'Brien, J.S., 1993. The consequences of demographic reduction and genetic depletion in the endangered Florida panther. Curr. Biol. 3, 340–349.

Rozhnov, V.V., Hernandez-Blanco, J.A., Naidenko, S.V., Lukarevskiy, V.S., Sorokin, P.A., Miquelle, D.G., Rybin, N.N., Kalinin, A.Y., Polkovnikova, O.N., 2014. Movements of an Amur tiger (*Panthera tigris altaica*) after release into the northwest portion of its range. In: Saaveleva, A.P., Seryodkin, I.V. (Eds.), Distribution, Migration, and Other Movements of Wildlife Reya. Pacific Geographical Society, Vladivostok, Russia, pp. 266–271.

Seal, U.S., 1994. A plan for genetic restoration and management of the Florida panther (*Felis concolor coryi*). Report to the Florida Game and Freshwater Fish Commission. Conservation Breeding Specialist Group, Apple Valley, Minnesota, USA, pp. 22.

Seddon, P.J., Griffiths, C.J., Soorae, P.S., Armstrong, D.P., 2014. Reversing defaunation: restoring species in a changing world. Science 345, 406–412.

Simón, M.A., Gil-Sánchez, J.M., Ruiz, G., Garrote, G., McCain, E., Fernández, L., López-Parra, M., Rojas, E., Arenas-Rojas, R., del Rey, T., GarcíaTardío, M., López, G., 2012. Reverse of the decline of the endangered Iberian lynx. Conserv. Biol. 26, 731–736.

Simón, M. et al., 2013. Ten years conserving the Iberian lynx. Consejería de Agricultura, Pesca y Medio Ambiente. Junta de Andalucía. Seville.

Solís, G., Peña J., Spørring, K., Boixader, J. Jiménez, I., 2014. Programa de funcionamiento del Centro Experimental de Cría de Yaguaretés en la Reserva Iberá. Version 3.0. pp. 78 http://www.proyectoibera.org/download/yaguarete/CECY_Programa_Funcionamiento.pdf

USFWS, 2008. Florida Panther Recovery Plan (*Puma concolor coryi*), Third Revision. United States Fish and Wildlife Service, Atlanta, GA, 217 pp.

Vargas, A., Sanchez, I., Martinez, F., Rivas, A., Godoy, J.A., Roldan, E., Simon, M.A., Serra, R., Perez, MaJ, Ensenat, C., Delibes, M., Aymerich, M., Sliwa, A., Breitenmoser, U., 2008. The Iberian lynx (*Lynx pardinus*) conservation breeding program. Int. Zoo Yearb. 42, 190–198.

Woodroffe, R., 2001. Strategies for carnivore conservation: lessons from contemporary extinctions. In: Gittleman, J.L., Funk, S.M., Macdonald, D., Wayne, R.K. (Eds.), Carnivore Conservation. Cambridge University Press, Cambridge, pp. 60–92.

Yerga J., Manteca, X., Vargas, A., Rivas, A. and Calzada, J., 2012. Etapas de la ontogenia del comportamiento del lince ibérico (*Lynx pardinus*) en cautividad. XIV Congreso Nacional y XI Iberoamericano de Etología (SEE). Sevilla.

SECTION V

Techniques and Technologies for the Study of a Cryptic Felid

CHAPTER

25

Snow Leopard Research: A Historical Perspective

Don Hunter, Kyle McCarthy**, Thomas McCarthy*[†]

*Rocky Mountain Cat Conservancy, Fort Collins, CO, USA

**Department of Entomology and Wildlife Ecology, University of Delaware, Newark, DE, USA

[†]Snow Leopard Program, Panthera, New York, NY, USA

IN THE BEGINNING

When India collided with Eurasia, it brought with it an abundance of wildlife, including many different cat species. As the two continental plates buckled and rose over 50 million years, the snow leopard (*Panthera uncia*) and its assemblage of wild ungulate prey remained atop what would become the highest mountains on earth. Indeed, seven major ranges emerged, forming the great arc of snow leopard habitat through Central Asia, where for millennia snow leopards lived free of human contact. Only in the last few thousand years have people moved upward from valleys and into their range. Early snow leopard accounts are usually of myth and mysticism and generally free of the disdain typically reserved for large carnivores. This is likely due to their elusive nature, beauty, and that they are especially wary of people and have never preyed on humans, although more than capable of doing so.

As explorers, intrepid naturalists, and hunters ventured further into the mountains of Central Asia, modern literature would begin to include the first details of the snow leopard. These accounts, as reviewed by Schaller (1977) and Sunquist and Sunquist (2002), led to the first targeted snow leopard study in the early 1970s. Koshkarev (1997) also notes the early Russian accounts of snow leopard. In this chapter we will cover a three-decade period, ~1970–2000, during which the science of snow leopard research moved from skilled observation (boots, binoculars, and notebook) to sensing animal movement via satellite; those engaged in snow leopard study would shift from primarily western scientists to well-trained scientists within the range countries; and the status of the snow leopard would move from a hunted species to one given full protection by all countries.

A historic review of the first 30 years of snow leopard research must begin with a perspective on the conditions particular to this cat. Its vast

Snow Leopards
http://dx.doi.org/10.1016/B978-0-12-802213-9.00025-0

range, extreme habitat, and elusive nature make snow leopards exceedingly difficult to study. To enter the cat's domain, researchers, especially westerners, require permits and provisions from countries that often have ongoing civil unrest or political instability. Border disputes, sensitive areas, military patrols, poor quality maps, and fickle bureaucrats often presented barriers greater than the mountains, and western researchers' field plans were often stymied by suspicious bureaucracies. Maps, especially in the 1970s and 1980s, were rarely available or considered "top secret." US Defense Mapping Agency 1:1,000,000 Operational Navigation Charts (ONC) were often the only means of navigation in the mountains and for mapping snow leopard distribution. At the beginning of this period China, the Soviet Union, and various allies (e.g., Mongolia) were relatively closed countries that required special arrangement to access remote areas. Thus, in the early years, the approach to snow leopard study was driven more by opportunity for access than any overarching, systematic scheme of range-wide scientific assessment.

Pioneering researcher George Schaller would cast a long shadow across snow leopard range that continues to the present. A renowned scientist before entering the mountains of Central Asia, Schaller became the archetypal snow leopard biologist, characterized by compassion for the species combined with exceptional observation skills and unrelenting perseverance in the field. He succeeded in opening the secret world of the snow leopard through his unique ability to take meticulous field observations and transcribe them into scientific literature and popular outlets. He used a comfortable writing style that engaged armchair naturalists worldwide, helping to make the snow leopard an iconic symbol of Central Asian wildness. Further, his work and words inspired talented scholars to follow in his footsteps.

Presently, the snow leopard is often labeled as "charismatic megafauna," a characterization that carries no scientific meaning, but animals deserving of the title do attract the people and funds needed for research. However, prior to 1970 the "charisma" of the snow leopard was little known. Schaller (1971) introduced the snow leopard to the world when he penned an engaging account of his first encounter with the cat. Jackson and Hillard (1986) would add to that with another *National Geographic* piece. But perhaps no publication brought the snow leopard more firmly into the general public's eye than Peter Matthiessen's *The Snow Leopard* (1978). Matthiessen, an author, naturalist, and Buddhist acolyte, accompanied Schaller on a trek into the Dolpo region of western Nepal, ostensibly to study blue sheep, but buoyed also with the hope of seeing a snow leopard. The poignant account of that journey, on which he never saw a snow leopard, won him a national book award and launched the snow leopard into global recognition.

A year before Schaller and Matthiessen's famous trek, two cats taken from the wild in the Soviet Republic of Kirghizia (now Kyrgyzstan) were sent to Seattle's Woodland Park Zoo where they came under the watchful eye of Helen Freeman, a volunteer docent. Filled with passion and energy, and recognizing how little was being done to understand or conserve the cats in the wild, Freeman started the International Snow Leopard Trust (ISLT) in 1981. Five years later this NGO hosted an international snow leopard symposium in India. Three more would be held in range countries, the last in Islamabad, Pakistan, in 1995. These symposia became critical forums for field researchers to share results, for range countries to represent their interest and showcase accomplishments, and for colleagues to interact. From its inception in 1981 through 2000, ISLT grew and became the international hub for snow leopard people and information. The Trust, as it came to be known, raised money to support field studies, hired in-country staff to conduct projects, and regularly published news about snow leopards as well as symposia proceedings. Importantly, it helped fund many of the needs identified during the symposia, including the development of common survey and monitoring protocol.

STEADY MARCH OF SCIENCE

Field Surveys

Of the first documented field studies, Schaller's observation of a snow leopard in Chitral Gol in northern Pakistan (Schaller, 1972, 1980) set the stage for a type of research that would become the most common in early snow leopard literature. Typically cited as "field survey," "status report," or "observations," these approaches collected information through a variety of methods such as scanning a hill slope for snow leopard (seldom actually seen) or prey, interviewing local people, walking transects noting scats or scrapes, and transcribing local records of predation. Observational studies of snow leopard sign and prey were the most practical given the expense and difficulty of live animal research (radio-collaring). They answered the most basic questions of presence/absence and distribution, while adding incrementally to life history information. However, the disparity of these mixed methods made it difficult to infer results for the leopard population in total (see The Need for Standard Methods, later).

Such field studies would continue through the ensuing three decades, covering much of the species range. From Pakistan in the early 1970s, Schaller would turn his attention to western China beginning in the mid-1980s (Schaller, 1998; Schaller et al., 1987, 1988a,b). In Mongolia, Bold and Dorjzunduy (1976) carried out an early study and Mallon (1984a) described the cat's range followed by others (McCarthy and Munkhtsog, 1997; Schaller et al., 1994). Ahmad et al. (1997) would return to northern Pakistan; Koshkarev (1989, 1997) surveyed portions of Kyrgyzstan, Eastern Siberia and Western Mongolia; Mallon (1984b, 1991), Fox et al. (1988, 1991), and Chundawat and Rawat (1994) surveyed Ladakh; and Jackson et al. (1994), Jackson and Ahlborn (1988, 1989), Oli et al. (1993, 1994) and Oli (Oli, 1994a,b) would venture into the Himalayan range of Nepal. Fox (1989, 1994) authored range-wide status reviews and perspectives.

Live Animal Research

Field surveys provide only limited information about a species. Radio telemetry is an essential tool to learn about animal movement, home range, feeding habits, and day-to-day activity. Further, with several animals radio-collared, it is possible to study social behavior, mating, and territoriality. By the 1970s, radio telemetry was a proven method for tracking mountain lion (*Felis concolor*), a similar-sized cat to snow leopard, in remote central Idaho (Hornocker, 1969). Schaller and Mel Sunquist made the first attempts to capture snow leopard in Chitral Gol in 1974 using box traps but were unsuccessful after 1.5 months of effort.

Eight years later, Rodney Jackson first captured and radio-collared a snow leopard in west Nepal (Jackson and Ahlborn, 1989). This seminal study lasted 4 years with a total of five animals collared. It produced a revealing corpus of data on snow leopard far beyond what could be learned by field surveys alone. Jackson collected data on food habits, habitat preferences, activity patterns, home range, abundance, density, and importantly, marking patterns. His findings suggested that sign – scats, tracks, scrape, and feces – may be reliable indicators of snow leopard presence and relative abundance. The relatively high sign and cat density in his study area was eventually used by researchers to infer comparative densities in other regions and in different habitat types. Density is a key metric for estimating the total number of snow leopards that exist in the wild – a pervasive question posed to researchers.

Jackson's research had to overcome many obstacles including an extremely remote study site that required a 200-mile small plane ride from Kathmandu and then 10 days walking over two high passes. With no examples to follow, he carried in equipment he could only hope would work, including leg hold snares (Novak, 1980) and VHF radio collars. This technology, advanced for its time, required multiple locations to triangulate a location. Signal bounce off the rocky broken terrain made it difficult to pinpoint the direction to the collar, a dilemma that would plague all future VHF-based snow leopard collaring studies. Once the source was established, direction was taken by handheld compass and later plotted on topographical maps to determine the location of the cat and home-range size determined by rudimentary minimum convex polygon methods. In extreme terrain at high altitudes, location points of the five leopards he eventually collared were hard earned. Over the 4 years that cats were followed they utilized home ranges of between 11.7 km^2 and 38.9 km^2 with substantial overlap both within and between sexes. From Jackson's first snow leopard capture in Nepal, through the late 1990s, only a few researchers would again attempt live animal studies.

While studying mountain ungulates in Ladakh, Chundawat et al. (1988) captured an adult male snow leopard in a cage trap made of local material. It slipped the collar, providing only minor information on movement and activity during 2.5 months in late winter. During that period the cat's home range was estimated to be 19.0 km^2. Nepal was again the site of snow leopard capture and collaring in 1990 when three cats were fitted with VHF transmitters by Oli (1997). Cats were tracked for only 1–2 months (late December–late February). Home range sizes were reported as 13.8–22.3 km^2 with substantial overlap, much as reported by the five cats tracked by Jackson and Ahlborn (1989).

In 1990, Schaller et al. (1994) initiated an ecological study of snow leopards in the Altai Mountains of southwest Mongolia and VHF-collared a single adult male. Over a 41-day period in November and December they located the cat on 36 different days and report a home range size of 12 km^2. The study was suspended for nearly 3 years due to economic and political uncertainty as Mongolia adjusted to the freedom from decades of communist rule.

In 1993, the study was reinitiated under the leadership of Tom McCarthy, and over the ensuing 5 years five more snow leopards were collared (McCarthy, 2000). The first four of those cats were again fitted with VHF collars, but in contrast to all previous studies, initial home range estimates for some of the cats were quite large, in excess of 140 km^2. McCarthy speculated that they could be even larger, given that the cats could not be relocated for long periods by ground-based telemetry. By this time, Argos satellite PTT collars had proven effective for tracking wildlife (Fancy et al., 1988). In February 1996, McCarthy placed an Argos collar on a female snow leopard (all three authors collaborated in the study). As expected, the satellite telemetry data showed that female utilized an area far greater than previous VHF estimates, exceeding 1500 km^2. With that finding alone, the era of VHF collaring of snow leopards came to an end.

As explained in Chapter 26, telemetry technology has moved ahead rapidly, and now even the Argos PTT is a relic with only one deployed on a snow leopard. Beyond eliminating the need to ground track snow leopards, through steep rugged terrain, at elevations up to 5000 m and in temperatures below −30°C, new technology means more accurate and more frequent locations of collared cats. Combined, McCarthy and Jackson recorded just 781 locations while VHF tracking 10 cats over 8 years. Today, a single GPS-collared snow leopard can easily provide more data points every few months.

The pioneers of early snow leopard research overcame many challenges and hardships to eke out the beginnings of the current scientific understanding of the species. In doing so, they exemplified snow leopard research as an enterprise of science and spirit.

Advances in the Lab to Support Work in the Field

Quality maps were a rarity in the early years of snow leopard research, and satellite imagery was only marginally useful (Prasad et al., 1991), as high mountains tend to cast shadows and have large areas of exposed rock, making image classification difficult. Yet most facets of snow leopard research are spatial in nature, from field surveys to delineation of potential protected areas. This led Hunter and Jackson (1997) to model snow leopard habitat across all 12 range countries, despite the shortcomings in satellite imagery. Using model parameters derived from snow leopard literature and expert input, they employed geographic information system (GIS) tools to produce maps of Total, Good, Fair, and Protected habitat (Fig. 25.1). The resultant visual representation of

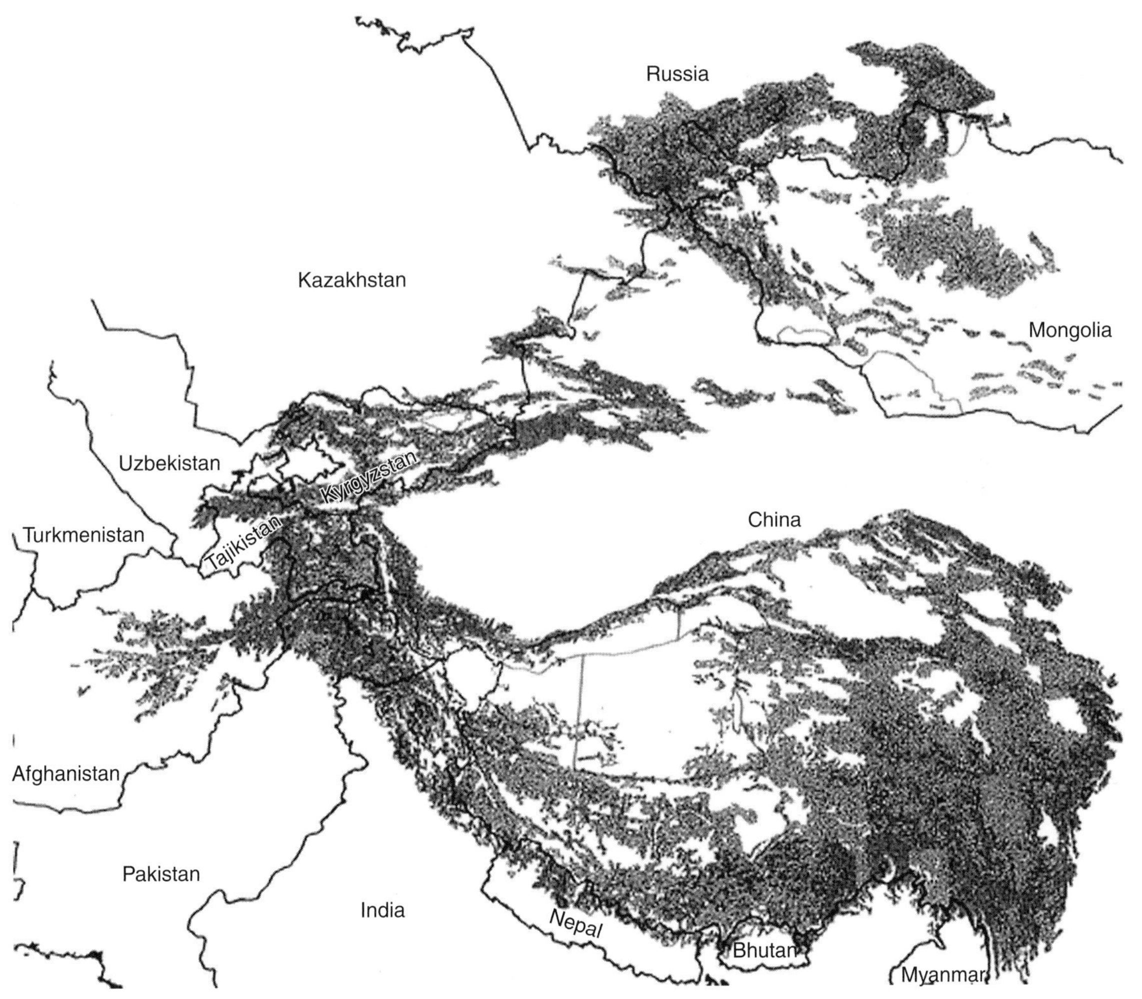

FIGURE 25.1 **Snow leopard range map from Hunter and Jackson (1997) including: good habitat, 549,706 km^2: fair habitat, 2,475,022 km^2; for a total of 3,024,728 km^2 of potential habitat.**

the snow leopard's vast and fragmented range reinforced the general consensus that the species is not in imminent danger of extinction, but rather threatened with localized extirpations. The new maps also showed the importance of China, with more than 60% of total snow leopard range, and provided the basis for new and more accurate estimates of density and total population size. The maps were an integral planning tool for snow leopard conservation efforts for more than a decade, when an expert knowledge mapping process in 2008 (Chapters 3 and 44) updated them with more current information.

The Need for Standard Methods

Helsinki Zoo sponsored the first international snow leopard symposium in 1978 with a focus on captive snow leopard care and breeding. Succeeding symposia continued the emphasis on zoo animals until the ISLT and Wildlife Institute of India hosted the fifth symposium in Srinagar, India, in 1986 with the primary topic of snow leopard conservation in the wild. Subsequent symposia became the de facto forum for presenting "status" reports by range country representatives, as well as survey data by individual scientists. Given the nature of early snow leopard research, reports provided abundance estimates extrapolated from a wide variety of methods, from actual field research to heuristic estimates by experts – a good starting point but far from scientifically valid or comparable. Jackson and Fox (1997) reviewed the variability of population estimates from several different surveys and cited the need for a common set of standard methods.

Spearheaded by ISLT, Project Snow Leopard (PSL) was presented at the snow leopard symposium in Xining, China (Freeman et al., 1994; Hunter et al., 1994). This was a strategy for uniting snow leopard range countries and other organizations to work toward shared goals for snow leopard conservation, modeled to some degree after India's successful "Project Tiger." As originally perceived, PSL would tackle many of the range-wide threats confronting snow leopard conservation. Targeted multinational workshops would focus on transboundary parks, travel corridors between parks, reducing livestock predation, improving reserve management, and curtailing international trade in snow leopard parts. Project Snow Leopard also introduced the Snow Leopard Information Management System (SLIMS; see next section), which included standard methods for surveys and a common database to store and share all types of snow leopard data.

On another level, PSL was presented as a means for promoting biodiversity conservation for all of Central Asia. By this time the snow leopard was recognized as an ecological apex species, or "indicator" species, for the high mountain ecosystems it occupied. Therefore, improvements in snow leopard conservation would benefit many other species, 15 of which were also endangered. Project Snow Leopard relied on several key elements for success: an unprecedented amount of cooperation among the 12 range countries, adequate funding and oversight, building an information network, and developing common methods vetted by field use and updated via periodic workshops or symposia.

Snow Leopard Information Management System

Delegates of the Xining symposium unanimously endorsed PSL and urged all countries to adopt and use the SLIMS. The concept of SLIMS began with an ISLT-sponsored workshop in 1990, attended by representatives from Pakistan, India, Nepal, China, Mongolia, Russia, and the US Fish and Wildlife Service (USFWS). From the workshop came also the Snow Leopard Survey and Conservation Handbook (Jackson and Hunter, 1996), the field survey portion of SLIMS. As envisioned, SLIMS was to meet PSL objectives by using common field methods to make survey results more consistent and robust; and, adopt a standardized computer database to store and share data. The handbook contained information about snow leopard ecology, data forms, and detailed instructions for conducting field surveys at two levels: First Order Surveys focused on sign (scats, tracks, scrapes) to establish snow leopard presence-absence, whereas Second Order Surveys sought to use sign density to arrive at relative snow leopard abundance. Field surveys also assessed prey diversity and abundance. Many of the handbook's data sheets and instructions were translated into several local languages, with the entire handbook translated into Chinese and Russian.

SLIMS software, developed by the USFWS, was designed for personal computers to be located in nodes in each country and connected via the emergent worldwide web with a central node at ISLT. The SLIMS software user interface was designed for easy entry of data collected as prescribed in the Handbook.

The SLIMS Handbook and software were envisioned as dynamic tools that could be improved upon with feedback from users. Training in field methods and software use became a priority for ISLT and its international partners. Workshops were held in China in 1993 and 1996; Pakistan in 1994 and 1995; Mongolia in 1994; Bhutan in 1997 and 2000; and in Nepal in 1999. Pakistan was the first to implement SLIMS on a countrywide basis.

SLIMS Discussion

SLIMS proved successful in many ways and yet fell short of its original goals. All of the in-country workshops were well received and let international experts interact with national biologists and park staff. These workshops elevated the importance of the snow leopard in the countries and brought attention to needs of wildlife departments often overlooked in national bureaucracy. SLIMS training introduced many scientific principles and the importance of meticulous data collection to both field biologists and upper level managers.

SLIMS aimed to be consistent across 12 countries and called for regular surveys in each country. These laudable and optimistic goals for SLIMS faced several insurmountable issues. Though approved by all countries, PSL and SLIMS were not front-funded and required ISLT and partners to continually seek funds for expensive field surveys. Placing a computer node in each country and finding the right people to train was also problematic. China's provinces proved too large and disconnected for a single national node. The First Order surveys were most easily learned by local staff; indeed, these simple presence-absence procedures were soon in use beyond the workshops. Second Order surveys were, however, not as easily learned and proved too difficult for park-level staff to comprehend. Thus, each country varied in its capability to implement SLIMS.

With the turn of the century, improved methods for estimating snow leopard abundance, such as camera traps and fecal genetics (Chapters 27 and 28, respectively), were emerging tools that moved beyond what SLIMS surveys and analyses could provide. Today, sign-based abundance estimates are generally discredited (McCarthy et al., 2008), yet basic SLIMS presence-absence surveys are still widely employed across the range, and in some cases are providing data for much advanced analytical methods such as occupancy modeling.

PSL and SLIMS were visionary for their time and brought snow leopard scientists and conservationists across the range together through the use of common methods and shared information. They set the stage, as this chapter does, for the scientific advances and progress in snow leopard conservation described in ensuing chapters.

References

Ahmad, I., Hunter, D.O., Jackson, R., 1997. A snow leopard and prey species survey in Khunjerab National Park, Pakistan. In: Jackson, R., Ahmad, A. (Eds.), Proceedings of the Eighth International Snow Leopard Symposium. Islamabad, Pakistan, International Snow Leopard Trust, Seattle, USA, and World Wildlife Fund-Pakistan, Islamabad, Pakistan, pp. 92–95.

Bold, A., Dorjzunduy, S., 1976. Information on the snow leopard *Uncia uncia* in the southern Gobi-Altai. Trudi Obshchei i Eksperimentalnoi Biologii (Ulan Bator) 11, 27–43 (in Mongolian).

Chundawat, R., Rawat, G., 1994. Food habits of snow leopard in Ladakh, India. In: Fox, J.L., Du Jizeng (Eds.), Proceedings of the Seventh International Snow Leopard Symposium, Xining, China. International Snow Leopard Trust, Seattle, and Northwest Plateau Institute of Biology, pp. 127–132.

Chundawat, R., Rogers, W.A., Panwar, H.S., 1988. Status report on snow leopard in India. In: Freeman, H. (Ed.), Proceedings of the Fifth International Snow Leopard Symposium, Srinagar, India. International Snow Leopard Trust and Wildlife Institute of India, pp. 113–121.

Fancy, S.G., et al., 1988. Satellite Telemetry: A New Tool for Wildlife Research and Management. US Fish and Wildlife Service, Washington, DC.

Fox, J.L., 1989. A review of the status of and ecology of snow leopard (*Panthera uncia*). International Snow Leopard Trust, Seattle, Oregon, 40 pp.

Fox, J.L., 1994. Snow leopard conservation in the wild – a comprehensive perspective on a low density and highly fragmented population. In: Fox, J.L., Du Jizeng (Eds.), Proceedings of the Seventh International Snow Leopard Symposium, Xining, China. International Snow Leopard Trust and Chicago Zoological Society, pp. 3–16.

Fox, J., Sinha, S., Chundawat, R., Das, P., 1988. A field survey of snow leopard presence and habitat use in northwestern India. In: Freeman, H. (Ed.), Proceedings of the Fifth International Snow Leopard Symposium, Srinagar, India. International Snow Leopard Trust and Wildlife Institute of India, pp. 99–111.

Fox, J.L., Sinha, S., Chundawat, R., Das, P., 1991. Status of snow leopard *Panthera uncia* in northwest India. Biol. Cons. 55, 283–298.

Freeman, H., Jackson, R., Hillard, D., Hunter, D.O., 1994. Project snow leopard: a multinational program spearheaded by the International Snow Leopard Trust. In: Fox, J.L., Du Jizeng (Eds.), Proceedings of the Seventh International Snow Leopard Symposium, Xining, China. International Snow Leopard Trust and Chicago Zoological Society, pp. 241–152.

Hornocker, M.G., 1969. Winter territoriality in mountain lions. J. Wildl. Mgmt. 33, 457–464.

Hunter, D.O., Jackson, R., 1997. A range-wide model of potential snow leopard habitat. In: Jackson, R., Ahmad, A. (Eds.), Proceedings of the Eighth International Snow Leopard Symposium, Islamabad, Pakistan. International Snow Leopard Trust, Seattle, USA, and World Wildlife Fund-Pakistan, Islamabad, Pakistan, pp. 51–56.

Hunter, D.O., et al., 1994. Project snow leopard – a model for conserving central Asian biodiversity. In: Fox, J.L., Du Jizeng (Eds.), Proceedings of the Seventh International Snow Leopard Symposium, Xining, China. International Snow Leopard Trust and Chicago Zoological Society, pp. 247–252.

Jackson, R., Ahlborn, G., 1988. Observations on the ecology of snow leopard in West Nepal. In: Freeman, H. (Ed.), Proceedings of the Fifth International Snow Leopard Symposium, Srinagar, India. International Snow Leopard Trust and Wildlife Institute of India, pp. 65–87.

Jackson, R., Ahlborn, G., 1989. Snow leopards (*Panthera uncia*) in Nepal – home range and movements. Natl. Geogr. Res. 5, 161–175.

Jackson, R., Fox, J. 1997. Snow leopard conservation: accomplishments and research priorities. In: Jackson, R., Ahmad, A. (Eds.), Proceedings of the Eighth International Snow Leopard Symposium, Islamabad, Pakistan. International Snow Leopard Trust, Seattle, USA, and World Wildlife Fund-Pakistan, Islamabad, Pakistan, pp. 128–145.

Jackson, R., Hillard, D., 1986. Tracking the elusive snow leopard. Natl. Geogr., 793–809.

Jackson, R., Hunter, D.O., 1996. Snow Leopard Survey and Conservation Handbook. International Snow Leopard Trust, Seattle, WA, 154 pp.

Jackson, R., Wang, Z., Lu, X., Chen, Y., 1994. Snow leopards in the Qomolangma Nature Preserve of the Tibet Autonomous Region. In: Fox, J.L., Du, J. (Eds.), Proceedings of the Seventh International Snow Leopard Symposium, Xining, China. International Snow Leopard Trust and Chicago Zoological Society, pp. 85–95.

Koshkarev, E., 1989. Snow leopard in Kirgizia: population, ecology, and conservation. Frunze: Academy of Sciences of Kirgizia (in Russian).

Koshkarev, E., 1997. Has the snow leopard disappeared from eastern Sayan and western Hovsogol? In: Jackson, R., Ahmad, A. (Eds.), Proceedings of the Eighth International Snow Leopard Symposium, Islamabad, Pakistan. International Snow Leopard Trust, Seattle, USA and World Wildlife Fund-Pakistan, Islamabad, Pakistan, pp. 96–107.

Mallon, D.P., 1984a. The snow leopard *Panthera uncia* in Mongolia. Int. Pedigree Book Snow Leopards 4, 3–9.

Mallon, D.P., 1984b. The snow leopard in Ladakh. Int. Pedigree Book Snow Leopards 4, 23–37.

Mallon, D.P., 1991. Status and conservation of large mammals in Ladakh. Biol. Conserv. 56, 101–119.

Matthiessen, P., 1978. The Snow Leopard. Viking Press, New York, 350 pp.

McCarthy, T.M., 2000. Ecology and conservation of snow leopards, Gobi brown bears, and wild Bactrian camels in Mongolia. PhD dissertation, University of Massachusetts, Amherst.

McCarthy, T., Munkhtsog, B., 1997. Preliminary assessment of snow leopard sign surveys in Mongolia. In: Jackson, R., Ahmad, A. (Eds.), Proceedings of the Eighth International Snow Leopard Symposium, Islamabad, Pakistan. International Snow Leopard Trust, Seattle, USA, and

World Wildlife Fund-Pakistan, Islamabad, Pakistan, pp. 57–64.

McCarthy, K.P., Fuller, T.K., Ming, M., McCarthy, T.M., Waits, L., Jumabaev, K., 2008. Assessing estimators of snow leopard abundance. J. Wildl. Manage. 72 (8), 1826–1833.

Novak, M., 1980. Chapman, J.A., Pursley, D. (Eds.), The foot-snare and leg-hold trap: a comparison, 3, Worldwide Furbearer Conference, Maryland, pp. 1671–1685.

Oli, M., 1994a. Ghost in the snow. BBC Wildlife 12 (8), 30–35.

Oli, M., 1994b. Snow leopards and blue sheep in Nepal: densities and predator:prey ratio. J. Mammal. 75, 998–1004.

Oli, M.K., 1997. Winter home range of snow leopards in Nepal. Mammalia 61 (3), 353–360.

Oli, M., Taylor, I., Rogers, M., 1993. Diet of snow leopard (*Panthera uncia*) in the Annapurna Conservation Area, Nepal. J. Zool. 231, 365–370.

Oli, M., Taylor, I., Rogers, M., 1994. Snow leopard *Panthera uncia* predation of livestock: an assessment of local perceptions in the Annapurna Conservation Area, Nepal. Biol. Cons. 68, 63–68.

Prasad, S.N., Chundawat, R.S., Hunter, D.O., Panwar, H.S., Rawat, G.S., 1991. Remote sensing snow leopard habitat in the trans-Himalaya of India using spatial models and satellite imagery: preliminary results. In: Proceedings Second International Symposium on Advanced Technology in Natural Resource Management, Washington, DC. American Society of Photogrammetry and Remote Sensing, pp. 519–523.

Schaller, G.B., 1971. Imperiled phantom of Asian peaks: first photographs of snow leopard in the wild. Natl. Geogr., 702–707.

Schaller, G.B., 1972. On meeting a snow leopard. Animal Kingdom (75–1), 7–13.

Schaller, G.B., 1977. Mountain Monarchs. University of Chicago Press, Chicago, Illinois, 425 pp.

Schaller, G.B., 1980. Stones of Silence. Viking Press, New York.

Schaller, G.B., 1998. Wildlife of the Tibetan Steppe. University of Chicago Press, Chicago, Illinois, 373 pp.

Schaller, G., Li, H., Talipu, H. Lu, Ren, J., Qiu, M., Wang, H., 1987. Status of large mammals in the Taxkorgan Reserve, Xinjiang, China. Biol. Conserv. 42, 53–71.

Schaller, G., Talipu, L.H., Ren, J., Qiu, M., 1988a. The snow leopard in Xinjiang, China. Oryx 22, 197–204.

Schaller, G., Ren, J., Qiu, M., 1988b. Status of snow leopard in Qinghai and Gansu Provinces, China. Biol. Conserv. 45, 179–194.

Schaller, G., Tserendeleg, J., Amarsanaa, G., 1994. Observations of snow leopard in Mongolia. In: Fox, J.L. and Du Jizeng (Eds.), Proceedings of the Seventh International Snow Leopard Symposium, Xining, China. International Snow Leopard Trust and Chicago Zoological Society, pp. 33–46.

Sunquist, M., Sunquist, F., 2002. Wild Cats of the World. University of Chicago Press, Chicago, Illinois, 452 pp.

CHAPTER

26

From VHF to Satellite GPS Collars: Advancements in Snow Leopard Telemetry

*Örjan Johansson**, *Anthony Simms***, *Thomas McCarthy*†

*Grimsö Wildlife Research Station, Swedish University of Agricultural Sciences, Riddarhyttan, Sweden

**Wildlife Conservation Society, Bronx, NY, USA

†Snow Leopard Program, Panthera, New York, NY, USA

INTRODUCTION

In terms of wildlife research, an ideal species for a biologist to study would be one that occurs at high densities in open habitat with a mild climate, which moves around at a steady pace and is easy for humans to follow, is not afraid of humans, and is active during daylight hours. In such a situation it would be easy to conduct field studies and generate robust data on most aspects of the species ecology. Unfortunately though, there are not many, if any, species like this. The snow leopard (*Panthera uncia*) is the complete opposite. The habitat where this species is found is remote and inaccessible, the climate harsh, and humans are poorly adapted to move around in it. Further, the species is highly elusive, its coat pattern camouflages perfectly with the surroundings, and it occurs in low densities. Together these factors make the animal almost impossible to detect. In such a situation, the only viable option to collect data on aspects of the species' ecology is to fit some kind of tracking device to a number of individuals. This in turn requires a safe and efficient method to capture and sedate the animals. In this chapter we describe the three different tracking technologies that have been used in snow leopard research to date and provide an overview of the studies that employed them.

VHF TELEMETRY – THE FIRST STUDIES

The Technology

Wildlife research entered a new era when the Craighead brothers fitted the first radio-collar to

Snow Leopards
http://dx.doi.org/10.1016/B978-0-12-802213-9.00026-2

a grizzly bear (*Ursus arctos horribilis*) in the early 1960s (Craighead and Craighead, 1965). This technology was a major breakthrough for studying wildlife because for the first time it allowed researchers to acquire data remotely without having to physically observe the animal (Fuller and Fuller, 2012). Since this first study, the basic technology has been improved but the principles remain the same: a transmitter fitted to the animal emits a signal that is picked up by an antenna-receiver combination, which is operated by the researcher. Commercial radio-collars send pulse signals broadcasted in very high frequency (VHF) radio waves, in bands between 30 and 300 megahertz (MHz) where each device/collar has a unique frequency, enabling identification of which animal is being tracked. The signal generated by the collar is transformed by the receiver into an audible, repetitive "beep." In general the closer the receiver gets to the transmitter, the stronger the beep. However, signal strength is also dependent on the type of antenna. There are two types of antenna: omnidirectional and directional. The first (omnidirectional) receives signals from all directions, while the latter picks up stronger signals in the direction of the transmitter or from where the signal bounces. Omnidirectional antennas are useful when searching for a collar's signal and can, for example, be fitted to a vehicle; however, a directional antenna is required to determine where the signal comes from.

VHF works best at relatively short ranges. The technique has its limitations, mainly that transmitted signals are easily blocked by landforms such as hills and mountains, or even vegetation. In addition, estimates of the collar location can be difficult to obtain or be erroneous as a result of signal "bounce." The extent of the error is difficult to assess but can be substantial. This occurs in almost all environments, but is particularly problematic in mountainous terrain such as snow leopard habitat. Signals in such areas are deflected by the topography and can result in errors of many kilometers in location estimates. The main tracking methods used when working with VHF collars are homing in, triangulation, and aerial telemetry. Homing in can be used to find an animal at close range, which is done by determining in which direction the signal is strongest and then moving there. It can be useful for making direct observations or to locate dead animals, dens, and kills. Triangulation is primarily used to determine an animal's approximate location from afar. With this technique, basically the researcher listens for signals from sites that can be identified on a map and determines the compass bearing to each signal (the collared animal). By repeating this process two or three times in close repetition, it is possible to approximate the animal's location remotely if the signals appear to be coming from the same place. Conversely, if multiple compass bearings are taken and the signals appear to be coming from different locations, then either bounce is occurring (White and Garrott, 1990) or too much time has elapsed between the bearings and the animal has moved a substantial distance. Aerial telemetry is usually the only viable option when working in areas that are too difficult to access by ground transport, or too remote, and for species that range over substantial areas (Miller et al., 2010).

Snow Leopard Studies

For large felids such as snow leopards, the advent of VHF telemetry provided a means of collecting data that had previously not been possible. The first telemetry study of snow leopards was conducted by Rodney Jackson in Nepal between 1982 and 1985 (Jackson and Ahlborn, 1989). This pioneering work was later followed by studies in India (Chundawat, 1990), Mongolia (Schaller et al., 1994), Nepal (Oli, 1997), and Mongolia (McCarthy, 2000) (Table 26.1). These early telemetry studies began to answer several fundamental ecological questions about the species, such as habitat use, home range size and overlap, movement patterns, etc. However, the limitations of VHF in mountainous terrain also made the research difficult because covering the distances required to obtain reliable locations for triangulation – or picking up any signal at all – while travelling by foot in snow leopard habitat is often impossible. As such, historic VHF studies report significant periods of time when the collared animals could not be located (Jackson and Ahlborn, 1989; Chundawat, 1990; Oli, 1997; McCarthy et al., 2005). This raised the question of whether the snow leopards were close by but the collar signals were blocked by landforms, or if the animals utilized larger home ranges than the tracking teams could cover, thereby putting the animals beyond the detection of the telemetry equipment. Furthermore, once the cats had been located, the extreme ruggedness and topographic variation caused unpredictable signal behavior (Oli, 1997). As a consequence, all VHF studies report rather small home ranges (Table 26.1) compared to recent GPS-based studies (Table 26.2). The largest home ranges were found in an area with relatively low prey densities (McCarthy et al., 2005). This could explain the larger home ranges, as range size and food abundance are negatively correlated (Tuqa et al., 2014). An alternative explanation could be that the mountainous terrain in Mongolia is gentler than in the Himalayas, which allowed the researchers to cover more ground, and that the collar signals were not as easily blocked in that study area as they were in the steeper Himalayan mountains (Fig. 26.1).

To illustrate the difficulties with VHF technology in snow leopard habitat, McCarthy et al.

TABLE 26.1 Telemetry Studies of Snow Leopards Using VHF Collars

Country	Year	Number of individuals collared	Days followed	Days located	Average # positions per individual	Average home range size (km^2)
Nepal[a]	1982–1984	5	370 (121–545)	122 (28–206)	142 (36–245)	19 (11–36)
India[b]	1989	1	70	28	28	19
Mongolia[c]	1991	1	41			12
Nepal[d]	1991	3	41 (27–60)		14 (10–16)	16 (10–22)
Mongolia[e]	1994	4	714 (530–985)	31 (17–61)	39 (23–84)	69 (14–142)

Note: Numbers presented are means and range. Home ranges were calculated using Minimum Convex Polygon 100%.
[a]*Jackson (1996).*
[b]*Chundawat (1990).*
[c]*Schaller et al. (1994).*
[d]*Oli (1997).*
[e]*McCarthy et al. (2005).*

TABLE 26.2 Telemetry Studies of Snow Leopards Using GPS Collars

Country	Year	Number of individuals collared	Days followed	Average # positions per individual	Average home range size (km^2)
Pakistan	2006	1		842	850
Mongolia	2008–present	19	9925	1715	503
Afghanistan	2012–present	4	1568	1903	509

Note: Numbers presented are means and range. Home ranges were calculated using Minimum Convex Polygon 100% for the Mongolia and Afghanistan data and using a Kernel for the Pakistan data.

FIGURE 26.1 **Tracking snow leopards using VHF radio telemetry is extremely difficult in the steep rocky terrain.** Figure (A) shows the area in Nepal where Rodney Jackson conducted his study (Jackson, 1996). Figure (B) shows Tom McCarthy in central Mongolia (McCarthy, 2000).

(2005) determined that one male snow leopard that had not been heard from for several months had crossed approximately 45 km of steppe to a different mountain system out of range of the telemetry equipment they were using. In addition, Chundawat (1990) collared and tracked one snow leopard for a 2.5-month period during a study in Ladakh, India. Substantial effort was invested in tracking the animal and monitoring its movements. However, as is common with VHF telemetry in mountainous habitat, much time and effort were spent acquiring reliable locations and the accuracy of tracking was severely compromised by signal bounce (Chundawat, 1990). These are not uncommon problems with VHF telemetry and by no means reflect poorly on those studies. One of the main ways to overcome such problems is through the use of aircraft to track the collared animals. However, in many cases no suitable aircraft were available in the study area countries, and the study areas were located far from the nearest airfield; and even if aircraft had been available, it would have been costly and quite a dangerous undertaking to aerial-track because of the generally remote and mountainous snow leopard habitat and its highly unpredictable weather.

To summarize, the early VHF studies provided insights into many aspects of snow leopard ecology that had previously been unknown. But all studies struggled to track the collared animals, which resulted in a small number and potentially inaccurate locations for each individual. Further, more than half of the studies collared few individuals and ran for only a short duration (Table 26.1).

ARGOS PTT TELEMETRY

The Technology

Argos satellite collars incorporate platform transmitter terminals (PTTs) that transmit signals to Argos satellites (Soutullo et al., 2007). As an Argos satellite passes by over a collar it measures the "Doppler shift" in the signal transmitted from the collar's PTT, where each PTT has a unique identification number (Soutullo et al., 2007). The data are then downlinked to earth stations where the collar's location is calculated. One location is calculated per satellite pass and at least four uplinks must be received to pinpoint the location (Wildlife Computers, 2015). Each location is given an accuracy class depending on signal strength. The estimated error can be substantial, as much as 59.6 km (Mate et al., 1997). Because locations can only be obtained when a satellite passes over the study area, the Argos PTT collars do not allow researchers to select what time of the day locations will be attempted. Satellite passes at any specific site are a function of latitude, with best coverage near the poles and worst near the equator (Mate et al., 1997; Soutullo et al., 2007; Tomkiewicz et al., 2010).

Snow Leopard Studies

During a snow leopard telemetry study in the Altai Mountains, Mongolia (McCarthy, 2000), it rapidly became clear to researchers that monitoring the cats using VHF ground telemetry was exceedingly difficult and the success rate for tracking locations was low (Table 26.3). Collared animals went unlocated for weeks at a time because it was not possible to search more than 50% of the study area in a single day, and movements and home range sizes of the snow leopards were clearly greater than what had been reported from previous studies (e.g., 11.7–38.9 km^2; Jackson and Ahlborn, 1989). In 1996, technological advances in radio-collar design made it possible to obtain satellite collars (Telonics ST-10 Argos PTTs) with a weight (800 g) and battery life (1 year) suitable for snow leopards. To maximize collar life, they were programmed to transmit daily for the first 30 days, and then afterwards, alternate days until the battery was depleted (estimated at 12–14 months after deployment). The collars were also fitted with a VHF transmitter to allow ground telemetry.

A female snow leopard was caught and fitted with an Argos PTT collar in the winter of 1996. The initial satellite dataset included 107 locations. A maximum acceptable area extent was defined for locations centered on the study area with an "*X* extent" of 1000 km and a "*Y* extent" of 700 km, assuming no snow leopard would range this far from its capture site and that any such location would represent telemetry error. Five locations that fell outside this range were deleted. Whether to include any of the remaining locations was decided by comparing each to its preceding and subsequent locations and making a subjective judgment on whether the snow leopard would likely have moved that much in the interval of time between transmissions. Also, if the preceding and subsequent locations were clustered, but the location itself was substantially removed from that cluster, it was assumed

TABLE 26.3 Radio-Telemetry History of Four Snow Leopards in the Saksai River Study Area, Gobi-Altai, Mongolia, 1994–1997

	Collared snow leopards					
	M-Red	**M-Blue**	**F-Green**	**F-Yellow**		
Capture date	**3/15/94**	**9/10/94**[a]	**3/28/96**	**2/16/96**		
				Ground	**Satellite**	**Total**
Total locations	**24**	**84**	**26**	**23**	**91**	**114**
Days located	22	61	17	22	79	94
Days attempted location	207	191	41	85	~199	300
Success rate (%)	10.6	31.9	41.5	25.9	39.7	31.3
Mean interval (days)	9.4	3.1	2.4	3.9	2.5	3.2
Consecutive days located	9	28	10			31
Last location	11/24/96	11/22/96	9/9/97	8/7/97	2/18/97	8/7/97

[a]*Subsequent captures on 9/15/94 and 5/10/95.*

that the location was in error. For any location that was questionable, the Argos-supplied data quality statistics for that point were taken into account. Eventually, 91 locations were accepted.

The home range identified from satellite telemetry data was unexpectedly large, much of it falling well outside the mapped study area. Home ranges of snow leopards in the study area calculated using minimum convex polygons for ground-based VHF telemetry locations, ranged from 14 km^2 to 142 km^2 (Table 26.3). For the Argos PTT-collared female, her VHF-based home range was approximately 58 km^2, but when all 91 of her plausible satellite locations were considered, her home range increased to 4530 km^2. When an even more conservative view of her satellite data was taken, and locations that represented single visits to sites well outside her core activity area were removed, the home range was about 1590 km^2, more than an order of magnitude greater than the largest home range that had been determined for any leopard using VHF telemetry.

In sum, even though it was difficult to determine the accuracy of the Argos PTT locations, the data clearly demonstrated that studying snow leopards would benefit greatly from a technology that was not dependent on a researcher's ability to follow the collared animals on the ground, as was the case with VHF telemetry.

GPS TELEMETRY

The Technology

The US Department of Defense began developing the global positioning system (GPS) in the late 1960s, and in 1993 the system became fully operational (McNeff, 2002). GPS utilizes low-orbiting satellites that send out constant "messages" containing information about the satellite's location and the current time. These messages can be received by a GPS device (e.g., a handheld GPS), which locates itself by triangulating signals and calculating the distance to each satellite via the time it takes the signals to reach it (Tomkiewicz et al., 2010). At least three satellites are needed to acquire a location, although for a "three-dimensional position," which increases accuracy, a minimum of four satellites are required (McNeff, 2002). The accuracy of GPS locations, or "fixes," is currently within a few meters; however, this can vary depending on time, location, habitat, number of satellites detected, among other factors. Besides being highly accurate, GPS has 24-h coverage, giving it a significant advantage over Argos PTT (Tomkiewicz et al., 2010).

The first wildlife studies employing GPS collars were conducted in the 1990s. At this time collars weighed about 1800 g and could only be fitted to large species such as moose (Rempel et al., 1995). But since then, the size and weight of collars has continually decreased and new features have been added. For instance, most manufacturers nowadays can equip their GPS collars with an accelerometer, which measures how much the collar is moving in two or three axes, providing detailed activity data. Mortality sensors can also be incorporated, which alert the researcher if the animal has not moved for a specified length of time. Thermometers are another common feature, measuring ambient temperature and helping the researcher to profile the animal's habitat preferences. It is also possible to fit collars with cameras, programmed to record either video or take photos at given intervals. In addition, collars can be programmed to automatically drop off the host animal at a specified date and time.

The most basic GPS collars have no means of communication, instead storing all their data in an internal memory to be later downloaded. This collar type is often referred to as "store-on-board" (Tomkiewicz et al., 2010). They weigh less and are often the cheapest of available models. However, if a store-on-board collar is lost, for example due to malfunction, the data it has collected are also lost. Therefore, as a safeguard to losing data, and to allow access to the data in "real time," GPS collars can be equipped with features that allow remote data transferal. The data can be transmitted through radio signals (UHF or VHF) to a handheld receiver via GSM (global system for mobiles), or through satellite systems such as Argos, Iridium, and Globalstar (Tomkiewicz et al., 2010). Similar to Argos PTT collars, GPS collars with remote data transfer do not require researchers to physically track and locate collared animals in the field. This frees up large amounts of time and other resources. Communication can be one-way, meaning that the collar will only send data; or two-way, meaning the collar can both send and receive data. Collars with one-way communication have to be programmed prior to deployment and cannot be changed once on an animal, whereas two-way systems enable the researcher to define a new program schedule for the collar, check battery status, activate the drop-off, and so on, while the collar is deployed on the host animal.

Of the collar types currently available, the most suitable for snow leopard studies are GPS collars with satellite communication. GSM is generally not appropriate because most parts of the snow leopard's range lack this service. Similarly, VHF and UHF download collars are generally not suitable because they only work at relatively short ranges (meaning that the collared animal has to be located and approached on foot as with traditional radio telemetry).

Satellite communications systems, such as Iridium and Globalstar, overcome the aforementioned limitations. Collars using the Iridium system can support both one-way and two-way communication and can "bundle" data and transmit it in short bursts, with up to 18 GPS locations in each bundle (Tomkiewicz et al., 2010). Bundling in this way can prolong a collar's battery life, but it also increases the risk of losing quantities of data if the transmission fails. At present, Globalstar collars only feature one-way communication; however, the data can be transmitted to satellite instantly as it is gathered (Fig. 26.2). Immediate upload like this minimizes the risk of data loss and offers a more real-time flow of information to the researcher. But it also consumes more battery power and thus reduces collar life.

FIGURE 26.2 **Adult snow leopard fitted with a GPS collar (Vectronic GPS Plus with Globalstar communication).**

Snow Leopard Studies

The first study to fit GPS collars on snow leopards took place in Chitral Gol National Park, Pakistan, in 2006 (McCarthy et al., 2007). This study saw a female snow leopard equipped with an Argos GPS collar (TGW 3481, Telonics) programmed to take three GPS locations per day. However, despite thorough initial testing of the collar, it did not uplink any positions once deployed on the host animal. The same snow leopard was recaptured approximately 2 months later and the collar replaced. The new collar uplinked all locations taken during the predeployment test period, but again, once fitted to the host snow leopard the uplinks ceased.

Similar problems were being faced by other wildlife researchers in the region. It was determined that unexplained "noise" in the Argos frequency range in parts of Europe and Asia (Fig. 26.3) was effectively drowning out the weak (0.5 W) signal of collars such as the one used in Pakistan (see sidebar). Since the Chitral collars worked well prior to deployment, it is assumed that the animal's body absorbed just enough of the signal to prevent it being detected over the background noise by the Argos satellite. The collars were equipped with an automatic drop-off and a VHF transmitter. The drop-off worked as planned (1 year after deployment), and once it had fallen off the host cat the signals again reached the Argos satellites successfully. Nonetheless, when retrieved it was found that the collars had collected 842 locations (collar 1:82 fixes, collar 2:760 fixes) and the calculated home range of the leopard was approximately 850 km^2, and showed her to cross into eastern Afghanistan. The number of location points obtained in this study, tracking one cat for a single year, nearly equaled all previous telemetry studies combined (Table 26.2). See McCarthy et al. (2007) for further details about the study.

IMPROVED CONDITIONS FOR LOW POWER ARGOS TAG TRANSMISSIONS

Fig. 26.3 illustrates the spatial extent and the amplitude of the wideband noise that has been interfering with the transmission of signals from Argos transmitters on the earth to those satellites carrying the Argos system, and how it has changed from May 2007 to May 2015. Fig. 26.3B shows that the spatial extent of the interference in 2015 is significantly less that it was in 2007 (Fig. 26.3A), particularly in the western Europe, northern Africa, and eastern Asia regions. Additionally, the measurements indicate that the amplitude of the noise in those regions where it was originally the strongest has been reduced by as much as 4–10 dBm, depending on the specific region. Consequently, Argos users are finding improved conditions in these regions for receiving more data from the low-power Argos tag transmissions generally associated with wildlife tracking.

In 2008, the first GPS collar with Globalstar communication was fitted to a snow leopard in Tost Mountains of South Gobi Province, Mongolia. This represented the start of the telemetry component in a long-term ecological study on the species. From August 2008 to the present, 20 individual snow leopards have been collared and 29 collars have been deployed (see Johansson et al., 2013, for details). North Star (King George, USA) collars were first used, during 2008–2009, but after several malfunctions, in 2010 the researchers switched to GPS-Plus

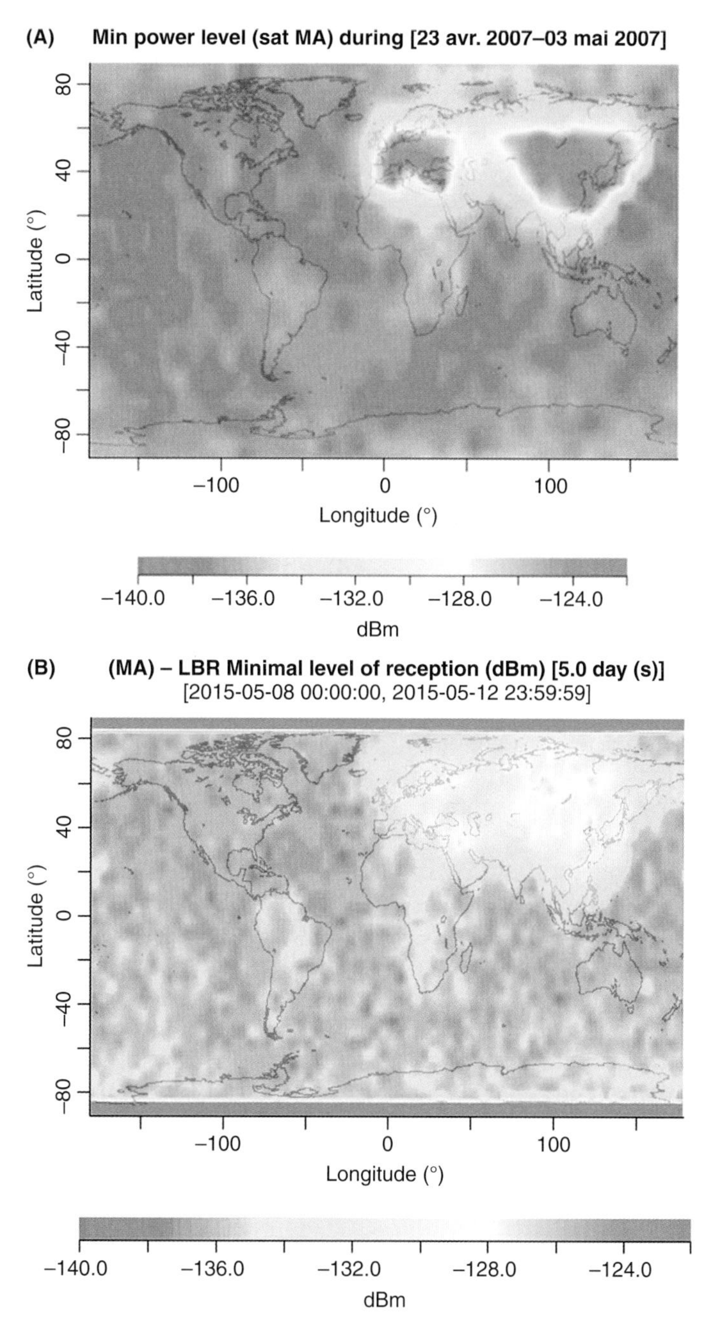

FIGURE 26.3 **Maps showing the interference (noise) associated with the Argos system and how it has decreased from 2007 to 2015.** (A) Measurements collected during the period April 23–May 3, 2007; (B) Measurements collected during the period May 8–12, 2015.

collars (Vectronic Aerospace, Berlin, Germany). The collars were programmed to take one GPS fix every 7 and 5 h for the North Star and the GPS-Plus collars, respectively. Since 2008, the collars have generated more than 30,000 GPS locations in total, and the study has yielded new insights into many aspects of snow leopard ecology. For instance, it had been deemed too difficult to locate kill sites from VHF-collared snow leopards (Jackson, 1996), however, the increased accuracy of GPS-collars and real-time uplinks have allowed kill sites to be identified (done when GPS points are clustered in one location over a period of days) and searched efficiently. Kill sites were visited between 2008 and 2013 and 249 prey eaten by the collared snow leopards were found (Johansson et al., 2015). Average home range for snow leopards that were collared for more than 3 months (n = 17) was 503 ± 286 km^2 (95% Minimum Convex Polygon). The collars also allowed the researchers to follow two young snow leopards that were dispersing from their collared mothers, and in 2011 the first dens were located and cubs counted and tagged with microchips (Johansson, unpublished data).

Similarly, a long-term telemetry study of snow leopards has been running since May 2012 in the Wakhan Corridor, Badakhshan Province, Afghanistan. To date, five Vectronic Aerospace GPS-Plus collars have been deployed on four snow leopards – two adult males and two adult females (one of the males was collared twice) – in the Hindu Kush Mountains. The collars were all programmed to take a GPS fix every 3 h for a period of 56 weeks (13 months), then automatically drop off. Over 9000 GPS locations have been obtained to date (Table 26.2). The study has provided the first ever insights to Afghanistan's snow leopards, such as predation habits, reproduction, and home range. It is also enabling researchers to develop sophisticated habitat preference models for the species and is helping the Afghan government and partners to develop conservation strategies for the species.

In addition to these studies, a number of snow leopards have been fitted with GPS collars in other places: one in Russia, one in Nepal, and three in Mongolia (outside the South Gobi study). These have not been long-term initiatives, and no published literature presently exists, and as such they are not discussed here.

In summary, GPS collars have for the first time provided researchers with an effective tool for studying the snow leopard. The sheer volume, quality, and variety of data this technology has yielded to date are significantly refining findings from earlier telemetry studies and broadening our understanding of the species.

CONCLUSIONS

Although the early VHF studies yielded insights into many previously unknown aspects of snow leopard ecology, it has become increasingly clear that for low-density, far-ranging species that inhabit inaccessible terrain, GPS-collars are far more effective, given the volume of data, accuracy of locations, and the ability to track numerous individuals simultaneously.

The reliability of collar communication can be a limitation with GPS telemetry. It is therefore recommended that researchers field-test each collar in the assigned study area for at least 24 h prior to deployment.

There have been several short-term GPS studies of snow leopards, which collared few (one or two) individuals. We believe it is far better to combine resources and attempt to study more animals for a longer time.

Finally, large felids depend on stealth and explosive rushes to catch prey, and for snow leopards this occurs in steep terrain. In such a situation it cannot be stressed strongly enough that collar weight and how well it is fitted can affect the host animal's survival. As such we recommend that collars not exceed 2% of the animal's body weight.

References

Chundawat, R., 1990. Habitat selection by a snow leopard in Hemis National Park, India. Int. Ped. Book of Snow Leopards 6, 85–92.

Craighead, F.C., Craighead, J.J., 1965. Tracking grizzly bears. BioScience 15 (2), 88–92.

Fuller, M.R., Fuller, T.K., 2012. Radio-telemetry equipment and applications for carnivores. In: Boitani, L., Powell, R.A. (Eds.), Carnivore Ecology and Conservation; A Handbook of Techniques. Oxford University Press, Oxford, UK.

Jackson, R., 1996. Home range, movements and habitat use of snow leopard in Nepal. Doctoral Dissertation. University of London, UK.

Jackson, R., Ahlborn, G., 1989. Snow leopards (*Panthera uncia*) in Nepal – home range and movements. Natl. Geogr. Res. 5 (2), 161–175.

Johansson, Ö., Malmsten, J., Mishra, C., Lkhagvajav, P., McCarthy, T., 2013. Reversible immobilization of free-ranging snow leopards (*Panthera uncia*) with a combination of medetomidine and tiletamine-zolazepam. J. Wildl. Dis. 49, 338–346.

Johansson, Ö., McCarthy, T., Samelius, G., Andrén, H., Tumursukh, L., Mishra, C., 2015. Snow leopard predation in a livestock dominated landscape in Mongolia. Biol. Conserv. 184, 251–258.

Mate, B.R., Nieukirk, S.L., Kraus, S.D., 1997. Satellite-monitored movements of the northern Right Whale. J. Wildl. Manage. 61, 1393–1405.

McCarthy, T. M., 2000. Ecology and conservation of snow leopards, Gobi brown bears, and wild Bactrian camels in Mongolia. Doctoral Dissertation. Paper AAI9960772. http://scholarworks.umass.edu/dissertations/AAI9960772

McCarthy, T.M., Fuller, T.K., Munkhtsog, B., 2005. Movements and activities of snow leopards in southwestern Mongolia. Biol. Conserv. 124, 527–537.

McCarthy, T., Khan, J., Ud-Din, J., McCarthy, K., 2007. The first study of snow leopards using GPS satellite collars underway in Pakistan. Cat News 46, 22–23.

McNeff, J.G., 2002. The global positioning system. IEEE T. Microw. Theory 50, 645–652.

Miller, C.S., Hebblewhite, M., Goodrich, J.M., Miquelle, D.G., 2010. Review of research methodologies for tigers: telemetry. Integr. Zool. 5, 378–389.

Oli, M.K., 1997. Winter home range of snow leopards in Nepal. Mammalia 61, 355–360.

Rempel, R.S., Rodgers, A.R., Abraham, K.S., 1995. Performance of a GPS animal location system under boreal forest canopy. J. Wildl. Manage. 59, 543–551.

Schaller, G.B., Tserendeleg, J., Amarsanaa, G., 1994. Observations on snow leopards in Mongolia. In: Fox, J.L., Du, J. (Eds.), Proceedings of the Seventh International Snow Leopard Symposium. Snow Leopard Trust and the Chicago Zoological Society, Chicago, USA.

Soutullo, A., Cadahia, L., Urios, V., Ferrer, M., Negro, J.J., 2007. Accuracy of lightweight satellite telemetry: a case study in the Iberian Peninsula. J. Wildl. Manage. 71, 1010–1015.

Tomkiewicz, S.M., Fuller, M.R., Kie, J.G., Bates, K.K., 2010. Global positioning system and associated technologies in animal behaviour and ecological research. Philos. T. Roy. Soc. B 365, 2163–2176.

Tuqa, J.H., Funston, P., Musyoki, C., Ojwang, G.O., Gichuki, N.N., Bauer, H., Tamis, W., Dolrenry, S., Van't Zelfde, M., de Snoo, G.R., de Longh, H.H., 2014. Impact of severe climate variability on lion home range and movement patterns in the Amboseli ecosystem, Kenya. Global Ecol. Conserv. 2, 1–10.

White, G.C., Garrott, R.A., 1990. Analysis of Wildlife Radio-Tracking Data, first ed. Academic Press, London.

Wildlife Computers, 2015. http://wildlifecomputers.com/learn/tracking_argos (accessed 24.02.2015).

CHAPTER

27

The Role of Genetics

OUTLINE

Snow Leopards
http://dx.doi.org/10.1016/B978-0-12-802213-9.00027-4

SUBCHAPTER

27.1

Conservation Genetics of Snow Leopards

*Anthony Caragiulo**, *George Amato**, *Byron Weckworth***
***American Museum of Natural History, Sackler Institute for Comparative Genomics, New York, NY, USA**
****Snow Leopard Program, Panthera, New York, NY, USA**

INTRODUCTION

The goal of conservation biology is to reduce the current rates of extinction and preserve biodiversity. The primary threat to many endangered species is small population size. At small population sizes, stochastic effects related to demographic, environmental, and genetic consequences increase extinction risks. Genetic diversity, along with species and ecosystem diversity, is recognized by the IUCN as one of the top three forms of biodiversity requiring conservation (Mcneely et al., 1990). Conservation geneticists apply genetic theory and techniques to preserve endangered species as dynamic entities, capable of coping with environmental change and thus minimizing their risk of extinction (Frankel and Soulé, 1981). The application of molecular tools to conservation research provides biologists, managers, and policy makers with insights into the drivers of extinction, which helps to inform appropriate management practices (Sarre and Georges, 2009).

Snow leopards (*Panthera uncia*) are an umbrella species of the high elevation regions of Central Asia, and a keystone for maintaining biodiversity within these fragile ecosystems amid the impacts of climate change and human perturbation (Li et al., 2014). Snow leopards are currently categorized as "Endangered" by the International Union for the Conservation of Nature (IUCN) and the Convention on International Trade in Endangered Species (CITES) Appendix I. This categorization is a result of declining snow leopard numbers and loss of habitat stemming from prey loss, conflict with humans, and poaching for hides and use of parts in traditional medicine. Snow leopards are believed to have been extirpated from as much as 15% of their historic range, and in some areas their numbers have declined by as much as 20% in the late twentieth century (Mccarthy and Chapron, 2003). Snow leopards are generally solitary and maintain stable home ranges delineated using markings such as scat, urine, and scrapes. A clear understanding of patterns of snow leopard population trends and genetic diversity is critical for guiding conservation initiatives that will ensure their long-term persistence. In this chapter we briefly review the most widely used genetic tools available for snow leopards, summarize the most important published studies of snow leopard genetics, and outline priority research questions and needs that would fill important gaps in our use and understanding of conservation genetics to support snow leopard conservation.

MAJOR GENETIC TOOLS AVAILABLE FOR SNOW LEOPARD CONSERVATION

Snow leopards are a cryptic species and naturally occur at low densities, making them difficult to study. These characteristics make it especially difficult to collect blood and tissue samples for genetic studies. Advances in molecular biology and genetics have made DNA collection much easier, as noninvasive molecular techniques have increasingly become the norm in studying large carnivores (Taberlet et al., 1996; Waits and Paetkau, 2005). *Noninvasive* is an umbrella term and refers to samples obtained without direct observation or handling of the target animal, with the most common sample types being hair and scat.

The benefit of noninvasive sampling is that the target species never has to be directly observed or handled, making it ideally suited for studying snow leopards. Noninvasive sampling also reduces the risks associated with immobilizing and handling rare and endangered species. The major detriment, however, is the poor DNA quality obtained from noninvasive samples. Advanced molecular techniques provide workarounds for obtaining reliable species and individual identifications from these samples, but extra care needs to be taken when analyzing data from noninvasive samples (Broquet et al., 2007; Miquel et al., 2006; Ruell and Crooks, 2007; Taberlet et al., 1999; Waits and Paetkau, 2005).

Hair samples can be collected through hair corrals, rub stations, and hair snares, and have been used to monitor a variety of carnivores including ursids, felids, canids, and mustelids (Kendall and Mckelvey, 2008). Scat has also been an effective means to monitor large carnivores, due to its ubiquity in nature and source of target DNA (Foran et al., 1997a). Additionally, scat provides the ability for dietary analysis, both through traditional methods of prey identification through undigested material (i.e., hair, bones) and molecular methods (Emmons, 1987; Farrell et al., 2000; see Chapter 27.2). Given their prevalence for marking predictable sites on the landscape with scat, and the cool, dry conditions that typify their habitat, noninvasive sampling for snow leopards is comparatively easy versus other large carnivores and may yield higher quality DNA for downstream analyses.

Hair and scat are a source of mitochondrial and nuclear DNA and can be used for both species and individual identification (Farrell et al., 2000; Foran et al., 1997b; Reed et al., 1997). Hair samples provide DNA in dried epithelial cells (dander) that cling to the shaft, and a higher proportion of DNA from cells within the hair follicle. Scat samples provide DNA from the epithelial cells lining the large intestine of the target species. The epithelial cells are sloughed off as feces move through the animal's colon and comprise the outer covering of the resultant scat.

Urine has also been utilized for noninvasive monitoring of populations, as DNA may be obtained from epithelial cells lining the urinary tract shed during urination, however, this method is less popular than hair and scat due to limitations in collection. Namely, urine has mostly been used for species inhabiting snowy areas and collected from urine-covered snow (Hedmark et al., 2004; Valiere and Taberlet, 2000; Van Der Hel et al., 2002), thereby limiting its ubiquity compared to hair and scat. However, when able to be collected, urine has been shown to provide quality genetic information about species and populations (Hedmark et al., 2004; Valiere and Taberlet, 2000). Additionally, scent spray marking, particularly in felids such as snow leopards, can also provide the raw genetic material from the field for identifying species and individuals (Caragiulo et al., 2015).

Noninvasive samples are an excellent source of mitochondrial DNA (mtDNA), which can be used for reliable species identification from numerous vertebrate species (Cronin et al., 1991; Farrell et al., 2000; Foran et al., 1997a; Kitano et al., 2007; Melton and Holland, 2007; Mukherjee et al., 2007; Paxinos et al., 1997). Mitochondria are numerous within a typical cell and contain small circular haploid DNA molecules that are maternally inherited and usually transmitted without recombination (Barr et al., 2005). Additionally, mtDNA evolves more rapidly than most nuclear DNA, resulting in the accumulation of differences between closely related species (Brown et al., 1979; Hebert et al., 2004). The abundance, ease of purification and sequencing, and interspecific sequence conservation make mtDNA ideal for identifying species from noninvasive samples (Foran et al., 1997a; Chaves et al., 2012). Mitochondrial DNA, specifically the *cytochrome oxidase I* gene region, provides DNA barcodes for the identification of samples from unknown species (Eaton et al., 2010; Hebert et al., 2003), and interspecific variation can be resolved from short DNA sequences, which is an important aspect when dealing with highly degraded DNA from noninvasive samples. Nuclear DNA is less commonly used for species identification and is less efficient for species identification than mtDNA markers (Rastogi et al., 2007); however, some nuclear markers (e.g., internal transcribed spacer regions) have shown promise for species identification (Schoch et al., 2012). Purification of DNA from noninvasive samples has advanced over the years, but the fact remains that it is still highly degraded and fragmented due to its source and the environmental conditions to which it is subjected prior to collection.

Mitochondrial DNA holds utility for discriminating between species, but analysis of nuclear DNA is necessary for identifying individuals and delineating populations. The most commonly used molecular marker is the microsatellite: short, tandemly repeated (usually between one and five base pair repeats) nuclear sequences that operate with traditional Mendelian inheritance (Jarne and Lagoda, 1996). They are selectively neutral and noncoding, making them ideal for examining population structure and gene flow (Rannala and Mountain, 1997), and many loci have been developed for snow leopards (e.g., Janečka et al., 2008; Waits et al., 2007).

Advances in sequencing technology have opened the door for single nucleotide polymorphisms (SNPs) to also be used similarly to microsatellites. SNPs are the most common type of genetic variability in most genomes and offer the potential for genome-wide scans of selectively neutral or adaptive variation with simple mutation models, powerful analytical methods, and application to noninvasive and historic DNA (Morin and Mccarthy, 2007; Morin et al., 2004; Morin et al., 2009). SNPs hold an advantage over microsatellites in that they are less prone to amplification error due to the single nucleotide nature versus longer sequences of tandem repeats in microsatellites, SNP error can be more easily quantified, and there is no user bias in the scoring of SNP genotypes. The disadvantage of SNPs is that panels of informative SNP loci have not been developed for many vertebrate species and this may need to be done *de novo* for a target species, whereas microsatellite panels exist for many vertebrate taxa.

MOLECULAR MARKERS FOR DETERMINING POPULATION STRUCTURE, CONNECTIVITY, AND PATTERNS OF GENE FLOW

All the previously mentioned molecular markers can be used to assess population structure, connectivity, and patterns of gene flow from noninvasively collected samples, and each addresses questions at different temporal scales. For example, mtDNA can be used to examine matrilineal lineages and pedigrees within species. Additionally, mtDNA is most commonly used in phylogeographic studies to examine the spatial distribution and divergence of populations (Avise et al., 1987; Avise, 1992). Haplotypic variation among populations and the pattern of mutational accumulation often provide insight into the evolutionary history of a species, as they have for numerous wild felid species such as pumas (Caragiulo et al., 2014a; Culver et al., 2000), cheetahs (Charruau et al., 2011), tigers (Driscoll et al., 2009; Luo et al., 2004; Luo et al., 2010), jaguars (Eizirik et al., 2001), leopards (Uphyrkina et al., 2001), clouded leopards (Kitchener et al., 2006), and lions (Barnett et al., 2006), but not yet on snow leopards. These analyses examine shared haplotypes amongst populations and stepwise mutational processes to look at the progression and divergence of the species on a spatial and geographic scale.

The major drawback to all of these molecular markers is that they are not interchangeable in their application. In order for meaningful comparisons between studies and species, the same markers need to be used. For instance, two studies on snow leopards must use the same microsatellite loci, or mtDNA markers, or SNP panel for comparison or combination of datasets for large-scale applications. Snow leopard conservation requires a coordinated effort regarding all genetic work to allow comparison between studies that use the same genetic markers to link local populations to amass as close to a range-wide dataset as possible.

GENETIC ANALYSIS OF NONNEUTRAL (ADAPTIVE) GENETIC VARIATION AND IMPLICATIONS FOR ADAPTATION TO A CHANGING ENVIRONMENT

The aforementioned measures of gene flow and connectivity are all based on selectively neutral molecular markers. These methods provide information about microevolutionary factors such as gene flow and can prioritize conservation units, but do not inform the potential for adaptation to changing environments. For this, one of the most common markers used is the major histocompatibility complex (MHC), which is a large cluster of loci involved in recognizing pathogen antigen molecules and regulating immune response (Frankham et al., 2010). Selection at the MHC favors heterozygotes to maintain high levels of genetic diversity; however, diversity is commonly lost by drift in small populations and this leads to a decrease in the ability to combat disease and potential extirpation (Frankham et al., 2010). Examining the MHC for select populations can indicate their ability to combat introduced or emergent pathogens For example, low diversity at the MHC is commonly cited as the cause of Tasmanian devil (*Sarcophilus harrisii*) decline in Australia due to a transmissible clonal facial tumor (Siddle et al., 2010). Genetic diversity in the MHC may be particularly critical for snow leopards as climate change hastens the emergence and spread of new diseases (Altizer et al., 2003), most imminently at the high altitudes they inhabit.

CONSERVATION GENETICS AND MOLECULAR ECOLOGY IN SNOW LEOPARDS TO DATE

Despite its high profile of a charismatic carnivore, snow leopard information is scarce and difficult to obtain due to its cryptic nature and remote habitat. Additionally, snow leopard range largely encompasses areas of political turmoil, yielding additional layers of complexity in their study, not the least of which includes difficulties for genetic work requiring transporting biological material for study. A literature survey on snow leopards yields surprisingly few studies for an animal with a such a large ecological footprint; nonetheless, we attempt to characterize and summarize what is known regarding snow leopards from a genetic standpoint, and address the importance of this work as a foundation for future snow leopard research.

One of the first studies to characterize snow leopard genetics was Zhang et al. (2007), which developed snow leopard-specific primers to amplify a section of the control region in the mitochondrial genome. The control region is highly variable, and species-specific primers are often necessary to characterize intraspecific variation. While Zhang et al. (2007) used skin and hair samples from 12 snow leopards from China, the primers developed are also useful in wildlife forensics, species identification, and population genetic studies using noninvasive snow leopard samples. They identified six haplotypes and six parsimony-informative sites within their 411 base pair (bp) sequence, making this primer set a good candidate for examining mitochondrial diversity and phylogeographic patterns of snow leopards in other portions of their range. Wei et al. (2009) went a step further and amplified the entire mitochondrial genome of the snow leopard, allowing researchers to design primers to examine any mitochondrial gene region for phylogeography and evolutionary studies of the Pantherines. As expected, the snow leopard mitochondrial genome structure is similar to other felids. Wei et al. (2009) performed a phylogenetic analysis using 10 felid species and approximately 4000 bp of mitochondrial DNA spanning seven genes and resolved the snow leopard as sister to lions (*Panthera leo*). In contrast, results from Caragiulo et al. (2014b), which used entire mitochondrial genomes for every Pantherine, identified snow leopards as sister to the monophyletic clade of leopards (*Panthera pardus*) and lions. A study by Davis et al. (2010), used a supermatrix of nuclear and mitochondrial genes with sex chromosome sequences, and contradicted both of these relationships, firmly placing snow leopards as sister species to tigers (*Panthera tigris*). The Davis et al. (2010) study compared the most comprehensive amount of genetic data, and the relationship between snow leopards and tigers was corroborated by an analysis using both genetic and morphological data by Tseng et al. (2013). The study of snow leopard evolutionary history was enhanced through the development of mitochondrial primers, but has yet to be used to examine the demographic history of snow leopards, a vital aid toward their conservation and management.

Whereas mtDNA provides information at the population and species level, finer-scale genetic analysis through individual identification is an essential advancement for studying snow leopards through noninvasive genetics. Individual identification of snow leopards was made possible through the development of a panel of polymorphic microsatellite loci, originally developed for domestic cats (*Felis catus*) (Waits et al., 2007). They reported 48 polymorphic microsatellite loci in snow leopards with 2–11 alleles per locus. Additionally, they identified 10 loci with significant power to discriminate among individuals. Identification of individuals through multiple microsatellite loci is used with mark-recapture statistics to estimate population numbers or the number of individuals in a given study area. Minimum numbers of individuals and population estimates of snow leopards through noninvasive genetic methods is incredibly useful given the elusiveness of snow leopards and the difficulty of direct observations in the wild, thereby potentially increasing the number of individuals that can be sampled.

The reporting of polymorphic microsatellite loci was a first step in snow leopard conservation genetics, and it allowed for the identification of individuals from noninvasively collected samples. This work was advanced by further refinement of primers for seven of the loci identified in Waits et al. (2007) to be snow leopard–specific (Janečka et al., 2008). The snow leopard-specific microsatellite primers had a higher success rate when used with scat samples because they amplified shorter DNA regions and primed to a more specific sequence, rather than to the closely related sequence of the domestic cat. The study by Janečka et al. (2008) was one of the first to examine wild snow leopards in different portions of their range (northwest India, central China, and southern Mongolia) using noninvasive genetic techniques. They surveyed a total transect length of 16.2 km and genetically identified 85 scat samples across those three geographic areas. Of the 85 genetically identified scat samples, 31 were snow leopards and equated to 10 individuals (2–4 males and 2 females in southwestern India; 1 of unknown gender in central China; and 3–5 males and 2 females in southern Mongolia); however, the individual from China yielded poor quality DNA and was not included in further analysis. Overall, they found low genetic diversity among the nine analyzed individuals, with 2.5–3 mean alleles per locus. Analysis of four mtDNA regions showed little genetic variability, with two haplotypes exhibited in the control region. All individuals sampled in southwestern India exhibited the two haplotypes, while central China and southern Mongolia samples exhibited the same singular haplotype. Due to small sample sizes, the study was merely a descriptive one, but provided the first baseline genetic information on the species.

Janečka et al. (2008) was followed by the only other noninvasive genetics study on snow leopards (Karmacharya et al., 2011), which focused on Shey Phoksundo National Park (SPNP) and Kanchenjunga Conservation Area (KCA) in Nepal. A total of 71 scat samples were collected, with 19 of them genetically identified as snow leopard using mitochondrial DNA genes. From these, genotypes were generated using six microsatellite loci described in Janečka et al. (2008) for 10 samples, revealing nine individuals (three males, six females). The Karmacharya et al. (2011) study was also purely descriptive and provided baseline genetic information on snow leopards in the area. The authors suggested that their data be used to design a more in-depth population survey to estimate snow leopard abundance at the study sites.

Advances in next-generation sequencing technology have increased the scope of genetic questions. Recently, 109 gigabases of total snow leopard sequence data with 40x coverage was compared to the genomes of the tiger and lion (Cho et al., 2013). This comparison highlighted a unique amino acid change in snow leopards that is consistent with adaptation to high altitudes. Specifically, snow leopards exhibited a specific genetic determinant in the *EGLN1* gene (Met39 > Lys39), which is a human homolog for mediating high altitude adaptation. Amino acid changes in the *EGLN1* gene, as well as the *EPAS1* gene, account for hypoxia tolerance in naked mole rats and led Cho et al. (2013) to hypothesize the amino acid change observed in snow leopards confers an adaptive advantage to high altitude. Next-generation sequencing gives scientists the ability to look deeper into the genomes of organisms and understand complex relationships between genes and potential adaptations. Additionally, next-generation sequencing amplifies short stretches of DNA, which makes noninvasive samples well suited for this methodology because their DNA is already fragmented by its very nature.

A few studies have compared the use of noninvasive genetic techniques to more traditional carnivore survey methods. McCarthy et al. (2008) compared in Kyrgyzstan and China the accuracy of Snow Leopard Information Management System (SLIMS) sign surveys to other abundance estimators, one of which was genetic analysis of snow leopard scat. The study found that noninvasive genetic methods of snow leopard abundance did not correspond to SLIMS sign surveys, with more snow leopards generally detected through genetic analysis. The authors concluded that none of the abundance estimators agree and estimating snow leopard numbers with confidence requires greater effort and better documentation. These findings are similar to those in Janečka et al. (2011) in which they estimated 5–6 individuals/100 km^2 based on noninvasive genetics compared to 1.5 individuals/100 km^2 based on camera trapping in the Gobi Desert of Mongolia.

As demonstrated already, a major void remains in our understanding of snow leopard genetics. The published molecular work on snow leopards has almost entirely focused on the phylogenetic relationship of snow leopards among other cat species (e.g., Davis et al., 2010) or survey techniques in species and individual identification for diet and occupancy analyses (e.g., Janečka et al., 2008; Janečka et al., 2011; Mccarthy et al., 2008). The Janečka et al. (2008) and Karmacharya et al. (2011) studies exemplify the information gap with regard to snow leopard genetic studies. They are, to our knowledge, the only two population genetic studies on snow leopards, and the small sample sizes and distance between all sampling locales make them inappropriate to build a comprehensive genetic network. The latter point speaks to the naturally low densities of snow leopards and the extreme difficulty in surveying their mountainous habitat. The solution to this problem requires more extensive sampling across a greater number of areas.

To date, no published studies describe snow leopard phylogeography or population and landscape genetics, and the number of published snow leopard-specific genetic studies remains far below that of the other imperiled big cats. There is clearly an urgent need to initiate conservation genetic research to begin filling the gaps in our understanding of the molecular ecology of endangered snow leopards.

SUBCHAPTER

27.2

Diet Reconstruction of Snow Leopard Using Genetic Techniques

Wasim Shehzad

Institute of Biochemistry & Biotechnology, University of Veterinary & Animal Sciences, Lahore, Pakistan

INTRODUCTION

Predation is a central interspecific relationship that can be studied by multidisciplinary approaches involving ecological, evolutionary, and behavioral sciences, leading to an understanding ecosystem function. Studies of carnivore diets can help evaluate resources used within an ecosystem (Mills, 1992) by characterizing prey selection with regards to prey availability, which in turn can provide an indicator of ecosystem stability. Accurate diet information can also be useful in understanding human-wildlife conflicts, such as those between large carnivores and human populations reliant on livestock (e.g., Bagchi and Mishra, 2006; Inskip and Zimmermann, 2009).

Livestock depredation is a challenge throughout the range of the snow leopard (see Chapter 5), which results in hostility toward the animal from local communities (Mishra, 1997; Mishra et al., 2006; Inskip and Zimmermann, 2009) and retribution killings of snow leopards (Hussain, 2003; Bagchi and Mishra, 2006). To date, the diet of the snow leopard has been analyzed using a variety of classical methods. Inference from field surveys, questionnaires, and interviews with local people can give an assessment of snow leopard predation and has been effectively used in some studies (e.g., Mishra, 1997; Namgail et al., 2007). But such studies may represent opinions and lack scientific rigor. Radio telemetry allows the study of snow leopard movements, home range, pattern of habitat utilization, and social organization (McCarthy, 2000), although locating the remains of killed prey in high, steep terrain is extremely difficult (Jackson, 1996). Johansson et al. (2015) assessed spatiotemporal variation in predation patterns of snow leopards and their kill rates in a GPS collaring study of snow leopards in Mongolia. However, that study is unique in its success and is not easily replicated across the snow leopard's broad range.

Examining feces may then represent the most readily available and easily collected source of diet information (Putman, 1984); this technique has been used extensively to study snow leopard diets (Oli, 1994; Lovari et al., 2009; Anwar et al., 2011). Such diet analysis requires the identification of undigested remains, bones, teeth, or hair in feces. There are two potential problems with fecal examination to assess snow leopard diets. The first relates to the accurate identification of snow leopard feces in the field, while the second deals with the limitations of accurately identifying the prey taxa.

LIMITATIONS OF CLASSICAL METHODS

Snow leopard feces are identified in the field mostly on the basis of size, shape, location, and associated sign such as pugmarks, scrapes, or the remains of prey species near the feces (Bagchi and Mishra, 2006). Yet carnivore feces are quite similar in their morphological characteristics, and it is often difficult to differentiate the feces of sympatric carnivores (Hansen and Jacobsen, 1999; Spiering et al., 2009). Diet assessment based on such erroneous fecal identification may have far-reaching consequences in terms of conservation planning for snow leopards. Some studies (e.g., Long et al., 2007; Vynne et al., 2011) have used detection dogs trained to locate the feces of specific carnivores in the field. In another study, scat detection dogs were trained to distinguish snow leopard feces from other nontarget feces *ex situ* (Snow Leopard Trust, unpublished data), thus eliminating the cost and complications of bringing a dog to the field. Neither method is now in common use for snow leopards and may not be cost effective.

Even in accurately identified feces, analysis of contents has many hurdles. Large bones and teeth are generally fragmented and therefore difficult to identify (Oli, 1993) from feces. Hair is commonly identified by comparisons with mounted reference specimens. However, this method is laborious and time consuming. Hairs from the same animal may also vary in structure according to their location on the body. Similarly, hair from several related species may possess similar characteristics (Oli, 1993). Finally, a reference collection that lacks specimens of all possible prey items can prohibit accurate identification of remains.

GENETIC METHODS

Genetic methods may address both the problem of identifying the species that deposited the material and the prey remains they contain. Feces collected in the field are now easily validated by genetic analysis (Davison et al., 2002). Yet in several studies (e.g., Lovari et al., 2009; Anwar et al., 2011; Wang et al., 2014), snow leopard feces have been genetically validated and then prey content determined using conventional microscopic analysis, hence accurate and complete prey species identification remained a potential problem. DNA-based approaches are particularly suitable to overcoming this problem.

Using feces as a source of DNA from food (Fernandes et al., 2008; King et al., 2008; Corse et al., 2010) provides a method for studying the feeding ecology of elusive and secretive animals such the snow leopard. The method uses universal primers, which requires no *a priori knowledge* of the diet to amplify prey DNA. The approach has been successfully implemented for herbivore (Valentini et al., 2009), yet the analysis of carnivore diets presents challenges, as predator DNA can be simultaneously amplified with prey DNA (Deagle et al., 2007). Furthermore, prey fragments might be rare in the fecal DNA extract, and consequently they have the tendency of being missed during the early stages of PCR, resulting in a PCR product containing almost exclusively the dominant sequence of predators (Green and Minz, 2005; Jarman et al., 2006).

So far, two PCR-based approaches exist that could be used to study the snow leopard diet. In the first approach, the PCR amplification primers are designed for a particular prey species and successful PCR amplification product from a sample indicates the presence of DNA from the target; such an approach is referred as the targeted approach. To our knowledge, the PCR-based specific approach has not been used to study snow leopard diets.

The second approach employs primers that bind to DNA regions conserved in a broad range of prey items and the PCR products amplified with these conserved primers are subsequently characterized; this can be thought of as the exploratory approach (see Shehzad et al., 2012a, b). The method uses a combination of a general primer for vertebrates (*12SV5*) that targets the 12rRNA gene of mitochondria for prey identification, and a blocking oligonucleotide (*UnciB*) sequence targeting the same gene and is highly specific to snow leopard DNA with a carbon 3 (C3) spacer at 3′end. The method has the potential to amplify all prey items present in the snow leopard diet, while addition of the blocking oligonucleotide limits the PCR amplification of snow leopard DNA present in a mixture of fecal DNA extract. To observe the rarest prey items in the diet the concentration of snow leopard blocking oligonucleotide may need to be as much as 20x the concentration of general primers for vertebrates.

Such an approach requires a next-generation sequencing platform, because there is the potential to amplify hundreds of thousands of sequences of both predator and prey as required to elaborate all pray taxa present in the feces (Shehzad et al., 2012a). Investment in or use of such equipment can be costly, but the methodology can produce a far more detailed and accurate assessment of the snow leopard diet than previously possible.

CONCLUSIONS

An unambiguous understanding of an endangered snow leopard's diet is crucial for conservation planning for this species. To date, despite numerous studies, the diet of the snow leopard has been inadequately assessed due to misidentifying feces as being those of snow leopards, and the inherent inaccuracies of classic macro- or microhistological examination of fecal content. The methods described by Shehzad et al. (2012a) can address both these shortcomings through genetic techniques. A better understanding of the diet of the snow leopard will allow a more accurate assessment of the level of conflict between the cats and pastoralists who rightly or wrongly attribute their livestock depredation losses to snow leopards. Mitigating measures can then be designed that address a real, as opposed to a perceived, conflict. The techniques can also be employed to help assess opportunities to increase snow leopard numbers in areas where they have been reduced. Knowledge of diet composition and prey availability in such instances would help conservationists determine whether adequate wild prey is available to support the hoped-for increase in snow leopard populations. This would help avoid situations where increasing leopard numbers only result in escalating conflicts with livestock and humans, dooming the effort to failure. Conversely, where conflict is already high and conservation efforts focus on reducing livestock depredation (predator-proof corrals, better guard dogs, etc.), an accurate assessment of diet composition and wild prey availability would help avert unintended stress to snow leopards already facing inadequate food supplies to sustain their existing numbers.

References

Altizer, S., Harvell, D., Friedle, E., 2003. Rapid evolutionary dynamics and disease threats to biodiversity. Trends Ecol. Evol. 18, 589–596.

Anwar, M., Jackson, R., Nadeem, M., Janečka, J., Hussain, S., Beg, M., Muhammad, G., Qayyum, M., 2011. Food habits of the snow leopard *Panthera uncia* (Schreber, 1775) in Baltistan, Northern Pakistan. Eur. J. Wildl. Res. 57, 1077–1083.

Avise, J.C., 1992. Molecular population structure and the biogeographic history of a regional fauna: a case history with lessons. Oikos 63, 62–76.

Avise, J.C., Arnold, J., Ball, R., Bermingham, E., Lamb, T., Neigel, J., Reeb, C., Saunders, N., 1987. Intraspecific phylogeography: the mitochondrial DNA bridge between population genetics and systematics. Ann. Rev. Ecol. Syst. 18, 489–522.

Bagchi, S., Mishra, C., 2006. Living with large carnivores: predation on livestock by the snow leopard (*Uncia uncia*). J. Zool. 268, 217–224.

Barnett, R., Yamaguchi, N., Barnes, I., Cooper, A., 2006. The origin, current diversity and future conservation of the modern lion (*Panthera leo*). P.Roy. Soc. Lond. B Bio. 273, 2119–2125.

Barr, C.M., Neiman, M., Taylor, D.R., 2005. Inheritance and recombination of mitochondrial genomes in plants, fungi and animals. New Phytol. 168, 39–50.

Broquet, T., Ménard, N., Petit, E., 2007. Noninvasive population genetics: a review of sample source, diet, fragment length and microsatellite motif effects on amplification success and genotyping error rates. Conserv. Genet. 8, 249–260.

Brown, W.M., George, M., Wilson, A.C., 1979. Rapid evolution of animal mitochondrial DNA. P. Natl. Acad. Sci. 76, 1967–1971.

Caragiulo, A., Dias-Freedman, I., Clark, J.A., Rabinowitz, S., Amato, G., 2014a. Mitochondrial DNA sequence variation and phylogeography of Neotropic pumas (*Puma concolor*). Mitochondrial DNA 25, 304–312.

Caragiulo, A., Dougherty, E., Soto, S., Rabinowitz, S., Amato, G., 2014b. The complete mitochondrial genome structure of the jaguar (*Panthera onca*). Mitochondrial DNA 0, 1–2.

Caragiulo, A., Pickles, R.S.A., Smith, J.A., Smith, O., Goodrich, J., Amato, G., 2015. Tiger (*Panthera tigris*) scent DNA: a valuable conservation tool for individual identification and population monitoring. Conserv. Genet. Resour. 7 (3), 681–683.

Charruau, P., Fernandes, C., Orozco Ter Wengel, P., Peters, J., Hunter, L., Ziaie, H., Jourabchian, A., Jowkar, H., Schaller, G., Ostrowski, S., Vercammen, P., Grange, T., Schlötterer, C., Kotze, A., Geigl, E.M., Walzer, C., Burger, P.A., 2011. Phylogeography, genetic structure and population divergence time of cheetahs in Africa and Asia: evidence for long-term geographic isolates. Mol. Ecol. 20, 706–724.

Chaves, P.B., Graeff, V.G., Lion, M.B., Oliveira, L.R., Eizirik, E., 2012. DNA barcoding meets molecular scatology: short mtDNA sequences for standardized species assignment of carnivore noninvasive samples. Mol. Ecol. Resour. 12, 18–35.

Cho, Y.S., Hu, L., Hou, H., Lee, H., Xu, J., Kwon, S., Oh, S., Kim, H.-M., Jho, S., Kim, S., 2013. The tiger genome and comparative analysis with lion and snow leopard genomes. Nat. Commun. 4, 2433.

Corse, E., Costedoat, C., Chappaz, R., Pech, N., Martin, J.F., Gilles, A., 2010. A PCR-based method for diet analysis in freshwater organisms using 18S rDNA barcoding on faeces. Mol. Ecol. Resour. 10, 96–108.

Cronin, M.A., Palmisciano, D.A., Vyse, E.R., Cameron, D.G., 1991. Mitochondrial DNA in wildlife forensic science: species identification of tissues. Wildl. Soc. Bull. 19, 94–105.

Culver, M., Johnson, W., Pecon-Slattery, J., O'brien, S.J., 2000. Genomic ancestry of the American puma (*Puma concolor*). J. Hered. 91, 186–197.

Davis, B.W., Li, G., Murphy, W.J., 2010. Supermatrix and species tree methods resolve phylogenetic relationships within the big cats, Panthera (Carnivora: Felidae). Mol. Phylogenet. Evol. 56, 64–76.

Davison, A., Birks, J.D.S., Brookes, R.C., Braithwaite, T.C., Messenger, J.E., 2002. On the origin of faeces: morphological versus molecular methods for surveying rare carnivores from their scats. J. Zool. 257, 141–143.

Deagle, B.E., Gales, N.J., Evans, K., Jarman, S.N., Robinson, S., Trebilco, R., Hindell, M.A., 2007. Studying seabird diet through genetic analysis of faeces: a case study on macaroni penguins (*Eudyptes chrysolophus*). PLoS One 2, e831.

Driscoll, C.A., Yamaguchi, N., Bar-Gal, G.K., Roca, A.L., Luo, S., Macdonald, D.W., O'brien, S.J., 2009. Mitochondrial phylogeography illuminates the origin of the extinct Caspian tiger and its relationship to the Amur tiger. PLoS One 4, e4125.

Eaton, M., Meyers, G., Kolokotronis, S., Leslie, M., Matrin, A., Amato, G., 2010. Barcoding bushmeat: molecular identification of Central African and South American harvested vertebrates. Conserv. Genet. 11, 1389–1404.

Eizirik, E., Kim, J.-H., Menotti-Raymond, M., Crawshaw, Jr., P., O'brien, S.J., Johnson, W., 2001. Phylogeography, population history and conservation genetics of jaguars (*Panthera onca*, Mammalia, Felidae). Mol. Ecol. 10, 65–79.

Emmons, L.H., 1987. Comparative feeding ecology of felids in a Neotropical rainforest. Behav. Ecol. Sociobiol. 20, 271–283.

Farrell, L., Roman, J., Sunquist, M., 2000. Dietary separation of sympatric carnivores identified by molecular analysis of scats. Mol. Ecol. 9, 1583–1590.

Fernandes, C.A., Ginja, C., Pereira, I., Tenreiro, R., Bruford, M.W., Santos-Reis, M., 2008. Species-specific mitochondrial DNA markers for identification of non-invasive samples from sympatric carnivores in the Iberian Peninsula. Conserv. Genet. 9, 681–690.

Foran, D., Crooks, K., Minta, S., 1997a. Species identification from scat: an unambiguous genetic method. Wildl. Soc. Bull. 25, 835–839.

Foran, D., Minta, S., Heinemeyer, K., 1997b. DNA-based analysis of hair to identify species and individuals for population research and monitoring. Wildl. Soc. Bull. 25, 840–847.

Frankel, O.H., Soulé, M.E., 1981. Conservation and Evolution. Cambridge University Press, Cambridge.
Frankham, R., Ballou, J.D., Briscoe, D., 2010. Introduction to Conservation Genetics. Cambridge University Press, Cambridge, UK.
Green, S.J., Minz, D., 2005. Suicide polymerase endonuclease restriction, a novel technique for enhancing PCR amplification of minor DNA templates. Appl. Environ. Microbiol. 71, 4721–4727.
Hansen, M.M., Jacobsen, L., 1999. Identification of mustelid species: otter (*Lutra lutra*), American mink (*Mustela vison*) and polecat (*Mustela putorius*), by analysis of DNA from faecal samples. J. Zool. 247, 177–181.
Hebert, P.D.N., Cywinska, A., Ball, S.L., Dewaard, J.R., 2003. Biological identifications through DNA barcodes. P. Roy. Soc. London Biol. 270, 313–321.
Hebert, P.D., Stoeckle, M.Y., Zemlak, T.S., Francis, C.M., 2004. Identification of birds through DNA barcodes. PLoS Biol. 2, e312.
Hedmark, E., Flagstad, Ø., Segerström, P., Persson, J., Landa, A., Ellegren, H., 2004. DNA-based individual and sex identification from wolverine (*Gulo gulo*) faeces and urine. Conserv. Genet. 5, 405–410.
Hussain, S., 2003. The status of the snow leopard in Pakistan and its conflict with local farmers. Oryx 37, 26–33.
Inskip, C., Zimmermann, A., 2009. Human-felid conflict: a review of patterns and priorities worldwide. Oryx 43, 18–34.
Jackson, R.M., 1996. Home Range, Movements and Habitat Use of Snow Leopard (*Uncia uncia*) in Nepal, Ph.D. thesis. University of London, p. 233.
Janečka, J., Jackson, R., Yuquang, Z., Diqiang, L., Munkhtsog, B., Buckley-Beason, V., Murphy, W., 2008. Population monitoring of snow leopards using noninvasive collection of scat samples: a pilot study. Anim. Conserv. 11, 401–411.
Janečka, J.E., Munkhtsog, B., Jackson, R.M., Naranbaatar, G., Mallon, D.P., Murphy, W.J., 2011. Comparison of noninvasive genetic and camera-trapping techniques for surveying snow leopards. J. Mammal. 92, 771–783.
Jarman, S.N., Redd, K.S., Gales, N.J., 2006. Group-specific primers for amplifying DNA sequences that identify Amphipoda, Cephalopoda, Echinodermata, Gastropoda, Isopoda, Ostracoda and Thoracica. Mol. Ecol. Notes 6, 268–271.
Jarne, P., Lagoda, P.J.L., 1996. Microsatellites, from molecules to populations and back. Trends Ecol. Evol. 11, 424–429.
Johansson, O., McCarthy, T., Samelius, G., Andren, H., Tumursukh, L., Mishra, C., 2015. Snow leopard predation in a livestock dominated landscape in Mongolia. Biol. Conserv. 184, 251–258.
Karmacharya, D.B., Thapa, K., Shrestha, R., Dhakal, M., Janecka, J.E., 2011. Noninvasive genetic population survey of snow leopards (*Panthera uncia*) in Kangchenjunga Conservation Area, Shey Phoksundo National Park and surrounding buffer zones of Nepal. BMC Res. Notes 4, 516.
Kendall, K.C., McKelvey, K.S., 2008. Hair collection. In: Long, R.A., MacKay, P., Zielinksi, W.J., Ray, J.C. (Eds.), Noninvasive Survey Methods for Carnivores. Island Press, Washington, DC.
King, R.A., Read, D.S., Traugott, M., Symondson, W.O.C., 2008. Molecular analysis of predation: a review of best practice for DNA-based approaches. Mol. Ecol. 17, 947–963.
Kitano, T., Umetsu, K., Tian, W., Osawa, M., 2007. Two universal primer sets for species identification among vertebrates. Int. J. Legal Med. 121, 423–427.
Kitchener, A.C., Beaumont, M.A., Richardson, D., 2006. Geographical variation in the clouded leopard, *Neofelis nebulosa*, reveals two species. Curr. Biol. 16, 2377–2383.
Li, J., Wang, D., Yin, H., Zhaxi, D., Jiagong, Z., Schaller, G.B., Mishra, C., McCarthy, T.M., Wang, H., Wu, L., 2014. Role of Tibetan Buddhist monasteries in snow leopard conservation. Conserv. Biol. 28, 87–94.
Long, R.A., Donovan, T.M., MacKay, P., Zielinski, W.J., Buzas, J.S., 2007. Effectiveness of scat detection dogs for detecting forest carnivores. J. Wildl. Manage. 71, 2007–2017.
Lovari, S., Boesi, R., Minder, I., Mucci, N., Randi, E., Dematteis, A., Ale, S.B., 2009. Restoring a keystone predator may endanger a prey species in a human-altered ecosystem: the return of the snow leopard to Sagarmatha National Park. Anim. Conserv. 12, 559–570.
Luo, S.J., Kim, J.H., Johnson, W.E., Walt, J., Martenson, J., Karanth, U.K., 2004. Phylogeography and genetic ancestry of tigers (*Panthera tigris*). PLoS Biol. 2, 2275–2293.
Luo, S.J., Kim, J.H., Johnson, W.E., Smith, J.L., O'brien, S.J., 2010. What is a tiger? Genetics and phylogeography. In: Tilson, R., Nyhus, P.J. (Eds.), Tigers of the World: The Science, Politics, and Conservation of *Panthera tigris*. second ed. Academic Press, London, UK.
McCarthy, T., 2000. Ecology and conservation of snow leopards, Gobi Brown bears and wild Bactrian camels in Mongolia. University of Massachusett, Amherst, p. 134.
McCarthy, T., Chapron, G. (Eds.), 2003. Snow Leopard Survival Strategy. International Snow Leopard Trust & Snow Leopard Network, Seattle.
McCarthy, K.P., Fuller, T.K., Ming, M., McCarthy, T., Waits, L.P., Jumabaev, K., 2008. Assessing estimators of snow leopard abundance. J. Wildl. Manage. 72, 1826–1833.
McNeely, J.A., Miller, K.R., Reid, W.V., Mittermeier, R.A., Werner, T.B. 1990. *Conserving the World's Biological Diversity,* IUCN, World Resources Institute, Conservation International, WWF-US and the World Bank, Washington, DC.
Melton, T., Holland, C., 2007. Routine forensic use of the mitochondrial 12S ribosomal RNA gene for species identification. J. Forensic Sci. 52, 1305–1307.

Mills, M.G.L., 1992. A comparison of methods used to study food habits of large African carnivores. In: McCullough, D.R., Barrett, H. (Eds.), Wildlife 2001: Populations. Elsevier Science Publishers, London.

Miquel, C., Bellemain, E., Poillot, C., Bessière, J., Durand, A., Taberlet, P., 2006. Quality indexes to assess the reliability of genotypes in studies using noninvasive sampling and multiple-tube approach. Mol. Ecol. Notes 6, 985–988.

Mishra, C., 1997. Livestock depredation by large carnivores in the Indian trans-Himalaya: conflict perceptions and conservation prospects. Environ. Conserv. 24, 338–343.

Mishra, C., Madhusudan, M.D., Datta, A., 2006. Mammals of the high altitudes of western Arunachal Pradesh, eastern Himalaya: an assessment of threats and conservation needs. Oryx 40, 29–35.

Morin, P.A., McCarthy, M., 2007. Highly accurate SNP genotyping from historical and low-quality samples. Mol. Ecol. Notes 7, 937–946.

Morin, P.A., Luikart, G., Wayne, R.K., 2004. SNPs in ecology, evolution and conservation. Trends Ecol. Evol. 19, 208–216.

Morin, P.A., Martien, K.K., Taylor, B.L., 2009. Assessing statistical power of SNPs for population structure and conservation studies. Mol. Ecol. Resour. 9, 66–73.

Mukherjee, N., Mondol, S., Andheria, A., Ramakrishnan, U., 2007. Rapid multiplex PCR based species identification of wild tigers using non-invasive samples. Conserv. Genet. 8, 1465–1470.

Namgail, T., Fox, J.L., Bhatnagar, Y.V., 2007. Carnivore-caused livestock mortality in Trans-Himalaya. Environ. Manage. 39, 490–496.

Oli, M.K., 1993. A key for the identification of the hair of mammals of a snow leopard (*Panthera uncia*) habitat in Nepal. J. Zool. 231, 71–93.

Oli, M.K., 1994. Snow leopards and blue sheep in Nepal: densities and predator:prey ratio. J. Mammal. 75, 998–1004.

Paxinos, E., McIntosh, C., Ralls, K., Fleischer, R., 1997. A noninvasive method for distinguishing among canid species: amplification and enzyme restriction of DNA from dung. Mol. Ecol. 6, 483–486.

Putman, R.J., 1984. Facts from feces. Mammal Rev. 14, 79–97.

Rannala, B., Mountain, J.L., 1997. Detecting immigration by using multilocus genotypes. P. Natl. Acad. Sci. USA 94, 9197–9201.

Rastogi, G., Dharne, M.S., Walujkar, S., Kumar, A., Patole, M.S., Shouche, Y.S., 2007. Species identification and authentication of tissues of animal origin using mitochondrial and nuclear markers. Meat Sci. 76, 666–674.

Reed, J., Tollit, D., Thompson, P., Amos, W., 1997. Molecular scatology: the use of molecular genetic analysis to assign species, sex and individual identity to seal faeces. Mol. Ecol. 6, 225–234.

Ruell, E.W., Crooks, K.R., 2007. Evaluation of noninvasive genetic sampling methods for felid and canid populations. J. Wildl. Manage. 71, 1690–1694.

Sarre, S.D., Georges, A., 2009. Genetics in conservation and wildlife management: a revolution since Caughley. Wildl. Res. 36, 70–80.

Schoch, C.L., Seifert, K.A., Huhndorf, S., Robert, V., Spouge, J.L., Levesque, C.A., Chen, W., Consortium, F.B., 2012. Nuclear ribosomal internal transcribed spacer (ITS) region as a universal DNA barcode marker for Fungi. Proc. Natl. Acad. Sci., 6241–6246.

Shehzad, W., Riaz, T., Nawaz, M.A., Miquel, C., Poillot, C., Shah, S.A., Pompanon F., Coissac, E., Taberlet, P., 2012a. Carnivore diet analysis based on next-generation sequencing: application to the leopard cat (*Prionailurus bengalensis*) in Pakistan. Mol. Ecol. 21, 1951–1965.

Shehzad, W., McCarthy, T.M., Pompanon, F., Purevjav, L., Coissac, E., Riaz, T., Taberlet, P., 2012b. Prey Preference of Snow Leopard (*Panthera uncia*) in South Gobi, Mongolia. PLoS One 7, e32104.

Siddle, H.V., Marzec, J., Cheng, Y., Jones, M., Belov, K. 2010. MHC gene copy number variation in Tasmanian devils: implications for the spread of a contagious cancer. Proc. Biol. Sci. 277 (1690): 2001–2006.

Spiering, P.A., Gunther, M.S., Wildt, D.E., Somers, M.J., Maldonado, J.E., 2009. Sampling error in non-invasive genetic analyses of an endangered social carnivore. Conserv. Genet. 10, 2005–2007.

Taberlet, P., Griffin, S., Goossens, B., Questiau, S., Manceau, V., Escaravage, N., Waits, L.P., Bouvet, J., 1996. Reliable genotyping of samples with very low DNA quantities using PCR. Nucleic Acids Res. 24, 3189–3194.

Taberlet, P., Waits, L.P., Luikart, G., 1999. Noninvasive genetic sampling: look before you leap. Trends Ecol. Evol. 14, 323–327.

Tseng, Z.J., Wang, X., Slater, G.J., Takeuchi, G.T., Li, Q., Liu, J., Xie, G. 2013. Himalayan fossils of the oldest known pantherine establish ancient origin of big cats. Proc. Biol. Sci. 281 (1774): 20132686.

Uphyrkina, O., Johnson, W., Quigley, H., Miquelle, D., Marker, L., Bush, M., O'brien, S., 2001. Phylogenetics, genome diversity and origin of modern leopard, *Panthera pardus*. Mol. Ecol. 10, 2617–2633.

Valentini, A., Miquel, C., Nawaz, M.A., Bellemain, E., Coissac, E., Pompanon, F., Gielly, L., Cruaud, C., Nascetti, G., Wincker, P., Swenson, J.E., Taberlet, P., 2009. New perspectives in diet analysis based on DNA barcoding and parallel pyrosequencing: the trnL approach. Mol. Ecol. Resour. 9, 51–60.

Valiere, N., Taberlet, P., 2000. Urine collected in the field as a source of DNA for species and individual identification. Mol. Ecol. 9, 2150–2152.

Van Der Hel, O.L., Van Der Luijt, R.B., Bas Bueno De Mesquita, H., Van Noord, P.A., Slothouber, B., Roest,

M., Van Der Schouw, Y.T., Grobbee, D.E., Pearson, P.L., Peeters, P.H., 2002. Quality and quantity of DNA isolated from frozen urine in population-based research. Anal. Biochem. 304, 206–211.

Vynne, C., Skalski, J.R., Machado, R.B., Groom, M.J., Jacomo, A.T.A., Marinho, J., Neto, M.B.R., Pomilla, C., Silveira, L., Smith, H., Wasser, S.K., 2011. Effectiveness of scat-detection dogs in determining species presence in a tropical savanna landscape. Conserv. Biol. 25, 154–162.

Waits, L.P., Paetkau, D., 2005. Noninvasive genetic sampling tools for wildlife biologists: a review of applications and recommendations for accurate data collection. J. Wildl. Manage. 69, 1419–1433.

Waits, L., Buckley-Beason, V., Johnson, W., Onorato, D., McCarthy, T., 2007. A select panel of polymorphic microsatellite loci for individual identification of snow leopards (*Panthera uncia*). Mol. Ecol. Notes 7, 311–314.

Wang, J., Laguardia, A., Damerell, P.J., Riordan, P., Shi, K., 2014. Dietary overlap of snow leopard and other carnivores in the Pamirs of Northwestern China. Chinese Sci. Bull. 59, 3162–3168.

Wei, L., Wu, X., Jiang, Z., 2009. The complete mitochondrial genome structure of snow leopard *Panthera uncia*. Mol. Biol. Rep. 36, 871–878.

Zhang F., Jiang, Z., Zeng, Y., McCarthy, T., 2007. Development of primers to characterize the mitochondrial control region of the snow leopard (*Uncia uncia*). Mol. Ecol. Notes 7, 1196–1198.

CHAPTER

28

Camera Trapping: Advancing the Technology

Wai-Ming Wong, Shannon Kachel***

*Department of Field Programs, Panthera, New York, NY, USA

**School of Environmental and Forest Resources, University of Washington, Seattle, WA, USA

CONSERVATION AND RESEARCH APPLICATIONS OF CAMERA TRAPPING

Since the pioneering efforts of Jackson et al. (2006), camera trapping has become a workhorse of snow leopard (*Panthera uncia*) research and monitoring (e.g., Janečka et al., 2011; McCarthy et al., 2008; Sharma et al., 2014). Noninvasive camera trapping can provide unambiguous, economical data regarding snow leopard abundance, occupancy, distribution, population dynamics, and even resource selection. Furthermore, camera trap photographs help raise awareness and promote local and global interest and investment in snow leopard conservation and stewardship.

Unfortunately, many small-scale studies are prone to analytical issues arising from insufficient data – only partially due to low snow leopard densities – which limit the inferential value of results to uncorrected abundance indices. Limited resources, low densities, low detection rates, and sparse data are all common issues in snow leopard camera trapping. Indeed, among 35 reported or published camera trapping studies compiled by the Snow Leopard Network (2014), only 14 reported error-bound inferential estimates compared to effort-based minimum abundance indices, and of those, only a single study obtained the sufficiently large multiyear samples needed to estimate demographic vital rates – survival, immigration, and emigration – in a wild snow leopard population (Sharma et al., 2014).

For all its virtues, snow leopard camera trapping clearly remains a substantial undertaking, requiring considerable resource investment and fieldwork in logistically and physically demanding conditions. Given the effort required, we believe that conservationists and researchers should, at a minimum, adhere to the standard reporting recommendations of Meek et al. (2014). Better yet, they should strive to maximize the value of their hard-earned data by allocating sufficient resources in advance and adhering to

Snow Leopards
http://dx.doi.org/10.1016/B978-0-12-802213-9.00028-6

design approaches that facilitate empirical analysis and comparisons in space and time. Investigators should scale up efforts to enable error-bound inferential estimation whenever possible. Though resource-intensive, such a philosophical shift in allocation of energy and resources could enhance the rigor of snow leopard research and conservation efforts, and enable more precise, unbiased detection of population trends and more thoughtful ecological investigations of this enigmatic species.

ANALYTICAL ADVANCES AND STUDY DESIGN PROGRESS

Establishing Clear Objectives

Researchers typically turn to camera trapping to confirm snow leopard presence and to estimate populations and monitor trends. Explicitly defined objectives help to clarify study design and maximize the value of the data to be collected. Data and model outputs are only useful within the context of specific science or management objectives, and do not alone justify camera trapping (Nichols et al., 2011). A single estimate of abundance is not inherently informative to research or conservation. Its value lies in its comparison with other estimates in space or time, the validity of which hinges on rigorous study design based on explicit objectives. Haphazard design may yield interesting parameter estimates, yet amount to little more than an anecdotal snapshot with limited relevance to larger scientific and conservation efforts. Good study design facilitates direct investigation of hypotheses and comparisons with other work. We do not mean to suggest that a single design prescription exists to treat all objectives. Indeed, many possible objectives exist, such as understanding underlying processes affecting snow leopard distribution, detecting patterns in population dynamics, or establishing robust baseline estimates. Instead, we echo Nichols et al. (2011) and argue that future camera trapping efforts need to carefully address some common deficiencies and, if necessary, acquire additional resources or consider alternative (i.e., molecular) methods when necessary in order to adequately address relevant objectives.

Regarding Noninvasive Genetics

Genetic methods can address many of the same research objectives as camera trapping, at generally lower costs, though other trade-offs arise (Janečka et al., 2011). Camera trapping requires higher initial costs and a greater concentration of field effort, yet equipment is reusable and data analysis is far less specialized and uncomplicated by customs and border control issues (Janečka et al., 2011). On the other hand, genetic sampling can boost detection probabilities and sample extensive areas rapidly. (For further discussion of molecular methods in snow leopard ecology, see Chapter 27). Overall, compared to camera trapping, molecular methods appear particularly well suited to large-scale regional studies (e.g., Russell et al., 2012) and investigations of low-density populations, and may thus be preferable in many cases.

Nontarget Species

Camera traps frequently provide ancillary data regarding the occurrence of livestock, poachers, prey species, and sympatric carnivores. Although such data are often relevant to research objectives, investigators should carefully assess potential sampling biases before drawing broad inferences from these data (Sollman et al., 2013). Even though modeling solutions exist to estimate density and abundance from camera trapping of unmarked populations (e.g., Rowcliffe et al., 2008), model assumptions are unlikely to be met if not accounted for in the study design. Whenever possible, researchers interested in populations of prey and sympatric carnivores should select species-specific targeted sampling

FIGURE 28.1 **Unique spotting patterns of two different snow leopards at a camera trap location in the Pamir Mountains, Tajikistan.** *Source: Photo courtesy of S. Kachel, Panthera, Tajikistan Academy of Sciences.*

methods or a comprehensive multispecies approach.

ANALYTICAL ADVANCES AND FUTURE DIRECTIONS

Over the past decade, advances in analytical methods for camera trapping data have kept pace with advances in technology. Two statistical frameworks – occupancy (MacKenzie et al., 2006) and spatial capture–recapture (SCR) modeling (Royle et al., 2014) – are particularly robust. Where SCR concerns detectability of individuals, occupancy modeling focuses on detectability of presence in a given survey patch. Both approaches incorporate spatial variability and uncertain detectability to arrive at error-bound estimates of target parameters, and accordingly require careful forethought regarding study design. Camera trapping data designed to meet the assumptions of SCR approaches may be suitable for occupancy modeling if sampling is representative of the landscape of interest. However, large-scale snow leopard occupancy can be investigated much more efficiently via sign or genetic surveys and thus we do not anticipate wide application of camera trap–based occupancy models. In contrast, given that snow leopards are individually identifiable (Fig. 28.1), the capture–recapture approach is ideal for investigation of density and vital rates (Jackson et al., 2006; Janečka et al., 2011; McCarthy et al., 2008; Sharma et al., 2014). SCR overcomes many of the limitations of the traditional capture–recapture approach and enables simulation of design alternatives and investigation of space use and resource selection in relation to underlying landscape covariates without intensive, invasive telemetry methods (Royle et al., 2014). For comprehensive treatment of these subjects, we refer readers to the relevant literature (Efford, 2011; Royle et al., 2014).

DESIGN CONSIDERATIONS AND PITFALLS

Precise, unbiased SCR estimation of population parameters is dependent upon the total number of snow leopards detected, recaptured and spatially recaptured at multiple locations, the total coverage and distribution of traps relative to home range size, and assumptions of population closure (Royle et al., 2014). Each of these needs imposes constraints and trade-offs in study design and increases the importance of allocating sufficient resources to camera trap studies. Based on simulations and literature review, Tobler and Powell (2013) advocated that camera trapping resources for jaguars (*Panthera onca*) would be best focused on large-scale studies covering larger areas with far more cameras than typically employed. For snow leopards this could mean study areas of 1000 km^2 or more and at least 30–40 camera stations, considerably more effort than most studies to date.

Camera Spacing

Adequate data rely at least in part on the spatial coverage of the area sampled by camera trapping. For example, the Snow Leopard Network (2014) found a median survey area of 1000 km^2 among studies reporting error-bound density estimates, compared to 158 km^2 among those reporting minimum abundance. To balance the competing model demands for maximizing spatial recaptures as well as unique individuals, cameras should be deployed over a sufficiently large area that nonetheless includes multiple cameras in most potential home ranges. In low-density populations, the polygon delineating a camera trap array should ideally cover multiple male home ranges (Tobler and Powell, 2013), which may be 500 km^2 or greater (McCarthy et al., 2010). Simultaneously, to ensure recaptures, most home ranges should contain multiple camera traps. Assuming a minimum female home range of 50–75 km^2, 4–5 km spacing is likely appropriate in most areas (Sharma et al., 2014).

Camera Placement

Very low densities, large home ranges, and challenging terrain all contribute to the low detection rates typical of snow leopard studies. Investigators can maximize detection by identifying travel routes (ridgelines, narrow gullies, and cliff bases) and marking sites (saddles, prominent boulders or outcroppings, and drainage constrictions) that snow leopards predictably visit (Jackson et al., 2006). Site characteristics should be included as covariates on detection parameters if sample sizes allow. To facilitate identification, investigators should capture images of each animal from multiple angles, by placing either paired cameras on opposite sides of trails or single cameras in locations where animals can be expected to linger (Sharma et al., 2014), or using lure to hold the animal's attention and encourage investigation. Cameras set to capture multiple images, with no delay between triggers, further increase the chance of capturing multiple identifiable markings.

Lures and Baits

Lures and baits are contentious in camera trapping and thus generally avoided in snow leopard studies. Attractants increase the precision of carnivore density estimates by increasing total captures and recaptures, but their influence on accuracy is unknown (Gerber et al., 2012). They may bias estimates by inducing trap responses among subsets of the population or by attracting animals from beyond the trap array. Baiting, which provides a food reward, is particularly problematic, given its high cost and clear potential to modify behavior, and should probably be avoided.

In contrast, lures, such as cologne, catnip oil, or an anchored clump of feathers, are cheap and seemingly unlikely to attract animals from great distances, which could violate geographic closure assumptions. Lure may still induce response biases and heterogeneous detection rates, but it may also help homogenize detection by compensating for natural trap site variability. Lure can increase detection in low-density populations where it is otherwise difficult to obtain adequate data for modeling, or when minimum abundance is the target parameter. The trade-off between precision and unknown accuracy bias should be weighed carefully and conservatively, but given that many snow leopard studies report minimum abundance by necessity, not choice, lure use probably deserves greater consideration.

OVERVIEW OF CAMERA TRAP TECHNOLOGY

Development of Equipment

Observing animals in the wild has been a long fascination to hunters, scientists, and wildlife enthusiasts. The ability to do so has been greatly enhanced with the development of photographic technology. In the 1890s, photographer and conservationist, George Shiras, pioneered the field of camera traps. Shiras used trip wires and a flash bulb to capture photographs of rarely seen animals, which were subsequently published in *National Geographic* (Shiras, 1906).

The first purely scientific use of camera traps was in the 1920s when ornithologist, Frank M. Chapman, used trip wires and bait to survey the species present on Panama's Barro Colorado Island, labeling it a "census of the living" (Chapman, 1927). However, due to technological difficulties, including battery power, equipment size, and trigger mechanism, camera trapping was considered a specialist activity unique to intrepid enthusiasts rather than mainstream researchers for decades (Kucera and Barrett, 1993).

In the late 1980s, remote cameras became popular amongst deer hunters who used them to scout potential hunting grounds for trophy bucks (Kays and Slauson, 2008). Manufacturers scrambled to meet this new demand by combining 35 mm cameras with active and passive infrared motion sensors. Remote cameras quickly transitioned from being bulky, complicated, and expensive to small, simple, and affordable. Biologists finally rediscovered remote cameras in the 1990s having realized that camera trap data can be combined with statistical analyses to answer fundamental ecological questions (Griffiths and van Schaik, 1993; Karanth, 1995; Mace et al., 1994).

Today, camera trapping continues to be transformed by novel technologies. Advanced trigger mechanisms such as miniaturized heat and motion sensors have replaced wires and pressure pads. Invisible infrared flash units provide nighttime black-and-white images without the startling effect of conventional flash. Modern batteries and low power microelectronics allow these devices to operate unsupervised for extended periods in remote locations. The shift from analog/film cameras to digital systems has enabled much greater storage capacity for data, instant viewing of images, and the ability to record metadata that comes with the images. The newest models of digital camera traps are now being integrated with wireless communication networks, such as cellular or satellite, allowing for real-time transmission of camera trap images taken in some of the most remote regions of the world (Olliff et al., 2014).

Assessment of Future Directions

Technological innovations in camera trapping are improving conservation capabilities around the world by facilitating real-time monitoring (Kays et al., 2011), wildlife surveillance (Gula et al., 2010), and public engagement (www.edgeofexistence.org/instantwild/). With global biodiversity in rapid decline, advances in technology are becoming ever more important in helping us solve complex conservation challenges. Understanding this need, conservation organizations are working with a range of experts to develop innovative and integrated technological tools that will revolutionize numerous facets of wildlife conservation ranging from biological monitoring to law enforcement. Through

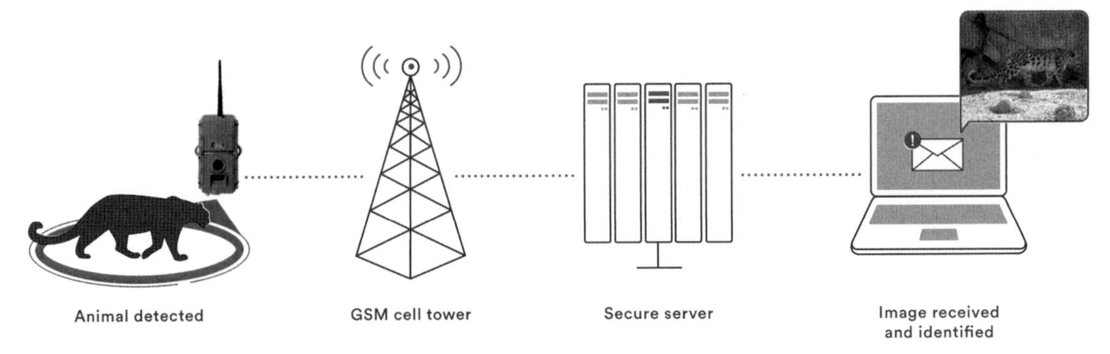

FIGURE 28.2 **Flow diagram showing the data transmission of a photographic image from a GSM camera trap to the recipient.** *Source: Photo courtesy of Snow Leopard Conservancy, India Trust.*

developing and using technology, we can more efficiently and effectively collect, process, and analyze data at different spatial and temporal scales. This in turn can better enable us to identify threats, develop mitigation strategies, and test their effectiveness.

GSM-Based Cameras

The retrieval of camera trap data is often arduous and costly. Threatened species of interest, like the snow leopard, live in remote areas that are difficult to access. It may take several days of driving to reach a village closest to the study area. From there, reaching actual field sites can require additional days of travel on foot.

Certain emerging camera trap models now feature a Group Special Mobile (GSM) module integrated directly into the device. These GSM camera traps require a SIM card from a GSM provider and use existing cell towers to transmit photographs via a cellular network, and a secure online web server to a designated recipient, much like a smartphone (Fig. 28.2). Most GSM camera traps send the photograph roughly 60 s after a picture is taken to an e-mail address or cell phone. A number of GSM camera trap models are available on the market such as the Covert Code Black 3G, Reconyx SC950c, and the HCO Blackout ScoutCam. While GSM camera traps have the potential to revolutionize many aspects of wildlife conservation and monitoring, these cameras have a number of advantages and disadvantages (Table 28.1).

For conservationists in the middle of a global poaching epidemic, GSM cameras may prove to be an invaluable law enforcement tool by enabling park rangers to respond to real-time photographs of intruders in a protected area. While commercially available GSM cameras act as forest sentries by transmitting images of any passing animal or human being, the indiscriminate transmission of all images poses a number of concerns. Park rangers may be inundated by large numbers of photographs, leading to data management issues. Additionally, indiscriminate image transmission can quickly reduce camera battery life while increasing the associated data charges. To address this issue, Panthera, a nonprofit organization dedicated to the conservation of the world's 37 wild cat species, has developed the PoacherCam, a GSM enabled camera trap. The PoacherCam features a human-detection algorithm that enables the camera to differentiate between humans and animals. In areas with GSM coverage, the PoacherCam allows for real-time transfer of human images, for example, after detecting illegal entry of people in a protected area (Olliff et al., 2014). This technology can facilitate law enforcement and anti-poaching efforts by enabling timely and

TABLE 28.1 Advantages and Disadvantages of Using GSM Camera Traps

Advantages	Disadvantages
Cellular coverage is already widespread throughout the world and constantly growing	Telecommunications companies tend not to build cellphone towers in remote wildlife areas
A single tower can provide a circular area of coverage with a radius of up to 35 km under ideal conditions	The signal can be weakened by a variety of factors including terrain (e.g., mountains) and weather (e.g., precipitation)
Cheap hardware costs and data rates	Limited use in areas with low GSM coverage
Cameras can be configured remotely using GSM network	Energy expenditure is high
Real-time image transfer allows for rapid response to incoming information (e.g., poachers)	Network outages can be a common occurrence in developing regions

targeted response to potential threats. Additional algorithms to detect vehicles and boats will further aid law enforcement efforts. Furthermore, algorithms developed to detect specific species of wildlife could potentially mitigate human-wildlife conflict by alerting villagers to the presence of the conflict species. Users of the PoacherCam will have the ability to adjust the settings to fit their specific needs.

Limitations

The use of GSM camera traps as a real-time monitoring and anti-poaching tool may be restricted depending on geography. In areas with poor or no cellular coverage, the GSM camera trap will have the same functionality as a traditional camera trap, which is more affordable. One solution is to create a cellular network in areas where GSM coverage is limited. A number of companies are developing technologies to facilitate cost-effective expansion of cellular coverage into remote areas. For example, NuRAN, a Canadian-based company, is a leading supplier of GSM solutions. Their energy-efficient hardware is encapsulated in a small, rugged box that can be installed in a tall tree or on a hill/mountaintop and run entirely off solar power. NuRAN's technology creates a private cellular network that is simple enough to be managed by individuals with limited technical skills and can be connected to the global GSM system if desired.

GSM Deployment Protocol

When using GSM cameras for remote monitoring and anti-poaching work, it is crucial to understand the cellular signal distribution and strength in the study area. A single cellular tower can provide a signal radius of up to 35 km, and many protected areas have varying degrees of coverage provided by towers located in nearby settlements. Locals who frequently enter the protected area (e.g., park rangers) are an incredible resource for learning about coverage distribution, because they often know which areas have higher signal quality. After determining the areas with the best coverage, it is best to visit those areas with the GSM camera trap to conduct *in situ* trials. Once it is confirmed that the camera has sufficient signal to reliably send an image, it can be placed along a trail and camouflaged. It is important to note that signal quality can fluctuate dramatically and is affected by numerous factors including weather, topography, and foliage. Even if a camera is able to reliably send an image during testing, subsequent sending attempts may fail if the signal weakens.

Satellite-Based Cameras

The Zoological Society of London has developed the world's first satellite-enabled camera trap – the Instant Wild CT1X (http://www.zsl.org/instant-wild-ct1x), which allows near real-time image transmission from virtually anywhere on earth. Designed as an alerting system to strengthen anti-poaching operations and improve the efficiency of biodiversity monitoring in remote locations, this system allows researchers to place cameras in sites that lack cellular coverage.

The cameras are preprogrammed to communicate with a dedicated central satellite node placed within a 1 km range. Up to 10 cameras can connect to one node. Following image capture and transmission from camera to node, images are prioritized and sent via satellite to a secure server to be assessed through dedicated user software. Images are attributable to specific cameras and nodes allowing the user to determine exactly where and when activity occurs. Furthermore, the system combines simple traditional ground sensors using magnetic and seismic activity to trigger alarms, warning of the proximity of humans. These alerts not only provide an early warning mechanism, but are also capable of determining the direction of travel of the poacher and potentially establishing the number of personnel in the hunting party.

With integrated seismic and magnetic sensors and near real-time data transmission, the Instant Wild CT1X has proven to be an effective alerting system and instrumental reconnaissance tool to help tackle animal poaching in East Africa. They have also been used to monitor wildlife in remote and difficult terrain as the satellite transmission together with improved battery life could reduce maintenance visits to just once a year.

Limitations

The satellite-based camera's ability to function in some of the most remote regions of the world is desirable to scientists and park rangers because it overcomes the significant limitation of the GSM camera traps. However, data transmission rates can be prohibitively expensive if sending large amounts of data. Due to high costs, consumer satellite technology is not nearly as popular as consumer cellular technology (Table 28.2). Consequently, there has been significantly less time and money directed toward the commercialization of the hardware needed for satellite communication. Commercially available hardware options are still expensive and bulky, and therefore not feasible for large-scale monitoring studies.

CAMERA TRAP DATA MANAGEMENT

Management of camera trap data is growing increasingly laborious due to the impressive storage capabilities of modern digital cameras (potentially up to 500 gigabytes, depending on the memory card). Despite the accelerated pace in the development of digital image capture, researchers still lack adequate software solutions to process and manage the increasing amount of information in a cost-effective way. Camera trap studies can generate huge quantities of images, usually only some of which are used by investigators. However, it is advised that all photographic data are recorded for all nontarget species, including humans, because this information can be extremely valuable to management and can potentially serve as auxiliary data to predict the target species' presence, distribution, or abundance. Not only are the data likely to be useful to examine interspecific interactions or impacts of human use, but a complete database will also make later analyses much easier, as researchers will not have to labor through the original photographs (Sunarto et al., 2013). Furthermore, a complete database enhances the ability to compare across sites or share data and contribute information to other projects

TABLE 28.2 Comparison Between GSM and Satellite-Based Cameras

Camera Type	Cost	Coverage	Hardware	Range	Availability
GSM	Inexpensive hardware and data rates due to widespread use of cellular networks	GSM coverage is lacking in remote areas, which can be extended, but the process can be complicated	Small cellular module for wireless communication easily integrates directly into the camera	Up to 35 km from cell tower under ideal conditions, but obstacles may shorten range dramatically	Several commercially available models
Satellite	Expensive hardware and data rates as consumer satellite market is still small	Several satellite companies offer global or near-global coverage	Current satellite communication hardware is bulky and must be housed separately from camera; requires either a wired or wireless connection to camera	Only requires a relatively clear view of the sky	Very new technology that is not yet readily available on the market

interested in different species. Consequently, there is a real need for data management systems and robust analytical methods to turn the many images generated into scientifically valid conclusions.

Developments in Image Data Storage and Processing

Digital photographs can already be annotated with metadata such as time, date, and geographical coordinates. Popular storage formats for digital images, such as the Joint Photographic Expert Group (JPEG), also support storage of custom metadata through open standards like Exchangeable Image File Format (EXIF) and Extensible Metadata Platform (XMP). This means all kinds of nonvariable annotations (e.g., species in a photograph) can be stored within the image file itself. However, retrieval of annotated information will require reading and parsing all the files in a collection, which is generally significantly slower than data retrieval from an external database.

File management, data annotation, and data extraction are key components of camera trap data management (Harris et al., 2010; Sundaresan et al., 2011). Traditionally, researchers have developed their own camera trap databases using a spreadsheet application such as Microsoft Excel or Access. However, documenting camera trap data this way is often slow and error-prone resulting in inconsistent labeling and tagging that can complicate data retrieval and sharing (Chaudhary et al., 2010; Harris et al., 2010; Maydanchik, 2007). Furthermore, these solutions are difficult to extend and customize, and lack attributes targeted to a particular study. Consequently, these basic frameworks are limited and cumbersome, and are not designed for targeted ecological camera trap studies, such as capture–recapture type surveys. Realizing these issues, researchers are starting to develop protocols to manage camera trap data in a more efficient way.

Harris et al. (2010) offer a generic method of managing camera trap data by organizing photos into specific folders, labeling species and number of animals captured in each photo, and then generating a text file that lists all file names and the directories within which they are contained. Camera trap data come with a large amount of ancillary data, such as date, time, species, individual ID, and gender, and generic methods, such as those described by Harris et al. (2010), do not offer a way to capture this critical information (Sundaresan et al., 2011). To address these issues, researchers are working with technologists to develop single software programs to make it easier to manage camera trap data and produce information in a cost-efficient way that is relevant to critical conservation questions.

A camera trap data management software program should serve two primary functions: organize files and manage the metadata associated with those individual files (Krishnappa and Turner, 2014). Examples of existing databases specifically formatted to manage camera trap data include DeskTEAM (Fegraus et al., 2011), WWF-Malaysia Camera-Trap Database (2012) and more recently Aardwolf (Krishnappa and Turner 2014). Single software programs such as these greatly reduce data tagging inconsistencies and irrecoverable data losses by managing entire data workflow from file management to data extraction. While these camera trap data management software systems have the ability to format data in a cost-efficient way, it also is important to address the analysis that follows data extraction specific to a study. Currently, more advanced software systems are being developed that have the capability to manipulate the data in order to produce data input files for various statistical analyses.

Camera Base (Tobler, 2010) is an existing tool that helps biologists manage the complete data from multiple camera trap surveys and provides tools for different types of data analysis including capture–recapture, occupancy, activity patterns, and diversity. However, a significant limitation is that Camera Base is not portable, such that it only runs on the Windows operating system with a database backend restricted to Microsoft Access, which has limited file capacity (Krishnappa and Turner, 2014). Panthera is developing a similar tool called the Camera Trap Research System (CTRS). In CTRS the image attribute data are stored as metadata in the header of the image files. Through the use of the .XMP standard, the investigator has complete flexibility in defining the image attributes of interest. Data security and access permissions are a key part of the system architecture. Utilities are provided to import the image data from the cameras, define independent capture events, tag the image attributes, and assist in individual identification using WildID pattern matching.

Once tagging is complete, the survey is published on the CTRS centralized server, which has unparalleled file storage capacity. On the server the XMP metadata are read from the image files and stored in a database along with references to the original image. The user can then perform arbitrary database queries or execute a script to create the input files necessary for a specific analysis, with spatially explicit capture recapture and SPACECAP to estimate animal densities and population size, as an example. A web interface provides third parties access the data in the CTRS database with all of the same security and permissions employed by the CTRS user itself. An important goal for any camera trap database management system is to standardize the collection and storage of data ultimately permitting analyses across surveys and provide an automated set of analysis tools reducing the workload of anyone leading a camera trap survey.

FUTURE DIRECTIONS IN TECHNOLOGY

Today, cameras are smaller, lighter, more energy efficient and affordable than their predecessors, ultimately making camera trapping more environmentally friendly and logistically feasible. With such developments, it is likely that camera traps will become better integrated with other data collection tools to record more detailed biological, climatic, and other environmental parameters. Another potential development is the use of three-dimensional imaging with multiple lenses (Moynihan, 2010). Theoretically, a single camera with multiple lenses linked to the unit via a wireless connection would enable an animal to be photographed from different angles at the same time. This would allow the user to create a three-dimensional model of the image subject, thus facilitating the identification of species, individuals, and/or physical condition.

Many of the technological advances in the gaming and cellular world can be applied to camera trap sensors, allowing the human body to be scanned and its movements recorded and analyzed. Similar technologies used in combination with existing databases and software suites might enable future camera traps to automatically identify species, individuals, and gender, measure body mass, describe general physiological condition, and characterize movement. Eventually such new technologies will become more accessible and economical. The last few years have also seen the development in camera trapping-related software for data management (Harris et al., 2010; Krishnappa and Turner, 2014; Tobler, 2010), including individual identification (Kelly, 2001; Hiby et al., 2009) and analytical software (Gopalaswamy et al., 2012).

While the developments in camera trap technology and camera trap data management systems are significantly improving our ability to monitor wildlife and mitigate threats, they also come with limitations that may have an impact on study designs. It is essential to choose the relevant equipment to collect the data needed, because not all technology is suitable for a specific research project (Trolliet et al., 2014). Subsequently, the choice of camera trap equipment should be appropriate for the purposes of the study and the data need to be managed appropriately.

References

Chapman, F.M., 1927. Who treads our trails? Natl. Geogr. 52, 330–345.

Chaudhary, V.B., Walters, L.L., Bever, J.D., Hoeksema, J.D., Wilson, G.W.T., 2010. Advancing synthetic ecology: a database system to facilitate complex ecological meta-analyses. ESA Bull. 91, 235–243.

Efford, M.G., 2011. Secr: Spatially explicit capture-recapture models in R. Development Core Team, R., Package, 2.0.0., Edition, R., Foundation for Statistical Computing, Vienna, Austria,

Fegraus, E.H., Lin, K., Ahumada, J.A., Baru, C., Chandra, S., Youn, C., 2011. Data acquisition and management software for camera trap data: a case study from the TEAM Network. Ecol. Informatics 6, 345–353.

Gerber, B.D., Karpanty, S.M., Kelly, M.J., 2012. Evaluating the potential biases in carnivore capture-recapture studies associated with the use of lure and varying density estimation techniques using photographic-sampling data of the Malagasy civet. Popul. Ecol. 54, 43–54.

Gopalaswamy, A.M., Royle, J.A., Hines, J.E., Singh, P., Jathanna, D., Kumar, N.S., Karanth, K.U., 2012. Program SPACECAP: software for estimating animal density using spatially explicit capture-recapture models. Methods Ecol. Evol. 3, 1067–1072.

Griffiths, M.G., van Schaik, C.P., 1993. Camera-trapping: a new tool for the study of elusive rain forest animals. Trop. Biodivers. 1, 131–135.

Gula, R., Theuerkauf, J., Rouys, S., Legault, A., 2010. An audio/video surveillance system for wildlife. Eur. J. Wildl. Res. 56, 803–807.

Harris, G., Thompson, R., Childs, J.L., Sanderson, J.G., 2010. Automatic storage and analysis of camera trap data. Bull. Ecol. Soc. Am. 91, 352–360.

Hiby, L., Lovel, P., Patil, N., Kumar, N.S., Gopalaswamy, A.M., Karanth, K.U., 2009. A tiger cannot change its stripes: using a three-dimensional model to match images of living tigers and tiger skins. Biol. Lett. 5, 383–386.

Janečka, J.E., Munkhtsog, B., Jackson, R.M., Naranbaatar, G., Mallon, D.P., Murphy, W.J., 2011. Comparison of noninvasive genetic and camera-trapping techniques for surveying snow leopards. J. Mammal. 92, 771–783.

Jackson, R.M., Roe, J.D., Wangchuk, R., Hunter, D.O., 2006. Estimating snow leopard population abundance using photography and capture-recapture techniques. Wildl. Soc. Bull. 34, 772–781.

Karanth, K.U., 1995. Estimating tiger *Panthera tigris* populations from camera-trap data using capture recapture models. Biol. Conserv. 71, 333–338.

Kays, R.W., Slauson, M., 2008. Remote cameras. In: Long, R.A., MacKay, P., Zielinski, W.J., Ray, J.C. (Eds.), Noninvasive Survey Methods for Carnivores. Island Press, Washington, DC, pp. 110–140.

Kays, R., Tilak, S., Kranstauber, B., Jansen, P.A., Carbone, C., Rowcliffe, J.M., Fountain, T., Eggert, J., He, Z., 2011. Monitoring wild animal communities with arrays of motion sensitive camera traps. Intl. J. Res. Rev. Wireless Sensor Net. 1, 19–29.

Kelly, M.J., 2001. Computer-aided photograph matching in studies using individual identification: An example from Serengeti cheetahs. J. Mammal. 82, 440–449.

Krishnappa, Y.S., Turner, W.C., 2014. Software for minimalistic data management in large camera trap studies. Ecol. Informat. 24, 11–16.

Kucera, T.E., Barrett, R.H., 1993. In my experience: the trailmaster camera system for detecting wildlife. Wildl. Soc. Bull. 21, 505–508.

Mace, R.D., Minta, S.C., Manley, T.L., Aune, K.E., 1994. Estimating grizzly bear population size using camera sightings. Wildl. Soc. Bull. 22, 74–82.

MacKenzie, D.I., Nichols, J.D., Royle, J.A., Pollock, K.H., Bailey, L.L., Hines, J.E., 2006. Occupancy Estimation and Modeling: Inferring Patterns and Dynamics of Species Occurrence. Elsevier, Amsterdam.

Maydanchik, A., 2007. Data quality Assessment. Technics Publications, Bradley Beach, NJ, USA.

McCarthy, K.P., Fuller, T.K., Ming, M., McCarthy, T.M., Waits, L., Jumabaev, K., 2008. Assessing estimators of snow leopard abundance. J. Wildl. Manage. 72, 1826–1833.

McCarthy, T., Murray, K., Sharma, K., Johansson, O., 2010. Preliminary results of a long-term study of snow leopards. Cat News 53, 15–19.

Meek, P.D., Ballard, G., Claridge, A., Kays, R., Moseby, K., O'Brien, T., O'Connell, A., Sanderson, J., Swann, E., Tobler, M., Townsend, S., 2014. Recommended guiding principles for reporting on camera trapping research. Biodivers. Conserv. 23, 2321–2343.

Moynihan, T., 2010. Fujifilm unveils its second-generation 3D camera. In: Fox, S. (Ed.), PC World. International Data Group (IDG), Boston, USA.

Nichols, J.D., Karanth, K.U., O'Connell, A.F., 2011. Science, conservation, and camera traps. In: O'Connell, A.F., Nichols, J.D., Karanth, K.U. (Eds.), Camera Traps in Animal Ecology: Methods and Analysis. Springer, New York, pp. 45–46.

Olliff, E.R.R., Cline, C.W., Bruen, D.C., Yarchuk, E.J., Pickles, R.S.A., Hunter, L., 2014. The Panthercam: a camera trap optimized for the monitoring of wild felids. Wild Felid Monitor 7, 21–23.

Rowcliffe, J.M., Field, J., Turvey, S.T., Carbone, C., 2008. Estimating animal density using camera traps without the need for individual recognition. J. Appl. Ecol. 45 (4), 1228–1236.

Royle, J.A., Chandler, R.B., Sollman, R., Gardner, B., 2014. Spatial Capture-Recapture. Academic Press, New York.

Russell, R.E., Royle, J.A., DeSimone, R., Schwartz, M.K., Edwards, V.L., Pilgrim, K.P., McKelvey, K.S., 2012. Estimating abundance of mountain lions from unstructured spatial sampling. J. Wildl. Manage. 76, 1551–1561.

Sharma, K., Bayrakcismith, R., Tumuruskh, L., Johansson, O., Sevger, P., McCarthy, T., Mishra, C., 2014. Vigorous dynamics underlie a stable population of the endangered snow leopard *Panthera uncia* in Tost Mountains, South Gobi, Mongolia. PLoS ONE 9, 1–10.

Shiras, G., 1906. One seasons' game bag with a camera. Natl. Geogr. 17, 366–423.

Snow Leopard Network, 2014. Snow Leopard Survival Strategy revised 2014 Version. Snow Leopard Network, Seattle. Available from: http://www.snowleopardnetwork.org/

Sollman, R., Mohamed, A., Samejima, H., Wilting, A., 2013. Risky business or simple solution – relative abundance indices from camera-trapping. Biol. Conserv. 159, 405–412.

Sunarto, Sollmann, R., Mohamed, A., Kelly, M.J., 2013. Camera trapping for the study and conservation of tropical carnivores. Raffles B. Zool. 28, 21–42.

Sundaresan, S.R., Riginos, C., Abelson, E.S., 2011. Management and analysis of camera trap data: alternative approaches (response to Harris et al., 2010). Bull. Ecol. Soc. America 92, 188–195.

Tobler, M., 2010. Camera Base. San Diego Zoo Institute for Conservation Research, San Diego, USA. www.atrium-biodiversity.org/tools/camerabase (accessed 28.12.2014).

Tobler, M.W., Powell, G.V.N., 2013. Estimating jaguar densities with camera traps: problems with designs and recommendations for future studies. Biol. Conserv. 159, 109–118.

Trolliet, F., Huynen, M.-C., Vermeulen, C., Hambuckers, A., 2014. Use of camera traps for wildlife studies. A review. Biotech. Agron. Soc. Environ. 18, 446–454.

WWF-Malaysia Camera-Trap Database 2012. Toolbox update 5 Camera trapping database. http://myrimba.org/2012/01/05/toolbox_update_5/ (accessed 10.02.2015).

CHAPTER

29

Landscape Ecology: Linking Landscape Metrics to Ecological Processes

Hugh S. Robinson, Byron Weckworth***

*Landscape Analysis Laboratory, Panthera, New York; College of Forestry and Conservation, University of Montana, Missoula, MT, USA

**Snow Leopard Program, Panthera, New York, NY, USA

"Biological conservation requires a rigorous scientific foundation, which landscape ecology should seek to provide." –*John A. Wiens (2002)*

INTRODUCTION

If ecology is the study of distribution and abundance (sensu Krebs, 2009), landscape ecology is simply the addition of a spatial or geographic component to the study of distribution and abundance. In essence, landscape ecology is a search for patterns, as well as the causes and consequence of those patterns (Fortin and Dale, 2005). Research into the landscape ecology of snow leopards is, essentially, still in its infancy. At the time when the first cougars were being radio-collared in Idaho (Seidensticker et al., 1973), *National Geographic* (1971) was just publishing the first photographs ever taken of a snow leopard (*Panthera uncia*). The nascent state of snow leopard research, as well as the inherent logistical challenges in accessing snow leopard range, has precluded the advancement of our understanding of snow leopard ecology, as compared to other large felids. As such, much may be learned by considering the techniques and findings of research conducted on other cat species such as leopards and cougars, as well as other large carnivores.

Conservation biology, historically, was largely focused at the species level; however, biodiversity is best managed at broad spatial (i.e., landscape) and temporal (i.e., generational) scales where processes such as disturbance, dispersal, and evolution itself occur (Hobbs, 1998). Both conservation biology and landscape ecology have evolved to embrace this idea. The bridge between single species programs and holistic ecosystem level efforts is provided by focal species such as large carnivores

Snow Leopards
http://dx.doi.org/10.1016/B978-0-12-802213-9.00029-8

(Caro, 2010; Lambeck, 1997). The natural history characteristics of snow leopards and other large carnivores make them simultaneously prone to extinction and excellent ambassadors for ecosystem or landscape level conservation interventions as keystone or umbrella species (Cardillo et al., 2004; Ray et al., 2005). This duality makes the study of large carnivores important not only to singular species conservation efforts, but also the protection of the ecosystems within which they live (Cardillo et al., 2004). Landscape ecology methods then provide a scientific line of inquiry with which to address these issues.

Several texts describe landscape ecology and spatial analysis. For a more detailed discussion, readers may refer to the expansive texts of Fortin and Dale (2005) or Turner et al. (2001). Additionally, MacArthur and Wilson (1967) provide an understanding of what are arguably the roots of landscape ecology. For the purposes of this chapter, we provide a review of basic considerations and methods specific to our understanding of landscape ecology and how they can be applied to snow leopards.

SCALE

We cannot discuss anything spatial without first addressing the issue of scale. "Scale" can mean different things to different people and is often used without much forethought. We are all familiar with the scale bar on a map. In that sense, scale refers to the ratio of distance on a map to distance on the ground (i.e., 1:50,000). The term "broad scale" is often used concerning geographically large studies; in this sense scale is a synonym for extent. Some may refer to pixel size of a GIS layer as scale (i.e., Landsat image pixels are 30 m × 30 m, or 900 m^2), however, this is more accurately described as resolution or grain. In landscape ecology, scale refers to the geographic or temporal resolution of an ecological process, and the data collected to describe that process; in this sense it is characterized by both grain and extent (Wiens and Milne, 1989) (Fig. 29.1). In the study of snow leopards, we often work on a hierarchy of scales ranging from a single home range, to a protected reserve, or even the entire distribution of the species within a range state or region.

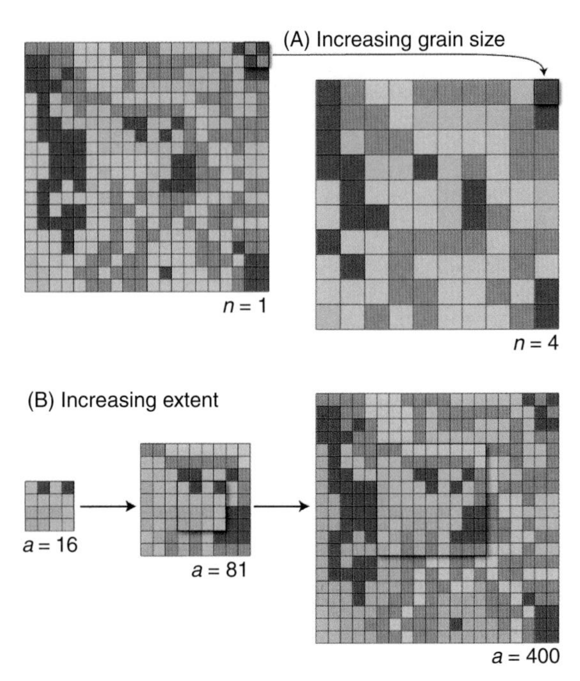

FIGURE 29.1 **The relationship of grain and extent in determining scale.** *Adapted from Turner et al. (1989).*

Choosing the Correct Scale

Spatial patterns vary with scale (Turner et al., 1989). While it is easy to imagine this in terms of spatial configurations (Fig. 29.1), it is perhaps less common for researchers to examine how temporal scale may affect our measurement of ecological processes. For example, most ecologists would suggest that migration is a seasonal movement where the animal returns to its starting location. Migration is differentiated from dispersal where individuals establish a new home range and do not return. If the sample unit is the gene rather than an individual,

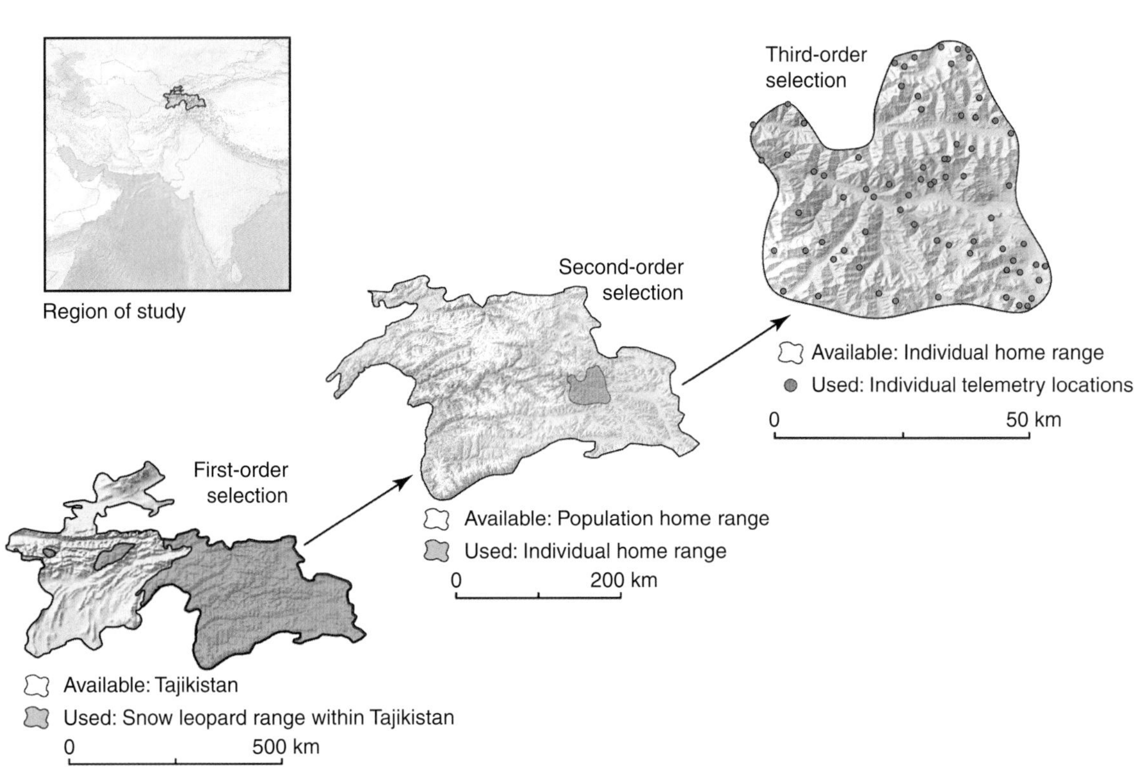

FIGURE 29.2 **Graphical depiction of selection scales following Johnson (1980).** *Adapted from Decesare et al. (2012).*

emigration of an animal in its lifetime becomes migration when the temporal scale is increased to account for gene flow and the return of genetic material to the original population. Study at this scale is called landscape genetics (Manel et al., 2003), where analysis combines principles from landscape ecology and population genetics (see Chapter 27.1). Time and space are both important considerations when describing the "natural" mosaic of disturbance observed on the landscape. For more on temporal scale, the reader may wish to see Bissonette and Storch (2007).

In a spatial context, Johnson (1980) provided four basic scales of resource selection that provide a great starting point when considering analysis and study design. He suggested a "natural ordering of selection"; first-order selection is the geographical range of a species, second-order is location of an individual's home range, third-order describes usage within an individual's home range, and fourth-order describes use at the patch or foraging level. These scales should suffice for most ecological applications; moreover, any analysis should identify scale at one of these widely accepted levels. Figure 29.2 depicts these levels as they may pertain to snow leopards.

Choosing the correct scale is not as simple as looking at the dimensions of your study area. Careful consideration needs to be taken regarding the proper temporal and spatial scale at which to measure the ecological question of interest (Bissonette, 2013). The scale of an analysis for snow leopards can be anything from a home range within the Tost Mountains of Mongolia, to the extent of the cat's range across Asia, but it must be defined *a priori*, and it must match the management objective or ecological question being asked. A mismatch of scale will lead to spurious results, hindering, rather than helping, conservation. The following example demonstrates how confusing scale and hypothesis can lead to incoherent conclusions.

A recent paper on wolf depredations in western North America suggested that destroying wolves that killed livestock increased conflict (Wielgus and Peebles, 2014). If true, this would have major management implications as the standard policy of removing conflict wolves would appear to exacerbate the problem. However, a closer inspection of the geographic scale used by Wielgus and Peebles (2014) calls their results into question. Their conclusion was based on the number of wolves killed in year x and the number of livestock depredated in year $x + 1$; however, they conducted their analysis at two levels, first by individual state and second by pooling across all states analyzed (Montana, Idaho, and Wyoming, an area of approximately 850,000 km^2). Both levels reflect a landscape scale (second-order) analysis, when one might expect influences of wolf control to be limited to a home range scale (third-order). Conducting the analysis at a landscape versus home range scale meant that the authors were expecting wolf removal in Idaho to reduce conflict in Wyoming. This is akin to suggesting that snow leopard-livestock conflict measures prescribed in Ladakh, India, would impact snow leopard conflict in the Pamirs of Tajikistan.

Levin (1992) states, "That there is no single correct scale or level at which to describe a system does not mean that all scales serve equally well or that there are not scaling laws." However, it is mostly a matter of carefully considering the behavior or system you are trying to quantify. An excellent example is provided by Squires et al.'s (2013) examination of Canada lynx (*Lynx canadensis*) dispersal corridors. They combined an analysis of home range characteristics (second-order resource selection) and step selection functions (third-order resource selection) to predict connectivity in western Montana. For further readings on how choice of scale may affect your research, see Turner et al. (2001), Dungan et al. (2002), and perhaps most importantly Wu et al. (2002).

Scaling Up

Most wildlife data we collect are point data. Tracks, scats, camera-trap photos, and telemetry points all represent a single location at a single point in time. These single events are of little use alone and we therefore aggregate them in order to elucidate meaningful patterns. Although methods exist for using presence data alone (e.g., MAXENT, Phillips et al., 2006), normally presence observations are compared to areas where animals were not located, or went unobserved, in order to make inferences regarding the impact of a spatial attribute on a behavior, individual, or an entire population. This type of comparison is at the root of many of the tools used by landscape ecologists. For instance, in spatially explicit Cox survival models (Hosmer et al., 2008; Cox, 1972), as well as resource selection and occupancy modeling (Manly et al., 2002; Mackenzie et al., 2006), spatial covariates where animals are observed are compared to areas where animals were not observed. A powerful tool for such comparisons is multiple logistic regression (Hosmer and Lemeshow, 2000).

Used and Unused Points – Resource Selection Functions

Logistic regression is the comparison of a binary condition, such as on/off, alive/dead, or used/unused (Hosmer and Lemeshow, 2000). Choosing the correct scale of analysis is never more important than with the use of logistic regression. Used point locations are simply those where the animal was observed. However, how random unused locations are placed on the landscape for comparison is less clear and will dramatically affect the model's output (Boyce, 2006; Johnson et al., 2006; Northrup et al., 2013). Referring back to Johnson's hierarchy of spatial scales (Fig. 29.2), to test a hypothesis regarding landscape-level resource selection of snow leopards you could compare individual used locations to available locations randomly generated within a study area polygon. If you are considering the use of telemetry data for modeling habitat use, the publications by Aarts et al. (2008), Beyer et al. (2010), Fieberg et al. (2010), and Barbet-Massin et al. (2012), as well as those mentioned earlier, are excellent references.

Grids and Occupancy Models

A second method of "scaling up" is by aggregating spatial covariates within a specified area. An example popular in the study of snow leopards is the use of camera trap data for occupancy modeling. As with telemetry data, we must aggregate to scale up, however, in this case the single point is instead considered representative of a snow leopard being present within a given area or cell. A grid is placed over the study area and spatial covariates within each grid cell are summarized. The assumption is that some characteristic within that grid cell is causal to the animal's presence or absence, and characteristics of cells where the animal was present are contrasted with those where the animal was absent, again using logistic regression.

Occupancy modeling works best for species with discrete spatial extents or units (i.e., a pond in the case of amphibians or a tree cavity in the case of a cavity-nesting tree species) that will act as the grid cell. In such cases, occupancy can reflect species distribution and also population trends calculated as site colonization and extinction rates in multiseason surveys (Mackenzie et al., 2003). In the study of snow leopards, and other highly mobile large carnivores, no specific geographic feature equates to the aforementioned ponds; where the sample unit is an entire population. While ponds are ephemeral and populations of amphibians may vary greatly between years, tracking populations of snow leopards using occupancy modeling would need a longer time period and a much broader scale (it would be daunting, but interesting, to combine presence/absence data from across snow leopard range in a grand occupancy model).

Several assumptions must be met when using occupancy models, two of which complicate their use for snow leopards and other large carnivores: (i) occupancy status at each site does not vary over the survey period (i.e., the site is closed), and (ii) detection of the species and detection histories at each location are independent (i.e., the same individual cannot occupy more than one site) (Mackenzie et al., 2006). To meet these assumptions when studying highly mobile animals, we impose a grid over our area of interest, with the grid cell now replacing the cavity or the pond. To account for our assumption that animals are not overlapping between grid cells, we try to make one grid cell equate to a single animal by making the grid cell size the same as a home range or larger, and limiting the study period (Bailey et al., 2014; Hines et al., 2010). The few published telemetry studies of snow leopards suggest complex spatial use with considerable overlap among individual home ranges, as well as periodic long-distance movements (Jackson and Ahlborn, 1989; Mccarthy et al., 2005). The likelihood that our grid cells will exactly match the boundaries of inviolate home ranges, thus satisfying the cell independence assumption, is low. Hence, as animals move across our grid we are now measuring intensity of use rather than occupancy. By violating the occupancy closure assumption, which is likely in snow leopards, we end up measuring habitat use and lose the ability to track the population, but can still determine what habitat covariates influence occupancy.

QUANTIFYING SPATIAL COVARIATES

The collection and quantification of spatial covariates is going through an exciting period with new sources of data coming in at both ends of the spatial scale. This is particularly true for the study of snow leopards whose habitat is exceedingly hard to access. In February 2013, the National Aeronautics and Space Administration (NASA) launched the Landsat program's eighth satellite, whose imagery became publicly available that year. The free dissemination of spectral data collected from the Landsat and MODIS (Moderate Resolution Imaging Spectroradiometer) programs, combined with new developments like Google Earth Engine, is making wide-scale analyses more feasible (e.g., Hansen et al., 2013). At the other end of the spectrum of scale, drones can collect highly detailed data over small areas. For information on acquisition and use of remotely sensed data, see Kerr and Ostrovsky (2003), Evans et al. (2005), Rose et al. (2015) and especially Pettorelli et al. (2014) announcing the launch of the journal *Remote Sensing in Ecology and Conservation*.

Deciding which sources you should look to for your study relies on the question you're asking, scale (again), and how transferable you wish your results to be. For instance the European Space Agency's GlobCover project (http://due.esrin.esa.int/page_globcover.php) provides a standard land cover map of the entire planet. A map of this scale is full of inaccuracies, but its use of 63 standardized land cover classes allows for easy comparison to future studies and/or studies from different geographic zones (Rondinini et al., 2011). The loss of accuracy at a local scale is offset by the ability to make comparisons across a broad scale (think Afghanistan to Nepal). An even better approach may be to avoid categorical land classifications all together (see subsequent section).

Matching Precision to Scale

Many researchers believe that for spatial data, fine grain size and high resolution are best. With 2.5-m^2 resolution, SPOT imagery (Airbus Defense and Space, Toulouse France) is beautiful to look at, but for most ecological applications, especially those centered on large carnivores such as snow leopards, that level of precision is unnecessary. Few if any conservation actions occur at that fine of scale. Very high-resolution imagery is only justified at very small scales (i.e., Johnson's fourth-order), too high of resolution wastes resources both in terms of finance and computing power. Just as wildlife species should be managed at the proper scale, the resolution of your data should match the resolution of your question.

If the precision of your data is low, increasing the scale of your analysis can help account for the "noise" or uncertainty in your data. An excellent example is the North American breeding bird survey conducted each June (Sauer et al., 2013) where citizen scientists record observations at specified intervals. Although data are collected at a grain size of 16 ha, analysis is conducted at a landscape or even continental scale (e.g., Bled et al., 2013; Karanth et al., 2014). This increases the number of replicates per sample unit, reduces the effect of erroneous or missed observations (Tyre et al., 2003) and compensates for a substantial variability in observer experience, skill, and reliability. This is particularly relevant to the study of snow leopards as use of citizen scientists (from Buddhist monks to ecotourists) grows. The effect of observer biases can be reduced by increasing the scale of an analysis to a landscape scale.

Categorical Versus Continuous Variables

Another consideration when quantifying spatial data is that of categorical versus continuous variables. Although it is easier to convey to a general audience that an animal favors a specific habitat type (e.g., pine forests or desert), it is likely better biologically to avoid categorical variables. Imposing a subjective habitat classification scheme can lead to false conclusions of preference or avoidance if animals are actually responding to something not captured in our categorization. Also, we run the risk of making discrete something that is a gradient (Mcgarigal and Cushman, 2005). For example, rather than three categorical levels of forest (open, closed, pine), percent canopy closure (0–100) yields a continuous covariate that is objective and allows greater flexibility in the data.

APPLYING LANDSCAPE ECOLOGY TO SNOW LEOPARDS

Research into the landscape ecology of snow leopards is in its infancy. Published literature focuses on food habits, human conflicts, and what Nichols and Williams (2006) called surveillance monitoring; namely monitoring programs that are not driven by *a priori* hypotheses. Counting snow leopards and monitoring their population trends is extremely important for conservation; however, to truly inform conservation we require a better understanding of the mechanisms that determine these population trends (reproduction, mortality, and dispersal) and the landscapes that shape these processes.

Camera trap data from multiyear studies can be used to quantify survival in snow leopards. Yet, in spite of the near ubiquitous use of camera traps in snow leopard research, to our knowledge only Sharma et al. (2014) have published an analysis of population parameters (i.e., survival, emigration, and population growth). Population models of other cat species with similar reproduction rates show population growth is most sensitive to adult female survival. To maintain a stable population, an adult female survival rate of approximately 0.8 or above is required (Robinson et al., 2014; Chapron et al., 2008). Survival estimates are possible using a robust mark-recapture design; however, confidence intervals are large and calculate only *apparent* survival, not *true* survival probability, and are not able to describe causes of mortality or delineate hazardous landscapes.

Cause-specific mortality is arguably the most valuable data for conservation programs, but is only obtained from known fate data (i.e., radio collar data), ideally based on cumulative incidence functions (Pintilie, 2006). Murray (2006) provides an excellent review of telemetry-based survival estimation. Ruth et al. (2011) combined known fate and landscape data to produce a "survival surface," mapping the probability of adult female cougar survival in the greater Yellowstone ecosystem. A similar exercise could prove invaluable for snow leopard conservation to help reduce human conflicts and ultimately snow leopard mortality.

Dispersal is a fundamental process in the expansion and persistence of wildlife species (Clobert et al., 2001; Bullock et al., 2002). Examples of dispersal and movements between habitats of varying quality, and the effects of these source-sink dynamics on population persistence, exist for a wide variety of species, but have not been addressed for snow leopards. Data on the dispersal patterns of snow leopards are a fundamental requirement for modeling source-sink and metapopulation dynamics.

Metapopulation theory as applied to large cats has been examined for African lions (*Panthera leo*) (Dolrenry et al., 2014), tigers (*Panthera tigris*) (Sharma et al., 2013; Wikramanayake et al., 2004), leopards (Dutta et al., 2013), and cougars (Andreasen et al., 2012) and could serve as good models for snow leopard investigations. Empirical works fall into categories of either genetic or telemetry-based data. Telemetry studies can provide data on core habitat use and dispersal (Fattebert et al., 2015) as well as the survival and fecundity estimates required for population modeling. Genetic data can provide estimates of gene flow and genetic drift. The combination of these data describe the rapidly expanding field of landscape genetics, which seeks to understand how landscape features affect spatial genetic structure and functional connectivity (Manel et al., 2003). An understanding of snow leopard landscape genetics is particularly relevant as they dwell in high mountain ecosystems that are naturally fragmented and often only narrowly connected. The added fragmentation of their habitat through anthropogenic causes may further impede important corridors for gene flow and thus the maintenance of adequate connectivity and genetic diversity.

Several papers on snow leopard food habits indicate considerable overlap between snow leopards and other carnivores (Jumabay-Uulu et al., 2014; Wang et al., 2014). The work of Johansson et al. (2015) on snow leopard predation is an invaluable contribution. As they state in their introduction "Knowledge of how [various factors] influence predation on natural prey or livestock can allow for more efficient mitigation efforts by directing them to the time periods and areas where predation is most likely to occur."

Snow leopards respond to prey availability as all carnivores do. In turn, their ungulate prey respond to primary productivity (i.e., photosynthesis), which is easily observed through satellite remote sensing products such as the enhanced vegetation index (EVI), normalized difference vegetation index (NDVI), and net primary productivity (NPP) – all MODIS products freely available (http://earthobservatory.nasa.gov/Features/MeasuringVegetation/). Given the barren nature of most snow leopard habitat, it may be tempting to conclude that plant growth provides little in terms of delineating their habitat preferences or spatial movements. On the contrary, exploration of spatial and temporal variation of this limiting resource could provide invaluable insight into the distribution of snow leopards and their prey, both natural and domestic (Pettorelli et al., 2005; Kerr and Ostrovsky, 2003).

CONCLUSIONS

Across their range, snow leopard basic ecology and behavior has only begun to emerge. Although attempts at describing the local distributions of the cat have been ongoing since the latter half of the nineteenth century (e.g., Gee, 1968), only in the past couple of decades have scientists begun to describe snow leopard ecology (first published telemetry study, Jackson and Ahlborn, 1989). The number of published works on snow leopards is increasing, but most studies are focused on diet (e.g., Lyngdoh et al., 2014; see Chapter 27.2), human-wildlife conflict (e.g., Suryawanshi et al., 2014; see Chapter 5), and climate change (e.g., Forrest et al., 2012; see Chapter 8). This research contributes to the body of knowledge that will enable successful conservation endeavors; however, a lack of understanding of snow leopard landscape ecology remains. Initiating conservation research is necessary in order to begin filling the knowledge gaps of population dynamics and landscape effects of snow leopards.

Acknowledgments

The authors wish to thank Nathaniel Robinson, University of Montana, for production of the figures for this chapter.

References

Aarts, G., Mackenzie, M., McConnell, B., Fedak, M., Matthiopoulos, J., 2008. Estimating space-use and habitat preference from wildlife telemetry data. Ecography 31, 140–160.

Andreasen, A.M., Stewart, K.M., Longland, W.S., Beckmann, J.P., Forister, M.L., 2012. Identification of source-sink dynamics in mountain lions of the Great Basin. Mol. Ecol. 21, 5689–5701.

Bailey, L.L., Mackenzie, D.I., Nichols, J.D., 2014. Advances and applications of occupancy models. Methods Ecol. Evol. 5, 1269–1279.

Barbet-Massin, M., Jiguet, F., Albert, C.H., Thuiller, W., 2012. Selecting pseudo-absences for species distribution models: how, where and how many? Methods Ecol. Evol. 3, 327–338.

Beyer, H.L., Haydon, D.T., Morales, J.M., Frair, J.L., Hebblewhite, M., Mitchell, M., Matthiopoulos, J., 2010. The interpretation of habitat preference metrics under use-availability designs. Philos. T. Roy. Soc. B. 365, 2245–2254.

Bissonette, J.A., 2013. Scale in wildlife management: the difficulty with extrapolation. In: Krausman, P.R., Cain, J.W.

(Eds.), Wildlife management and conservation: Contemporary principles and practices. The John Hopkins University Press, Baltimore, MY.

Bissonette, J.A., Storch, I., 2007. Temporal dimensions of landscape ecology: wildlife responses to variable resources. Springer, New York, NY.

Bled, F., Sauer, J., Pardieck, K., Doherty, P., Royle, J.A., 2013. Modeling trends from North American breeding bird survey data: A spatially explicit approach. PloS One 8, 14.

Boyce, M.S., 2006. Scale for resource selection functions. Divers. Distrib. 12, 269–276.

Bullock, J.M., Kenward, R.E., Hails, R.S. (Eds.), 2002. Dispersal Ecology: The 42nd Symposium of the British Ecological Society held at the University of Reading, 2–5 April 2001. first ed. Blackwell, Malden, MA.

Cardillo, M., Purvis, A., Sechrest, W., Gittleman, J.L., Bielby, J., Mace, G.M., 2004. Human population density and extinction risk in the world's carnivores. PloS Biol. 2, 909–914.

Caro, T.M., 2010. Conservation by proxy: indicator, umbrella, keystone, flagship, and other surrogate species. Island Press, Washington, DC.

Chapron, G., Miquelle, D.G., Lambert, A., Goodrich, J.M., Legendre, S., Clobert, J., 2008. The impact on tigers of poaching versus prey depletion. J. Appl. Ecol. 45, 1667–1674.

Clobert, J., Danchin, E., Dhondt, A.A., Nichols, J.D., 2001. Dispersal. Oxford University Press, Oxford; New York.

Cox, D.R. 1972. Regression models and life-tables. J. R. Stat. Soc. Ser. B Stat. Methodol. 34, 187–220.

Decesare, N.J., Hebblewhite, M., Schmiegelow, F., Hervieux, D., McDermid, G.J., Neufeld, L., Bradley, M., Whittington, J., Smith, K.G., Morgantini, L.E., Wheatley, M., Musiani, M., 2012. Transcending scale dependence in identifying habitat with resource selection functions. Ecolog. Applicat. 22, 1068–1083.

Dolrenry, S., Stenglein, J., Hazzah, L., Lutz, R.S., Frank, L., 2014. A Metapopulation Approach to African Lion (*Panthera leo*) Conservation. PloS One 9, 9.

Dungan, J.L., Perry, J.N., Dale, M.R.T., Legendre, P., Citron-Pousty, S., Fortin, M.J., Jakomulska, A., Miriti, M., Rosenberg, M.S., 2002. A balanced view of scale in spatial statistical analysis. Ecography 25, 626–640.

Dutta, T., Sharma, S., Maldonado, J.E., Wood, T.C., Panwar, H.S., Seidensticker, J., 2013. Gene flow and demographic history of leopards (*Panthera pardus*) in the central Indian highlands. Evolution. Applicat. 6, 949–959.

Evans, K.L., Warren, P.H., Gaston, K.J., 2005. Species-energy relationships at the macroecological scale: a review of the mechanisms. Biolog. Rev. 80, 1–25.

Fattebert, J., Robinson, H.S., Balme, G., Slotow, R., Hunter, L., 2015. Structural habitat predicts functional dispersal habitat of a large carnivore: how leopards change spots. Ecolog. Applicat. 25, 1911–1921.

Fieberg, J., Matthiopoulos, J., Hebblewhite, M., Boyce, M.S., Frair, J.L., 2010. Correlation and studies of habitat selection: problem, red herring or opportunity? Phil. Trans. R. Soc. B 365, 2233–2244.

Forrest, J.L., Wikramanayake, E., Shrestha, R., Areendran, G., Gyeltshen, K., Maheshwari, A., Mazumdar, S., Naidoo, R., Thapa, G.J., Thapa, K., 2012. Conservation and climate change: assessing the vulnerability of snow leopard habitat to treeline shift in the Himalaya. Biol. Conserv. 150, 129–135.

Fortin, M.-J.E., Dale, M.R.T., 2005. Spatial analysis: a guide for ecologists. Cambridge University Press, New York.

Gee, E.P., 1968. Occurrence of the snow leopard, *Panthera uncia* (Schreber), in Bhutan. J. Bombay Nat. Hist. Soc. 64, 552–553.

Hansen, M.C., Potapov, P.V., Moore, R., Hancher, M., Turubanova, S.A., Tyukavina, A., Thau, D., Stehman, S.V., Goetz, S.J., Loveland, T.R., Kommareddy, A., Egorov, A., Chini, L., Justice, C.O., Townshend, J.R.G., 2013. High-resolution global maps of 21st-century forest cover change. Science 342, 850–853.

Hines, J.E., Nichols, J.D., Royle, J.A., Mackenzie, D.I., Gopalaswamy, A.M., Kumar, N.S., Karanth, K.U., 2010. Tigers on trails: occupancy modeling for cluster sampling. Ecolog. Applicat. 20, 1456–1466.

Hobbs, R.J., 1998. Managing ecological systems and processes. In: Peterson, D.L., Parker, V.T. (Eds.), Ecological Scale: Theory and Applications. Columbia University Press, New York.

Hosmer, D.W., Lemeshow, S., 2000. Applied Logistic Regression. Wiley, New York.

Hosmer, D.W., Lemeshow, S., May, S., 2008. Applied Survival Analysis: Regression Modeling of Time-to-Event Data. Wiley-Interscience, Hoboken, NJ.

Jackson, R., Ahlborn, G., 1989. Snow leopards (*Panthera uncia*) in Nepal – home range and movements. Natl. Geogr. Res. 5, 161–175.

Johnson, D.H., 1980. The comparison of usage and availability measurements for evaluating resource preference. Ecology 61, 65–71.

Johnson, C.J., Nielsen, S.E., Merrill, E.H., McDonald, T.L., Boyce, M.S., 2006. Resource selection functions based on use-availability data: theoretical motivation and evaluation methods. J. Wildl. Manage. 70, 347–357.

Johansson, O., McCarthy, T., Samelius, G., Andren, H., Tumursukh, L., Mishra, C., 2015. Snow leopard predation in a livestock dominated landscape in Mongolia. Biol. Conserv. 184, 251–258.

Jumabay-Uulu, K., Wegge, P., Mishra, C., Sharma, K., 2014. Large carnivores and low diversity of optimal prey: a comparison of the diets of snow leopards *Panthera uncia* and wolves *Canis lupus* in Sarychat-Ertash Reserve in Kyrgyzstan. Oryx 48, 529–535.

Karanth, K.K., Nichols, J.D., Sauer, J.R., Hines, J.E., Yackulic, C.B., 2014. Latitudinal gradients in North American avian species richness, turnover rates and extinction probabilities. Ecography 37, 626–636.

Kerr, J.T., Ostrovsky, M., 2003. From space to species: ecological applications for remote sensing. Trends Ecol. Evol. 18, 299–305.

Krebs, C.J., 2009. Ecology: The Experimental Analysis of Distribution and Abundance. Pearson Benjamin Cummings, San Francisco, CA.

Lambeck, R.J., 1997. Focal species: A multi-species umbrella for nature conservation. Conserv. Biol. 11, 849–856.

Levin, S.A., 1992. The problem of pattern and scale in ecology. Ecology 73, 1943–1967.

Lyngdoh, S., Shrotriya, S., Goyal, S.P., Clements, H., Hayward, M.W., Habib, B., 2014. Prey preferences of the snow leopard (*Panthera uncia*): regional diet specificity holds global significance for conservation. PloS One 9, 11.

MacArthur, R.H., Wilson, E.O., 1967. The Theory of Island Biogeography. Princeton University Press, Princeton, NJ.

MacKenzie, D.I., Nichols, J.D., Hines, J.E., Knutson, M.G., Franklin, A.B., 2003. Estimating site occupancy, colonization, and local extinction when a species is detected imperfectly. Ecology 84, 2200–2207.

Mackenzie, D.I., Nichols, J.D., Royle, J.A., Pollock, K.H., Bailey, L.L., Hines, J.E., 2006. Occupancy estimation and modeling: inferring patterns and dynamics of species occurrence. Elsevier, Amsterdam; Boston.

Manel, S., Schwartz, M.K., Luikart, G., Taberlet, P., 2003. Landscape genetics: combining landscape ecology and population genetics. Trends Ecol. Evol. 18, 189–197.

Manly, B.F.J., McDonald, L.L., Thomas, D.L., McDonald, T.L., Erickson, W.P., 2002. Resource selection by animals: statistical design and analysis for field studies. Kluwer Academic Publishers, Dordrecht; Boston.

McCarthy, T.M., Fuller, T.K., Munkhtsog, B., 2005. Movements and activities of snow leopards in Southwestern Mongolia. Biol. Conserv. 124, 527–537.

McGarigal, K., Cushman, S.A., 2005. The gradient concept of landscape structure. In: Wiens, J.A., Moss, M. (Eds.), Issues and Perspectives in Landscape Ecology. Cambridge University Press, Cambridge.

Murray, D.L., 2006. On improving telemetry-based survival estimation. J. Wildl. Manage. 70, 1530–1543.

Nichols, J.D., Williams, B.K., 2006. Monitoring for conservation. Trends Ecol. Evol. 21, 668–673.

Northrup, J.M., Hooten, M.B., Anderson, C.R., Wittemyer, G., 2013. Practical guidance on characterizing availability in resource selection functions under a use-availability design. Ecology 94, 1456–1463.

Pettorelli, N., Olav Vik, J., Mysterud, A., Gaillard, J.-M., Tucker, C., Chr. Stenseth, N., 2005. Using the satellite-derived NDVI to assess ecological responses to environmental change. Trends Ecol. Evol. 20, 503–510.

Pettorelli, N., Nagendra, H., Williams, R., Rocchini, D., Fleishman, E., 2014. A new platform to support research at the interface of remote sensing, ecology and conservation. Remote Sens. Ecol. Conserv. 1, 1–3.

Phillips, S.J., Anderson, R.P., Schapire, R.E., 2006. Maximum entropy modeling of species geographic distributions. Ecol. Model. 190, 231–259.

Pintilie, M., 2006. Competing Risks: A Practical Perspective. John Wiley & Sons, Chichester, England; Hoboken, NJ.

Ray, J.C., Redford, K.H., Steneck, R.S., Berger, J., 2005. Large Carnivores and the Conservation of Biodiversity. Island Press, Washington.

Robinson, H.S., Desimone, R., Hartway, C., Gude, J.A., Thompson, M.J., Mitchell, M.S., Hebblewhite, M., 2014. A test of the compensatory mortality hypothesis in mountain lions: a management experiment in West-Central Montana. J. Wildl. Manage. 78, 791–807.

Rondinini, C., Di Marco, M., Chiozza, F., Santulli, G., Baisero, D., Visconti, P., Hoffmann, M., Schipper, J., Stuart, S.N., Tognelli, M.F., Amori, G., Falcucci, A., Maiorano, L., Boitani, L., 2011. Global habitat suitability models of terrestrial mammals. Phil. Trans. R. Soc. B 366, 2633–2641.

Rose, R.A., Byler, D., Eastman, J.R., Fleishman, E., Geller, G., Goetz, S., Guild, L., Hamilton, H., Hansen, M., Headley, R., Hewson, J., Horning, N., Kaplin, B.A., Laporte, N., Leidner, A., Leinagruber, P., Morisette, J., Musinsky, J., Pintea, L., Prados, A., Radeloff, V.C., Rowen, M., Saatchi, S., Schil, S., Tabor, K., Turner, W., Vodacek, A., Vogelnaann, J., Wegmann, M., Wilkie, D., 2015. Ten ways remote sensing can contribute to conservation. Conserv. Biol. 29, 350–359.

Ruth, T.K., Haroldson, M.A., Murphy, K.M., Buotte, P.C., Hornocker, M.G., Quigley, H.B., 2011. Cougar survival and source-sink structure on greater Yellowstone's northern range. J. Wildl. Manage. 75, 1381–1398.

Sauer, J.R., Link, W.A., Fallon, J.E., Pardieck, K.L., Ziolkowski, Jr., D.J., 2013. The North American breeding bird survey 1966-2011: Summary analysis and species accounts. North Am. Fauna 79, 2–32.

Seidensticker, J.C., Hornocker, M.G., Wiles, W.V., Messick, J.P., 1973. Mountain lion social organization in the Idaho Primitive Area. Wildl. Monogr. 35, 3–60.

Sharma, S., Dutta, T., Maldonado, J.E., Wood, T.C., Panwar, H.S., Seidensticker, J., 2013. Forest corridors maintain historical gene flow in a tiger metapopulation in the highlands of central India. Proc. R. Soc. Lond. B 280, 9.

Sharma, K., Bayrakcismith, R., Tumursukh, L., Johansson, O., Sevger, P., McCarthy, T., Mishra, C., 2014. Vigorous dynamics underlie a stable population of the endangered snow leopard *Panthera uncia* in Tost Mountains, South Gobi, Mongolia. PloS One 9, 10.

Squires, J.R., Decesare, N.J., Olson, L.E., Kolbe, J.A., Hebblewhite, M., Parks, S.A., 2013. Combining resource selection and movement behavior to predict corridors for

Canada lynx at their southern range periphery. Biol. Conserv. 157, 187–195.

Suryawanshi, K.R., Bhatia, S., Bhatnagar, Y.V., Redpath, S., Mishra, C., 2014. Multiscale factors affecting human attitudes toward snow leopards and wolves. Conserv. Biol. 28, 1657–1666.

Turner, M.G., O'Neill, R.V., Gardner, R.H., Milne, B.T., 1989. Effects of changing spatial scale on the analysis of landscape pattern. Landscape Ecol. 3, 153–162.

Turner, M.G., Gardner, R.H., O'Neill, R.V., 2001. Landscape Ecology in Theory and Practice: Pattern and Process. Springer, New York.

Tyre, A.J., Tenhumberg, B., Field, S.A., Niejalke, D., Parris, K., Possingham, H.P., 2003. Improving precision and reducing bias in biological surveys: Estimating false-negative error rates. Ecol. Appl. 13, 1790–1801.

Wang, J., Laguardia, A., Damerell, P.J., Riordan, P., Shi, K., 2014. Dietary overlap of snow leopard and other carnivores in the Pamirs of Northwestern China. Chinese Sci. Bull., 3162–3168.

Wielgus, R.B., Peebles, K.A., 2014. Effects of wolf mortality on livestock depredations. PloS One 9, e113505.

Wiens, J.A., 2002. Central concepts and issues of landscape ecology. In: Gutzwiller, K.J. (Ed.), Applying Landscape Ecology in Biological Conservation. Springer-Verlag, New York.

Wiens, J.A., Milne, B.T., 1989. Scaling of 'landscapes' in landscape ecology, or, landscape ecology from a beetle's perspective. Landscape Ecol. 3, 87–96.

Wikramanayake, E., McKnight, M., Dinerstein, E., Joshi, A., Gurung, B., Smith, D., 2004. Designing a conservation landscape for tigers in human-dominated environments. Conserv. Biol. 18, 839–844.

Wu, J.G., Shen, W.J., Sun, W.Z., Tueller, P.T., 2002. Empirical patterns of the effects of changing scale on landscape metrics. Landscape Ecol. 17, 761–782.

SECTION VI

Snow Leopard Status and Conservation: Regional Reviews and Updates

CHAPTER

30

Central Asia: Afghanistan

Zalmai Moheb, Richard Paley

Afghanistan Program, Wildlife Conservation Society, Kabul, Afghanistan

INTRODUCTION: HISTORICAL RECORDS AND PAST CONSERVATION EFFORTS

One of the first scientific observations of the snow leopard (*Panthera uncia*) in Afghanistan comes from the Second Danish Expedition (1898–1899) in the Pamirs. Olufsen (1899) wrote in his expedition record that "*The long-haired, light-grey panther with black spots is very common here, and is much hunted. One specimen had a height of 70 centimeters, was 130 centimeters from snout to base of tail, and had a tail a meter long.*" Nearly 70 years later, Cullman (1965, cited in Hassinger, 1973) reported seeing a snow leopard skin in Kabul, which allegedly came from Wakhan. The presence of snow leopards in Wakhan was later confirmed by Petocz (1978), and subsequently by a slew of records since the beginning of this century (Fitzherbert and Mishra, 2003; Habibi, 2003; Mishra and Fitzherbert, 2004; Schaller, 2004; Simms et al., 2011; Simms et al., 2013; UNEP, 2003; Habib, 2008; Moheb et al., 2012). The second area where snow leopards have been recorded is in the eastern province of Nuristan (approximately 100 km south of Wakhan) where Petocz and Larson (1977) reported observing a snow leopard feeding on an ibex (*Capra sibirica*) carcass in winter. In 1966, Gaisler et al. (1968 cited in Hassinger, 1973) purchased a snow leopard skin, which allegedly came from an animal in Nuristan. It is interesting to note that Stevens et al. (2011) did not find any conclusive evidence of snow leopard in Nuristan in 2006 and 2007; though they acknowledged that this might have been due to the inexperience of their survey team and the fact they did not access remote areas owing to the security situation. However data from satellite telemetry showed that in summer 2007, a collared female snow leopard moved from Chitral Gol National Park in Pakistan to Nuristan (McCarthy et al., 2007), which would indicate that Nuristan provides suitable habitat for snow leopard and it might therefore support a population.

Conservation efforts in Afghanistan go back to the 1960s and 1970s when international conservation organizations, with the

Snow Leopards
http://dx.doi.org/10.1016/B978-0-12-802213-9.00030-4

encouragement of the Afghan government, conducted countrywide surveys for wildlife. This resulted in a proposal by the Food and Agriculture Organization (FAO) that 14 locations across Afghanistan should be designated as protected areas. Four of these lay within snow leopard range in northeastern Afghanistan. However, in the aftermath of the 1979 Soviet invasion, and the protracted period of conflict that followed, these recommendations remained unrealized. It was not until the collapse of the Taliban regime in 2001 that conservation efforts were resumed. Consequently sustained and coherent snow leopard conservation interventions are a relatively new phenomenon in Afghanistan.

PRESENT STATUS OF SNOW LEOPARDS IN AFGHANISTAN

Assessment of Existing and Potential Snow Leopard Geographical Range

A number of assessments of snow leopard range in Afghanistan, both contemporary and historical, have been conducted over the past two decades (see Habibi, 2003; Hunter and Jackson, 1997; McCarthy and Chapron, 2003). However, the most recent and comprehensive was undertaken by WCS under the GEF-funded Program of Work on Protected Areas (POWPA) in 2008. Using both previous survey records and priority zone analysis covering the whole of Afghanistan, the first tentative distribution map for snow leopard was produced, which has since been enhanced by data from later reports and field observations. The resulting map (Fig. 30.1) shows those areas of preferred habitat in Afghanistan where the presence of snow leopards has been confirmed through direct or indirect observation (11,648.55 km^2), where interviews with local communities suggest it is likely (3,013.51 km^2), and where their presence is as yet unconfirmed (37,385.06 km^2). The current geographical model shows that the snow leopard overall distribution range lies within 10 provinces in north and northeastern Afghanistan, however, the current confirmed areas are in Badakhshan and Nuristan provinces (Fig. 30.1).

At an international snow leopard conservation conference in 2008 in China (see Chapter 3), experts designated the Wakhan District as a priority Snow Leopard Conservation Unit and later it was designated as a priority area in the GSLEP (2013). Fitzherbert and Mishra (2003) had already identified the area between Ishkashim and Qala-e Panja as the main stronghold for the species, but the presence of snow leopards has also been confirmed in the Big and Little Pamir mountain ranges, which fall within the same district (Simms et al., 2011; Simms et al., 2013; Moheb et al., 2012).

More recent field surveys, conducted by the Wildlife Conservation Society (WCS), have recorded evidence of snow leopards in Zebak and Ishkashim districts (Ismaily, S., Simms, A. Large mammal survey of Zebak and Ishkashim districts 2012 & 2013. Wildlife Conservation Society, unpublished report, 2013), which are west and southwest of and adjacent to Wakhan District, and further afield in the Darwaz Region of Badakhshan Province (Moheb and Mostafawi, 2012; Moheb and Mostafawi, 2013). Based on statements from local communities in the 1970s, Habibi (2003) also reported the presence of snow leopards in the eastern provinces of Laghman and Nuristan, and the Ajar Valley of Bamyan Province in central Afghanistan.

Estimates of Snow Leopard Population in Afghanistan

No accurate estimate of snow leopard population in Afghanistan is currently available. Nevertheless, based on assessments of appropriate habitat and the density of the species

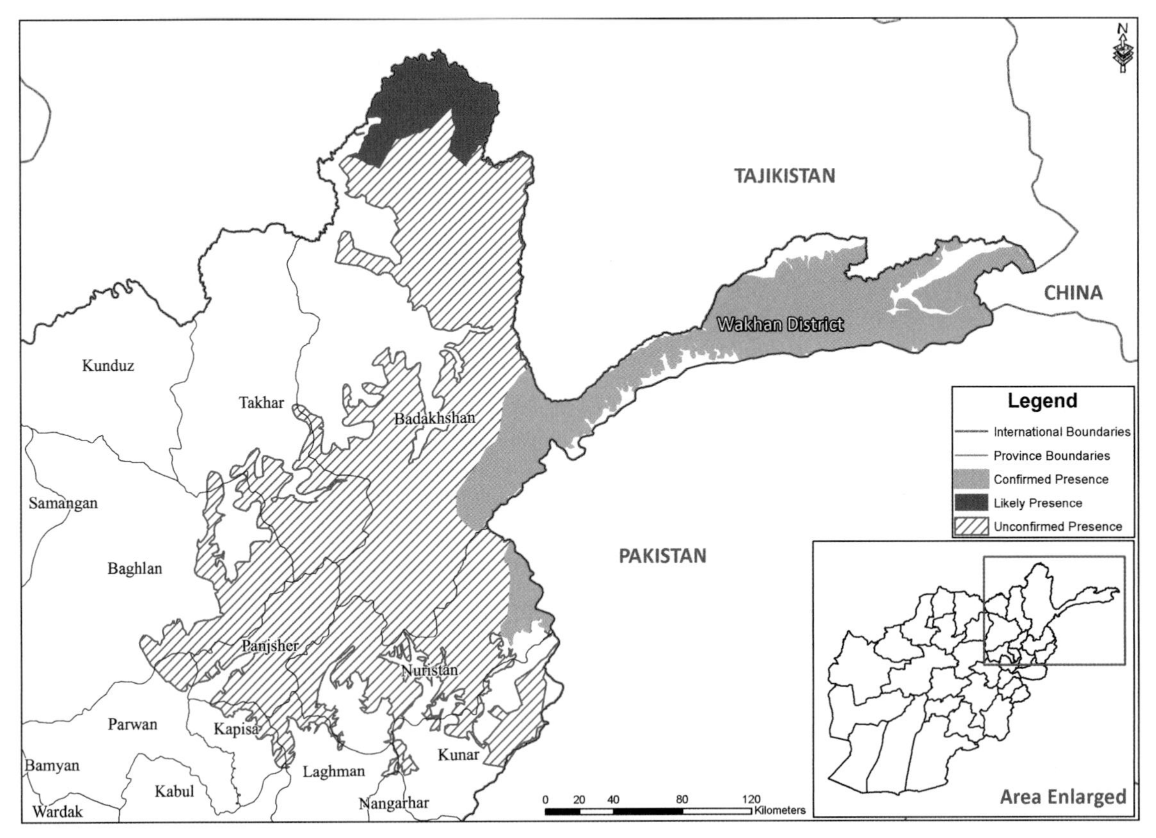

FIGURE 30.1 **Area of preferred habitat for snow leopard in Afghanistan.**

elsewhere in its range, the population in the country has been variously estimated at 50 to 200 individuals (Snow Leopard Working Secretariat, 2013; Simms et al., 2013). Furthermore, no valid measurements indicate that the snow leopard population in northeast Badakhshan (Fig. 30.1) has decreased over the past decade, results of WCS questionnaire surveys carried out in 2006–2007 in Nuristan (Karlstetter, M. Wildlife surveys and wildlife conservation in Nuristan, Afghanistan, including scat and small rodent collection from other sites, WCS unpublished report, 2008) and in 2011–2012 in northern Badakhshan (Moheb and Mostafawi, 2012; Moheb and Mostafawi, 2013) suggest an overall decline in snow leopard populations over the past 1–3 decades.

CURRENT THREATS TO SNOW LEOPARD POPULATIONS

The major threats to the survival of the snow leopard in Afghanistan are declines in prey species, brought about by competition with domestic livestock, hunting for meat and the fur trade, and the killing of snow leopard in retaliation for livestock predation (Habib, B. Status of large mammals in Proposed Big Pamir Wildlife Reserve, Wakhan, Afghanistan, WCS, unpublished report, 2006). Although data on prey species are lacking for most snow leopard range, recent community-based surveys in the north of Badakhshan Province suggest that prey populations have shrunk since the 1970s (Moheb and Mostafawi, 2011; Moheb and Mostafawi, 2012; Moheb and Mostafawi, 2013). A decrease in availability of prey is likely to have negatively affected individual snow leopard survival and reproductive rates, and local communities claim that predation of domestic livestock has increased accordingly. Habib (2008) reported 0.65% total livestock mortality in the Wakhan Corridor due to snow leopard predation during the winter of 2007. The resulting economic loss, though limited across all communities, has significant implications for the individual families affected, and consequently generates negative attitudes toward large predators in general, and especially the snow leopard, frequently leading to calls for retaliatory killing. Mishra and Fitzherbert (2004) relayed reports from local communities of 10 snow leopard killings in Wakhan for the period 1989–2002.

Snow leopards are a valuable commodity in the wildlife trade, and though evidence from the Wakhan suggests they are hunted primarily in retaliation, rather than for commercial reasons (Fitzherbert and Mishra, 2003), pelts are sold for significant sums of money in Afghan wildlife markets. A complete pelt may sell for as much as USD 300–1500 in Kabul (Johnson and Wingard, 2010). Rodenburg (1977) estimated that 50–80 skins were sold in Afghan markets per year around that time. Recent estimates are lacking, but visits to the fur markets of Kabul by WCS staff in 2014 show that pelts are still very much available, though the provenance of the pelts is uncertain. Some traders state their origin is Badakhshan in Afghanistan, while others claim they come from neighboring snow leopard range countries.

MEASURES TO CONSERVE SNOW LEOPARD IN AFGHANISTAN

Research on Snow Leopard and Prey Species

Since 2006 WCS, in coordination with the Afghan National Environmental Protection Agency and the Ministry of Agriculture, has been conducting research on the snow leopard and its prey species. The work was initially conducted in Badakhshan and Nuristan provinces, but due to growing insecurity in the latter, it is now focused on Badakhshan Province exclusively. A wide array of methodologies has been employed to gain a better understanding of the status and ecology of snow leopards in Afghanistan, including camera trapping, satellite telemetry of collared cats, monitoring of prey species, and interview-based approaches.

The first camera trap photographs of snow leopards were taken in the Hindu Kush Mountains of the western Wakhan District in 2009, with the aim of collecting sufficient data to make population estimates using photo capture-recapture techniques. More than 40 camera traps were deployed between 2011 and 2013 and the geographical scope has been extended to areas of the Big and Little Pamir mountains to the north and east. To date, more than 5000 photographs of snow leopards have been taken and though the data has yet to be fully analyzed, initial results suggest that Wakhan District supports a healthy population of snow leopards (Simms et al., 2011).

Satellite collaring operations began in Wakhan District in 2012. Since then four snow leopards have been collared (one of them twice) using Vectronic GPS collars (Vectronic Aerospace, Berlin, Germany) on three separate expeditions. In June 2012 two male snow leopards were collared followed by a female that was accompanied by two cubs in September. The following year in September a male and a female were collared. Each collar was programmed to remain in place for 13 months, and the purpose of the research was to provide data on snow leopard movement, home range, and habitat preferences (see Rahmani, 2014), as well as eventually yielding a population estimate throughout its range in Afghanistan.

Wild ungulates such as Siberian ibex (*Capra sibirica*), markhor (*Capra falconeri*), urial (*Ovis vignei*), and Marco Polo sheep (*Ovis ammon polii*) along with small mammals such as marmots (*Marmota caudata*), hares (*Lepus* spp.), and rodents constitute the major prey species for snow leopards in northeast Afghanistan. Comprehensive surveys of selected prey species have been conducted since 2006 in the Hindu Kush and Pamir ranges of Wakhan District and primary surveys conducted in Zebak, Ishkashim, and Gharan districts (adjacent to Wakhan). Reconnaissance surveys have also been conducted further afield in the Darwaz and Shah-r Buzurg areas of north and northwest Badakhshan Province, respectively. These surveys have yielded information on the presence of Siberian ibex and markhor. Table 30.1 presents the number of prey species counted in different areas of Badakhshan over the past 30 years.

The Legal and Management Frameworks

The snow leopard receives protection under a number of policy and legal instruments in Afghanistan. These include the Environmental Law, National Biodiversity Strategy and Action Plan, National Protected Area System Plan, Presidential Decree No. 25 banning all hunting, and Afghanistan's Protected Species List. The draft Wildlife Management and Hunting Law will provide additional protection when ratified by the National Assembly. Afghanistan is also a signatory to several international conventions linked to snow leopard conservation such as the Convention on International Trade in Endangered Species (CITES), Convention on Biological Diversity (CBD), and the Convention on Migratory Species (CMS).

TABLE 30.1 The Number of Prey Species Counted in Different Areas of Badakhshan over Past 30 Years

Prey spp.	Observer	Observation area	Total counts	Population estimate
Marco Polo sheep	Petocz et al. (1978)	Afghan Pamirs	1260	2500
	Schaller (2004)		620	1000
	Habib (2008)	Western Big Pamir	85	211
	Winnie and Harris (2007)			244
	Harris et al. (2010)			172 F
Ibex	Petocz (1978)	Big Pamir	919	
	Schaller (2004)	Upper Istimoch Valley	37	
	Habib (2008)	Western Big Pamir	162	
	Winnie and Harris (2007)		118	
Markhor	Moheb and Mostafawi (2011)	Shahr-e Buzurg District	4	>20
	Moheb and Mostafawi (2012)	Darwaz Region	6	>80

On March 30, 2014, the Government of the Islamic Republic of Afghanistan declared Wakhan District (10,984.99 km^2) as the country's second national park. The park covers more than 70% of Afghanistan's confirmed snow leopard habitat and will be managed as an IUCN Category VI protected area. However, certain designated areas within its boundaries such as the proposed Big Pamir (576.64 km^2) and Teggermansu (248.51 km^2) wildlife reserves will receive stricter protection under IUCN Category I.

Threat Mitigation Efforts

As management of these protected areas grows in effectiveness, conservation agencies are better able to address the direct threats to snow leopards through monitoring of illegal hunting and application of the law. Though law enforcement presents many challenges in a country experiencing continuing conflict and where the rule of law is not universal, increasing cooperation between communities, protected area staff, local government officials, and border police in Wakhan District is yielding results in terms of crime detection and follow-up.

At the national level, the issue of illegal trade in wildlife is receiving increasing attention. WCS has implemented programs to raise awareness among police and customs officials of Afghanistan's wildlife-related laws and policies, as well as its commitments to international conventions. It has also provided parallel training to US military personnel, both in Afghanistan and the United States, in an effort to reduce the demand for furs and other wildlife products in the markets of US and coalition bases.

Having identified potential hotspots for human-snow leopard conflict through predation surveys, WCS is endeavoring to mitigate conflict by training shepherds on improved techniques for guarding their flocks and through facilitating the building of more than 30 predator-proof corrals across Wakhan District since 2010. To date, these predator-proof corrals, which are communally owned and managed, have achieved a 100% level of protection for animals corralled in them. The hope is that as more corrals are built and predation of livestock is further reduced, a corresponding downward trend in retaliatory killing of snow leopards will be observed.

These attempts at direct mitigation efforts have been underpinned by a broader effort to promote wildlife conservation in Afghanistan, with the snow leopard as one of several flagship species. Capacity development and awareness programs across snow leopard range have involved government officials, spiritual leaders, community representatives, and teachers and students, and employed a variety of tools including workshops, posters, brochures, and environmental celebration days and participation in on-the-ground conservation activities. At the national level, government ministries, international NGOs, and donor agencies have been targeted through formal training programs, university curricula, and support for oversees study.

Community-Based Conservation

Since launching their snow leopard conservation initiatives in 2006, both the Afghan government and WCS have acknowledged that local community participation is an essential ingredient of success. Though some of the most immediate threats to the survival of snow leopards emanate from the behavior of local people, those same individuals and communities can be converted into the snow leopard's staunchest allies.

Central to the process of bringing about this conversion is the creation of a legally registered community governance institution, the

Wakhan Pamir Association (WPA). The WPA is led by an elected board comprised of community representatives from across Wakhan District. The board makes important decisions relating to conservation and natural resource management on behalf of its constituent communities. Among these are decisions supporting the establishment and management of several protected areas in snow leopard range. Furthermore, WPA is increasingly assuming management responsibility for the 55 community rangers who, trained and equipped by WCS, currently conduct various activities assisting snow leopard conservation, such as camera trapping, field surveys, and the monitoring and reporting of illegal hunting. The WPA is also directing its efforts toward improving tourism facilities and experience with the intention of attracting larger numbers of international visitors to the Wakhan District and thereby increasing the benefits to both communities and wildlife.

Transboundary Initiatives

The belief that the Pamir Mountains are a critical haven for snow leopards, and that their management as a holistic landscape is most likely to achieve species conservation goals, has prompted a number of initiatives aimed at enhancing transboundary cooperation. Indeed satellite telemetry has shown that snow leopards have no respect for international boundaries and cross them at will (McCarthy et al., 2007; Zahler and Schaller, 2014). In 2006, WCS held a workshop in Urumqi, China, at which senior government participants from Afghanistan, Tajikistan, Pakistan, and China committed to a range of recommendations for regional conservation cooperation, including exploring the idea of a transboundary park comprised of adjacent areas of the Pamir Mountains in all four countries.

This initiative has been followed by two tripartite initiatives between Afghanistan, Tajikistan, and Pakistan. In 2011, WCS implemented a project focusing on the management of diseases common to livestock and wildlife, which are known to be a significant threat to key snow leopard prey species such as Marco Polo sheep, markhor, urial, and ibex. This was followed in 2013 by a Climate Change Vulnerability Assessment in the Pamir region, which aimed to better understand how climate change is affecting mountain communities in these three countries.

CONCLUSIONS

Since 2006, the Afghan government and its international partners have developed a sustained and coherent program for conserving snow leopards. The approach as described in this chapter has been a holistic one based on sound science and encompassing a broad range of conservation tools, which include legal protection, threat mitigation, and crucially, community engagement.

Afghanistan is also an active participant in the World Bank Global Snow Leopard Ecosystem Protection Program, developing first its own National Snow Leopard Ecosystem Protection (NSLEP) Plan (NEPA, 2013) and contributing to the preparation of the Global Snow Leopard and Ecosystem Protection Program (GSLEP). The future actions outlined for Afghanistan in the GSLEP conform to existing conservation strategies. This includes a commitment by the Afghan government to continue field monitoring and satellite collaring of snow leopards to gain further understanding of their ecology and population size; more rigorous enforcement of the Environment Law and its protected species lists in order to provide effective protection for the snow leopard and its prey species; and full implementation of the National Protected Area System Plan, specifically the designation of the Wakhan District as a protected area (now declared).

Acknowledgment

The work described in this chapter is the result of the joint efforts of Afghanistan's National Environmental Protection Agency, the Ministry of Agriculture, Irrigation and Livestock, and the Wildlife Conservation Society between June 2006 and December 2014. Most of the work was funded through the generous support of the US Agency for International Development, with additional support from the National Geographic Society (the first satellite collaring of a snow leopard in Afghanistan) and the World Food Program (predator-proof corral construction).

References

Fitzherbert, A., Mishra, C., 2003. Afghanistan Wakhan Mission Technical Report. United Nations Environment Program and Food and Agriculture Organization of the United Nations. 101 pp.

Habibi, K., 2003. Mammals of Afghanistan. Zoo Outreach Organization, India.

Habib, B., 2008. Status of Mammals in Wakhan Afghanistan. Afghanistan Wildlife Survey Program. Wildlife Conservation Society (WCS), New York.

Harris, R.B., Winnie, J., Amish, S.J., Beja-Pereira, A., Godinho, R., Costa, A., Luikart, G., 2010. Argali abundance in the Afghan Pamir using capture-recapture modeling from fecal DNA. J. Wildl. Manag. 74, 668–677.

Hassinger, J., 1973. A survey of the mammals of Afghanistan resulting from the 1965 Street Expedition. Fieldiana Zool. 60, 156–158.

Hunter, D.O., Jackson, R., 1997. A range-wide model of potential snow leopard habitat. In: Jackson, R., Ahmad, A. (Eds.). Proceedings of the 8th International Snow Leopard Symposium, Islamabad, November 1995. International Snow Leopard Trust, Seattle and WWF-Pakistan, Lahore. pp. 51–56.

Johnson, M.F., Wingard, J.R., 2010. Wildlife Trade in Afghanistan. Wildlife Conservation Society, Kabul.

McCarthy, T.M., Chapron, G., 2003. Snow Leopard Survival Strategy. ISLT and SLN, Seattle, USA.

McCarthy, T., Khan, J., Ud-Din, J., McCarthy, K., 2007. The first study of snow leopards using GPS satellite collars underway in Pakistan. Cat News 46, 22–23.

Mishra, C., Fitzherbert, A., 2004. War and wildlife: a post-conflict assessment of Afghanistan's Wakhan Corridor. Oryx 38, 102–105.

Moheb, Z., Mostafawi, N.S., 2011. Biodiversity Reconnaissance Survey in Shahr-e Buzurg District, Badakhshan Province, Afghanistan, unpublished report. Wildlife Conservation Society (WCS), New York.

Moheb, Z., Mostafawi, S.N., 2012. Biodiversity Reconnaissance Survey in Darwaz region, Badakhshan Province, Afghanistan, unpublished report. Wildlife Conservation Society (WCS), New York.

Moheb, Z., Mostafawi, S.N., 2013. Biodiversity Reconnaissance Survey in Maymai Disrtict, Badakhshan Province, Afghanistan, unpublished report. Wildlife Conservation Society (WCS), New York.

Moheb, Z., Mostafawi, N., Noori, H., Rajabi, A.M., Ali, H., Ismaily, S., 2012. Urial survey in the Hindu Kush Range in the Wakhan Corridor, Badakhshan Province, Afghanistan, unpublished report. Wildlife Conservation Society (WCS), New York.

NEPA, (2013). National Snow Leopard Ecosystem Priority Protection (NSLEP) for Afghanistan. National Environmental Protection Agency, Government of the Islamic Republic of Afghanistan.

Olufsen, O., 1899. Through the unknown Pamirs: The Second Danish Pamir Expedition (1898–99), Translations accessed from http://www.iras.ucalgary.ca/~volk/sylvia/Pamir1.htm#one

Petocz, R., 1978. Report on the Afghan Pamirs, Part 1 Ecological Reconnaissance. UNEP and FAO of the United Nations.

Petocz, R., Larson, J.Y., 1977. Ecological Reconnaissance of Western Nuristan with Recommendations for Management. FAO.

Petocz, R.G., Habibi, K., Jamil, A., Wassey, A., 1978. Report on the Afghan Pamirs. Part 2 Biology of Marco Polo sheep (*Ovis ammon polii*), Food and Agriculture Organization of the United Nations.

Rahmani, H., 2014. Snow Leopard Habitat Preference Modeling in the Wakhan Corridor using Satellite Telemetry Data. MSc thesis, University of Leeds.

Rodenburg, W.F., 1977. The Trade in Wild Animal Furs in Afghanistan. FAO.

Schaller, G.B., 2004. The Status of Marco Polo Sheep in the Pamir Mountains of Afghanistan. National Geographic Society (Grant No.EC-0182-04) and WCS.

Simms, A., Moheb, Z., Salahudin, I., Ali, H., Ali, I., Wood, T., 2011. Saving threatened species in Afghanistan: snow leopards in the Wakhan Corridor. Int. J. Environ. Stud. 68, 299–312.

Simms, A., Ostrowski, S., Ali, H., Rajabi, A.M., Noori, H., Ismaili, S., 2013. First radio-telemetry study of snow leopards in Afghanistan. Cat News 58, 29–31.

Snow Leopard Working Secretariat, 2013. Global Snow Leopard and Ecosystem Protection Program (GSLEP). Snow Leopard Working Secretariat; Bishkek, Kyrgyz Republic.

Stevens, K., Dehgan, A., Karlstetter, M., Rawan, F., Tawhid, M.I., Ostrowski, S., Ali, J.M., Ali, R., 2011. Large mammals surviving conflicts in the eastern forests of Afghanistan. Oryx 45, 265–271.

UNEP, 2003. Afghanistan: Post-conflict Environmental Assessment. United Nations Environment Program, Nairobi, Kenya.

Winnie, J., Harris, R., 2007. Marco Polo argali research in the Big Pamir Mountains of Afghanistan: year-end summary, unpublished report. Wildlife Conservation Society (WCS), New York.

Zahler, P., Schaller, G.B., 2014. Saving More Than Just Snow Leopards. The New York Times. Available from: http://www.nytimes.com/2014/02/02/opinion/saving-more-than-just-snow-leopards.html?_r=0

CHAPTER

31

Central Asia: Kyrgyzstan

Askar Davletbakov, Tatjana Rosen**, Maksat Anarbaev†,‡, Zairbek Kubanychbekov‡,§, Kuban Jumabai uulu¶, Jarkyn Samanchina††, Koustubh Sharma‡‡*

*Institute for Biology and Soil Sciences, National Academy of Sciences of the Kyrgyz Republic, Bishkek, Kyrgyz Republic

**Tajikistan and Kyrgyzstan Snow Leopard Programs, Panthera, Khorog, Gorno-Badakhshan, Tajikistan

†National Center for Mountain Regions Development, Bishkek, Kyrgyz Republic

‡Snow Leopard Program, Panthera, Bishkek, Kyrgyz Republic

§Panthera Foundation Kyrgyzstan, Bishkek, Kyrgyz Republic

¶Snow Leopard Trust/Snow Leopard Foundation in Kyrgyzstan, Bishkek, Kyrgyz Republic

††Fauna & Flora International, Bishkek, Kyrgyz Republic

‡‡GSLEP Secretariat, Bishkek, Kyrgyz Republic

SNOW LEOPARD HABITAT AND DISTRIBUTION

Snow leopards (*Panthera uncia*) in Kyrgyzstan inhabit about 89,000 km^2 (Undeland, 2005) including portions of the northern, southern, and western Tien Shan and the Pamir-Alai range. The northern Tien Shan includes the Kyrgyz and Chu-Ili ranges; the Kungey and Terskey Ala-Too ranges that border Djetim Bel, from the river Uzungush and east to the Kakshal range. The southern Tien Shan consists of the Ferghana and Moldo-Too ranges, the southern side of the Djetim range and to the south, the Kakshal range. The western Tien Shan is bordered by the Ferghana range on the east; by the rivers Kara-Kulzha and Kara-Darya to the south; the Talas range to the north; and the Pskem range to the northwest. The Pamir-Alai system includes the Turkestan, Alai, and Trans-Alai ranges. Snow leopard presence has been confirmed for all of these areas.

A genetic study conducted in 2009 in the Sarychat-Ertash reserve showed a minimum of 18 snow leopards in a 1,341 km^2 area (Jumabay-Uulu et al., 2013), whereas preliminary results from systematic camera trapping in 2014 revealed a minimum of 15 snow leopards in the same area

Snow Leopards
http://dx.doi.org/10.1016/B978-0-12-802213-9.00031-6

(Jumabay-Uulu et al., unpublished data). Preliminary camera trapping in Naryn shows the presence of at least 5 snow leopards using a study area of 200 km^2. New surveys are in progress in Naryn, Sarychat, and three nascent community-based conservancies in the Alai valley.

Koshkarev and Vyrypaev (2000) estimated the number of snow leopards for the whole of Kyrgyzstan to be 150–200 individuals, attributing the decline from an earlier estimate of ca. 650 to widespread poaching in the 1990s following the breakup of the Soviet Union. More recent population estimates are closer to 350–400 individuals for the whole country (National Academy of Sciences of Kyrgyzstan, unpublished data), though some sources still put the numbers at 300–350 individuals (Snow Leopard Network, 2014). About 44% of the land area suitable for snow leopards (around 89,000 km^2) is used as pasture for livestock (Undeland, 2005). Snow leopards occur in eight nature reserves and three national parks: Besh-Aral State Reserve (632 km^2), Kara-Buura SR (615 km^2), Karatal-Japyryk SR (364 km^2), Padysha-Ata SR (305 km^2), Kulun-Ata SR (274 km^2), Naryn SR (910 km^2), Sarychat-Ertash SR (1341 km^2), Sary Chelek Biosphere Reserve (238 km^2), Ala Archa NP (194 km^2), Kara-Shoro NP (120 km^2), Chong-Kemin (120 km^2), and Karakol National Park (382 km^2).

In 2013, the Issyk-Kul Oblast State Administration decided to establish Khan Tengri National Park (more than 1870 km^2) in order to implement a Decision of the Parliamentary Committee of the Kyrgyz Republic. This proposed site directly borders Kazakhstan and China, and links Naryn and Sarychat-Ertash State Reserves and Karakol National Park in Kyrgyzstan with Tomur Reserve in Xinjiang, China.

STATUS OF SNOW LEOPARD PREY

Key snow leopard prey in Kyrgyzstan includes argali (*Ovis ammon*), Siberian ibex (*Capra sibirica*), and red deer (*Cervus elaphus*). Marmots (*Marmota* spp.), pikas (*Ochotona* spp.) and snowcock (*Tetraogallus* spp.) also constitute important prey. In Kyrgyzstan, according to the latest data, there are 47,668 ibex (Davletbakov, 2010), of which 902 are in the Turkestan range; 4,116 in the Ferghana, Chatkal, and Pskem ranges; 15,884 on the Terskey, Kungey Ala-Too, and Kakshal ranges; 12,516 in the Kakshal and At-Bashy ranges; 6,103 in the Alai and Za-Alai ranges; 3,480 in the Talas range; and 4,667 in the Kyrgyz range. Across Kyrgyzstan, there are approximately 16,500 argali (Davletbakov, 2010), of which 10,500 are in the Kungey Ala-Too and Kakshal ranges; 5,000 in the Kakshal and At-Bashy ranges; 680 in the Ferghana range; 200 in the Alai and Za-Alai ranges; 70 in the Talas range; and 50 in the Kyrgyz range.

Under a new hunting law established in 2014, hunting of ungulates is allowed only in assigned areas and hunters have to obtain permits from the area managers (Almaz Musaev, personal communication, 2015). The new law and involvement of NGOs has encouraged the establishment of the first community-based hunting conservancies in Chon Kemin, Aksu (Issyk-kul), and the Alay valley, aimed at rehabilitating ungulate populations while providing financial incentives to the members through trophy hunting. For a description of the law and the work of the community-based hunting conservancies in Kyrgyzstan and Tajikistan, see Chapter 16.3.

LEGAL PROTECTION

Hunting, possession, and trade of snow leopard are prohibited in Kyrgyzstan through the Law on the Animal World (1999). Hunting of snow leopards has been prohibited since 1948 and the species has been listed in the national Red Data Book of the Kyrgyz SSR since 1985. The snow leopard is listed as "critically endangered" in the second edition of the Red Book of the Kyrgyz Republic (Government of Kyrgyzstan, 2006). The fine for harming a snow leopard has been increased from 199,640 Kyrgyz som (USD 3,992) to 500,000 som (USD 8,400). Species listed in the Red Book are generally protected, but can be taken from nature based on special decisions by the government.

THREATS TO SNOW LEOPARDS IN KYRGYZSTAN

Key threats identified include killing of snow leopards for their skins and other parts, intentional and incidental trapping, poaching of the prey base, and habitat fragmentation. Although legal protection may have contributed to a decrease in trapping and killing of snow leopards, it is still documented in many parts of the country. Based on interviews, for example, in the Alai valley alone, 4–5 snow leopards are reportedly trapped each year or imported from neighboring Tajikistan to meet what is described as a regional demand for snow leopards pelts as ornaments in the house and as symbols of power and wealth (authors' personal observation, 2014). An occupancy-based study conducted in the Alai valley seems to indicate high levels of local extinctions of snow leopards. (Taubmann et al., 2015), a possibility indicated by other studies (Izumiama et al., 2009; Watanabe et al., 2010). According to interviews with local people in the Alai in 2014, there could be fewer than 20 snow leopards left in the entire Pamir-Alai range.

Poaching of the prey base, especially argali and ibex, that indirectly contributes to lower snow leopard densities, is still rampant despite efforts such as community-based hunting conservancies and other community-based conservation programs; incentive programs to reward rangers and citizens who apprehend poachers may progressively address this issue. Some poaching is done by local people, mostly for sustenance, whereas some is also reportedly done by border officers in areas adjacent to international borders.

NATIONAL ACTION PLAN, THE NSLEP, AND MANAGEMENT PLANS FOR PROTECTED AREAS

The drafting process for the National Action Plan was initiated in 2012. The final version of the national plan was approved by the Prime Minister in August 2013 but has not yet been published. In 2013, the Kyrgyz National Snow Leopard and Ecosystem Protection Plan was adopted during the Snow Leopard Forum in Bishkek (Snow Leopard Working Secretariat, 2013). In June 2014 during the National Focal Points Action Planning, Leadership and Capacity Building Workshop, the Sarychat landscape (13,200 km^2) was identified as part of the Global Snow Leopard and Ecosystem Protection Program (GSLEP) goal of securing at least 20 landscapes by 2020. Another transboundary landscape, Alai-Gissar (approximately 30,000 km^2), was also identified by the officials from Kyrgyzstan and Tajikistan to be secured for the snow leopard and mountain ecosystem.

TRANSBOUNDARY CONSERVATION INITIATIVES

In 2005, the Global Environment Facility (GEF) West Tien Shan project (2005–2009) was launched with the goal of improving and increasing cooperation between five protected areas, all of which hold snow leopards: Chatkal State Reserve (Uzbekistan), Sary-Chelek and Besh-Aral SRs (Kyrgyzstan), and Aksu-Djabagly SR (Kazakhstan). The objectives also include strengthening institutional capacity and national policies, supporting regional cooperation, and enhancing income generation within the protected areas.

The Tien Shan Ecosystem Development Project, also funded by GEF, was launched in 2009 to support management of protected areas and sustainable development in Kazakhstan and Kyrgyzstan. About the same time, the Pamir-Alai Transboundary Conservation Area project (PATCA), funded by the European Union, looked at options of establishing a transboundary protected area between Kyrgyzstan and Tajikistan. A biological database was assembled (Asikulov, 2008; Doempke, 2008; Murray, 2008; Sagimbaev, 2007; Toropova, 2007), although PATCA is yet to be implemented.

In 2013, the "Mountains of Northern Tien Shan" project (2013–2016) was launched, with the assistance of the government of the Federal Republic of Germany and the Nature and Biodiversity Conservation Union (NABU). A transboundary protected area is planned at the junction of Chon-Kemin (Kyrgyzstan), Chu-Or Nature Reserve, and Almaty reserves (Kazakhstan).

Over the years, several initiatives have aimed at stimulating cross-border collaboration between scientists and conservation officials across the region. Some involved joint surveys, such as the one organized on the border between Kazakhstan and Kyrgyzstan by scientists from both counties (FFI, 2007). GIZ coordinated training and exchanges between staff of forest and hunting departments of Kyrgyzstan and Tajikistan and also brought members from nascent community-based hunting conservancies in Kyrgyzstan to similar established hunting conservancies in Tajikistan. The US-based NGO Panthera brought hunting conservancy members from Tajikistan to nascent hunting conservancies in Kyrgyzstan. A technical workshop brought together experts from Tajikistan, Uzbekistan, Kazakhstan, and the Kyrgyz Republic to identify transboundary snow leopard landscapes in Central Asia, as well as stimulate cross-border collaboration (FFI and CMS, 2015).

NGOs WORKING IN KYRGYZSTAN ON THE CONSERVATION OF SNOW LEOPARDS

NABU

NABU is a German organization, established in 1899, that has been active since 2001 in snow leopard conservation projects in Kyrgyzstan. In 2002, NABU established the "Gruppa Bars," an anti-poaching unit of four people, aimed at stopping illegal hunting and trade in wildlife protected under the Red Book of Kyrgyzstan. The unit has the power to arrest suspects and seize live animals, skins, weapons, and other evidence. Since its establishment, Gruppa Bars has captured 180 poachers and confiscated many snow leopard pelts as well as furs from other endangered animals. They have also confiscated and destroyed hundreds of traps to date. In 2013, the unit also started deploying camera traps on different mountain ranges.

NABU is also active in many outreach activities. Every year, they conduct campaigns to promote the conservation of difference species. They also manage a rehabilitation center in the Issyk-Kul lake district, where three snow leopards are kept in captivity. One of NABU's most recent initiatives is the Snow Leopard Conservation Forum, launched in 2013 by Kyrgyz President Atambaev.

WWF

WWF has been active in Kyrgyzstan since 1999, initially through the Econet project, a platform for developing protected nature areas of different status and territories with different sustainable use of natural resources. In 2009, WWF started working on developing a concept for the conservation of the central Tien Shan Mountains and wildlife. In 2013, this culminated in the approval of a GEF/UNDP project to establish Khan-Tengri National Park. WWF is also actively supporting Sarychat-Ertash State Reserve through technical support and capacity-building activities. WWF also supports a festival called "The home of the Snow Leopard," which brings together people who live in the proximity of the Sarychat-Ertash reserve. It also supports a fund for the development of the villages Ak-Shiyrak and Enilchek (outside Sarychat-Ertash). Money from the fund goes towards anti-poaching activities. A DNA survey conducted in Sarychat-Ertash in 2011 by WWF confirmed the presence of approximately 20 snow leopards (WWF, personal communication, 2015).

Snow Leopard Trust (SLT) in Partnership With the Snow Leopard Foundation in Kyrgyzstan (SLFK)

SLT has been active in Kyrgyzstan since 2003. It has mainly worked in the Sarychat-Ertash State Reserve as well as the villages of Ak-Shiyrak and Enilchek. It has provided training to the rangers, led sign and camera-trap surveys, and engaged the two villages through the Snow Leopard Enterprises program. This provides training and equipment that enables women from participating villages to make handicrafts from the raw wool of their livestock. SLT purchases the finished items at mutually agreed upon prices, and sells the items through their online stores. The program has drastically increased the value of each herding family's raw wool, and in turn enhanced their tolerance to livestock losses to wild predators. In order for a community to participate and earn this additional income, each member must sign a conservation agreement. Every year, these participating communities review and sign an agreement that requires each person to protect the snow leopards and wild prey species living in their area from poaching. If a participating community fulfills their collective conservation agreements, an additional cash bonus is awarded at the end of each year. However, if any poaching takes place during that time, the entire community loses the bonus.

In 2014, another program was initiated by SLFK/SLT in collaboration with Interpol and the Protected Areas Department of Kyrgyzstan and with support from the UK government's Illegal Wildlife Trade Challenge Fund. In 2014, the program was first piloted in Sarychat and Naryn, rewarding rangers and community members who successfully stopped illegal hunting, before being launched across the 19 protected areas in 2015. A series of training programs for the front line as well as senior officials are planned in the next several years to improve their efficiency. In 2015, with support from the Christensen Foundation, SLFK has also initiated a project to address the issues of biodiversity and cultural erosion by beginning to amass a repository of biocultural folklore and creating an educational strategy for children living in and around snow leopard habitats in Kyrgyzstan. In 2014, SLT/SLFK conducted a camera-trap survey in the Sarychat-Ertash strict nature reserve. Fifteen snow leopards have been identified and the analysis of data using spatially explicit mark-recapture puts the estimated population close to earlier works. Annual camera-trap surveys are planned in the reserve and some adjoining areas for the next 10 years.

Fauna & Flora International (FFI)

FFI started its work with Sarychat-Ertash Reserve in 2004, aimed at staff capacity building, improvement of the technical and material base, strengthening anti-poaching activities, conducting biodiversity survey with specialists from the Kyrgyz National Academy of Sciences, and developing a management plan for the reserve. Under this project, in 2005–2006, FFI also carried out a project in the three villages situated in the immediate proximity of Sarychat-Ertash – Enilchek, Ak-Shiyrak, and Karakolka – in order to develop alternative income-generating activities and reduce poaching of snow leopard and its prey base. The Sarychat-Ertash management plan was finalized and submitted to the State Agency for Environmental Protection and Forestry in November 2014 to await approval.

KAIBEREN (in Partnership With the National Academy of Sciences of Kyrgyzstan, Shinshu University, and Panthera)

Kaiberen is a project on the research and conservation of carnivores and their mountain ungulate prey launched in 2008 under the umbrella of the National Center for the Development of Mountain Regions of the Kyrgyz Republic. Initially supported only by the National Academy of Sciences of Kyrgyzstan, Shinshu University in Japan, and the Japanese Ministry of Education and Science, since 2014 it is also now a partner of Panthera.

Between 2009 and 2013, Kaiberen deployed satellite collars on argali in Sarychat-Ertash to monitor their migration movements. In 2014, in cooperation with Panthera, it conducted the preliminary camera-trap survey in the Naryn State Reserve described earlier, as well as a full scale camera-trap survey. With Panthera, it is also leading the process of supporting three different community-based hunting conservancies in the Alai valley. During the winter of 2014–2015, it launched exploratory camera-trap surveys in the hunting conservancies. In the fall of 2015, more extensive surveys will be conducted. In the spring of 2015, Kaiberen and Panthera started a long-term ecological study in Besh Moinok, in the Issyk-Kul region, on the border with China with the objective of understanding spatial and behavioral ecology of sympatric snow leopards and wolves. The study will involve the collaring of snow leopards, wolves, and argali.

FUTURE NEEDS

Scope research in Kyrgyzstan is large. Most of the camera-trapping surveys have been conducted in Sarychat-Ertash, Naryn, and Kemin Reserves. It will be vital to expand to other areas, including community-based conservation areas and hunting concessions to gain a better understanding of snow leopard distribution and densities, and in particular the factors that contribute to higher densities. As already described, some studies and monitoring efforts that are proposed for launch in 2015 will shed some new information on the ecology of snow leopards.

Greater efforts to support the establishment of community-based hunting conservancies in snow leopard habitat before it is too late are also needed. Community-based hunting conservancies have been established in Chong-Kemin and Aksu as described earlier, and three newly established community-based hunting conservancies are awaiting assignment of hunting grounds. Plans for community-based hunting areas near the Sarychat-Ertash reserve and in the Jalalabad region, where intense poaching and illegal trade in wildlife parts is threatening the survival of snow leopards and their prey, are also underway. As described in Chapter 16.3, communities motivated by the opportunity to use regulated subsistence hunts and trophy hunts to sustainably harvest the mountain ungulate prey of the snow leopard are more likely to protect them.

References

Asikulov, T., 2008. PATCA expert reports: social and economic issues in the Alay region. Available from: http://patca.zerofive.co.uk/

Davletbakov, A., 2010. National Academy of Sciences Report. Bishkek, Kyrgyz Republic.

Doempke, S., 2008. Socio-economic mission report of the Pamir-Alai Transboundary Conservancy Area (PATCA). Available from: http://patca.zerofive.co.uk/

Government of the Kyrgyz Republic, 2006. Red Book of the Kyrgyz Republic.

FFI, 2007. Central Asia snow leopard workshop, Bishkek, June 2006. Meeting report. Fauna & Flora International, Cambridge, UK.

FFI and CMS, 2015. Aspects of transboundary snow leopard conservation in Central Asia.

Izumiama, S., Anarbaev, M., Watanabe, T., 2009. Inhabitation of larger mammals in the Alai valley of the Kyrgyz Republic. Geogr. Stud. 84, 14–21.

Jumabay-Uulu, K., Wegge, P., Mishra, C., Sharma, K., 2013. Large carnivores and low diversity of optimal prey: a comparison of the diets of snow leopards *Panthera uncia* and wolves *Canis lupus* in Sarychat-Ertash Reserve in Kyrgyzstan. Oryx 48, 529–535.

Koshkarev, E.P., Vyrypaev, V., 2000. The snow leopard after the break-up of the Soviet Union. Cat News 32, 9–11.

Murray, M., 2008. Support to the establishment of the Pamir-Alai Transboundary Conservancy Area between Kyrgyzstan and Tajikistan (PATCA project): Biodiversity Surveys of 2007. Available from: http://patca.zerofive.co.uk/

Sagimbaev, S., 2007. PATCA expert reports: the mammals of the Alay Valley. Available from: http://patca.zerofive.co.uk/

Snow Leopard Network., 2014. Snow Leopard Survival Strategy. Revised version 2014.1. Snow Leopard Network. Available from: www.snowleopardnetwork.otg

Snow Leopard Working Secretariat., 2013. Global Snow Leopard & Ecosystem Recovery Program (GSLEP). Bishkek, Kyrgyz Republic.

Taubmann, J., Sharma, K., Zhumabai Uulu, K., Hines, J.E., Mishra, C., 2015. Status assessment of the endangered snow leopard *Panthera uncia* and other large mammals in the Kyrgyz Alay, using community knowledge corrected for imperfect detection. Oryx. 50 (2), 220–230.

Toropova, V., 2007. PATCA expert reports: the use of resources Alay and conservation of biodiversity. Available from: http://patca.zerofive.co.uk/

Undeland, A., 2005. Kyrgyz Livestock Study: Pasture Management and Use. World Bank, Bishkek.

Watanabe, T., Izumiyama, S., Gaunavinaka, L., Anarbaev, M., 2010. Wolf depredation on livestock in the Pamir. Geogr. Stud. 85, 26–35.

CHAPTER

32

Central Asia: Kazakhstan

Oleg Loginov

Snow Leopard Fund, Ust-Kamenogorsk, Kazakhstan

INTRODUCTION

Snow leopard or irbis (*Panthera uncia*) is one of the rarest large mammals of Kazakhstan. In 1978, simultaneously with the Red Book of the USSR, the species appeared in the Red Book of Kazakhstan, which has been republished four times. In all editions, the snow leopard is listed in category III as "a rare species, whose area and number are reduced" (Grachev, 1996). It makes sense to raise the protected status of the snow leopard in Kazakhstan to reflect its peripheral, fragmented range and low numbers. The snow leopard is also the national symbol of Kazakhstan as presented by the president of the Republic of Kazakhstan, Nursultan Nazarbaev, in a message to the people, "Strategy – 2030." Until now, no special document regulated measures for the protection of this animal in Kazakhstan. The National Strategy for snow leopard conservation in Kazakhstan (Loginov, 2011) could become such a document. The development of this strategy was carried out by a Snow Leopard Network (SLN) small grant, on the initiative the Snow Leopard Fund together with "Conservation and sustainable development of biodiversity in the Kazakhstan part of Altai-Sayan eco-region" a joint project of the government, United Nations Development Program (UNDP), and the Global Environment Facility (GEF). The strategy was approved by the Scientific-Technical council of the Forest and Hunting Committee of the Ministry of Agriculture of Republic Kazakhstan on August 4, 2011.

DISTRIBUTION

Snow leopards are found in the extreme east, southeast, and south of Kazakhstan, along the borders with Russia, China, Kyrgyzstan, and Uzbekistan. The distribution is fragmented and covers about 2.7% of the global total, but is nonetheless important. For example, the Boro-Horo, Zhongar Alatau, and Saur ranges form a natural corridor connecting the Tien Shan and Altai-Sayan mountains. The distribution of snow leopard in Kazakhstan practically coincides with that of Siberian ibex (*Capra sibirica*), its basic prey species.

Snow Leopards
http://dx.doi.org/10.1016/B978-0-12-802213-9.00032-8

POPULATION SIZE

In 1920–1950, the snow leopard was still quite common in the mountains of Central Asia and Kazakhstan. In the early 1980s, there were 180–200 individuals in ~50,000 km^2 of suitable habitat according to Fedosenko (1982). Now it is estimated that there are approximately 100–120 snow leopards in Kazakhstan (Loginov, 2011; Zhiryakov and Baidavletov, 2002).

The current distribution of snow leopard in Kazakhstan remains as it was in earlier years. But unfortunately, the lack of detailed research and information from some regions does not allow precise estimates of population size. There are five subpopulations of the snow leopard in Kazakhstan.

Western Tien Shan

Located in South Kazakhstan and Dzhambul (Zhambylsky) oblasts, the snow leopard occupies several mountain ranges adjoining Kyrgyzstan and Uzbekistan: the Talass Alatau, Ugam, and part of Karzhantau. In the Ugam range, snow leopards occur in upper courses of the rivers Ugam, Sairam, Ularsaj, and Chilhursaj; and in Talass Alatau, in Aksu-Zhabagly Nature Reserve in the upper parts of the Zhabagly, Kshikaindy, Balabaldarbek, and Aksu valleys. Rough number of irbis here are 8–10 individuals (Loginov, 2011). The snow leopard is also found within the Kazakhstan part of the Kyrgyz Alatau in the northern Tien Shan. There are no protected areas in the Kazakh part of Kyrgyz Alatau, nor even rough estimates of numbers.

North Tien Shan

Here the snow leopard occupies ranges bordering Kyrgyzstan and China, including Transili Alatau, eastern part of Kungei-Alatau, and Ketmen (or Uzynkara). There are three protected areas in Transili Alatau, including Almaty Reserve and Ile-Alatau National Park near the city of Almaty with a population of more than 1 million people. The number of snow leopards in Almaty Reserve was estimated at 30–35 individuals by Zhiryakov and Baidavletov (2002).

Zhongar Alatau

The most important ridges here are Toksanbai, Borohoro (adjoining China), and the Central Zhongar ridge or Zhetysu Alatau. Snow leopards are only found in the eastern half of the ridge from the Aksu River in the west, to Chindal in the east. Zhongar Alatau is one of the major snow leopard areas in Kazakhstan, connecting the Tien Shan with the Saur-Tarbagatay and Altai populations. Snow leopards also occur in the higher part the Altyn-Emel and Kojandytau ridge, on the southern side of Zhongar Alatau. This grouping may total about 45–55 snow leopards.

There have been reports of snow leopards at lower elevations in this region. For example, approximately 20 years ago in winter the leopard was observed near a poultry farm in Malinovka, east of the city of Tekeli at about 1000 m. (R. Minibaev, Tekeli, personal communication). There are also sightings of snow leopards in in semidesert habitat of the Alakol hollow, to the east of Zhongar Alatau, that indicate movement from these mountains toward Tarbagatay and Saur. In January 2010 a snow leopard was observed in deep snow in Korinsky canyon at approximately 1600 m in coniferous forest. It was possible to photograph the animal from 1.5 m (Figs 32.1 and 32.2).

Saur-Tarbagatay

These two ranges lie in eastern Kazakhstan, close to the border with China. The range Saur rises to 3723 m at Muztau peak. Tarbagatay is lower, but more extensive. Records of snow leopard mainly occur near Muztau. The population of Siberian ibex on Saur is considered to be stable. In Tarbagatay, snow leopards have not

FIGURE 32.1 **Snow leopard photographed in deep snow in Zhongar Alatau.**

FIGURE 32.2 **Snow leopard in coniferous forest in Zhongar Alatau.**

been sighted for a long time, despite the fact that argali migrate through this range. Tarbagatay Wildlife Sanctuary protects the argali, and plans are in the works to transform it into a national park.

South Altai

This population of snow leopards living in the Kazakhstan part of Altai-Sayan ecoregion borders Russia and China, and its presence is limited to the South Altai, Sarymsakty, and Katun ridges. The snow leopard is rare here and numbers are unlikely to exceed 10 individuals (Baidavletov, 1999). In November 2010, snow leopard signs were found in Archaty in Katon-Karagay National Park. In December 2014, the first image of a snow leopard there was taken by camera trap near to Archaty village within Katon-Karagay National Park, as part of the Snow Leopard Fund/UNDP/GEF project (Fig. 32.3).

An adult male snow leopard was caught in a trap in the winter of 1983 on the northern slope of Kamenistaya Mountain near Cheremshanka (80 km from Ust-Kamenogorsk). Unfortunately, this animal, a male, died. The same year, one more snow leopard was trapped on the Linejsky ridge in the present West Altai Game Reserve. On Ivanovsky ridge in March 1980, a group of tourists of observed a snow leopard on top of a ridge, eating a rock ptarmigan (*Lagopus muta*). On seeing the people, the snow leopard showed no aggression and slowly departed (Zinchenko and Loginov, 2011).

THREATS

The main threats to the snow leopard in Kazakhstan are poaching, decrease in number of prey, degradation and fragmentation of habitats, and lack of effective protection.

FIGURE 32.3 **First photo of a snow leopard in Katon-Karagay NP, Altai.**

SNOW LEOPARD CONSERVATION IN KAZAKHSTAN

Legal

Protection of snow leopards in Kazakhstan is regulated by the criminal code of the Republic of Kazakhstan (RK), specifically the laws "On protection, reproduction and fauna use" and "On specially protected natural territories." Article 288 concerning "Illegal hunting" stipulates large fines or jail sentences of up to two years, and more if the crimes are carried out by organized groups or especially large groups. Article 289, "Infringement of rules of protection of fauna," and Article 290 also contain high penalties or imprisonment up to two years.

Protected Areas

There are 15 protected areas of different types that are known to harbor snow leopard populations (Table 32.1). Since Kazakhstan's independence, two projects have addressed snow leopard habitat: the Government/UNDP/GEF project "Conservation and sustainable development of biodiversity in the Kazakhstan part of Altai-Sayan ecoregion" and "Conservation in-situ of mountain agrobiodiversity in Kazakhstan." Together these resulted in the creation of six new national parks and expanded the area of some other reserves. The highest level of protection is provided by State Reserves (Zapovednik), which are strict nature reserves (category Ia in the IUCN classification). Snow leopards occur in four of these "strict reserves." There are also six state national natural parks (IUCN category II) and four wildlife sanctuaries (category IV), which were created during the Soviet period.

To improve conservation of the snow leopard and its habitats, as well as other biodiversity of Kazakhstan, additional protected areas are needed. The following are proposed:

1. A Game Reserve (or National Park) in Zhongar-Alatau: this should adjoin the existing Zhongar-Alatau NP on the west, and including Verhnekoksu Wildlife Sanctuary and most of the Toksanbai Ridge.
2. Muztau Game Reserve in the Saur range: in the eastern part of the northern slope of the Saur range (a proposal regarding creation of this game reserve has been submitted to the Forest and Hunting Committee).
3. National park in the Kyrgyz Alatau: this should be established in the western part of the Kyrgyz Alatau range, from the Aspara River, including Merkensky canyon, and the western extremity of the ridge, approximately up to Lugovoye village.

TABLE 32.1 Protected Areas in Kazakhstan Where the Snow Leopard Has Been Recorded

	Name/Region	IUCN category	Area (km^2)
	West Tien Shan		
1	Aksu-Zhabagly Game Reserve	(Ia)	1281.18
2	Sairam-Ugam National Park	(II)	1500
	Northern Tien Shan		
3	Almaty Game Reserve	(Ia)	717
4	Ile-Alatau National Park	(II)	1992.92
5	Almaty Wildlife Sanctuary	(IV)	5124
6	Kolsai Kolderi National Park	(II)	1610.45
	Zhongar Alatau		
7	Altyn-Emel National Park	(II)	5200
8	Verhnekoksuisky Wildlife Sanctuary	(IV)	2400
9	Toktinsky Wildlife Sanctuary	(IV)	1870
10	Lepsinsky Wildlife Sanctuary	(IV)	2580
11	Zhongar-Alatau National Park	(II)	3560.22
	Saur-Tarbagatay		
12	Tarbagatay Wildlife Sanctuary	(IV)	2400
	Altai		
13	Katon-Karagay National Park	(II)	6434.17
14	Markakol Game Reserve	(Ia)	1030
15	West Altai Game Reserve	(Ia)	5607.8

4. National park in Narynkolsky region: the proposed protected area is in the southwest part of Almaty region, bordering Kyrgyzstan and China. It should include the spurs of Saryjaz (and the highest point of Kazakhstan), Khan Tengri, and the Kazakhstan part of the Terskei Alatau.
5. Ecological corridors: systematic research is needed to delineate ecological corridors connecting core protected areas in snow leopard habitat to form a uniform chain. Because many sites that are natural migratory "stepping stones" are on boundaries with China and Kyrgyzstan, mechanisms of transboundary cooperation will also be needed. Especially important ridges are Saur, Dzungarian Alatau, and Ketmen (or Usynkara) on the border with China and the northern Tien Shan ranges bordering Kyrgyzstan.

ACTION PLAN FOR SNOW LEOPARD CONSERVATION IN KAZAKHSTAN

A draft summary of the objectives recommended for the snow leopard conservation action plan:

1. Perform scientific research and constant monitoring of the status of all snow leopard populations in Kazakhstan.
2. Conduct a large-scale ecological awareness campaign to highlight the need for the conservation of the snow leopard as a rare animal and national symbol of Kazakhstan.
3. Strengthen measures to counter snow leopard poaching and sale of its fur and body parts for traditional medicine.
4. Expand and develop international transboundary cooperative measures to protect the snow leopard and its habitat.
5. Perfect measures of protection and reproduction of all wild ungulates, which are prey of snow leopard.
6. Establish a system of compensation for damage caused by an attack of a snow leopard on livestock.
7. Develop ecological tourism and other forms of employment for local communities in and around snow leopard habitat.
8. Enhance the protected areas network within Kazakhstan by expanding existing reserves and national parks, and creating new protected areas and ecological corridors.
9. Improve conditions for breeding of snow leopards in zoos of Kazakhstan and create a rehabilitation center to maintain and breed snow leopards near to its natural habitat.

References

Baidavletov, R. Zh., 1999. Report on the Kazakh part of the international programme "Ensuring long-term conservation of the Altay-Sayan Ecoregion," section "current status of the snow leopard population in the Kazakh Altai". WWF grant 435/RU0074.01/GLP. Almaty, Kazakhstan (in Russian).

Fedosenko, A.K., 1982. Snow leopard. Mammals of Kazakhstan Academy of Sciences, Alma-Ata, Kazakhstan, pp. 222–240. (in Russian).

Grachev, Yu.A., 1996. Snow leopard. Red Book of the Kazakh SSR Academy of Sciences, Alma-Ata, Kazakhstan, pp. 246–247.

Loginov, O.V., 2011. Strategy for the conservation of the snow leopard in Kazakhstan. Selevinia, 7–30, (in Russian).

Zhiryakov, V.A., Baidavletov, R.Zh., 2002. Ecology and behavior of the snow leopard in Kazakhstan. Selevinia (1-4), 184–199, (in Russian).

Zinchenko, Yu.K., Loginov, O.V., 2011. On an encounter with the snow leopard in western Altai. Selevinia, 188, (in Russian).

CHAPTER

33

Central Asia: Tajikistan

Abdusattor Saidov, Khalil Karimov**, Zayiniddin Amirov[†], Tatjana Rosen[††]*

*Academy of Sciences, Dushanbe, Tajikistan

**Tajikistan Snow Leopard Program, Panthera, Khorog, GBAO, Tajikistan

[†]Institute of Zoology and Parasitology, Academy of Sciences, Dushanbe, Tajikistan

[††]Tajikistan and Kyrgyzstan Snow Leopard Programs, Panthera, Khorog, Gorno-Badakhshan, Tajikistan

SNOW LEOPARD HABITAT IN TAJIKISTAN

The mountain ecosystem in Tajikistan is of great importance for the survival of the snow leopard (*Panthera uncia*). Tajikistan is located in the center of the distribution, and its mountains have a key connecting role in the entire range. The presence of snow leopards and their wild prey are indicators of a healthy mountain ecosystem of unique ecological, economic, aesthetic and spiritual significance. The total habitat of the snow leopard in Tajikistan covers about 85,700 km^2, which represents 60% of the total area of the country and about 2.8% of the current global range of the species. Snow leopards are widely distributed in the Pamir and Pamir-Alai systems; density is highest in the Rushan, Yazgulem, Vanch, Shugnan, Ishkashim, Sarykol, and Trans-Alai ranges. Muratov (2004) estimated total numbers at 200–220. While localized camera trap studies have been conducted, there is as yet no reliable country-wide population estimate, but rather an educated guess of 250–280 snow leopards in Tajikistan, mostly concentrated in the Pamirs. The Pamir and Pamir-Alai ranges are the main link between the southeastern part of the global range of the species (particularly the Hindu Kush and the Karakoram ranges) and the Tien Shan system and the northern part of the range. Field research in June–August 2014 in the Gissar range confirmed the presence of snow leopards and a preliminary estimate of four individuals (Tara Meyer, Yale University, Panthera, and Tajik Academy of Sciences, unpublished data).

The optimal habitat of the snow leopard in almost all parts of the country is located at an altitude of 2000–4000 m. However, in some areas the terrain and the availability of prey drive snow leopards to lower elevations, even as low as 1000 m. While snow leopards inhabit alpine and subalpine zones with rugged relief, steep

Snow Leopards
http://dx.doi.org/10.1016/B978-0-12-802213-9.00033-X

slopes, and deep gorges across most of its range, the eastern Pamirs are characterized by high-elevation plateaus and snow leopards there use alpine meadows abutting cliffs and other rocky formations.

STATUS OF KEY PREY SPECIES

The main prey species of the snow leopard in Tajikistan include Siberian ibex (*Capra sibirica*), Marco Polo sheep (*Ovis ammon polii*), markhor (*Capra falconeri*), marmot (*Marmota caudata*), Tolai hare (*Lepus tolai*), pika (*Ochotona roylei*), chukar partridge (*Alectoris chukar*), Himalayan snowcock (*Tetraogallus himalayensis*), and Tibetan snowcock (*Trimeresurus tibetanus*). Of these species, ibex, Marco Polo sheep, and markhor are legally harvested, and urial may be hunted in the future. Sustainable use, through hunting tourism, can be an incentive to conserve these species, thus ensuring the availability of prey for the snow leopard and the integrity of the ecosystem and promoting socioeconomic development of local communities in the snow leopard range.

PROTECTED AREAS WHERE SNOW LEOPARDS OCCUR

In Tajikistan snow leopards are found in several protected areas, including Tajik National Park (26,116 km^2), Zorkul State Reserve (SR) (877 km^2), Romit SR (161 km^2), Dashtijum SR (534 km^2), two natural parks (Shirkent, Sarikhosor), and eight reserves with regulated natural resource use. A camera-trap study conducted during the summer of 2011 in Zorkul SR (Diment et al., 2012) showed the presence of at least four different snow leopards (Mallon and Diment, 2014). Marco Polo sheep, ibex, and marmots constitute the main prey for snow leopards in this reserve. Poaching and illegal trophy hunting, especially of Marco Polo sheep, persist although the reserve is taking steps to curb it through the use of camera traps.

Community-Based and Private Conservancies

In the Pamirs, including the westernmost edges in the Darvaz range, and in the Hazratisho range in Shuroabad, community and family-based conservancies have developed trophy hunting programs for ibex, argali, and markhor that provide financial incentives to limit poaching of wild prey species and reduce livestock overgrazing. Implementation of such programs has reversed local declines in ibex, argali, markhor, and snow leopard populations.

Surveys using both camera trap and fecal genetic methods have been conducted in several of these conservancies, providing baseline estimates on snow leopard numbers and showing high snow leopard densities in some areas. In the Ravmeddara conservancy (500 km^2) in the Bartang valley, a survey conducted in 2012 showed the presence of 6 snow leopards; a 2012 survey in the Murghab hunting concession in Jarty Gumbez (1000 km^2) in the eastern Pamirs showed 19 snow leopards (23 snow leopards according to the genetic analysis), and 6 snow leopards (16 snow leopards according to the genetic analysis) in Pshart and Madiyan valleys (1000 km^2) also in the eastern Pamirs. A 2013 survey in Darshaydara (500 km^2) showed also 6 snow leopards and 1 snow leopard in nearby Zong (500 km^2). A 2013 survey in the Zighar conservancy (40 km^2) in the Darvaz Range showed 6 snow leopards. A survey in 2014 in the Alichur conservancy (900 km^2) in the eastern Pamirs showed 3 snow leopards (Panthera, unpublished data).

Community-based conservancies in Tajikistan that have implemented the trophy hunting programs for mountain ungulates received the 2014 International Council for Game and Wildlife Conservation (CIC) Markhor Award. The award showcases the effectiveness that community-based conservancies can have in Central Asia.

THREATS TO SNOW LEOPARDS IN TAJIKISTAN

Snow leopards face many threats to their survival, ranging from retaliatory killing as a result of livestock-snow leopard conflict, illegal trade in snow leopards and their parts, and decline in the prey base.

Decline in Snow Leopard Prey

The decline in snow leopard prey occurs mainly because of poaching, which is the main limiting factor in the number of snow leopards. The reduction in populations of mountain ungulates in many parts of the country is suspected to have dramatically affected the population of snow leopards. Previously, during Soviet Union times, the population of Marco Polo sheep decreased from 70,000 animals in the 1960s to 25,000 animals in the early 1980s due to intensive hunting and poaching, including from members of geological expeditions working each year in the Pamirs (Tajik Academy of Sciences, Institute of Zoology and Parasitology, unpublished data). The civil war (1992–1997) caused a further sharp decline in the population of mountain ungulates (ibex, markhor, urial, Marco Polo sheep) given the general availability of weapons. In recent years, poaching has decreased in some places thanks to protected areas and hunting concessions that actively increased the interest in the conservation and sustainable use of Marco Polo sheep and ibex. This has led to a partial recovery in the populations of argali and ibex. A survey conducted in 2009 counted 23,700 Marco Polo sheep (Michel and Muratov, 2010), and this increase is expected to have had positive effects on the snow leopard.

However, in many places poaching still prevents the recovery of mountain ungulates, leading to a further reduction in the number, range, and distribution of the different ungulate species. The urial sheep, after the death of the last known specimen in 2013, has likely disappeared from the Tajik Wakhan and Badakhshan; populations in other parts of the country continue to decline and thus have little nutritional value for the snow leopard. Ibex have also witnessed a decline in many parts of Badakhshan, but in some parts of the Pamir far away from human settlements the population is stable. Markhor are found in the southwestern part of the Pamir range, in the Darvaz and Hazratishoh ranges along the border with Afghanistan. Conservation efforts in community-based and other hunting conservancies have resulted in a notable increase in markhor numbers. In April 2012, 1018 markhor were counted (Michel et al., 2015), and the latest census, conducted in April 2015, produced an estimate of 1200 (Institute of Zoology and Parasitology, unpublished data). An annual quota of six animals for trophy hunting has been set since 2012.

Degradation and Fragmentation of Habitats

The past 20 years have witnessed increasing human pressure on mountain ecosystems and biodiversity in Tajikistan. Overgrazing; intensive use of mountain land for farming, construction of new settlements, and growth of existing mountain villages; and construction of roads and new power lines increase erosion of mountain slopes and create the preconditions for the degradation and fragmentation of snow leopard habitat, including that of its prey.

Reduction in the Prey Base as a Result of Competition with Livestock

The reduction in prey is also due to competition with livestock as the number of domestic herds and lands allocated to pasture use increase. This is particularly the case for Marco Polo sheep and urial. Overgrazing and haying on the alpine meadows are thought to deprive Marco Polo sheep and ibex from access to grazing grounds, especially in the winter, and significantly reduce their survival and reproduction.

Decrease in Prey Availability for the Snow Leopard as a Result of Collection of Wild Plants for Fuel

The main natural resource used by local people in the Pamirs is teresken (*Ceratoides papposa*), which is widely used as fuelwood. Intensive uprooting of teresken year after year degrades the high steppe ecosystem and pastures (Breckle and Wucherer, 2006). This causes shortages of winter forage and general land degradation. The most affected areas seem to be those where argali are already absent due to poaching and grazing, but as the most accessible teresken stands are already overused the pressure increases in areas that overlap with argali and ibex habitats.

Poaching in Connection with Illegal Trade in Snow Leopard Skins, Bones, and Derivatives

Snow leopard skin is a valuable commodity and in great demand. Specialized local poachers use traps and rifles to kill snow leopards for their skin and derivatives. More than 20 traps were confiscated in 2014 in the eastern Pamirs alone. Often attacks on livestock are used as an excuse for illegally selling snow leopard parts. Anecdotal information suggests that each year 4–5 snow leopards are killed for their skin and other parts. Demand for snow leopard bones comes mainly from outside Tajikistan, and with increasing trade relationships with neighboring countries, border patrols lack the capacity to address the illegal trade of snow leopard parts going out of the country. According to local sources, snow leopards are sometimes trapped, tranquilized, and then released for a hunter to shoot. One snow leopard was also confiscated after being "purchased" by a hunting outfitter for just such a trophy hunt (Panthera, unpublished data).

Panthera has assisted in the establishment of informal anti-poaching networks that have been critical in acquiring information on demand for snow leopards and their parts.

Retaliatory Killing as a Result of Attacks on Livestock

Attacks and retaliatory killings are observed in the Pamirs. Many of the snow leopard attacks occur in late fall and winter. In most cases, the attacking snow leopard enters the corral from an opening in the roof and cannot get out. The snow leopard then kills all the animals in the corral and in response the farmer kills the cat. In such cases the farmer may then try to sell the skin to compensate for the livestock loss. Predator-proofing of corrals (see Chapter 14.1) has been an effective means of reducing conflicts with snow leopards.

LEGAL PROTECTION

Snow leopards have been included in the Red Book of Tajikistan as "a rare species, decreasing in number," but the assessment requires updating and its reconciliation with the global assessment of the species on the IUCN Red List. The conservation and use of rare and endangered species of flora and fauna, such as the snow leopard, are included in the Red Book of Tajikistan and regulated under the laws on "Environmental Protection" (1993), "Animal World" (2007), "Protected Areas" (2012), and relevant regulations. Illegal hunting of snow leopard is punished with a penalty of at least 4000 somoni (approximately 1,000 USD) up to 240,000 somoni (approximately 50,000 USD).

In 2000, Tajikistan ratified the Convention on the Conservation of Migratory Species of Wild Animals (CMS), where snow leopards are listed under Appendix I. As of March 30, 2016, Tajikistan is a member of the Convention on International Trade in Endangered Species of Fauna and Flora (CITES).

The Snow Leopard Action Plan

In 2010, the Institute of Zoology and Parasitology of the Academy of Sciences of Tajikistan, supported by Fauna & Flora International and Panthera, initiated a process to develop a national action plan. The draft plan is still awaiting final approval by the government.

NSLEP 2014–2020

In 2013, the Tajik government endorsed the National Snow Leopard Ecosystem Protection Priorities (NSLEP) plan, developed in the context of the Global Snow Leopard and Ecosystem Protection Program (Chapter 45).

FUTURE NEEDS AND PRIORITIES

Priorities include reducing human-snow leopard conflicts through the use of predator-proof corrals, livestock guard dogs, and improved husbandry practices. Therefore, financial resources need to be identified to help communities in these actions. Addressing the threats to the key snow leopard prey (Marco Polo sheep, ibex, and markhor, in particular) from poaching is also critical and is being accomplished by the community conservancies and the anti-poaching networks. These need to be expanded to have a greater impact across the range in Tajikistan. The capacity of protected area staff also has to be increased, including by providing equipment and vehicles to patrol the areas. It also necessary to establish collaboration and partnerships across different ministries (such as Security and Customs) to increase the ability to track and combat illegal trade in snow leopards and their parts, which includes technical support.

References

Breckle, S.W., Wucherer, W., 2006. Vegetation of the Pamir (Tajikistan): land use and desertification problems. In: Spehn, E.M., Liberman, M., Korner, C. (Eds.), Land Use Change and Mountain Biodiversity. CRC Press, Boca Raton, FL, pp. 227–239.

Diment, A., Hotham, P., Mallon, D., 2012. First biodiversity survey of Zorkul reserve, Pamir Mountains, Tajikistan. Oryx 46, 13–14.

Mallon, D., Diment, A., 2014. Biodiversity Survey of Zorkul Nature Reserve – Summer 2011. Fauna & Flora International, Cambridge, UK.

Michel, S., Muratov, R., 2010. Survey on Marco Polo Sheep and other mammal species in the Eastern Pamir (Republic of Tajikistan, GBAO). Nature Protection Team, Khorog, and Institute of Zoology and Parasitology of the Academy of Sciences of the Republic of Tajikistan, Dushanbe.

Michel, S., Rosen Michel, T., Saidov, A., Karimov, Kh., Alidodov, M., Kholmatov, I., 2015. Population status of Heptner's markhor *Capra falconeri heptneri* in Tajikistan: challenges for conservation. Oryx 49, 506–513.

Muratov, R. Sh., 2004. On the condition of the population of snow leopard in Tajikistan. In Fauna of Tajikistan, pp. 228–230. Academy of Sciences, Dushanbe, Tajikistan (in Russian).

CHAPTER

34

Central Asia: Uzbekistan

*Alexander Esipov**, Elena Bykova**, Yelizaveta Protas†, Bakhtyor Aromov‡*

*Institute of the Gene Pool of Plants and Animals, Uzbek Academy of Sciences, Tashkent

**Institute of the Gene Pool of Plants and Animals, Uzbek Academy of Sciences, Tashkent, Uzbekistan

†Independent, Jackson Heights, New York, NY, USA

‡Gissar State Nature Reserve, Gosbiokontrol, Tashkent, Uzbekistan

SNOW LEOPARD STATUS

Uzbekistan is the range state with the smallest snow leopard (*Panthera uncia*) population, on the westernmost edge of the range. At this time, it remains relatively unstudied, likely fragmented, and in many ways data deficient. The Uzbekistan population is nonetheless an important population, as a healthy peripheral population serves as an indicator of the health of the species as a whole.

In Uzbekistan the snow leopard is found in the West Tien Shan (Ugam, Maidantal, Talass Alatau, Pskem, and Chatkal ranges) and the Pamir-Alay system (Turkestan, Zarafshan, and Gissar ranges). The total area of snow leopard habitat in Uzbekistan is about 10,000 km^2, or 0.5% of the global range (Biocontrol of Uzbekistan, 2014).

The total snow leopard population in Uzbekistan is variously estimated at 30–50 individuals based on questionnaires and yearly counts from reserves (Kreuzberg-Mukhina et al., 2004; Azimov, 2009), or 80–120 individuals based on expert estimates from unpublished estimation methods (Biocontrol of Uzbekistan, 2014), that is, 1–3% of the global population. The populations in West Tien Shan and the Pamir-Alay systems are 10–15 (or 30–40) individuals and 20–30 (or 50–80) individuals, respectively. The population size also varies seasonally due to transboundary movements.

However, accurate data on snow leopard population size are lacking because targeted research has not been conducted in Uzbekistan in the last decade. The available data are sporadic, cover a limited geographic area, and are not species-specific, having been mostly collected during yearly counts of multiple species in Chatkal and Gissar Reserves. No data exist for some parts of the range including Zaamin Reserve, Jizzakh region, Gissar range outside of

Snow Leopards
http://dx.doi.org/10.1016/B978-0-12-802213-9.00034-1

Gissar Reserve, and the Tupalang and Surkhandarya river basins.

Snow leopards typically inhabit elevations of 2200–4500 m, preferring areas with rugged topography, such as rocky gorges and crags interspersed with small plateaus and alpine vegetation. This landscape provides good cover and is inhabited by Siberian ibex (*C. sibirica*), its most important prey species. In Gissar, snow leopards usually stay in the open juniper (*Juniperus*) forest zone and higher. In winter, snow leopards migrate vertically, following the ungulates to lower elevations, usually not descending below the juniper zone, about 2500–2800 m, except in winters with heavy snow.

PREY SPECIES

In Uzbekistan, the Siberian ibex constitutes the key prey species. Wild boar (*Sus scrofa*), voles (*Alticola argentatus, Microtus* spp.), red pika (*Ochotona rutila*), red marmot (*Marmota caudata*), Menzbier's marmot (*M. menzbieri*), tolai hare (*Lepus tolai*), snowcock (*Tetraogallus himalayensis*), and partridge (*Alectoris chukar*) constitute prey of secondary importance (Ishunin, 1961; B. Aromov, Gissar Reserve, unpublished data). In Gissar Reserve, the red marmot constitutes a significant prey species, particularly in spring/early summer. Occasional depredation of domestic livestock (cattle, goats, sheep, horses, and donkeys) does occur, but does not seem to be a major food source. Kill site inspections documented in Gissar Reserve since 1981 include 65 kills of ibex, 11 wild boar, 45 marmot, (hunted either by lying in wait or digging), and 118 partridge and snowcock kills (B. Aromov, Gissar Reserve, unpublished data).

Siberian Ibex

The Siberian ibex is widespread in both the West Tien Shan and Pamir-Alay, and its geographic range largely coincides with that of the snow leopard (Fig. 34.1). Males are more frequently preyed upon than females (Table 34.1). The ibex population is estimated at more than 2000 individuals (Bykova and Esipov, 2006). Approximately 1200–1300 reside in the West Tien

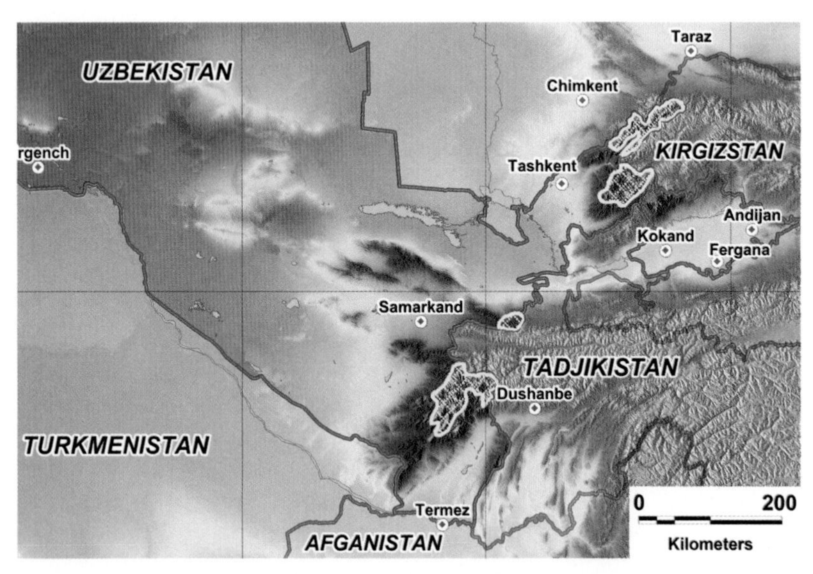

FIGURE 34.1 **Current distribution of snow leopard and Siberian ibex in Uzbekistan.**

TABLE 34.1 Demographics of Killed Ibex from Snow Leopard Kills Discovered in Gissar Reserve, 1981–2014 (Gissar Reserve, Uzbekistan, unpublished data)

Age	Total	Male	Female
1	0		
2	0		
3	4	1	3
4	2	1	1
5	7	5	2
6	3	1	3
7	12	10	4
8	12	8	4
9	5	4	1
10	15	12	3
11	2	1	1
12	2	2	–
13	1	1	–
Total	65	44	21

Shan (Dyakin, 2002), of which approximately 400 are in Chatkal Reserve (Kashkarov, 2002b). Gissar Reserve in the Pamir-Alay contains up to 1000 individuals. The smallest population, 35–40 individuals, is in Zaamin Reserve.

Sympatric Carnivores

The gray wolf (*Canis lupus*) and Eurasian lynx (*Lynx lynx isabellinus*) might compete with the snow leopard for prey, but further study is needed. Wolf counts have not been conducted in a long time, but population size for Uzbekistan overall was estimated at 1500 including 1200 individuals in the mountain regions and 300 in the deserts (according to unpublished government figures in 1992). The population has likely decreased since then, but more recent studies conclude that the wolf is a common species (Kashkarov, 2002b; Mitropolsky, 2005). The wolf population in Western Tien Shan is estimated at 230–250 individuals, present at all elevations (Dyakin, 2002). Wolf status in Gissar Range is not well known.

FIGURE 34.2 **Lynx on camera traps in Gissar Reserve, 2014.** *Source: Photo courtesy of Y. Protas, Panthera, WWF Central Asia Program, Uzbek Biocontrol Agency, Gissar Nature Reserve.*

The lynx is locally listed in the Red Data Book of Uzbekistan. It has been sporadically reported in Tashkent Region, extending into neighboring Kazakhstan (Mitropolsky 2005; B. Tuichiev, Burichmulo Forestry Farm, and A. Mokh, personal communications). It was recently photographed by camera traps in the Chatkal range (Fig. 34.2), altering our understanding of its distribution in the Western Tien Shan (Esipov et al., 2014, 2015). In Gissar, the lynx has been recently confirmed in all available habitat types (Y. Protas and B. Aromov, Gissar Reserve, Uzbekistan, unpublished data).

EXISTING PROTECTED AREAS AND THEIR EFFECTIVENESS (CHATKAL, GISSAR, ZAAMIN)

In Uzbekistan, the snow leopard is protected in three Strict Reserves: Chatkal, Gissar, and Zaamin; and two National Parks (NP): Ugam-Chatkal and Zaamin (Table 34.2). Together, these encompass approximately 65% of the total snow

TABLE 34.2 Snow Leopard Habitat in Uzbekistan Coverage by Protected Areas

Name of PA (established year)	Location	Area (km^2)	Administered by	Geographical placement	IUCN category	Snow leopard presence	
						Year round	Seasonal
STRICT RESERVES							
Chatkal Biosphere Reserve (1947)	Tashkent Region	357	Tashkent City Government	West Tien-Shan Boshkyzylsai section: western Chatkal range; Maidantal section: north-facing slope of Chatkal range	I	+	+
Gissar Biosphere Reserve (1983)	Kashkadarya Region	810	State Committee of the Republic of Uzbekistan for Nature Protection	Pamir-Alay West-facing slope of Gissar range	I	+	+
Zaamin mountain and juniper Reserve (1926, 1960)	Jizzakh Region	268	Ministry of Agriculture and Water Resources of the Republic of Uzbekistan	Pamir-Alay North-facing slope of the Turkestan range	I		+
NATIONAL PARKS							
Ugam-Chatkal National Park (1990)	Tashkent Region	5746	Tashkent City Government	West Tien-Shan Pskem River basin, western and northern slopes of Chatkal range	II	+	
Zaamin National Park (1976)	Jizzakh Region	241	Ministry of Agriculture and Water Resources of the Republic of Uzbekistan	Pamir-Alay North-facing slope of the Turkestan range	II		+

leopard habitat. However, only about 6% of the total area is inside Strict Reserves. The upper elevations of Gissar, Pskem, Zaamin, and Tupalang ridges are classified as closed, security-protected border zones.

The Ugam-Chatkal NP covers most of the West Tien Shan mountain range and directly borders Sairam-Ugam NP and Aksu-Jabagly Reserve in Kazakhstan, and Besh Aral Reserve in Kyrgyzstan; the new Padysha-Ata Reserve in Kyrgyzstan is also located close by, thus greatly increasing overall protection that allows animals to migrate along existing ecological corridors. However, no neighboring protected areas exist near Gissar Reserve.

Snow leopard population density is likely close to maximum inside reserves and is much higher than in neighboring unprotected territories (Mitropolskaya, 2009). However, protected areas (Pas) do not afford complete protection, and about 35% of snow leopard habitat is outside protected areas, with a high level of poaching and anthropogenic disturbance all over. Where the protected areas are small and do not include seasonal habitats (e.g., ibex winter range lies outside Chatkal Reserve), the ungulate population can migrate and experience increased seasonal poaching (Mitropolskaya, 2009).

Chatkal Biosphere Reserve

The snow leopard population is estimated at 2–3 individuals and was likely negatively affected by the administrative expiration of the wildlife sanctuary formerly located in the Akbulok River basin bordering on the reserve.

Gissar Biosphere Reserve

The largest Strict Reserve in the country contains the highest snow leopard population and density. The most complete data on Uzbekistan snow leopard ecology comes from here. The latest population estimate is a maximum of 25, varying seasonally with movements across the border. Good snow leopard habitat inside the reserve totals about 500 km^2, in the Kizilsu and Tanhaz river valleys and the mountains around them. The snow leopards mainly remain in the alpine and subalpine meadows in the summer, coming down into the open juniper forest in the winter, shifting between 2200 m and 4300 m elevation. In the Gilan sector in the north, snow leopards are present in the warmer months, migrating to the warmer south-facing slopes across the border in Tajikistan in winter.

TABLE 34.3 Estimated Snow Leopard Population in Gissar Biosphere Reserve (800 km^2) 1981–2014

1981–1990	2–13
1991–2000	11–15
2001–2010	16–24
2011–2014	23–25

The population has been steadily increasing for all the years data have been available (Table 34.3). This may be due to improved protection measures implemented since the reserve's inception in 1975, lack of tourists, relocation of two villages, and more recently, seasonal emigration of human population for temporary work. Data have been gathered since 1981 with regular twice-in-a-year sign counts along permanent transect routes in the reserve, allowing comparison over several decades. Visual encounters occasionally take place, sometimes including cubs. Fourteen visual encounters with cubs have been recorded between 1981 and 2014.

In 2013, a camera trap study was launched in Gissar Reserve. Although inadequate data were obtained to estimate population size, presence of snow leopards was confirmed photographically for the first time (Fig. 34.3), with at least two individuals and six capture events over three field seasons in 2013 and 2014. Snow leopard population status has not been studied in areas near Gissar Reserve where it might be present, such as the Tupalang River basin or Surkhandarya province of Uzbekistan (Western Gissar).

FIGURE 34.3 **Camera trap photo of snow leopard in the Tanhaz River basin, Gissar Reserve.** *Source: Photo courtesy of Y. Protas, Panthera, WWF Central Asia Program, Uzbek Biocontrol Agency, Gissar Nature Reserve.*

Zaamin Nature Reserve

In recent years, the Zaamin Reserve has not conducted research or sign counts, because this protected area is entirely inside a designated border zone and is closed to all visitors. The small staff lacks the ability to conduct regular monitoring. If a population exists, it is likely small, perhaps just 2–3 individuals.

In 1992, the upper Pskem Forestry Division, which used to be subjected to heavy grazing, was added to Ugam-Chatkal National Park, and the entire territory of Pskem Valley from Karabuloq upwards was declared a border zone, with entry forbidden without a special pass. Anthropogenic activity from the border post to the national border was limited to protected area employees. The two protected areas, Zaamin and Ugam-Chatkal, combined protect an estimated 15–20 snow leopards.

PLANNED PROTECTED AREA EXPANSION

The National Biodiversity Strategy and Action Plan (National Biodiversity Strategy Project Steering Committee, 1998) calls for extending the protected area network and increasing the total area covered to 10% of the country. The strategy and plan are currently being reworked to take into account the value of ecosystem services and ecosystem adaptations to climate change. The new Pskem Reserve, which will be located inside the Ugam-Chatkal National Park as a core zone, is currently in the development stage. However, there is even greater potential to increase territorial protection for snow leopards.

We recommend the following steps to optimize protection and expand the protected area network:

1. Expand Chatkal Reserve by adding the Akbulak River Basin and several territories on the southern slopes of Chatkal range, specifically the south-facing slopes of Kurgantash Mountain, which serves as a wintering area for ibex and snow leopards; create an ecological corridor between Boshkyzylsay and Maidantal sections of the reserve.
2. Expand Zaamin Reserve and Zaamin National Park by annexing the north-facing slopes of the Turkestan Range.
3. Create a national park in the middle and eastern part of the Gissar Range.

THREATS

A combination of anthropogenic and other factors present direct or indirect threats to snow leopards.

Direct Threats

Traditional Snow Leopard Hunting

Historically, snow leopards were hunted for skins, both for clothing and status. Killing a snow leopard was considered prestigious and demonstrated the hunter's skill. In the Soviet era, redlisting and hunting bans were enacted, though poaching still occurred. In the post-Soviet era, there is one known report of a skin on sale for USD 1000 in 2004.

Live Capture of Cubs and Adults

Although cubs were not captured from this region during the Soviet era, it later became popular to capture young cubs for private zoos. By the early 2000s there were multiple documented cases of illegal captures. The cost of a cub was about USD 1,000 and of a live adult about USD 5,000-10,000. Current information on the capture and selling of individuals is not available.

Conflicts with Local Herders

A widespread belief is that snow leopards greatly impact the domestic animal population. In reality, this is greatly exaggerated, and snow leopards attack domestic livestock only occasionally. Goats and sheep are depredated the most, less frequently cattle, horses, and donkeys. A total of 82 cases of snow leopards attacking domestic animals have been recorded for the Gissar Range outside of the reserve. In one attack, 35 individual domestic animals were killed, although the average is 5.5 per case, and retaliatory killing after such incidences does occur (Bykova et al., 2004). Such incidents may be related to a diminished natural prey base, as a result of disease, overhunting, and other factors, which forces snow leopards to switch to other prey types. Predation on livestock increases in winter, when preying upon ibex becomes harder due to heavy snow.

In summer, livestock herds are often left unguarded in the daytime, or guarded by children, who cannot always protect the animals. Poor construction of summer corrals contributes to this problem, since they are easy to penetrate (low walls, no roof), but at the same time prevent livestock from escaping. In the winter, snow leopards may penetrate pens through roofs covered only with reeds. Villages near reserves also pose a problem; for example Gissar Reserve is surrounded by 11 Uzbek and Tajik villages. Conflict between the reserve and the mostly unemployed residents of Tashkurgan and Chopukh villages was finally resolved through forced relocation of the ~500 villagers into low-lying farmland in the 2000s.

Armed Human Conflict

Minefields, which are a relic of the civil war in neighboring Tajikistan, are located along the border near Gissar Reserve where animals cross during seasonal migrations. There are known cases of brown bear (*Ursus arctos*), ibex, and other large animals triggering mines.

Natural Mortality

The West Tien Shan receives deep snow in winter, and avalanches are common. Local residents say at least three snow leopards have been killed by avalanches while hunting ibex in the Pskem River valley in recent years.

Indirect Threats

Loss of wild ungulates is a significant threat. The ibex is intensely poached along its entire range. On average, 30 ibex are poached each winter just from Pskem Village near Chatkal Reserve (Kashkarov, 2002a). The Pamir-Alay ibex are vulnerable even inside protected areas. Despite large sections of the Turkestan and Gissar ranges being classified as border regions with highly restricted access, ibex populations are heavily poached near the Tajik border. Disease periodically causes prey populations to decrease. For example, an outbreak of sarcoptic mange caused a steep decline of ibex populations in the 1970s.

Subsistence Hunting

Traditional hunting by local residents for meat and skins is illegal and not easily measured. Target species are ibex and wild boar collected for meat and skins, and marmots, collected for purported medicinal properties. Regulations limiting travel and tourism in mountain border regions, and the ban on rifle possession by civilians, are intended to stabilize or increase wild ungulate numbers. However, the low quality of life in the remote mountain villages and loose law enforcement result in continued poaching. Residents of remote villages usually possess limited understanding of hunting laws and poorly understand the objectives of protected areas. Wildlife is perceived as anyone's free, unlimited resource. People are not concerned with sustainable use because they have short-term goals.

Sport Hunting

Hunting for sport is practiced mainly by city residents, who usually hunt legally, although unlicensed hunts have been reported. Birds are usually the main object, with ungulates taking second place. Legal hunting has resulted in ungulate population declines. For example, in the late 1980s, commercial wild boar hunting in West Tien Shan led to a disruption of breeding female numbers and a steep decline in overall population size. Today, that population is on the increase, but has still not recovered to its former size. Wild boar is not the preferred snow leopard prey, but this demonstrates the effects of overexploitation on a formerly numerous species.

VIP Hunting

VIP hunts are conducted by locally influential officials and military personnel. This category of hunter is active at any time of year, targeting any type of game, including bear and snow leopard, and attempts to control their actions are ineffective. Snow leopard hunting by law enforcement officials has been reported in the past (Bykova et al., 2004; Mitropolsky, 2005).

Disturbance Factors: Human Land Use, Collection of Natural Products

In the past, alpine regions were frequently visited by pastoralists, geologists, tourists, hunters, religious pilgrims, and residents collecting herbs, nuts, and fruit. More recently visitor disturbance has decreased in some alpine regions due to tighter border protection for security reasons (Pskem, Gissar, Turkestan ranges). Transport such as cars or helicopters is becoming prohibitively expensive, and partial cessation of geological research and extraction (strip-mining of barium ore near Chatkal Biosphere Reserve ceased in 2002) have all led to a decrease in human disturbance in some areas.

Recreational activity in Uzbekistan is largely associated with the mountain regions, which contain numerous resorts, vacation homes, children's camps, and so on. Recreational activity creates ever-increasing pressure on mountain territories, due to urban population growth, hiking, skiing, paragliding, rafting, and other sports take place, particularly in the Tien Shan Range around Tashkent, and less so in the Pamir-Alay Mountains. Building and expansion of mountain villages, road building and electrification, use of mountain sides for food cultivation (fruit and nut trees, grains, potato, tobacco), and tree cutting, all contribute significantly to wildlife habitat loss and mountain soil erosion.

Grazing Herds: Competition Between Wild Prey and Livestock

Use of high elevation alpine meadows for seasonal grazing is widespread in the country. In the Soviet era, these pastures were used by both local pastoralists, as well as those from nearby republics. Today herd sizes have shrunk due to economic hardship. Use of pastures is allowed only for Uzbekistan nationals, and only if they have a license granted by the appropriate agency. Such a license stipulates type and number of animals to be grazed.

Until 1990, the Pskem River basin was under active agricultural use. All alpine territories were used for grazing from March to September, up to 4500 m elevation, mostly by Kyrgyz herders under long-term lease agreements. Such use stopped in the early 1990s when the Central Asian Republics gained independence. The Chatkal and Gissar ranges, with higher human populations, still receive substantial pressure from agriculture, particularly pastoralists.

HISTORY OF THE SNOW LEOPARD NATIONAL STRATEGY AND ACTION PLAN

The Snow Leopard Conservation Strategy and Action Plan was prepared in 2004 (Kreuzberg-Mukhina et al., 2004), based on prior conservation plans by the IUCN Cat Specialist Group (Nowell and Jackson, 1996) and the Snow Leopard Survival Strategy (McCarthy and Chapron, 2003). National strategy priorities recognize the problems of species survival in the modern world, develop guiding principles for decision making on conservation, build an information network for collecting and use of data on snow leopard populations, and develop a platform for continued cooperation between stakeholders and international action plans.

In 2013, a review of the National Strategy and Action Plan was conducted. Following this, the National Priorities for Snow Leopard Ecosystem Protection in Uzbekistan 2014–2020 (NSLEP) was developed (Snow Leopard Working Secretariat, 2013). This document is part of the Global Snow Leopard and Ecosystem Protection Program, drafted at the Global Snow Leopard Conservation Forum meeting on October 22–23, 2013, in Bishkek, Kyrgyz Republic (Chapter 45).

The NSLEP Priority Actions are:

- Human-snow leopard conflict reduction through improved livestock pens
- Effective guard dog training and use
- Improved agriculture methods
- Reduction in snow leopard threats, such as poaching, habitat degradation from overgrazing, human overpopulation, and village growth
- Improving equipment and skills capacity of protected areas for effective snow leopard protection
- Stimulating local resident participation in conservation of snow leopard and its prey
- Expanding snow leopard research and monitoring programs
- Conducting an education program using mass media and educational institutions
- Increasing the capacity of key ministries and agencies and their employees for detecting and preventing illegal trade in live animals and their parts
- Technical assistance for illegal trade detection through conducting training seminars for border guards and customs officers
- Increased collaboration and transboundary cooperation with neighboring countries of Kazakhstan, Kyrgyzstan, and Tajikistan

References

Azimov, Zh. (Ed.), 2009. The Red Book of the Republic of Uzbekistan. Volume II. AnimalsChinor-ENK, Tashkent.

Biocontrol of Uzbekistan, 2014. Snow Leopard in Uzbekistan. Econews, http://econews.uz.

Bykova, E., Esipov, A., 2006. Modern status of ungulate game species in Uzbekistan. Selevinia, 194–197.

Bykova, E., Esipov, A., Aromov, B., Kreuzberg, E., 2004. Questionnaire as method of collecting data on endangered species such as snow leopard. Protected areas of Central Asia. Tashkent, pp. 208–214.

Dyakin, B., 2002. Game species resources in Western Tien Shan. Biodiversity of Western Tien Shan: protection and sustainable use. Tashkent, pp. 90–97.

Esipov, A., Golovtsov, D., Bykova, E., 2014. Camera trapping experience in Chatkal Biosphere Nature Reserve, Uzbekistan. Environment and Natural Resource Management. Tyumen, Russia, pp. 96–98.

Esipov, A., Golovtsov, D., Bykova, E., 2015. Fauna of mammals and birds of Western Chatkal ridge by camera trapping. Vestnik of Tyumen University.

Ishunin, G., 1961. Mammals (Carnivores and Ungulates). Fauna of Uzbek SSR. V.3. Tashkent, p. 230 (in Russian).

Kashkarov, R., 2002a. On the fauna of mammals (Carnivora and Artiodactyla) in Pskem river basin. Selevinia, 1–4, 150–158.

Kashkarov, R., 2002b. Modern status and resources of carnivorous mammals (Canidae, Ursidae, Mustelidae). Biodiversity of Western Tien Shan: protection and sustainable use. Tashkent, pp. 115–121.

Kreuzberg-Mukhina, E., Bykova, E., Esipov, A., Aromov, B., Vashetko, E., 2004. Strategy and Action Plan for Conservation of the Snow Leopard in Uzbekistan. Tashkent, Uzbek Zoological Society and State Committee of Nature Protection.

McCarthy, T., Chapron, G. (Eds.), 2003. Snow Leopard Survival Strategy. International Snow Leopard Trust & Snow Leopard Network, Seattle, USA.

Mitropolskaya, Y., 2009. The ratio of protected areas of Uzbekistan and areas of the most important mammal species and communities. Actual Problems of Zoological Science. Tashkent, 10–12.

Mitropolsky, O., 2005. Biodiversity of Western Tien Shan. Data on the mammals and birds in Chirchik and Akhangaran rivers basins, Uzbekistan. Kazakhstan. Tashkent-Bishkek, 166.

National Biodiversity Strategy Project Steering Committee, 1998. Biodiversity Conservation National Strategy and Action Plan. Government of the Republic of Uzbekistan, Tashkent.

Nowell, K., Jackson, P. (Eds.) 1996. Wild Cats: Status Survey and Conservation Action Plan. IUCN/SSC Cat Specialist Group. IUCN, p. 382.

Snow Leopard Working Secretariat, 2013. Global Snow Leopard and Ecosystem Protection Program (GSLEP): Annex. Snow Leopard Working Secretariat, Bishkek, Kyrgyz Republic.

CHAPTER

35

South Asia: Bhutan

Tshewang R. Wangchuk, Lhendup Tharchen***

*Bhutan Foundation, Washington, DC, USA

**Jigme Dorji National Park, Damji, Bhutan

INTRODUCTION

Bhutan lies in the Eastern Himalayas where precipitation is higher (Baillie and Norbu, 2004) than other parts of the snow leopard (*Panthera uncia*) range. Vegetation, therefore, grows at much higher altitudes here, with trees often growing up to 4750 m (Miehe et al., 2007) on north-facing slopes. Forest connectivity from the southern border on the Indian plains to alpine meadows and high Himalayas adjoining the Tibetan border allow for the movement of lowland species such as tiger (*Panthera tigris*) to over 4200 m (Jigme and Tharchen, 2012).

Locally, snow leopard is known as *chen, chengo, sa,* or *tsagay* in the different regions of northern Bhutan. However, in the national language, Dzongkha, it is popularly known as *gangzig* (literally "leopard of snow-capped mountains"). Photographs, genetic evidence, and encounters with herders have confirmed the presence of snow leopards in three protected areas: Jigme Khesar Strict Nature Reserve (JKSNR; 609.5 km^2; previously Torsa Strict Nature Reserve), Jigme Dorji National Park (JDNP; 4316 km^2), and Wangchuck Centennial National Park (WCNP; 4914 km^2) (Tharchen, 2013; T.R. Wangchuk, unpublished data). Although no confirmed evidence of the species has been recorded from Bumdeling Wildlife Sanctuary (BWS; 1564 km^2) so far, based on anecdotal evidence and herders' accounts of predation on livestock, its occurrence here is plausible. However, prey density (especially blue sheep *Pseudois nayaur*) is low in this area. There is also relatively less livestock husbandry in the alpine pastures here as compared to the western parts of northern Bhutan. Scat samples we collected from BWS in 2008 turned out to be all from leopard cat (*Prionailurus bengalensis*) and red fox (*Vulpes vulpes*). Confirming that snow leopards are present in BWS would provide for a better understanding of possible snow leopard conservation units and define its eastern limit for Bhutan. This could be rigorously done in an occupancy framework (such as MacKenzie et al., 2002) using noninvasive techniques (either camera traps or fecal DNA sampling).

Snow Leopards
http://dx.doi.org/10.1016/B978-0-12-802213-9.00035-3

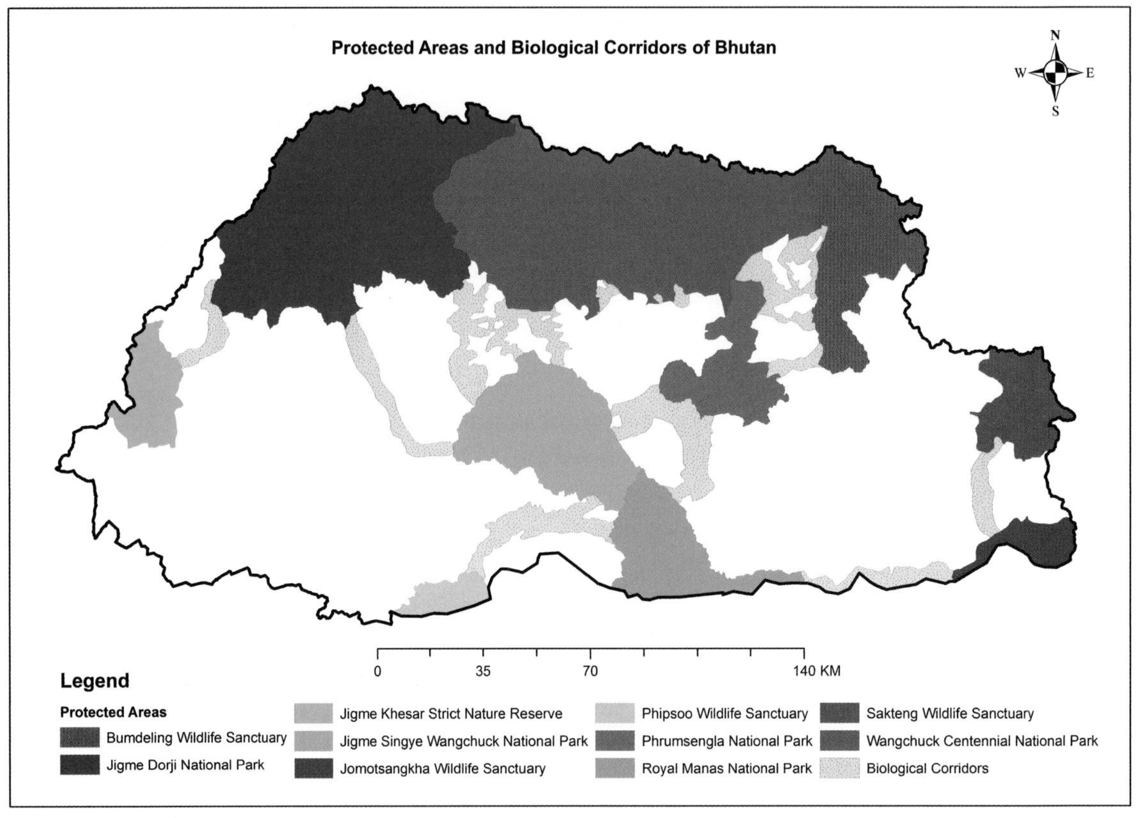

FIGURE 35.1 **Protected area network of Bhutan.** Snow leopard confirmed in Jigme Khesar Strict Nature Reserve, Jigme Dorji National Park, Wangchuck Centennial National Park. It is expected (but not confirmed) in Bumdeling Wildlife Sanctuary.

SNOW LEOPARD HABITAT DISTRIBUTION IN BHUTAN

A seemingly contiguous landscape with no obvious breaks in habitat connectivity comprising JKSNR, JDNP, WCNP, and BWS, and their biological corridors, encompass Bhutan's possible snow leopard habitat (Fig. 35.1). Toward the north, this habitat extends further into the Tibetan Autonomous Region of China after crossing the Bhutan Himalayas. However, it needs to be confirmed whether the Kurichhu River, originating in Tibet, poses a potential break in snow leopard habitat connectivity as it flows through BWS. Within the ca. 9000 km^2 general range in Bhutan, the snow leopard mainly restricts itself to an elevation belt of 3500–5500 m with occasional forays into lower habitats such as alpine valley floors or subalpine forests, usually traversed in transit. At the lower limits, snow leopard often overlaps with tiger and leopard (*Panthera pardus*), making Bhutan the only place in the world for this unique phenomenon where these three large cats occur in the same habitat. Inhospitable barren rocks, snowfields, and permanent ice lie above the upper limit.

Snow leopard habitat in Bhutan overlaps with that of five other felids – golden cat (*Catopuma temminckii*), leopard cat, leopard, Pallas's cat (*Otocolobus manul*), and an occasional tiger (Thinley, 2013; Wangchuk et al., 2004; WWF, 2012). However, these cats

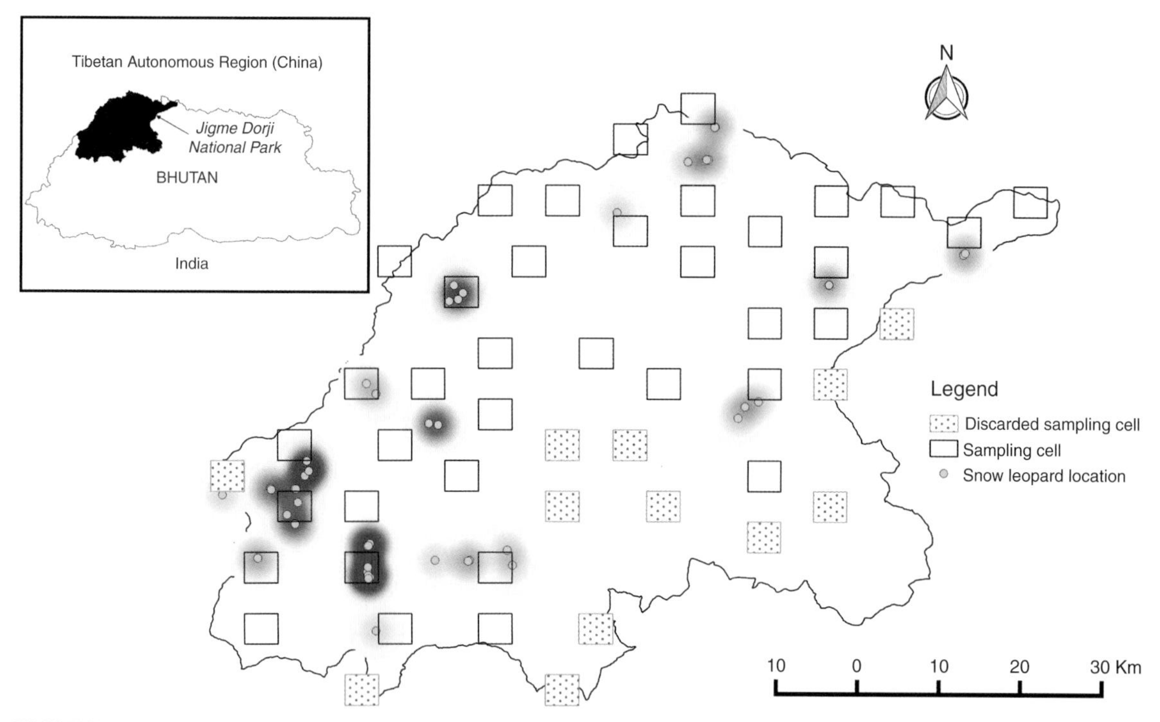

FIGURE 35.2 **Relative snow leopard density in Jigme Dorji National Park.** Based on 64 confirmed samples comprising 40 unique individuals identified from genetic sampling (from a total of 283 putative samples). Dark red indicates higher density; each sampling cell is 4 km × 4 km.

restrict themselves to the vicinity of forested areas, whereas snow leopard is mostly confined to higher, open, and rocky areas. Canids present in snow leopard habitat include red fox, wild dog (*Cuon alpinus*), and wolf (*Canis lupus*). In addition, Himalayan black bears (*Ursus thibetanus*) and the Himalayan weasel (*Mustela sibirica*) are also found in the same areas. The most common prey for snow leopard in Bhutan is blue sheep. Smaller animals such as marmot (*Marmota himalayana*), Royle's pika (*Ochotona roylei*), and various phasianids have been known to contribute to snow leopard diet in similar habitat in Nepal (Oli et al., 1993), and from evidence in our scat samples, we believe this to be true for Bhutan as well. Yak is the most common livestock species in snow leopard habitat, and young yaks often contribute to snow leopard diet.

So far, snow leopard presence has been confirmed from numerous camera trap photos, DNA samples, and direct sightings from JKSNR, JDNP, and WCNP. In many instances groups of three and even four snow leopards (most likely groups with mother and grown cubs) have been photographed or sighted by park staff and herders. In JDNP alone, systematic genetic sampling yielded a total of 40 unique individuals from 64 confirmed samples (T.R. Wangchuk et al., unpublished data). The western part of the park, in the Jomolhari and Lingzhi regions, showed a much higher relative density of snow leopards in comparison to the rest of the park (Fig. 35.2). In one such location, near Bongtey La Pass, a herder sighted a group of four snow leopards (K. Dorji, personal communication, 2014). Groups of snow leopards with grown cubs indicate that they are breeding and surviving into maturity.

THREATS

Direct Threats

Habitat degradation and persecution by poachers and angry herders are major causes of snow leopard population decline globally. Such threats common elsewhere (poaching, retaliatory killings, habitat destruction, and prey depletion) are minimal in Bhutan. While livestock predation by snow leopard is recorded all across its range in Bhutan, retributive killing is rare. This mainly stems from a general aversion toward killing animals influenced by Buddhist values, and fear of the law. There is general awareness that it is illegal to kill a snow leopard or blue sheep. However, attitudes of herders who lose livestock to snow leopards could potentially change from tolerance to anger, this shift often propagated by the introduction of the hitherto unknown concept of "human-wildlife conflict." Although some levels of predation were always tolerated by herders traditionally, the foreign concept of "conflict" (often introduced by conservationists based on experiences from Africa, the American west, and India where predators were persecuted in retaliation for predation), coupled with ineffective processes of monetary compensation, are primarily responsible for this shift in attitude. For Bhutan, preventing this shift in attitude is of utmost importance for snow leopard conservation. If livestock mortality is the main cause of contention and probable retaliation in the future, we propose alternative methods of solving this problem without going down the spurious path of ineffective monetary compensation, which is flawed for two reasons. First, the amount generally allotted for compensation cases is far less than the cost of a yak. Secondly, and more importantly, tedious verification requirements make the process lengthy and bureaucratic. Both these reasons cause herders to raise their expectations only to be quashed with inaction by authorities. In some areas (e.g., Soe), we found that herders lose an equal number or more yaks to *gid* disease (also called sturdy or stagger) than to predation. Treating yaks and dogs (which are the definitive hosts for the parasite responsible, *Taenia multiceps*) and thus reducing yak mortality from *gid* disease appears to be a more practical and effective way to help herders maintain healthy numbers of yaks rather than to only give attention when herders lose their cattle to predation, which is inevitable.

Indirect Threats

Because predator populations depend directly upon prey species, sustaining a prey base is essential for the long-term survival of the snow leopards. Snow leopard habitats are also growing grounds for the highly valued caterpillar-fungus association (*Ophiocordyceps*), which is valued for its medicinal properties (see Chapter 10.4). Although the extent has not been quantified, disturbance of the habitat by *Ophiocordyceps* collectors in summer will most likely have a negative impact on the health of the prey species, which will ultimately affect snow leopard survival. With recent changes in the economy (e.g., large income from *Ophiocordyceps*) and social structure (children moving away to schools and rural-to-urban migration), there is reason to believe that the yak herding economy is also changing, and people in many areas are giving up yak herding, especially in JKSNR. The absence of human and livestock presence in the alpine meadows tend to encourage unpalatable woody vegetation such as rhododendron shrubs to take over open meadows, thereby reducing habitat for important snow leopard prey such as blue sheep. The extent and effect of this recent phenomenon is poorly understood and urgently needs to be studied, along with quantifying potential yak-blue sheep competition in other areas where livestock numbers may be increasing.

Climate Change

Although the Himalayan region in general bears the brunt of climate change through decreasing snow mass and increased frequencies of unpredictable freak weather patterns, it is still unclear how climate change might affect snow leopards (see Chapter 8). Using coarse resolution 500 m habitat covariate maps, Forrest et al. (2012) paint a bleak future whereby Bhutan could lose up to 55% of snow leopard habitat under a scenario of increased emissions. The model considers snow leopard habitat contraction from possible upward shift of forests, but fails to account for corresponding range expansion from upward movement of alpine meadows and snow leopard adaptation to hypoxia, all appreciably long-term phenomena that cannot be ruled out. Such alarmist messages based on models simulated on coarse resolution and scant time-series data run the risk of distracting focus away from important present threats.

LEGAL STATUS OF SNOW LEOPARD

In Bhutan, the snow leopard is listed in Schedule I (totally protected species) of the Forest and Nature Conservation Act of Bhutan 1995, which treats it on a par with species in Appendix I of the Convention on International Trade in Endangered Species of Wild Fauna and Flora (CITES). The Act prohibits "taking, injuring, destroying, killing, shooting, capturing, trade and/or use of snow leopard parts and products" (RGoB, 1995). In 2013 the government revised penalties to further strengthen protection. Fines starting at Nu. 0.5–1.0 million (USD 8,000–16,000), or imprisonment of up to 10 years, or both, may be imposed on poachers attempting to kill or killing a snow leopard. The Forest and Nature Conservation Rules of Bhutan (DoFPS, 2006) and subsequent revisions guide law enforcement against killing snow leopards and other wildlife. Bhutan's constitution mandates that it shall maintain at least 60% of country under forest cover in perpetuity. Presently, more than 50% of the country has been designated under the category of protected areas and biological corridors, and this encompasses all snow leopard habitat in the country. Overall, Bhutan presents a unique opportunity for understanding snow leopard ecology in an area of minimal current threats where ecological conditions within the range have not changed much for the last century.

SNOW LEOPARD CONSERVATION IN BHUTAN

As an apex predator, conserving snow leopards allows for the conservation of a myriad species living with it. The entire snow leopard habitat in Bhutan is a treasure trove of alpine flora and fauna including many medicinal herbs. Income from collection of medicinal herbs contributes to livelihood of highland communities, while the plants themselves provide health benefits to many people who use traditional medicine. Snow leopard habitat serves as the main pasture and grazing ground for yak-herding communities. It is also the source and main catchment area for the major river systems of Bhutan. Almost all the rivers of the country originate in snow leopard country.

For Bhutan, a country that generates more than 40% of its GDP from the sale of hydropower, conserving and protecting the watersheds in the snow leopard habitat are of utmost importance. Therefore, investments in snow leopard and habitat conservation have the added benefit of not only helping sustain upland communities, but also continued generation of national income through hydropower production from its fast-flowing rivers.

Snow leopard habitat, by virtue of occupying the highlands of Bhutan, is also important for spiritual Buddhist practitioners. Many places with religious and cultural significance fall in these areas and are used by hermits and ascetics in their spiritual pursuits. The following list includes major initiatives undertaken by the government, communities, and conservation partners that contribute toward snow leopard conservation:

1. In protecting the livelihood of the people and conserving wild biodiversity, managing human wildlife interactions has been a priority. To address potential retaliation on snow leopards by communities for predating on livestock, awareness creation among the communities on the importance of the species and the ecosystem is being undertaken by park officials through frequent village meetings. The Department of Forests and Park Services (DoFPS) also collaborates with schools in the respective parks in creating awareness of the species and on landscape conservation initiatives.
2. To minimize grazing pressure from livestock, the government has initiated livestock intensification initiatives by introducing high-value cattle breeds. Efforts are also under way to improve livestock breeds to increase their productivity.
3. A monetary compensation program for herders who lost cattle to predators had been started in 2003, but has since run out of funds.
4. As communities are instrumental for long-term conservation, JDNP and WCNP have started community-based conservation initiatives. In these two parks, several yak herders have been modestly compensated for livestock losses through direct cash payment, solar lighting facilities, improved pasture development, and community-based ecotourism ventures. Conservation committees have been formed within the communities, and they have been trained in wildlife monitoring skills to assist park staff. This approach is gaining momentum as it promotes better partnership with local community members in conservation programs.

Park authorities have instituted the Jomolhari Mountain Festival in JDNP, and the Nomad Festival in WCNP to boost tourism, increase awareness, and garner public support in conservation.

Currently, a snow leopard conservation and highlanders' livelihood upliftment program in northern Bhutan, covering most of the areas falling within the snow leopard range, is under implementation by the government. This program is implemented in partnership with WWF-Bhutan Program, the Bhutan Foundation, the Bhutan Trust Fund for Environment Conservation, Global Environmental Facility, and the Snow Leopard Conservancy.

CHRONOLOGY OF SNOW LEOPARD CONSERVATION EFFORTS IN BHUTAN

1. First Snow Leopard Information and Management System (SLIMS) training in Bhutan was held in May 1997 to provide hands-on training in detecting snow leopard sign and counting blue sheep.
2. SLIMS workshop and training was conducted in 2000 and 2007, funded by WWF Bhutan, International Snow Leopard Trust, and Royal Government of Bhutan.
3. In 2005, the government, in collaboration with WWF, organized the consultation workshop to develop a regional strategy and action plan to conserve the snow leopard in the Himalayas. It brought together international experts and those from snow leopard range countries who helped prepare the strategy. It identified four main areas to address snow leopard conservation in Bhutan and neighboring countries: (i) protecting snow leopard habitat and ecosystems, (ii) management of human wildlife conflicts, (iii) management of illegal wildlife trade in snow leopard body parts, and (iv) transboundary initiatives and cooperation.

4. Systematic and opportunistic snow leopard surveys were carried out in JDNP using fecal DNA in 2008. Results are currently being finalized for publication.
5. Opportunistic and systematic field surveys using camera traps have been conducted in WCNP and JDNP from 2009.
6. Bhutan participated in the discussions on global snow leopard conservation held in Bishkek, Kyrgyzstan, in 2012 and has been actively involved with this initiative.
7. JDNP initiated a participatory community assessment to establish a community-based snow leopard conservation program for the Jomolhari region in the park involving the communities of Soe Yutoed and Yaksa.
8. In 2014, DoFPS initiated plans for a national snow leopard survey. Preliminary surveys are currently underway in the protected areas within snow leopard range in Bhutan.

FUTURE PLANS

In order to guide evidence-based snow leopard conservation, a few information gaps have to be closed. These include:

1. Establishing baseline data on population estimates and distribution of snow leopards and prey for all the protected areas with snow leopards to feed into a parkwide and national snow leopard population trend monitoring program. A key first step would be to confirm the eastern limit of snow leopards in Bhutan.
2. Establishing how habitat covariates explain snow leopard and prey distribution, and testing for functionality of existing biological corridors.
3. Understanding movement patterns of snow leopards to better define home range sizes, to refine population metrics, and to understand movement behavior.
4. Mapping critical habitats and areas of high livestock predation to better explain causes of livestock mortality in order to properly guide appropriate countermeasures.
5. Understanding the effects of human and livestock presence and absence on vegetation communities in mountain pasture systems.
6. Monitoring for poaching and illegal trade of snow leopard.
7. Developing appropriate education and awareness materials on snow leopard conservation, and using it for herders, school children, *Ophiocordyceps* collectors, enforcement personnel, and the general public.
8. Initiating snow leopard tourism in snow leopard habitat to increase awareness and generate revenue for locals so benefits from conservation flow to communities.
9. Engaging yak-herding communities in helping the park staff monitor snow leopard and prey populations through snow leopard conservation committees.
10. In the long term, contributing toward better understanding of population genetics and phylogeography of snow leopards globally to inform conservation.

CONCLUSIONS

While the conservation of endangered large cats remains a race against time in most countries across its range, Bhutan presents a unique opportunity for proactive and pre emptive conservation. Strong political will for conservation is exemplified by a constitutional mandate of maintaining at least 60% of the country under forest cover at all times, and having set aside over half of the country as protected areas and biological corridors. Core Buddhist beliefs of compassion toward all sentient beings and tolerance enable an environment that is conducive for humans to coexist with many other species.

Frequent encounters with female snow leopards with two or three cubs allude to a bright future for snow leopards in Bhutan. There is minimal direct threat to the cat and its prey that roam undisturbed in our mountains. Nevertheless, potential threats could arise from a shift in attitude by herders who lose yaks to snow leopards. When the scale of tolerance is tipped toward retaliation, it might be too late. For that matter, it is of utmost importance that the snow leopard be seen as an asset whose conservation would provide direct benefits to communities that share the mountains with them. A combination of science and pragmatism, a better understanding of the human-livestock-snow leopard interface, and prompt action informed by evidence will ensure that Bhutan continues to play an important role in snow leopard conservation.

References

Baillie, I.C., Norbu, C., 2004. Climate and other factors in the development of river and interfluve profiles in Bhutan, Eastern Himalayas. J. Asian Earth Sci. 22, 539–553.

DoFPS, 2006. Forest and Nature Conservation Rules, 2006. Department of Forest and Park Services. Ministry of Agriculture and Forests, Royal Government of Bhutan.

Forrest, J.L., Wikramanayake, E., Shrestha, R., Areendran, G., Gyeltshen, K., Maheshwari, A., Mazumdar, S., Naidoo, R., Thapa, G.J., Thapa, K., 2012. Conservation and climate change: Assessing the vulnerability of snow leopard habitat to treeline shift in the Himalaya. Biol. Conserv. 150, 129–135.

Jigme, K., Tharchen, L., 2012. Camera-trap records of tigers at high altitudes in Bhutan. Cat News 56, 14–15.

MacKenzie, D.I., Nichols, J.D., Lachman, G.B., Droege, S., Andrew Royle, J., Langtimm, C.A., 2002. Estimating site occupancy rates when detection probabilities are less than one. Ecology 83, 2248–2255.

Miehe, G., Miehe, S., Vogel, J., Co, S., La, D., 2007. Highest treeline in the northern hemisphere found in Southern Tibet. Mt. Res. Dev. 27, 169–173.

Oli, M.K., Taylor, I.R., Rogers, M.E., 1993. The diet of the snow leopard (*Panthera uncia*) in the Annapurna Conservation Area, Nepal. J. Zool. (London) 231, 365–370.

RGoB (1995). Forest and Nature Conservation Act of Bhutan. Royal Government of Bhutan.

Tharchen, L., 2013. Protected Areas and Biodiversity of Bhutan. CDC Publishers, Kolkatta, Bhutan.

Thinley, P., 2013. First photographic evidence of a Pallas's cat in Jigme Dorji National Park, Bhutan. Cat News 58, 27–28.

Wangchuk, T., Thinley, P., Tshering, K., Tshering, C., Yonten, D., Pema, B., 2004. A field guide to the mammals of Bhutan. Department of Forestry and Park Services, Ministry of Agriculture and Forests, Royal Government of Bhutan.

WWF., 2012. Near threatened Pallas's cat found in WCP. http://www.wwfbhutan.org/?206453/Near-threatened-Pallas-Cat-found-in-WCP (accessed 16.01.2015).

CHAPTER

36

South Asia: India

Yash Veer Bhatnagar[*,**], *Vinod Bihari Mathur*[†], *Sambandam Sathyakumar*[†], *Abhishek Ghoshal*[*,†], *Rishi Kumar Sharma*[‡], *Ajay Bijoor*[*,**], *Rangaswamy Raghunath*[*], *Radhika Timbadia*[*,**], *Panna Lal*[†]

[*]Nature Conservation Foundation, Mysore, Karnataka, India

[**]Snow Leopard Trust, Seattle, WA, USA

[†]Wildlife Institute of India, Dehradun, Uttarakhand, India

[‡]World Wide Fund for Nature – India, New Delhi, India

SNOW LEOPARD RANGE IN INDIA

The Himalaya extend over an arc of ca. 2500 km along the north and northeastern boundary of India, stretching across the five Indian states of Jammu & Kashmir (JK), Himachal Pradesh (HP), Uttarakhand (UK), Sikkim (SK), and Arunachal Pradesh (AP). The alpine tracts of the Himalaya and the arid marginal mountains of the Tibetan Plateau in the rain-shadow of the main Himalayan range known as Trans-Himalaya (Rodgers and Panwar, 1988; Rodgers et al., 2000) together constitute ca. 100,000 km^2 of snow leopard (*Panthera uncia*) habitat in India (Fig. 36.1). These tracts show a gradient of increasing aridity from east to west and south to north. Snow leopards mostly occur in the more arid, nonforested tracts in India between 3200 m and 5200 m (Chundawat, 1992; Fox et al., 1991), and as low as 2700 m in Kedarnath WLS in UK (Green, 1979). Recent reports from UK (Sathyakumar unpublished) suggest they may use the lower forests tracts (<3200 m) occasionally.

The high altitudes of the Indian Himalaya support a diverse and unique assemblage of wild flora and fauna (Polunin and Stainton, 1984; Schaller, 1977), owing to distinct biogeographic characteristics (Mani, 1974; Rodgers et al., 2000). Some parts of the region are even classified as global biodiversity hotspots (Myers et al., 2000; Olson and Dinerstein, 1998). Asiatic ibex (*Capra sibirica*) and bharal (*Pseudois nayaur*) constitute the primary prey of snow leopards (Lyngdoh et al., 2014; Schaller, 1977).

All the major rivers of northern India that support millions of people and their livelihoods originate in the snow leopard range. The Himalaya thus, in addition to having immense

Snow Leopards
http://dx.doi.org/10.1016/B978-0-12-802213-9.00036-5

TABLE 36.1 Recent Density Estimates Available From Various Sites in India Based on Camera Trapping (CT) [Using the ½ Mean of Maximum Distances Moved (MMDM) or Spatially Explicit Capture Recapture (SECR)] or Genetic Analysis (GA)

Method and density model	Survey year	No. of snow leopards	Density estimate	Standard error
CT, 1/2 MMDM (Jackson et al., 2006)	2004	6	4.45	0.16
CT, 1/2 MMDM (Jackson et al., 2006)	2003	6	8.49	0.22
CT, SECR (Sathyakumar et al., 2014)	2009-11	4	4.77	1.81
GA, N/Catchment area (Suryawanshi et al., NCF-SLT, unpublished data)	2010	4	1.66	NA
		1	0.45	NA
		2	0.74	NA
		4	1.17	NA
		7	1.94	NA
		7	3.30	NA
CT, ML SECR (Sharma et al., NCF-SLT, unpublished data)	2011	7	0.21	0.10
		10	1.03	0.38
		15	0.47	0.13

biological value, also provide invaluable ecosystem services (MEA, 2005; World Bank, 2013).

The snow leopard has been accorded the highest level of legal protection under Schedule 1 of the Indian Wildlife Protection Act, 1972. Given the endangered status of the snow leopard, its vast interface with humans, and the unique habitat, the government of India recently developed a national strategy for its effective protection – Project Snow Leopard (PSL, 2008). In this chapter, we briefly describe the state of knowledge on the species ecology and threats, revise the abundance guesstimate of snow leopard using robust density figures, and finally suggest the way forward for effective conservation.

STATE OF KNOWLEDGE

India has relatively rich natural history records from snow leopard range spanning over a century (see references in Schaller, 1977), but systematic surveys began in the 1980s, covering parts of the western Himalaya such as Ladakh (Chundawat and Qureshi, 1999; Fox et al., 1991; Mallon, 1991), Himachal Pradesh (Bhatnagar et al., 2008; Vinod and Sathyakumar, 1999; A. Ghoshal, NCF-SLT, unpublished data), UK mountains (Sathyakumar, 1993), and more recently in parts of Sikkim (Anonymous, 2010; Sathyakumar et al., 2014). Ecological studies in Ladakh (Chundawat and Rawat, 1994), Spiti (Bhatnagar, 1997; Mishra, 1997; Mishra et al., 2004; Suryawanshi et al., 2013) and UK (Bhattacharya et al., 2012; Kandpal and Sathyakumar, 2010; Sathyakumar, 1994) have also added to the understanding of snow leopard and prey ecology, as well as to interactions between people and wildlife. Various studies exploring snow leopard abundance using modern tools have provided some robust estimates (Table 36.1).

We analyzed approximately 100 papers and reports to understand the gaps in knowledge to help guide future research and conservation efforts. We divided the entire potential snow leopard range in India (3200–5200 m) into 311 subcatchments (mostly at tertiary level) and plotted the level of knowledge as (3) *"good"*: robust ecological studies on snow leopard and prey have been carried out (e.g., Upper Spiti and Hemis) versus

(2) *"moderate"*: where some structured surveys on snow leopard and prey have been carried out (e.g., rest of Spiti or western SK) versus (1) *"poor"*: where general surveys on prey have been carried out with notes on snow leopard occurrences (e.g., eastern Uttarakhand), or (0) *"no information"*: areas yet to receive research attention (e.g., Pir Panjal range of JK and most of AP) (Fig. 36.1).

In India, a substantial third of the snow leopard range (ca. 100,146 km^2) has not yet received research attention (Fig. 36.1). This area includes most of the Pir Panjal Range and the rest of Kashmir Himalaya, Nubra region of Ladakh, the Dhauladhar range in HP, and most of western Uttarakhand (Fig. 36.1). Almost the entire AP Mountains remain unexplored. About half of the potential snow leopard range has poor information (56%), while small portions of the range have moderate (7%) and good (4%) information (Fig. 36.1). From the 90 years preceding 1990, we found only 25 papers on snow leopard in India, while in the three decades since then there are close to 75 papers, showing an improvement in both extent and quality of research. We however emphasize that for a large country such as India where more than 95% of snow leopard range remains largely unstudied, there is a need for more concerted and cooperative efforts for understanding snow leopard abundance, resource use, and threats so that meaningful conservation in larger landscapes can be achieved.

REVISING SNOW LEOPARD POPULATION ESTIMATES FOR INDIA

In the 1980s, a guesstimate of snow leopard population in India was presented along with other range countries (Fox, 1989). Of the estimated 4000–7500 snow leopards globally, 400–700 were estimated for India. There is an urgent need to make a revised estimate of snow leopard abundance using improved methodology, especially since reliable density estimates are now available (see Table 36.1). The challenges to monitoring snow leopards and estimating population parameters are discussed in Chapter 25. Given the constraints, we have attempted to provide a revised estimate of snow leopard population in India using a two-step approach.

We first evaluated the 311 subcatchments mentioned earlier to score habitat quality as "good" or "relatively poor" ("good": e.g., Hemis, with known high-quality areas for snow leopard and prey populations; and "relatively poor": e.g., Lahaul, where surveys suggest otherwise) (Fig. 36.2). Then we used snow leopard density estimates from a large (>3000 km^2) area in Spiti, HP, which included a gradient of habitat qualities, to derive representative density estimates (Table 36.2). The snow leopard density estimates for the "relatively poor" and "good" quality habitats in Spiti were 0.21 snow leopard/100 km^2 and 1.03/100 km^2, respectively (Table 36.2). The estimate for a larger region spanning a gradient of habitat qualities was 0.47/100 km^2 (Table 36.2). A few good areas are known for snow leopard in the eastern Himalaya, but one study from a relatively small and isolated but good block in the Kanchendzonga NP suggests values of 4.25 (SE 2.55/100 km^2) (Sathyakumar et al., 2014), which has also been used for the representative sites in Sikkim. Finally, we extrapolated the snow leopard density estimates to the classified snow leopard habitat in India (Table 36.3). We clarify that our approach uses the only available snow leopard density estimates emerging from a large-scale and long-term study in the Trans-Himalaya (R. Sharma, NCF-SLT, unpublished data). Because climate and habitat characteristics in other parts of the range, such as the Greater Himalaya, NE Himalaya, and the more rolling Tibetan Plateau, may be different, our extrapolated estimate should be treated with caution and as an evolving one.

Based on this analysis, about two-thirds of the snow leopard range (ca. 63,000 km^2) was found to be "relatively poor," while 37,000 km^2 was considered "good" habitat (Table 36.3,

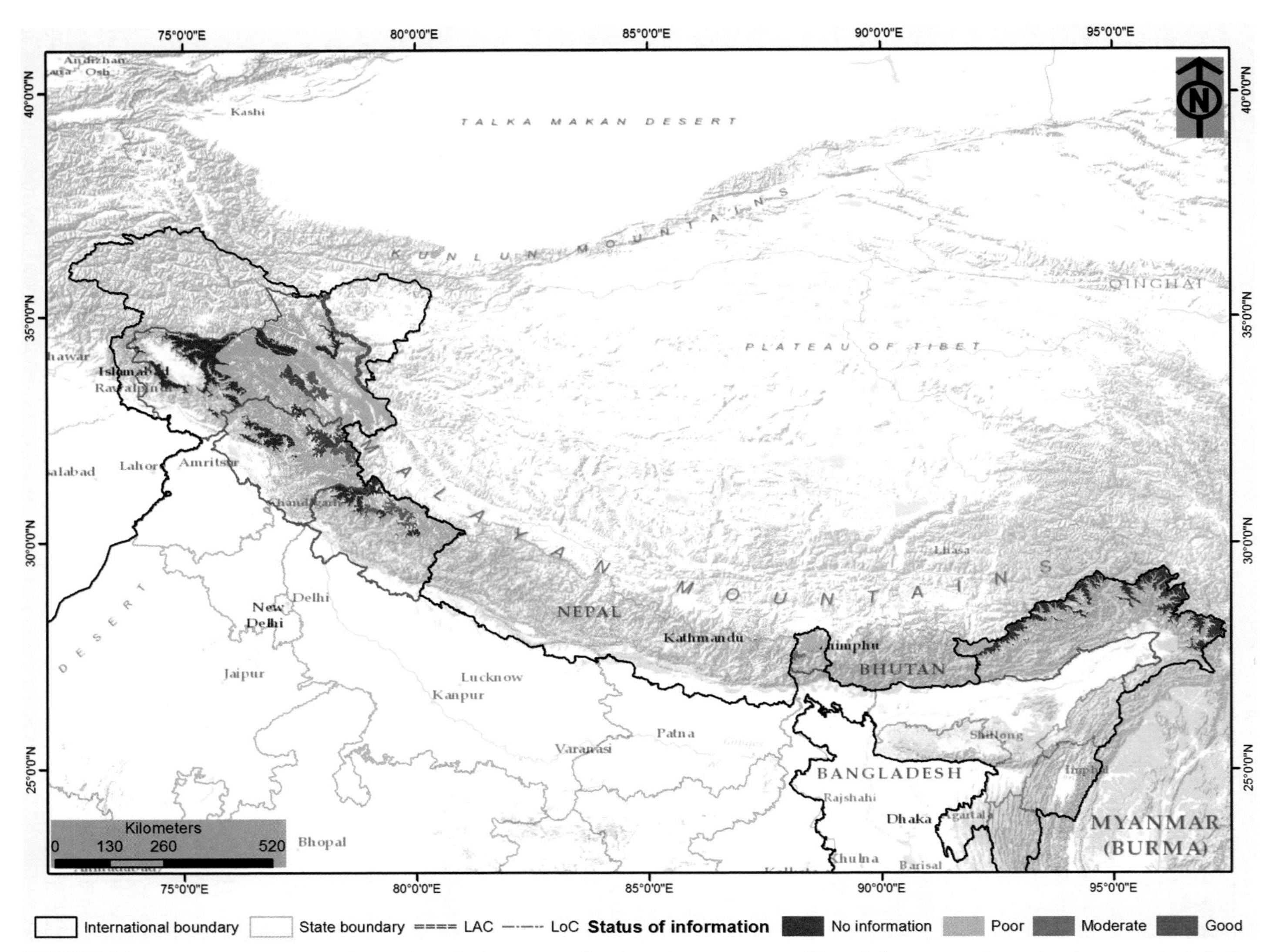

FIGURE 36.1 **Potential snow leopard range in India (3200–5200 m).** Information gaps: status of knowledge on snow leopard and prey species in India as assessed by attributing values to 311 tertiary catchments within snow leopard range ranked as good (3), moderate (2), poor (1), and no information (0). See text for details. Data on catchments used for this analysis are available from the corresponding author. The international borders are indicative and may not be accurate.

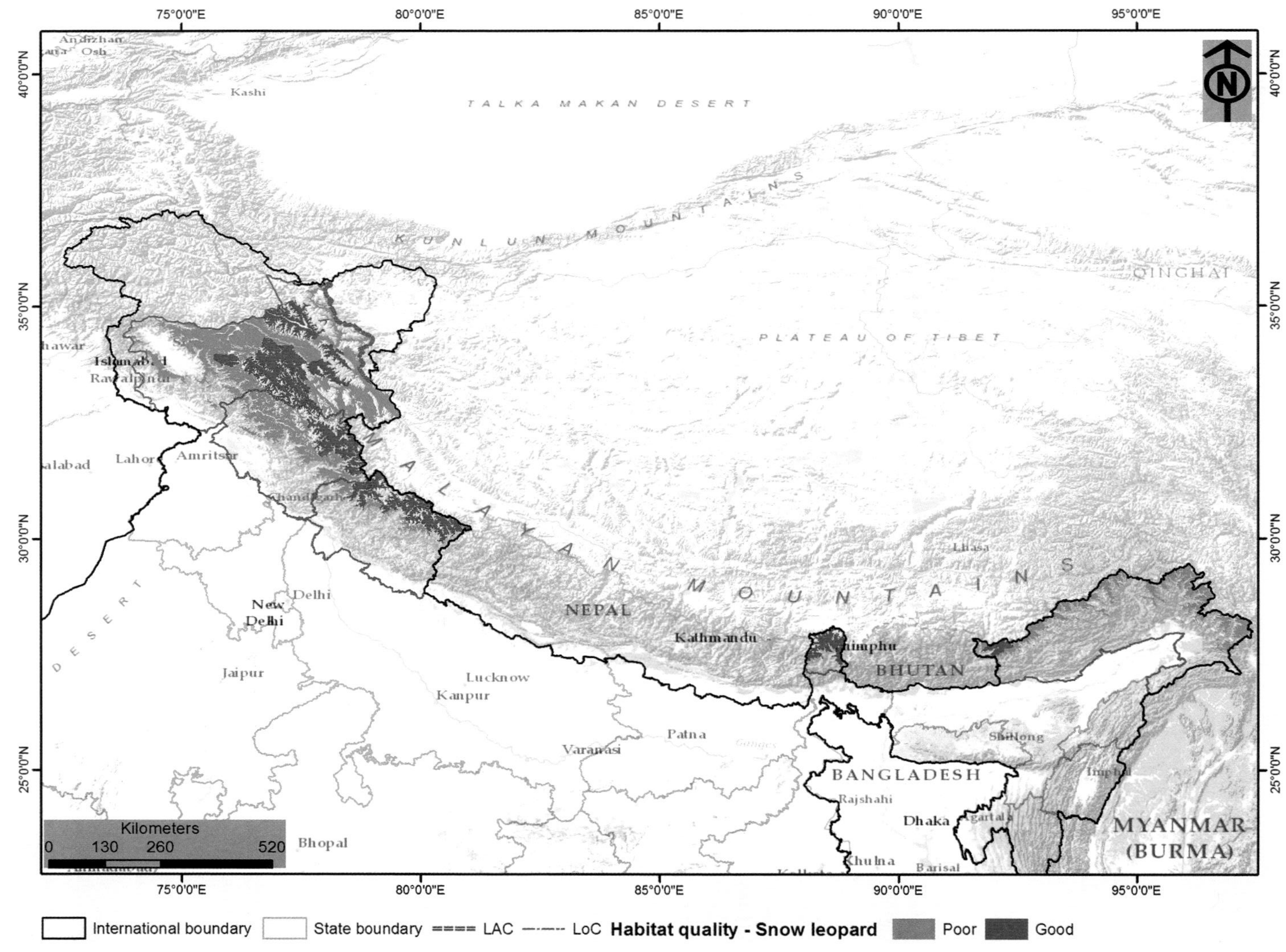

FIGURE 36.2 **Snow leopard habitat quality.** Status of habitat for snow leopard and prey species in India as assessed by attributing values to 311 tertiary catchments within snow leopard range ranked as good (2) and relatively poor (1). See text for details. The international borders are indicative and may not be accurate.

TABLE 36.2 Snow Leopard Density Estimates from Different Quality Habitats Derived Using Spatially Explicit Capture–Recapture (SECR) Models from Study Sites in Spiti, Himachal Pradesh, India (Sharma, R., et al., NCF, Unpublished Data)

S. No.	No. of trap sites	MCP	N	C	ML-ESA	Density (nos./km^2)	SE	Note
1	15	650	7	21	3387	0.21	0.10	Low-density areas
2	15	206	10	43	983	1.03	0.38	High-density areas
3	30	954	15	66	3234	0.47	0.13	Overall landscape scale density

Low density areas are characterized by low wild prey densities. High density areas are characterized by high wild prey densities. Overall landscape scale density includes a gradient of poor, medium, and high-quality habitats over a large landscape.
Abbreviations: No. of trap sites, Number of camera trap locations used in estimating snow leopard density; MCP, minimum convex polygon, indicating the area (in square kilometres) enclosed by the outmost camera trap locations; N, number of adult individual snow leopards captured; C, number of independent snow leopard photo captures; ML-ESA, maximum-likelihood effective sampling area derived from the SECR model; SE, standard error.

Fig. 36.2). The snow leopard population in India is thus likely to be ca. 516 (238–1039, using the SE in Table 36.2). An alternative estimate for Sikkim will take the national estimate to 524. Our value is similar to the earlier one (Fox, 1986), but is based on more robust density estimates and is replicable, as more information on habitat quality and density estimates emerge from future studies.[1]

THREATS TO SNOW LEOPARDS

A detailed assessment of global and countrywide threats was recently made (SLN, 2014; World Bank, 2013). In this section, we provide a brief summary of the context of threats and recent issues of national relevance.

Native people in the snow leopard range of India are agropastoralists and are mostly Tibetan Buddhists and *Pahadis* of different ethnic backgrounds (Bhatnagar and Singh, 2011; Mishra et al., 2006). Numerous transhumant tribes graze livestock, mainly sheep-goats, in the snow leopard range during summer (Bhasin, 2011; Saberwal, 1996). Most of these people depend on local biomass extraction for fuel, fodder, and nontimber produce and also collect dung from their livestock for fuel and manure. The people pervasively use the mountains, including areas within protected areas (PA), for sustenance. Given that wildlife is pervasive in the snow leopard habitat in India, there is an extensive human–wildlife interface. This creates a wide range of situations where human activities can negatively impact wildlife, and vice versa.

Livestock Grazing

As mentioned earlier, resident and migratory livestock grazing has resulted in decimation and poor population performance in the wild prey of snow leopard in the Himalaya (Bagchi et al., 2004; Bhatnagar et al., 2000; Mishra et al., 2004; Suryawanshi et al., 2010). With increasing herds, the migratory herders are now penetrating more arid parts of the Trans-Himalaya in search of pastures, as seen in the Upper Spiti Valley. An ongoing study in

[1]It is hoped that researchers should be able to update information as knowledge on occurrences and newer snow leopard densities become available. The detailed methodology followed and data files (Excel and Shape) can be obtained from the corresponding author.

TABLE 36.3 Snow Leopard Population Estimates in India Using Extrapolation of Densities Typical of Relatively Poor and Good Quality Habitats

Habitat quality	Area (km^2)	Snow leopard estimate
Using the estimate of "relatively poor" and "good" habitats		
Relatively poor	62,924	133
Good	37,222	383
Total estimate		**516**
Using overall estimates encompassing both "relatively poor" to "good" habitats		
Overall area	100,146	**470**
Statewise estimates		
AP		**42**
Relatively poor	13,946	29
Good	1,235	13
HP		**90**
Relatively poor	13,745	29
Good	5,916	61
JK		**285**
Relatively poor	30,437	64
Good	21,421	221
SK		**13**
Relatively poor	2,029	4
Good[a]	816	8
UK		**86**
Relatively poor	2,767	6
Good	7,834	81

An overall estimate using average snow leopard densities encompassing poor, medium and good quality habitats is also given. See text for explanation. Relatively poor, 0.21/100 km^2, SE 0.10; good quality, 1.03/100 km^2, SE 0.38; and overall area, 0.47/100 km^2, SE 0.13. The extrapolated values of snow leopard population at national and state levels are indicative.

[a] *Using the value of 4.25 (SE 2.55) based on Sathyakumar et al., 2014, this estimate will go up to 35 for the "good" areas, 39 for Sikkim State, and 542 for the nation.*

HP indicates poor availability of wild prey in vast areas of the Lahaul valley potentially due to intense grazing (A. Ghoshal, NCF-SLT, unpublished data). Investigations in Nanda Devi region in UK reveal a substantial diet overlap between wild and domestic ungulates (Bhattacharya and Sathyakumar, 2011; Bhattacharya et al., 2012). Berger et al. (2013) further suggest cashmere or *pashmina* production is becoming an important threat to the snow leopard due to competition from increasing herds of the cashmere goats.

Human-Snow Leopard-Wild Prey Interaction

Livestock is often lost to wild carnivores, including snow leopards, in the entire range. In some areas the problem is intense with high economic losses and negative attitudes (Bhatnagar et al., 1999; Suryawanshi et al., 2014). Crop losses, including valuable cash crops (USL, 2011), and even a perception of pasture degradation due to wild ungulates such as kiang compound the issue (Bhatnagar et al., 2006). Occasionally, the remains of retaliatory killing of wildlife, including snow leopard, may even find their way into the illegal wildlife trade market (Theile, 2003; see Chapter 7).

Emerging Threats

Infrastructure development, including hydropower (Rajvanshi et al., 2012), especially in the areas bordering China and Pakistan, has increased over the past few decades. Additionally, development of tourism infrastructure is escalating in the snow leopard habitat (PSL, 2008; World Bank, 2013). These activities generally involve an influx of relatively large numbers of outsiders, especially laborers, within a short time period into landscapes that usually support only low human densities. The laborers exacerbate local biomass usage through poaching and collecting other valuable natural resources from the region. Increased garbage produced by tourism facilities, outsiders, and local people (especially livestock carcasses), along with garbage mismanagement in the expanding villages, are facilitating population increase of free-ranging dogs that have emerged as a serious threat to wildlife and livestock due to depredation (Ghoshal, 2011; Hughes and Macdonald, 2013; Suryawanshi et al., 2013; USL, 2011). The development drive, while important for the region, underscores the enhanced responsibility of both infrastructure development and conservation agencies to seek a balance toward convergence with biodiversity management.

CONSERVATION EFFORTS IN INDIA

Of the approximately 700 PAs in India, 53 include areas that lie within potential snow leopard range. Many PAs include the less-used forests below 3200 m and/or permafrost areas above 5200 m (Appendix 36.1). The combined notified area for the 53 PAs is about 36,000 km^2, with an average size of 693 km^2, with few PAs covering >1000 km^2. For calculating the actual area of snow leopard range under protection, we used the polygon for the PAs from the database in the Wildlife Institute of India and the potential snow leopard range within it (3200–5200 m). However, polygons were available for only 39 PAs, as many states are still in the process of digitizing the PA boundaries. We recognized issues regarding boundaries of certain PAs (e.g., Changthang WLS – calculated area ca. 18,000 km^2 vs. notified area 4,000 km^2; Karakorum WLS – calculated area 19,000 km^2 vs. notified area 5,000 km^2; Kibber WLS – calculated area 960 km^2 vs. notified area 2,200 km^2). Based on the available 39 PA polygons (estimated area of PAs = 65,580 km^2) the area of snow leopard habitat within PAs is about 33,900 km^2 or 33% of the potential snow leopard range in India. If we use the notified area for Changthang WLS (4000 km^2) and Karakorum WLS (5000 km^2), and correct it with 56% snow leopard habitat (the average for the 39 PAs), the total area stands at 22,200 km^2, or about 22% of snow leopard range in India.

Even at 22%, a substantial proportion of snow leopard range in India appears to be under legal protection, where all forms of consumptive use are restricted under the Indian Wildlife (Protection) Act (Anonymous, 2002). However, given the pervasiveness of snow leopard and people's dependence, enforcing exclusionary PAs can be counterproductive for both conservation and livelihoods. Additionally, the staff strength and capacity of the primary agency mandated for conservation, the State Forest or Wildlife Departments, is often inadequate (PSL, 2008; USL, 2011). This leads to inefficient monitoring and inability to engage with stakeholders for conservation activities (World Bank, 2013). Nearly 80% of snow leopard range is still unprotected (excluding the buffers of the three biosphere reserves) and studies indicate that many wildlife strongholds remain in such regions (Chundawat and Qureshi, 1999; PSL, 2008). With successful recent experiments using participatory conservation in the Trans-Himalaya including socially fenced grazing-free reserves (Mishra et al., 2003) where the community agrees to stop or reduce damaging use (see Chapter 14.2), conflict resolution, and livelihood support (Jackson and Wangchuk, 2004; Jackson et al., 2010), the MoEFCC and the Himalayan States, with assistance from conservation agencies, developed an innovative participatory, knowledge-based and landscape-level program for conservation planning and action in the snow leopard range – Project Snow Leopard (PSL, 2008).

THE WAY FORWARD

Under PSL, each state is identifying at least one landscape of 2000–5000 km^2. The landscape will typically include a mosaic of high-quality habitats (PAs and socially fenced areas) interspersed in a matrix of lower-quality habitat for snow leopard under multiple-use. We envisage that these nonexclusionary reserves will serve as source populations that will maintain or augment local wildlife populations (Bhatnagar and Mishra, 2014; Mishra et al., 2010; PSL, 2008). Each village cluster will formulate and implement proactive protection and mitigation mechanisms to offset people–wildlife conflicts. Conservation agencies will work in tandem with local government and civil society organizations to boost local incomes based on indigenous enterprises. There will be new organizations at the state, landscape, and village cluster levels to facilitate participatory planning and implementation of conservation interventions. The capacity of forestry staff and other facilitators will constantly be enhanced. The emphasis will be on ecological, social, and institutional knowledge (especially because nearly 95% of snow leopard range remains unstudied), capacity building, coordination through convergence of activities among agencies, and concerted and sustained action.

Under the PSL, even amidst persisting and burgeoning challenges, management of the Upper Spiti Landscape in Himachal Pradesh (USL, 2011) and most of Ladakh are in progress and showing promising results. We remain hopeful that through these innovative approaches, India will be able to achieve effective conservation of snow leopard in its entire range by 2030.

References

Anonymous, 2002. The Wildlife (Protection) Act 1972 (as amended up to 2002). Natraj Publishers, Dehradun, p. 158.

Anonymous, 2010. Non-Invasive Monitoring to Support Local Stewardship of Snow Leopards and their Prey. Final report submitted to Critical Ecosystems Partnership Fund and Ashoka Trust for Research in Ecology and the Environment, Nature Conservation Foundation, Mysore and The Mountain Institute-India.

Bagchi, S., Mishra, C., Bhatnagar, Y.V., 2004. Conflicts between traditional pastoralism and conservation of Himalayan ibex (*Capra sibirica*) in the Trans-Himalayan mountains. Anim. Conserv. 7 (2), 121–128.

Berger, J., Buuveibaatar, B., Mishra, C., 2013. Globalization of the cashmere market and the decline of large mammals in Central Asia. Conserv. Biol. 27 (4), 679–689.

Bhasin, V., 2011. Pastoralists of Himalayas. J. Human Ecol. 33 (3), 147–177.

Bhatnagar, Y.V., 1997. Ranging and habitat use by Himalayan Ibex (*Capra ibex sibirica*) in Pin Valley National Park. PhD dissertation, Saurashtra University, Rajkot.

Bhatnagar, Y.V., Mishra, C., 2014. Conserving without fences. Project Snow Leopard. In: Rangarajan, M., Madhusudhan, M.D., Shahabuddin, G. (Eds.), Nature Without Borders. Orient Blackswan, New Delhi, pp. 157–177.

Bhatnagar, Y.V., Singh, N.J., 2011. Nomadism in the Indian Changthang: Changes and Implications on Society and Biodiversity. Global Change, Biodiversity and Livelihoods in Cold Desert Region of Asia. Bishen Singh Mahendra Pal Singh, Dehradun, pp. 135–146.

Bhatnagar, Y.V., Stakrey, R.W., Jackson, R., 1999. A survey of depredation and related wildlife-human conflicts in Hemis National Park, Ladakh, Jammu & Kashmir, India. International Snow Leopard Trust, Seattle, Washington, USA.

Bhatnagar, Y.V., Rawat, G.S., Johnsingh, A.J.T, Stüwe, M., 2000. Ecological separation between ibex and resident livestock in a Trans-Himalayan protected area. In: Richard, C, Basent

K., Sah, J.P. Raut, Y. (Eds.), Grassland Ecology and Management in Protected Areas of Nepal, vol. 2. Technical and status papers on grasslands of mountain protected areas. Royal Bardia National Park, Thakurwara, Bardia, Nepal, March 15–19, 1999. ICIMOD, Kathmandu, pp. 71–84.
Bhatnagar, Y.V., Wangchuk, R., Prins, H.H., Van Wieren, S.E., Mishra, C., 2006. Perceived conflicts between pastoralism and conservation of the kiang *Equus kiang* in the Ladakh Trans-Himalaya, India. Environ. Manag. 38 (6), 934–941.
Bhatnagar, et al., 2008. Exploring the Pangi Himalaya: A preliminary wildlife survey in the Pangi region of Himachal Pradesh. Technical Report. The Wildlife Wing, Himachal Pradesh Forest Department & Nature Conservation Foundation, Mysore.
Bhattacharya, T., Sathyakumar, S., 2011. Natural resource use by humans and response of wild ungulates: a case study from Bedini-Ali, Nanda Devi Biosphere Reserve, India. Mt. Res. Dev. 31 (3), 209–219.
Bhattacharya, T., Kittur, S., Sathyakumar, S., Rawat, G.S., 2012. Diet overlap between wild ungulates and domestic livestock in the Greater Himalaya: Implications for management of grazing practices. Proceedings of the Zoological Society (vol. 65, No. 1, pp. 11–21), Springer-Verlag.
Chundawat, R.S., 1992. Ecological studies on the snow leopard and its prey species in Hemis NP, Ladakh. PhD hesis submitted to the University of Rajasthan, Jaipur.
Chundawat, R.S., Qureshi, Q., 1999. Planning wildlife conservation in Leh and Kargil districts of Ladakh, Jammu and Kashmir. Draft Report submitted to the Wildlife Institute of India, Dehradun.
Chundawat, R.S., Rawat, G.S., 1994. Food habits of snow leopard in Ladakh, India. In: Fox, J.L., Juzeng, D. (Eds.), Proceedings of the seventh International snow leopard symposium. International Snow Leopard Trust, Seattle, pp. 127–132.
Fox, J.L., 1989. A review of the status and ecology of the snow leopard. *Panthera uncia*. International Snow Leopard Trust, Seattle, Pp. 40.
Fox, J.L., Sinha, S.P., Chundawat, R.S., Das, P.K., 1991. Status of the snow leopard *Panthera uncia* in north-west India. Biol. Conserv. 55, 283–298.
Ghoshal, A., 2011. Impact of Urbanization on Winter Resource Use and Relative Abundance of a Commensal Carnivore, the Red Fox (*Vulpes vulpes*). MSc Dissertation. Forest Research Institute University, Dehradun, India.
Hughes, J., Macdonald, D.W., 2013. A review of the interactions between free-roaming domestic dogs and wildlife. Biol. Conserv. 157, 341–351.
Jackson, R., Roe, J.D., Wangchuk, R., Hunter, D.O., 2006. Estimating snow leopard population abundance using photography and capture–recapture techniques. Wildl. Soc. Bull. 34, 772–781.
Jackson, R.M., Wangchuk, R., 2004. A community-based approach to mitigating livestock depredation by snow leopards. Hum. Dimens. Wildl. 9 (4), 1–16.
Jackson, R.M., Mishra, C., McCarthy, T.M., Ale, S.B., 2010. Snow leopards: conflict and conservation. In: Macdonald, D.W., Loveridge, A.J. (Eds.), Biology and Conservation of Wild Felids. Oxford University Press, UK, pp. 417–430.
Kandpal, V., Sathyakumar, S., 2010. Distribution and relative abundance of mountain ungulates in Pindari Valley, Nanda Devi Biosphere Reserve, Uttarakhand, India. Galemys 22 (1), 277–294.
Lyngdoh, S., Shrotriya, S., Goyal, S.P., Clements, H., Hayward, M.W., Habib, B., 2014. Prey preferences of the snow leopard (*Panthera uncia*): regional diet specificity holds global significance for conservation. PLoS One 9 (2), e88349.
Maheshwari, A., Sharma, D., Sathyakumar, S., 2013. Snow leopard (*Panthera uncia*) surveys in the Western Himalayas, India. J. Ecol. Nat. Environ. 5 (10), 303–309.
Mallon, D.P., 1991. Status and conservation of large mammals in Ladakh. Biol. Conserv. 56 (1), 101–119.
Mani, M.S., 1974. Ecology and Biogeography in India. Dr. W. Junk B.V. Publishers, The Hague.
MEA, 2005. Mountain systems. In: Millennium Ecosystem Assessment. Ecosystems and Human Well-Being: Current Status and Trends Assessment. Island Press, Washington, DC. 681–716 pp.
Mishra, C., 1997. Livestock depredation by large carnivores in the Indian Trans-Himalaya: conflict perceptions and conservation prospects. Environ. Conserv. 24 (04), 338–343.
Mishra, C., Allen, P., McCarthy, T.O.M., Madhusudan, M.D., Bayarjargal, A., Prins, H.H., 2003. The role of incentive programs in conserving the snow leopard. Conserv. Biol. 17 (6), 1512–1520.
Mishra, C., Van Wieren, S., Ketner, P., Heitkonig, I.M.A., Prins, H.H.T., 2004. Competition between domestic livestock and wild bharal *Pseudois nayaur* in the Indian trans-Himalaya. J. Appl. Ecol. 41, 344–354.
Mishra, C., Madhusudan, M.D., Datta, A., 2006. Mammals of the high altitudes of western Arunachal Pradesh, eastern Himalaya: an assessment of threats and conservation needs. Oryx 40 (01), 29–35.
Mishra, C., Bagchi, S., Namgail, T., Bhatnagar, Y.V., 2010. Multiple use of Trans-Himalayan rangelands: reconciling human livelihoods with wildlife conservation. In: du Toit, J.T., Kock, R., Deutsch, J.C. (Eds.), Wild Rangelands: Conserving Wildlife While Maintaining Livestock in Semi-Arid Ecosystems. first ed. Blackwell Publishing, pp. 291–311.
Myers, N., Mittermeier, R.A., Mittermeier, C.G., Da Fonseca, G.A., Kent, J., 2000. Biodiversity hotspots for conservation priorities. Nature 403 (6772), 853–858.
Olson, D.M., Dinerstein, E., 1998. The Global 200: a representation approach to conserving the Earth's most biologically valuable ecoregions. Conserv. Biol. 12 (3), 502–515.
Polunin, O., Stainton, A., 1984. Flowers of the Himalaya. Oxford University Press, Oxford, UK.

PSL, 2008. The Project Snow Leopard. Ministry of Environment & Forests, Government of India, New Delhi.

Rajvanshi, A., Arora, R., Mathur, V.B., Sivakumar, K., Sathyakumar, S., Rawat, G.S., Johnson, J.A., Ramesh, K., Dimri, N.K., Maletha, A., 2012. Assessment of cumulative impacts of hydroelectric projects on aquatic and terrestrial biodiversity in Alaknanda and Bhagirathi Basins, Uttarakhand. Wildlife Institute of India, Technical Report.

Rodgers, W.A., Panwar, H.S., 1988. Planning a Wildlife Protected Area Network in Indiavols. I & IIWildlife Institute of India, Dehradun.

Rodgers, W.A., Panwar, H.S., Mathur, V.B., 2000. Wildlife Protected Area Network in India: A Review (Executive Summary). Wildlife Institute of India, Dehradun.

Saberwal, V.K., 1996. Pastoral politics: Gaddi grazing, degradation, and biodiversity conservation in Himachal Pradesh, India. Conserv. Biol. 10 (3), 741–749.

Sathyakumar, S., 1993. Status of Mammals in Nanda Devi National Park. Scientific and Ecological Expedition on Nanda Devi. Wildlife Institute of India, Dehradun, pp. 5–15.

Sathyakumar, S., 1994. Habitat ecology of major ungulates in Kedarnath Musk deer Sanctuary, Western Himalaya. PhD Thesis submitted in Saurashtra University, Rajkot, Gujarat.

Sathyakumar, S., Bashir, T., Bhattacharya, T., Poudyal, K., 2011. Assessing mammal distribution and abundance in intricate eastern Himalayan habitats of Khangchendzonga, Sikkim, India. Mammalia 75 (3), 257–268.

Sathyakumar, S., Bhattacharya, T., Bashir, T. and Poudyal, K., 2014. Developing a monitoring programme for mammals in Himalayan Protected Areas: A case study from Khangchendzonga National Park and Biosphere Reserve, Sikkim, India. Parks, 20.2, pp. 35–48.

Schaller, G.B., 1977. Mountain Monarchs: Wild Sheep and Goats of the Himalaya. University of Chicago Press, Chicago, USA.

SLN, 2014. Snow Leopard Survival Strategy. Revised 2014 Version Snow Leopard Network, Seattle, Washington, USA.

Suryawanshi, K.R., Bhatnagar, Y.V., Mishra, C., 2010. Why should a grazer browse? Livestock impact on winter resource use by bharal *Pseudois nayaur*. Oecologia 162 (2), 453–462.

Suryawanshi, K.R., Bhatnagar, Y.V., Redpath, S., Mishra, C., 2013. People, predators and perceptions: patterns of livestock depredation by snow leopards and wolves. J. Appl. Ecol. 50, 550–560.

Suryawanshi, K.R., Bhatia, S., Bhatnagar, Y.V., Redpath, S., Mishra, C., 2014. Multiscale factors affecting human attitudes toward snow leopards and wolves. Conserv. Biol. 28 (6), 1657–1666.

Theile, S., 2003. Fading Footprints: The Killing and Trade of Snow Leopards. TRAFFIC International.

USL, 2011. Management Plan for Upper Spiti Landscape, Including the Kibber Wildlife Sanctuary. Himachal Pradesh Forest Department, Shimla and Nature Conservation Foundation, Mysore.

Vinod, T.R., Sathyakumar, S., 1999. Ecology and Conservation of Mountain Ungulates in Great Himalayan National Park, Western Himalaya. Wildlife Institute of India.

World Bank, 2013. Global snow leopard and ecosystem protection program: Conference Document for Endorsement (71 pages) and Annex Document of Complete National Programs (234 pages). Washington, DC.

APPENDIX 36.1 Protected areas (PAs) across the snow leopard range (3200–5200 m) in India covering the five states of Jammu & Kashmir, Himachal Pradesh, Uttarakhand, Sikkim, and Arunachal Pradesh

State	S. No.	PAs (national park - NP; wildlife sanctuary – WLS; wetland – WL)	Notified geographic area (km^2)	Estimated geographic area (km^2)	% Snow leopard habitat
Jammu and Kashmir	1	Baltal Thajwas WLS	211	–	–
	2	Changthang Cold Desert WLS	4000	18,792	56
	3	Dachigam NP	171	170	44
	4	Gulmarg WLS	139	196	60
	5	Hanle/Chushul WL	–	–	–
	6	Hemis NP	3350	5100	81
	7	Hirapora WLS	115	–	–
	8	Hygam WL	7	–	–
	9	Karakoram (Nubra Shyok) WLS	5000	19,387	34
	10	Kishtwar NP	400	–	–
	11	Lachipora WLS	94	–	–
	12	Limber WLS	44	42	32
	13	Malgam WL	5	–	–
	14	Norrichain (Tsokar) WL	2	–	–
	15	Overa Aru WLS	511	430	78
	16	Rajparian WLS	20	–	–
	17	Tsomoiri (Ramsar Site) WL	120	–	–
Subtotals			14,188	44,117	22
Himachal Pradesh	1	Great Himalayan NP	755	831	79
	2	Pin Valley NP	675	720	60
	3	Chandra Tal WLS	39	37	79
	4	Churdhar WLS	66	–	–
	5	Daranghati WLS	172	37	6
	6	Dhauladhar WLS	944	1009	78
	7	Gamgul Siahbehi WLS	109	112	28
	8	Kanawar WLS	61	68	26
	9	Kais WLS	14	13	29
	10	Kibber WLS*	2220	968	36
	11	Kugti WLS	379	355	87
	12	Lippa Asrang WLS	31	29	100
	13	Manali WLS	32	36	32
	14	Nargu WLS	278	265	6

APPENDIX 36.1 Protected areas (PAs) across the snow leopard range (3200–5200 m) in India covering the five states of Jammu & Kashmir, Himachal Pradesh, Uttarakhand, Sikkim, and Arunachal Pradesh *(cont)*

State	S. No.	PAs (national park - NP; wildlife sanctuary – WLS; wetland – WL)	Notified geographic area (km^2)	Estimated geographic area (km^2)	% Snow leopard habitat
	15	Raksham Chitkul WLS	304	330	81
	16	Rupi Bhaba WLS	738	718	63
	17	Sainj WLS	90	97	73
	18	Sechu Tuan Nala WLS	103	503	86
	19	Talra WLS	40	–	–
	20	Tirthan WLS	61	66	58
	21	Tundah WLS	64	371	62
Subtotals			7174	6564	77
Uttarakhand	1	Askot Musk Deer WLS	600	477	8
	2	Gangotri NP	2200	2441	51
	3	Govind NP	472	376	59
	4	Govind Pashu Vihar WLS	481	786	59
	5	Kedarnath WLS	975	1054	52
	6	Nanda Devi NP**	625	545	45
	7	Valley of Flowers NP**	88	75	86
Subtotals			5440	5754	49
Sikkim	1	Khangchendzonga NP	1784	2474	55
	2	Kyongnosla Alpine WLS	31	28	100
	3	Pangolakha WLS	128	134	47
	4	Shingba (Rhododendron) WLS	43	45	93
Subtotals			1986	2682	56
Arunachal Pradesh	1	Dibang WLS	4149	4589	75
	2	Kamlang WLS	783	–	–
	3	Namdapha NP	1808	1875	8
	4	Yordi-Rabe Supse WLS	492	–	–
Subtotals			7231	6464	55
Totals			36,019	65,580	35

Area of PAs and percentage of snow leopard habitat within each PA have been calculated using available polygons from the Wildlife Institute of India database (2006, updated on October 26, 2014). Area of snow leopard habitat within the Changthang Cold Desert WLS, Karakoram WLS, and Kibber WLS has been corrected (see text for details). All figures have been rounded off.

* *The polygon size available is smaller than notified area and had a potential snow leopard habitat of 344 km^2.*

** *ca. 2500 km^2 of Trans-Himalayan region around the two PAs form the buffer zone of Nanda Devi Biosphere Reserve, and is mostly snow leopard habitat.*

CHAPTER

37

South Asia: Nepal

Som Ale[*,**], *Karan B. Shah*[†], *Rodney M. Jackson*[**]

*Biological Sciences, University of Illinois, Chicago, IL, USA

**Snow Leopard Conservancy, Sonoma, CA, USA

†Natural History Museum, Tribhuvan University, Swoyambhu, Kathmandu, Nepal

DISTRIBUTION, ABUNDANCE, AND POPULATION STATUS

Snow leopards are distributed in the main Himalayan range in Nepal, mostly within alpine and subalpine zones. Their presence has been confirmed in 10 protected areas (PAs) (Table 37.1, Fig. 37.1) and outside PAs in Humla, Jumla, Mugu, Manang, and Sankhuwasabha districts. Snow leopard range is divided into five large, relatively contiguous habitat blocks: Western, Rolwaling, Sagarmatha, Makalu-Barun, and Kanchenjunga areas (WWF-Nepal, 2009).

Throughout Nepal, snow leopard density varies. Based on radio-telemetry, Jackson and Ahlborn (1989) reported a density of 10–12 snow leopards per 100 km^2 in the uninhabited Langu Valley. Lovari et al. (2009) used genetic analysis to estimate four adults in Sagarmatha National Park (1148 km^2). Karmacharya et al. (2011) used noninvasive techniques to estimate four adults in Shey-Phoksundo and five in Ghunsa and Yagma regions in Kangchenjunga. Wegge et al. (2012) identified six adults in 125 km^2 of Phu Valley (Manang) using fecal DNA. A minimum of three adult snow leopards inhabited Jomosm, Lubra, and Thini areas of Mustang in 2011, based on camera trapping in ca. 75 km^2 (Ale et al., 2014). Subsequent camera trapping in 2012–2013 revealed two additional individuals making a total of five.

Jackson and Ahlborn (1990) projected 300–500 snow leopards in approximately 27,432 km^2, based upon expert opinion, local interviews, and a simple GIS model. The Department of National Parks and Wildlife Conservation (DNPWC), in partnership with the Department of Forests, National Trust for Nature Conservation (NTNC), and WWF-Nepal, reassessed the snow leopard population in 2009, using a model based on relationships between sign surveys, genetic analyses, and the extent of potentially suitable habitat, and estimated 301–400 snow leopards residing in ca. 13,000 km^2 (Government of Nepal, 2013; Shrestha et al., 2010; WWF-Nepal, 2009). Some 28% of potential snow leopard habitat in Nepal is located outside PAs.

The snow leopard's altitudinal range in the country extends from 2700 m to 5600 m. Dispersing individuals may cross higher

http://dx.doi.org/10.1016/B978-0-12-802213-9.00037-7

TABLE 37.1 Protected Areas in Nepal Harboring Snow Leopards

		Snow leopard presence through				
Protected area	Area (km²)	Radio-collar	Genetics	Camera-trap/photo	Sign-survey	Local interview
Langtang National Park (LNP)	1710				*	*
Api Nampa Conservation Area (ANCA)	1903				*	*
Gaurishankar (GCA)	2179				*	*
Shey-Phoksundo (SPNP)	3555	*	*			*
Annapurna Conservation Area (ACA)	7629			*	*	*
Makalu-Barun National Park (MBNP)	2233					*
Kangchenjunga Conservation Area (KCA)	2035	*		*	*	*
Manaslu Conservation Area (MCA)	1663				*	*
Sagarmatha (Mt. Everest) National Park (SNP)	1148			*	*	
Dhorpatan Hunting Reserve (DHR)	1325				*	*

mountain passes; for example, Shah and Baral (2012) reported signs along Tashi-Lapcha-La pass (5755 m) in the Everest region.

Blue sheep (*Pseudois nayaur*) and Himalayan tahr (*Hemitragus jemlahicus*) are the principal prey in Nepal (Ale and Brown, 2009; Lovari et al., 2009; Oli, 1994; Wegge et al., 2012). Other species such as Tibetan argali, (*Ovis ammon hodgsoni*), kiang (*Equus kiang*), musk deer (*Moschus* spp.), marmot (*Marmota* spp.), woolly hare (*Lepus oiostolus*), pika (*Ochotona* spp.), and several pheasant species (Phasianidae) constitute its supplementary diet (Shah and Baral, 2012, and references therein).

Oli et al. (1993) reported that livestock comprised almost 30% of snow leopard diet in Manang. A more detailed study in Sagarmatha (Lovari et al., 2009) indicated that Himalayan tahr, musk deer (*Moschus chrysogaster*), and livestock (*Bos* spp.) were the most frequent prey. With the decline in Himalayan tahr recruitment in Sagarmatha, livestock increased in snow leopard diet by as much as 29%, making it a significant alternative prey. An even higher amount (42%) of livestock in the diet was reported from Nar-Phu, Manang, by Wegge et al. (2012).

CONSERVATION THREATS AND CHALLENGES

Although human population density is relatively low in northern Nepal, land use practices are pervasive within and outside PAs. The welfare of snow leopards therefore hinges upon coexistence with pastoralists and farmers (Ale et al., 2010; Jackson et al., 2010; Shrestha and Wegge, 2008). Livestock depredation varies widely but could exceed 12% of livestock holdings in parts of Nepal (Jackson et al., 1996). High

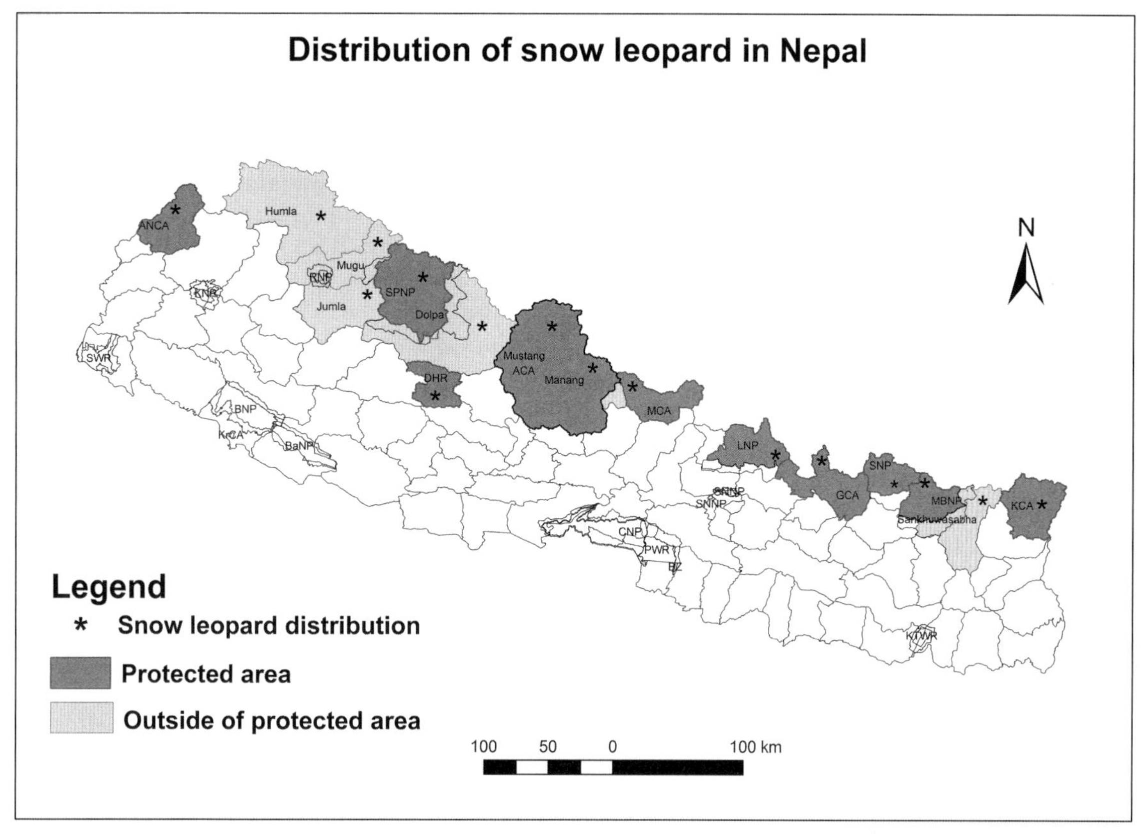

FIGURE 37.1 **Protected areas (PA) and areas outside PAs with snow leopard presence.** *Map: Tej Thapa, Tribhuvan University, Nepal.*

depredation rates can at least partly be attributed to human disturbance, less vigilant guarding practices, and increased livestock numbers. Herders are especially angered by events of "surplus killing" where a snow leopard may take 50 or more sheep and goats in a single instance. This loss is equivalent to USD 7500 (at 2014 market value) – a significant sum given per capita annual incomes of USD 250–400 (Jackson et al., 2010).

Oli et al. (1993) reported the local people's negative attitude toward snow leopards in the Manang area. Out of 102 respondents, more than 50% recommended elimination of snow leopard as the only solution to livestock depredation. Occasionally the carcass of snow leopard used to be paraded through villages by hunters to collect a reward; this practice has ceased, at least in Annapurna. As recently as 2012, two snow leopards were poisoned in Marpha, Mustang, and the culprits were later sentenced to jail. In 2011, a snow leopard carcass was located by game scouts at 4700 m in the Everest region that was apparently poisoned by local herders.

In addition to human-snow leopard conflict, loss of prey such as in Sagarmatha (Everest) (Ale, 2007; Lovari et al., 2009) poses a challenge. Recently the collection of *Cordyceps* and other non-timber forest products, especially in west Nepal (e.g., Shey-Phoksundo, Annapurna), has not only caused disturbance to snow leopard and its prey but also increased the chance of snow leopard poaching by outsiders (see also Chapter 10.4).

STRATEGIES TO MITIGATE CONSERVATION THREATS

Strategies to mitigate conservation threats to snow leopard utilize both coarse-filter and fine-filter approaches (Noss, 1987). The former aims to preserve entire communities of flora and fauna by protecting large extents of habitat (e.g., by establishing protected areas), while the latter focuses on protecting species.

Snow Leopard Conservation in Protected Areas (Ecosystem Approach)

Snow leopard conservation in Nepal began with the establishment of protected areas targeting species in the 1960s. In 1984, Shey-Phoksundo was established as a high-altitude national park to protect the pristine area, and snow leopard became a conservation target for Shey-Phoksundo and received support from WWF-Nepal working in partnership with DNPWC.

In Sagarmatha NP, an ecosystem approach is being applied following the return of snow leopards after an absence of nearly four decades (Ale and Boesi, 2005; Ale and Brown 2009; Lovari et al., 2009). These individuals may have begun dispersing to adjoining valleys, such as Rolwaling (Ale et al. 2010).

ICDP (Integrated Conservation-Development Project) With Community-Based Ecosystem Approach

The real focus on snow leopard conservation in Nepal occurred in the early 1990s with what Heinen and Shrestha (2006) termed the *"dawn of social conservation."* In accordance with the Third Amendment to the 1973 Act (HMG, 1973), several Conservation Areas were designated; these are reserves managed for integrated conservation and development, corresponding to IUCN Category VI (managed resource or extractive) reserves.

The first local Snow Leopard Conservation Committee (SLCC) comprising seven members was established in Annapurna, the country's first Conservation Area (Ale, 1997; Ale and Karki, 2002). The committee operated under Annapurna Conservation Area Project's (ACAP) legal village level unit (Conservation Area Management Committee, or CAMC). Actions such as hiring communal herders and patrolling snow leopard habitat to deter poachers were initiated. In Mustang district, from 2001 to 2006, the USD 2.3 million Upper Mustang Biodiversity Conservation Project was launched under the ICDP approach that targeted snow leopard conservation (Ale and Karki, 2002). A massive awareness program was launched, dozens of corrals were predator-proofed, and an income-generation program established to compensate people for livestock lost to predators.

Nepal has established five conservation areas where participatory approaches are employed for snow leopard conservation. The formation of local committees like the SLCC, Women's Groups, and anti-poaching units along with the Buffer Zone Council represented by locals has been effective in furthering snow leopard conservation.

Protection Beyond Protected Areas

Few PAs in Nepal are large enough to contain viable populations of snow leopards. More than 65% of Nepal's snow leopards (Jackson and Ahlborn, 1990) and as much as 28% of their habitat may fall outside PAs (WWF-Nepal, 2009). Each national park maintained a headquarters and a base for an army unit (consisting of 1200 soldiers) whose responsibility included anti-poaching patrols. The government introduced Community Forestry in 1990 to improve management in areas outside the PA. The community forestry program helps create and restore potential corridors for snow leopard, with pastures managed by local community forest user groups based on officially approved operational plans. As of 2013, there were 17,809 user groups managing more than 1.6 million hectares of forest and adjacent pasture involving 2.3 million households (Pokharel, 2013).

Species Approach in Snow Leopard Conservation

Nepal has targeted this goal through the National Parks and Wildlife Conservation Act of 1973 (and subsequent amendments), together with the Rare, Endangered Wildlife and Plants Trade Act of 2002 (Heinen and Shrestha, 2006) and action plans.

Snow Leopard Action Plan

The government of Nepal endorsed the National Snow Leopard Conservation Action plan in 2005 (Government of Nepal, 2005). The eight objectives included an assessment of snow leopard status, establishment of habitat monitoring, human-snow leopard conflict resolution, pasture improvement, ecological studies, promoting conservation awareness, reducing poaching activities, and encouraging collaboration with national and international conservation organizations. Altogether 44 activities were identified to meet these objectives, requiring USD 2.92 million over a period of 10 years. Only a small proportion of recommended activities have been implemented, largely due to lack of funding. The action plan was revised in 2012 to include emerging threats such as climate change and the proliferation of rural roads (Government of Nepal, 2013).

Savings and Credit Program

The US-based Snow Leopard Conservancy (SLC), in partnership with the NTNC and the Sagarmatha Buffer Zone Management Committee, has embarked on a community-based initiative in the Everest region. The project aims at minimizing people-wildlife conflict and responding to poverty, an important driving factor. Community-based savings-and-credit (SAC) micro-finance schemes were established in four settlements in 2011 and are now functioning smoothly. Each SAC group, consisting of at least 80% of a community's households, was formed following a week-long workshop and each has received multiple training.

As of July 2013, these groups with more than 200 members from four villages own more than Rs 1,700,000 (ca. USD 20,000). In 2014, the fund increased to more than USD 27,000. Within three years of its implementation, the original fund had quadrupled in large part from community contributions and fund-raising actions. The SAC use their funds to provide loan programs to their members. This program's novel aspect is that 25% of net profits support local snow leopard conservation (e.g., patrolling, partially compensating livestock losses, and undertaking awareness and education-related activities through local schools). The SAC program has been extended to communities in Annapurna. Also in Annapurna, NTNC (supported by SLC) has predator-proofed more than 40 livestock corrals across Mustang district.

Livestock insurance program

A community-managed livestock insurance scheme was successfully piloted in Gunsa valley of Kangchenjunga Conservation Area in 2005 (Gurung et al., 2011), under the leadership of DNPWC. An endowment fund (USD 16,000) was established in the first year, which has grown since then as members insure their livestock and pay yearly premiums. The rate was set at USD 0.73 per yak calf (1.1% of yak compensation value) per annum. During the pilot phase, compensation was set at USD 33.30 per calf or 50% of the current value of a young yak. The Snow Leopard Conservation Committee (SLCC) Members have been trained to verify depredation events and make relief payments. Retaliatory killing of snow leopards has declined since the program was initiated. This program has now been introduced into Langtang and Shey- Phoksundo NPs.

Species to Ecosystem to Landscape Approaches to Snow Leopard Conservation

Nepal has moved from the "fences and fines" approach of national parks in the 1970s to the

1990s social conservation (conservation areas and buffer zones) and the current landscape conservation strategy (Heinen and Shrestha, 2006) pioneered by a joint initiative of government and NGOs (Bajimaya, 2002; HMGN/MFSC, 2005; Wikramanayake et al., 2011).

Declaration of the Sacred Himalayan Landscape (SHL) Strategic Plan (2006–2016) in 2006, the SHL Interim Plan (2010–2014) in 2010, and the design of Kailash Sacred Landscape in 2013 indicate Nepal's interest in landscape-level conservation to protect this and other wide-ranging species. The 2010 memorandum of understanding between China and Nepal and a resolution between India and Nepal the same year are important steps towards transboundary conservation of this species.

The Role of Religion and Traditional Social Norms in Snow Leopard Conservation

Nature historically has been considered sacred in Nepal. The Buddhist concept of dependent origination envisages the human species having a strong unity of interdependence with all other living beings and encourages humans to protect the fabric of nature. People in Manang strongly believe that snow leopards and domestic cats are born to remove their sins from past lives; therefore killing these creatures means having their sins transferred to your own life. Another folktale describes the snow leopard as a "fence" for crops, meaning that in the absence of snow leopards, livestock would be free-ranging and likely invade crop fields. So, folk wisdom suggests that the presence of the snow leopard is an indicator of a decent livelihood and healthy ecosystem. In these herding communities, roasting meat over fire is not allowed or the mountain god will send its dog, that is, snow leopard, and one will have to suffer livestock losses (Ale and Karki, 2002).

The legend in Lubra in Mustang has it that Bon-Po and Buddhist deities, disguised as snow leopards, travel from one settlement to another guarding them from demons and natural calamities. So killing a snow leopard may also mean angering one's ancestral spirits (Ale et al., 2014).

Snow leopards are venerated as "god" of the mountains in some high-altitude enclaves, while in other regions they are considered "dog" of the mountain god. Khumbu (Everest) and Rolwaling (Dolakha) contain *beyuls* (the fabled Shangri-la or Shambala), valleys that locals consider are hidden from evil and protected from the world by the mountain gods (Ale et al., 2010). Should a person with an ill intention try to reach a *beyul*, it is believed that the person will be attacked and driven away by snow leopards at mountain passes.

In Dolpo, there is a story of a great incarnated lama frequently making trips to Tibet in the guise of a snow leopard, in search of rare medicinal herbs. In Langu valley, Mugu district, almost one third of open forests and scrubland has been under religious protection initiated by local lamas. These sacred places are a refuge for wildlife including snow leopard, where hunting, felling of trees, and clearing juniper bushes are not allowed. In most mountain enclaves, snow leopards enjoy a legendary place in local folklore and are not as widely persecuted as their wild cousins such as the tiger.

Snow Leopard Population Monitoring

Monitoring Through Remote Cameras

In Annapurna, in partnership with the NTNC, SLC introduced camera monitoring of snow leopards in 2012 (Fig. 37.2). The program trains every year a few dozens of school youths (Snow Leopard Scouts) and herders – as citizen scientists – on snow leopard (Fig. 37.3 its prey, and habitat monitoring in parts of Mustang and Manang (Ale et al., 2014). To date, more than 100 Snow Leopard Scouts have been trained in three different parts of Annapurna and one in Everest. In Kangchenjunga, members of SLCC, under the supervision of DNPWC and WWF-Nepal, have been monitoring snow leopards through cameras and field signs.

FIGURE 37.2 An adult snow leopard self-captured in 2013 in a remote camera strategically located in Lubra in Annapurna, Nepal.

Snow Leopard Monitoring Using Snow Leopard Information Management System (SLIMS)

Nepal has developed "Snow Leopard Monitoring Guidelines" (Thapa, 2007), building on the earlier work of Jackson and Hunter (1996). Snow leopard monitoring using systematically established sign transects is ongoing in Kangchenjunga and Shey-Phoksundo, Langtang, and Sagarmatha (Shah and Thapa, 2007).

CONCLUSIONS AND NEXT STEPS FORWARD

Nepal witnessed a ground-breaking, VHF radio-tracking study in the 1980s (Jackson, 1996). In 2013, DNPWC, WWF-Nepal, and NTNC embarked on satellite tracking of snow leopards in Kangchenjunga, with one snow leopard collared so far. Snow leopard conservation is challenging because of Nepal's rugged and often roadless terrain that makes law enforcement and the implementation of resource management strategies taxing. The situation has further deteriorated for the last three decades or so due to political turmoil.

Because the snow leopard shares its habitat with China and India, transboundary cooperation and protected areas become vital. The future of Nepal's snow leopards ultimately rests in the hands of local communities including livestock herders who share their habitat. We need locally focused, community-driven projects to alleviate existing people-wildlife conflicts, but also policies and legislation to deter poaching and other harmful activities to snow leopard, its prey, and habitat. Large or small-scale projects will only work if they bring about fundamental, long-term behavioral change.

FIGURE 37.3 Snow Leopard Scouts (2013 batch) from Annapurna, Nepal.

References

Ale, S.B., 1997. The Annapurna Conservation Area Project: a case study of an integrated conservation and development project in Nepal, pages 155-169 In: R. Jackson and A. Ahmad (Eds.). Proceedings of the 8th International Snow Leopard Symposium, Islamabad, November 1995. International Snow Leopard Trust, Seattle, and WWF-Pakistan, Lahore, Pakistan.

Ale, S.B., Shrestha, B., Jackson, R., 2014. On the status of snow leopard *Panthera uncia* (Schreber, 1775) in Annapurna, Nepal. J. Threat. Taxa 6, 5534–5543.

Ale, S. B., 2007. Ecology and conservation of the snow leopard and the Himalayan tahr in Sagarmatha (Mt. Everest) National Park, Nepal. PhD Thesis. University of Illinois-Chicago.

Ale, S. B., Karki, B. S., 2002. Observations on conservation of snow leopards in Nepal. In: Proceedings of the International Snow Leopard Survival Summit, pp 1–8. International Snow Leopard Trust. Seattle, USA.

Ale, S.B., Boesi, R., 2005. Snow leopard sightings on the top of the world. CAT News 43, 19–20.

Ale, S.B., Brown, J.S., 2009. Prey behavior leads to predator: a case study of the Himalayan tahr and the snow leopard in Sagarmatha (Mt. Everest) National Park, Nepal. Israel J. Ecol. Evol. 55, 315–327.

Ale, S.B., Thapa, K., Jackson, R.M., Smith, J.L.D., 2010. The fate of snow leopards in and around Mt. Everest. CAT News 53, 19–21.

Bajimaya, S., 2002. Sharing Nepal's experience in transboundary conservation endeavors. Wildlife Nepal 6, 8–13.

Government of Nepal, 2005. Snow Leopard Conservation Action Plan for Kingdom of Nepal (2005–2010). Government of Nepal, Ministry of Forest and Soil Conservation. Department of National Parks and Wildlife Conservation, Babarmahal, Kathmandu, Nepal, pp.18.

Government of Nepal, 2013. Snow Leopard Conservation Action Plan for Nepal, 2005–2015 (Revised 2012). Government of Nepal, Ministry of Forest and Soil Conservation. Department of National Parks and Wildlife Conservation, Babarmahal, Kathmandu, Nepal, p. 17.

Gurung, G.S., Thapa, K., Kunkel, K., Thapa, G.J., Kollmair, M., Boeker, U.M., 2011. Enhancing herders' livelihood and conserving the snow leopard in Nepal. CAT News 55, 17–21.

Heinen, J.T., Shrestha, S.K., 2006. Evolving policies for conservation: an historical profile of the protected area system in Nepal. J. Environ. Plann. Man. 49, 41–58.

HMG, 1973. National Park and Wildlife Conservation Act 2029. Ministry of Law and Justice, Nepal Gazette, 2029-11-28 B.S.

HMGN/MFSC, 2005. Proceedings of the National Consultation on Sacred Himalayan Landscape in Nepal. Printed with support from ICIMOD, TMI, and WWF-Nepal Program.

Jackson, R., 1996. Home range, movements and habitat use of snow leopard (*Uncia uncia*) in Nepal. PhD dissertation, University of London, External Programme.

Jackson, R., Ahlborn, G., 1989. Snow Leopards in Nepal – home range and movements. Nat. Geog. Res. 5, 161–175.

Jackson, R., Ahlborn, G., 1990. The role of protected areas in Nepal in maintaining viable population of snow leopards. Intl. Ped. Book of Snow Leopard. 6, 51–69.

Jackson, R., Hunter, D. O., 1996. Snow leopard survey and conservation handbook. International Snow Leopard Trust, Seattle, and US Geological Survey, Biological Resources Division, 154 pages.

Jackson, R. M., Ahlborn, G., Gurung, M., Ale, S., 1996. Reducing livestock depredation losses in the Nepalese Himalaya. In: Timm, R. M. and Crabb, A.C. (Eds.). Proceedings of the 17th Vertebrate Pest Conference. University of California, Davis, pp. 241–247.

Jackson, R.M., Mishra, C., McCarthy, T.M., Ale, S.B., 2010. Snow leopards: conflict and conservation. In: Macdonald, D.W., Loveridge, A.J. (Eds.), Biology and Conservation of Wild Felids. Oxford University Press, UK.

Karmacharya, D. B., Thapa, K., Shrestha, R., Dhakal, M., Janecka, J.E., 2011. Noninvasive genetic population survey of snow leopards (*Panthera uncia*) in Kangchenjunga conservation area, Shey Phoksundo National Park and surrounding buffer zones of Nepal. BMC Research Notes 4, p. 516.

Lovari, S., Boesi, R., Minder, I., Mucci, M., Randi, E., Dematteis, A., Ale, S.B., 2009. Restoring a keystone predator may endanger a prey species in a human-altered ecosystem: the return of the snow leopard to Sagarmatha National Park. Anim. Conserv. 12, 559–570.

Noss, R.F., 1987. From plant communities to landscapes in conservation inventories: a look at the Nature Conservancy (USA). Biol. Conserv. 41, 11–37.

Oli, M.K., 1994. Snow leopards and blue sheep in Nepal: densities and predator–prey ratio. J. Mammal. 75 (4), 998–1004.

Oli, M.K., Taylor, I.R., Rodgers, M.E., 1993. Diet of the snow leopard (*Panthera uncia*) in the Annapurna in the Annapurna Conservation Area, Nepal. J. Zool. (London) 23, 365–370.

Pokharel, R.K., 2013. Forestry: Community. In: Jha, P.K., Neupane, F.P., Shrestha, M.L., Khanal, I.P. (Eds.), Biological Diversity and Conservation. Nepal Academy of Science and Technology, Khumaltar, Lalitpur, pp. 347–352.

Shah, K. B., Baral, H. S., 2012. Conservation of Snow Leopard in Nepal. A booklet in Nepali. Himalayan Nature, Kathmandu, Nepal, p. 102.

Shah, K.B., Thapa, T. B., 2007. Establishment of New SLIMS Program and Assessment of Existing Ones in the Sacred Himalayan Landscape (SHL) and North Mountains (NM), Nepal. An unpublished final report submitted to the WWF-Nepal, Kathmandu, Nepal, p. 70.

Shrestha, R., Wegge, P., 2008. Wild and domestic sheep in Nepal Trans–Himalaya: competition or resource partitioning? Environ. Conserv. 35, 125–136.

Shrestha R., Thapa K., Thapa G., Wikramanayake E., Bajimaya S., Pradhan N., Bhatta S., Janecka J., Wegge P., 2010. Estimating snow leopard (*Panthera uncia*) populations in the Nepal Himalaya. 2010. A paper presented on 23rd International Congress for Conservation Biology, Society for Conservation Biology, Beijing, China.

Thapa, K., 2007. Snow leopard monitoring guideline. Shah, K. B., Bajimaya S. (Eds.), p. 60.

Wegge, P., Shrestha, R., Flagstad, Ø., 2012. Snow leopard *Panthera uncia* predation on livestock and wild prey in a mountain valley in northern Nepal: implications for conservation management. Wildl. Biol., 18131–18141.

Wikramanayake, E., Dinerstein, E., Seidensticker, J., Lumpkin, S., Pandav, B., Shrestha, M., Mishra, H., Ballou, J., Johnsingh, A.J.T., Chestin, I., Sunarto, S., Thinley, P., Thapa, K., Jiang, G., Elagupillay, S., Kafley, H., Pradhan, N.M.B., Jigme, K., Teak, S., Cutter, P., Aziz, M.A., Than, U., 2011. A landscape-based conservation strategy to double the wild tiger population. Conserv. Lett. 4, 219–227.

WWF-Nepal. 2009. Estimating Snow Leopard Population in the Nepal Himalaya. WWF-Nepal program, Baluwatar, Kathmandu, Nepal.

CHAPTER

38

South Asia: Pakistan

OUTLINE

Snow Leopards
http://dx.doi.org/10.1016/B978-0-12-802213-9.00038-9

SUBCHAPTER

38.1

Snow Leopard Conservation in Pakistan: A Historical Perspective

Ahmad Khan

Department of Geographical Sciences, University of Maryland, College Park, MD, USA

THE SNOW LEOPARD'S PLACE AND PRESENTATION

The snow leopard (*Panthera uncia*), an emblematic species of Asia's high peaks, is sometimes referred to as "Queen of the High Mountains." This beautiful but endangered animal, cryptic and elusive, roams the elevations above 2000 m, where its survival is challenged by ecological fragmentation, climatic change, social resistance and behavior, development initiatives and economic needs, retaliation for economic losses it inflicts by killing livestock, and hunting for its pelts and body parts for trade.

Pakistan is one of the 12 countries in Asia where this magnificent cat is known to inhabit. Occurring in the Himalayas, Karakoram, and Hindu Kush mountains, the country's snow leopard population estimates are as uncertain as those of any other range state. In the 1970s, George Schaller estimated a total of 250 snow leopards in Pakistan (Schaller, 1977). However, over the next decade new information, resulting from increased attention to the species and enhanced capacity of the provincial wildlife departments to monitor populations, along with the application of modern technologies such as satellite image interpretation, indicated that this initial estimate was likely low.

Upon its creation in 1947, Pakistan inherited the Indian Forest Act of 1927, which was adopted to address conservation of forests needs in the country. The first legislation that addressed the wildlife conservation in Pakistan was the Wildlife Protection Ordinance 1959. However, the snow leopard was not covered under this ordinance and did not appear on the list of protected or unprotected animals. The reason for this omission might have been that most snow leopard habitat was in the independent states of Swat, Dir, Chitral, Azad Jammu and Kashmir, and Gilgit Baltistan, hence little or no information existed regarding the species' occurrence. The independent states had their own rules and traditions that governed protection and conservation of natural resources. These varied from state to state and from ruler to ruler, but were mostly focused on protection of hunting areas for the entertainment of the ruling class.

Evolving conservation efforts in Pakistan focused more on a utilitarian approach toward natural resources until the 1970s. Regulation of hunting through permits and maintenance of hunting reserves was the primary focus, rather than endangered species, habitat, or prey conservation. Ungulates and birds, which could be hunted for sport, were given the highest priority, while predators, including snow leopard, remained the lowest priority for conservation.

In the 1970s, when most of the provincial legislation was promulgated, the snow leopard was included in the list of protected animals. The first legislation mentioning snow leopards was the Pakistan Wildlife Ordinance of 1971. Legislation promulgated later and covering snow leopards includes the Punjab Wildlife Act of 1974, Khyber Pakhtunkhwa Wildlife Act of 1975, Gilgit Baltistan Wildlife Preservation Act of 1975, and Azad Jammu and Kashmir Wildlife Act of 1975. Although the snow leopard was declared a protected animal by these provincial legislative actions and it had Endangered status on the IUCN Red List, no special emphasis was placed on its protection, other than a generalist approach that was extended to all wildlife species under the wildlife laws in the 1970s and 1980s.

The snow leopard's current legal status doesn't allow them to be killed for any reason except in self-defense where a human life is at stake. According to the legal provisions, possession of a dead or live snow leopard is illegal for any purpose, whether as a pet or for body parts or trade. Any violator found with a snow leopard or its body parts can be punished with fines and/or imprisonment as prescribed under the law. Along with snow leopards, the provincial laws have given protected status to most of its prey species including the markhor (*Capra falcomeri*), Marco Polo sheep (*Ovis ammon polii*), blue sheep (*Pseudois nayaur*), Himalayan ibex (*C. sibirica*), musk deer (*Moschus* spp.), snowcock (*Tetraogallus* spp.), and Monal pheasant (*Lophophorus impejanus*). The provincial laws also have provisions to establish protected areas, including national parks, wildlife sanctuaries, game reserves, and private game reserves. The protected area network covering snow leopard range in Pakistan includes 10 national parks with a combined area of ~15,000 km^2, 4 wildlife sanctuaries totaling ~1000 km^2, 11 game reserves with an area of about 2380 km^2, and 8 community game reserves covering more than 2200 km^2 (Fig. 38.1.1). Looking at the distribution of protected areas by province we find that Gilgit–Baltistan has 4 national parks, 5 wildlife sanctuaries and 9 game reserves. Khyber-Pakhtunkhwa has 4 national parks, 1 wildlife sanctuary, 6 game reserves, and 10 community game reserves within snow leopard range. Finally, Azad Jammu and Kashmir has 1 national park, where the occurrence of snow leopard is reported.

MILESTONES IN SNOW LEOPARD CONSERVATION IN PAKISTAN

Snow leopards received little attention until the International Snow Leopard Trust (now known as Snow Leopard Trust or SLT) turned its attention to Pakistan in the late 1980s, and thus elevated national interest in the cat. At about the same time, WWF-Pakistan pioneered a project on conservation and sustainable use of wild ungulates in the Bar Valley, focused on community management of the flare-horned markhor and Himalayan ibex. This early community-based conservation of the primary prey species of the snow leopard brought a new dimension to conservation in Pakistan by ensuring that 80% of the trophy hunting license went to the local communities (for more detailed information see Chapter 16.1).

The snow leopard conservation efforts in Pakistan picked up in 1994 when SLT initiated the training of Pakistani biologists and wildlife conservationists in the Snow Leopard Information Management System (SLIMS), a tool designed to collect information about snow leopard, its prey species, and habitat. Leading the effort, Rodney Jackson, the then Director of Conservation at SLT and Don Hunter from the

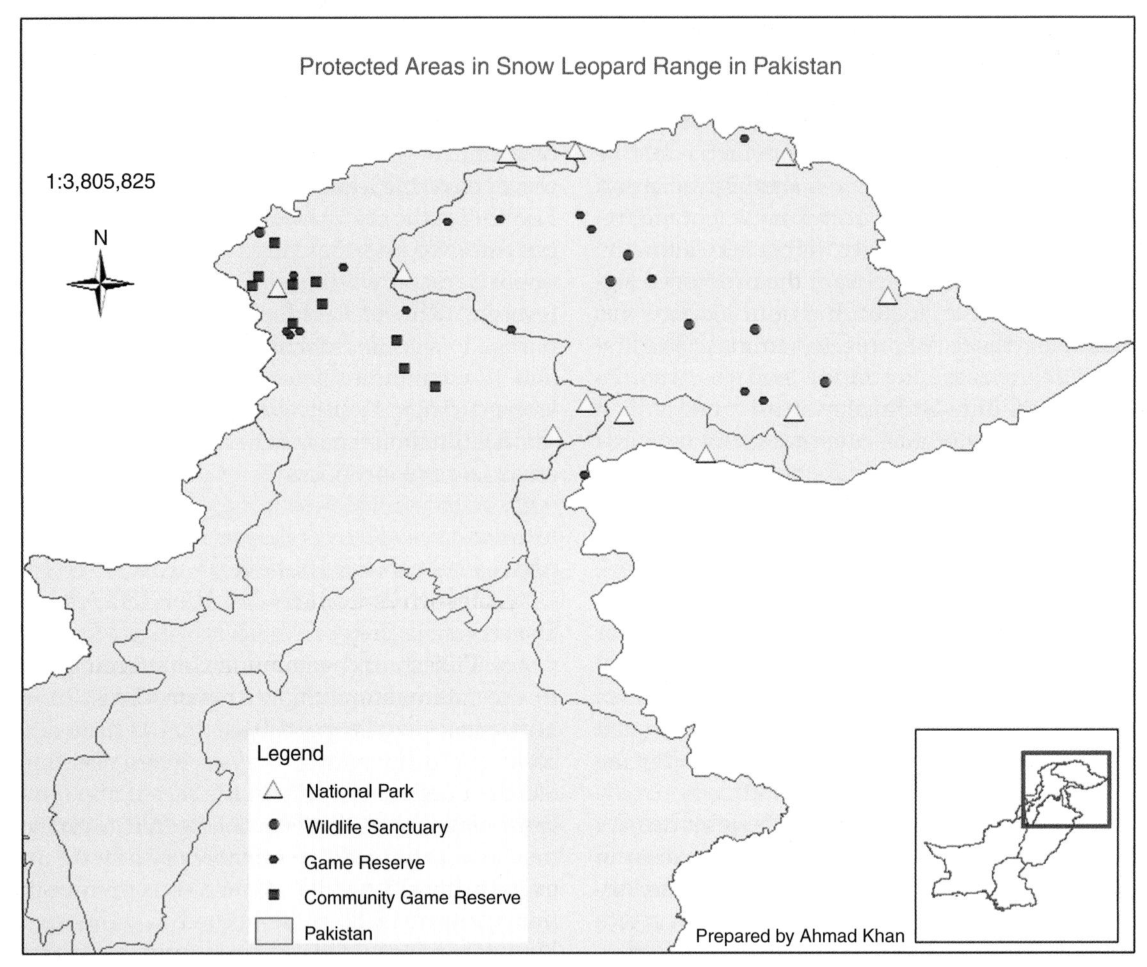

FIGURE 38.1.1 **Map of protected area type and location in snow leopard habitat of Pakistan.**

US Fish and Wildlife Service, conducted snow leopard surveys in Chitral Gol National Park and Khunjerab National Park, two key protected areas. The information collected from these areas revealed a population that might be larger than the prevailing estimates (Ahmad et al., 1997). These efforts were followed by the organization of the Eighth International Snow Leopard Symposium in Islamabad, where snow leopard conservationists from Pakistan and other range states, joined international experts from academia, research institutions, and nongovernmental organizations to present their work on snow leopard, its prey species, and habitat, and work together to address the conservation issues faced by the species in Pakistan and elsewhere.

In 1998, SLT, in partnership with WWF-Pakistan, appointed this author as the first in-country representative in Pakistan to work exclusively on snow leopard conservation, conducting surveys under SLIMS, building capacity of the park staff in SLIMS and launching a snow leopard conservation education program. The joint program resulted in the finding of new habitats in Tirich Valley, documenting snow leopard presence on a regular basis in protected areas such as Chitral Gol National Park and Tooshi Game Reserve in Chitral, raising interest and awareness of local communities to address snow leopard conservation, and the development of a community incentive program based on local handicrafts. The first community in which the enterprise was established in the late 1990s, Goleen Gol, has been working actively with WWF-Pakistan and SLT after their capacity was built in making value-added woolen handicrafts. This economic incentive brings much needed income to the local community and ensures their engagement in snow leopard conservation.

The Snow Leopard Conservancy, formed by snow leopard conservationist Rodney Jackson and his wife Darla Hillard, a conservation education specialist, in partnership with Shafqat Hussain of Yale University, launched a livestock insurance scheme in Gilgit-Baltistan in 1999. The program relied on payments to herders for livestock losses to dissuade them from taking retributive actions against snow leopards that kill livestock (see Chapter 13.3). To meet the goals of the Snow Leopard Survival Strategy (McCarthy and Chapron, 2003) WWF-Pakistan undertook the initial development of Pakistan's Snow Leopard Conservation Strategic Action Plan in 2004, which was endorsed by the then Ministry of Environment in 2007 (Khan, 2008). A major technological milestone for both Pakistan and the global snow leopard community occurred in 2006 when the first GPS-satellite collar was put on a snow leopard in Chitral Gol National Park. The project, implemented jointly by WWF-Pakistan and SLT, revealed interesting information on snow leopard home range and territorial behavior (McCarthy et al., 2006). The data documented movements of the cat across the political border between Pakistan and Afghanistan. This suggests that conserving of the species in one country is not sufficient and pointed to the need for international collaboration on both research and conservation of the species.

JOINING HANDS TO CONSERVE THE CAT

Modern techniques and technological applications, along with community participation to support government efforts, have proved successful ways to conserve a unique resource – the snow leopard. Lessons learned from the Bar Valley experiment became the basis for other programs in Pakistan such as Community Based Biodiversity Conservation Project, Mountain Areas Conservancy Project, and community managed game reserves (Edwards, 2006; Hunnam et al., 2003). Today, the snow leopard in Pakistan remains the focus of collaborative conservation efforts among local communities and nongovernmental and government organizations. As a result, this large predator, previously killed by people for a variety of reasons, is much less commonly persecuted (Hayward and Somers, 2009; Khan, 2014). People have started reporting trapped animals to conservation organizations for safe rescue, although some losses of snow leopards caught killing livestock are still reported. Though all is not glittering, the success that comes with our continuing efforts suggests an improved future for this magnificent cat in Pakistan.

SUBCHAPTER

38.2

The Current State of Snow Leopard Conservation in Pakistan

Jaffar Ud Din, Hussain Ali*, Muhammad Ali Nawaz*,***

***Snow Leopard Foundation, Islamabad, Pakistan**
****Department of Animal Sciences, Quaid-i-Azam University, Islamabad, Pakistan**

PAKISTAN'S SNOW LEOPARDS

"*Suddenly I saw the snow leopard. Wisps of clouds moved between us, and she became a ghost creature, appearing and disappearing as if in a dream.*" These thoughts were expressed by naturalist and wildlife biologist George B. Schaller (1971) after the world's first human encounter with the majestic cat in the wild in the Hindu Kush in northern Pakistan. This coincidence led to the snow leopard being dubbed the "Imperiled Phantom of Asian Peaks" (Schaller, 1971) and served as a catalyst in sensitizing range countries and experts to the fate of the cat in its entire range, while conceding that the snow leopard range in Pakistan was geoecologically and culturally unparalleled.

The Inimitable Range

Snow leopards are reported to occur in 12 countries of South and Central Asia, including Pakistan where the cat's range covers about 80,000 km^2 spread across five major mountain ranges; the Hindu Kush and Hindu Raj in the Khyber Pakhtunkhwa (KP) province and the Pamir, Karakorum, and Himalaya in Gilgit-Baltistan (GB) and Azad Jammu and Kashmir (AJK).

The Mighty Mountains

Pakistan's snow leopard range falls mainly in the northern part of the country at the intersection of the Hindu Kush, Karakorum, and Himalaya (Fig. 38.2.1). These magnificent mountain ranges are home to 5 of the world's 14 peaks higher than 8000 m, including K-2, the second highest peak in the world (8611 m), Gasherbrum I, Gasherbrum II, Broad Peak, and Nanga Parbat. These mountain ranges also boast more than 160 peaks in the 7000 m category and another 700 peaks in the 6000 m category (Government of Pakistan and IUCN, 2003).

Rich Geoecological Profile

The unique geoecological status of the snow leopard range in Pakistan cannot be overstated. High altitudes and subzero temperatures mean that these mountainous regions are among the

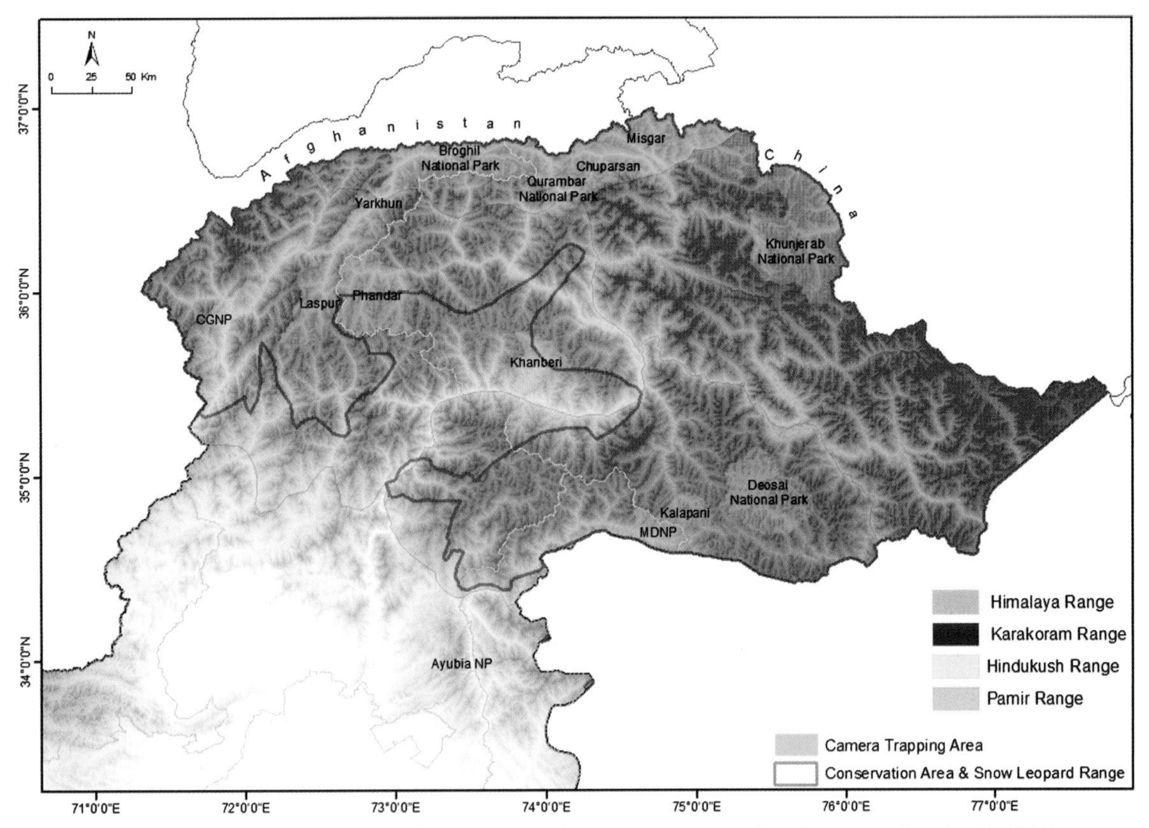

FIGURE 38.2.1 **Snow leopard range and Snow Leopard Foundation research and conservation sites in Pakistan.**

most heavily glaciated parts of the world. Three of the world's longest glaciers outside the polar regions, namely Biafo Glacier, Baltoro Glacier, and Batura Glacier, are found in GB. In addition, several high-altitude lakes form the head of the Indus River – a lifeline for Pakistan's economy.

Snow leopard range in Pakistan harbors rich biological diversity with Palearctic affinities. The majority of the country's forest cover and medicinal plants occur here, in addition to various wild sheep and goats, including the flare-horned markhor (*C. falconeri*), Himalayan ibex (*C. ibex*), Ladakh urial (*O. orientalis vignei*), blue sheep (*P. nayaur*), Marco Polo sheep (*O. ammon polii*), and musk deer (*Moschus sagristurus*) (Government of Pakistan, 2000). Mountain passes, including those on the famous Silk Route, are still used for trade.

Land of Culture and Traditions

Snow leopard habitat displays unique biodiversity and ecosystem functions and services, due to its remoteness, rich history, and comparatively little development. Functions and services include watershed protection, genetic resources and wild crop cultivars, traditional knowledge, customary laws, and spiritual and cultural values. More than 9 million people from different mountain tribes have dwelt in these remote valleys for centuries. There are at least a dozen local dialects, most of which stem from ancient civilizations. Examples of historic linkages include the Kalash tribe from three small hamlets in the Chitral district of KP who claim to be the descendants of Alexander the Great, and the Buddha sculptures in districts Gilgit and Diamer in GB, which are thought to have been created in the ninth millennium B.C.E. (roughly the late Stone Age)[1].

Snow Leopards and Fables

Old folk tales and songs from the region reference wild cats, and people often name their children "Purdoom" (snow leopard). This is an excellent example of the intimate relationship between local people and snow leopards (Din, 2014). Another cultural connection is the Indus River, which is said to resemble a snow leopard's mouth. Similarly, local stories often speak of a snow leopard "in the old days" that hung over a ledge to allow a hunter to pass by. Traditionally, however, snow leopard hunting remained a sign of bravery and dignity for the Rajas, Maha Rajas, Walis, and Mehtars (tribal heads) who displayed pelts as a matter of pride and prestige for almost two decades (Government of Pakistan, 2013).

THE DIMMING MOUNTAINS: THREATS AND CHALLENGES

Unfortunately, snow leopard populations are thought to be declining due to increasing anthropogenic pressures, despite the snow leopard's socioecological importance (Din and Nawaz, 2011; Jackson et al., 2008). Major threats are human induced and economically driven and can be divided into direct and indirect threats.

Direct Anthropogenic Threats

A major source of human-snow leopard conflict stems from predation on domestic livestock, which frequently results in retaliatory killing of the cat. Annual predation pressure on livestock in snow leopard range is estimated to be 1.3 animals per household, which translates into about 30% of total household income, or an annual economic loss of USD 119 per household; monthly incomes are generally about USD 50 (Government of Pakistan, 2013). Equally problematic is poaching for pelts that fetch handsome sums on the black market.

Indirect Anthropogenic Threats

According to the "National Snow Leopard Ecosystem Protection (NSLEP) priorities" (Government of Pakistan, 2013), major threats affecting snow leopard survival include habitat loss and degradation, poaching of prey, foraging competition between wild prey and livestock, wildlife diseases, lack of awareness, poor law enforcement, and climate change.

SAVING THE ARK: ONGOING CONSERVATION PRACTICES

The historical context of snow leopard research and conservation in Pakistan was covered earlier in this chapter, and we see that efforts between 1962 and 2008 provided a platform for a more focused, long-term approach. Snow leopard conservation and management have gained momentum recently, and many government and nongovernment organizations (NGOs) are now actively involved. A major breakthrough was the establishment of the Snow Leopard Foundation (SLF) in 2008. It was set up as a dedicated institution to increase the scale of snow leopard conservation work in Pakistan.

Research and Monitoring

The efforts of organizations such as SLF, SLT, WWF, Panthera, IUCN, and the Snow Leopard Conservancy (SLC) have contributed significantly to our understanding of snow leopards in Pakistan. Nonetheless, the elusive nature

[1] http://www.dawn.com/news/629659/basha-dam-threatens-thousands-of-ancient-rock-carvings

of the species and its remote habitat mean that research is still in its nascent stage. Schaller (1976) estimated 250 cats remained in Pakistan, Roberts (1997), quoting Jackson (1993) estimated 300, and Hussain (2003) estimated 300–420 snow leopards in the country. Din and Nawaz (2011) believe there are about 36 snow leopards in Chitral district alone. Estimates are based on anecdotal reports and sign surveys collected from smaller sites and then extrapolated to the entire Pakistan range. Recent studies into snow leopard diet include that of Anwar et al. (2011), who studied the biomass consumption of snow leopards in Baltistan and reported that livestock constituted 70% of snow leopard diet while wild prey accounted for the remaining 30%. WWF-Pakistan has been involved in sign and questionnaire surveys in some parts of Hunza-Nagar district. These studies were more focused on understanding snow leopard distribution and human-snow leopard conflict (Khan et al., 2014).

SLF has been studying the status of snow leopards in Pakistan using human-snow leopard interaction surveys (Schaller, 1976), occupancy modeling (MacKenzie and Nichols, 2004), camera traps (Karanth and Nichols, 1998), and genetic analysis of fecal DNA (Aryal et al., 2014) since 2010. Total snow leopard habitat in Pakistan covers about 8000 km^2, and SLF has confirmed its presence over large landscapes, starting from Gahriat Gol and Chitral Gol in the west, to the Hindu Kush ranges in the east, including the Torkhow, Laspur, Mastuj, and Broghil valleys in Chitral district. The highest photo-capture rate of the species was recorded in the Khunjerab, Misgar, and Shimshal valleys. Results of genetic analyses of more than 800 suspected snow leopard feces, including population estimates, were not available at the time of writing.

Conservation Measures

Conservation programs are aimed at offsetting human-induced threats to snow leopards in the country. Project Snow Leopard (PSL) is implementing a livestock insurance scheme in three valleys of the Baltistan region (Rosen et al., 2012). WWF-Pakistan is carrying out community mobilization, education, and awareness efforts in KP and GB through a US Agency for International Development (USAID)-funded project (Government of Pakistan, 2013). SLF concurrently began implementing the Snow Leopard Conservation Strategy using 3-year planning cycles and expanded the program geographically and thematically. SLF has linked livelihoods with conservation by developing integrated conservation tools such as the Snow Leopard Friendly Livestock Vaccination Program, Snow Leopard Enterprises, predator-proof corrals, and livestock insurance schemes. The latter has reached about 5000 households in 22 valleys in the Hindu Kush and Karakoram-Pamir mountain ranges, covering about 20% of snow leopard range in Pakistan (unpublished SLF reports).

Education, Advocacy, and Capacity Building

A lack of awareness, support, skills, and the expertise required to understand snow leopard conservation needs are major threats to the survival of snow leopards in Pakistan. Organizations such as WWF and WCS are implementing community sensitization programs in snow leopard habitats. SLF is working with schools and youth in conservation program sites by establishing nature clubs and developing and helping implement action plans. Capacity-building efforts by SLF in carnivore assessment techniques through theoretical and field training are also underway. More than 25 research students and 200 government and NGO field staff are involved.

Evaluation of Successes

Ongoing community-based conservation programs and research endeavors are vital but inadequate, considering the size of snow leopard range in Pakistan and the magnitude of conservation needs. Increasing efforts from the current limited villages and valleys to a wider area is essential to ensure a reversal in population trend.

THE WAY FORWARD

A holistic conservation approach is required. This may be achieved by national and international collaborations to tap resources and help improve our understanding of snow leopards. The Global Snow Leopard & Ecosystem Protection (GSLEP) program is one such initiative. SLF has helped the government of Pakistan formulate the NSLEP and prioritize large landscapes in the Hindu Kush (10,541 km^2), Karakoram-Pamirs (25,498 km^2), and Himalayas (4,659 km^2). The implementation of the NSLEP is expected to ensure the survival of viable snow leopard populations. There is little doubt that a thriving snow leopard population is a prerequisite for maintaining Pakistan's unique mountain ecology and heritage.

References

Ahmad, I., Hunter, D.O., Jackson, R., 1997. A snow leopard and prey species survey in Khunjerab National Park, Pakistan. In Jackson, R., Ahmad, A. (Eds.). Proceedings of 'Eighth International Snow Leopard Symposium, Islamabad, Pakistan. International Snow Leopard Trust, Seattle. pp. 92–95.

Anwar, M.B., Jackson, R., Nadeem, M.S., Janecka, J.E., Hussain, S., Beg, M.A., Muhammad, G., Qayyum, M., 2011. Food habits of the snow leopard *Panthera uncia* (Schreber, 1775) in Baltistan, Northern Pakistan. Eur. J. Wildl. Res. 57, 1077–1083.

Aryal, A., Brunton, D.H., Ji, W., Karmacharya, D.B., McCarthy, T.M., Bencini, R., Raubenheimer, D., 2014. Multi-pronged strategy including genetic analysis for assessing conservation options for the snow leopard in the central Himalaya. J.Mammal. 95 (4), 871–881.

Din, Z., 2014. Hunters become guards of snow leopard in Chitral. Dawn.com. http://www.dawn.com/news/1152206 (accessed 27.12.2014).

Din, J.U., Nawaz, M.A., 2011. Status of snow leopard and prey species in Torkhow Valley, District Chitral, Pakistan. J. Anim. Plant Sci. 21 (4), 836–840.

Edwards, S.R., 2006. Saving Biodiversity for Human Lives in Northern Pakistan. Mountain Areas Conservancy Project. The World Conservation Union (IUCN), Pakistan Country Office, Karachi, Pakistan, 38 p.

Government of Pakistan, World Wide Fund for Nature, Pakistan, IUCN, Pakistan, 2000. Biodiversity Action Plan for Pakistan.

Government of Pakistan, IUCN, 2003. Northern Areas Strategy for Sustainable Development. IUCN Pakistan, Karachi.

Government of Pakistan, 2013. Pakistan National Snow Leopard Ecosystem Protection Priorities. Climate Change Division, Islamabad.

Hayward, M.W., Somers, M.J. (Eds.), 2009. Reintroduction of Top-Order Predators. Wiley-Blackwell and ZSL. Conservation Sciences and Practice No. 5. p. 167.

Hunnam, P., Brodnig, G., Khawar, H., Khan, M.M., 2003. Mountain Areas Conservancy Project, Pakistan: Mid Term Evaluation Report (Final Draft). IUCN, Pakistan. 61 p.

Hussain, S., 2003. The status of the snow leopard in Pakistan and its conflict with local farmers. Oryx 37 (1), 26–33.

Jackson, R., Mallon, D., McCarthy, T., Chundaway, R.A., Habib, B., 2008. Panthera uncia. The IUCN Red List of Threatened Species, version 2014.3. www.iucnredlist.org (accessed 24.12.2014).

Karanth, K.U., Nichols, J.D., 1998. Estimation of tiger densities in India using photographic captures and recaptures. Ecology 79 (8), 2852–2862.

Khan, A.A., 2008. Draft Strategic Plan, Snow Leopard Conservation in Pakistan. Ministry of Environment, Government of Pakistan, Islamabad, p. 27.

Khan, M.I., 2014. Pakistan Snow Leopard to have new home BBC News, Islamabad. http://www.bbc.co.uk/news/world-asia-30381441 (accessed 08.12.2014).

Khan, M.Z., Awan, S., Khan, B., Abbas, S., Ali, A., 2014. A review of behavioural ecology and conservation of large predators inhabiting the Central Karakoram National Park (CKNP). J. Biodivers. Environ. Sci., 439–446, 5.3.

MacKenzie, D.I., Nichols, J.D., 2004. Occupancy as a surrogate for abundance estimation. Anim. Biodivers. Conserv. 27 (1), 461–467.

McCarthy, T.M., Chapron, G. (Eds.), 2003. A Snow Leopard Survival Strategy. International Snow Leopard Trust, Seattle, USA, 125 p.

McCarthy, T.M., Khan, J., Din, J.U., McCarthy, K.M., 2006. First study of snow leopards using GPS-satellite collars in Pakistan. Cat News 46, 222–223.

Roberts, T.J., 1997. Mammals of Pakistan. Oxford University Press, Karachi.

Rosen, T., Hussain, S., Mohammad, G., Jackson, R., Janecka, J.E., Michel, S., 2012. Reconciling sustainable development of mountain communities with large carnivore conservation. Mt. Res. Dev. 32 (3), 286–293.
Schaller, G.B., 1971. Imperiled phantom of Asian peaks. Natl. Geogr. Magazine 140, 701–706.
Schaller, G.B., 1976. Mountain mammals in Pakistan. Oryx 13 (4), 351–356.
Schaller, G.B., 1977. Mountain Monarchs: Wild Sheep and Goats of the Himalaya. University of Chicago Press, Chicago.

CHAPTER

39

Northern Range: Mongolia

Bariushaa Munkhtsog, Lkhagvajav Purevjav**, Thomas McCarthy†, Rana Bayrakçısmith†*

*Institute of General and Experimental Biology, Mongolian Academy of Sciences, Irbis Mongolian Centre, Ulaanbaatar, Mongolia

**Snow Leopard Conservation Fund, Ulaanbaatar, Mongolia

†Snow Leopard Program, Panthera, New York, NY, USA

INTRODUCTION

Mongolia's snow leopard (*Panthera uncia*) population is second in size only to that of China's and stands at about 1000 adult cats. They are distributed along the mountain system of Mongolian Altai, Gobi Altai, Trans Altai Gobi, Khangai, and Khovsgol mountains. The status of snow leopards in Mongolia had been discussed prior to 1989 (Bannikov, 1954; Bold and Dorjzunduy, 1976; Mallon, 1984; O'Gara, 1988; Zhirnov and Ilyinsky, 1986) and most information concerned distribution, abundance, and basic food habits. Bannikov (1954) provided the first account and distribution map of snow leopards in Mongolia, summarizing the information on population status. Schaller et al. (1994) estimated a population of 1500–1700 cats inhabiting 107 counties of 10 provinces of Mongolia. The species is estimated to occur at densities of <1–1.1 individuals per 100 km^2 (McCarthy and Chapron 2003) across the rugged mountain habitats of Mongolia. Snow leopard range in Mongolia is estimated to cover between 90,000 (Schaller et al., 1994) and 130,000 km^2 (Mallon, 1984). McCarthy (2000) produced a more refined estimate of approximately 103,000 km^2 (Fig. 39.1). The cats are not uniformly distributed within that range; high-density areas include the Gobi Altai, Mongolian Altai, and the northern Altai ranges.

The habitat of snow leopards in Mongolia occurs within the elevation range of 600–4200 m above sea level, which is generally lower than snow leopards are found in the majority of their range (McCarthy and Chapron, 2003). Optimal habitat is found in broken and moderately broken mountains with clearly defined ridge lines and cliffs. Siberian ibex and other ungulates inhabit these areas and are the snow leopard's main prey. The cats may also hunt on more open slopes and forest edges for roe

Snow Leopards
http://dx.doi.org/10.1016/B978-0-12-802213-9.00039-0

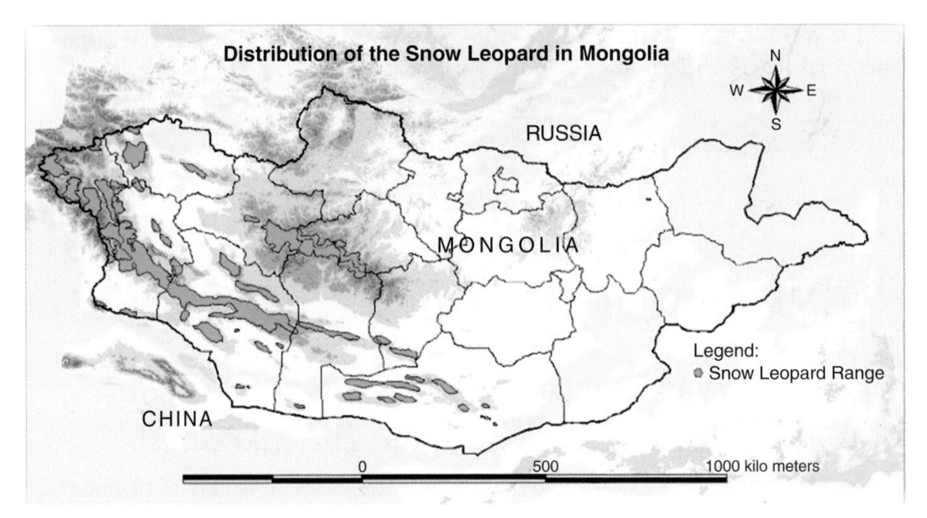

FIGURE 39.1 **Current snow leopard range in Mongolia.**

deer, wild boar, hare, and other prey. Marmots, chukar partridge, and snow cock are additional food sources. Because it is a top predator of the mountain ecosystem of Mongolia and Central Asia, it is an umbrella species for conservation in those ecosystems. These same areas have been used for livestock pasture for thousands of years and support the nomadic herding culture in Mongolia. For example, the sale of cashmere provides herders of 21 provinces with more than USD 200 million each year, and about one-third of all cashmere comes from areas where snow leopards occur.

STATUS AND THREATS

Snow leopards are faced with many anthropogenic threats in Mongolia, with the primary ones being (i) loss of prey base due to competition with livestock for pasture and water sources, (ii) loss of habitat due to increased livestock numbers, (iii) development of mining and transportation infrastructure, and (iv) direct poaching. We examine specific threats more closely here.

Decreases in Snow Leopard Prey

Pastoralism has been the principal livelihood in Mongolia for thousands of years, but livestock were privatized in the 1990s and numbers rapidly reached ~40 million, leading to pasture degradation from overgrazing. The number of herder families also increased and pushed deeper into snow leopard habitat. Herder families that once survived on 400 head of sheep and goats, increased herd sizes to 1000 or more, in many cases to increase income from cashmere. These changes degraded the range, putting added pressure on the wild ungulate prey of the snow leopard.

A reduction in wild ungulate numbers is one of the most important factors that may lead to declines in snow leopards. In 1990s and early 2000s, excessive poaching of ibex is believed to have led to more snow leopard attacks on livestock. Since 2005 mountain ungulate numbers have remained stable or are increasing throughout their range in the country. With larger herd sizes, people have been less reliant on illegal trapping or hunting of wildlife. At the same time, the penalty for hunting of animals has increased and includes confiscation of hunting equipment and vehicles as added deterrents.

Snow Leopard Killing Due to Livestock Depredation

The killing of snow leopards by herders due predation on livestock is a serious threat, resulting from a combination of increased livestock numbers, reduced wild prey population, and poor guarding practices. While livestock depredation occurs across the range in Mongolia, it seems to be most prevalent southern and western Mongolia where livestock and herders spend winter months in snow leopard habitat. Domestic sheep and goats are killed as they come and go to their pastures, as well as inside their relatively poorly constructed night corrals. Free roaming yaks and horses are also easy prey for the cats.

Infrastructure Development and Habitat Degradation

Mining is increasing and is now a primary contributor to the Mongolian economy (see Chapter 10.2). The rapid rise in mineral development, such as the Oyu Tolgoi copper and gold deposit and many coalmines, has caused a large increase in truck traffic to China, cutting through populations of endangered wildlife and habitats (see Chapter 10.3). In November 2010, the Parliament authorized construction of a 1766 km long railroad network to link Mongolia's major coal and copper mines with China.

Road construction in snow leopard habitat can lead to reduction of wild prey through disturbance. Mining can also lead to the localized destruction of key snow leopard habitat. The extent of the impact mining may be having in Mongolia is not yet known, but a mining license map shows the potential for extensive fragmentation within snow leopard habitat. Development of mining infrastructure also results in increased disturbance to and poaching of ungulates, or even snow leopards.

Poaching of Snow Leopards and Their Prey

Poorly managed and illegal hunting is an additional threat and may lead to decreased number of wild snow leopard prey including marmot, Siberian ibex, argali, and snow cock), which may in turn lead to increased depredation by snow leopard attacks on domestic livestock and ultimately may lead to retaliatory killing of snow leopards.

Snow leopard bones and other body parts are sometimes used as substitutes for tiger in traditional Asian medicine. The cat's pelt is highly sought after for luxury fur garments. Snow leopard skins are confiscated at Mongolian customs and by anti-poaching teams, which indicates the high level of threat from this source. Anti-poaching teams documented 12 cases of illegal snow leopard hunting and trade since 2001, including confiscation of four snow leopard skins during one inspection (see Section "Wildlife Law Enforcement"). In another case in Bayan-Olgii Province, 15 snow leopard skins were confiscated while in transit to Russia. Local residents may receive as little as USD 200 per skin by traders from larger cities. High prices were reportedly frequently offered by Chinese traders for derivatives of snow leopards, musk deer, and other species.

THE HISTORY OF SNOW LEOPARD CONSERVATION IN MONGOLIA

In 1994, the Mongolian Association for Conservation of Nature (MACNE) started the first snow leopard conservation project in Mongolia, which was initially focused on Gobi-Altai province. It was supported in part through a partnership with the Wildlife Conservation Society (WCS) and Dr George Schaller. This collaboration led to the first radio collaring study and country-wide survey for snow leopards in the country, funded by WCS for the years 1993–1998. A side project became the community-based conservation program originally called Irbis Enterprises (see Chapter 13.2). In 1997, WWF-Mongolia implemented a snow leopard conservation project in western Mongolia. The UNDP/GEF funded projects that were initiated in Mongolia in the mid-1990s placed conservation of the snow leopard and its habitat among its key components. In 2000, the International Snow Leopard Trust (SLT) became formally involved in conservation in Mongolia and adopted the fledgling Irbis Enterprises project (now known as Snow Leopard Enterprises, or SLE) and expanded it over time to include villages and herding families throughout the cat's range in the country (see Chapter 13.2).

From this list, it is clear that many, or most, snow leopard conservation projects between the early 1990s and today were funded and undertaken primarily by international and national NGOs. These included those named already as well as the Snow Leopard Conservancy, Irbis Mongolia, Snow Leopard Conservation Fund (a partner organization of SLT), and several UNDP/GEF funded environment projects. The National Snow Leopard and Ecosystem Priorities (NSLEP) of Mongolia (Snow Leopard Working Secretariat, 2013) lists several successful outcomes of more than two decades of work including:

- Landscape-based conservation strategies approved and implemented at the province level (Khuvsgol, Khovd, Bayan-Olgii, and Uvs) through UNDP Altay Sayan Project.
- Many new protected areas were established in potential snow leopard habitat. Today 20 state protected areas contain snow leopards and key habitats.
- Two transboundary nature reserves established in important snow leopard habitats at the border of Russia and Mongolia (Uvs Lake SPA and Siilhem NP).
- Two interagency "Irbis" anti-poaching teams established and conducting regular patrols in snow leopard habitat. Poaching incidents in five western provinces decreased rapidly.
- Snow leopard monitoring program using advanced techniques (camera trapping, genetic analysis) done with support of WWF-Mongolia and Institute of Biology/ Irbis Mongolian Center to monitor key snow leopard populations in state protected areas.
- More than 400 herding families in seven provinces participate in the SLE handicrafts project generating sustainable income with a commitment to nonpoaching.
- WWF Mongolia initiated a livestock insurance program in three model sites in western Mongolia, which delivered sheep to the herders who had lost livestock to snow leopards.
- Long-term snow leopard study in Tost Uul led to establishment of a local protected area in this key part of Mongolian snow leopard range.
- Livestock insurance program was introduced in Tost Mountain of South Gobi Province in 2009 involving more than 30 households to mitigate human–wildlife conflict in the area.
- The Nature Conservancy assessed mining impacts for the southern Mongolian ecoregion and recommended areas for better protection, which included snow leopards.

In addition, and during the same time period, the following conservation plans for protection of snow leopards in Mongolia were developed and implemented:

- National Biodiversity Conservation Policy of Mongolia, 1996
- Management plan of Gobi Gurvansaikhan National Park, 1997
- Snow leopard conservation management plan of Mongolia, 1999
- Snow leopard conservation management plan of Uvs province, western Mongolia, 2000
- Management plan of Uvs Lake Strictly Protected Areas, western Mongolia, 2002
- Snow leopard conservation policy of Mongolia, 2005
- Snow leopard conservation policy of Uvs province, western Mongolia, 2011

- Conservation program of rare and very rare wildlife species of Mongolia, 2011
- Mongolia National Snow Leopard and Ecosystem Protection Priorities (NSLEP), 2013

SNOW LEOPARDS IN LAW AND POLICY

The snow leopard is listed in the current Mongolian Red Data Book (2013) and protected as a "very rare" species under the Mongolian Law of Wildlife (2012). Since 1972, it has been illegal to kill snow leopards in Mongolia, although licensed trophy hunting was allowed until 1993. In 2011, the government approved the National Program on Conservation of Very Rare and Rare Wildlife Species. Key regulations concerning the conservation and use of wildlife, including snow leopards, and their habitats are contained in conservation laws, the majority of which were updated and endorsed by the Mongolian parliament in 2012. The primary piece of legislation regulating wildlife conservation is the State Wildlife Law of Protected Areas, which mandates sustainable use of wildlife, habitat conservation and restoration to ensure biological diversity, and law enforcement. Numerous sublegislative and agency-level regulatory acts provide the foundation of wildlife management and law enforcement agencies' conservation activities, regulation of the use of rare and threatened species, and habitat protection, as well as a regulatory mechanism with reasonably well-defined jurisdiction and distinctions between federal and regional government agencies.

However, the effectiveness of the regulatory management system is significantly reduced both by the absence of an effective enforcement policy and the presence of multiple regulatory, legal, and methodological loopholes. Wildlife-related environmental impact assessments (EIAs) are only mandated for protected areas. Outside of protected areas, development and even large-scale infrastructure projects do not require EIAs, and there is no legal basis forbidding such activity even if it has the potential to negatively affect endangered species or their habitat.

TRANSBOUNDARY INITIATIVES

Preservation of the transboundary area at the intersection of Russia, Mongolia, and China is of particular importance to the conservation of snow leopards at the northern edge of its distribution because it connects populations in western Mongolia and northwestern China to a remnant snow leopard population in Russia. Mongolia and Russia have an excellent history of collaboration on transboundary snow leopard conservation, including establishment of the transboundary Uvs Lake UNESCO World Heritage Site, existence of a government-level MoU for communication and cooperation, and bicountry research teams regularly collaborating on both sides of the border. The priority is to maintain existing snow leopard populations along the Russian-Mongolian border at Tsagaan Shuvuut and Siilhem B-Chihachev ridges. Continued assessments are needed to determine the importance of other potential transboundary snow leopard corridors to support the recovery of the Russian populations; initial surveys have been completed around the Tavan Bogd Uul area. Potential corridors along the boundaries of Russia, Mongolia, China, and Kazakhstan are Tavan Bogd Uul Ridge, Southern Altai Ridge, and the mountain ridges to the north of Khuvsgol Lake.

The NSLEP of Mongolia (Snow Leopard Working Secretariat, 2013) lists several transboundary cooperative actions that are needed, including:

- Develop and adopt a snow leopard conservation action plan for the Russian–Mongolian transboundary zone as well as Mongolian–Chinese border.
- Expand coverage of international transboundary protected areas along the

Russian–Mongolian border, including the Siilhem B, Chihachev, Tsagaan Shuvuut, and Tunkinsky ridges and mountains north of Khuvsgol Lake.
- Expand the "Golden Mountains of Altai" UNESCO World Heritage site to encompass all contiguous transboundary Altai Mountain range in Russia, Mongolia, China, and Kazakhstan.
- Coordinate actions between these four countries to curtail illegal wildlife trade in protected species. This collaboration should involve coordination, monitoring, and information exchange among the concerned countries' customs agencies regarding wildlife trade, corresponding governments, and international structures such as the Convention on International Trade in Endangered Species of Wild Fauna and Flora (CITES) and INTERPOL.
- Coordinate between specialists from Russia, Mongolia, China, and Kazakhstan to develop collaborative research programs on the snow leopard and its prey, especially monitoring snow leopard populations in the Russian-Mongolian transboundary area.
- Develop transboundary ecotourism in the habitats of snow leopards and other rare species focused on creating alternative income sources for local communities in Russia and Mongolia. The first step in this direction was made in 2010 by WWF and the UNDP/GEF "Biodiversity Conservation in the Russian Altai-Sayan Ecoregion: the Land of the Snow Leopard" project to develop ecotourism in local communities in snow leopard habitat in Altai, Tuva, and western Mongolia.

RESEARCH, MONITORING, AND CAPACITY BUILDING

A. G. Bannikov (1954) first summarized information on the status and distribution of snow leopards in Mongolia, and since the 1970s, scientists from the Institute of Biology, Mongolian Academy of Sciences, have studied snow leopards. In 1976, Dr A. Bold published a map of snow leopard distribution in southern Mongolia, assessed the population, and recommended ways to solve snow leopard depredation of livestock. In the 1980s, G. Amarsanaa studied the species and published articles on its biology and ecology. Modern snow leopard research and conservation in Mongolia was initiated in the 1990s by MACNE, bringing to Mongolia cutting-edge technology and partnering with international researchers such as George Schaller and Thomas McCarthy, assisted by Mongolian scientist B. Munkhtsog. Additionally, long-term snow leopard surveys were funded by Snow Leopard Conservation Foundation/Snow Leopard Trust.

In 2008, a Snow Leopard Research Center named after the famous Mongolian conservationist J. Tserendeleg was established and has been functioning in the Tost Local Protected Area. The center hosted a long-term research program developed and funded by Panthera and the Snow Leopard Trust. In that study, 19 snow leopards were radio-collared and the population was monitored via camera trapping. The field camp is providing training for field biologists, rangers, and students. Similarly, the Jargalant research camp was established in western Mongolia funded by WWF Mongolia and could be expanded into another long-term comprehensive research center to study and monitor endangered species including the snow leopard. From the Institute of Biology, doctorate, masters, and bachelor students frequently join scientists in the field on these research programs to learn, collect, and analyze data for their theses.

Since 1997, monitoring of key snow leopard populations in western Mongolia has been done by staff from Uvs Lake SPA, Altai Tavan Bogd, and Khar Us Lake National Park with the support of WWF Mongolia and UNDP/GEF. Park biologists were trained in monitoring methodology in 1998, 2004, and 2006. The software program "Biosan" was developed by the WWF

Mongolia and endorsed by the Ministry of Environment and Green Development in 2007 to aggregate data collected by park staff.

WILDLIFE LAW ENFORCEMENT

Between 1997 and 2013, there were a number of cases prosecuting poachers for killing snow leopards, and state inspectors discovered 19 cases (2 in Bayankhongor Province, 4 in Gobi-Altai, 1 in South Gobi, 7 in Khovd, 4 in Bayan-Olgii, and 1 in Ulaanbaatar) of illegal hunting and trade of skin of snow leopards. Guilty parties were sentenced up to 1.6 years in prison. In several instances between 2000 and 2011, a number of snow leopard pelts were smuggled to Russia's Altai Republic from Mongolia and the violators were prosecuted.

However, state and local nature protection agencies and inspectors have extremely limited or nonexistent funding, staff, and/or equipment to effectively patrol or monitor border posts. Additionally, local residents are known to possess a significant number of illegal and unregistered weapons used for poaching. It is imperative that additional funding from the central budget be allocated to law enforcement to ensure effective work by state and provincial nature protection agencies in the fight against illegal hunting in snow leopard habitat. It is also necessary to devote more attention to the fight against the illegal trade in endangered species and their parts.

Cooperation between conservation and enforcement agencies is urgently needed to address illegal trade in snow leopards and other rare species. WWF's extensive experience in creating and supporting interagency anti-poaching brigades can be used to advance such initiatives. Snow leopard conservation enforcement is insufficient in most protected and unprotected areas in Mongolia. There is also a need to increase cooperative patrolling in transboundary areas between Russia and Mongolia. Additionally, expansion of local protected areas with effective management and enforcement plans should be financed by the local government and central budget or these areas will remain as protected areas only on paper.

LEGAL FRAMEWORK TO EMPOWER COMMUNITIES TO COMANAGEMENT WILDLIFE AND HABITAT

The project "Nature Conservation and Sustainable Management of Natural Resources, Gobi Component," funded by GTZ (1999–2006), formed herder communities to increase herders' livelihoods through collaborative community-based natural resource management. After 2001, the number of donor-supported development projects in Mongolia, especially in the countryside, increased considerably. And based on the success in the Gobi, almost every development project has encouraged and supported herders in establishing herder associations to deliver interventions, encourage pasture management and livestock production, promote income-generating activities such as value-added processing of livestock products or livelihood diversification into nonlivestock-related activities, or arrange joint marketing of dairy products.

The Mongolian government legalized "community" as an officially recognized rural institution through the amendment of the Environmental Protection Law (2006). Under this concept, the herder community organizations are allowed to designate Community Responsible Areas for managing natural resources within their responsible territories.

In order to increase the effectiveness of local communities' conservation efforts it is necessary to cultivate strategies to develop community-managed resource use of local protected areas, and improve conditions for economic development by attracting funding to develop tourism, small businesses, and service-related alternative employment. Development of sustainable pasture use management plans that account for the needs of wild ungulate species in snow leopard habitat should be a high priority. To reduce livestock losses that lead to conflict, interventions such as predator-proof corrals (see Chapter 14.1) and methods to improve livestock guarding are needed. The highly successful Snow Leopard Enterprises model should be expanded to offer incentives and help compensate herders for loss of livestock. More community-based inspection teams are needed to actively patrol and protect rare species by engaging local residents who reside in snow leopard habitat. And more emphasis should be placed on developing sustainable ecotourism opportunities to observe snow leopards, their habitat, and sign, by providing guiding services, accommodation, transport, food, and so on.

FUTURE NEEDS TO MITIGATE SNOW LEOPARD THREATS

In this chapter we have shown the evolution of snow leopard research and conservation in Mongolia, a country that is critical to the survival of the species. Yet there is much still to do. We close by listing several actions we feel are urgently needed to ensure a long future for *irbis*, Mongolia's snow leopard.

First, the state and local protected area network must be expanded, and the capacity of the staff and the administration must be elevated. More scientific research is needed on both snow leopards and their prey. Although some steps have been taken, more actions are needed to reduce competition between livestock and wild ungulates in snow leopard areas. To reduce conflict with herders, such antidepredation actions as predator-proof corrals need to be expanded. To alleviate the high cost of depredation on livestock, more support is needed for sustainable alternative income generation for herding families in snow leopard habitat. Environmental education and awareness raising need to be conducted at many levels, from villages in snow leopard habitat to the broader national public. Snow leopards are a national treasure and people need to have them firmly in their consciousness. Laws pertaining to illegal hunting of snow leopard prey need much stronger enforcement. And lastly, a more thorough review of the potential impacts of mineral extraction on snow leopards and their ecosystem needs to be urgently undertaken.

Acknowledgments

The three lead authors of this chapter have more than 50 cumulative years of experience working on snow leopard research and conservation issues in Mongolia. However, for this chapter we also drew extensively from the National Snow Leopard and Ecosystem Protection Priorities (NSLEP) of Mongolia (Snow Leopard Working Secretariat, 2013), which brought in a great range of knowledge from individuals, NGOs, and government on the history as well as the current status of snow leopards in this country.

References

Bannikov, A.G., 1954. Mammals of the Mongolian People's Republic. Academy of Sciences, Moscow (in Russian).

Bold, A., Dorjzunduy, S., 1976. Report on snow leopards in the southern spurs of the Gobi Altai. Proceedings of the Institute of General and Experimental Biology–Ulaanbaatar 11, 27–43, (in Mongolian with Russian abstract).

Mallon, D., 1984. The snow leopard, *Panthera uncia*, in Mongolia. International Pedigree Book of Snow Leopards. Helsinki Zoo, Finland, vol. 4, 3–10.

McCarthy, T.M., 2000. Ecology and Conservation of Snow Leopards, Gobi Brown Bears, and Wild Bactrian Camels in Mongolia. PhD thesis. University of Massachusetts. 134 pp.

McCarthy, T.M., Chapron, G. (Eds.), 2003. Snow Leopard Survival Strategy. International Snow Leopard Trust and Snow Leopard Network, Seattle, USA.

O'Gara, B., 1988. Snow leopards and sport hunting in the Mongolian People's Republic. Proceedings of the Fifth International Snow Leopard Symposium 5, 215–225.

Schaller G. B., Tserendeleg J., Amarsanaa G., 1994. Observations on snow leopards in Mongolia. In: Fox, J.L., Du, Jizeng. (Eds.). Proceedings of the Seventh International Snow Leopard Symposium (Xining, Qinghai, China, July 25–30, 1992),International Snow Leopard Trust, Seattle, pp. 33–42.

Snow Leopard Working Secretariat, 2013. Global Snow Leopard and Ecosystem Protection Program (GSLEP). Snow Leopard Working Secretariat; Bishkek, Kyrgyz Republic.

Zhirnov, L.V., Ilyinsky, V.O., 1986. The Great Gobi National Park – A refuge for Rare Animals of the Central Asian Deserts. Academy of Sciences, Moscow (in Russian).

CHAPTER

40

Northern Range: Russia

Mikhail Paltsyn, Andrey Poyarkov**, Sergei Spitsyn†, Alexander Kuksin‡, Sergei Istomov§, James P. Gibbs*, Rodney M. Jackson¶, Jennifer Castner††, Svetlana Kozlova‡‡, Alexander Karnaukhov**, Sergei Malykh**, Miroslav Korablev**, Elena Zvychainaya**, Vyacheslav Rozhnov***

*Department of Environmental and Forest Biology, State University of New York, College of Environmental Science and Forestry, Syracuse, NY, USA

**Laboratory of Behavior and Behavioral Ecology of Mammals, A.N. Severtsov Institute of Ecology and Evolution, Russian Academy of Sciences, Moscow, Russia

†Science Department, Altaiskiy State Nature Biosphere Reserve, Yailu, Russia

‡Science Department, Ubsunurskaya Kotlovina State Nature Biosphere Reserve, Kyzyl, Russia

§Science Department, Sayano-Shushensky State Nature Biosphere Reserve, Shushenskoe, Russia

¶Snow Leopard Conservancy, Sonoma, CA, USA

††The Altai Project, East Lansing, MI, USA

‡‡Independent Consultant on Results-Based Management, Syracuse, NY, USA

INTRODUCTION

Snow leopards (*Panthera uncia*) in Russia represent the northernmost segment of the species' range, where its current distribution is limited to the Altai-Sayan Ecoregion. Areas currently inhabited by snow leopards in the Ecoregion do not exceed 20,000 km^2. These are home to likely 70–90 snow leopards (1–2% of the total wild population) (Paltsyn et al., 2012). The few stable snow leopard populations in

Snow Leopards
http://dx.doi.org/10.1016/B978-0-12-802213-9.00040-7

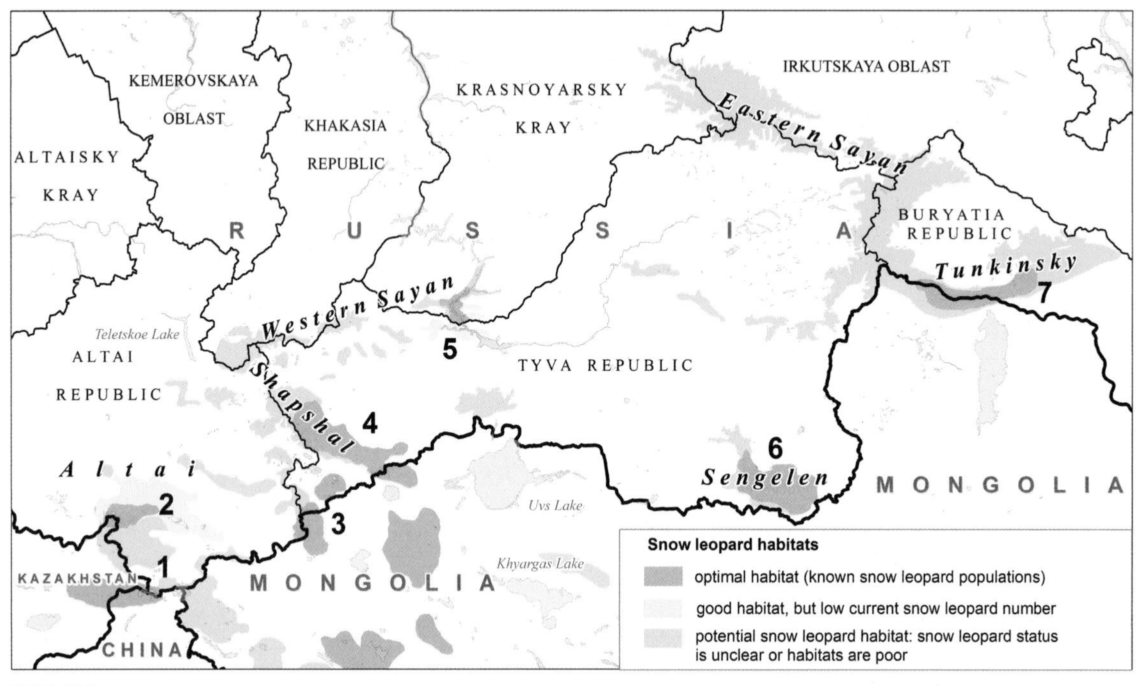

FIGURE 40.1 **Snow leopard habitat and known populations in Russia.** Locations of snow leopard populations: 1, South Altai Ridge; 2, Argut River Watershed; 3, Chikhachev Ridge and Mongun-Taiga Massif; 4, Tsagan-Shibetu and Shapshal Ridges; 5, Sayano-Shushensky Nature Reserve and its buffer zone; 6, Sengelen Ridge; 7, Tunkinsky Ridge.

Russia occupy no more than 12,000 km^2 in the mountains of Altai, Tyva, and Western and Eastern Sayan (Fig. 40.1).

Since 1998, the snow leopard has been a flagship species for the WWF Program in the Altai-Sayan Ecoregion (WWF, 2012) and UNDP/GEF Project "Biodiversity Conservation in the Russian portion of the Altai-Sayan Ecoregion" (UNDP/GEF, 2005). Considerable financial resources have been dedicated to snow leopard conservation aimed at mitigation of two major threats: poaching and retaliatory killing. About 90% of project efforts were implemented in key habitats: Argut River watershed, Chikhachev Ridge, Tsagan-Shibetu and Shapshal Ridges, and Sayano-Shushensky Nature Reserve (Fig. 40.1).

Here we emphasize lessons learned over the past 15 years in snow leopard conservation in Russia. Sharing experiences and insights is an extremely valuable tool for advancing conservation, regionally and range-wide. Yet, paradoxically, it is widely recognized that conservation practitioners generally do not share their experiences in published form (Sunderland et al., 2009). Therefore a huge volume of collective conservation experience is lost for future projects. Moreover, unexpected outcomes and project failures regularly occur in conservation projects given the complex socioeconomic and uncertainty issues associated with species conservation. Yet these are rarely reported despite their high practical value for others working in similar situations (Sunderland et al., 2009). For these reasons we highlight a number of lessons learned in Russia while working to mitigate snow leopard poaching and retaliatory killing during 2000–2014.

SNOW LEOPARD-HERDER CONFLICT MITIGATION PROJECTS

Insurance of Livestock Against Snow Leopard Depredation in Western Tyva

This project was implemented by WWF-Russia in cooperation with RESO-Garantia insurance company in three districts of Tyva Republic in 2000–2003. The overall goal was to protect the snow leopard population in western Tyva from retaliatory killings by decreasing conflicts between herders and snow leopards. Local herders in snow leopard habitats were provided the opportunity to insure their livestock against snow leopard predation. WWF paid the full premium to RESO-Garantia. Every herder willing to participate received a simple camera with a roll of film to photograph livestock killed by snow leopards as well as special forms to record details of any attack. Each participating herder was expected to report any cases of snow leopard depredation to the project coordinator as soon as possible. The project coordinator and expert zoologist were expected to investigate on-site every reported case of livestock depredation to determine whether the livestock was killed by a snow leopard. Every case approved had to be covered by RESO-Garantia in full. In turn, herders participating in the program had to tolerate snow leopards killing their livestock and not pursue and kill any of the cats that may have been involved (Chimed Ochir, 2003). As a result, a total of 54,841 head of livestock were insured in 2000, 21,105 in 2001, and 24,048 in 2002 (Kynyraa, 2001). Twenty reports of snow leopard attacks on livestock were collected in 2000–2003 claiming 233 livestock kills. Only seven cases were investigated and four cases (56 head of livestock killed) were covered by insurance. In July 2003 the project was terminated.

Project successes:

- This was the first initiative in Russia to mitigate conflicts between snow leopard and local herders. It was implemented at the right time – after collapse of the Soviet system of collective livestock management – when herders were left without any support from the government and relied only on their livestock to survive. Any livestock losses, including to wild predators, are economically damaging to herders; therefore, the insurance program was accepted by local communities with great enthusiasm.
- Before the project, Tyvan herders such as those elsewhere regarded snow leopards only as a pest and few if any knew the species to be globally endangered and in need of protection (Chimed Ochir, 2003; Jackson and Wangchuk, 2001; Kynyraa, 2001). This represented the first time in Tyvan history that the issues of snow leopard predation on livestock and necessity of leopard conservation were raised at regional and national levels and discussed in the media. Thanks to an information campaign, most Tyvan herders became willing to tolerate livestock losses due to snow leopard predation if these were compensated by insurance.
- The project collected valuable information about locations and intensity of snow leopard predation on livestock in western Tyva.
- The fact that four local herders received compensation for livestock killed by snow leopards inspired other herders to tolerate attacks for 2 or 3 more years after the project was terminated. Herders continued to report snow leopard attacks until 2008 because they still hoped to receive compensation and desisted from killing snow leopards involved in livestock depredation.

Project shortcomings:

- The insurance program was planned "top–down" by WWF-Russia and RESO-Garantia without consulting key stakeholders, especially local herders (Chimed Ochir, 2003). Moreover, the project did not have clear success indicators, and therefore it was difficult to measure whether the insurance program reduced the level of retaliatory killing of snow leopards.

- The project was based on limited data collected in 1998–1999 by external experts during three short visits. Obviously, these data were insufficient to clearly define the project area. Thus, Ovur District, where not a single snow leopard attack on livestock was documented in 1998–2003, was included in the project area. Poor understanding of the spatial pattern of snow leopard attacks resulted in a huge project area (about 8,000 km^2) spread over a remote mountain area resulting in a high number of insured livestock (up to 55,000 head in 2000) and considerable premiums paid to RESO-Garantia (e.g., total insurance premium of USD 61,866 in 2000) (Kynyraa, 2001).
- Small technical issues generated problems. Herders were not trained in the proper use of cameras. Most images were blurred and inappropriate as evidence. Also, the herders had only one roll of 12-exposure film each, too little to collect the information needed. Moreover, most herders had limited knowledge of Russian and could not record data correctly on the forms provided.
- Many herders lived in remote areas and had no means of communication. It took them at least 1 day on horseback to reach the district center and report attacks to the project coordinator, who lived 500 km away in the capital of Tyva Republic. At best, it required 2–3 days for the project coordinator and expert to reach the site of a snow leopard attack and make an assessment. In many cases 3–4 days were sufficient to enable scavengers to damage carcasses and destroy all signs of snow leopard near dead livestock. No local experts were trained by the project to help with documentation or to offer expertise. As a result only seven cases from 20 reported were investigated by experts and only four were approved for insurance payments.
- RESO-Garantia was only able to pay insurance coverage to herders via bank accounts. This was impractical for local people because there were no bank branches in Mongun-Taiga District. A herder had to travel 500–600 km to Kyzyl (capital of Tyva Republic), open a bank account, and then return again to actually withdraw money. The considerable time and money required for impoverished herders to obtain compensation made the project's utility dubious.

Protection of Livestock Corrals from Snow Leopards in Western Tyva

After the conclusion of the livestock insurance project, WWF and UNDP-GEF Project invested significant funds in field surveys of snow leopards and analyzing predation on livestock in western Tyva. We learned that 57% of attacks occurred on Tsagan-Shibetu and 36% on Shapshal Ridges. Also 94% of all cases of livestock depredation (n = 104) occurred in open pastures and only 6% in corrals (n = 6). However, just a few snow leopard attacks in corrals were responsible for 56% (n = 260) of all livestock killed by snow leopards in western Tyva in 2000–2011 (Paltsyn et al., 2012), a pattern similar to Ladakh, India (Jackson and Wangchuk, 2001). Whereas a snow leopard only kills 1–3 animals in an attack in open landscape, it is capable of killing and wounding dozens of sheep and goats (up to 80 head) in a corral. Corrals also become traps when the snow leopard is often not able to jump out through the roof and is killed by the herder. Between 2000 and 2008 we learned of six snow leopards killed by herders during livestock attacks, four of which were inside corrals. Tyvan herders usually tolerated small losses, but after catastrophic losses inside corrals the snow leopard was killed in most cases.

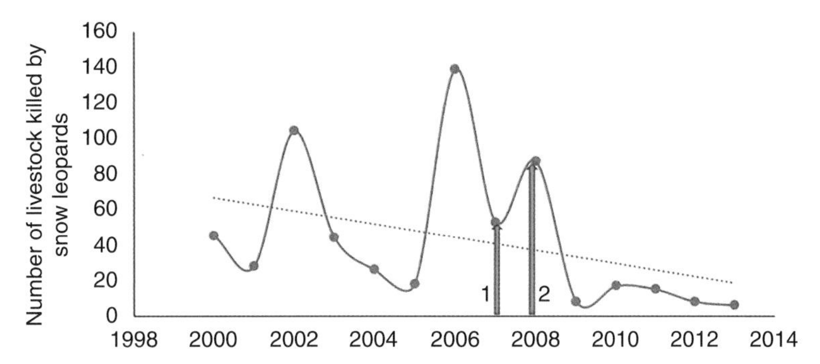

FIGURE 40.2 **Number of livestock killed by snow leopards in western Tyva (Chikhachev Ridge, Mongun-Taiga Massif, Tsagan-Shibetu, and Shapshal Ridges), Russia, in 2000–2013 based on reports by local herders.** Three big spikes in 2002, 2006, and 2008 indicate massive killing of sheep and goats inside corrals. Green arrows: 1–21 corrals are reinforced with metal mesh in snow leopard habitat on Tsagan-Shibetu Ridge in 2007; 2–25 corrals are protected against snow leopards on Chikhachev, Mongun-Taiga, and Shapshal Ridges in 2008.

As a result, we developed the idea of covering ventilation holes in the corral roof and windows with metal mesh to prevent snow leopards from entering, and in October 2007 all winter corrals in 21 herder camps in Tsagan-Shibetu Ridge (a hot spot of snow leopard predation on livestock) were strengthened. In 2009 corrals were improved in 25 herder camps in Shapshal, Mongun-Taiga, and Chikhachev Ridges. More than 70 herders participated in the project and learned about this method for corral protection. Since 2008 not one snow leopard attack on livestock inside corrals has been recorded in western Tyva. This simple low-cost effort eliminated 50–60% of livestock loss due to snow leopard and decreased losses to more tolerable levels (no more than 20 livestock killed annually) during 2009–2014 (Fig. 40.2).

Project successes:

- The project was based on substantial information on snow leopard distribution, location and intensity of livestock predation events, and on grazing patterns in western Tyva.
- The project had clearly defined goals, objectives, and measurable results. Specifically, the 46 protected corrals in the key snow leopard habitats resulted in significant project outcomes: 50–60% decrease in the number of livestock killed by snow leopards and at least 50% decrease in mortality of snow leopards. The expected project impact – increase of snow leopard population in western Tyva – was measurable as well, through local snow leopard populations monitoring nearly every year since 2004. In 2004, density of snow leopard tracks on Tsagan-Shibetu Ridge was 3.6 tracks/100 km of routes walked and no cub tracks were found; in 2010 – 4.6 tracks/100 km walked and two cub tracks; and in 2011 – 5.6 tracks/100 km walked and three cub tracks. We cannot attribute this increase in the population only to the reinforced corrals, but this intervention likely contributed positively.
- Local herder communities and authorities took great interest in the project and were actively involved in planning and implementation. Herders participated in reinforcing corrals, contributing horses, vehicles, and labor. Community capacity building to decrease livestock loss from snow leopards was part of the initiative. Lastly, every herder who participated signed an agreement with Ubsunurskaya Kotlovina Nature Reserve pledging to protect snow leopards, and in 2012 some of the herders were involved in camera trapping.
- The project was cheap and logistically simple.

Project shortcomings:

- The project did not suggest any solutions for the protection of free-ranging livestock. These losses are more or less tolerable now, but could increase with growing snow leopard numbers in the future. An attempt to protect pastured livestock using electric fences in 2009 was unsuccessful due to difficulties in setting up and maintaining long electric fences in the rugged terrain.

LESSONS LEARNED FROM ANTI-POACHING PROJECTS

Protecting Snow Leopards from Poaching in the Argut River Watershed, Altai Republic, 2005–2008

In December 2005, WWF-Russia and Game Management Department (GMD) of Russian Federal Agricultural Control Agency (Rosselkhoznadzor) launched the first anti-poaching campaign aimed at protection of snow leopards in Argut River watershed and the surrounding area. According to the "Strategy for Snow Leopard Conservation in Russia" (Poyarkov et al., 2002), this area was believed to have the largest snow leopard population in the country, totaling 30–40 cats. WWF funded four patrol groups (12 inspectors), including fuel, car parts, and field equipment, to organize regular anti-poaching patrols in Shavla Wildlife Refuge, a key protected area in the Argut River watershed. The total area is about 328,811 ha of rugged and remote terrain. In 2006–2007 GMD groups patrolled every month (a level of protection four to six times higher than in 2004–2005) with active participation of the police. Intensive patrolling resulted in 30 poachers, mainly local hunters, being fined, and 25 firearms and 100 snares confiscated (in 2004–2005 only 5 poachers were discovered and punished in the same area). At the end of 2007 the project was halted, because Rosselkhoznadzor lost its authority to fight poaching, and this was transferred to the new regional Wildlife Protection Committee of Altai Republic. It appears that the project contributed to a temporary and significant decrease in poaching in the lower and middle Argut River valley: in February 2004 we found 35–40 signs of poaching in the area (snares, traps, skins of killed animals, etc.) but only two in March 2007.

Project successes:

- GMD was a federal agency with an adequate number of professional staff (35 inspectors) and relatively good salaries. Every patrol group had its own vehicle. All they needed was fuel, parts for vehicles, and equipment (not provided sufficiently by the federal government).
- Formal competition was established among inspector groups, measured in numbers of citations and confiscated firearms. The best groups received annual financial bonuses.

Project shortcomings:

- The project was based on the false assumption that in 2005–2007 the Argut snow leopard population still had 30–40 individuals concentrated in the middle of Argut River watershed. Detailed sign and camera trap surveys in 2004–2011 did not find any signs of permanent snow leopard presence in that part of the watershed: camera trapping in 2010–2011 proved that all signs of big cats in the area belonged to Eurasian lynx (*Lynx lynx*) (Paltsyn et al., 2012). The only obvious presence of snow leopards was discovered in 2004 and 2012 in a remote part of the upper Argut, where three to five snow leopards survived. Thus, anti-poaching raids were conducted in an area where no or few snow leopards remained, and no patrolling occurred in the upper Argut watershed that still supported leopards. Therefore, despite considerable anti-poaching efforts, the project was of little help to the remaining Argut snow leopard population, although it likely generally benefited wildlife in the area.

- About 90% of all cases of poaching discovered by anti-poaching brigades in 2005–2007 were located in easily accessible areas near roads or significant trails, but not in the remote habitat so important for snow leopards. Anti-poaching activities in remote habitat requires a lot of time and labor and usually results in a maximum of 1–3 poachers caught and often none at all for 2 weeks of hard work, whereas patrolling relatively accessible areas may result in 10–12 discovered poaching cases within a week. Thus, brigades achieve better results by patrolling accessible places. But anti-poaching brigades thus missed high concentrations of snares in more remote areas where snow leopards might still live. Also in such cases inspectors miss snare-poachers – a small segment of the poacher population that presents the most danger to snow leopards. Therefore, seemingly good results offered little real benefit to snow leopards.
- Conservationists in Altai-Sayan persistently underestimated snare poaching until it recently became evident that the Argut snow leopard population was devastated by snares and that the other significant population in the Sayano-Shushensky Nature Reserve decreased similarly for the same reason (Paltsyn et al., 2012). Snare removal – a critical aspect of snow leopard conservation – was not part of the 2005–2007 anti-poaching campaign.

Snare Removal Campaigns in Argut River Watershed and Sayano-Shushensky Nature Reserve, 2008–2014

Snare poaching is the most serious and widespread threat contributing to the decline of at least five of seven known stable snow leopard populations in the Altai-Sayan (Fig. 40.3) (Paltsyn et al., 2012). Local residents who overwinter in snow leopard habitat are the main poachers. High prices for derivative products of snow leopards, musk deer, and other species are the main drivers of snaring. Animal parts are one of the few income sources for local residents living in remote villages and herder camps where unemployment can reach 80–90%.

Snare removal in the Argut River watershed and Sayano-Shushensky NR started in 2008–2009. Based on lessons learned from the anti-poaching project in 2005–2007, the snare removal campaign used different tactics than previously:

- The geographic focus was narrow: 600 km^2 of optimal snow leopard habitat in Argut River watershed in 2008–2009 and 500 km^2 in Sayano-Shushensky NR and its buffer zone. Only habitats currently inhabited by snow leopards were patrolled to increase the effectiveness of anti-poaching activities.
- Patrol groups in both areas had not only inspectors but also included snow leopard researchers and local people who knew the area well and had good prior knowledge about snaring.
- Success was measured by the number of snares found and removed and the number of snare poachers fined and held criminally liable, instead of the total number of poachers fined and weapons confiscated.
- During snare removal, field groups remained in the habitat for at least 2 weeks and also collected data on snow leopard presence and prey species abundance. Camera traps were deployed to monitor the species after the team left the field.
- Snare removal activities occurred during the key snaring period of November–March (frozen rivers increase access and pelage is most valuable). Raid intensity increased from two patrols per snaring season in 2008–2009 to six in Argut and four per season in Sayano-Shushensky NR in 2010–2014.

As a result of the snare removal campaign in Sayano-Shushensky NR, the number of snares in snow leopard habitats decreased from more

FIGURE 40.3 **Snow leopard female with a wire snare around its neck in Sayano-Shushensky Nature Reserve.** The cat was caught in a snare in spring 2013 when regular snare removal patrolling was terminated by the Reserve's new administration. This is the final picture of this snow leopard recorded. In fall 2013, the cat and its three cubs disappeared. *Photo credit S. Istomov.*

than 800 in 2008 to only 5 in 2012. Snow leopard numbers were relatively stable in 2008–2012 and estimated at 7–9 individuals. But in 2013–2014 snare removal activities were ceased completely in the Reserve when the new administration fired nearly all inspectors who had participated in these activities and hired new inspectors without snare removal experience. The absence of snare removal over one winter resulted in a dramatic increase in snare numbers to several hundred and an abrupt decrease in the local snow leopard population to only 2–3 individuals by spring 2014.

In the Argut area the number of snares in the patrolled habitats decreased from more than 600 in 2008 to 0 in winter 2013–2014. Two snow leopards – Kryuk (male) and Vita (female) – found in 2012 successfully survived until the present and even produced two cubs in 2013. One more snow leopard was camera trapped, and tracks of 2–3 others were found in the upper Argut valley in 2013–2014. Therefore, 7–8 snow leopards now dwell in the Argut River watershed and give hope for the restoration of a healthy population of 25–30 individuals within 10 years. The success of the Argut snare removal project was achieved not only by regular patrols, but also due to the following:

- The project has been long term with adequate, stable funding provided by a range of public and private donors.
- In addition to snare removal, our group deployed camera traps. Local hunters did not know the locations of the cameras, and many of them thought that they had been hidden to take pictures of poachers. Therefore, some poachers stopped snaring

in the project area in a misguided attempt to avoid prosecution. Ironically, in winter 2014 camera traps were actually used by Sailyugem National Park inspectors to identify and prosecute a poacher.

- To increase effectiveness of anti-poaching raids, a satellite-based poacher detection system developed by Wildlife Intel (2014) was deployed in two Argut cabins used by local hunters. The system was triggered by a heat sensor when a fire was lit in a stove. An alert message was sent in real time, through satellite messaging, to mount a response from a patrol group of the Wildlife Protection Committee. In winter 2012–2013 the system revealed 3–4 visits of poachers, but unfortunately none of the alerts was responded to due to the limited staff and resources of the Wildlife Protection Committee. Nevertheless, the systems helped collect data about the frequency of poacher visits and this information was used to direct anti-poaching raids by Sailyugem National Park staff in 2014.
- Two professional snow leopard poachers living in the project area were contracted to protect and monitor snow leopards in the upper Argut River watershed in 2013. These people voluntarily agreed to cooperate on restoring the Argut snow leopards and to give up snaring. They signed a contract and received several camera traps with the task of monitoring snow leopards and removing snares. As a result, three more previously unknown snow leopards were identified and the key leopard habitat remained snare-free in 2013–2014. The men lost their illegal income from snaring; in compensation they began receiving incentives for snow leopard conservation as well as horse rental fees and per diem expenses for participating in fieldwork. Currently their legal annual income is about twice their previous illegal income from snaring. To make this possible WWF-Russia organized a special fundraising campaign amounting to about USD 40,000 in 2014. Additionally support from Disney Worldwide Conservation Fund has provided further incentives. Ex-poachers will receive their annual fee for conservation of snow leopards (about USD 3,000 each) as long as they prove with camera trap photographs that snow leopards still persist. At the same time these two ex-poachers were presented to the general public as protectors of Argut snow leopards and one of them, Mergen Markov, received a Disney Conservation Hero Award in 2014. These men are proud of their new profile and livelihood obtained by helping to restore the snow leopard population in their homeland.

We believe that the ex-poachers/donor agreement offers a promising and ethical tool for conservation of critical populations of other endangered species. It is similar to a payment for ecosystem services. It provides local people in remote areas who depend on poaching and snaring a more attractive source of income and a profitable alternative. This allows people to use their traditional knowledge as hunters, does not require them to switch to unknown activities to generate income, does not require the considerable time for development that small business would, and can be used in remote and disadvantaged communities where small business development is not feasible. Moreover, the income of local ex-poachers is directly connected to the well-being of snow leopard and other endangered species. Camera trapping, with individual identification of snow leopards, offers a sound tool for verifying the program's conservation targets. Of course, we do not suggest ex-poachers/donor agreements as an option for all alternative income generation and community-based conservation mechanisms, but they can be an effective crisis tool where immediate action is required.

Development of Alternatives to Poaching Income in Altai Republic, 2010–2014

Beginning in 2010, WWF and Citi Foundation implemented a small business development project for local communities living near snow leopard habitat and protected areas in Altai. The main goal of the project has been to decrease poaching by providing local communities with alternative sources of income such as tourism, souvenirs, and felt production as well as small-scale agriculture. The project trained more than 1000 local people in basic small business skills and provided financial support totaling USD 170,000 to 129 local entrepreneurs in 2010–2014 (WWF-Russia 2014). As a result of the project, 230 new jobs for local people were established in Altai. The project has had a positive socioeconomic effect on local communities, but its value for snow leopard conservation is questionable due to following factors:

- The project mainly targeted tourist enterprises, but these are mainly summer activities. Therefore, in winter, when poaching is most viable, local people have no occupation or income and may continue snaring and poaching.
- Small and remote communities, such as Argut village, are not promising for the development of ecotourism or souvenir production because they are rarely visited by any tourists and have no facilities.
- None of the known snow leopard poachers have been involved in the project. Snaring poachers live in remote places, are not aware of alternative income projects, and typically have no interest in them. Often they follow a traditional herder and hunter lifestyle and earn greater income from illegal wildlife trade thanks to low levels of government control of poaching. Finally, locals may not be eager to switch livelihoods, especially when there are few comparable economic incentives to do so.
- There is no direct link between support for project enterprises that benefit local people and the well-being of snow leopards. For example, tourism in mountain areas does not need the presence of snow leopards as a guarantee for income generation. Thus, local communities do not consider snow leopards as a necessary economic resource nor do snow leopards necessarily benefit from such projects.

Online Public Awareness Campaign to Prohibit Legal Snare Poaching

In December 2013, Ministry of Natural Resources and Environment (MNRE) of Russia adopted decree # 581, "On changes in hunting regulations," which allowed snaring for wolves in some regions of Russia, including Krasnoyarsky Krai, where the most northern Russian snow leopard population survives. This legal amendment practically legalized snaring in critical snow leopard habitat adjacent with Sayano-Shushensky NR. A. Poyarkov organized a petition opposing this decree to Ministry of Natural Resources and Environment of Russia (MNRE) and Russian President Vladimir Putin with an appeal to completely ban snaring in the habitats of snow leopard and Amur tiger. The petition was published on the public website, www.change.ru, and was signed by 88,000 Russian citizens. In August 2014 in response to the petition MNRE prohibited snaring in the habitats of snow leopard and other species in Krasnoyarsky, Khabarovsky, and Zabaikalsky Krais (MNRE, 2014).

This positive experience clearly demonstrates that public awareness campaigns can be effective in improving legislation in endangered species protection. On the other hand, given extremely limited funding of wildlife law enforcement in Russia, we cannot assume these changes will result in any positive outcomes for snow leopards.

CONCLUSIONS

Conservation of snow leopard is challenging due to remoteness and limited accessibility of the snow leopards' habitats, extreme poverty of local communities, the low economic development of most snow leopard range countries, limited funding options for the conservation of this species, and lack of government enforcement (Snow Leopard Secretariat, 2013). Often conservationists must act despite limited background information on snow leopard populations, threats, and within unstable socioeconomic and political situations. In such conditions, success is impossible without trial and error. Lessons learned from the shortcomings of conservation projects in different parts of the snow leopard's range can have tremendous value for the rapid improvement of overall effectiveness in snow leopard protection.

Therefore, we reiterate that planning and implementation of programs for snow leopard and any other endangered species should incorporate learning from previous projects both regionally and across the species' range. This enables such programs to adapt over time based on lessons learned – to practice adaptive management in the face of uncertainty of natural systems and processes – and thereby make conservation more effective and efficient in the long run (Allen et al., 2011; Parma et al., 1998; WWF, 2012). We hope that lessons learned in Russia will encourage other snow leopard conservation practitioners to disseminate their experiences broadly to enable more effective conservation measures that secure a sustainable future for this charismatic wild cat.

References

Allen, C.R., Fontaine, J.J., Pope, K.L., Garmestani, A.S., 2011. Adaptive management for a turbulent future. J. Environ. Manag. 92 (5), 1339–1345.

Chimed Ochir, B., 2003. Livestock insurance campaign against snow leopard predation on livestock in the Republic of Tyva. Evaluation Report, WWF-Russia.

Jackson, R., Wangchuk, R., 2001. Linking snow leopard conservation and people-wildlife conflict resolution: grassroots measures to protect the endangered snow leopard from herder retribution. Endangered Species UPDATE 18 (4), 138–141.

Kynyraa, M.M., 2001. Report on activities of WWF RU 0074.01 project in the framework of voluntary insurance of livestock from snow leopard predation. WWF Archive, (in Russian).

Ministry of Natural Resources and Environment of Russia, 2014. Ministry of Natural Resources and Environment of Russia will initiate a ban of wolf snaring in the habitats of snow leopard and Amur tiger (in Russian). Available from: http://www.mnr.gov.ru/news/detail.php?ID=134933.

Paltsyn, M.Y., Spitsyn, S.V., Kuksin, A.N., Istomov, S.V., 2012. Conservation of snow leopard in Russia. WWF, Krasnoyarsk. Available from: http://wwf.ru/resources/publ/book/eng/599.

Parma, A., Amarasekare, P., Mangel, M., Moore, J., Murdoch, W.W., Noonburg, E., Pascual, M.A., 1998. What can adaptive management do for our fish, forests, food, and biodiversity? Integr. Biol. 1, 16–26.

Poyarkov, A.D., Lukarevsky, V.S., Subbotin, A.E., Zavatsky, B.P., Prokofyev, S.M., Kelberg, G.V., Malkov, N.P., 2002. Strategy for snow leopard conservation in Russia. WWF, Moscow. Available from: http://wwf.ru/resources/publ/book/eng/7

Snow Leopard Secretariat, 2013. Global Snow Leopard and Ecosystem Protection Program (GSLEP). Snow Leopard Working Secretariat; Bishkek, Kyrgyz Republic.

Sunderland, T., Sunderland-Groves, J., Shanley, P., Campbell, B., 2009. Bridging the gap: how can information access and exchange between conservation biologists and field practitioners be improved for better conservation outcomes. Biotropica 41(5), 549–554. Available from: http://onlinelibrary.wiley.com/doi/10.1111/j.1744-7429.2009.00557.x/pdf

UNDP/GEF, 2005. Biodiversity Conservation in the Russian Portion of the Altai-Sayan Ecoregion – Phase I. Full size project document. Available from: http://www.thegef.org/gef/project_detail?projID=1177

Wildlife Intel, 2014. Cutting-edge technologies for monitoring trespass in remote areas. Available from: http://wildlifeintel.com/index.php/how-it-works/how-it-works.

WWF, 2012. WWF Standards of Conservation Project and Programme Management (PPMS). Practical Guide.

WWF-Russia, 2014. Sustainable small business development in the Altai helps to conserve the nature values. Available from: http://wwf.ru/resources/news/article/eng/12620

CHAPTER

41

China: The Tibetan Plateau, Sanjiangyuan Region

Yanlin Liu, Byron Weckworth**, Juan Li*[†]*,*
Lingyun Xiao[‡]*, Xiang Zhao*, Zhi Lu**[*,‡]

*Shan Shui Conservation Center, Beijing, China
**Snow Leopard Program, Panthera, New York, NY, USA
[†]Department of Environmental Science, Policy and Management, University of California, Berkeley, CA, USA
[‡]Center for Nature and Society, College of Life Sciences, Peking University, Beijing, China

The Tibetan Plateau is the largest and highest plateau in the world. It extends over 2,500,000 km^2 at an average elevation in excess of 4500 m. Our research is focused primarily in the Sanjiangyuan region of the Tibetan Plateau in Qinghai, China (Fig. 41.1). The Sanjiangyuan, or "Three Rivers Region," is an area of 395,000 km^2 that encompasses the headwaters of three of the world's major rivers, the Yangtze, Yellow, and Mekong – rivers that provide water to more than a billion people downstream. The Sanjiangyuan Region is also noted as home to China's second largest protected area, the Sanjiangyuan National Nature Reserve (SNNR, 152,000 km^2). Sanjiangyuan's high elevation grasslands are punctuated with rugged mountain ranges that are home to not only the charismatic snow leopard (*Panthera uncia*), but also carnivores such as gray wolf (*Canis lupus*), Tibetan brown bear (*Ursus arctos*), Eurasian lynx (*Lynx lynx*), red fox (*Vulpes vulpes*), Tibetan fox (*V. ferrilata*), Chinese mountain cat (*Felis bieti*), and Pallas's cat (*Otocolobus manul*). The human inhabitants across the region number about 1 million, 90% are Tibetan, and the majority of those practice some form of pastoralism (e.g., yak, sheep, goat, and horses).

When George Schaller finished his pioneering study on giant panda (*Ailuropoda melanoleuca*) in 1984, China's State Forestry Administration (SFA) invited him to survey snow leopards in western China. From 1984 to 1986, he travelled extensively in the Sanjiangyuan Region to survey the elusive species (Schaller et al., 1988).

After 30 years, research and conservation of giant panda has greatly improved. The fourth national giant panda survey indicates 1864 individuals in the wild, 268 more than the estimation in the third survey in 2003, and 67 established nature reserves to protect them, with 27 of them established after 2003 (WWF, 2015). The giant panda is now

Snow Leopards
http://dx.doi.org/10.1016/B978-0-12-802213-9.00041-9

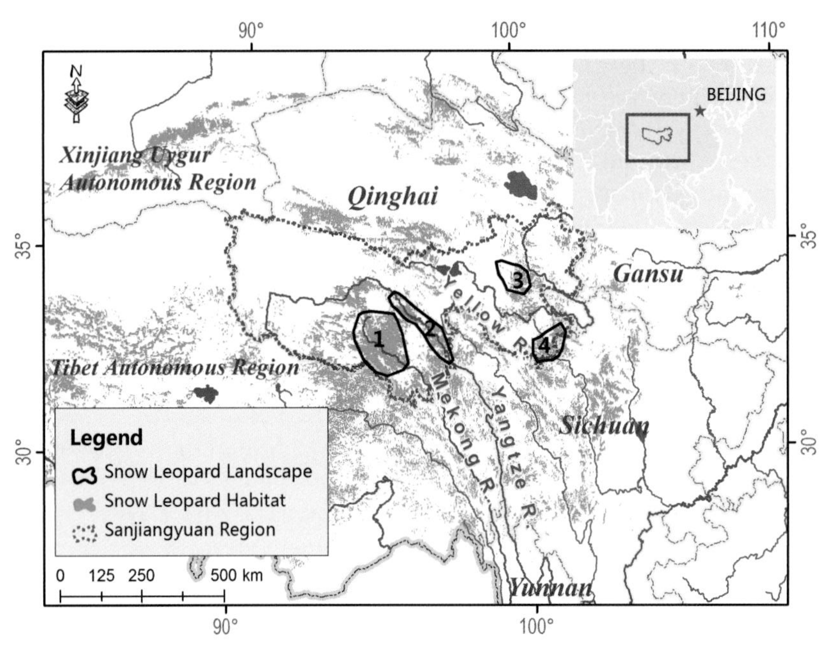

FIGURE 41.1 **Sanjiangyuan Region locates in eastern part of Tibetan plateau.** Snow leopard habitat prediction is showed in dark green, and the four important landscapes are sketched: (i) the source region of the Mekong, (ii) an area along the Upper Yangtze, (iii) Anemaqen, and (iv) Nyanpo Yutze.

considered a "national treasure" and the conservation symbol of China. In contrast, there remain substantial information gaps on snow leopards, and no nature reserves exist to fill their needs.

Is it possible to replicate the success of giant panda conservation for snow leopards in the Sanjiangyuan Region? What can we learn from the rich history of panda conservation? What is the pathway to ensure the long-term survival of snow leopards given the complexities of modern social and ecological trajectories?

STATUS OF SNOW LEOPARD IN SANJIANGYUAN REGION

The Sanjiangyuan region contains some of the best snow leopard landscapes across China. The habitats in this region are large and connected. Using MaxEnt modeling and presence points, Li et al. (2014) predicted 65,000 km^2 of suitable habitat in this region, 15% of all snow leopard habitat across China (Fig. 41.1). Li (2012) defined "important habitat" as continuous habitat that can support at least 50 individuals. In Li et al. (2014) snow leopard density was calculated as $2.7 \pm 0.7/100$ km^2, thus the minimum area for an important habitat is estimated to be 1900 km^2. Thirty-four of these important habitats were identified, of which eight are located in or around the Sanjiangyuan region. The largest patch was found in the source region of the Mekong River. Extending this habitat prediction model outside the region is problematic, given fewer presence points available. Nevertheless, the prediction provides a baseline estimate from the best available data and is useful for indicating general patterns on a coarse scale. The prediction model is continuously updated, and ongoing field surveys verify that snow leopards indeed roam beyond the predicted habitats.

The habitats are believed to be well connected. Connectivity analysis using Circuitscape revealed abundant connections between each pair of "important habitat" patches, and no pinch points (Li, 2012). Although there are 430 km of railroads in the west, there are no fenced expressways, and a low road density (0.02 km/km^2). Thus far, there has been no indication that roads are barriers to snow leopard movements.

Across the Sanjiangyuan Region, blue sheep (*Pseudois nayaur*) are the major prey for snow leopards, with Himalayan marmot (*Marmota himalayana*) as well (Li, 2012). After suffering heavy hunting from the 1960s to the 1980s (Schaller et al., 1988), blue sheep have slowly recovered over the last two decades. There is no range-wide census information available, although several recent surveys have begun to describe blue sheep numbers at individual sites. In 2012, Xiao et al. (2012) surveyed four sites along the upper Yangtze River, employing block count and double-observers methods. Within the 564 km^2 of snow leopard habitat covered, 100 km of transects were walked in 15 days, and 239 herds that included 11,638 individual blue sheep were recorded, yielding an average density of 30 head/km^2. In Nyanpo Yutze, along the Yellow River, local monks censused 487 households by interview in 2010; from the interviews, local herders estimated a total of 8000 blue sheep within an area of 2000 km^2 (Zhaxi Sange, personal communication). This rough estimate suggests that blue sheep have reached a relatively high density compared to other countries within blue sheep range (Snow Leopard Network, 2015). However, it is likely that there were more blue sheep in the 1950s, and the increased livestock grazing is predicted to reduce the number of blue sheep that can persist in the landscape.

Perhaps not coincidental to the blue sheep recovery, snow leopard numbers are also believed to have increased over the past two decades. Schaller et al. (1988) estimated 650 snow leopards in Qinghai Province, with about two-thirds of those in the Sanjiangyuan Region. Qinghai Forestry estimated at least 1000 individuals across the province, based on camera trapping surveys in the Sanjiangyuan Region. Li (2012) estimated 40 individuals across the 1500 km^2 habitat in Suojia Township near the Yangtze River. Shan Shui maintains ongoing camera trap surveys at two other sites in the Sanjiangyuan Region. In 2013, 6–8 individuals were identified within 300 km^2 of habitat in Yunta village, and in 2014, in the Zhaqing Township north of the Mekong River, 20 individuals were identified in 950 km^2 of habitat (Shan Shui, unpublished data). During the same interview survey in Nyanpo Yutze in 2010, 40 individuals were estimated (Zhaxi Sange, founder of Nyanpo Yutze Environmental Protection Association, personal communication).

Snow leopards are known to be threatened by both local and outside factors. However, a rigorous assessment of these threats is lacking. Poaching and retaliatory killing for livestock depredation are documented region-wide, but may not be the primary factors for endangerment of the species (Li et al., 2014). Gun confiscation in the early 2000s resulted in a significant reduction in hunting. Emergent disease via transmission from other carnivores may be a mounting problem, particularly given the increasing number of feral dogs. However, almost no information is available on this potentially critical issue. Although blue sheep densities are believed to be lower now than what they were in the 1950s, prey depletion due to competition with livestock or hunting does not seem dire, as relatively high densities of blue sheep were detected in several sites, and government-led conservation programs are devoted to limit livestock numbers. In contrast, infrastructure expansion, mining, and hydro development are the most concerning threats as activities that will destroy or fragment current habitats.

THE BIG BROTHER RULES?

Government oversight dominates both economic development and environmental conservation in China. Government investment was key for successful panda conservation. In the Sanjiangyuan Region prior to 2003, government priorities focused on improving the livelihood of herders by increasing livestock number and providing better services in health care and education. However, since then, conservation issues have begun to occupy increasing resources and attention. The catalyst for this change in focus is the degradation of the grasslands, which is believed to endanger the livelihood both of local herders and people downstream.

Since 2003, three conservation milestones in this region have had varied impact on snow leopards and other wildlife. First was the establishment of the SNNR in 2003 (Li et al., 2012). The reserve is composed of 18 patches distributed across the region. The SNNR includes 38,000 km^2 of snow leopard suitable habitat, 25,000 km^2 of which is deemed "important habitat" (Li, 2012). However, the reserve is more like the "conservation area" defined by IUCN, which is used by both humans and wildlife, instead of a strictly protected area without people. The land within the reserve was divided and contracted to herders in the late 1990s, and livestock grazing has continued after the establishment of the reserve. When the SNNR was established, 21 conservation stations were set up to manage the reserve, however, insufficient capacity and funding constraints have limited their effectiveness. The SNNR Administration has cooperated with NGOs to promote community-based conservation programs within the nature reserve, and has achieved success in some areas (Shen and Tan, 2012).

The second milestone was the initiation of the Sanjiangyuan Ecological Protection and Construction Program in 2005. This program, in combination with the rebuilding of Yushu Prefecture after the 2010 earthquake, has changed the Sanjiangyuan Region profoundly. The program aims to reverse grassland degradation by relocating large numbers of herders, constructing large-scale fencing, employing widespread pika (*Ochotona* spp.) poisoning, and using artificial precipitation. These programs were implemented at a cost of 7.5 billion RMB (~1.25 billion USD) from 2005 to 2013 for the First Phase, and 25 billion RMB (~4.2 billion USD) was approved by the central government in 2014 for the Second Phase, 2013–2019 (NDRC – National Development and Reform Commission, 2013). One third to half of the herders were relocated to settlements in towns or cities at the early stage of the program (Du, 2012; Foggin, 2008).

In addition to these government-mandated relocations, more herders are moving to towns and cities for better education and health care, which contributes to the increasing depopulation of the grasslands. In some remote villages, 70% of herders have moved since 2005 (Chen Song, Leader of Zhaqing Township Government, personal communication). Livestock numbers are likely decreasing in some areas where the collection of caterpillar fungi, for use in traditional medicine, has provided a more lucrative source of income (Sulek, 2011) (see Chapter 10.4). Wild ungulates may benefit from such change, although no evidence of this is yet available. Pika poisoning is an increasingly controversial practice that has been carried out for more than half a century. The current policy continues even though scientific evidence has demonstrated its lack of effectiveness (Smith et al., 2006). Fencing large tracts of grassland to prevent grazing is widely adopted to reverse degradation, but the fences are also problematic for native ungulates species, such as Tibetan antelope (*Pantholops hodgsonii*) and Tibetan wild ass (*Equus kiang*) (Zhaduo, Founder and Director of The Snowland Great Rivers Environmental Protection Association, personal communication). The evaluation of the program in 2013 showed that grassland coverage has been improved, but there was no information on its impacts on wildlife (NDRC, 2013; Shao et al., 2010).

Finally, in 2011, the State Council of China announced its approval of the National Comprehensive Experimental Zone for Ecological Conservation in the Sanjiangyuan Region, a special zone to experiment with policy reforms. The specific targets of the Experimental Zone, as stated by the State Council, is to demonstrate "Ecological Civilization" through recognizing Tibetan cultural values toward conservation and the central role of pastoralists, while also initiating a scheme of payment for ecosystem services (or eco-compensation). Following the approval of the experimental zone, two policies were implemented in 2012, the "Eco-compensation" and the "Conservation Guard." The eco-compensation scheme required the herders to restrain livestock numbers, and funds were transferred directly to the accounts of individual households based on the area of pastureland contracted. In 2014, 3.45 billion RMB (~0.6 billion USD) was paid to herders in this region. The Conservation Guard policy hired herders to take care of both their land and the wildlife it contains. Nearly 60,000 herders were hired in 2014 and paid at a monthly rate of 1,400 RMB (~230 USD).

However, the conservation outcomes cannot be guaranteed. Where overstocking is clearly causing damage, Harris (2010) argued that *"we lack sufficient understanding of current socio-ecological systems to identify ultimate and proximate drivers of pastoralist behavior, and thus policy initiatives aimed at sustainability may well fail."* According to our field observation, even the monetary incentive is not always effective in reducing livestock numbers. More importantly, conservation in such landscapes is not possible through individual households, but can only be accomplished by the collective actions in a grazing group or a village. Transferring funding to individual households may create obstacles for collective actions (Li, 2014). Another challenge is the proper training for herders, who typically do not have the skills to do more than simple patrolling.

These conservation programs have the potential to create a culture and atmosphere that could benefit wildlife conservation. Tourism based on wildlife observation or hunting in North America and Africa has provided large amounts of funding to sustain the wildlife management mechanism. In western China, while wildlife tourism or trophy hunting is poorly developed, eco-compensation from the central government could be a driver for wildlife conservation. However, this system in China is far from efficient or effective, with a lack of incentive mechanisms and technical support hindering progress. Other stakeholders, such as local Tibetan communities, monasteries, and NGOs should be involved for successful policy making and implementation.

The success of these conservation plans are also jeopardized by the Qinghai provincial government's programs for increased infrastructure and hydro development, mining, and tourism. The monetary investment into these programs vastly outpaces the investment into conservation. This agenda also threatens the current conservation landscape. For example, in 2012 a proposal was put forth to shrink the SNNR to allow for expansion in mining and hydro development. The proposal was ultimately rejected by the central government in 2014, but it is a clear indication of where the challenges lie at the provincial level. Ongoing dilemmas also include two highways currently under construction that penetrate the region. One of these highways will bisect the Anemaqen region of the SNNR, a holy place for Tibetan Buddhists. It is difficult for the SNNR or Qinghai Forestry to argue against such planning within the government system.

As in most societies, in China the government has the most resources and authority, and thus the potential to be both a villain and a savior for conservation. Other sectors, such as conservation groups, must optimize the opportunities provided through effective policies, while enduring and mitigating the impacts of those that act in opposition to conservation goals. In much the same way as the giant panda, snow leopards could be used to evaluate the effectiveness of current government-funded programs, with monitoring and conservation made part of the routine of government-hired guards. Shan Shui

Conservation Center and its partners have been working in this region to train local herders in effective monitoring techniques to document snow leopards and blue sheep. These monitoring efforts are introduced and integrated at the local scales of townships and counties in an attempt to legitimize the philosophical concepts of the Experimental Zone.

THE MAKING OF A FLAGSHIP SPECIES

Large, charismatic mammals are often used as flagship species. Their high profile and ability to evoke empathy toward wildlife make them ideal for drawing financial support and raising broader environmental awareness for conservation endeavors (Sergio et al., 2008). Although the strategy has been criticized (Entwistle and Dunstone, 2000), the giant panda is widely used to promote conservation, and even provides the symbol for one of the world's largest conservation NGOs, World Wide Fund for Nature (WWF). This strategy serves to help both the species and the landscapes on which they dwell. A true flagship species garners recognition from the government as well as popularity among laypeople. How did the giant panda evolve to become a "national treasure"? Studying and following the lessons learned from giant panda conservation can help inform and lead the effort on successfully doing the same for snow leopards.

The snow leopard has great potential through both ecological and cultural aspects to be a flagship species for the Tibetan Plateau. As a top predator, snow leopards play an integral ecological role in maintaining a healthy alpine mountain ecosystem. Tibetan herders value the species as the pets of mountain gods (Li et al., 2014), and Qinghai Forestry Department is keen to promote snow leopard as the second "Provincial Ecological Name Card," after the Tibetan antelope, to publicize its conservation achievements (Zhang Li, Director of International Cooperation Officer of Qinghai Forestry, personal communication). As an important area for the snow leopard, it will be valuable to promote the snow leopard as a flagship species for the Sanjiangyuan region.

Scientific research is one driver for panda conservation. After George Schaller's study in the early 1980s, his Chinese collaborators continued research to deepen our understanding of the behavior and ecology of giant pandas. From 1985 to 1995, Pan Wenshi and his team studied the giant panda in Qingling (Pan et al., 2014). The research provided in-depth knowledge of the species and contributed toward winning recognition by the central government. Snow leopard surveys by Chinese researchers only started in 2002 in Xinjiang (Wang et al., 2012), and in the Tibetan Plateau, with the first comprehensive study beginning in 2009 (Li, 2012). The collaborative team of Shanshui Conservation Center and Peking University, supported by Panthera and the Snow Leopard Trust, has been working on snow leopard research and conservation in this region since 2009. Since that time, information has been gathered, publicity increased, and the interaction between different groups promoted. The continuing work of the team holds great hopes.

Mass media plays an important role in promoting species conservation (Blewitt, 2010). In China, the conservation of Tibetan antelope is a primary example. The species was heavily poached in the 1980s and 1990s (Schaller, 1998). The Wildyak Team is a volunteer-based group that was organized in 1994 to control poaching in the Kekexili area, particularly on Tibetan antelope. National mass media reported extensively on the fate of the species and the heroism of the team. In 2000, in part as a result of the media turning the Tibetan antelope's plight into a high-profile conservation issue, the area was transitioned into the Kekexili Nature Reserve. The reserve took over patrolling, but the administration cooperated well with the Wildyak Team and other NGOs to involve volunteers and to keep the Tibetan antelope in the public consciousness. This success translated into the Tibetan antelope becoming one of the five mascots of Beijing Olympics in 2008. Furthermore, and most importantly, it is estimated that the Tibetan antelope population has doubled since the mid-1990s (Liu, 2009).

Snow leopards have only begun to attract domestic attention in the past decade and a half. The first Chinese newspaper report on snow leopards was not published until 2000, and in 2014 there was an increase to more than 250 (Huike websearch, 2014, www.wisers.com). Nature films are an increasingly common and useful vehicle for promoting conservation. In recent times, several film groups have shown interest in capturing the snow leopard on camera. China Central Television produced the documentary "Snow Leopard" with Shan Shui in 2014, which will be released sometime in 2015. Disney Nature has been filming snow leopards in Suojia Township since 2014, with an estimated release in April 2016. The increasing exposure of snow leopards and their conservation issues in the public media and in upcoming documentary films is an essential step to elevate the species to flagship designation.

A key difference between the conservation of snow leopards versus giant pandas and Tibetan antelope is the absence of an immediate extinction crisis. The extinction crisis of giant pandas in 1980s (Pan et al., 2014) and of Tibetan antelope in 1990s (Schaller, 1998) greatly promoted their recognition by both government and people. Snow leopards are distributed over much larger landscapes and at lower density, making it unlikely across the vast Tibetan plateau that they would experience a similar dramatic crisis as the giant pandas or Tibetan antelope. Yet, we know that snow leopard numbers will never be high (perhaps 2000–3000 in China currently), so precisely because of their naturally low densities, we must seize the opportunity to protect them and their habitat and prevent future crisis situations.

COEXISTENCE IN THE "SNOW LEOPARD LANDSCAPE"

China is at the turning point where near-term action will lead to either the loss or preservation of its wild west (Harris, 2008). The Sanjiangyuan region is an indispensable part of the valuable natural heritage. There are yet healthy populations of snow leopard, wolf, and brown bear roaming across the region, which is indicative of a healthy and "wild" ecosystem. Coexistence between carnivores and people is paramount (Chapron et al., 2014) if we want to retain the Wild West.

To ensure the long-term survival of snow leopards in the Sanjiangyuan Region, a landscape-level conservation approach should be adopted. The Global Snow Leopard and Ecosystem Protection Program (Snow Leopard Working Secretariat, 2013) called for a joint effort to identify and secure 20 snow leopard landscapes across the snow leopard's range by 2020, and general guidelines for landscape management planning was also developed. Based on the predicted "important habitats" (Li, 2012), and information from prey surveys and camera trapping, we propose four snow leopard landscapes in the Sanjiangyuan Region, which require further survey and conservation intervention. They are the source region of the Mekong, an area along the Upper Yangtze, Anemaqen, and Nyanpo Yutze (Fig. 41.1).

Many knowledge gaps must be filled throughout all four of these landscapes. Although coarse evidence demonstrates an intact system full of top carnivores, and the prey base to maintain them, conditions can change rapidly. Rigorous surveys and scientific research are fundamental to properly evaluating the status of, and threats to, snow leopards, their prey, and habitat. How snow leopards and a rich guild of sympatric carnivores maintain an ecological balance and each of their ecological roles are poorly understood. The exploration of this question will not only deepen our understanding of the ecosystem, but also clarify arguments for conserving snow leopards as keystone species. The impacts of anthropogenic interventions should also be understood, including both current and planned infrastructure, different policies on grassland conservation, and restoration.

For conservation interventions, landscape level conservation should be adopted, by involving multiple stakeholders to secure important "Snow Leopard Landscapes." Protected area networks should integrate nature reserves managed by government with sacred lands protected culturally (see Chapter 15.2), especially where these two have significant overlap. Wildlife-human conflicts need to be mitigated, and poaching controlled. Wildlife monitoring and conservation at the level of grazing group or village need to fit into the eco-compensation and conservation guard policies in order to make the best of large-scale governmental funding. The needs of snow leopards and other wildlife should be weighed during development planning. When securing the four important landscapes, we must consider important corridors to maintain connectivity among them.

The Sanjiangyuan Region is a representative snapshot of the Tibetan Plateau. Across the Chang Tang, Kunlun, Himalayan, and Qilian ranges, snow leopards live in human-dominated landscapes. In many cases, as with the Sanjiangyuan Region, large nature reserves have been established, such as Chang Tang NR, Arjin Shan NR, Mid-Kunlun NR, West Kunlun NR, and Chomolungma; large-scale government-funded conservation programs are conducted, such as the State Ecological Safe Shelter Zone Program in Tibet (Zhong et al., 2006), and Qilian Mountain Program in Gansu and Qinghai. Without a doubt, the human economic impulse for development is high. The experiences of Sanjiangyuan Region, success or failure, may serve as a guide and portent for snow leopard conservation across the entire Tibetan Plateau.

References

Blewitt, J., 2010. Media, Ecology and Conservation: Using the Media to Protect the World's Wildlife and Ecosystems. Green Books, Cambridge, UK.

Chapron, G., Kaczensky, P., Linnell, J.D.C., 2014. Recovery of large carnivores in Europe's modern human-dominated landscapes. Science 346 (6216), 1517–1519.

Du F., 2012. Ecological resettlement of Tibetan herders in the Sanjiangyuan: a case study in Madoi County of Qinghai. Nomadic Peoples 16 (1), 116–133.

Entwistle, A., Dunstone, N. (Eds.), 2000. Priorities for the Conservation of Mammalian Diversity: Has the Panda Had Its Day?. Cambridge University Press, Cambridge, UK.

Foggin, J.M., 2008. Depopulating the Tibetan grasslands: national policies and perspectives for the future of Tibetan herders in Qinghai Province, China. Mt. Res. Dev. 28 (1), 26–31.

Harris, R.B., 2010. Rangeland degradation on the Qinghai-Tibetan plateau: a review of the evidence of its magnitude and causes. J. Arid Environ. 74 (1), 1–12.

Harris, R.B., 2008. Wildlife Conservation in China: Preserving the Habitat of China's Wild West. M.E. Sharpe, New York.

Li, J., Wang, D.J., Yin, H., Zhaxi, D.J., Jiagong, Z.L., Schaller, G.B., Mishra, C., McCarthy, T.M., Wang, H., Wu, L., Lu, Z., 2014. Role of Tibetan Buddhist monasteries in snow leopard conservation. Conserv. Biol. 28 (1), 87–94.

Li, J., 2012. Ecology and conservation strategy of snow leopard (*Panthera unica*) in Sanjiangyuan Area on the Tibetan Plateau. Doctoral thesis, Peking University.

Li, S., 2014. Community-Based Conservation and External Intervention. Peking University Press, Beijing, China.

Li, X., Brierley, G., Shi, D., Xie, Y., Sun, H., 2012. Ecological protection and restoration in Sanjiangyuan National Nature Reserve, Qinghai Province, China. In: Higgit, D. (Ed.), Perspectives on Environmental Management and Technology in Asian River Basins. Springer, Netherlands, pp. 93–120.

Liu, Wulin, 2009. Tibetan Antelope. Chinese Forestry Press, Beijing, China.

NDRC, 2013. The Plan of Project of Qinghai Sanjiangyuan Ecological Protection and Construction--Phase II. http://www.sdpc.gov.cn/zcfb/zcfbghwb/201404/W020150203319341893483.pdf (accessed 20.05.2015).

Pan, W., Lu, Z., Zhu, X., Wang, D., Wang, H., Long, Y., Fu, D., Zhou, X.A., 2014. Chance for Lasting Survival: Ecology and Behavior of Wild Giant Pandas. Smithsonian Institution Scholarly Press, Washington DC, USA.

Schaller, G.B., 1998. Wildlife of the Tibetan Steppe. University of Chicago Press, Chicago, USA.

Schaller, G.B., Junrang, R., Mingjiang, Q., 1988. Status of the snow leopard (*Panthera uncia*) in Qinghai and Gansu Provinces, China. Biol. Conserv. 45 (3), 179–194.

Sergio, F., Caro, T., Brown, D., Clucas, B., Hunter, J., Ketchum, J., McHugh, K., Hiraldo, F., 2008. Top predators as conservation tools: ecological rationale, assumptions, and efficacy. Annu. Rev. Ecol. Evol. Syst. 39, 1–19.

Shao, Q., Zhao, Z., Liu, J., Fan, J., 2010. The characteristics of land cover and macroscopical ecology changes in the Source Region of Three Rivers in Qinghai-Tibet plateau during last 30 years. Geoscience and Remote Sensing Symposium (IGARSS), IEEE International, pp. 363–366.

Shen, X., Tan, J., 2012. Ecological conservation, cultural preservation, and a bridge between: the journey of Shanshui Conservation Center in the Sanjiangyuan region, Qinghai-Tibetan Plateau, China. Ecol. Soc. 17 (4), 38.

Smith, A.T., Zahler, P., Hinds, L.A., 2006. Ineffective and unsustainable poisoning of native small mammals in temperate Asia: a classic case of the science–policy divide. In: McNeely, J.A., McCarthy, T.M., Smith, A.T., Olsvig-Whittaker, L., Wikramanayake, E.D. (Eds.), Conservation Biology in Asia. Society for Conservation Biology and Resources Himalaya Foundation, Kathmandu, Nepal, pp. 285–293.

Snow Leopard Network, 2015. Snow Leopard Survival Strategy. Revised version 2014.1. www.snowleopardnetwork.org (accessed 02.072015).

Snow Leopard Working Secretariat, 2013. Global Snow Leopard and Ecosystem Protection Program (GSLEP). Snow Leopard Working Secretariat; Bishkek, Kyrgyz Republic.

Sulek, E., 2011. Disappearing sheep: the unexpected consequences of the emergence of the caterpillar fungus economy in Golok, Qinghai, China. HIMALAYA 30 (1), 9.

Wang, Y., Ma, M., Turghan, M.A., 2012. Bibliometric evaluation of research on snow leopard. Chinese J. Ecol. 31 (3), 766–773.

WWF, 2015. Wild Giant Panda Population Increases Nearly 17%. <http://www.worldwildlife.org/press-releases/wild-giant-panda-population-increases-nearly-17 (accessed 01.07.2015).

Xiao, L., Liu, Y., Dou, X., 2012. Blue sheep. Rapid biodiversity assessment report in the Sanjiangyuan Region. Shanshui Conservation Center, technical report, Beijing, China.

Zhong, X., Liu, S., Wang, X., Zhu, W., Li, X., Yang, L., 2006. A research on the protection and construction of the state ecological safe shelter zone on the Tibet Plateau. J. Mt. Sci. 24 (2), 129–136.

CHAPTER

42

China: Current State of Snow Leopard Conservation in China

Philip Riordan, Kun Shi***

*Department of Zoology, University of Oxford, Oxford, UK

**The Wildlife Institute, Beijing Forestry University, Beijing, China

China holds more snow leopards (*Panthera uncia*) than any other country, possibly more than half the global population. The unprecedented rate of socioeconomic change in China over the past 30 years places unique pressures on wildlife and ecosystems within its borders. Of the large predators in China, the snow leopard has arguably suffered least as a consequence of these changes. Their occurrence in remote mountain regions, where there are few people, has contributed to their continued survival. The cultural prominence of snow leopard through Chinese history has been less pronounced than for other species, such as tiger, and their avoidance of people has reduced the attention they have received. As a consequence, people's awareness of the snow leopard across China is relatively poor and building support for conservation efforts has had to overcome this initial hurdle.

Deng Xiaoping's key policies in the late 1970s of "opening up" to the rest of the world and a market model for economic reform, known within China as 改革开放 (*Gǎigé kāifàng*), set the nation on its current trajectory of economic development. Following the Maoist era, the economic growth of China in the past three decades led to the country becoming the second largest economy in the world. Within China, the impacts of this unprecedented growth have stretched across its almost 10 million km^2 of land area, nearly equal to the entirety of Europe. Major land reforms such as the collectivization and subsequent decollectivization of agriculture in 1949 and the early 1980s, respectively, have had significant impact on China's natural ecosystems, leading to more recent policy attempts to repair damage to key habitats.

A BRIEF HISTORY OF THE SNOW LEOPARD IN CHINA

The environmental history of China is a long one, with frequent attempts by ruling authorities to tame the wilderness (Elvin, 2004). Social policies, including movement of people into remote mountainous regions, have bought changes to

Snow Leopards
http://dx.doi.org/10.1016/B978-0-12-802213-9.00042-0

previously isolated communities. The cultural transfer from the incoming populations has altered the way in which indigenous peoples view their environment. Increasing urbanization and the provision of free schooling in towns and cities to children from poor rural areas have drawn people away from traditional farming and ancestral land. The traditional transhumant forms of livestock management are changing in response to larger-scale market forces than previously important local trade and commerce routes, with livestock being increasingly more carefully managed through central authorities.

Cultural (In)Significance

In the earlier dynasties that shaped imperial China, such as the Zhou Dynasty (ca. 1046–256 BCE), wildlife was regarded as often dangerous or destructive and needed to be controlled. Forests and swamps were cleared and drained to make way for agriculture and support the increasing human population. By contrast, the mountains of ancient China had few people and were seldom visited, being home to magic and supernatural creatures. Possible accounts of snow leopards are given in the ancient texts describing these mysterious lands. In the "Yizhoushu" (逸周书) account of the Western Zhou Dynasty (1046–771 BCE), Zun Er (拵栭), resembled a tiger or leopard, with a tail of nearly one meter in length. In the seminal geographical history of the pre-Qin Dynasty (256 BCE), entitled *Classics of Mountains and Seas* (山海经), Meng Ji (孟極) was described with black spotted forehead and a white, leopard-shaped body and secretive nature. Zhu Jian (驺虞) was described as *"living alone in the mountain, making no noise when walking on the vegetation, leopard shape with longer tail."* Other accounts include "Zou Yu" (邹瑜), characterized as a mysterious and beneficent hunter who help local people by killing pest animals such as wild boar and deer. Later reports in the Ming Dynasty identified four further features of Zou Yu: a tiger's body and lion's head; white fur with black spots, and an extremely long tail; graceful movements; and superb agility, running like wind. Perhaps the most popular account is that of Pi Xiu (貔貅), first recorded in "Shih Chih." Pi and Xiu were identified as different gendered beasts that can earn money (male) and not squander it (female). The snow leopard was also recorded as Ai Ye Bao (艾叶豹) and He Ye Bao (荷叶豹) in the *Compendium of Materia Medica* (本草纲目) published during the Ming Dynasty (1368–1644 CE).

Most of the early references are accounts previously handed down as oral tradition. These traditions will have many subtle differences and peculiarities depending on the region. Importantly, many mountain regions were beyond the reach of early Chinese writers and so there is likely to be much traditional knowledge that is unrecorded. While early Chinese dynasties came and went in the east, other empires and cultures were in similar procession in what is now the northwest of China, with their own mythologies and interpretations of nature. Conversations between the authors and many mountain community dwellers living in Xinjiang and elsewhere in China have revealed that written accounts are indeed rare. It is also becoming apparent that, with the social and economic changes in China, much traditional knowledge is being lost as younger members of communities move to the cities and away from their ancestral homes (Riordan and Shi, unpublished data).

Political and Legislative Developments

Since the 1950s, the snow leopard has been identified as a notable species requiring scientific investigation in its natural habitats. Its status was formalized in 1988 with the adoption of "The Law of the People's Republic of China on the Protection of Wildlife" at the Fourth Meeting of the Standing Committee of the Seventh National People's Congress (November 8, 1988). Upon enactment of the law in 1989, the snow leopard, along with other priority species, was included as National Class I Protected Wildlife. Following regional surveys and ecological research, the snow leopard was listed as a priority species for the first national survey of terrestrial wildlife resources, which reported in 1995. The second national survey is currently underway. Meanwhile, China began to insist on monitoring and evaluation for key snow leopard populations and habitats, understanding threats, and providing a reliable scientific basis for snow leopard conservation (Shi et al., unpublished data).

Under the 1988 Law on Protection of Wildlife, key measures for the protection of endangered wildlife were enshrined, including articles to combat poaching and illegal trade of snow leopard skin and bones. Enforcement of these crimes was strengthened through improved patrolling, port inspection, and market inspection, effectively curbing the momentum of illegal activities against snow leopards. Also, while not initiated at national level, in Tibet, Qinghai, Yunnan, and Gansu provinces, damage caused by snow leopard has received government compensation to ease conflicts with local people and to promote support for snow leopard conservation. In the wake of the 1988 act, publicity and education concerning snow leopard and other wildlife conservation were started. Public warnings against buying snow leopard products and encouragement to report illegal activity or accidentally captured animals were strengthened by setting up telephone hotlines and e-mails.

Since 2000, significant ecological construction projects have been undertaken in snow leopard areas. These include China's natural forest resources protection, returning farmland to forests, returning grazing land to grassland, wildlife conservation, and construction of new nature reserves. China has established 27 nature reserves in snow leopard distribution range with a total area of more than 600,000 square kilometers and covering more than 50% of the snow leopard habitat, constructing the most basic network for snow leopard conservation and monitoring. Chinese government regards PA establishment and management as extremely important for maintaining snow leopard populations and their habitats.

CURRENT SNOW LEOPARD DISTRIBUTION IN CHINA

In November 2012, a meeting held in Beijing bought together local experts from across the snow leopard range in China. The goal was to develop the National Action Plan and obtain a current update on the status of the snow leopard. The details from that meeting and also from communications among the members of the China Cat Specialist Group follow.

The snow leopard's range in China occurs across 1.1 million km^2 of the provinces in the west of the country, coinciding with more than 10,000 km of national border adjoining 10 neighboring and one possible snow leopard range states. China contains about 60% of the potential habitat available to snow leopards, and an estimated population of 2000–2500 individuals accounts for one-third to one-half of the total global population in the wild (McCarthy and Chapron, 2003).

Full population assessments across China have not been carried out, though this work is underway. The majority of previous estimates are essentially guesswork, with a handful based on reliable and repeatable methodologies. A few more recent efforts have started to yield improved estimates. What is rapidly becoming apparent, through the use of new technologies such as remote camera surveillance, is that there are more snow leopards in China than was previously thought. Reassessment of the species' range in China during the 2012 meeting highlighted several discrepancies with many published range maps. Participants identified most closely with the potential distribution included in the SLSS, though some disagreed about the distinction between "good" and "fair" habitats

(McCarthy and Chapron, 2003). Using this range map, the estimated area of habitat available to snow leopards in China is 2.08 million km^2. The map published on the IUCN Red List (2008 reassessment, version 2012.2), which is attributed to *International Snow Leopard Trust and the Wildlife Conservation Society 2008,* was deemed highly inaccurate, reducing the species' range in China by 68.6% to 0.65 million km^2. Participants rejected the validity of this map, claiming that many areas known to have snow leopard were excluded.

Given the uncertainty about space use and social structure of snow leopard populations, overall density estimates are difficult to make. The most recent version of the SLSS (Snow Leopard Network, 2014) makes this clear, with the review of populations estimates ranging from 0.15 to 8.49 per 100 km^2. Discussions between Chinese experts in 2012 considered the minimum density estimate published at that time (0.2 animals per 100 km^2) to be low. Nonetheless, at this density the overall snow leopard population of China is 4500 individuals, twice that of previous estimates. Of course, snow leopards do not occur evenly across their entire range, but experts thought this was a helpful conservative overall figure from which to proceed. Future refinement of this estimate currently in preparation may put the population even higher.

Provinces and Autonomous Regions

Snow leopards are known to occur in eight provinces and Autonomous Regions: Gansu, Qinghai, Sichuan, Tibet (Xizang), Xinjiang, Yunnan, Inner Mongolia, and Ningxia. Details provided by participants from each province that attended the 2012 meeting are provided in Table 42.1. Here we consider the areas within each province that are important for snow leopard, although Tibet and Qinghai are dealt with in detail in Chapter 41. All sites confirmed by Ma et al. (2002) were identified as still having populations of snow leopard.

Gansu Province: Present in the Qilian Shan range along the border with Qinghai and in the Die Shan along the border with Sichuan. Previously thought to have been extirpated from the outlying ranges along the Gansu-Inner Mongolia boundary, local evidence suggests there may be at least low level use of these areas by snow leopards, possibly by dispersing individuals. The Qilian Shan National Nature Reserve (>20,000 km^2) and Yanchiwan National Nature Reserve (5000 km^2) have apparently healthy populations of prey and of snow leopards.

Inner Mongolia Autonomous Region: Records suggest that snow leopards once occupied many of the larger high-elevation outcrops in the desert ranges on the Inner Mongolia-Ningxia border. These populations were believed to have been extirpated by the 1990s; however recent surveys and incidental accounts suggest snow leopard still used these areas in the early 2000s and single snow leopards were captured and relocated in 2011 and in 2013 (Shi and Riordan, unpublished data). These mountain ranges likely served as one of several important linkages for the snow leopard population and understanding the current use of these areas is a priority in China.

Qinghai Province: Qinghai is situated on the Qinghai-Tibetan Plateau and forms a mixture of high-elevation grasslands, rangelands, wetlands, and mountainous outcrops. The latter are important for snow leopard and include Arjin Shan, which borders Xinjiang, the semi-connected Qilian Shan, which borders Gansu, and the Kunlun Shan, which stretches west along the southern edge of the Tarim Basin. The core areas of the Sanjiangyuan-Kekexili Reserve complex, which borders Tibet, include more than 10,000 km^2 of snow leopard habitat.

Sichuan Province: Liao and Tan (1988) listed 10 counties where snow leopard have been reported, including Yaan, Baoxing, Jinchaun, and Xiaojin, along with Aba, Garze, Dege, and Batang. Its presence has been confirmed in Giant Panda Reserves such as in Wolong and it is

TABLE 42.1 Summary Details and Guiding Objectives for Snow Leopard Conservation across China's Provinces

	Gansu	Nei Mongol	Ningxia	Qinghai	Sichuan	Xinjiang	Xizang	Yunnan
SL population (estimate)	168	18	4	1039	505	872	1797	99
95% CI	48	5	1	297	144	249	513	28
Habitat area km^2	77,585	8095	1904	479,620	233,287	402,583	829,563	45,822
MAJOR PREY SPECIES[a]								
Bharal	G_WL_D	G_RL_D	G_RL_D	G_WL_D	G_WL_D	G_WL_D	G_WL_D	G_RL_D
Ibex	n/a	n/a	n/a	n/a	n/a	G_R	n/a	n/a
Argali	G_RL_S	G_RL_S	n/a	G_RL_S	G_RL_S	G_RL_S	G_RL_S	n/a
Marmot	G_WL_D	G_RL_D	G_RL_D	G_WL_D	G_WL_D	G_WL_D	G_WL_D	G_RL_D
Galliformes	G_WL_D	G_WL_D	G_RL_D	G_WL_D	G_WL_D	G_WL_D	G_WL_D	G_WL_D
Key locations	Qilianshan; Yanchiwan; Gobi region	Urad Houqi; Helan Shan Gobi region	Helan Shan; Gobi region	Sanjanyuan; Kekexili; Zadoi Yushu Arjin Shan	Ganzi; Gonggashan Baoxing-Wolong complex	Tian Shan; Karakorum; Kunlun; Alai; Alatau; Altai	Zanda; Qomo-langma; Xigaze	Hengduan Shan
SL PA number	22	3	2	22	120	21	26	14
SL PA area	17,832	2341	1649	62,296	91,434	106,547	443,685	2793
% SL area	23%	29%	87%	13%	39%	26%	53%	6%
Transboundary opportunities[b]	Mongolia	Mongolia	Mongolia (i)	Mongolia (i)	Myanmar (i) India (i)	Kazakhstan Kyrgyzstan Russia Mongolia Tajikistan Afghanistan Pakistan India	Myanmar India Bhutan Nepal	Myanmar India (i)
Joint province action[b]	Nei Mongol Ningxia Qinghai Sichuan Xinjiang	Gansu Ningxia Xinjiang (i)	Gansu Nei Mongol	Gansu Sichuan Xinjiang Xizang	Gansu Qinghai Xizang Yunnan	Gansu Nei Mongol (i) Qinghai Xizang	Qinghai Sichuan Xinjiang Yunnan	Sichuan Xizang

[a] *Prey species distributions defined as geographically widespread (G_W) or geographically restricted (G_R) living in locally dense (L_D) or locally scarce populations (L_S).*
[b] *Transboundary and joint provincial opportunities specified as indirect (i) when not immediately bordering, but where participation in regional cooperation strategies would be advantageous.*

present in low numbers in various areas above the timberline (Schaller, 1998). Distribution and status in Sichuan Province are poorly understood, and field surveys are needed to establish the current distribution.

Tibet Autonomous Region (TAR): Political sensitivities in Tibet and the scale of the Autonomous Region have made assessments of snow leopard difficult. Tibetan government participants to the meeting in Beijing in 2012 reported snow leopard occurring in several areas consistently. Key prefectures were highlighted: Xigazê (日喀则市); Qamdo (昌都市); Nyingchi (林芝地区); Lhoka (山南地区); and Ngari (阿里地区). Currently, focus is on Qomolangma National Nature Reserve, home to the world's highest mountain peak. Other areas in the Himalayas are likely to hold good snow leopard populations and also offer opportunities for transboundary planning with India, Nepal, Bhutan, and possibly Myanmar.

Xinjiang Autonomous Region: Xinjiang has a mixed topography, with vast desert basins (Tarim and Dzungaria), surrounded by chains of mountains. The spatial structure of the connected mountain ranges in Xinjiang against the background of economic development and simultaneous regional tensions provides the greatest challenge for maintaining connectivity between different parts of the species' range. Snow leopards are reported from the Tien Shan mountains stretching from Kyrgyzstan to Mongolia; along the border with Mongolia and north to Russia in the Altay; in the Dzungarian Alatau bordering Kazakhstan; the Arjin Shan and Kun Lun ranges along the northern edge of the Tibetan Plateau; the Pamirs along the Tajikistan-Afghanistan border; and the Karakorum mountains along the Pakistan border.

Yunnan Province: Snow leopards reportedly occur in the Hengduan Shan region of northwest Yunnan near the borders with the Tibet Autonomous Region, Sichuan, and Myanmar (Yunnan Forestry Administration, unpublished data). Local rangers have reported sightings and sign in areas around Baimaxue Shan. Further surveys are needed to more accurately determine population status.

Protected Areas

At the end of 2011, China had established 2640 nature reserves, covering 1.49 million km^2, representing 14.93% of the total land area (the global average is 12%). The amount of suitable snow leopard habitat within protected areas is 28%, almost five times higher than the overall figure of 6% calculated for the entire range by Hunter and Jackson (1995). The largest protected areas within the snow leopard range in China are Qiangtang Nature Reserve in Tibet and the adjoining Sanjiangyuan Nature Reserve in Qinghai, with a combined area of approximately 600,000 km^2. Aerjinshan Nature Reserve in Xinjiang and Kekexili Nature Reserve in Qinghai also abut these reserves forming a protected area complex of 740,000 km^2. Of this total area, 375,300 km^2 (50%) is designated as suitable snow leopard habitat, with the majority of this (258,300 km^2) in Qinghai. Areas of suitable habitat within protected areas are not contiguous and it is unclear the degree to which the interstitial unsuitable habitats are impermeable to snow leopards, presenting significant obstacles to dispersal.

Key Regions for Long-Term Survival

Animals are seldom distributed evenly across their entire range, with individuals needing different areas according to various life-history stages, such as breeding, foraging, and dispersing. These functions need not, and often do not, occur in the same areas within an individual's range, so defining a sufficient range for survival and at an appropriate spatial scale is not a trivial problem. In attempting to distill these complexities down into areas that might support a breeding population and those through which snow leopards might pass to get from one breeding area to another, we must confront the problem that there is little knowledge to draw on to help us. Information about prey resources is sparse and uneven, as are data about the quality of the ecosystems that support the food chain headed by snow leopards. We are therefore left with few options but to mathematically model the possibilities using the scant information available (Riordan et al., 2015).

Breeding Populations

We sadly have precious little knowledge about the areas in which snow leopard successfully breed. The anecdotal evidence from the 2012 meeting is that they appear to breed across their entire range. Camera trap data from our own studies have revealed breeding in Qilianshan, Taxkurgan, Bortala, Qomolangma, and Sanjiangyuan. One compelling argument against the current population density estimates for snow leopard is that individuals would be so widely dispersed that they would not be able to find each other to breed at a sufficient rate to have maintained the population thus far.

Linking Habitats

To examine connectivity across the snow leopard range, we constructed models to identify areas for movement and dispersal across their vast global range in fragmented landscapes (Riordan et al., 2016). These highlight locations that may support infrequent, but critical, dispersal movements between parts of the range. Although individuals may not permanently reside in these areas, such as the small stepping-stone patches and narrow corridors, their existence as connections between established populations may be an essential service for the maintenance of healthy snow leopard populations.

Predicting connectivity across the snow leopard range highlights three patterns. First, the west Himalayan-Karakorum-Pamir region of Tibet and Xinjiang is predicted to have moderate to high levels of movement of dispersing snow leopards across a wide area. Second, in contrast, the Dzungarian region is predicted to have two relatively restricted corridors, with low levels of movement around the Dzungarian Basin. This would be a valuable place to monitor movement, to determine if snow leopards are crossing this zone and if they are using the predicted corridors. Limited movement is known to occur; for example, in 2009 a male snow leopard was found dead, having apparently attempted to cross a section of the Dzungarian Basin approximately 50 km outside of the predicted snow leopard range (Ablimit Abdukadkir, personal communication, 2009).

The Gobi Desert region in China presents a third pattern of predicted connectivity. Specifically, potential stepping-stones are highlighted within the desert. Snow leopard populations historically occurred in some of these patches, declining since the 1940s (Schaller, 1998; Wang and Schaller, 1996). Connectivity through this area would have important implications for population structure, genetic diversity, and conservation planning. Assessing movement and gene flow through this area and use of these stepping-stones is a priority as urgent protection may be required. Snow leopards have historically been reported occasionally in the Mongolia-China border area (Schaller, 1998), and also moving across open steppe between isolated hills in Mongolia (McCarthy, 2000). A dead snow leopard was found in open desert 30 km south of Sevrei in southern Mongolia (B. Munkhtsog, personal communication, 2007), and snow leopard occurrence up to 600 km from the nearest mountains have been reported in Russia (Heptner and Sludskii, 1972). More recently, in March 2013 livestock herders captured, and subsequently released, a young male snow leopard in Inner Mongolia, China, approximately 40 km from Helan Shan, an area where snow leopard have not been reported for over 30 years (Ningxia and Inner Mongolia Forestry Administrations, unpublished data).

NATIONAL STRATEGIES FOR SNOW LEOPARD PROTECTION

In 2013, the *China Snow Leopard Conservation Action Plan (2013–2020)* was drafted by the Wildlife Institute at Beijing Forestry University with support from the State Forestry Administration of China. This document will form the basis for national policy development in China's central government. The plan acknowledged the scale of the snow leopard range, the unique biodiversity of the mountains, and their role in water conservation, carbon exchange reserves, climate regulation, and so forth. Snow leopard action plans for each province are being developed on the basis of the overarching national document.

CONSERVATION GOALS

Measures are set out to improve surveys and monitoring of snow leopard population and habitat in the field. Understanding the basic situation will allow better scientific planning, protection, and support systems, including habitat restoration and ecological corridor construction. Further measures include combating illegal activities, increasing publicity and education, as well as expanding international cooperation and support local sustainable economical developments. By 2020, the encroachment of agricultural into snow leopard habitat should be reduced to a rate 70% below current expansion identified under current climate change (Riordan et al., 2016), and habitat improvements should be optimized and extended, with corridors between populations identified (Riordan et al., 2015) and initiated.

PRIORITY AREAS

To realize the snow leopard conservation objectives of China, according to snow leopard conservation strategy, China will take conservation actions in the following priority areas:

1. Investigating and monitoring snow leopard population and habitat dynamics, strengthening basic research and conservation planning.
2. Improve management systems for protection and enhance the protection of habitats.
3. Coordination of snow leopard conservation and the local community's social and economic development.
4. Strengthening law enforcement, cracking down on poaching, smuggling and illegal trade of snow leopard and other wildlife products.
5. Expanding international cooperation for global snow leopard conservation.

WORKING TOGETHER: OPPORTUNITIES AND ACTIONS FOR INTERNATIONAL COOPERATION

Snow leopards' primarily inhabit areas along over 10,000 km of sensitive political boundaries creates significant problems for their conservation, but it also offer benefits. Many areas of international conjunction occur at key sites for range connectivity, such as the Himalayan region (Tibet–Bhutan; Nepal; India); Karakorum and Pamirs (Xinjiang–India; Pakistan; Afghanistan; Tajikistan); and Tian Shan (Xinjiang–Tajikistan; Kyrgyzstan; Kazakhstan).

China was one of the founding states of the United Nations in 1945, with the People's Republic of China taking its seat on the Security Council in 1971. As such, China's international role has a long history, placing it at the center of the developments of key agreements of relevance to the nation's biodiversity, including the snow leopard. China was among the first signatories to the Convention on Biological Diversity (CBD) in 1993 and the Convention on the International Trade in Endangered Species (CITES), as an accession member since 1981. China is a signatory to the United Nations Framework

Convention on Climate Change (UNFCCC), since 1992. These agreements tend to be rather dry and framed in dense impenetrable legal jargon. They do, however, indicate an overarching desire and for transboundary collaborations.

WHAT FUTURE FOR SNOW LEOPARD IN CHINA?

It is, of course, impossible to predict the future. But that has never stopped anyone trying. The first step would be to predict the future economic, political, and social trends on which the fate of the snow leopard hangs. If we conduct a thought experiment where people had no interest in either them or their landscapes, then they would probably be just fine, barring any random natural disasters. However, at this time the snow leopard's future trajectory is irrevocably entwined with our own. It appears that most people around the world and also within China would not like to see snow leopard and other charismatic species become extinct because of our actions. This is a good start. The government of China is also concerned not to lose a key species, not least because they realize their importance as the principal country that will ensure the snow leopard's survival.

But how much sway do these voices have, amid the noise of international and domestic politics and commerce? Equally, how concerned are people really? If the answer to both these questions is "not much," then we lose the snow leopard while the majority of people are looking the other way.

SYNONYMS AND LOCAL NAMES FOR SNOW LEOPARD USED IN CHINA

Chinese synonyms: Ai Ye Bao; Cao Bao; He Ye Bao; Xue Bao

Other languages and dialects: Pis (Tajik); Sha (Tibetan); Yiletes (Uiygur); Yilibus (Mongolian)

References

Elvin, M., 2004. The Retreat of the Elephants: An Environmental History of China. Yale University Press, USA.

Heptner, V.G., Sludskii, A.A., 1972. Mammals of the Soviet Union, Volume II, Part 2 Carnivora (Hyaenas and Cats). Vysshaya Shkola, Moscow. English translation 1992. E.J. Brill, Leiden, The Netherlands.

Hunter, D.O., Jackson, R., 1995. A range-wide model of potential snow leopard habitat. In: Jackson, R., Ahmad, A. (Eds.), In: Proceedings of the 8th International Snow Leopard Symposium, Islamabad, November. International Snow Leopard Trust, Seattle and WWF-Pakistan, Lahore, pp. 51–56, 1997.

Ma, J.Z., Zou, H.F., Cheng, K., 2002. The distribution status of snow leopard (*Panthera uncia*) in China. Northeast Forestry University, Beijing, China.

McCarthy, T.M., 2000. Ecology and conservation of snow leopards, Gobi brown bears, and wild bactrian camels in Mongolia. PhD thesis. University of Massachusetts, Massachusetts, USA.

McCarthy, T.M., Chapron, G., 2003. Snow Leopard Survival Strategy. International Snow Leopard Trust, Seattle, USA.

Riordan, P., Cushman, S.A., Mallon, D., Shi, K., Hughes, J.R., 2015. Predicting global population connectivity and targeting conservation action for snow leopard across its range. Ecography 38, 1–8.

Riordan, P., Wang, J., Hughes, J.R., Shi, K., 2016. Seasonal shifts in livestock grazing in mountainous China in response to climate change and the implications for conservation of snow leopard. Climatic Change.

Schaller, G.B., 1998. Wildlife of the Tibetan Steppe. Chicago University Press, Chicago, USA.

Snow Leopard Network, 2014. Snow Leopard Survival Strategy. Version 2014.1. Snow Leopard Network, www.snowleopardnetwork.com

Wang, X.M., Schaller, G.B., 1996. Status of large mammals in western Inner Mongolia, China. J. East China Normal Univ. Nat. Sci. 12, 93–104.

SECTION VII

The Future of Snow Leopards

CHAPTER

43

Sharing the Conservation Message

Rana Bayrakçısmith[*,‡], *Sibylle Noras*[**], *Heather Hemmingmoore*[†]

[*]Snow Leopard Program, Panthera, New York, NY, USA
[**]Snow Leopard Network, Melbourne, Australia
[†]Snow Leopard Network, London, UK
[‡]Snow Leopard Network, Seattle, WA, USA

INTRODUCTION

The methods and tools used to communicate snow leopard (*Panthera uncia*) conservation messages have changed substantially in recent years through improved speed and ability to engage with larger audiences in unprecedented ways. Global social media fundamentally enhances conservation organizations' capacity to reach the public and changes unidirectional communication into real time engagement. Meanwhile, mobile phone technology expansion provides the opportunity to reach more remote areas and improve connections with residents of snow leopard range countries. We are now accustomed to instantaneous communication via email and social media, and news feeds available on personally tailored websites. Less than 25 years ago, snow leopard conservationists relied heavily on postal mail and expensive, time-consuming, and rare in-person meetings to communicate with stakeholders. Fax machines in the 1980s initiated rapid document sharing, while today a professional's work day could include emails with fellow scientists around the world, global online meetings enabled by Voice over Internet Protocol (VoIP) technology, a local school presentation, and posting snow leopard photos for fans to share over social media.

Although communication options have proliferated and messages can reach an ever-widening audience, barriers remain. English is ever more commonly used in international communication, yet often scientists, field assistants, and the public in snow leopard range countries are not fluent in English and therefore unable to access widely available information. Conversely, not all information coming out of snow leopard range country is available in English. Communication access remains an issue for some, and while mobile phone service is expanding, many remote mountain villages still lack service.

Snow Leopards
http://dx.doi.org/10.1016/B978-0-12-802213-9.00043-2

Effective conservation programs require understanding, input, and active support from the general public, scientific community, and government sectors. The majority of conservation messages come from the snow leopard conservation community, comprised of NGOs, scientists, academics, and conservationists. Building a constructive dialogue between stakeholders allows for the sharing of information and ideas, especially concerning solutions and failures. It is essential to share knowledge gained from failed efforts, rather than to deemphasize such projects (see Chapter 40).

Communication should encourage collaboration and involvement in conservation programs targeted at saving snow leopards. How that message is communicated varies depending on who is conveying the information and who is intended to receive it. Recipients drive the key messages, methodology, and style of communication, and they include the general public, scientific community, and government.

COMMUNICATING CONSERVATION MESSAGES WITH THE GENERAL PUBLIC

The general public is the largest and most diverse group of stakeholders that includes interested people both within and outside snow leopard range countries. They provide the largest pool of potential financial support for conservation NGOs and the capacity to influence governmental policy through lobbying and activism in some countries. Public support can make or break a conservation program. Fostering feelings of environmental and social responsibility is essential to instill a culture of conservation that people carry throughout their lives. Engagement and education of youth is important to facilitate a generational shift in cultural attitudes toward conservation, ensuring that iconic species and ecosystems will persist for further generations.

Methods of Communication with the Public

Traditional media, including television, newspapers, and radio still provide a common source of conservation information for the general public. Yet social media is becoming ever more important to individuals, NGOs, and news organizations who often distribute conservation stories that way, reaching far more people than traditional media forms would allow.

Social Media

Social media provides the opportunity for individuals to engage with conservation organizations by sharing information while overlaying their own opinions. Many NGOs have embraced social media as an effective means of distributing information through their already-engaged member base. These followers further disseminate materials within their own networks, allowing conservation messages to spread more widely than traditional media. Such sharing can potentially reach audiences that would not otherwise seek out conservation content, thereby educating and engaging new followers. Snow leopard conservation lends itself to this model because pictures and videos of the iconic cat captivate social media users.

Social media comprises more than personal online profiles and tweeted quotes; tools such as crowd sourcing provide opportunities for large numbers of people to contribute information or funds to a specific campaign or project.

Other Online Tools

Conservation information is also shared online via blogs, podcasts, and websites. Through blogs and podcasts, individuals or organizations share their messages with more targeted audiences than traditional broad spectrum media. In all media formats, from large news outlets to individuals, accidental or intentional misinformation is always a risk. Because blogs are not held to the same reporting rigor as traditional journalism, they may rely on biased stories or be biased themselves, and individual bloggers are not constrained by editorial policy or advertisers.

Websites and email updates from conservation organizations can be powerful resources to inform the public and effectively share successes and stories from the field. These websites are searched out by individuals and therefore reach a highly engaged audience, provide detailed and potentially more accurate information, and often contain specific calls to action including financial contributions. Users are often able to sign up for, or unsubscribe from, email alerts and newsletters, but NGOs must be careful not to overload recipients, and risk numbing recipients to their cause.

Popular Media

Popular books and documentaries help capture the public's imagination regarding snow leopards. Peter Matthiessen's classic book *The Snow Leopard* (1978) established the mystique of the elusive cat and remains in print and available electronically. It continues to be one of the most commonly mentioned sources of information on the species by the general public. The BBC's *Planet Earth* series (2006) reached millions of viewers with breathtaking first footage of a snow leopard hunt, while educating the public on biodiversity in general. Wildlife conservation-oriented documentaries and films are an excellent method of engaging the public. Such films even command their own festivals such as the annual international Jackson Hole Wildlife Film Festival (USA), the Wildscreen Festival in Bristol, UK, and others around the globe.

Challenges in Communicating with the Public

The ultimate outcome of effective conservation messaging is inspiring the public to actively participate in snow leopard conservation in various ways, including volunteering with conservation organizations, donating money, or participating in ecotourism by visiting homestays in snow leopard range areas. However, snow leopard conservation is a complicated topic involving multiple parties: NGOs, universities, governments, and people living in range countries. Each group has different, overlapping, or sometimes opposing agendas and priorities. It is critical that communicators clarify their specific conservation message to avoid ambiguity and set themselves apart from other groups.

Challenges in communicating conservation messages to the press will always exist, especially when they contain complex scientific concepts. To mitigate this danger, press releases could be vetted by a nonscientist before distribution and strive to be free of jargon. The public must be able to understand and engage with a news story, while at the same time respect the science-based information and opinions represented.

There is a place for factual, scientific news stories, and for advocacy communications that use more passionate language. With social media, enlisting fans to help distribute a story by sharing is important, but runs the risk of "slacktivism" wherein people perform a token action and therefore feel their civic duty is finished without taking concrete action (Rotman et al., 2011). Social media also poses the risk that users only superficially educate themselves before losing interest and moving on to the next emotive cause.

It is vital to convey snow leopard conservation messages to the public within snow leopard range countries to not only educate, but to foster feelings of national pride in the presence of this species. This task proves especially challenging as access to media can be difficult due to bureaucracy, legal restrictions, censorship, language and cultural barriers, and lack of infrastructure. Availability of the Internet is increasing, allowing greater connections with the general public, although many rural areas still remain out of reach. Language barriers are being reduced online as free translation services become more dependable and widely used, but are not wholly reliable.

Key Conservation Messages for the Public

Conservation messages for the general public have two distinct goals:

- Educating the public regarding the role of the snow leopard in its Asian mountain ecosystems
- Instilling a sense of responsibility toward the species and environment as a whole

As a charismatic and iconic species, snow leopards easily capture the public's attention. People generally appreciate large and attractive animals, and can often be moved by emotive imagery and information. Snow leopards have benefited greatly by this status. They sustain two species-specific NGOs in the United States (the Snow Leopard Trust [SLT] and Snow Leopard Conservancy [SLC]), as well a dedicated program at Panthera, a felid-specific NGO, among others. In addition to species-specific conservation, snow leopards are frequently used alongside giant pandas, polar bears, and other appealing species by global conservation NGOs such as World Wildlife Fund for Nature to inspire the public to action and to raise funds. Using snow leopards to convey more general conservation goals, these organizations work to instill a culture of conservation and convey the message that endangered species conservation is everyone's responsibility.

Positive Messaging

Studies by psychologists and conservation researchers have explored the efficacy of diverse communication approaches. Most now suggest that messages focused purely on environmental destruction are counterproductive because people become overwhelmed, apathetic, and uninspired. These studies suggest the message is more powerful, and more likely to encourage engagement and action, when doomsday stories of species extinction and habitat loss are balanced with those of hope and success (Harre, 2011; Swaisgood and Sheppard, 2010). However, optimistic narratives should not ignore the threats thereby justifying the status quo. Rather, like the Common Cause for Nature Report (Blackmore and Holmes, 2013) suggests, conservation communications should be clear about both problems and solutions so people understand why action is important. The report suggests the most effective communications illustrate the intrinsic value of nature, highlight actions that can be taken at both individual and community levels, encourage caring for other people, and inspire creativity in how they show support for conservation. By sharing stories with a positive focus on local people and incorporating inspiring photographs and videos, the public is able to see that their active participation and support can make a difference. The past 20 years produced many positive snow leopard stories, including the use of remote camera trapping technology that has documented higher numbers of snow leopards than expected in many study areas. Other positive news includes the success of NGOs encouraging people in snow leopard habitat to see benefits in living harmoniously with the animals. As the Snow Leopard Survival Strategy 2014 (Snow Leopard Network, 2014) shows, there is still real and significant work to be done on snow leopard research and conservation, however organizations should not be tempted to inflate bad news in order to attract funding and support for needed research and conservation activities.

COMMUNICATING CONSERVATION MESSAGES WITHIN THE SCIENTIFIC COMMUNITY

The scientific component of the snow leopard community includes professional scientists and conservationists, whether independent or associated with an NGO, university, or government agency. It may also include students in postgraduate programs. Their communications often pertain to:

- Current status of the species in all range countries, including threats and protection levels
- Ongoing research and conservation projects
- Knowledge and data sharing

Communication within this group is critical for the timely exchange of ideas and research techniques, to ensure that limited funding is utilized most effectively, to generate support for research and conservation initiatives, to facilitate collaboration, and to support advocacy in government.

Traditional methods of interaction within the scientific community, aside from direct personal communication, occur predominantly via publications and conferences. Publications allow data transfer between professionals who keep current via journals (hard copy and electronic), online newsletter subscriptions, university affiliations, and web-based alerts and networks. Conferences stimulate idea exchange and encourage future personal communication, and they have been especially successful for the professional snow leopard community.

Conferences

A series of eight international snow leopard symposia were conducted between 1978 and 1995. They were held in Finland (1978), West Germany (1980 and 1984), the United States (1982), India (1986), the USSR (1989), China (1992), and Pakistan (1995). The proceedings from four symposia were published in the International Pedigree Book of Snow Leopards (Blomqvist, 1978, Blomqvist, 1980, Blomqvist, 1982, Blomqvist, 1989) and the rest were published strictly as conference proceedings (Blomqvist, 1984; Fox, 1994; Freeman, 1986; Jackson and Ahmad, 1997).

The ninth symposium, the Snow Leopard Survival Summit, was held in Seattle, Washington, USA, in May 2002, with a record 65 participants from 17 countries. In contrast, just 14 participants from 6 countries were able to attend the very first symposium (L. Blomqvist, 2015, Nordens Ark, Sweden, personal communication). The 2002 Summit was hosted by the International Snow Leopard Trust (ISLT) and Woodland Park Zoo. The meeting led to two significant and innovative conservation outcomes: The first was the Snow Leopard Survival Strategy (McCarthy and Chapron, 2003), a document that identified threats to snow leopards and strategies to combat them through research, conservation actions, and national action plans. Second, the Snow Leopard Network (SLN), a global community of professionals and organizations that facilitates communication between snow leopard researchers and conservationists, was created (more information follows).

The conference on Range-wide Conservation Planning for Snow Leopards took place in Beijing in 2008. More than 100 people from 17 countries attended the conference with the goal of initiating range-wide conservation planning for snow leopards (see Chapter 44) and using expert knowledge to update range maps (see Chapter 3). The exercise clearly demonstrated where knowledge of snow leopard status is available or missing, allowing a more comprehensive and targeted focus to future research and conservation efforts (Williams, 2008).

The Global Snow Leopard Forum was held in October 2013 in Bishkek, Kyrgyz Republic (see Chapter 45), hosted by the World Bank's Global Tiger Initiative to specifically engage government officials from the range countries in addition to scientists and conservationists. This conference resulted in the joint Bishkek Declaration between all range country governments committing to collaborative transboundary conservation policies as articulated in the Global Snow Leopard and Ecosystem Protection Program (GSLEP) (Snow Leopard Working Secretariat, 2013).

The Snow Leopard Network (SLN)

SLN is an international consortium of experts, practitioners, and organizations dedicated to facilitating "sound scientifically-based conservation of the endangered snow leopard through networking and collaboration between individuals, organizations, and governments" (http://snowleopardnetwork.org). A primary aim of SLN involves updating and supporting the implementation of the Snow Leopard Survival Strategy (McCarthy and Chapron, 2003; Snow Leopard Network, 2014), and it facilitates research and communication to this end. The updated SLSS (2014) is now available as a living document on the SLN website.

SLN members include scientists, government officials, conservation organizations, and members of the public from around the world. It is run by volunteers and one part-time staff coordinator. Membership is free and has grown from 65 founding members to more than 600 individuals and organizations. A 7-member Steering Committee elected every 3 years provides leadership. In addition, SLN operates the Snow Leopard Conservation Grant Program, which supports education, research, or conservation projects on snow leopards that meet the needs identified in the SLSS. The SLN also hosts an extensive bibliography of articles, a news feed, and reports from grantees. Much of the website is available to the general public, but also includes a members-only area. All web content is available in English and many pages are available in the range country languages of Chinese, Russian, Nepali, and Mongolian.

Challenges in Communicating with the Scientific Community

Even with the efforts of individual professionals and a targeted organization such as the SLN, challenges still exist in communicating within the scientific community. Language boundaries can present challenges in sharing information, and yet it is imperative to make crucial and timely data accessible to non-English-speaking professionals working within range countries. Professional rivalries, refusal to share data, and differences in opinion over scientific rigor may occur, which could result in the misinterpretation or distortion of information. Sharing information in a professional manner between scientists is critical without isolating or targeting individuals or organizations.

COMMUNICATING THE CONSERVATION MESSAGE WITH GOVERNMENT

Maintaining communications with policymakers and various government bodies is essential to help inform and influence conservation policy and to maintain working relationships that enable further research.

Methods of Communicating with Government

Professionals working in snow leopard conservation communicate directly with governments of the 12 snow leopard range countries through personal connections, informal channels, and via intergovernmental agreements such as the Global Snow Leopard and Ecosystem Protection Program (GSLEP). Communication also takes place between international NGOs and governments to ensure that officials receive current information regarding snow leopard status and conservation best practices. NGOs and government agencies also regularly form active conservation and research partnerships.

Because range country governments are ultimately responsible for conservation policy within their borders, it is essential for snow leopard researchers and conservation NGOs to maintain strong relationships with governments to help facilitate appropriate development, implementation, enforcement, and improvement of conservation policy. Scientists and NGOs can also provide input into the creation of protected areas and cooperation with adjacent countries.

Methods for communicating conservation messages to government depend on the topic and the setting. Conferences such as the 2013 Global Snow Leopard Forum in Bishkek allow for discourse between governments and the scientific community and private sector. Ideally, these forums not only inform official policy, but also allow for ad-hoc and one-on-one meetings with government officials in attendance.

Challenges in Communicating with Governments

Establishing dialogue with governments regarding snow leopard conservation can be challenging. Some officials may not have the necessary background or knowledge to adequately assess conservation threats and proposed projects. When political agendas do include conservation programs, governments must contend with the competing needs of many species and ecosystems. Where conservation is not a priority, governments may focus on interests that often conflict with conservation goals, such as road construction or mining. Additionally, the variety of political structures and turnover of government officials in range countries can alter the political landscape so dramatically that conservation work supported by one administration may be rejected by the next. Ongoing engagement is essential to ensure collaboration with and between governments and effective conservation policy within range countries.

CONCLUSIONS

Looking forward, communication through technological channels should continue to improve and language barriers be further reduced, making it possible to affect ever-growing audiences. However, reaching more people is only part of the goal: the desired outcomes of snow leopard conservation messages are often about changing long-held perceptions and behavior, in terms of engendering a sense of pride and responsibility to the snow leopard and the ecosystem in which it resides. Behavioral changes are often based on personal engagement and trust. Therefore, it is important to continue to engage with politicians, scientists, individuals, and communities in snow leopard range and in the broader international community. Although challenges to effective communication will continue, the snow leopard conservation community should expect and encourage greater engagement with the general public, scientific community, and governments, resulting in increasingly effective conservation programs facilitating the survival of the snow leopard.

References

Blackmore, E., Holmes, T. (Eds.), 2013. Common Cause for Nature: Values and Frames in Conservation. http://valuesandframes.org/initiative/nature (accessed 15.11.2014).

Blomqvist, L. (Ed.), 1978. First International Snow Leopard Conference, March 7–8, 1978, Helsinki. Proceedings published in International Pedigree Book of Snow Leopards, *Panthera uncia*, vol. 1, Helsinki Zoo, Helsinki.

Blomqvist, L. (Ed.), 1980. Second International Snow Leopard Conference, October 9–10, 1980, Zurich Zoo. Proceedings published in International Pedigree Book of Snow Leopards, *Panthera uncia*, vol. 2, Helsinki Zoo, Helsinki.

Blomqvist, L. (Ed.), 1982. Third International Snow Leopard Symposium, June 22–25, 1982, Woodland Park Zoo, Seattle. Proceedings published in International Pedigree Book of Snow Leopards, *Panthera uncia*, vol. 3, Helsinki Zoo, Helsinki.

Blomqvist, L. (Ed.), 1984. Fourth International Snow Leopard Symposium, September 20–21, 1984, Krefeld Zoo, West Germany. Helsinki Zoo, Helsinki.

Blomqvist, L. (Ed.), 1990. Sixth International Leopard Symposium, October 2–7, 1989, Alma-Ata, Kazakhstan, USSR. Proceedings published in the International Pedigree Book of Snow Leopards, *Panthera uncia*, vol. 6, Helsinki Zoo, Helsinki.

Fox, J. (Ed.), 1994. Seventh International Snow Leopard Symposium, July 25–30, 1992, Xining, Qinghai, Peoples Republic of China. Proceedings published by International Snow Leopard Trust, Seattle, WA, USA and Chicago Zoological Society, Chicago.

Freeman, H. (Ed.), 1986. Fifth International Snow Leopard Symposium, October 1986, Srinagar, India. International Snow Leopard Trust, Seattle.

Harre, N., 2011. Psychology for a Better World: Strategies to Inspire Sustainability. http://www.psych.auckland.ac.nz/uoa/home/about/our-staff/academic-staff/niki-harre/psychologyforabetterworld (accessed 15.11.2014).

Jackson, R., Ahmad, A., 1997. Eighth International Snow Leopard Symposium, November 12–16, 1995, Islamabad, Pakistan. Proceedings published by International Snow Leopard Trust, Seattle and WWF-Pakistan, Islamabad.

Matthiessen, P., 1978. The Snow Leopard. Viking Press, New York.

McCarthy, T.M., Chapron, G. (Eds.), 2003. Snow Leopard Survival Strategy. Snow Leopard Trust, Seattle, 125 pages.

Planet Earth: Mountains, 2006, television series, British Broadcasting Company Natural History Unit, United Kingdom.

Rotman, D., Vieweg, S., Yardi, S., Chi, E., Preece, J., Shneiderman, B., Pirolli, P., Glaisyer, T., 2011. From Slacktivism to Activism: Participatory Culture in the Age of Social Media. Proceedings of the 2011 Annual Conference on Human Factors in Computing Systems extended abstracts, New York, pp. 819–822.

Snow Leopard Network, 2014. Snow Leopard Survival Strategy, Revised 2014 Version. http://www.snowleopardnetwork.org/docs/Snow_Leopard_Survival_Strategy_2014.1.pdf (accessed 02.07.2015).

Snow Leopard Working Secretariat, 2013. Global Snow Leopard and Ecosystem Protection Program (GSLEP). Snow Leopard Working Secretariat; Bishkek, Kyrgyz Republic.

Swaisgood, R., Sheppard, J.K., 2010. The culture of conservation biologists: show me the hope! BioScience 60 (8), 626–630.

Williams, N., 2008. International conference on range-wide conservation planning for snow leopards: saving the species across its range. Cat News 48, 33–34.

CHAPTER

44

Global Strategies for Snow Leopard Conservation: A Synthesis

Eric W. Sanderson, David Mallon**, Thomas McCarthy†, Peter Zahler*, Kim Fisher**

*WCS Asia Program, Wildlife Conservation Society, New York, NY, USA

**Division of Biology and Conservation Ecology, Manchester Metropolitan University, Manchester, UK

†Snow Leopard Program, Panthera, New York, NY, USA

INTRODUCTION

The idea of a strategy derives from the ancient Greek concept of "strategia," or the art of the troop leader; that is, strategies are matter of command and generalship (Thesarus Linguae Graecae, 2011). Generals concern themselves with high-level decision making to obtain a specified objective, especially under conditions of uncertainty and limited resources where not every option can be pursued simultaneously. In military terms, strategy is distinguished from tactics, which are the subset of skills required to meet the objective, such as logistics or intelligence. The term first came into use in the Eastern Roman Empire around the sixth century and was only translated into the Western vernacular in the nineteenth century. Similar ideas have been articulated by great leaders in other parts of the world as well. In more recent times, the idea of strategy has left a purely military realm and entered politics, business, sports, and even conservation.

A good strategy includes articulating a vision, setting one or more time-bound goals, determining actions to achieve that goal, and mobilizing resources to execute those actions (c.f. Groves et al., 2003). The strategy therefore describes how the ends are achieved by the means, and enables leaders to deploy resources wisely and effectively

Snow Leopards
http://dx.doi.org/10.1016/B978-0-12-802213-9.00044-4

while coordinating among partners. Sun Tzu, the famous Chinese strategist, wrote in the *Art of War* (2007) that unity, not size, is a source of strength. Mintzberg et al. (2002) defined strategy as "a pattern in a stream of decisions" not simply planning, while Mckeown (2012) argues that "strategy is about shaping the future," achieving "desirable ends with available means."

Conservation of biological diversity is a desirable end, but it is sadly limited in its available means, especially given the rather ambitious goal of preserving the diversity of life on Earth for future generations (Wilson, 2003). Because conservation leaders in governments and other institutions constantly face decisions in a climate of uncertainty and limited resources, strategy is critical to the practice of conservation. Strategic planning often occurs as a mechanism for building consensus, enthusiasm, and support for conservation of a species (Sanderson et al., 2002). In the past, strategic planning efforts have led to renewed investments for tigers (ExxonMobil, 2000) and jaguars (Society for Conservation Biology, 2002).

Most species are not so lucky. The vast majority of species lack explicit strategies for their conservation (Species Conservation Planning Task Force, 2008). As a result, conservation goals for most of biodiversity lack specificity beyond the implicit goal suggested by the IUCN Red List categories; that is, for taxa to be listed as "Species of Least Concern" and not something else (IUCN SSC, 2000). Least Concern species are not endangered, threatened, or vulnerable because they do not face a high risk of extinction in the wild.

Avoiding imminent extinction does not seem to be a high bar for conservation (c.f. Westwood et al., 2014; Neel et al., 2012; Sanderson, 2006), but the finality of species extinction has successfully mobilized policy tools such as the US Endangered Species Act and the IUCN Red Listing process and continues to be an ideological mainstay of conservation (Neel et al., 2012; Ladle and Jepson, 2008). Snow leopards are currently listed as endangered under the IUCN Red List process (Jackson et al., 2008), though one might question this listing, given that the species remains widely distributed in multiple populations (see Chapter 3).

Although most conservationists would agree that avoiding species extinction is an immediate goal, it should not be the end goal for conservation (Sanderson, 2006). Redford et al. (2011), for example, suggest a more affirmative definition for successful species conservation. They write that any species that is successfully conserved will have the following characteristics: (i) be self-sustaining demographically and ecologically, (ii) be genetically robust, (iii) have healthy populations, (iv) have representative populations distributed across the historical range in ecologically representative settings, (v) have replicate populations within each ecological setting, and (vi) be resilient (i.e., able to continue to express key demographic, genetic, behavior, and ecological attributes even when disturbed by climate change or other factors) across the range.

Snow leopards (*Panthera uncia*) are a species where considerable effort has been expended on strategic thinking by governments, conservation organizations, academic institutions, and development agencies, and where many of the components of the Redford et al. (2011) definition of successful conservation might be usefully applied. Just since 2000, four global strategies have been developed for the species: the Snow Leopard Survival Strategy (McCarthy and Chapron, 2003); the Snow Leopard Range-wide Assessment and Conservation Planning meeting in Beijing, China, in 2008 (Williams, 2008: McCarthy et al., 2009; see Chapter 3); the Global Snow Leopard and Ecosystem Protection Program (Snow Leopard Working Secretariat, 2013; see Chapter 45); and the Snow Leopard Survival Strategy, Revised Version 2014.1 (Snow Leopard Network, 2014). Hereafter these will be referred to as SLSS 2003, the SLRAC 2008, GSLEP 2013, and SLSS 2014, respectively. Many of these strategies pick up elements of the Redford et al. (2011)

prescription for successful conservation, while the exact elements and their construction into strategic formulations shows how snow leopard conservationists have evolved their strategic thinking over the past two decades.

This chapter identifies areas of convergence and highlights points of divergence among these strategies by considering their goal/vision statements, especially those that address the "why," "where," and "how" of snow leopard conservation. "Why" refers to why people should invest effort in conserving the species; answers to "why snow leopard conservation" thus express the values the snow leopard strategists hold for snow leopards. Convergence of values among governments, scientists, conservationists, and communities suggests a basis for cooperation. "Where" refers to geographic representations of snow leopard range and conservation sites in the four conservation strategies. Understanding the distribution of snow leopard across an immense range (>3,000,000 km^2), and showing how that range intersects with nation states and ecological settings, suggests where conservation efforts should be optimally distributed from a strategic perspective. "How" refers to the conservation activities necessary to fulfill the vision of conservation of the species. If why and where are strategic elements, then how brings in the logistics, including the kinds of expertise necessary for conservation of this species. More importantly, bringing the why, where, and how of snow leopard conservation together suggests an emerging strategic synthesis for snow leopard conservation against which future progress can be measured.

SNOW LEOPARD STRATEGIES

Snow Leopard Survival Strategy (SLSS 2003)

In 2002, approximately 58 snow leopard specialists, including researchers, advocates, and conservationists from range states and elsewhere met in Seattle, USA, for the Snow Leopard Survival Summit. In a foreword to the first SLSS, Urs and Christine Breitenmoser wrote "that across mountain ridges and deserts, national borders and cultural barriers – the common values for which we fight, far outweigh the few differences that separate us." What those values were was not clearly expressed, but the Executive Summary emphasizes that the SLSS is necessary to "save the endangered snow leopard" and the title of the volume (a "survival strategy") suggests the main driver was avoiding extinction. This focus made sense at the time, because the snow leopard was listed as endangered again in 2002, just as it had been in 1986, 1988, 1990, 1994, and 1996 (Baillie and Groombridge, 1996; Groombridge, 1994; IUCN, 1990; IUCN Conservation Monitoring Centre, 1988; IUCN Conservation Monitoring Centre, 1986). As of July 2015, the snow leopard was still considered endangered. The latest assessment by Jackson et al. (2008) listed the species as endangered because the species was "suspected to have declined by 20% over the past two generations (16 years) due to habitat and prey base loss, and poaching and persecution," and the global snow leopard effective population size was suspected to be fewer than 2500 individuals (Jackson et al., 2008).

Given the IUCN status of the species, the SLSS 2003 set out the following tactical objectives (note, in this chapter, we distinguish between goals, which refer to the desired state of the species at some point in time, and objectives, which are activities to help achieve a desired state, which is somewhat different terminology that used by the SLSS 2003; for further discussion of the confusing terminology of conservation planning, see Sanderson, 2006, or Groves et al., 2003):

- Assess and prioritize threats to snow leopards on a geographic basis.
- Define and prioritize appropriate conservation, education, and policy measures appropriate to alleviate threats.

FIGURE 44.1 **Ecological settings defined for potential snow leopard range in 2008 at the Snow Leopard Range-wide Assessment and Conservation Planning (Williams 2008).** Ecological settings are portions of the range where snow leopards have a distinct relationship to the prey and ecosystems where they live. Potential range is defined in Chapter 3.

- Prioritize subjects for snow leopard research and identify viable and preferred research methods.
- Build a network of concerned scientists and conservationists to facilitate open dialogue and cross-border cooperation.
- Gain consensus on a fundamental Snow Leopard Survival Strategy document that will be made available to the range states in conservation planning at national and local levels.

The SLSS document itself satisfied the first three and the fifth objectives upon publication. And the process of writing the SLSS 2003 was vital to creation of the Snow Leopard Network, which remains an active organization, satisfying the fourth objective (see Chapter 43). On its own terms, the SLSS 2003 was an immediate success.

Geographically, the SLSS 2003 divided and assessed snow leopard range by political units (i.e., range states, or the countries where snow leopards have been reported). Status assessments were provided for 12 range states and one possible (Myanmar) (Fig. 44.1), and the legal status was reported for those 12 states. Threats and research needs were evaluated in four snow leopard regions, which were agglomerations of nation states or parts of nation states. The SLSS 2003 thus set the precedent for thinking about snow leopard populations in biogeographic groupings that crossed international boundaries. These regions included the Himalaya (abbreviated HIMLY), including the Tibetan Plateau and other parts of southern China, India, Nepal, and Bhutan; the Karakorum and Hindu Kush Range (KK/HK), including Afghanistan, Pakistan, and southwest China; the Commonwealth of Independent States and western China (CISWC), including Uzbekistan, Tajikistan, Kyrgyzstan, Kazakhstan, and Xinjiang Province in China; and the northern snow leopard range (NRANG), which included China's Altai and Tien Shan Mountains, Mongolia, and Russia.

SLSS 2003 included a short section on "potential actions to address threats" that followed an extensive analysis of the threats to snow leopard conservation. The potential actions considered were grazing management, income generation (for people living near snow leopards), cottage industries, an ungulate trophy hunting program (recognizing the competition between human hunters and snow leopards for the same prey), reducing poaching and trade in snow leopard parts, reducing livestock depredation by snow leopards, animal husbandry (for livestock), and conservation education and awareness. Each action type was addressed at the policy and local community level. The remainder of the strategy was given to identification of research and information needs and a short section range state action planning.

Snow Leopard Survival Strategy, Revised Version 2014.1 (SLSS 2014)

The Snow Leopard Network, which formed after the first SLSS meeting, published an updated SLSS in 2014. Interestingly, in the update the values for snow leopards were recast in a more positive light and to give a wider perspective, in keeping with the GSLEP 2013 (Global Snow Leopard and Ecosystem Protection Program; see following), which was being developed concurrently. Endangerment is not mentioned; rather SLSS 2014 emphasizes the iconic nature of the cat and its ecological dependency on prey species and healthy rangelands. The passage is worth quoting in full:

> The iconic snow leopard is the least known of the "big cats" due to its elusive nature, secretive habits and the remote and challenging terrain it inhabits. As an apex predator, its survival depends on healthy populations of mountain ungulates, the major prey; these in turn depend on the availability of good-quality rangeland minimally degraded by concurrent use from livestock and humans. The snow leopard has a large home range size, so viable populations can be secured only across large landscapes. The snow leopard therefore represents the ideal flagship and umbrella species for the mountain ecosystems of Asia.
>
> Snow leopards share their range with pastoral communities who also require healthy rangelands to sustain their livestock and livelihoods. Moreover, these high altitude mountains and plateaus provide invaluable ecosystem services through carbon storage in peat lands and grasslands, and serve as Asia's 'water towers', providing fresh water for hundreds of millions of people living downstream in Central, East and South Asia.

From this basis, SLSS 2014 reiterated the first three action-oriented goals of SLSS 2003 for the 2014 revision. No meeting was held because the process was conducted in parallel with the GSLEP 2013 strategy, which included a large international meeting in October 2013 of many of the same people and institutions. The revised SLSS provided updated status assessments for 12 countries and mentioned Myanmar as a possible, but unlikely, range state. Threats and research needs were addressed as a series of chapters on a range-wide scale, though some chapters and the three appendices included data by nation and subnation (e.g., regions within China), including a valuable list of camera trapping studies and a list of protected areas where snow leopards are known to occur. In general the approach to snow leopard geography emphasized political boundaries based on range states.

SLSS 2014, like its predecessor, included a review of threats to the species, along with a revised set of action items for each threat. Chapters were dedicated to explaining and suggesting action-oriented responses to livestock competition, illegal trade, climate change, and large-scale infrastructure development, including mining, electrical power infrastructure, and linear barriers (e.g., railroads, highways, fences). A separate chapter titled "Conservation Actions" focused on community-based conservation efforts to help local people see the presence of wild snow leopards as beneficial or at least neutral. These include handicrafts, savings and credit programs, corral improvements, livestock insurance, veterinary assistance, ecotourism, and education and awareness-raising.

Snow Leopard Range-wide Assessment and Conservation Planning (2008)

In 2008, a similar but not identical group of researchers, conservationists, and government officials met in Beijing, China, at a meeting organized by Panthera, the Snow Leopard Network, Snow Leopard Trust (SLT), and the Wildlife Conservation Society (WCS) (Williams, 2008; McCarthy et al., 2009; see Chapter 3). The purpose of this meeting was to develop a range-wide assessment and conservation vision for the snow leopard, using procedures described by the SSC (2008) and in related planning efforts for large wild felids, including jaguars (Sanderson et al., 2002), lions (IUCN SSC Cat Specialist Group, 2006), cheetahs (Durant, 2007), and tigers (Sanderson et al., 2006; Dinerstein et al., 2007).

Working together, this group prepared a conservation vision statement for the snow leopard, explicitly linked back to datasets developed through the range-wide assessment:

> A world where snow leopards and their wild prey thrive in healthy mountain ecosystems across all major ecological settings of their entire range, and where snow leopards are revered as unique ecological, economic, aesthetic and spiritual assets.

A vision like this one represents an aspirational state, which may never be realized, but whose expressed values guide conservation efforts. In this vision statement, snow leopards, their wild (as opposed to domesticated) prey, and healthy mountain ecosystems are considered a joint unit for conservation. Multiple instances of this unit are valued across different "ecological settings" that collectively encompass the entire range. In other words, one example population, or even several in the same ecological circumstances, would not satisfy the vision statement. Rather, variable combinations of snow leopards, prey, and "healthy mountain ecosystems" are desired. At the Beijing meeting, these were described as regions where snow leopards display ecological characteristics or interactions not found in other parts of the range. These could include different habitat usage patterns, different prey base, different home ranges or dispersal patterns, or different behavioral repertoires. To bring even greater clarity, the vision statement was directly linked to a map that described seven ecological settings that collectively span the potential range of the species. The experts assembled at that meeting created consensus maps displaying snow leopard conservation units (SLCUs), ecological settings, and the potential range of the species (Fig. 44.2) (see Chapter 3).

The ecological values of snow leopards were explicitly coupled with other kinds of values, including both material (ecological and economic) and nonmaterial (aesthetic and spiritual) benefits to people. The language is unusual, coupling a spiritual concept, "reverence," with an economic one, "assets."

After consideration of a conservation vision for the species, the research community was joined by representatives from most of the range states to develop country-specific action plans. Those action plans highlighted a wide variety of conservation actions for snow leopards as indicated in Table 44.3.

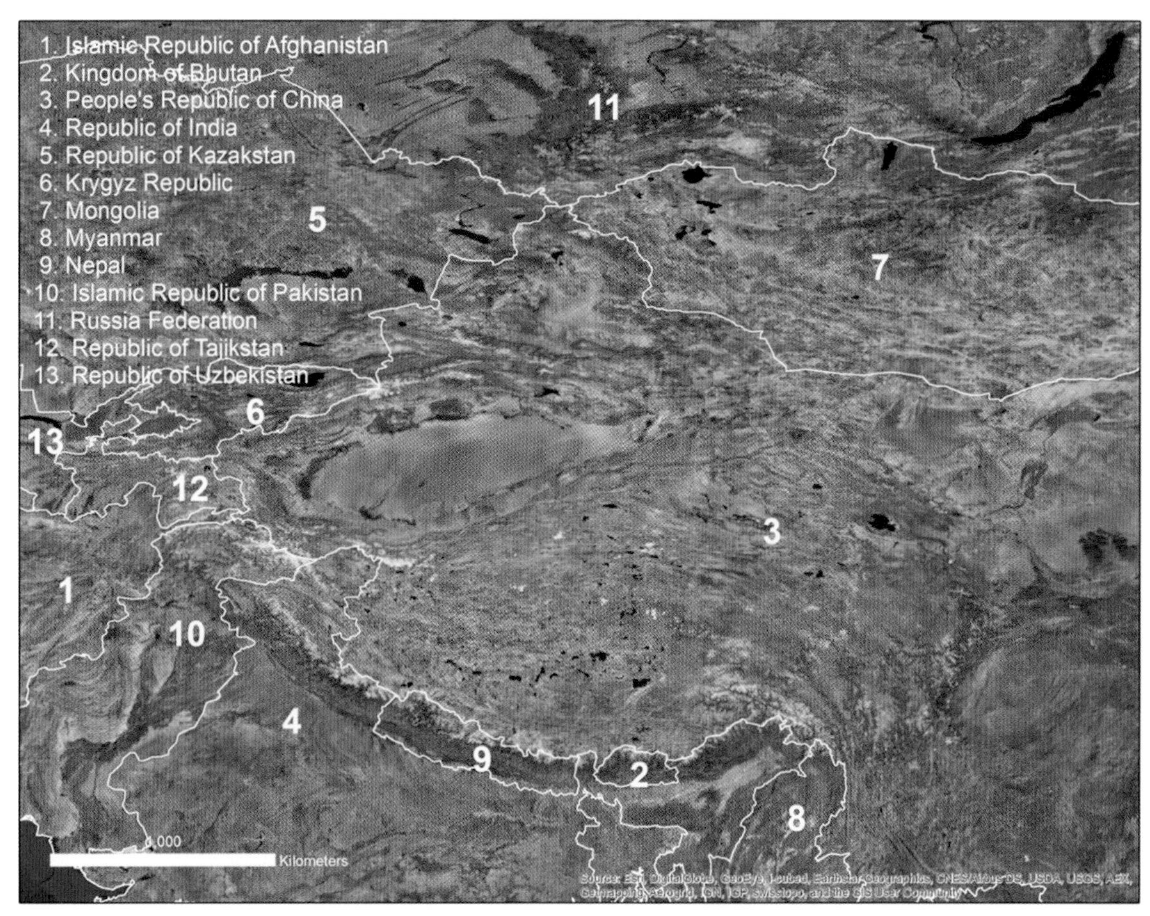

FIGURE 44.2 **Snow leopard range states.** Range states are countries that overlap the potential distribution of snow leopards as shown in Figure 44.1.

Global Snow Leopard and Ecosystem Protection Program (GSLEP) (2013)

In October 2013, representatives of 12 range state nations and a variety of snow leopard conservation and ecosystem protection partner organizations (similar but not identical to the groups that produced the SLSS 2003 and the SLRAC 2008) met in Bishkek, Kyrgyz Republic, to launch a new international effort to save the snow leopard and conserve high-mountain ecosystems (see Chapter 45). This process was aimed at creating a policy framework for snow leopard conservation among range state governments. To that end, an important output of the meeting was the jointly agreed Bishkek Declaration on the Conservation of Snow Leopards, which acknowledged that snow leopards were:

> ". . . an irreplaceable symbol of our nation's natural and cultural heritage and an indicator of the health and sustainability of mountain ecosystems" and recognized that mountain ecosystems inhabited by snow leopards provide "essential ecosystem services, including storing and releasing water . . ., sustaining pastoral and agricultural livelihoods … ; and offering inspiration, recreation, and economic opportunities."

Thus the Bishkek Declaration creates a symbolic relationship between the conservation of snow leopards and the ecosystem services provided by snow leopard habitat.

The Bishkek Declaration was published with an extensive report describing the new program. The first chapter expresses the same idea as the SLRAC 2008 that snow leopards, their wild prey, and their ecosystems form a joint unit for conservation. It goes on to elaborate on the various ecosystem services provided to people in Central and South Asia by this unit, including cultural services, water services, biodiversity, medicine, agro-pastoralism, carbon sequestration and storage, and recreation and economic opportunities. The report argues that "many of the threats to snow leopards and to their prey and ecosystems have the potential to degrade the provisioning of these ecosystem services."

To address where to conserve snow leopards, the GSLEP described an explicit goal called "20 by 2020," which states:

> The goal of GSLEP is for the 12 range countries, with support from interested organizations, to work together to identify and secure 20 snow leopard landscapes across the big cat's range by 2020.

Secure snow leopard landscapes are defined as those that (i) contain at least 100 breeding age snow leopards conserved with the involvement of local communities, (ii) support adequate and secure prey populations, and (iii) have functional connectivity to other snow leopard landscapes, some of which cross international boundaries (Snow Leopard Working Secretariat, 2013). A proposed set of 23 landscapes were described on a map that collectively ensure that the 12 nations that have supported the Bishkek Declaration each have at least one area important for snow leopards. Conservation of these areas is described under the slogan: "Secure 20 by 2020."

Actually securing snow leopard landscapes of course requires actions on the ground. The Bishkek Declaration recognized this by describing a set of five direct impact activities and three enabling activities for snow leopard conservation, namely:

1. Engaging local communities in conservation, including promoting sustainable livelihoods, and addressing human-wildlife conflict
2. Managing habitat and prey based upon monitoring and evaluation of populations and range areas
3. Combatting poaching and illegal trade
4. Transboundary management and enforcement
5. Engaging industry
6. Building capacity and enhancing conservation policies and institutions
7. Research and monitoring
8. Building awareness

These eight activities were further elaborated in the GSLEP 2013 as a set of "good practices for snow leopard, prey, and habitat conservation" and a further set of activity "portfolios."

WHY CONSERVE SNOW LEOPARDS?

The four strategies describe a set of broadly shared values for snow leopards, which have evolved over the first decade and a half of the twenty-first century (Table 44.1). In the SLSS 2003, the values of snow leopard conservation were largely implicit. A group of people who already shared in an interest in conservation and feared for the future of the species gathered together to make a strategy for conservation. Their efforts were largely tactically focused: What does the conservation community know about snow leopards? What are the most important threats and places? How do we cooperate better to save the species? Values and geographies were not expressed explicitly.

It is not surprising therefore that the SLRAC 2008 planning meeting attempted to fill gaps in the earlier work by making an explicit value statement through a vision statement, and a more exact description of the geography

TABLE 44.1 Comparison of Values for Snow Leopards in Four Twenty-First-Century Conservation Strategies

Conservation value	SLSS 2003	SLRAC 2008	GSLEP 2013	SLSS 2014
Avoid extinction; existence value	[X]	[X]	[X]	[X]
Ecological functionality of species		X		
Ecosystem services of habitat		[X]	X	X
Cultural significance		X	X	X
Representation on an ecological basis		X		
Representation on a national basis	[X]	[X]	X	X

[X], Implicit values in the strategies, as interpreted by the authors; X, Explicit values stated in the strategies.

through a well-established and globally recognized range-wide assessment process (e.g., Sanderson et al., 2002). This process carried with it an emphasis on ecological, as opposed to political, geography. This idea can trace its origins back to work on tigers in the 1990s (e.g., Dinerstein et al., 1997), which was replicated and refined through work on other large cats (e.g., jaguars – Sanderson et al., 2002; lions – Nowell and Bauer, 2006). The Beijing vision statement asserts the importance of the shared conservation units comprised of snow leopards and their prey and the mountain ecosystems where they live. Multiple instances of these units are desired for conservation across the range, an idea later expressed more generally for all species by Redford et al. (2011).

The GSLEP 2013 and SLSS 2014 further build and expand on the idea of a conservation unit by placing snow leopard conservation in the context of ecosystem services. The "reverence" for snow leopards as "assets" in the SLRAC 2008 vision has been complemented by a carefully articulated argument for ecosystem services delivered both to people who share snow leopard range and the millions that benefit from mountain "water towers" downstream. Nearly one-third of the world's current human population draws power, irrigation, industrial waters, and fishery benefits from snow leopard habitat (GSLEP, 2013).

SLSS 2014 represents the leading expression of this concept to date by chaining the ideas together logically. The snow leopard as a predator depends on wild prey (largely grazing and browsing ungulates), which in turn depend on functioning mountain ecosystems. Functioning mountain ecosystems also provide carbon storage, water, and agro-pastoral opportunities for local, often impoverished, communities.

It is clear from this logic, as elaborated in the four strategies and other chapters of this book, that snow leopards are dependent on functioning mountain ecosystems in Asia. It is perhaps less clear that the ecosystems are dependent on snow leopards as such (see relevant discussion in Ray et al., 2005). No one claims that snow leopards directly provide water or sequester much carbon. What depends on snow leopards in particular are the cultural values of the species, whether we describe these as "cultural services" or "revered assets" or "existence values." Snow leopards are beautiful, valuable, and iconic in their own right, and they are also symbolic of the beautiful, valuable, iconic ecosystems where they live.

WHERE TO CONSERVE SNOW LEOPARDS?

The strategies express two different ways of framing the geography of species conservation: conservation by nation state or conservation by ecological setting (Table 44.2). Three of the four strategies considered here adopt contemporary

TABLE 44.2 Geographic Treatment of Snow Leopard Range in Four Twenty-First-Century Conservation Strategies

Geographic description	SLSS 2003	SLRAC 2008	GSLEP 2013	SLSS 2014
By nation states	13*	11**	12***	13
By ecological region or setting	4 regions	9 ecological settings	None	None
Number of units, areas or landscapes	None	69 snow leopard conservation units	20 snow leopard landscapes	186 protected areas
Extent of potential range (km^2)	3,024,728	3,256,840	Not given	1,200,000–3,000,000+
Extent of occupied habitat (km^2)	1,846,000	1,230,881****	1,766,000	Various

* *Afghanistan, Bhutan, China, India, Kazakhstan, Kyrgyz Republic, Myanmar, Mongolia, Nepal, Pakistan, Russia, Tajikistan, and Uzbekistan.*
** *Kazakhstan and Myanmar were not represented.*
*** *Myanmar was not represented.*
**** *Definitive plus probable range (see Chapter 3).*

political geography as the frame for conserving a species that evolved over 2 million years (see Chapter 1). A political perspective dominates because human decision making about conservation is in fact political and state specific; no country has no formal influence on what happens beyond its boundaries except through international agreements between sovereign states (such as the Bishkek Declaration). Political cooperation is necessary because snow leopards often inhabit transboundary mountains, where high ridges demarcate polities as well as watersheds. Moreover cooperation is necessary, across nations and across sectors, also because the limited resources for snow leopard conservation and the immensity of the task require pooling resources together.

The main difficulty with politically defined goals for conservation is that they may be hindered by other aspects of the political process that have nothing to do with conservation (Leader-Williams et al., 2010; Czech et al., 1998). Other political priorities may and do intervene, leading to science being sidelined or ignored, and/or limiting participation. The GSLEP 2013 landscapes, for example, identify only a few marginal areas in China, despite the fact that science clearly indicates that broad swaths of China are important snow leopard habitat. The opposite can also occur. No one knows the scientific status of the small corner of potential range in Myanmar where the species might occur, so that country did not participate in the GSLEP process.

An ecological lens on snow leopard geography is complementary to a politically oriented system of conservation priorities and less subject to political factors that have nothing to do with conservation. Ecological conservation focuses on what makes a particular conservation unit (e.g., snow leopards, prey, ecosystem characteristics) distinctive in terms consistent with long-term co-evolution of the species and other aspects of the ecosystem (Dinerstein et al., 1997; Sanderson et al., 2002). Distinctiveness can be defined in terms of predator-prey relations, habitat use, behavioral differences, or for ecosystems, differences in provisioned ecosystem services (c.f. Hidasi-Neto et al., 2015). Once one has the concept of ecological distinctiveness within a species, and mechanisms to define it, then one can also express goals for conservation in terms of the ecology of the species. These logical interconnections are all deeply satisfying to scientifically minded conservationists. For example, in the SLRAC 2008, that expression is a vision to conserve areas important for snow leopards within the seven major ecological settings defined across the entire potential range of the species. The implicit rationale is that conservation of these areas will conserve snow leopards, evolutionary potential, ecological functionality, and

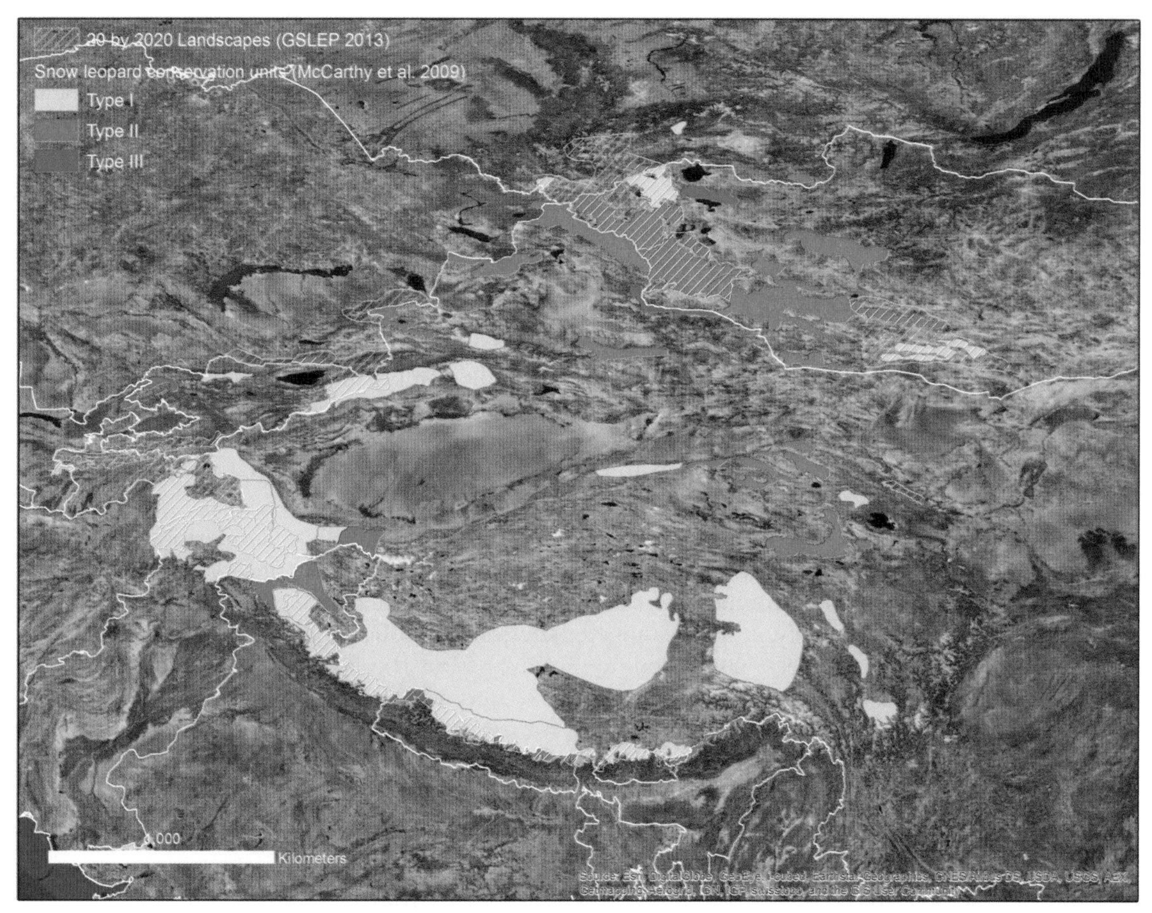

FIGURE 44.3 **Geographic overlap between the "20 by 2020 snow leopard landscapes" defined GSLEP 2013 (Snow Leopard Working Secretariat, 2013) and the snow leopard conservation units defined by SLRAC 2008 (Williams, 2008).** A snow leopard conservation unit was defined as a Type I if it contains a population of resident snow leopards large enough to be potentially self-sustaining over the next 100 years (note this implies a stable prey base by definition); Type II if it contains fewer snow leopards than Type I (i.e., not self-sustaining for 100 years), but with adequate habitat such that snow leopard numbers could increase if threats were alleviated; or Type III, if snow leopards are definitely extant in the area, but have inadequate habitat or wild prey, or the population is too small or too threatened to be considered viable over the long-term without large investments to reduce threats in the short term (see Chapter 3).

provide the widest possible diversity of ecosystem services. The disadvantage of the ecological approach to conservation is that no one person or organization has responsibility for snow leopards in, for example, the Himalayas. Cooperation is required.

The best of all possible worlds is to find overlap between ecological and political conservation priorities. Comparison of the snow leopard conservation units produced in Beijing and the snow leopard landscapes produced in Bishkek shows that there is substantial overlap, especially through the mountain ranges that ring the range (Fig. 44.3). There are many more SLRAC snow leopard conservation units (69) than the GSLEP's 23 "20 by 2020 landscapes," but the landscapes make up for that by crossing or encompassing multiple conservation units. The only area where there is not substantive agreement is in China, where the GSLEP process analysis shows only three landscape, while the conservation units from the SLRAC 2008 are numerous, varied, and in the case of the Tibetan Plateau, geographically enormous.

HOW TO CONSERVE SNOW LEOPARDS?

There is no one way to conserve snow leopards; there are lots of ways. One might expect a diversity of conservation approaches given the vast geography of the species, encompassing not only many different ecologically distinct mountain ranges, but also many different countries and cultures. The art of the conservation strategist is picking the right approach in the right place and the right time (often in an environment of constrained resources). Fortunately, a wide range of different activities have been suggested to conserve snow leopards – a virtual panoply of twenty-first-century conservation strategies – and many of them have been tried in different places, at different times, and by different organizations. Table 44.3 combines the suggested conservation activities from these four strategies into a coherent framework. Only three approaches to snow leopard conservation are shared across all four strategies: local community involvement, capacity building for snow leopard conservation, and improved implementation of treaty obligations (e.g., CITES, CMS) on behalf of signatory snow leopard range states. Fourteen other approaches are endorsed by three of the strategies. Detailing these shared approaches is beyond the scope of this chapter, but they are discussed in other chapters of this book (e.g., Chapters 13, 14, and 16).

A STRATEGIC SYNTHESIS

Given the evolving expressions of values for snow leopard conservation and the significant overlap in conservation geography among the four strategies, the snow leopard conservation community is well placed for a strategic synthesis. Table 44.1 shows how the values for snow leopard conservation have evolved over time, to the current emphasis on the cultural values of snow leopards and the ecosystem services of the mountain ecosystems they represent. Table 44.2 indicates the many opportunities for snow leopard conservation, which can be framed in overlapping political and ecological geographies. Table 44.3 suggests agreement on a small number of approaches to range-wide conservation (community engagement, capacity building, international agreements), and many other approaches that are recommended across multiple strategies.

On the basis of these similarities, we believe that the snow leopard conservation community is poised for a global strategic synthesis. We suggest the following vision statement, which draws from the four strategies and represents a shared aspirational state of the species in the future:

> A world where snow leopards and their wild prey thrive in healthy mountain ecosystems across the full extent of snow leopard's potential range and where those interconnected ecosystems provide ecological, economic, aesthetic and cultural benefits to people in the mountains and downstream watersheds.

As a goal, the show leopard conservationists and range states should commit to:

> By 2025, to have successfully conserved snow leopard populations with their wild prey in nationally and ecologically representative, healthy mountain landscapes that ensure range-wide connectivity and that provide abundant ecosystem services to local and downstream populations.

Expressing the goal in the context of the four strategies enables conservation "generals" working in governments, conservation organizations, and civil society to divide and conquer the conservation problem. Terms are now well defined. Everyone agrees that there should be

TABLE 44.3 Comparison of Suggested Actions for Snow Leopard Conservation in Four Twenty-First-Century Conservation Strategies

Actions to conserve snow leopards	SLSS 2003	SLRAC 2008	GSLEP 2013	SLSS 2014
Engaging local communities in conservation				
Use participatory approach, integrate community needs into snow leopard conservation	X	X	X	X
Employ community-based conflict mitigation/resolution		X	X	
Improve livestock husbandry, including monitoring of depredation events	X			X
Livestock insurance/compensation and vaccination programs linked to snow leopard conservation	X		X	X
Support construction of predator proof corrals	X		X	X
Encourage alternative livelihoods for local communities other than livestock grazing, including savings and credit programs and ecotourism	X		X	X
Work on climate change adaptation strategies with communities				X
Managing habitat and prey based on scientific monitoring				
Establish long-term monitoring of snow leopard, prey, and habitat		X	X	
Create/enhance/expand protected area networks		X	X	
Create management and monitoring plans for existing protected areas			X	
Support sustainable pasture and grazing management	X			X
Support restoration in degraded landscapes to attain snow leopard conservation		X		
Manage ungulate trophy hunting programs in a sustainable way compatible with snow leopard needs	X			
Also see section: Build capacity and enhance conservation policies and institutions.		X	X	X
Combatting poaching and illegal trade				
Improve/enforce laws on conservation, hunting/poaching, trade	X	X	X	
Education and outreach to relevant communities (e.g., development agencies, military, tourists)			X	
Regularly monitor markets for illegal snow leopard and prey parts	X			X
Also see section: Engaging local communities activities.		X	X	X
Also see section: Transboundary management and enforcement.		X	X	X
Also see section: Build capacity and enhance conservation policies and institutions.	X	X	X	X

(Continued)

TABLE 44.3 Comparison of Suggested Actions for Snow Leopard Conservation in Four Twenty-First-Century Conservation Strategies *(cont.)*

Actions to conserve snow leopards	SLSS 2003	SLRAC 2008	GSLEP 2013	SLSS 2014
Transboundary management and enforcement				
Initiate/enhance transboundary cooperation, including bilateral and multilateral agreements		X	X	
Foster a landscape level approach to conservation, including no net-loss policies for biodiversity		X	X	X
Capacity building for border and customs officials			X	
Establish transboundary protected areas			X	
Engaging industry in snow leopard conservation				
Use snow leopards as indicator species of impacts of development			X	
Invite industrial officials to relevant snow leopard conservation events			X	
Consult with development-oriented ministries regarding snow leopard conservation priority areas, including through the Environmental and Social Assessment process			X	X
Build capacity and enhance conservation policies and institutions				
Improve treaty (e.g., CITES, CMS) implementation	X	X	X	X
Improve and strengthen national laws regarding snow leopard conservation	X		X	X
Build national capacity for research, monitoring and enforcement, including workshops and seminars	X	X	X	
Develop/implement National Snow Leopard Action Plan		X	X	
Establish rescue centers for rehabilitation of orphan cubs, injured adults		X		
Also see section: Build awareness of snow leopards and snow leopard conservation efforts.	X			
Research and monitoring				
Assess status of snow leopards, prey, and habitat relative to conservation goals	X	X	X	
Assess impact of climate change on snow leopards, prey, and habitat		X	X	X
Build awareness of snow leopards and snow leopard conservation efforts				
Establish awareness campaigns to facilitate snow leopard conservation at all levels	X	X		X
Provide educational materials to schools about snow leopard conservation	X		X	X
Also see section: Engage local communities activities.				

enough snow leopards to ensure long-term population viability of at least representative populations on a national and ecological basis and across the entire range (see Chapter 3). Moreover, successfully conserved snow leopard prey populations should be abundant enough so that snow leopards do not depend on domesticated animals, but can maintain themselves on wild prey alone (see Chapter 4). Successful conservation includes enforced legal protections that prohibit snow leopard hunting and trade (see Chapter 18), and manage harvest of prey species sustainably (see Chapter 16). Successfully conserved healthy mountain landscapes provide ecosystem services such as water provision, carbon sequestration, agro-pastoral opportunities (consonant with snow leopard conservation), and recreational opportunities. Every range state has a part to play, and all ecological settings, as defined in the SLRAC 2008, will receive conservation effort, which further ensures that the widest range of ecosystem services are provided. Finally, expressing snow leopard conservation in these terms parallels the Redford et al. (2011) definition of successful conservation, which moves the conversation beyond a focus on avoided extinction to a proactive definition of why and where snow leopards must be conserved.

Beyond these agreed terms, however, we see even greater opportunities. There appear to be enough populations in enough places that truly range-wide conservation of the species can be imagined. That is why we set the vision for the long-term conservation at the range-wide scale, with an emphasis not only on individual populations in individual places, but on interconnectivity between them.

As Groves et al. (2003) describes, a good strategy includes articulating a vision, setting one or more time-bound goals, determining actions to achieve that goal, and mobilizing resources to execute those actions. The four snow leopard strategies outlined over the past decade-plus, in combination, articulate a way forward for the larger conservation community – describing how the ends are achieved by the means, and enabling leaders to deploy resources wisely and effectively while coordinating among partners. Generals require strategies, but so do foot soldiers. Snow leopard conservation will be a long fight. Coordination and cooperation are key, but so are putting our values forward, so that we can enlist greater help and find greater success mobilizing the resources to enact our vision of a world where snow leopards and people coexist in perpetuity.

References

Baillie, J., Groombridge, B. (Eds.), 1996. 1996 IUCN Red List of Threatened Animals. International Union for Conservation of Nature and Natural Resources, Gland, Switzerland.

Czech, B., Krausman, P.R., Borkhataria, R., 1998. Social construction, political power, and the allocation of benefits to endangered species. Conserv. Biol. 12, 1103–1112.

Dinerstein, E., Loucks, C., Wikramanayake, E., Ginsberg, J., Sanderson, E., Seidensticker, J., Forrest, J., Bryia, G., Heydlauff, A., Klenzendorf, S., Leimbruber, P., Mills, J., O'Brien, T.G., Shrestha, M., Simons, R., Songer, M., 2007. The fate of wild tigers. BioScience 57, 508–514.

Dinerstein, E., Wikramanayake, E., Robinson, J.G., Karanth, K.U., Rabinowitz, A.R., Olson, D.M., Matthew, T., Hedao, P., Connor, M., 1997. A Framework for Identifying High Priority Areas and Actions for the Conservation of Tigers in the Wild. World Wildlife Fund-US, Wildlife Conservation Society, and National Fish and Wildlife Foundation's Save the Tiger Fund, Washington, DC.

Durant, S., 2007. Range-wide conservation planning for cheetah and wild dog. Cat News 46, 13.

Groves, C., The Nature Conservancy, Hunter, M., 2003. Drafting a Conservation Blueprint: A Practitioner's Guide to Planning for Biodiversity, second ed. Island Press, Washington, DC.

ExxonMobil. 2000. ExxonMobil Strengthens Commitment to Tiger Conservation; Company Has Committed $9 Million to Tiger Conservation Save The Tiger Fund Conservation Grants Announced – Press Releases on CSRwire.com. http://www.csrwire.com/press_releases/25816-ExxonMobil-Strengthens-Commitment-to-Tiger-Conservation-Company-Has-Committed-9-Million-to-Tiger-Conservation-Save-The-Tiger-Fund-Conservation-Grants-Announced (accessed 27.07.2015).

Groombridge, B. (Ed.), 1994. IUCN Red List of Threatened Animals. International Union for Conservation of Nature and Natural Resources, Gland, Switzerland.

Hidasi-Neto, J., Loyola, R., Cianciaruso, M.V., 2015. Global and local evolutionary and ecological distinctiveness of terrestrial mammals: identifying priorities across scales. Divers. Distrib. 21, 548–559.

IUCN, 1990. 1990 IUCN Red List of Threatened Animals. IUCN, Gland, Switzerland, and Cambridge, UK.

IUCN Conservation Monitoring Centre. 1988. 1988 IUCN Red List of Threatened Animals. International Union for Conservation of Nature and Natural Resources/United Nations Environment Programme, Gland, Switzerland.

IUCN Conservation Monitoring Centre, 1986. 1986 IUCN Red List of Threatened Animals. International Union for Conservation of Nature and Natural Resources, Gland, Switzerland, and Cambridge, UK.

IUCN Species Survival Commission, 2000. IUCN Red List Categories and Criteria. Version 3.1, Second edition IUCN Council, Gland, Switzerland.

Jackson, R., D. Mallon, T. McCarthy, R.A. Chundaway, and B. Habib. 2008. *Panthera uncia* (Ounce, Snow Leopard). The IUCN Red List for Threatened Species, Verison 2015.4. http://www.iucnredlist.org/details/22732/0 (accessed 27.07.2015).

Ladle, R.J., Jepson, P., 2008. Toward a biocultural theory of avoided extinction. Conserv. Lett. 1, 111–118.

Leader-Williams, N., Adams, W.M., Smith, R.J. (Eds.), 2010. Trade-Offs in Conservation: Deciding What to Save. Wiley-Blackwell, Oxford, UK.

McCarthy, T.M., Chapron, G., 2003. Snow Leopard Survival Strategy. International Snow Leopard Trust and Snow Leopard Network, Seattle, WA.

McCarthy, T., Sanderson, E., Mallon, D., Fisher, K., Zahler, P., Hunter, L., 2009. Range-Wide Conservation Planning for the Snow Leopard. International Congress for Conservation Biology, Beijing, China.

Mckeown, M., 2012. The Strategy Book. Pearson, New York.

Mintzberg, H., Lampel, J.B., Quinn, J.B., Ghoshal, S., 2002. The Strategy Process: Concepts, Context, Cases. Prentice Hall, Upper Saddle River, NJ.

Neel, M.C., Leidner, A.K., Haines, A., Goble, D.D., Scott, J.M., 2012. By the numbers: how is recovery defined by the US Endangered Species Act? BioScience 62, 646–657.

Nowell, K., Bauer, H., 2006. Conservation Strategy for the Lion in Eastern and Southern Africa. IUCN SSC Cat Specialist Group, Gland, Switzerland.

Ray, J., Redford, K.H., Steneck, R., Berger, J. (Eds.), 2005. Large Carnivores and the Conservation of Biodiversity. Island Press, Washington, DC.

Redford, K.H., Amato, G., Baillie, J., Beldomenico, P., Bennett, E.L., Clum, N., Cook, R., Fonseca, G., Hedges, S., Launay, F., Lieberman, S., Mace, G.M., Murayama, A., Putnam, A., Robinson, J.G., Rosenbaum, H., Sanderson, E.W., Stuart, S.N., Thomas, P., Thorbjarnarson, J., 2011. What does it mean to successfully conserve a (vertebrate) species? BioScience 61, 39–48.

Sanderson, E.W., Redford, K.H., Chetkiewicz, C-L.B., Medellin, R.A., Rabinowitz, A.R., Robinson, J.G., Taber, A.B., 2002. Planning to save a species: the jaguar as a model. Conserv. Biol. 16, 58–72.

Sanderson, E.W., 2006. How many animals do we want to save? the many ways of setting population target levels for animals. BioScience 57, 911–922.

Sanderson, E., Forrest, J., Loucks, C., Ginsberg, J., Dinerstein, E., Seidensticker, J., Leimgruber, P., Songer, M., Heydlauff, A., O'Brien, T., Bryja, G., Klenzendorf, S., Wikramanayake, E., 2006. Setting priorities for the conservation and recovery of wild tigers: 2005-2015. The Technical Assessment. Wildlife Conservation Society, and World Wildlife Fund–US, and National Fish and Wildlife Foundation–Save the Tiger Fund, Washington, DC.

Society for Conservation Biology., 2002, Jaguar conservation spotty, Science Daily. www.sciencedaily.com/releases/2002/01/020128080643.htm.

Snow Leopard Network, 2014. Snow Leopard Survival Strategy. Revised Version 2014.1. Snow Leopard Network, Seattle, WA.

Snow Leopard Working Secretariat, 2013. Global Snow Leopard and Ecosystem Protection Program (GSLEP): A New International Effort to Save the Snow Leopard and Conserve High-Mountain Ecosystems. Snow Leopard Working Secretariat, Bishkek, Kyrgyz Republic.

Species Conservation Planning Task Force. 2008. Strategic Planning for Species Conservation: A Handbook. Version 1.0. IUCN Species Survival Commission (SSC), Gland, Switzerland. http://cmsdata.iucn.org/downloads/scshandbook_2_12_08_compressed.pdf (accessed 06.03.2015).

Thesarus Linguae Graecae. 2011. στρατηγὶια. The Thesarus Linguae Graecae Project, Irvine, CA. http://stephanus.tlg.uci.edu/lsj/#eid=99950&context=search (accessed 06.03.2015).

Tzu, S., 2007. The Art of War. Filiquarian, Minneapolis, MN.

Westwood, A., Reuchlin-Hugenholtz, E., Keith, D.M., 2014. Re-defining recovery: a generalized framework for assessing species recovery. Biol. Conserv. 172, 155–162.

Williams, N., 2008. International Conference on Range-wide Conservation Planning for Snow Leopards: saving the species across its range. Cat News 48, 33–34.

Wilson, E.O., 2003. The Future of Life. Vintage, New York.

CHAPTER

45

The Global Snow Leopard and Ecosystem Protection Program

Andrew Zakharenka, Koustubh Sharma**, Chyngyz Kochorov**, Brad Rutherford‡, Keshav Varma§, Anand Seth¶, Andrey Kushlin¶, Susan Lumpkin¶, John Seidensticker§§, Bruno Laporte***, Boris Tichomirow†††, Rodney M. Jackson‡‡‡, Charudutt Mishra§§§, Bakhtiyar Abdiev****, Abdul Wali Modaqiq††††, Sonam Wangchuk‡‡‡‡, Zhang Zhongtian§§§§, Shakti Kant Khanduri*****, Bakytbek Duisekeyev†††††, Batbold Dorjgurkhem‡‡‡‡‡, Megh Bahadur Pandey§§§§§, Syed Mahmood Nasir******, Muhammad Ali Nawaz††††††, Irina Fominykh‡‡‡‡‡‡, Nurali Saidov§§§§§§, Nodirjon Yunusov********

*Global Tiger Initiative Secretariat, World Bank, Washington, DC, USA

**GSLEP Secretariat, Bishkek, Kyrgyz Republic

‡Snow Leopard Trust, Seattle, WA, USA

§GTI Council, New Delhi, India

¶Independent Advisor, Washington, DC, USA

§§Smithsonian Conservation Biology Institute, Washington, DC, USA

***Leadership, Knowledge, Learning, LLC, Washington, DC, USA

†††Middle Asia Program, NABU, Berlin, Germany

‡‡‡Snow Leopard Conservancy, Sonoma, CA, USA

§§§Snow Leopard Trust and Nature Conservation Foundation, Mysore, Karnataka, India

****State Agency on Environment Protection and Forestry, Bishkek, Kyrgyz Republic

Snow Leopards
http://dx.doi.org/10.1016/B978-0-12-802213-9.00045-6

[††††]National Environmental Protection Agency, Kabul, Afghanistan
[‡‡‡‡]Ministry of Agriculture and Forests, Thimpu, Bhutan
[§§§§]Department of International Cooperation, State Forestry Administration, Beijing, China
[*****]Ministry of Environment and Forests, New Delhi, India
[†††††]Wildlife Department, Ministry of Agriculture, Astana, Kazakhstan
[‡‡‡‡‡]International Cooperation Division, Ministry of Environment and Green Development, Ulaanbaatar, Mongolia
[§§§§§]Department of National Parks and Wildlife Conservation, Ministry of Forest and Soil Conservation, Kathmandu, Nepal
[******]Ministry of Climate Change, Islamabad, Pakistan
[††††††]Department of Animal Sciences, Quaid-i-Azam University, Islamabad, Pakistan
[‡‡‡‡‡‡]Department of International Cooperation, Ministry of Natural Resources and Environment, Moscow, Russian Federation
[§§§§§§]State Agency of Natural Protected Areas, Dushanbe, Tajikistan
[*******]International Relations Department, State Committee for Nature Protection, Tashkent, Uzbekistan

GENESIS: HOW THE GLOBAL SNOW LEOPARD AND ECOSYSTEM PROTECTION PROGRAM AND THE SNOW LEOPARD INITIATIVE WERE FORMED

An elusive denizen of the mountains of Central and South Asia, the snow leopard (*Panthera uncia*) inhabits 12 countries. Inhabiting an estimated 1.8 million km^2 of area at elevations from 540 to more than 5000 m, snow leopards share landscapes with people who depend on various traditional forms of agro-pastoralism. The snow leopard is a culturally, ecologically, and economically important symbol of healthy mountain ecosystems and the communities living there, yet this cat is under threat of extinction across its range.

Value of the Snow Leopard and Its Landscapes

The snow leopard is an indicator of healthy high-mountain ecosystems, which support the cat, its prey, and a vast amount of biodiversity, as well as contribute to human well-being locally, regionally, and globally. Hundreds of millions of people depend on these landscapes for water, hydropower, agriculture, forage for livestock, food for themselves, mineral resources, medicinal products, cultural traditions and spiritual values, and inspiration (Snow Leopard Working Secretariat, 2013a). The cultural and aesthetic value of snow leopard ecosystems is immeasurable. The lifestyle, religious and spiritual beliefs, traditional agriculture, food, marriage systems, and governance of societies inhabiting these areas are all unique. The Himalayan ranges harbor many mystical and sacred linkages to several religions and beliefs, including many sacred mountains. Snow leopards in particular offer iconic representation of these areas and appear in the coats of arms and other symbols of some nations and cities in the snow leopard range. Ecosystems that support snow leopards are of immense economic value. An estimate of the economic value of some prominent services generated from snow leopard habitats in India is nearly USD 4 billion a year, the bulk of which comes from water provisioning, hydropower, livestock, agriculture, and tourism (Snow Leopard Working Secretariat, 2013a).

High-mountain ecosystems accumulate precipitation, regulate seasonal runoff, and provide water to the human population beyond regional borders, benefiting nearly one-third of the world's human population. Himalayan glaciers are the headwaters of ten major river systems in Asia that generate hydropower, maintain fisheries, support industry, and irrigate farmland (Snow Leopard Working Secretariat, 2013a). They are important to carbon storage and sequestration. For instance, carbon storage across the snow leopard range in China equates to as much as 14 gigatons of carbon, equivalent to almost half of the carbon stored in the forests of Asia (Snow Leopard Working Secretariat, 2013b, pp. 30–46). These ecosystems are extremely rich and diverse in plant and animal species. For example, in India alone, snow leopard habitat supports 350 species of mammals and 1200 species of birds, while the Altai mountains support nearly 4000 species of plants, 143 species of mammals, and 425 species of birds (Snow Leopard Working Secretariat, 2013b, pp. 47–62).

Apart from its intrinsic value, biodiversity underpins agriculture. More than 335 species of wild relatives of cultivated crops are found in the region. Wild relatives of all major domesticated livestock—cattle, horses, sheep, and goats also occur in this region. Thus, the biodiversity resources of the area not only provide life support to the people, but also serve as repositories of wild genetic material for plant and animal breeders (Snow Leopard Working Secretariat, 2013a).

A New Approach to Snow Leopard Conservation

Early in 2011, the government of the Kyrgyz Republic began spearheading an initiative that would comprehensively address high-mountain environmental issues using the conservation of snow leopards as a flagship. Then, President Roza Otunbayeva supported a proposal from Germany's Nature and Biodiversity Conservation Union (NABU) to host a global forum on snow leopard conservation in Bishkek, which was subsequently endorsed by Germany's Chancellor Angela Merkel in August 2011. In February 2012, the subsequent president of the Kyrgyz Republic, Almazbek Atambayev, solicited support for this initiative from the World Bank, asking then-President Robert Zoellick to help replicate the effort of the Global Tiger Initiative (GTI) for the conservation of the snow leopard.

At President Atambayev's request, the GTI's Secretariat at the World Bank, in technical partnership with NABU and the Snow Leopard Trust (SLT), offered its support and advice to guide the process of developing a Global Snow Leopard and Ecosystem Protection Program (GSLEP) with the participation of the 12 snow leopard range countries. Subsequently, the snow leopard range countries, with many partners, held a series of meetings and worked intensely to develop individual National Snow Leopard and Ecosystem Protection Priorities (NSLEPs). These NSLEPs are the core of the GSLEP. In addition, the international community developed Global Support Components (GSCs) to offer assistance when the issues to be addressed transcend national boundaries and go beyond the capacity of any one country to address alone. These also form part of the GSLEP.

While the GSLEP was being developed, the government of the Kyrgyz Republic prepared to host leaders in the governments of the snow leopard range countries at the first Global Snow Leopard Conservation Forum. At that Forum, held on October, 2013, the government leaders issued the Bishkek Declaration on the Conservation of Snow Leopards, and unanimously endorsed the GSLEP as the road map for achieving the declaration's goal.

FRAMEWORK: KEY PRINCIPLES, STRUCTURE, AND APPROACHES OF THE SNOW LEOPARD INITIATIVE

In this section, we explore the framework of the new approach that snow leopard range countries are taking.

Experience of the Global Tiger Initiative

The Snow Leopard Initiative emerged following the principles of the GTI and its approach of building collective engagement and actions around the tiger conservation challenges (GTRP, 2010).

National Governments Leading the Efforts

The first principle of the Snow Leopard Initiative is that range country governments determine the key directions and parameters of the initiative and program implementation. The involvement and cooperation of national governments at high levels is essential because the threats to snow leopards and their habitats originate either outside national boundaries or are beyond the regular influence of environment ministries and agencies. For instance, curbing demand for cashmere or for cat body parts and products is a regional and global economic issue; addressing organized wildlife crime requires the involvement of national and regional law enforcement agencies; planning environmentally friendly large infrastructure projects that takes into account habitat connectivity involves mining, oil and gas, and transport agencies.

Following the GTI model, a working group of snow leopard experts was formed to assist the countries to deliver the Global Forum, chart the political declaration, and develop the range-wide program that would bring various players together to work on the multisectoral challenges of conservation and development. Even though such ideas were not new among the scientists and practitioners, the challenge was to break out of sectoral and country-specific silos. The prospects for progress improved significantly by March 2012 when the then-president of the World Bank offered support. A series of governmental meetings among representatives of the 12 range countries and their partners followed (described subsequently).

Mutual Accountability of 12 Snow Leopard Range Countries and Partners

The second important principle of the program is mutual accountability among the range country governments and partners to deliver the program. It sets no single government or party above others; it enables conditions for equal respect and consensus-based decision making; and it establishes trust for monitoring performance and reporting results and issues to one another. This principle enabled the governments to open up a platform of collaboration and establish the trusting relationships required for dealing with both sovereign and shared issues.

Political leadership created an opportunity for technical specialists to work out mutually agreeable solutions to long-term challenges. It was not always easy for all players to accept new rules, but gradually – through consistent application of the initiative's principles – trust and working relationships developed. The GTI Secretariat with key global team members played an important role in facilitating these relationships, applying the principles and providing general services of support, coordination, collating working materials, and maintaining communication with and among the governments and partners.

Good Practices and Knowledge Exchange Taken to Scale

The third defining principle of the Snow Leopard Initiative is knowledge exchange that enables taking good and tested practices as well as knowledge to the scale required to achieve the GSLEP goal. This process involves collation of good practices in a standardized manner, briefly recording the specific issue that is addressed by each practice, its scope, the staff and annual budget required, and its verified and measurable results. Capturing good practices in a standardized way enables countries to select a needed practice and evaluate what it will take in terms of capacity and resources to scale up or emulate it. The comprehensive list of good practices for snow leopard conservation, created by May 2013, contained 24 good practices grouped into six broader themes and was used for developing and scaling up the portfolio of national and global support activities of the Global Snow Leopard and Ecosystem Protection Program (Snow Leopard Working Secretariat, 2013a).

PREPARATION STAGE AND MILESTONES OF THE GSLEP AND GLOBAL FORUM ON SNOW LEOPARD CONSERVATION

The envisioned high-level forum required preparation of a robust program document that would outline effective responses to the challenge of conserving and recovering the snow leopard and its habitats. Moreover, with the range of the snow leopard spanning 12 countries with different histories, cultures, and political systems, and no previous experience of range-wide collaboration of such scale, it was essential to first build a coalition among the governments and technical partners.

Early on, an *Approach Paper* was developed to prepare the national priorities and the global program itself. It was first discussed with the core group and then later shared with the governments to capture their concerns and establish ownership of the program, agree on a common approach and structure of national and global components, describe the context and the process, the steps, and associated timelines, as well as the proposed outlines of the NSLEPs and GSLEP. Most important, it served as an agreed-upon quality measure of each of the 12 NSLEPs, their standardized and rich structure and content to come down the road.

It was agreed that government representatives would have to be fully sponsored to ensure full attendance at the meetings, and that the quality of the meetings' outcomes would be ensured by consistent and structured preparation work for each. The meetings' objectives and expected outcomes were discussed and agreed with members of the core group of experts, at least 6 weeks in advance of any meeting. Then specific forms and templates were developed to solicit inputs from the range governments and partners and to provide working materials toward each of the objectives and expected outcomes.

The challenge was to establish lines of communication with the governments to ensure quality preparation work and to provide the technical help of trusted and respected national and international experts. This was done via regular teleconferences and email distributions aimed at establishing informal and then more formalized lists of the appropriate officials. A formal group of governmental officials was later established in each range country at three levels: the *Minister* or Director of the agency responsible for setting national snow leopard conservation agenda; the *Director General* or similar mid-level senior official responsible for overseeing the implementation; and *National Focal Point* who is responsible for keeping track of communications among the Secretariat, national governmental agencies, and partners. A list of credible technical experts and partners for each country was also developed and shared with the governments along with the forms and requested materials.

Keeping a single line of two-way communications between a designated member of the core group and the National Focal Points and partner representatives helped to establish trusting working relationships. An important element of such communications was to ensure that range governments' collective responsibility for developing their own national programs and sharing inputs, experience, and practices across boundaries. This was achieved through testing the forms and templates with a few countries, developing and sharing their exemplary-quality forms with the rest of the countries, and having regular group communications with status of each country's preparation milestones.

The deliberations and results of each meeting were captured in a summary report and shared with all participants. Working materials were also shared among the participants. Finally, an important element of the process was to agree at the end of each meeting on four or five next steps and those responsible for their completion by the agreed deadlines.

Bishkek Working Meeting, December 2012

The first meeting of representatives of all snow leopard range country governments was hosted by the government of Kyrgyz Republic in Bishkek during December 1–3, 2012. It was coorganized and cosponsored by SLT, Snow Leopard Network (SLN), NABU, and the Word Bank/Global Tiger Initiative. Its objectives were multiple: to consolidate the leadership of the range countries' governments, articulate the benefits and value of the snow leopard and its ecosystems, attempt to define concrete national and global goals of the future program, share proven good practices, invite and engage industry and donors, and agree on specific actions that both governments and partners would have to accomplish between the end of the workshop and the proposed forum.

Outcomes were rich and diverse. Countries presented early thinking on their national priorities and actions, and shared experience and concerns. The *Bishkek Recommendations on the Conservation of Snow Leopards and Their High-Mountain Ecosystems* were prepared and adopted during the workshop. The recommendations became the first document collectively developed and owned by the governments of snow leopard range countries. The path to the Global Forum was discussed at length and steps forward were identified and agreed upon, including mobilizing political support, defining national priorities and the global snow leopard conservation agenda, and enhancing partner support. A short list of specific near-term tasks was agreed upon, too. President Atambayev was informed about the workshop outcomes (Fig. 45.1).

Bangkok Planning Workshop, March 2013

The next meeting was held, together with the tiger range countries, on the invitation of the government of Thailand in Bangkok on March 9, 2013. Its objective was to assist countries with preparation of the 2013 Global Forum through a collaborative review and discussion of the "zero-draft" versions of their NSLEPs and making further improvements to ensure the consistency of the NSLEPs as the building blocks of the proposed GSLEP. Such a meeting and face-to-face discussion of the challenges and ways to overcome them was necessary to help each country complete its full draft NSLEP by the end of April 2013, as required by the Forum preparation timeline.

This was achieved through developing and sharing a nine-point reference document explaining what a high-quality NSLEP would look like. It was also agreed that key documents needed for the Global Forum would include high-quality NSLEPs, the GSLEP, and the Declaration. Teams were also identified to make it happen, including the National Focal Points,

FIGURE 45.1 **Representatives of the range countries and partner organizations.**

technical Country Support Teams of experts, the Secretariat in Bishkek, and the Global Support Team. It was agreed that NSLEPs would be revised and final drafts will be ready by the beginning of May 2013 and that a pre-forum drafting meeting would be held in Moscow in mid-May.

Beijing International Workshop on Snow Leopard Conservation in China, May 2013

Hosted by the State Forestry Administration of the People's Republic of China and the Wildlife Institute at Beijing Forestry University, the workshop was held in Beijing on May 26–27, 2013, this workshop was supported by the World Bank and GTI, SLT, and other organizations. The goal of this workshop was to coordinate preparation of the provincial programs on the conservation of snow leopard and its ecosystems, review inputs into the China's NSLEP, and to identify China's leading positions in regional and global processes related to snow leopard conservation.

China, being home to more than half of the snow leopard's global range and population, required more detailed attention to the NSLEP preparation at the national and province levels. The workshop brought together officials of forestry administrations from seven provinces:

Xinjiang, Tibet, Qinghai, Gansu, Sichuan, Inner Mongolia, and Yunnan. Each of the provinces shared its experience in snow leopard conservation, and China's draft NSLEP was revised and finalized based on these inputs.

Moscow Pre-Forum Drafting Meeting of Senior Officials, May 2013

The purpose of the Moscow meeting on 29–30 May was to review and finalize the drafts of key inputs for the Forum; discuss and agree on the final draft of the *Bishkek Declaration on the Conservation of Snow Leopards and their Ecosystems*; present and finalize in good quality all 12 NSLEPs; and review building blocks of the GSLEP. This was the final face-to-face meeting before the Forum, which was scheduled for October 2013.

Step by step, the quality of the NSLEPs had improved. Each country received an individual score card and specific suggestions to improve its NSLEPs, identifying lead countries in describing each of the components. The outline of the GSLEP, which was envisioned to be not a simple summation of the NSLEPs but rather a new document inspired by them, and the Global Drafting Team were also introduced. By the time of the Moscow meeting, trusting working relationships had been developed among the teams. After half-day deliberations and overnight work by the drafting team, the consensus draft of the *Bishkek Declaration on the Conservation of Snow Leopards* emerged on May 29, 2013.

Global Snow Leopard Conservation Forum, Bishkek, October 2013

The Forum led by the president of Kyrgyz Republic was possible thanks to the leadership of the Kyrgyz Republic and the World Bank that helped to convene the key players, the active engagement of range country representatives, the support and experience of the GTI and NABU teams, and immeasurable support from key donors and partners such as Global Environmental Facility (GEF), Snow Leopard Conservancy (SLC), SLN, SLT, United Nations Development Program (UNDP), United States Agency for International Development (USAID), and World Wildlife Fund (WWF). After fine-tuning its language, the range country leaders issued the *Bishkek Declaration on the Conservation of Snow Leopards* and endorsed the GSLEP as the road map for achieving its overarching goal (Fig. 45.2).

THE GLOBAL SNOW LEOPARD AND ECOSYSTEM PROTECTION PROGRAM

This section describes the GSLEP itself – its goal, the national priorities, global support components, factors for success, mechanism for GSLEP implementation and financing, and the expected outputs and enabling conditions. The GSLEP seeks to address high-altitude mountain issues using the conservation of the charismatic and endangered snow leopard as a flagship. The GSLEP is a range-wide effort that unites range-country governments, nongovernmental and intergovernmental organizations, local communities, and the private sector around a shared vision to conserve snow leopards and their valuable high-mountain ecosystems (Fig. 45.3).

The Common Goal

The snow leopard range countries and partners unanimously agree to the shared goal of the GSLEP for the 7 years through 2020. As stated in the Bishkek Declaration, "*The snow leopard range countries agree, with support from interested organizations, to work together to identify and secure at least 20 snow leopard landscapes across the cat's range by 2020 or, in shorthand – Secure 20 by 2020.*" Secure snow leopard landscapes are defined as those that contain at least 100 breeding-age snow leopards conserved with the involvement of local communities, support

FIGURE 45.2 **President Atambayev addresses participants of the Forum.**

adequate and secure prey populations, and have functional connectivity to other snow leopard landscapes, some of which cross international boundaries. Secure 20 by 2020 will lay the foundation to reach the ultimate goal: ensuring that snow leopards remain the living icon of mountains of Asia for generations to come.

NSLEPs and Global Support Components

The foundation of the GSLEP is 12 individual NSLEPs. Each NSLEP incorporates a set of priority, concrete project activities to be implemented to meet national goals and, collectively, the overarching global goal. The NSLEPs are buttressed by five Global Support Components prepared by international organizations to address issues that transcend national boundaries and go beyond the capacity of any one country to address alone. The GSCs aim to support and assist the range countries, as needed, in the areas of wildlife law enforcement; knowledge sharing; transboundary cooperation; engaging with industry; and research and monitoring. The activities of the countries and the international community are grouped under broad themes that correspond to the commitments of the Bishkek Declaration:

- Engaging local communities in conservation
- Managing habitats and prey
- Combating poaching and illegal trade
- Transboundary management and enforcement
- Engaging industry
- Research and monitoring
- Building capacity and enhancing conservation policies and institutions
- Building awareness

FIGURE 45.3 **Participants of the Global Snow Leopard Conservation Forum.**

The first five are direct impact activities; the last three are enabling ones to create conditions for successfully performing or improving the direct impact activities. Together, the portfolio of national activities, supported by the GSCs, will move the countries toward their national and global goals.

Success Factors

The GSLEP represents the first-ever comprehensive, coordinated effort to conserve snow leopards and their mountain habitats, moving from isolated interventions to collective impact initiatives that unify the efforts of countries and the global conservation community to achieve a shared vision and goal. The success of GSLEP implementation depends on scaling up known and tested key actions and good practices, which will require incremental domestic and external financing of about USD 150–250 million over the first 7 years of the program. Successful GSLEP implementation is being shaped by political support and joint collective actions by partner organizations. The GSLEP provides for regular information sharing, coordinated by a country-led Secretariat, to maintain momentum and high-level attention to progress toward the goal. Regular coordination and information sharing will also enable countries, partners, and donors to fine-tune their efforts to reflect changing circumstances and new knowledge. With about 90% of the program costs in national activities and most range countries reporting gaps in policy and institutional capacity, successful implementation of the program will require substantial political will, leadership, vision, and knowledge sharing to create effective institutional arrangements for national implementation, monitoring, and reporting purposes.

Options for financing the program will vary by range country but include official bilateral programs; multilateral development bank programs; Global Environment Fund programs; inter- and nongovernmental organizations; private sector social responsibility programs; and various forms of payment for ecosystem services schemes. A Secretariat has been established in Bishkek with national and international staff to coordinate the activities of the countries and the international community, and to help coordinate program implementation and use of funding.

Outputs and Enabling Conditions

The outputs of GSLEP implementation are designed to generate both enabling conditions for boosting protection and conservation efforts as well as to produce tangible results toward the common goal. Based on the national and global portfolios of activities, the following anticipated outcomes or expected areas of impact will contribute toward the program's goal; estimated costs and share of the total costs of each outcome are also shown.

1. *Engaging local communities and reducing human-wildlife conflict: USD 16.0 m/9%*
   - **a.** Reduction in livestock predation and mortality, decreased killing of snow leopards and prey.
   - **b.** Snow leopard numbers maintained or increased to form viable populations.
2. *Controlling poaching of snow leopards and prey: USD 41.4 m/24%*
   - **a.** Threats halted; populations of snow leopard and prey increased.
3. *Managing habitat and prey: USD 50.3 m/30%*
   - **a.** Extent of habitat protection, management, and connectivity documented and increased.
   - **b.** Gene flow between populations maintained or restored.
   - **c.** Prey numbers maintained or increased to support viable snow leopard populations.
4. *Transboundary management and enforcement: USD 4.6 m/2%*
   - **a.** Reduced degradation of transboundary landscapes and poaching and smuggling of snow leopard and prey and their products.
   - **b.** Increased capacity for and better transboundary coordination.
5. *Engaging industry: USD 7.2 m/4%*
   - **a.** Piloted approaches for mining and other industry involvement toward joint planning and conservation of snow leopard landscapes.
6. *Research and monitoring: USD 33.7 m/18%*
   - **a.** Major knowledge gaps studied; range, key reproduction sites, existing and potential connecting corridors for snow leopard populations identified and incorporated into landscape level-planning.
   - **b.** Setting of baselines to track progress and effectiveness of conservation programs.
   - **c.** Adaptive management of conservation programs, identification of priority areas for protection.
7. *Strengthening policies and institutions and strengthening capacity of national and local institutions: USD 21 m/7%*
   - **a.** Strengthened policy and institutional environment as well as law enforcement and protected area management, community-based conservation, and industry participation in landscape management.
   - **b.** Highly trained and equipped conservation practitioners; restructured roles and responsibilities among agencies; increased funding for snow leopard conservation.

8. *Awareness and communication: USD 2.6 m/1%*
   a. General public and target groups better equipped with knowledge about snow leopard ecosystems and values associated with them.
   b. Greater political and financial support for snow leopard and ecosystem conservation.

GSLEP LAUNCH, IMPLEMENTATION, AND INFORMATION SHARING

Following the Forum, the range countries met again at the Issyk-Kul Action Planning, Leadership and Capacity Development Global Workshop in the Kyrgyz Republic. The purpose was to (i) identify a minimum of 20 snow leopard landscapes in which to achieve the GSLEP's "Secure 20 by 2020" goal, (ii) define National Priority Activities (NPAs) and Global Priority Activities (GPAs) for the first 2-year implementation plan, (iii) agree on an approach for developing key performance indicators to measure progress toward the GSLEP goal and advance preparation of specific project proposals to funding partners, and (iv) enhance the capacity of the National Focal Points and Working Secretariat staff.

Making the Common Goal a Reality: Identifying Snow Leopard Conservation Landscapes

In total, 23 landscapes totaling more than 500,000 km^2 (25% of snow leopard habitat) have been identified by the range countries to be secured for snow leopard conservation by 2020 through better protection, community-based conservation efforts, and multisectorial cooperation to encourage wildlife conservation and green growth (Table 45.1 and Fig. 45.4).

A seven-point definition for a secured landscape was postulated during the meeting by a working group with representatives from range countries and international experts. The criteria agreed upon by the range countries to define secure landscapes were:

1. Snow leopard landscapes designated as "ecologically fragile" zones that have defined "values" and biodiversity-sensitive land-use and development planning for various zones within the landscape. Critical wildlife areas and corridors designated within the landscapes where damaging land use is minimized.
2. Stable or increasing population of snow leopards and sufficient prey populations maintained in the landscapes.
3. Sustainable and socially responsible development achieved through community-based efforts and business models to enhance livelihoods of local communities within the ecologically fragile zones (landscapes).
4. Industry encouraged to aid local communities in the multiple-use zones within the snow leopard landscapes (chipping in funds for conservation and livelihood activities).
5. Local community involvement in conservation planning and implementation through community-based conservation efforts, provisioning of economic and other incentives, and policy and legal support.
6. Policy initiatives and strengthening of laws to effectively address traditional and emerging threats including climate change.
7. Sustainability of GSLEP and NSLEPs through capacity building, technology, research, resource mobilization, multi-country information exchange, and cooperation among the range countries.

Monitoring efforts involve two groups of activities: impact and process-oriented activities.

TABLE 45.1 Snow Leopard Conservation Landscapes

Country	Landscape	Area (km^2)
Afghanistan	Wakhan National Park	10,951
Bhutan	Snow Leopard Habitat	12,110
China	Qilianshan	13,600
	Tuomuerfeng	2,376
	Taxkorgan	15,000
India	Hemis-Spiti	29,000
	Nanda Devi-Gangotri	12,000
	Kanchendzonga-Tawang	5,630
Kazakhstan	Zhetysu Alatau (Jungar Alatau)	16,008
	Northern Tien Shan	23,426
Kyrgyzstan	Sarychat	13,201
Mongolia	Altai	56,000
	South Gobi	82,000
	North Altai	72,000
Nepal	Eastern	9,674
	Central Complex	9,258
	Western	10,436
Pakistan	Hindu Kush	10,541
	Pamir	25,498
	Himalaya	4,659
Russia	Altai	48,000
Tajikistan	Pamir	92,000
Kyrgyzstan–Tajikistan	Alai–Gissar	30,000

National Priority Activities and 2-Year Implementation Plan

With the broad set of priorities defined in the NSLEPs, governments prepared a set of well-defined, measurable, and costed National Priority Activities (NPAs 2014–2015) that would be a basis for the GSLEP's 2-year implementation plan for 2014–2015. The NPAs were developed as first milestones to achieve priorities defined by the range countries in the NSLEPs and grouped by the themes of the Bishkek Declaration. Each of the NPAs describes the specific activities, measurable outcomes, expected costs in 2014 and 2015, including governmental budget allocations and donor funding, location in the priority landscapes, contributing donors, and implementing agencies.

In addition to the NPAs, partners supporting the governments in GSLEP implementation prepared five sets of Global Priority Activities (GPAs 2014–2015), one for each of the GSLEP's Global Support Components to launch global and regional support of national activities in collaboration with governments.

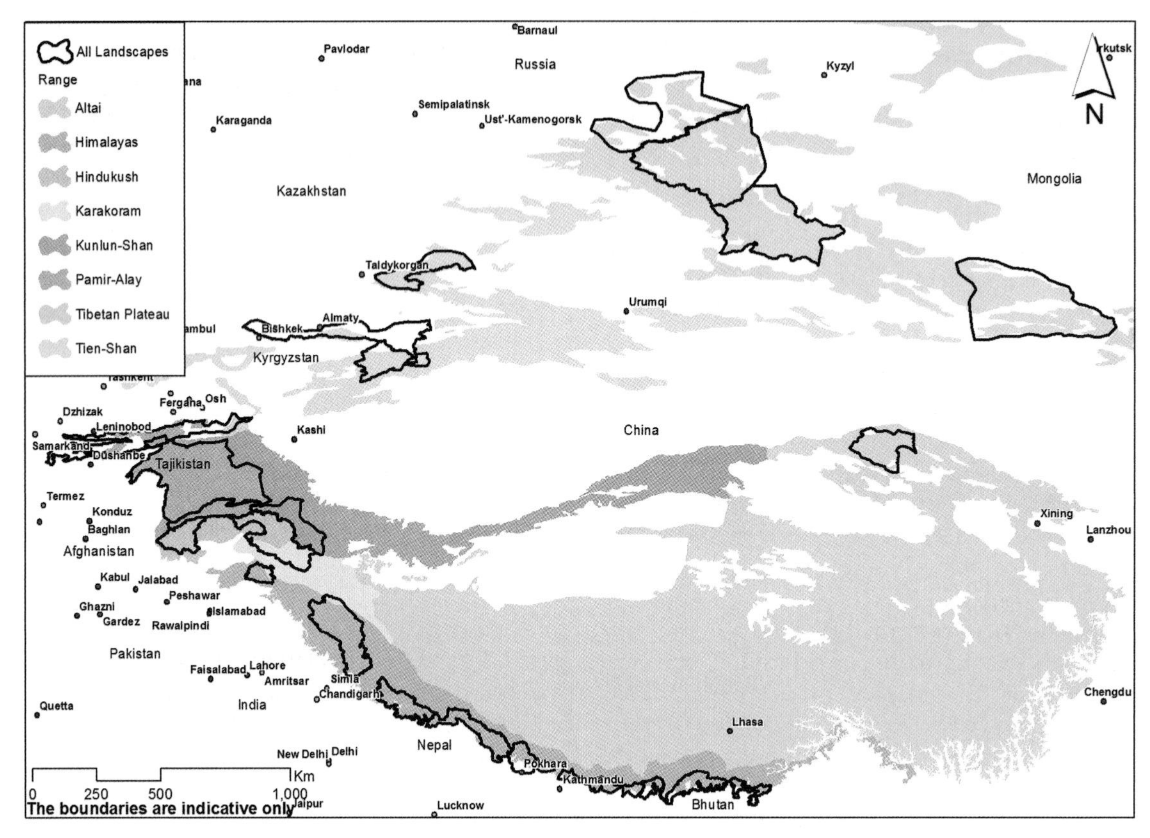

FIGURE 45.4 **Identified snow leopard landscapes.**

Leadership and Collective Engagement

The success of the Snow Leopard Initiative is due to a large extent to the engagement of leaders and champions who have escorted the process since the beginning. The leadership of President Atambayev was key to launching this initiative. He was able to call the attention of the global community to the plight of snow leopards and to mobilize the leadership of the 12 range countries. In addition to the high-level leadership that has created political space, champions have emerged throughout the process. Coalitions of change agents are starting to form and are driving the changes that are called for in each of countries and snow leopard landscapes.

A workshop in June 2014 placed focus on how to develop effective leadership teams to support national institutional arrangements for GSLEP implementation. Participants engaged in a series of collaborative exercises to equip the Secretariat and National Focal Points with a set of tools and insights to help understand the complexity of the biodiversity and conservation challenges they face on the ground, and to use that understanding to develop more targeted and robust national action plans and global priority actions.

GSLEP Steering Committee

The first ministerial-level meeting of the steering committee took place during March, 2015 in Village Koi Tash in the Kyrgyz Tien

Shan Mountains. The steering committee unanimously elected Mushahidullah Khan, Honorable Minister of Climate Change from Pakistan, as the chairperson, and Sabir Atadjanov, Director of State Agency on Environment and Forestry Protection of the Kyrgyz Republic, as its co-chair for the next 2 years. The meeting was attended by senior officials and representatives of the governments of 11 range countries, alongside international and national organizations.

The meeting resulted in formal adoption of the resolution establishing the steering committee membership, roles, and operational guidelines. The steering committee includes the environment ministers of all snow leopard range countries. The six major contributing partners, including GEF, Global Tiger Initiative Council, NABU, SLT, United Nations Development Program, and World Wildlife Fund for Nature, were considered as observer members of the steering committee to be reviewed and voted on again by the range country governments after 2 years. Management planning guidelines, jointly developed by a working group of National Focal Points and international experts were released during the meeting. The management plans will contain a situation analysis, including current and projected threats to the snow leopards and their ecosystems within the landscape, and provide direction to securing these as per the definitions agreed upon by range countries during the June 2014 meeting.

Vision and Next Steps

The future success of GSLEP implementation depends on the collective leadership of all 12 snow leopard countries and various organizations involved in GSLEP implementation. Sustaining such collective leadership requires coherent and continuous work by key players toward achieving the GSLEP goal by 2020.

Political attention from heads of states is necessary to ensure leadership and a cross-sectoral approach to conservation. Supporting such political space for GSLEP implementation creates momentum and eases coordination of national agencies dealing with environment, law enforcement, agriculture, transport, mining, and energy.

Effective guidance by the steering committee and coordination by the Secretariat are required for the 12-country program. Program coordination will be determined by the Secretariat's capacity to ensure regular meetings of all range countries and program partners to drive the implementation agenda forward.

Short-term action planning based on steering committee guidance and proven good practices will ensure continuous sharpening of the implementation agenda. Twelve sets of NPAs and GPAs comprising the first 2-year implementation plan are the first step forward. The next iteration of the NPAs and GPAs (2016–2017) would be likely linked to the key performance indicators (KPIs).

KPIs will ensure robust program monitoring. This includes actions on the ground; biological monitoring of the cats, their prey species, and habitats; institutional monitoring; and strategic monitoring of how we get closer to achieving the long-term GSLEP goal. KPIs will give the steering committee a comprehensive understanding of the current status and progress in GSLEP implementation and serve as a basis for making decisions, mobilizing political support, and financing.

References

Global Tiger Recovery Program (GTRP), 2010. Global Tiger Initiative Secretariat. The World Bank, Washington, DC.

Snow Leopard Working Secretariat, 2013a. Global Snow Leopard and Ecosystem Protection Program (GSLEP). Snow Leopard Working Secretariat; Kyrgyz Republic, Bishkek.

Snow Leopard Working Secretariat, 2013b. Global Snow Leopard and Ecosystem Protection Program (GSLEP): Annex. Snow Leopard Working Secretariat; Kyrgyz Republic, Bishkek.

CHAPTER

46

Joining up the Spots: Aligning Approaches to Big Cat Conservation from Policy to the Field

Urs Breitenmoser, Tabea Lanz*, Roland Bürki**, Christine Breitenmoser-Würsten**

***IUCN Cat Specialist Group, Bern, Switzerland**
****KORA, Bern, Switzerland**

INTRODUCTION

The Convention on Biological Diversity (www.cbd.int) defines biological diversity as the diversity within and between species, and of ecosystems. The convention aims to protect biological resources including genetic resources (genetic material of actual or potential value), organisms, populations, and other biotic components of ecosystems. A given species should hence not only by conserved as a part of the global biological diversity, but also as a large (and therefore variable) gene pool and as an actor in its ecosystem. Carnivores are important to maintain ecological processes through their influence on lower trophic levels and their high evolutionary (selective) potential because of the co-evolutionary relationships with their prey (Dawkins and Krebs, 1979; Ginsberg, 2001). For large cats, this implies that they should be conserved not only as a viable population, but as an important ecological player across their "original range." Although this is unfortunately an illusion for most *Panthera* species – for example, the tiger (*Panthera tigris*) has lost 93% of its original range over the past 100 years (Sanderson et al., 2006) – it is still possible for the snow leopard (*Panthera uncia*), which has experienced a much smaller range reduction, perhaps no more than 10%, (see Chapter 3) and is still widespread within its traditional distribution area. However, this implies coordinated conservation efforts over an area of almost 2,760,000 km^2, stretching 3500 km west-east and 3000 km north-south, across several climatic zones and habitats,

Snow Leopards
http://dx.doi.org/10.1016/B978-0-12-802213-9.00046-8

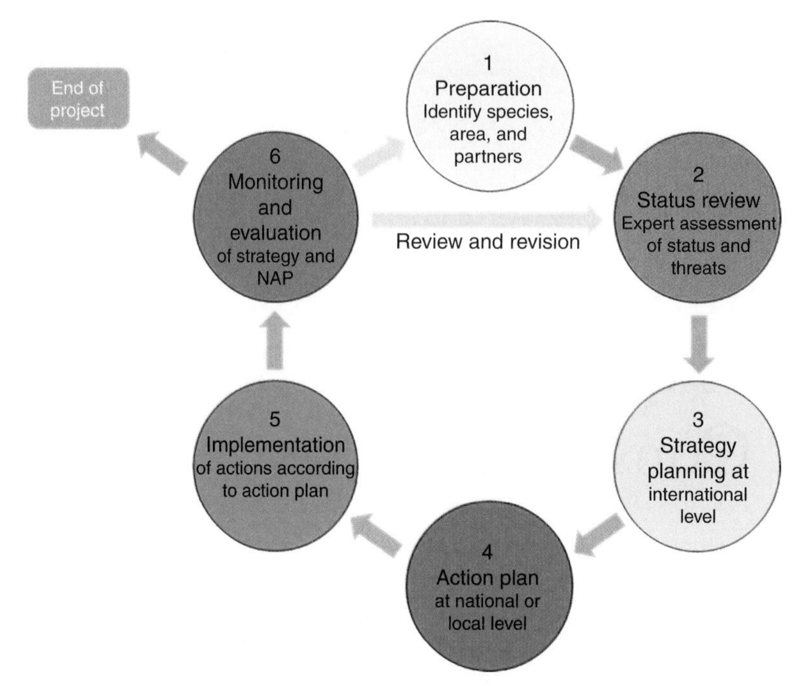

FIGURE 46.1 **Strategic planning cycle for species conservation projects.**

12 countries, and various cultures, languages, and economic systems.

Such an endeavor requires a carefully planned, well-coordinated, and transparent process allowing the combining of large-scale multilateral initiatives with on-the-ground conservation and facilitating dialogue and cooperation within and between national authorities, scientists and experts, (local) stakeholders, and the public. The question is: How can we develop an agreed-upon and comprehensive range-wide conservation plan and implement it through well-adapted, site-specific conservation activities with the involvement of local people? The challenges are how to (1) develop a common understanding of the situation, (2) develop goals and objectives matching at the global and local levels and achieve them through target-driven conservation activities, and (3) organize the international, intersectoral and intercultural cooperation. The Species Survival Commission (SSC) of the International Union for Conservation of Nature (IUCN) has developed guidelines for the strategic planning for species conservation (IUCN/SSC, 2008a, b), which have substantially informed this chapter. We outline here a strategic planning cycle for (cat) conservation projects (Fig. 46.1; IUCN/SSC Cat Specialist Group, 2015), which should, if applied repeatedly and adaptively at various levels and geographic scales and supported by an efficient information system, facilitate the integration of partners with different backgrounds at various levels and help organize their continuous cooperation. We then review strategic planning efforts for the conservation of the snow leopard considering the strategic planning cycle.

PLANNING APPROACH FOR THE CONSERVATION OF LARGE CATS

Strategic planning for species conservation according to IUCN/SSC (2008a, b) should be participative, transparent and informed by the

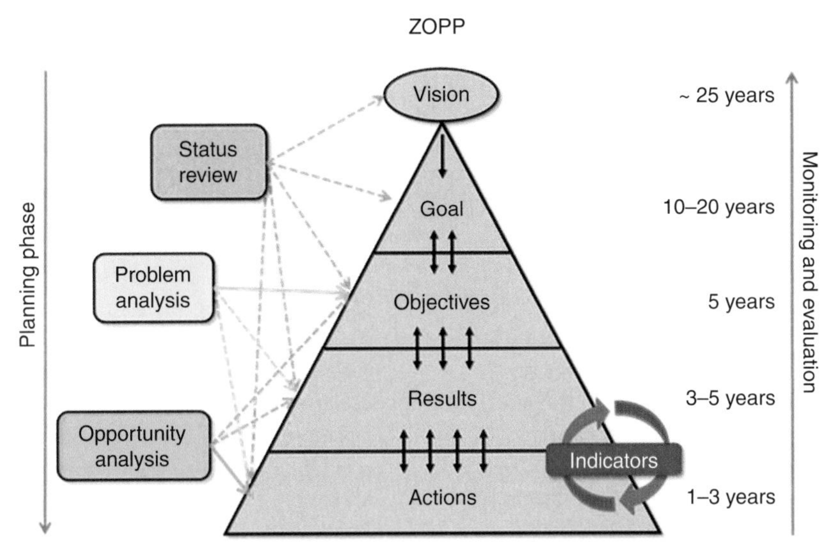

FIGURE 46.2 **The ZOPP pyramid for developing a Species Conservation Strategy (SCS).**

best available science. The planning process is based mainly on the Ziel-Orientierte Projekt-Planung (ZOPP, goal-oriented project planning) combined with the Logical Framework Approach (logframe and LFA), as it was developed by the German agency for international cooperation in the 1980s (GTZ, 1997). A ZOPP pyramid (Fig. 46.2) is typically developed in a workshop where all relevant interest groups participate. The result is a species conservation strategy (SCS) possibly with an integrated action plan (AP), possibly in the form of a logframe matrix, which is today widely used for project management and supervision. The purpose of such a careful planning process is of course the implementation of widely accepted and supported conservation measures. The effect of the activities will need to be monitored and the SCS and/or the AP may need to be adapted in a repeated planning process according to the results of the monitoring. The strategic planning cycle (Fig. 46.1) combines the different phases of a conservation project into six distinct steps, which should be repeated until the goal of the project is fulfilled:

1. *Preparation*: Before developing an SCS, the ground must be carefully prepared. The conservation unit (species, subspecies, or metapopulation) and the geographical scale are determined, partners are identified, and the support from relevant stakeholders is secured. The participation of relevant interest groups and the cooperation between key players is essential for the success of the planning process and the subsequent implementation of the conservation plan. Conservation is not only a rational and scientific procedure; it is as much an emotional and sociopolitical process. Partnership has to be carefully built through early involvement and consistent mutual information. Governmental institutions, experts, and relevant NGOs and stakeholders (including potential opponents) have to be integrated into the process, and they have to be informed about their different roles. A nongovernmental organization (NGO), which is often the initiator of a conservation project, should try to get a mandate from the relevant national

or international institutions. A mandate can considerably ease the process and the subsequent political endorsement of the SCS and the AP.

2. *Status Report*: In a second step, all information relevant for the planning process is compiled. Most important is a thorough assessment of the conservation status of the target species/unit within the target range, including an analysis of threats and use of the IUCN Red List assessment procedures (IUCN Standards and Petitions Subcommittee, 2014). The Status Report should however go beyond biological and ecological aspects and also provide background information to understand the threats, human dimension aspects and conflicts, socioeconomic issues, policy, and enabling conditions (Fig. 46.2). Compiling a status report is foremost a scientific and technical process done by experts. However, the more the (potential) partners and interest groups are involved (e.g., as contributors or reviewers), the more the conclusions of the report will be accepted and used in the following planning process. The status report will not only inform the strategic planning, but also serve as a reference point for the implementation and subsequent monitoring.
3. *Strategy (global level)*: When the scope and the mandate are clear, the partners and stakeholders (the participants) have been identified, and the status report is available, the strategic planning is done in a participatory process, most likely as a facilitated workshop. In a ZOPP approach (Fig. 46.2; IUCN/SSC Cat Specialist Group, 2015), a long-term vision and goal(s) are defined. The goal describes the concrete aim the entire conservation project or program is heading for. To reach the goal, the threats (identified in the status report and reviewed in a problem analysis during the workshop) must be overcome. To do this, objectives are identified, which directly address the threats. To achieve an objective, one to several concrete results (equivalent to targets in IUCN/SSC, 2008a, b), and actions are defined (Fig. 46.2). Results must be SMART (specific, measurable, attainable, relevant, and time-bound); their fulfilment will be monitored by means of indicators. Finally, objectives, results, actions, indicators, and additional parameters such as responsibilities, time lines, and budget frame are compiled in a logframe. For practical reasons, this is often done by a designated committee after the workshop, but then reviewed by all participants.

The ZOPP approach allows working stepwise from a vision to concrete actions. Hence the strategic planning and the action planning (Point 4) are not really separated steps. However, all large cats including the snow leopard are distributed over many countries in the form of transboundary populations, and we hence distinguish between the planning at global (range-wide) level and at national (sometimes even subnational) level (Fig. 46.3). To agree on strategic goals and priorities at a global (e.g., range-wide) level is more a sociopolitical than a scientific-technical issue and needs to be discussed with those directly concerned (e.g., the national authorities in charge of wildlife conservation). Internationally acting NGOs (and their national partners) are often the driving forces behind these processes and can also take organizational responsibilities. In practice, an organization committee consisting of international and local specialists and representatives of national institutions must be elected before or at the range-wide workshop. This committee will draft the SCS after the workshop and organize its review and endorsement in all participating countries and its implementation at the global level, but it can also assist in the translation of the SCS into (national) action plans.

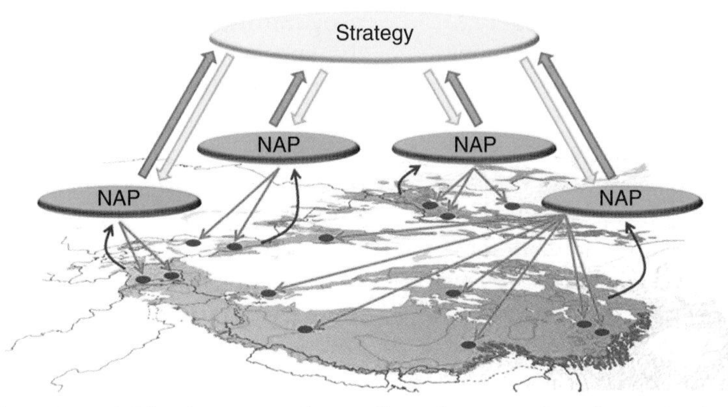

FIGURE 46.3 **Schematic model for the range-wide coordinated conservation of a species through a species conservation strategy (SCS), national action plans (NAPs), and in situ conservation projects (blue dots).** The plans (top–down) inform the *in situ* projects (yellow and green arrows), whereas the information collected during the monitoring process (bottom-up) help to evaluate and revise the NAPs and the strategy (purple and blue arrows).

4. *Action Plan (national or local level)*: Countries unite a section of the species' range under a common legal framework and are therefore the most appropriate units to implement conservation actions. An SCS for (large) cats is generally international and needs to be transformed into more concrete and more precise national action plans (NAP). Certain actions will have to be defined on the global level, but most activities (though often similar in each country) need to be adapted to the national conditions and implemented at national and local level. The NAPs are however informed by the SCS and describe the contributions of each country in solidarity with its neighbors to the overarching goal(s) and objectives. A NAP may be organized in a way similar to an SCS (including a status assessment and problem analysis, a strategic planning part, a logframe, etc.), but differs in three important aspects: (1) The development process must include all local interest groups (i.e., representatives from relevant GOs and NGOs, experts, and local stakeholders or people, who for practical reasons often cannot be integrated at the global level); (2) a NAP must be tailored to the national prerequisites (e.g., legislation, wildlife management and conservation systems, traditions, socioeconomic and human dimension aspects); (3) the NAP must be developed and made available in the national language(s). In large countries or in countries with a federal structure, such as China or India, respectively, a NAP for the snow leopard may even have to be split into several provincial action plans.

The process for developing a NAP is almost identical as for the SCS: Participatory, facilitated workshops including all partners and stakeholders, with a steering committee organizing the process are necessary. The development of a NAP is often more demanding than that of a SCS, because conflicts of interests occur usually more at the national level and because institutions involved in the development of the SCS at the global level are generally conservation-oriented. However, a sensible SCS – which defines clear global goals and principles, but leaves enough freedom for national discussions – can considerably support the development of NAPs.

5. *Implementation*: Conservationists regard the implementation of the actions as the "real conservation" and tend to forget the conceptual, planning, reporting, and monitoring parts of a project. But neglecting these tasks reduces the efficiency and sustainability of the project, leads to a loss of time and funding, and hinders the transfer of experience. The interface between the planning process and the implementation of the conservation actions is the logframe, which should be complemented by a work and time plan and a specific monitoring plan. Depending on the scale and complexity of a project, a kind of "adaptive project cycle" may even be beneficial at the project level or even adding a local cycle to the scheme outlined in Fig. 46.3.

6. *Monitoring and Evaluation*: Conservation projects – especially if they are large-scale and long-lasting programs such as the implementation of an SCS or NAP with many different partners – must be iterative and adaptive processes requiring a continuous thorough observation and evaluation of the performance. The plans must be regularly reviewed, revised, and updated. Monitoring, evaluation, and follow-ups must therefore be an integral part of every SCS and NAP. To ensure the monitoring quality, consistent and concise progress reporting is important. During the implementation of actions, the parameters as defined by the indicators are measured, analyzed, and reported, allowing judgment as to whether a defined result, the objectives, and finally the goal are achieved. The careful definition of SMART results and indicators is crucial for an effective monitoring.

External supervision can provide independent review and advice. Supervision, monitoring, and (terminal) evaluation of the implementation of an SCS or NAP must be agreed upon at the workshops where the plans are developed. IUCN's SSC's Species Conservation Planning Sub-Committee (www.iucn.org/about/work/programmes/species/who_we_are/about_the_species_survival_commission_/ssc_leadership/ssc_sub_committees/species_conservation_planning_sub_committee), or the relevant Specialist Groups can assist the review and evaluation of species conservation plans according to the IUCN standards.

If, according to the monitoring and evaluation report, the goal of the project is fulfilled, the activities may end here. However, both SCSs and NAPs are often documents with a timeline for evaluation and revision, but not a limited lifespan. Thus after the monitoring and evaluation, the strategic planning cycle starts again (Fig. 46.1).

During the implementation of the conservation activities, all project partners and the local community concerned are regularly informed about the progress. After the evaluation (at the end of the cycle), the larger audience is updated (e.g. through media coverage or scientific publications). An SCS and NAP should be public documents, and their revised version should also be released.

STRATEGIC PLANNING STIPULATIONS FOR THE CONSERVATION OF THE SNOW LEOPARD

Several essential elements of the strategic planning cycle (Fig. 46.1) have already been realized for the conservation of the snow leopard.

Species and area: Two subspecies of the snow leopard were described, *P. u. uncia* (Schreber, 1775) north of the Himalayas, and *P. u. uncioides* (Horsfield, 1855) in the Himalayas (Nowell and Jackson, 1997), based on morphology. For the status assessment (e.g., for the IUCN Red List), transboundary conservation planning, and the conservation breeding programs of zoos, the snow leopard was so far considered a monotypic species. The phylogenetic differentiation of the snow leopard should be addressed by genetic research. However, the question it is not urgent for *in situ* projects as long as no translocations are foreseen.

Initiators and partners: Snow leopard conservation was primarily initiated by the zoos when the first snow leopard studbook was released (Blomqvist, 1978a) and the first international conference was held at Helsinki Zoo, Finland, in 1978 (Blomqvist, 1978b). The following conferences, 1980 at Zurich Zoo, Switzerland, and at the Woodland Park Zoological Gardens, Seattle, USA, were still mainly focused on captive breeding programs, but with the perspective of potential releases into the wild (Eisen, 1982). *In situ* conservation became more prominent at the fourth conference at Krefeld Zoo, Germany (Blomqvist, 1984), but conservation breeding continued to be an important subject (e.g., Blomqvist, 1990). The next four snow leopard conferences were held in the range countries: 1986 in India (Freeman, 1988), 1989 in Kazakhstan (Blomqvist, 1990), 1992 in China (Fox and Du, 1994), and 1995 in Pakistan (Jackson and Ahmad, 1997). These meetings brought the first assessments of the *in situ* conservation status of the species (Freeman, 1988). In 2002, snow leopard conservationists met at the Snow Leopard Survival Summit in Seattle, USA, to develop the first Snow Leopard Survival Strategy (McCarthy and Chapron, 2003). This summit was also the start of a global expert coalition, the Snow Leopard Network (SLN), which now includes more than 600 individual members, but also GOs and NGOs (www.snowleopardnetwork.org). The SLN is open to all individuals and organizations, but over 70% of the 632 members (membership list in www.snowleopardnetwork.org, accessed June 23, 2015) are from NGOs or scientific institutions. NGOs (Table 46.1) are the most important actors with regard to the implementation of *in situ* activities and raising global and local awareness on snow leopard conservation.

An organization such as the Global Tiger Forum (www.globaltigerforum.com) or the Global Tiger Initiative (globaltigerinitiative.org), where the range countries play an important role, was lacking for the snow leopard. In October 2013, initiated by the government of the Kyrgyz Republic and assisted by the World Bank, the Global Snow Leopard and Ecosystem Protection Program (GSLEP) was launched with a conference in Bishkek (Snow Leopard Working Secretariat, 2013; see Chapter 45). The GSLEP united for the first time official representatives of all 12 snow leopard range countries, conveying a clear commitment for the conservation of the snow leopard (the "Bishkek Declaration," Snow Leopard Working Secretariat, 2013). Hence, the SLN and the GSLEP together nowadays should encompass all relevant partners – GOs, NGOs, and scientific institutions – for snow leopard conservation.

Status reviews and population monitoring: The population size has been estimated several times (overview in the Snow Leopard Survival Strategy, SLSS; Snow Leopard Network, 2014): 4510–7350 individuals (Fox, 1994); ca. 10,000 (Hunter and Jackson, 1997); 6000–8000 (McCarthy and Chapron, 2003); 4500–7500 (Jackson et al., 2010); 3920–6390 (GSLEP; Snow Leopard Working Secretariat, 2013). The estimated distribution range varied from 1,835,000 km^2 (Fox, 1994) to 3,024,728 (Hunter and Jackson, 1997) and 1,776,000 km^2 (Snow Leopard Working Secretariat, 2013). The review of snow leopard distribution at the 2008 Beijing workshop (Chapter 3) estimated an area of 2,758,213 km^2 as a sum of the confirmed, probable, and possible distribution range. From 1986 to 2008, the snow leopard was seven times assessed according to IUCN Red List protocols and each time listed as Endangered (EN). Since the introduction of the formal criteria, the assessment has always been based on an assumed continuous population decline, although the figures given above demonstrate how vague the population estimations always were. Until now, there has been no scientific, robust estimation available that would allow the size and trend of the snow leopard population to be assessed.

TABLE 46.1 List of Selected International and National NGOs and the Snow Leopard Range Countries in Which They Are Active According to Their Websites (Organizations are Listed in Alphabetical Order of the Abbreviations)

Country	International NGOs							National NGOs					
	FFI[a]	NABU[b]	Panthera[c]	SLC[d]	SLT[e]	WCS[f]	WWF[g]	NCF[h]	SLCF[i]	SLF[j]	SLFK[k]	SLFP[l]	SSCC[m]
Afghanistan						X							
Bhutan				X			X						
China			X		X	X	X						X
India			X	X	X		X	X					
Kazakhstan				X						X			
Kyrgyzstan	X	X	X		X	X	X				X		
Mongolia			X	X	X	X	X		X				
Nepal			X	X			X						
Pakistan			X	X	X	X	X					X	
Russia				X			X						
Tajikistan	X		X	(X)									
Uzbekistan			X			X							

[a]*Fauna & Flora International: http://www.fauna-flora.org/species/snow-leopard/ [29.01.2015]*
[b]*Nature and Biodiversity Conservation Union: http://www.schneeleopard.de [29.01.2015]*
[c]*Panthera: http://www.panthera.org/programs/snow-leopard/snow-leopard-program [29.01.2015]*
[d]*Snow Leopard Conservancy: http://snowleopardconservancy.org/ [29.01.2015]*
[e]*Snow Leopard Trust: http://www.snowleopard.org/ [29.01.2015] including Snow Leopard Enterprises*
[f]*Wildlife Conservation Society: http://www.wcs.org/saving-wildlife/big-cats/snow-leopard.aspx [29.01.2015]*
[g]*World Wildlife Fund: http://wwf.panda.org/ [29.01.2015]*
[h]*Nature Conservation Foundation: http://ncf-india.org/ [29.01.2015]*
[i]*Snow Leopard Conservation Foundation: https://www.facebook.com/pages/Snow-leopard-Conservation-Foundation-in-Mongolia/161537973927505 [29.01.2015]*
[j]*Snow Leopard Fund: http://www.slf.kz/en/ [29.01.2015]*
[k]*Snow Leopard Foundation, Kyrgyzstan*
[l]*Snow Leopard Foundation, Pakistan: https://www.facebook.com/snowleopard.pakistan?fref=ts [29.01.2015]*
[m]*Shan Shui Conservation Center*

Conservation strategies: The first comprehensive conservation plan was the Snow Leopard Survival Strategy, developed following the Snow Leopard Survival Summit in Seattle, USA, in May 2002 (McCarthy and Chapron, 2003). It contained most of the elements suggested for an SCS (IUCN/SSC, 2008a, b), including a review of the state of knowledge and the conservation status, a threat and gap analysis, a mission statement, goal and objectives, and recommendations on how to transfer the SLSS into NAPs. The Seattle Summit also established the Snow Leopard Network, which 12 years later released an updated version of the SLSS (Snow Leopard Network, 2014). The SLSS 2003 and 2014 were technical instruments reviewing not only the conservation status of the snow leopard, but also conservation measures and methodologies available and used, including up-to-date monitoring and population estimation methods for snow leopard and prey populations. The two SLSS were documents informed by experts from conservation NGOs and scientific institutions. They were based on "best knowledge" and "best practices," but lacked a policy part and buy-in from the range country governments.

In October 2013, the Global Snow Leopard Conservation Forum released the Global Snow Leopard and Ecosystem Protection Program (GSLEP; Snow Leopard Working Secretariat, 2013; see earlier section), to which many international, governmental, and nongovernmental organizations have contributed. All 12 range countries were represented at the conference, but also international organizations: Convention on International Trade in Endangered Species of Wild Fauna and Flora (CITES), Convention on the Conservation of Migratory Species of Wild Animals (CMS), Global Environment Facility (GEF), INTERPOL, United Nations Development Programme (UNDP), and the World Bank; several NGOs: Fauna & Flora International (FFI), Nature and Biodiversity Conservation Union (NABU), Snow Leopard Conservancy (SLC), Snow Leopard Trust (SLT), TRAFFIC, WildCRU, Wildlife Conservation Society (WCS), Panthera, WWF; and other organizations: Global Tiger Initiative (GTI), USAID, Snow Leopard Network. The GSLEP report defined core principles of snow leopard conservation such as integrating snow leopard conservation with local or regional economy, landscape-level transboundary conservation, and capacity needs for cross-sectoral responses. The GSLEP furthermore defined a concrete goal to be achieved by 2020, an activity portfolio, program management, and a portfolio for the National Snow Leopard and Ecosystem Protection Programs (NSLEP). Many of the aspects listed were already addressed by the 2003 and 2014 versions of SLSS; the main achievement of the GSLEP was that it had a strong involvement of the governments of the range countries and addressed topics that are important for the definition of global and local policies.

National action plans: A NAP has so far been developed in 9 of the 12 range countries (Snow Leopard Network, 2014; Table 46.2). Afghanistan, Bhutan, and China have no NAPs, the plans from Kyrgyzstan, Mongolia, and Tajikistan have not yet been approved by the authority in charge and not yet released (Snow Leopard Network, 2014). Table 46.2 compares the available six NAPs with the standards as proposed by IUCN/SSC (2008a, b). Most NAPs were drafted by a special commission, and the involvement of interest groups and local people was not considered or remained unclear. Only two NAPs refer to the SLSS 2003 as an overarching document. Most plans define objectives and actions, but elements facilitating the monitoring (e.g., results/targets and indicators) or a chapter explaining the implementation – including a time line, responsibilities, and a tentative budget – are mostly lacking.

The Annex to the GSLEP (Snow Leopard Working Secretariat 2013) contained, for each of the 12 range countries, a chapter mostly called "National Snow Leopard Ecosystem Protection Priority" (NSLEP). The NSLEPs address many elements of a NAP, such as habitat assessment, threats, monitoring, research, institutions involved, and so on. Most documents also present a list of objectives, activities, and a rough budget, but lack concrete indications on actors, implementation, approaches, and a time frame.

DISCUSSION AND RECOMMENDATIONS: STRATEGIC PLANNING REQUIREMENTS FOR CONSERVATION OF THE SNOW LEOPARD

Is a conservation program better organized bottom-up, from the needs and desires of local people and interest groups, or top-down, informed by expert knowledge? Is it in better hands with strongly committed NGOs or with an enduring governmental agency? Should it consider more the ecological situation or economic framework? Clearly, a sustainable and large-scale program such as the range-wide conservation (and not just survival) of the snow leopard would have to consider all of these aspects. Conservation specialists and representatives

TABLE 46.2 Overview of National Action Plans (NAP) of Snow Leopard Range Countries and Summary of Elements Suggested by IUCN/SSC (2008a, b)

					ZOPP and Logframe elements						Implementation factors			
Country	**Year**	**Drafting**	**SLSS**	**End**	**Vi**	**Go**	**Ob**	**Re**	**Ac**	**In**	**Imp**	**TL**	**Act**	**Bud**
Afghanistan	*No NAP available*													
Bhutan	*No NAP available*													
China	*No NAP available*													
India	2008	National Committee	+	X		X	X		X			X		X
Kazakhstan	2013	Snow Leopard Fund	(+)	X			X							
Kyrgyzstan	2013	Unknown		X										
Mongolia	2005	Not available		X										
Nepal	2013	Draft. and rev. teams	–	X			X	X	X	X	X			
Pakistan	2008	WWF–Pakistan	–	X			X		X		X	X	X	
Russia	2012	WWF–Russia	–	X					X					
Tajikistan	2010	Tajik Acad. Sc., int. NGOs	–	–										
Uzbekistan	2004	Working Group	–	X			X		X			X	X	

Year, Year of release or endorsement; drafting, group that developed the NAP; SLSS, Snow Leopard Survival Strategy 2003 (McCarthy and Chapron, 2003) considered; end, formally endorsed by parliament (Mongolia), the government or a ministry; ZOPP and Logframe elements: Vi, vision; Go, goal; Ob, objectives; Re, results or targets; Ac, actions; In, indicators, implementation factors; Imp, implementation addressed; TL, time line defined, act, actors defined; bud, budget for implementation outlined. For all range countries, an NSLEP was published as an appendix to the GSLEP (Snow Leopard Working Secretariat, 2013), which may partly have superseded the NAP listed here.

of conservation organizations (hence the main contributors and readers of this book) tend to overvalue the expert (mainly biological) parts of a conservation program and to neglect the socioeconomic and policy aspects. While in snow leopard conservation, (local) socioeconomic situations have been considered and addressed in specific projects (see Chapters 11 and 14), the transboundary and the transsectoral cooperation should still be strengthened. An improved strategic cooperation requires an agreed and standardized approach, hence merging and harmonizing the presently available strategic instruments, namely the SLSS, the GSLEP, and the NAPs and NSLEPs based on these strategic documents.

Perception and goals of nature conservation are also a matter of fashion and hence of occasional paradigm shifts. Mace (2014) identified four different (scientific) framings of conservation over the past 50 years: "Nature for itself" (up to 1970), "nature despite people" (until 1990), "nature for people" (until 2005), and most recently "people and nature," where "*science has moved fully away from a focus on species and protected areas and into a shared human-nature environment, where the form, function, adaptability, and resilience provided by nature are valued most highly.*" With regard to the role of an individual species such as the snow leopard, this modern framing seems to be consistent with the broad definition of biodiversity as already outlined. Beyond its ecological function, the snow leopard seems to be the ideal flagship species throughout its distribution range. It is a visible, charismatic, and well-known species causing relatively few conflicts and respected by the local societies.

Despite the various threats to the survival of the species (see overview in the SLSS 2014; Snow Leopard Network, 2014; see Chapters 5–10), the snow leopard has the highest potential among all large cat species to thrive across its original range as an ambassador for nature conservation together with people. Such an approach requires an integral, integrated, and range-wide strategic conservation planning (see Chapter 44) along the concept outlined previously, and implies that proper institutions are established to advance and secure snow leopard conservation. According to IUCN standards, strategic planning in species conservation is an integral and participatory process (IUCN/SSC, 2008 a, b). For species with a large distribution, it is however for practical reasons not possible to integrate national or even local stakeholders into a global planning process (IUCN/SSC Cat Specialist Group, 2015). It might therefore be needed to establish national structures analogous to the international arrangement. We here discuss only the possibilities at the global scale.

Cooperation: With the SLN (integrating mainly conservation NGOs and experts) and the GSLEP (uniting range countries and their GOs with international organizations), two organizational structures are already available that include all the relevant key players. These two structures should ideally work together or even build a common platform. The SLN hosts mainly the technical expertise, whereas the GSLEP unites the policy level. But neither of the bodies at the moment have the structure to guarantee that they can play their role in the future. The SLN is an informal network. The members do not share a single view with regard to snow leopard conservation, standards, or practical or scientific approaches. The GSLEP has been able to bring together the range countries under the lead of the World Bank – probably also with the hope for funding opportunities. But the World Bank is not a conservation institution with a long-term commitment to species conservation. Indeed, it announced in November 2014 its intention to phase out of the GTI by June 2015. This will be the same for the GSLEP. A high-level group or forum of decision makers integrating all range states is needed, but it should be hosted within an entity that has a lasting commitment in conservation and integrating all range countries (e.g., an international convention such as CMS or CITES).

It is crucial that all partners understand the importance, but also the differences of the roles of the key players: The range countries, GOs, and international institutions need to define the legal framework, agree on transboundary cooperation, and they are responsible for the management, law enforcement, and (in the long run) the monitoring of the populations. The conservation NGOs are best positioned for *in situ* project implementation, working with local people, and engaging in education, awareness, and capacity building. The scientific community needs to suggest robust monitoring schemes and carry out surveys and ecological and sociological research projects. These activities are of course not strictly separated; communication, cooperation, and constant mutual information are crucial.

Strategic planning: The SLSS and the GSLEP are both strategic documents for the range-wide conservation of the snow leopard. The two documents overlap, but also considerably differ. It is not good conservation practice to have two "global strategies" for the same species. The ideas and approaches of the SLSS and the GSLEP should hence be merged in a comprehensive range-wide species conservation strategy according to the IUCN guidelines (IUCN/SSC, 2008 a, b; IUCN/SSC Cat Specialist Group, 2015). The scientific-technical, socioeconomic, and political considerations should be integrated into one comprehensive conservation strategy, and, most important, should reflect the agreement of all institutions so far involved in snow leopard conservation planning. Hence a widely accepted body should be created to merge the two approaches. A global snow leopard conservation strategy needs to give a clear conceptual and methodological frame for the development of national action plans. Conservation action plans for individual snow leopard range countries now exist as nine national action plans (Table 46.2), 12 National Snow Leopard Ecosystem Protection Priority documents (Appendix to the GSLEP; Snow Leopard Working Secretariat, 2013), and several action plans developed by NGOs and experts. Again, the overlap between the different documents is large, but their structure and content is inconsistent, and most of them are not detailed and concrete enough to serve as an implementation tool. Each country should develop one NAP tailored to the specific situation and needs of the respective country, but considering the strategic specifications of the range-wide strategy. Each NAP then needs to be endorsed by the relevant GOs and implemented according to a practical logframe and adequate work plans.

Survey and monitoring: Baseline surveys of snow leopard and important prey species populations and a continuous monitoring of the populations are important prerequisites for the success of the SCS and the NAPs. Up to now, the assessments of the snow leopard population were not convincing, despite the numerous *in situ* surveys and modeling approaches. Experts and conservation organizations need to agree on a robust monitoring scheme based on the best practice and the modern methodology as, for example, outlined in the SLSS 2014 (Snow Leopard Network, 2014). All GOs and NGOs involved in snow leopard work need to adopt this scheme and to share data in order to provide more reliable information on snow leopard and prey populations. Finally, the conservation activities implemented according the SCS and the NAPs must be monitored, in order to evaluate their effectiveness and to allow a constant adjustment of snow leopard conservation actions to changing situations and needs.

References

Blomqvist, L. (Ed.), 1984. International Pedigree Book of Snow Leopards, *Panthera uncia*, vol. 4, Helsinki Zoo, Helsinki.

Blomqvist, L. (Ed.), 1990. International Pedigree Book of Snow Leopards, *Panthera uncia*, vol. 6, Helsinki Zoo, Helsinki.

Blomqvist, L., 1978a. First report on the snow leopard studbook, *Panthera uncia*, and 1976 world register. International Zoo Yearbook, vol. 18. pp. 227–231.

Blomqvist, L., 1978b. First international snow leopard conference in Helsinki, 7–8 March 1978. Int. Zoo News 25 (5), 5–6.

Dawkins, R., Krebs, J.R., 1979. Arms races between and within species. Proc. R. Soc. Lond. B 205, 489–511.

Eisen L., 1982. Symposium held on snow leopard. Woodland Park Zoological Gardens Newsletter, October/November 1982, pp. 1–3.

Fox J.L., Du Jizeng J., 1994. Proceedings of the Seventh International Snow Leopard Symposium (Xining, Qinghai, China, July 25–20, 1992). International Snow Leopard Trust, Seattle, Washington.

Fox, J.L., 1994. Snow leopard conservation in the wild – a comprehensive perspective on a low density and highly fragmented population. In: Fox, J.L., Du Jizeng, (Eds.). Proceedings of the Seventh International Snow Leopard Symposium. International Snow Leopard Trust, Seattle, pp. 3–15.

Freeman, H. (Ed.), 1988. Proceedings of the fifth international snow leopard symposium. International Snow Leopard Trust and Wildlife Institute of India. Conway, Bombay.

Ginsberg, J.G., 2001. Setting priorities for carnivore conservation: what makes carnivores different? In: Gittleman, J.L., Funk, S.M., Macdonald, D., Wayne, R.K. (Eds.), Carnivore Conservation. Cambridge University Press, Cambridge, pp. 498–523.

GTZ (Deutsche Gesellschaft für Technische Zusammenarbeit), 1997. Ziel Orientierte Projekt Planung – ZOPP. Roetherdruck, Darmstadt.

Hunter, D.O., Jackson, R., 1997. A range-wide model of potential snow leopard habitat. In: Jackson, R., Ahmad, A., (Eds.). Proceedings of the 8th International Snow Leopard Symposium, Islamabad, November 1995. International Snow Leopard Trust, Seattle and WWF-Pakistan, Lahore, pp. 51–56.

IUCN Standards and Petitions Subcommittee, 2014.Guidelines for Using the IUCN Red List Categories and Criteria. Version 11. Prepared by the Standards and Petitions Subcommittee. IUCN, Gland, Switzerland.

IUCN/SSC Cat Specialist Group, 2015. The Cat Conservation Compendium. Cat News Special Issue Nr. 10, in print.

IUCN/SSC, 2008a. Strategic Planning for Species Conservation: A Handbook. Version 1.0. IUCN Species Survival Commission, Gland, Switzerland.

IUCN/SSC, 2008b. Strategic Planning for Species Conservation: An Overview. Version 1.0. IUCN, Gland. Switzerland.

Jackson R., Ahmad A. (Eds.), 1997. Proceedings of the 8th International Snow Leopard Symposium, Islamabad, November 1995. International Snow Leopard Trust, Seattle and WWF Pakistan, Lahore.

Jackson, R.M., Mishra, C., McCarthy, T.M., Ale, S.B., 2010. Snow leopards: conflict and conservation. In: Macdonald, D.W., Loveridge, A.J. (Eds.), Biology and Conservation of Wild Felids. Oxford University Press, Oxford, pp. 417–430.

Mace, G., 2014. Whose conservation? Changes in the perception and goals of nature conservation require a solid scientific basis. Science 345 (6204), 1558–1560.

McCarthy, T.M., Chapron, G. (Eds.), 2003. Snow Leopard Survival Strategy. ISLT and SLN, Seattle.

Nowell, K., Jackson, P., 1997. Wild Cats. Status Survey and Conservation Action Plans. IUCN/SSC Cat Specialist Group, Gland, Switzerland.

Sanderson, E.W., Forrest, J., Loucks, C., Dinerstein, E., Ginsberg, J., Seidensticker, J., Leimgruber, P., Songer, M., Heydlauff, A., O'Brien, T., Bryja, G., 2006. Setting Priorities for the Conservation and Recovery of Wild Tigers: 2005-2015. The Technical Assessment. WCS, WWF, Smithsonian and NFWF-STF, Washington, DC, USA.

Snow Leopard Network, 2014. Snow Leopard Survival Strategy. Revised 2014 Version. Snow Leopard Network, Seattle.

Snow Leopard Working Secretariat. 2013. Global Snow Leopard and Ecosystem Protection Program (GSLEP). Snow Leopard Working Secretariat; Bishkek, Kyrgyz Republic.

CHAPTER

47

Future Prospects for Snow Leopard Survival

David Mallon, Thomas McCarthy***

*Division of Biology and Conservation Ecology, Manchester Metropolitan University, Manchester, UK

**Snow Leopard Program, Panthera, New York, NY, USA

The preceding 46 chapters represent the most comprehensive and up-to-date review of information on all aspects of the snow leopard ever assembled. Striking images carved on rocks millennia ago illustrate that the snow leopard has always provided a powerful source of inspiration. The role as an icon of the high mountains of Asia continues to inspire people to this day and is attested to by the many programs and the huge amount of attention devoted to it. The Snow Leopard Survival Summit, organized by the Snow Leopard Trust (SLT) in Seattle, 2002, was a seminal event in snow leopard conservation, leading to the development of the Snow Leopard Survival Strategy (McCarthy and Chapron, 2003; Snow Leopard Network, 2014) and the founding of the Snow Leopard Network (SLN, see Chapter 43). Since then, millions of dollars and hundreds of thousands of person-hours have been invested in the conservation of the snow leopard, its habitats, and its prey, though scientific research, field programs, local community engagement, capacity building, training, and anti-poaching. Education and awareness programs have raised the profile and contributed to the creation of more positive attitudes toward the cat by local people as well as government officials, and have led to many new protected areas being established. The recent application of advanced research techniques have sharply improved the quality of information available, bringing knowledge of the species more into line with other big cats, behind which the snow leopard has always lagged.

Livestock herding is the prevalent livelihood option across most of snow leopard range (Chapter 5), and local people depend on domestic animals – yaks, cattle, sheep, goats, camels, donkeys, horses – for meat and milk for consumption, wool for tents, ropes and clothing, as draft and transport animals, and to provide products for sale. Livestock share the mountain rangelands with snow leopards and their prey and usually exceed wild ungulates, both in number and biomass, many times over. The resulting competition for the best grazing, and

Snow Leopards
http://dx.doi.org/10.1016/B978-0-12-802213-9.00047-X

disturbance caused by the presence of herds and guard dogs, may force wild ungulates to use suboptimal areas while a further risk is transmission of disease (Chapter 6). Overgrazing reduces the quality and carrying capacity of rangelands, indirectly affecting the snow leopard through a reduced prey base. Snow leopards also prey on livestock and may be killed in retaliation. So there is a clear need for local people to have a meaningful stake in conservation through community projects, economic incentives, or alternative income generation, if snow leopards and their prey are to persist.

There is a notable range of innovative programs that engage local communities, as we learn in Chapter 13. One facet of the programs described here is that they are not "compensation" programs that simply pay herders for livestock losses to snow leopards and other predators. What sets them apart from direct compensation schemes is that built-in conditions require certain actions by participating individuals or communities, ranging from a portion of homestay proceeds going into community environment protection, to better livestock husbandry practices that help reduce losses to predators, and commitment to purchase own vaccines or maintain repairs to corrals. Snow Leopard Enterprises makes clear the ultimate condition by taking the added step of requiring a contract with the community that stipulates bonuses will be withheld in the event a snow leopard is killed within the village's "community responsible area."

In some instances, compensation programs have proven useful in mitigating losses to predators (Nyhus et al., 2003), but their track record in snow leopard range is not positive. Low compensation amounts, corruption, bureaucratic apathy in government-run programs (Mishra, 1997; Jackson and Wangchuk, 2004), along with donors becoming weary of funding NGO-managed compensation schemes, have all played a role in the low success rate of such efforts.

Killing of livestock by snow leopards (and other carnivores) impinges directly on livelihoods. Predation rates vary, as do the effects, according to the number and value of the animal killed (e.g., horse or yak versus sheep or goat). However, the most serious effects occur when a snow leopard enters a poorly constructed nighttime corral and kills many sheep or goats in a single attack, leading to economic hardship and stimulating negative attitudes toward the snow leopard. Schemes to improve the security of corrals have been undertaken in five of the 12 range countries – Afghanistan, India, Pakistan, Tajikistan, and Russia – as detailed in Chapter 14.1. Several hundred corrals have been protected so far and no subsequent predation events have been recorded at any of them. There is therefore no doubt that these initiatives have reduced livestock mortality and should be straightforward to replicate widely elsewhere.

While predation places a burden on pastoralists, many more livestock are lost annually to disease than to snow leopards, 1.5–5 times more in the example described in Chapter 14.3 from northern Pakistan. The response was to vaccinate livestock while agreeing on limits to herd sizes and securing an undertaking by livestock owners not to poach snow leopards. The result was reduced mortality of livestock and an increase in local incomes, as well as stable, rather than increasing, herd sizes. Vaccination involves relatively low initial inputs but must be repeated annually to cover newborn animals. A more ambitious initiative aims to reduce grazing pressure through negotiating a village reserve (grazing set-aside) in return for compensation payments (Chapter 14.2). Establishment of the reserve resulted in higher numbers of wild ungulates using the area and increased use by snow leopards.

In their *Guiding Principles on Trophy Hunting as a Tool for Creating Conservation Incentives*, the IUCN SSC states ". . . the wise and sustainable use of wildlife can be consistent with and contribute to conservation, because the social and economic benefits derived from use of species can provide incentives for people to conserve them and their habitats" (IUCN SSC 2012). Trophy hunting is such a form of wildlife use that can provide much needed income to people living in remote snow leopard range, and if well managed, provide strong economic incentives to protect snow leopards, their prey, and their habitat. As we see in Chapter 16, community-based trophy hunting programs (CTHPs) have been in practice or have been proposed in at least four snow leopard range countries. It has been practiced in some form in Pakistan since the late 1970s. Tajikistan CTHPs hosted their first hunters in 2012 and neighboring Kyrgyzstan has established several community conservancies but they have yet to conduct hunts. In contrast, in Mongolia it is argued that a well-managed, community-based trophy hunting program there could lead to sustainable management of ungulates while providing resources for conservation and building community support, but "thus far, developing such an approach has proved elusive." Although benefits have accrued in some instances, there are also lessons to be learned from the experiences of existing programs. A risk not specifically addressed in Chapter 16 is that of communities seeing snow leopards themselves as a potential trophy of great worth to foreign hunters. Jackson (2004) even cites a high-level Pakistan official asking, "What about a snow leopard trophy hunting program?" He goes on to point out that unanticipated consequences of CTHPs, such as a desire for even greater income, can have negative impacts for snow leopards.

The aforementioned examples point to the advances made toward addressing various aspects of human–wildlife conflict around the snow leopard and may be seen as a blueprint for success elsewhere. The challenge lies in scaling them up to cover substantial parts of the snow leopard's global range. Predator-proof corrals have already been implemented in several different localities, so may be the most straightforward to extend further (though the difficulty in transporting wire mesh and other materials long distances over difficult terrain should not be underestimated). Grazing set-asides contain great potential, but the reserve described in Chapter 14.2 covers only 5 km^2, so applying it to much wider areas would require substantially higher total payments, which have to be sourced from somewhere and may be unsustainable. As a conservation tool, community-managed trophy hunting has a foothold in snow leopard range, yet it will not be appropriate in all range states since cultural or regulatory obstacles may limit its applicability. For any community-based program to thrive, it will be essential to ensure the commitment by local people over the long term to assure the sustainability of these and similar programs.

Despite many clear successes, lest the reader be tempted to directly replicate any particular model described here, or elsewhere, it is advisable to heed the advice offered by Jackson et al. (2010) who point out that the mountain ranges of snow leopard habitat "*. . . are renowned for their varied climates, geography, biodiversity, and cultures that change dramatically over relatively short horizontal or vertical distances.*" It follows that any conservation initiative must address the particular set of threats and conditions, as well as the aspirations, of the community one seeks to work with. The community-based conservation incentive programs portrayed in this chapter all work, at least in part, because the instigators first sought to understand the conservation situation and then proceeded to develop highly participatory methods.

Although all the programs described here can surely be expected to have benefitted the snow leopard and its prey, the extent has not always been evaluated and confirmed. Strengthening of corrals has certainly reduced livestock mortality from snow leopard predation, but to what extent they have also prevented retaliatory killing is much more difficult to measure, as the authors indicate. Chapter 17 makes a similar point with regard to education programs, few of which have been assessed as to whether they resulted in changes in attitudes or behavior. Chapter 40 explicitly lists successes and failures of programs in Russia, making the valid point that failures are rarely reported, but all lessons learned, positive and negative, should be integrated into an adaptive management cycle.

Chapter 20 presents a viewpoint not often heard in our field, that of corporations engaged in resource extraction potentially mobilizing resources for biodiversity conservation and even providing "safe havens" for snow leopard and their prey. Such a case can be made for the Kumtor mine in Kyrgyzstan, which effectively guards one of the primary access routes to the Sarychat-Ertash Reserve, a key snow leopard landscape in the Central Tien Shan. In addition to restricting access, Kumtor has supported, both financially and logistically, NGOs conducting snow leopard research and conservation activities within and near the reserve. While mining has recently joined list of key threats to snow leopards, mitigation of such development may well provide a mechanism to actually strengthen and fund improvements in national policy and institutional capacity as we learned in Chapter 10.2.

The role of zoos in both in situ and ex situ conservation of snow leopards is highlighted in Chapters 21–23, including essays on how captive snow leopards serve as ambassadors of their wild kin. A pair of "ambassador" snow leopards made what is arguably the most significant impact ever on conservation of snow leopards in the wild, when zoo keeper Helen Freeman fell in love with Nicholas and Alexandra, Woodland Park Zoo's first snow leopards, and took up their wild cousin's cause in 1981 by establishing the International Snow Leopard Trust (ISLT).

Unlike every other entry in this book, Chapter 24.1 is not about snow leopards, yet the implications of this chapter should be clear to snow leopard conservationists. Each year we learn about snow leopards, both cubs and adults, that have been taken into captivity, "rescued," for one reason or another. Some are orphans, some are injured, some confiscated from perpetrators of illegal canned trophy hunts. The end result is the same – a snow leopard suffers in an inadequate facility while the international conservation community wrings its collective hands trying to decide what can or should be done. In the rare cases there is a positive outcome, such as Leo, a cub from Pakistan, who found a home in the Wildlife Conservation Society's Bronx Zoo (see Chapter 23.3). More often, the animals die, such as a captive cub held for over a year in a make-shift cage at the Pamir Biological Institute in Tajikistan, while government officials debated if the cat could be sent to a reputable zoo in England that had offered a home and transport. More recently, another captive orphaned cub in Pakistan generated serious debate between several NGOs and snow leopard scientists on the merits of a monitored soft release to the wild or the establishment of an educational facility within which to permanently house the cat. In the end it was decided to build the facility and not attempt a release. The editors invited the four case studies of this chapter which cover rescue, rehabilitation, and reintroduction of other large felids in Asia, Europe, North America, and South America to provide valuable insights and perspectives so that we may be better prepared to meet the growing challenge of snow leopards coming into captivity where resources for their care are lacking. There is at present no clear need for reintroductions of snow leopards into the wild, and current conservation efforts are rightly focused on ensuring the persistence of existing populations. This situation may change at some point in the future and these case studies present valuable insights into the complexities of such operations.

We devoted an entire section of the book to the rapidly advancing techniques and technologies available in the study of this highly cryptic cat, appropriately labeled the "ghost of the mountains." Chapter 25 provides a historic backdrop covering 30 years of work by several pioneers in our field. Today's tech-savvy practitioners may be somewhat amused to learn that earlier researchers had to hide their paper maps since such "top secret" items could land one in jail just for carrying them into parts of snow leopard range. Three chapters (and a few decades) later, we learn of camera traps so advanced they can identify a human and alert rangers to possible poachers via cellular or satellite uplinks in real time. Today, GPS-satellite collars (Chapter 26) track snow leopard movements at such a fine scale that locations obtained from a single cat in one year would likely exceed all VHF collar data points obtained over the first 20 years of collaring studies. While many cutting-edge technologies have been fully taken advantage of, the authors of Chapter 27.1 explain that in the case of genetic tools, snow leopard research is lagging behind. Whether to inform conservation actions, or to monitor their impact, noninvasive genetic methods are viable tools that are now underutilized. Given the cool, dry conditions that typify snow leopard habitat, and the predictable marking habits of the species, sampling snow leopard feces is considered by the authors to be comparatively easy versus other large carnivores and may yield higher quality DNA for analyses. As the cumulative advances in research techniques yield more and better data on snow leopard ecology, examining that at a landscape scale is the purview of spatial ecology. Echoing what we heard on genetics, Chapter 29 points to the rudimentary application of this science to the understanding of snow leopards. We opened this section of the book looking at where we came from, and close it looking forward. The tools and techniques are there for our use, we need only apply them more vigorously.

Despite this impressive body of work, our understanding of snow leopard biology and ecology remains incomplete. Among the gaps in knowledge are such fundamental parameters as the size of the global snow leopard population as well as reliable estimates of prey abundance. The snow leopard is secretive, has mainly crepuscular or nocturnal habits, and dwells in rugged, often remote, habitats that pose logistical challenges to researchers. As an apex predator in an environment with low basal productivity, snow leopards naturally live at relatively low densities so sample sizes tend to be low, making extrapolations problematic.

Population estimates to date have therefore largely been based on informed guesswork. However, some recent field studies cited here using advanced techniques of camera-trapping, satellite collaring, and fecal genetics report higher densities of 1–3 animals/100 km^2 in several sites (e.g., in China and Nepal), higher national population estimates, and confirm snow leopard presence at new sites and even its return to the Sagarmatha (Mt. Everest) region after a long absence. An implication of these findings is that the global population may be higher than current estimates of 4080–6590 (McCarthy and Chapron, 2003) and 4500–7500 (Jackson et al., 2010).

With hindsight, it anyway seems demographically implausible that only 4080 snow leopards, (taking the lowest figure, let alone fewer than the 2500 capable of reproduction that are cited in the current IUCN Red List assessment) were scattered across as much as 3.2 million km^2 of global range. If that were the case, it would raise the question of how enough individual snow leopards would manage to locate a mate. An alternative explanation, that viable subpopulations occur in small pockets within a matrix containing no snow leopards, is equally improbable, given that survey data indicates otherwise. Earlier population estimates may simply reflect the fallacy that "cryptic equals rare." Calculating a robust estimate of global population size and trends at global and regional levels should be seen as a research priority.

FIGURE 47.1 **Seemingly secure in its lofty, remote, and rugged home, it will take more than a cryptic lifestyle and superb camouflage to ensure a future for the magnificent snow leopard.** *Photo credit Raghu Chundawat.*

It is also becoming clearer that the global range appears relatively intact and snow leopards are able to disperse widely (see also Riordan et al., 2015). The northern and southern sectors of the global range may possibly be cut off from each other, but this is not established beyond doubt and some occasional exchange of individuals – and therefore genetic material – could still take place. There seems little to prevent an individual snow leopard in theory from moving from Eastern India or Bhutan along the entire Himalaya to Ladakh in the northwest and thence to the Karakoram, Hindu Kush, and Pamirs – and then potentially northeast along the Tien Shan or back eastward through China along the Kun Lun range.

The prospect that there may in reality be more snow leopards than previously believed, living in a practically contiguous and unfragmented range will be welcomed by all, as it greatly increases the chance of persistence. There are however some caveats. First, the situation is not uniform and in some locations, for example, the Russian Federation at the northwest edge of the distribution, the status is certainly more precarious. Second, new and old threats still have to be countered, so there is no ground for complacency or for slackening of conservation efforts. Traditional threats and constraints include retaliatory killing, illegal trade, reduced prey base, overuse of rangelands for grazing and fuelwood collection, lack of resources and other capacity; among the emerging threats are construction of linear barriers – of which border fences are likely the more serious – and mining. The dramatic upsurge in harvesting of caterpillar fungus, and concomitant degradation of pastures, illustrates the difficulty in predicting all possible future threats. The specter of climate change hovers over the mountains, but site-specific predictions are hindered by the lack of fine-scale data.

In fact the point of unavailable data applies to threats more widely. There is a lot of speculation over existing and emerging threats, but little quantified evidence to show the severity of their actual impact on snow leopards, directly or indirectly. For example, retaliatory killing by livestock herders is frequently cited as a major issue, but no figures for the number of snow leopards thus killed are available to permit a balanced judgment.

Nevertheless, the goal of snow leopard conservation must surely be to maintain the relatively intact and interconnected populations across most of the range over the long term, with the associated benefits for viable prey, the health of mountain ecosystems, and the services that these provide.

The range-wide priority-setting exercise that took place in Beijing in 2008 and the Global Snow Leopard and Ecosystem Protection Program (GSLEP) that aimed to secure government commitment widened the strategic framework. Unifying these and the SLSS into a common program as proposed in Chapter 46 would be a desirable aim and the vision and goal articulated in Chapter 44 contribute toward such a unifying theme.

The evidence presented in this volume reinforces what was said in the latest version of the Snow Leopard Survival Strategy (Snow Leopard Network, 2014): "*The snow leopard has a large home range size, so viable populations can only be secured across large landscapes ... the snow leopard therefore represents the ideal flagship and umbrella species for the mountain ecosystems of Asia.*" (Fig. 47.1)

References

IUCN SSC., 2012. IUCN SSC Guiding principles on trophy hunting as a tool for creating conservation incentives. Ver. 1.0. IUCN, Gland, Switzerland.

Jackson, R., 2004. Pakistan's Community-based trophy hunting programs and their relationship to Snow Leopard conservation, unpublished report. Snow Leopard Network; Sonoma, CA.

Jackson, R.M., Mishra, C., McCarthy, T., Ale, S.B., 2010. Snow leopards, conservation and conflict. In: Macdonald, D., Loveridge, A. (Eds.), The Biology and Conservation of Wild Felids. Oxford University Press, UK, pp. 417–430.

Jackson, R., Wangchuk, R., 2004. A community-based approach to mitigating livestock depredation by snow leopards. Hum. Dimens. Wildl. 9, 307–315.

McCarthy, T.M., Chapron, G., 2003. Snow Leopard Survival Strategy. Snow Leopard Network and International Snow Leopard Trust, Seattle. USA.

Mishra, C., 1997. Livestock depredation by large carnivores in the Indian trans-Himalaya: Conflict perceptions and conservation prospects. Environ. Conserv. 24, 338–343.

Nyhus, P., Fischer, F., Madden, F., Osofsky, S., 2003. Taking the bite out of wildlife damage: the challenge of wildlife compensation schemes. Conserv. Pract. 4, 37–40.

Riordan, P., Cushman, S., Hughes, J., Mallon, D., Shi, K., 2015. Predicting global population connectivity and targeting conservation action for snow leopard across its range. Ecography 38, 1–8.

Snow Leopard Network, 2014. Snow Leopard Survival Strategy. Revised Version 2014. Snow Leopard Network. www.snowleopardnetwork.org.

Subject Index

H

I

J

K

N

O